PROGRAMMABLE
CONTROLLERS

PROGRAMMABLE CONTROLLERS
THEORY AND IMPLEMENTATION

Second Edition

L. A. Bryan
E. A. Bryan

An Industrial Text Company Publication
Atlanta • Georgia • USA

Editor Stephanie Philippo
Art Director Gina Kory
Associate Editor Deborah Suwala

© 1988, 1997 by Industrial Text Company
Published by Industrial Text Company
All rights reserved
First edition 1988. Second edition 1997
Printed and bound in the United States of America
03 02 01 00 99 98   10 9 8 7 6 5 4 3

Requests for permission, accompanying workbooks, or
further information should be addressed to:
Industrial Text and Video Company
1950 Spectrum Circle
Tower A-First Floor
Marietta, Georgia 30067
(770) 240-2200
(800) PLC-TEXT

Library of Congress Cataloging-in-Publication Data

Bryan, L.A.
    Programmable controllers: theory and implementation/L.A. Bryan,
E.A. Bryan.—2nd ed.
      p.   cm.
    Includes index.
    ISBN 0-944107-32-X
    1. Programmable controllers.   I. Bryan, E.A.   II. Title.
TJ223.P76B795   1997
629.8'9—dc21   96-49350
                                          CIP

CONTENTS

Section 5   Advanced PLC Topics and Networks

PREFACE

Since the first edition of this book in 1988, the capabilities of programmable logic controllers have grown by leaps and bounds. Likewise, the applications of PLCs have grown with them. In fact, in today's increasingly computer-controlled environment, it is almost impossible to find a technical industry that does not use programmable controllers in one form or another. To respond to these phenomenal changes, we introduce the second edition of *Programmable Controllers: Theory and Implementation.*

This second edition, like the first, provides a comprehensive theoretical, yet practical, look at all aspects of PLCs and their associated devices and systems. However, this version goes one step further with new chapters on advanced PLC topics, such as I/O bus networks, fuzzy logic, the IEC 1131-3 programming standard, process control, and PID algorithms. This new edition also presents revised, up-to-date information about existing topics, with expanded graphics and new, hands-on examples. Furthermore, the new layout of the book—with features like two-tone graphics, key terms lists, well-defined headings and sections, callout icons, and a revised, expanded glossary—makes the information presented even easier to understand.

This new edition has been a labor-intensive learning experience for all those involved. As with any task so large, we could never have done it alone. Therefore, we would like to thank the following companies for their help in bringing this book to press: Allen-Bradley Company—Industrial Computer Group, ASI-USA, B & R Industrial Automation, Bailey Controls Company, DeviceNet Vendors Association, ExperTune Software, Fieldbus Foundation, Hoffman Engineering Company, Honeywell—MicroSwitch Division, LANcity—Cable Modem Division of Bay Networks, Mitsubishi Electronics, Omron Electronics, Phoenix Contact, PLC Direct, PMC/BETA LP, Profibus Trade Organization, Schaevitz Engineering Company, Siemens Automation, Square D Company, Thermometrics, and WAGO.

We hope that you will find this book to be a valuable learning and reference tool. We have tried to present a variety of programmable control operations; however, with the unlimited variations in control systems, we certainly have not been able to provide an exhaustive list of PLC applications. Only you, armed with the knowledge gained through this book, can explore the true limits of programmable logic controllers.

Stephanie Philippo
Editor

ABOUT THE AUTHORS

LUIS BRYAN

Luis Bryan holds a Bachelor of Science in Electrical Engineering degree and a Master of Science in Electrical Engineering degree, both from the University of Tennessee. His major areas of expertise are digital systems, electronics, and computer engineering. During his graduate studies, Luis was involved in several projects with national and international governmental agencies.

Luis has extensive experience in the field of programmable controllers. He was involved in international marketing activities, as well as PLC applications development, for a major programmable controller manufacturer. He also worked for a consulting firm, providing market studies and company-specific consultations about PLCs. Furthermore, Luis has given lectures and seminars in Canada, Mexico, and South America about the uses of programmable controllers. He continues to teach seminars to industry and government entities, including the National Aeronautics and Space Administration (NASA).

Luis is an active member of several professional organizations, including the Institute of Electrical and Electronics Engineers (IEEE) and the IEEE's instrument and computer societies. He is a senior member of the Instrument Society of America, as well as a member of Phi Kappa Phi honor society and Eta Kappa Nu electrical engineering honor society. Luis has coauthored several other books about programmable controllers.

ERIC BRYAN

Eric Bryan graduated from the University of Tennessee with a Bachelor of Science in Electrical Engineering degree, concentrating in digital design and computer architecture. He received a Master of Science in Engineering degree from the Georgia Institute of Technology, where he participated in a special computer-integrated manufacturing (CIM) program. Eric's specialties are industrial automation methods, flexible manufacturing systems (FMS), and artificial intelligence. He is an advocate of artificial intelligence implementation and its application in industrial automation.

Eric worked for a leading automatic laser inspection systems company, as well as a programmable controller consulting firm. His industrial experience includes designing and implementing large inspection systems, along with developing PLC-based systems. Eric has coauthored other publications about PLCs and is a member of several professional and technical societies.

HOW TO USE THIS BOOK

Welcome to *Programmable Controllers: Theory and Implementation*. Before you begin reading, please review the following strategies for using this book. By following these study strategies, you will more thoroughly understand the information presented in the text and, thus, be better able to apply this knowledge in real-life situations.

BEFORE YOU BEGIN READING

- Look through the book to familiarize yourself with its structure.

- Read the table of contents to review the subjects you will be studying.

- Familiarize yourself with the icons used throughout the text:

     Chapter Highlights

     Key Terms

- Look at the appendices to see what reference materials have been provided.

AS YOU STUDY EACH CHAPTER

- Before you start a chapter, read the Chapter Highlights paragraph at the beginning of the chapter's text. This paragraph will give you an overview of what you'll learn, as well as explain how the information presented in the chapter fits into what you've already learned and what you will learn.

- Read the chapter, paying special attention to the bolded items. These are key terms that indicate important topics that you should understand after finishing the chapter.

- When you encounter an exercise, try to solve the problem yourself before looking at the solution. This way, you'll determine which topics you understand and which topics you should study further.

WHEN YOU FINISH EACH CHAPTER

- At the end of each chapter, look over the list of key terms to ensure that you understand all of the important subjects presented in the chapter. If you're not sure about a term, review it in the text.

- Review the exercises to ensure that you understand the logic and equations involved in each problem. Also, review the workbook and study guide, making sure that you can work all of the problems correctly.

- When you're sure that you thoroughly understand the information that has been presented, you're ready to move on to the next chapter.

SECTION ONE

INTRODUCTORY CONCEPTS

- Introduction to Programmable Controllers
- Number Systems and Codes
- Logic Concepts

CHAPTER ONE

INTRODUCTION TO PROGRAMMABLE CONTROLLERS

I find the great thing in this world is not so much where we stand as in what direction we are moving.

—Oliver Wendell Holmes

CHAPTER HIGHLIGHTS Every aspect of industry—from power generation to automobile painting to food packaging—uses programmable controllers to expand and enhance production. In this book, you will learn about all aspects of these powerful and versatile tools. This chapter will introduce you to the basics of programmable controllers—from their operation to their vast range of applications. In it, we will give you an inside look at the design philosophy behind their creation, along with a brief history of their evolution. We will also compare programmable controllers to other types of controls to highlight the benefits and drawbacks of each, as well as pinpoint situations where PLCs work best. When you finish this chapter, you will understand the fundamentals of programmable controllers and be ready to explore the number systems associated with them.

1-1 DEFINITION

Programmable logic controllers, also called *programmable controllers* or *PLCs*, are **solid-state** members of the computer family, using integrated circuits instead of electromechanical devices to implement control functions. They are capable of storing instructions, such as sequencing, timing, counting, arithmetic, data manipulation, and communication, to control industrial machines and processes. Figure 1-1 illustrates a conceptual diagram of a PLC application.

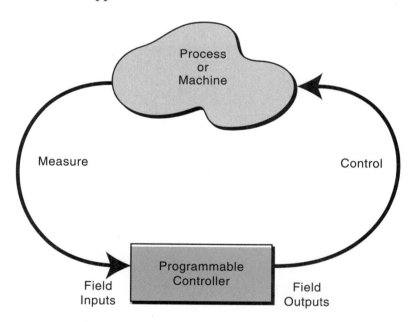

Figure 1-1. PLC conceptual application diagram.

Programmable controllers have many definitions. However, PLCs can be thought of in simple terms as industrial computers with specially designed architecture in both their central units (the PLC itself) and their interfacing circuitry to field devices (input/output connections to the real world).

As you will see throughout this book, programmable logic controllers are mature industrial controllers with their design roots based on the principles of simplicity and practical application.

1-2 A HISTORICAL BACKGROUND

The Hydramatic Division of the General Motors Corporation specified the design criteria for the first programmable controller in 1968. Their primary goal was to eliminate the high costs associated with inflexible, relay-controlled systems. The specifications required a solid-state system with computer flexibility able to (1) survive in an industrial environment, (2) be easily programmed and maintained by plant engineers and technicians, and (3) be reusable. Such a control system would reduce machine downtime and provide expandability for the future. Some of the initial specifications included the following:

- The new control system had to be price competitive with the use of relay systems.

- The system had to be capable of sustaining an industrial environment.

- The input and output interfaces had to be easily replaceable.

- The controller had to be designed in modular form, so that subassemblies could be removed easily for replacement or repair.

- The control system needed the capability to pass data collection to a central system.

- The system had to be reusable.

- The method used to program the controller had to be simple, so that it could be easily understood by plant personnel.

THE FIRST PROGRAMMABLE CONTROLLER

The product implementation to satisfy Hydramatic's specifications was underway in 1968; and by 1969, the programmable controller had its first product offsprings. These early controllers met the original specifications and opened the door to the development of a new control technology.

The first PLCs offered relay functionality, thus replacing the original hardwired **relay logic**, which used electrically operated devices to mechanically switch electrical circuits. They met the requirements of modularity, expandability, programmability, and ease of use in an industrial environment. These controllers were easily installed, used less space, and were reusable. The controller programming, although a little tedious, had a recognizable plant standard: the ladder diagram format.

In a short period, programmable controller use started to spread to other industries. By 1971, PLCs were being used to provide relay replacement as the first steps toward control automation in other industries, such as food and beverage, metals, manufacturing, and pulp and paper.

THE CONCEPTUAL DESIGN OF THE PLC

The first programmable controllers were more or less just relay replacers. Their primary function was to perform the sequential operations that were previously implemented with relays. These operations included ON/OFF control of machines and processes that required repetitive operations, such as transfer lines and grinding and boring machines. However, these programmable controllers were a vast improvement over relays. They were easily installed, used considerably less space and energy, had diagnostic indicators that aided troubleshooting, and unlike relays, were reusable if a project was scrapped.

Programmable controllers can be considered newcomers when they are compared to their elder predecessors in traditional control equipment technology, such as old hardwired relay systems, analog instrumentation, and other types of early solid-state logic. Although PLC functions, such as speed of operation, types of interfaces, and data-processing capabilities, have improved throughout the years, their specifications still hold to the designers' original intentions—they are simple to use and maintain.

TODAY'S PROGRAMMABLE CONTROLLERS

Many technological advances in the programmable controller industry continue today. These advances not only affect programmable controller design, but also the philosophical approach to control system architecture. Changes include both **hardware** (physical components) and **software** (control program) upgrades. The following list describes some recent PLC hardware enhancements:

- Faster scan times are being achieved using new, advanced microprocessor and electronic technology.

- Small, low-cost PLCs (see Figure 1-2), which can replace four to ten relays, now have more power than their predecessor, the simple relay replacer.

- High-density input/output (I/O) systems (see Figure 1-3) provide space-efficient interfaces at low cost.

- Intelligent, microprocessor-based I/O interfaces have expanded distributed processing. Typical interfaces include PID (proportional-integral-derivative), network, CANbus, fieldbus, ASCII communication, positioning, host computer, and language modules (e.g., BASIC, Pascal).

- Mechanical design improvements have included rugged input/output enclosures and input/output systems that have made the terminal an integral unit.

- Special interfaces have allowed certain devices to be connected directly to the controller. Typical interfaces include thermocouples, strain gauges, and fast-response inputs.

- Peripheral equipment has improved operator interface techniques, and system documentation is now a standard part of the system.

Courtesy of Mitsubishi Electronics, Mount Prospect, IL

Figure 1-2. Small PLC with built-in I/O and detachable, handheld programming unit.

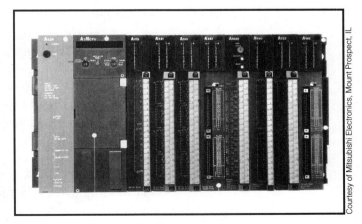

Courtesy of Mitsubishi Electronics, Mount Prospect, IL

Figure 1-3. PLC system with high-density I/O (64-point modules).

All of these hardware enhancements have led to the development of programmable controller families like the one shown in Figure 1-4. These families consist of a product line that ranges from very small "microcontrollers," with as few as 10 I/O points, to very large and sophisticated PLCs, with as many as 8,000 I/O points and 128,000 words of memory. These family members, using common I/O systems and programming peripherals, can interface to a local communication network. The family concept is an important cost-saving development for users.

Courtesy of Allen-Bradley, Highland, Heights, OH

Figure 1-4. Allen-Bradley's programmable controller family concept with several PLCs.

Like hardware advances, software advances, such as the ones listed below, have led to more powerful PLCs:

- PLCs have incorporated object-oriented programming tools and multiple languages based on the IEC 1131-3 standard.

- Small PLCs have been provided with powerful instructions, which extend the area of application for these small controllers.

- High-level languages, such as BASIC and C, have been implemented in some controllers' modules to provide greater programming flexibility when communicating with peripheral devices and manipulating data.

- Advanced functional block instructions have been implemented for ladder diagram instruction sets to provide enhanced software capability using simple programming commands.

- Diagnostics and fault detection have been expanded from simple system diagnostics, which diagnose controller malfunctions, to include machine diagnostics, which diagnose failures or malfunctions of the controlled machine or process.

- Floating-point math has made it possible to perform complex calculations in control applications that require gauging, balancing, and statistical computation.

- Data handling and manipulation instructions have been improved and simplified to accommodate complex control and data acquisition applications that involve storage, tracking, and retrieval of large amounts of data.

Programmable controllers are now mature control systems offering many more capabilities than were ever anticipated. They are capable of communicating with other control systems, providing production reports, scheduling production, and diagnosing their own failures and those of the machine or process. These enhancements have made programmable controllers important contributors in meeting today's demands for higher quality and productivity. Despite the fact that programmable controllers have become much more sophisticated, they still retain the simplicity and ease of operation that was intended in their original design.

PROGRAMMABLE CONTROLLERS AND THE FUTURE

The future of programmable controllers relies not only on the continuation of new product developments, but also on the integration of PLCs with other control and factory management equipment. PLCs are being incorporated, through networks, into computer-integrated manufacturing (CIM) systems, combining their power and resources with numerical controls, robots, CAD/CAM systems, personal computers, management information systems, and hierarchical computer-based systems. There is no doubt that programmable controllers will play a substantial role in the factory of the future.

New advances in PLC technology include features such as better operator interfaces, graphic user interfaces (GUIs), and more human-oriented man/machine interfaces (such as voice modules). They also include the development of interfaces that allow communication with equipment, hardware, and software that supports artificial intelligence, such as fuzzy logic I/O systems.

Software advances provide better connections between different types of equipment, using communication standards through widely used networks. New PLC instructions are developed out of the need to add intelligence to a controller. Knowledge-based and process learning–type instructions may be introduced to enhance the capabilities of a system.

The user's concept of the flexible manufacturing system (FMS) will determine the control philosophy of the future. The future will almost certainly continue to cast programmable controllers as an important player in the factory. Control strategies will be distributed with "intelligence" instead of being centralized. Super PLCs will be used in applications requiring complex calculations, network communication, and supervision of smaller PLCs and machine controllers.

1-3 PRINCIPLES OF OPERATION

A programmable controller, as illustrated in Figure 1-5, consists of two basic sections:

- the central processing unit
- the input/output interface system

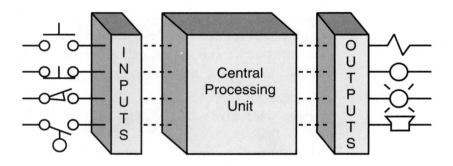

Figure 1-5. Programmable controller block diagram

The central processing unit (CPU) governs all PLC activities. The following three components, shown in Figure 1-6, form the CPU:

- the processor
- the memory system
- the system power supply

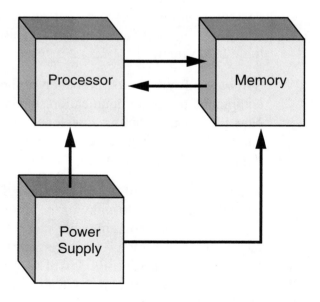

Figure 1-6. Block diagram of major CPU components.

The operation of a programmable controller is relatively simple. The **input/ output (I/O) system** is physically connected to the field devices that are encountered in the machine or that are used in the control of a process. These field devices may be discrete or analog input/output devices, such as limit switches, pressure transducers, push buttons, motor starters, solenoids, etc. The I/O interfaces provide the connection between the CPU and the information providers (inputs) and controllable devices (outputs).

During its operation, the CPU completes three processes: (1) it **reads**, or accepts, the input data from the field devices via the input interfaces, (2) it **executes**, or performs, the control program stored in the memory system, and (3) it **writes**, or updates, the output devices via the output interfaces. This process of sequentially reading the inputs, executing the program in memory, and updating the outputs is known as **scanning**. Figure 1-7 illustrates a graphic representation of a scan.

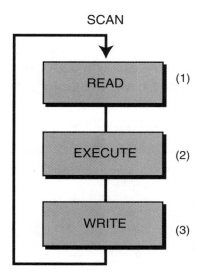

Figure 1-7. Illustration of a scan.

The input/output system forms the **interface** by which field devices are connected to the controller (see Figure 1-8). The main purpose of the interface is to condition the various signals received from or sent to external field devices. Incoming signals from sensors (e.g., push buttons, limit switches, analog sensors, selector switches, and thumbwheel switches) are wired to terminals on the input interfaces. Devices that will be controlled, like motor starters, solenoid valves, pilot lights, and position valves, are connected to the terminals of the output interfaces. The system **power supply** provides all the voltages required for the proper operation of the various central processing unit sections.

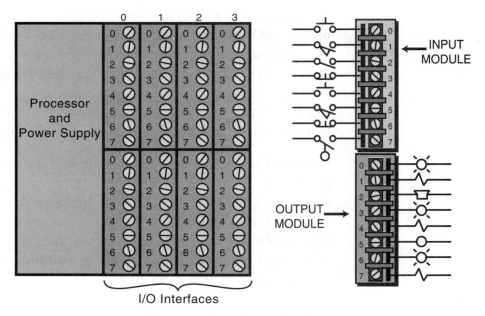

Figure 1-8. Input/output interface.

Although not generally considered a part of the controller, the **programming device**, usually a personal computer or a manufacturer's miniprogrammer unit, is required to enter the control program into memory (see Figure 1-9). The programming device must be connected to the controller when entering or monitoring the control program.

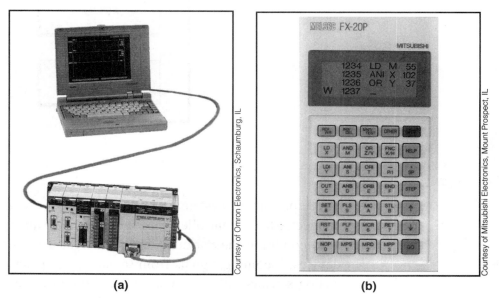

(a)   **(b)**

Figure 1-9. (a) Personal computer used as a programming device and **(b)** a mini-programmer unit.

Chapters 4 and 5 will present a more detailed discussion of the central processing unit and how it interacts with memory and input/output interfaces. Chapters 6, 7, and 8 discuss the input/output system.

1-4 PLCs VERSUS OTHER TYPES OF CONTROLS

PLCs VERSUS RELAY CONTROL

For years, the question many engineers, plant managers, and original equipment manufacturers (OEMs) asked was, "Should I be using a programmable controller?" At one time, much of a systems engineer's time was spent trying to determine the cost-effectiveness of a PLC over relay control. Even today, many control system designers still think that they are faced with this decision. One thing, however, is certain—today's demand for high quality and productivity can hardly be fulfilled economically without electronic control equipment. With rapid technology developments and increasing competition, the cost of programmable controls has been driven down to the point where a PLC-versus-relay cost study is no longer necessary or valid. Programmable controller applications can now be evaluated on their own merits.

When deciding whether to use a PLC-based system or a hardwired relay system, the designer must ask several questions. Some of these questions are:

- Is there a need for flexibility in control logic changes?

- Is there a need for high reliability?

- Are space requirements important?

- Are increased capability and output required?

- Are there data collection requirements?

- Will there be frequent control logic changes?

- Will there be a need for rapid modification?

- Must similar control logic be used on different machines?

- Is there a need for future growth?

- What are the overall costs?

The merits of PLC systems make them especially suitable for applications in which the requirements listed above are particularly important for the economic viability of the machine or process operation. A case which speaks for itself, the system shown in Figure 1-10, shows why programmable controllers are easily favored over relays. The implementation of this system using electromechanical standard and timing relays would have made this control panel a maze of large bundles of wires and interconnections.

Courtesy of Omron Electronics, Schaumburg, IL

Figure 1-10. The uncluttered control panel of an installed PLC system.

If system requirements call for flexibility or future growth, a programmable controller brings returns that outweigh any initial cost advantage of a relay control system. Even in a case where no flexibility or future expansion is required, a large system can benefit tremendously from the troubleshooting and maintenance aids provided by a PLC. The extremely short cycle (scan) time of a PLC allows the productivity of machines that were previously under electromechanical control to increase considerably. Also, although relay control may cost less initially, this advantage is lost if production downtime due to failures is high.

PLCs Versus Computer Controls

The architecture of a PLC's CPU is basically the same as that of a general purpose computer; however, some important characteristics set them apart. First, unlike computers, PLCs are specifically designed to survive the harsh conditions of the industrial environment. A well-designed PLC can be placed in an area with substantial amounts of electrical noise, electromagnetic interference, mechanical vibration, and noncondensing humidity.

A second distinction of PLCs is that their hardware and software are designed for easy use by plant electricians and technicians. The hardware interfaces for connecting field devices are actually part of the PLC itself and are easily connected. The modular and self-diagnosing interface circuits are able to pinpoint malfunctions and, moreover, are easily removed and replaced. Also, the software programming uses conventional relay ladder symbols, or other easily learned languages, which are familiar to plant personnel.

Whereas computers are complex computing machines capable of executing several programs or tasks simultaneously and in any order, the standard PLC executes a single program in an orderly, sequential fashion from first to last instruction. Bear in mind, however, that PLCs as a system continue to become more intelligent. Complex PLC systems now provide multiprocessor and multitasking capabilities, where one PLC may control several programs in a single CPU enclosure with several processors (see Figure 1-11).

Courtesy of Giddings & Lewis, Fond du Lac, WI

Figure 1-11. PLC system with multiprocessing and multitasking capabilities.

PLCs VERSUS PERSONAL COMPUTERS

With the proliferation of the personal computer (PC), many engineers have found that the personal computer is not a direct competitor of the PLC in control applications. Rather, it is an ally in the implementation of the control solution. The personal computer and the PLC possess similar CPU architecture; however, they distinctively differ in the way they connect field devices.

While new, rugged, industrial personal computers can sometimes sustain midrange industrial environments, their interconnection to field devices still presents difficulties. These computers must communicate with I/O interfaces not necessarily designed for them, and their programming languages may not meet the standards of ladder diagram programming. This presents a problem to people familiar with the ladder diagram standard when troubleshooting and making changes to the system.

The personal computer is, however, being used as the programming device of choice for PLCs in the market, where PLC manufacturers and third-party PLC support developers come up with programming and documentation systems for their PLC product lines. Personal computers are also being employed to gather process data from PLCs and to display information about the process or machine (i.e., they are being used as graphic user interfaces, or GUIs). Because of their number-crunching capabilities, personal computers are also well suited to complement programmable controllers and to bridge the communication gap, through a network, between a PLC system and other mainframe computers (see Figure 1-12).

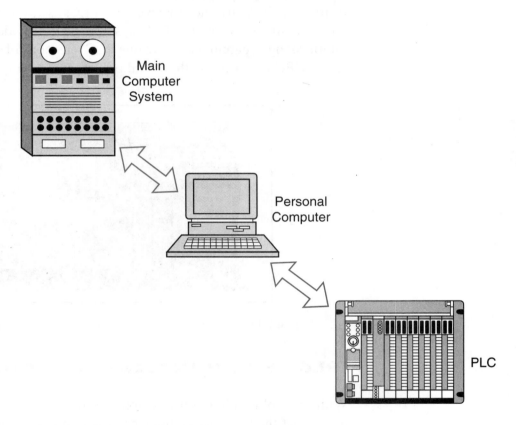

Main
Computer
System

Personal
Computer

PLC

Figure 1-12. A personal computer used as a bridge between a PLC system and a main computer system.

Some control software manufacturers, however, utilize PCs as CPU hardware to implement a PLC-like environment. The language they use is based on the International Electrotechnical Commission (IEC) 1131-3 standard, which is a graphic representation language (sequential function charts) that includes ladder diagrams, functional blocks, instruction lists, and structured text. These software manufacturers generally do not provide I/O hardware interfaces; but with the use of internal PC communication cards, these systems can communicate with other PLC manufacturers' I/O hardware modules. Chapter 10 explains the IEC 1131-3 standard.

TYPICAL AREAS OF PLC APPLICATIONS

Since its inception, the PLC has been successfully applied in virtually every segment of industry, including steel mills, paper plants, food-processing plants, chemical plants, and power plants. PLCs perform a great variety of control tasks, from repetitive ON/OFF control of simple machines to sophisticated manufacturing and process control. Table 1-1 lists a few of the major industries that use programmable controllers, as well as some of their typical applications.

CHEMICAL/PETROCHEMICAL
Batch process
Finished product handling
Materials handling
Mixing
Off-shore drilling
Pipeline control
Water/waste treatment

GLASS/FILM
Cullet weighing
Finishing
Forming
Lehr control
Packaging
Processing

FOOD/BEVERAGE
Accumulating conveyors
Blending
Brewing
Container handling
Distilling
Filling
Load forming
Metal forming loading/unloading
Palletizing
Product handling
Sorting conveyors
Warehouse storage/retrieval
Weighing

LUMBER/PULP/PAPER
Batch digesters
Chip handling
Coating
Wrapping/stamping

MANUFACTURING/MACHINING
Assembly machines
Boring
Cranes
Energy demand
Grinding
Injection/blow molding
Material conveyors
Metal casting
Milling
Painting
Plating
Test stands
Tracer lathe
Welding

METALS
Blast furnace control
Continuous casting
Rolling mills
Soaking pit

MINING
Bulk material conveyors
Loading/unloading
Ore processing
Water/waste management

POWER
Burner control
Coal handling
Cut-to-length processing
Flue control
Load shedding
Sorting
Winding/processing
Woodworking

Table 1-1. Typical programmable controller applications.

Because the applications of programmable controllers are extensive, it is impossible to list them all in this book. However, Table 1-2 provides a small sample of how PLCs are being used in industry.

AUTOMOTIVE

Internal Combustion Engine Monitoring. A PLC acquires data recorded from sensors located at the internal combustion engine. Measurements taken include water temperature, oil temperature, RPMs, torque, exhaust temperature, oil pressure, manifold pressure, and timing.

Carburetor Production Testing. PLCs provide on-line analysis of automotive carburetors in a production assembly line. The systems significantly reduce the test time, while providing greater yield and better quality carburetors. Pressure, vacuum, and fuel and air flow are some of the variables tested.

Monitoring Automotive Production Machines. The system monitors total parts, rejected parts, parts produced, machine cycle time, and machine efficiency. Statistical data is available to the operator anytime or after each shift.

Power Steering Valve Assembly and Testing. The PLC system controls a machine to ensure proper balance of the valves and to maximize left and right turning ratios.

CHEMICAL AND PETROCHEMICAL

Ammonia and Ethylene Processing. Programmable controllers monitor and control large compressors used during ammonia and ethylene manufacturing. The PLC monitors bearing temperatures, operation of clearance pockets, compressor speed, power consumption, vibration, discharge temperatures, pressure, and suction flow.

Dyes. PLCs monitor and control the dye processing used in the textile industry. They match and blend colors to predetermined values.

Chemical Batching. The PLC controls the batching ratio of two or more materials in a continuous process. The system determines the rate of discharge of each material and keeps inventory records. Several batch recipes can be logged and retreived automatically or on command from the operator.

Fan Control. PLCs control fans based on levels of toxic gases in a chemical production environment. This system effectively removes gases when a preset level of contamination is reached. The PLC controls the fan start/stop, cycling, and speeds, so that safety levels are maintained while energy consumption is minimized.

Gas Transmission and Distribution. Programmable controllers monitor and regulate pressures and flows of gas transmission and distribution systems. Data is gathered and measured in the field and transmitted to the PLC system.

Pipeline Pump Station Control. PLCs control mainline and booster pumps for crude oil distribution. They measure flow, suction, discharge, and tank low/high limits. Possible communication with SCADA (Supervisory Control and Data Acquistion) systems can provide total supervision of the pipeline.

Oil Fields. PLCs provide on-site gathering and processing of data pertinent to characteristics such as depth and density of drilling rigs. The PLC controls and monitors the total rig operation and alerts the operator of any possible malfunctions.

Table 1-2. Examples of PLC applications.

GLASS PROCESSING

Annealing Lehr Control. PLCs control the lehr used to remove the internal stress from glass products. The system controls the operation by following the annealing temperature curve during the reheating, annealing, straining, and rapid cooling processes through different heating and cooling zones. Improvements are made in the ratio of good glass to scrap, reduction in labor cost, and energy utilization.

Glass Batching. PLCs control the batch weighing system according to stored glass formulas. The system also controls the electromagnetic feeders for infeed to and outfeed from the weigh hoppers, manual shut-off gates, and other equipment.

Cullet Weighing. PLCs direct the cullet system by controlling the vibratory cullet feeder, weight-belt scale, and shuttle conveyor. All sequences of operation and inventory of quantities weighed are kept by the PLC for future use.

Batch Transport. PLCs control the batch transport system, including reversible belt conveyors, transfer conveyors to the cullet house, holding hoppers, shuttle conveyors, and magnetic separators. The controller takes action after the discharge from the mixer and transfers the mixed batch to the furnace shuttle, where it is discharged to the full length of the furnace feed hopper.

MANUFACTURING/MACHINING

Production Machines. The PLC controls and monitors automatic production machines at high efficiency rates. It also monitors piece-count production and machine status. Corrective action can be taken immediately if the PLC detects a failure.

Transfer Line Machines. PLCs monitor and control all transfer line machining station operations and the interlocking between each station. The system receives inputs from the operator to check the operating conditions on the line-mounted controls and reports any malfunctions. This arrangement provides greater machine efficiency, higher quality products, and lower scrap levels.

Wire Machine. The controller monitors the time and ON/OFF cycles of a wire-drawing machine. The system provides ramping control and synchronization of electric motor drives. All cycles are recorded and reported on demand to obtain the machine's efficiency as calculated by the PLC.

Tool Changing. The PLC controls a sychronous metal cutting machine with several tool groups. The system keeps track of when each tool should be replaced, based on the number of parts it manufactures. It also displays the count and replacements of all the tool groups.

Paint Spraying. PLCs control the painting sequences in auto manufacturing. The operator or a host computer enters style and color information and tracks the part through the conveyor until it reaches the spray booth. The controller decodes the part information and then controls the spray guns to paint the part. The spray gun movement is optimized to conserve paint and increase part throughput.

MATERIALS HANDLING

Automatic Plating Line. The PLC controls a set pattern for the automated hoist, which can traverse left, right, up, and down through the various plating solutions. The system knows where the hoist is at all times.

Table 1-2 continued.

Storage and Retrieval Systems. A PLC is used to load parts and carry them in totes in the storage and retrieval system. The controller tracks information like ⌐ lane numbers, the parts assigned to specific lanes, and the quantity of parts in a particular lane. This PLC arrangement allows rapid changes in the status of parts loaded or unloaded from the system. The controller also provides inventory printouts and informs the operator of any malfunctions.

Conveyor Systems. The system controls all of the sequential operations, alarms, and safety logic necessary to load and circulate parts on a main line conveyor. It also sorts products to their correct lanes and can schedule lane sorting to optimize palletizer duty. Records detailing the ratio of good parts to rejects can be obtained at the end of each shift.

Automated Warehousing. The PLC controls and optimizes the movement of stacking cranes and provides high turnaround of materials requests in an automated, high-cube, vertical warehouse. The PLC also controls aisle conveyors and case palletizers to significantly reduce manpower requirements. Inventory control figures are maintained and can be provided on request.

METALS

Steel Making. The PLC controls and operates furnaces to produce metal in accordance with preset specifications. The controller also calculates oxygen requirements, alloy additions, and power requirements.

Loading and Unloading of Alloys. Through accurate weighing and loading sequences, the system controls and monitors the quantity of coal, iron ore, and limestone to be melted. It can also control the unloading sequence of the steel to a torpedo car.

Continuous Casting. PLCs direct the molten steel transport ladle to the continuous-casting machine, where the steel is poured into a water-cooled mold for solidification.

Cold Rolling. PLCs control the conversion of semifinished products into finished goods through cold-rolling mills. The system controls motor speed to obtain correct tension and provide adequate gauging of the rolled material.

Aluminum Making. Controllers monitor the refining process, in which impurities are removed from bauxite by heat and chemicals. The system grinds and mixes the ore with chemicals and then pumps them into pressure containers, where they are heated, filtered, and combined with more chemicals.

POWER

Plant Power System. The programmable controller regulates the proper distribution of available electricity, gas, or steam. In addition, the PLC monitors powerhouse facilities, schedules distribution of energy, and generates distribution reports. The PLC controls the loads during operation of the plant, as well as the automatic load shedding or restoring during power outages.

Energy Management. Through the reading of inside and outside temperatures, the PLC controls heating and cooling units in a manufacturing plant. The PLC system controls the loads, cycling them during predetermined cycles and keeping track of how long each should be on or off during the cycle time. The system provides scheduled reports on the amount of energy used by the heating and cooling units.

Table 1-2 continued.

Coal Fluidization Processing. The controller monitors how much energy is generated from a given amount of coal and regulates the coal crushing and mixing with crushed limestone. The PLC monitors and controls burning rates, temperatures generated, sequencing of valves, and analog control of jet valves.

Compressor Efficiency Control. PLCs control several compressors at a typical compressor station. The system handles safety interlocks, startup/shutdown sequences, and compressor cycling. The PLCs keep compressors running at maximum efficiency using the nonlinear curves of the compressors.

Pulp and Paper

Pulp Batch Blending. The PLC controls sequence operation, ingredient measurement, and recipe storage for the blending process. The system allows operators to modify batch entries of each quantity, if necessary, and provides hardcopy printouts for inventory control and for accounting of ingredients used.

Batch Preparation for Paper-Making Processing. Applications include control of the complete stock preparation system for paper manufacturing. Recipes for each batch tank are selected and adjusted via operator entries. PLCs can control feedback logic for chemical addition based on tank level measurement signals. At the completion of each shift, the PLC system provides management reports on materials use.

Paper Mill Digester. PLCs control the process of making paper pulp from wood chips. The system calculates and controls the amount of chips based on density and digester volume. Then, the percent of required cooking liquors is calculated and these amounts are added to the sequence. The PLC ramps and holds the cooking temperature until the cooking is completed.

Paper Mill Production. The controller regulates the average basis weight and moisture variable for paper grade. The system manipulates the steam flow valves, adjusts the stock valves to regulate weight, and monitors and controls total flow.

Rubber and Plastic

Tire-Curing Press Monitoring. The PLC performs individual press monitoring for time, pressure, and temperature during each press cycle. The system alerts the operator of any press malfunctions. Information concerning machine status is stored in tables for later use. Report generation printouts for each shift include a summary of good cures and press downtime due to malfunctions.

Tire Manufacturing. Programmable controllers are used for tire press/cure systems to control the sequencing of events that transforms a raw tire into a tire fit for the road. This control includes molding the tread pattern and curing the rubber to obtain road-resistant characteristics. This PLC application substantially reduces the space required and increases reliability of the system and the quality of the product.

Rubber Production. PLCs provide accurate scale control, mixer logic functions, and multiple formula operation of carbon black, oil, and pigment used in the production of rubber. The system maximizes utilization of machine tools during production schedules, tracks in-process inventories, and reduces time and personnel required to supervise the production activity and the shift-end reports.

Plastic Injection Molding. A PLC system controls variables, such as temperature and pressure, which are used to optimize the injection molding process. The system provides closed-loop injection, where several velocity levels can be programmed to maintain consistent filling, reduce surface defects, and shorten cycle time.

Table 1-2 continued.

1-5 PLC PRODUCT APPLICATION RANGES

Figure 1-13 graphically illustrates programmable controller product ranges. This chart is not definitive, but for practical purposes, it is valid. The PLC market can be segmented into five groups:

1. micro PLCs

2. small PLCs

3. medium PLCs

4. large PLCs

5. very large PLCs

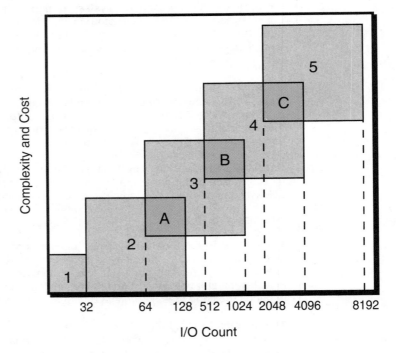

Figure 1-13. PLC product ranges.

Micro PLCs are used in applications controlling up to 32 input and output devices, 20 or less I/O being the norm. The micros are followed by the small PLC category, which controls 32 to 128 I/O. The medium (64 to 1024 I/O), large (512 to 4096 I/O), and very large (2048 to 8192 I/O) PLCs complete the segmentation. Figure 1-14 shows several PLCs that fall into this category classification.

The A, B, and C overlapping areas in Figure 1-13 reflect enhancements, by adding options, of the standard features of the PLCs within a particular segment. These options allow a product to be closely matched to the application without having to purchase the next larger unit. Chapter 20 covers, in detail, the differences between PLCs in overlapping areas. These differences include I/O count, memory size, programming language, software functions, and other factors. An understanding of the PLC product ranges and their characteristics will allow the user to properly identify the controller that will satisfy a particular application.

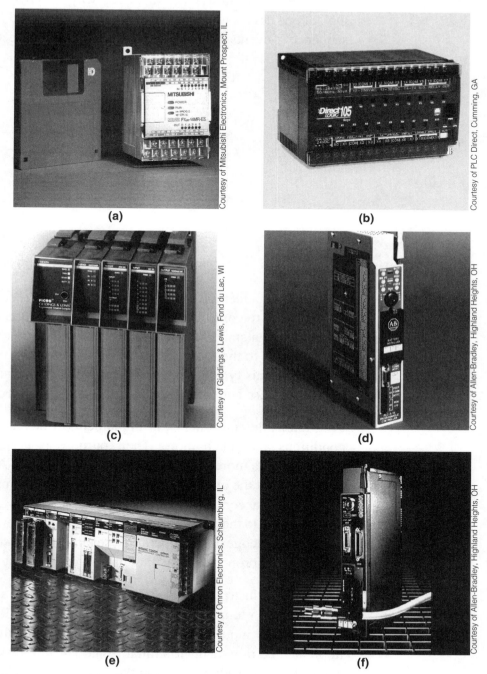

Figure 1-14. (a) Mitsubishi's smallest print size PLC (14 I/O), **(b)** PLC Direct DL105 with 18 I/O and a capacity of 6 amps per output channel, **(c)** Giddings & Lewis PIC90 capable of handling 128 I/O with motion control capabilities, **(d)** Allen-Bradley's PLC 5/15 (512 I/O), **(e)** Omron's C200H PLC (1392 I/O), and **(f)** Allen-Bradley's PLC 5/80 (3072 I/O).

1-6 LADDER DIAGRAMS AND THE PLC

The **ladder diagram** has and continues to be the traditional way of representing electrical sequences of operations. These diagrams represent the interconnection of field devices in such a way that the activation, or turning ON, of one device will turn ON another device according to a predetermined sequence of events. Figure 1-15 illustrates a simple electrical ladder diagram.

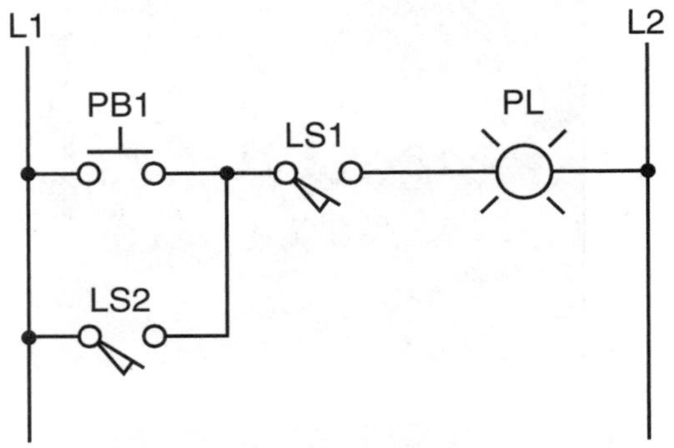

Figure 1-15. Simple electrical ladder diagram.

The original ladder diagrams were established to represent hardwired logic circuits used to control machines or equipment. Due to wide industry use, they became a standard way of communicating control information from the designers to the users of equipment. As programmable controllers were introduced, this type of circuit representation was also desirable because it was easy to use and interpret and was widely accepted in industry.

Programmable controllers can implement all of the "old" ladder diagram conditions and much more. Their purpose is to perform these control operations in a more reliable manner at a lower cost. A PLC implements, in its CPU, all of the old hardwired interconnections using its software instructions. This is accomplished using familiar ladder diagrams in a manner that is transparent to the engineer or programmer. As you will see throughout this book, a knowledge of PLC operation, scanning, and instruction programming is vital to the proper implementation of a control system.

Figure 1-16 illustrates the PLC transformation of the simple diagram shown in Figure 1-15 to a PLC format. Note that the "real" I/O field devices are connected to input and output interfaces, while the ladder program is implemented in a manner, similar to hardwiring, inside the programmable controller (i.e., *softwired* inside the PLC's CPU instead of *hardwired* in a panel). As previously mentioned, the CPU reads the status of inputs, energizes the corresponding circuit element according to the program, and controls a real output device via the output interfaces.

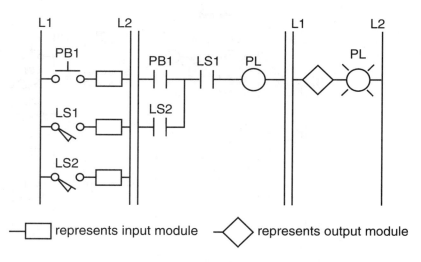

—☐ represents input module   —◇ represents output module

Figure 1-16. PLC implementation of Figure 1-15.

As you will see later, each instruction is represented inside the PLC by a reference **address**, an alphanumeric value by which each device is known in the PLC program. For example, the push button PB1 is represented inside the PLC by the name PB1 (indicated on top of the instruction symbol) and likewise for the other devices shown in Figure 1-16. These instructions are represented here, for simplicity, with the same device and instruction names. Chapters 3 and 5 further discuss basic addressing techniques, while Chapter 6 covers input/output wiring connections. Example 1-1 illustrates the similarity in operation between hardwired and PLC circuits.

EXAMPLE 1-1

In the hardwired circuit shown in Figure 1-15, the pilot light PL will turn ON if the limit switch LS1 closes <u>and</u> if either push button PB1 <u>or</u> limit switch LS2 closes. In the PLC circuit, the same series of events will cause the pilot light—connected to an output module—to turn ON. Note that in the PLC circuit in Figure 1-16, the internal representation of contacts provides the equivalent power logic as a hardwired circuit when the referenced input field device closes or is pushed. Sketch hardwired and PLC implementation diagrams for the circuit in Figure 1-15 illustrating the configurations of inputs that will turn PL ON.

SOLUTION

Figure 1-17 shows several possible configurations for the circuit in Figure 1-15. The highlighted blue lines indicate that power is present at that connection point, which is also the way a programming or monitoring device represents power in a PLC circuit. The last two configurations in Figure 1-17 are the only ones that will turn PL ON.

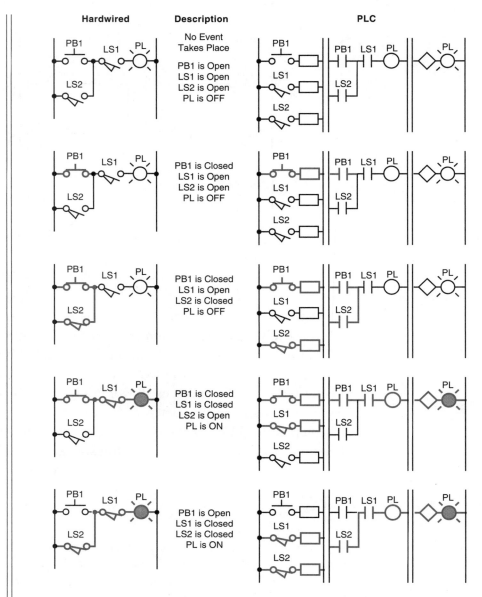

Figure 1-17. Possible configurations of inputs and corresponding outputs.

1-7 ADVANTAGES OF PLCs

In general, PLC architecture is modular and flexible, allowing hardware and software elements to expand as the application requirements change. In the event that an application outgrows the limitations of the programmable controller, the unit can be easily replaced with a unit having greater memory and I/O capacity, and the old hardware can be reused for a smaller application. A PLC system provides many benefits to control solutions, from reliability and repeatability to programmability. The benefits achieved with programmable controllers will grow with the individual using them—the more you learn about PLCs, the more you will be able to solve other control problems.

Table 1-3 lists some of the many features and benefits obtained with a programmable controller.

Inherent Features	Benefits
Solid-state components	• High reliability
Programmable memory	• Simplifies changes • Flexible control
Small size	• Minimal space requirements
Microprocessor-based	• Communication capability • Higher level of performance • Higher quality products • Multifunctional capability
Software timers/counters	• Eliminate hardware • Easily changed presets
Software control relays	• Reduce hardware/wiring cost • Reduce space requirements
Modular architecture	• Installation flexibility • Easily installed • Reduces hardware cost • Expandability
Variety of I/O interfaces	• Controls a variety of devices • Eliminates customized control
Remote I/O stations	• Eliminate long wire/conduit runs
Diagnostic indicators	• Reduce troubleshooting time • Signal proper operation
Modular I/O interface	• Neat appearance of control panel • Easily maintained • Easily wired
Quick I/O disconnects	• Service without disturbing wiring
System variables stored in memory data	• Useful management/maintenance • Can be output in report form

Table 1-3. Typical programmable controller features and benefits.

Without question, the "programmable" feature provides the single greatest benefit for the use and installation of programmable controllers. Eliminating hardwired control in favor of programmable control is the first step towards achieving a flexible control system. Once installed, the control plan can be manually or automatically altered to meet day-to-day control requirements without changing the field wiring. This easy alteration is possible since there are no physical connections between the field input devices and output devices (see Figure 1-18), as in hardwired systems. The only connection is through the control program, which can be easily altered.

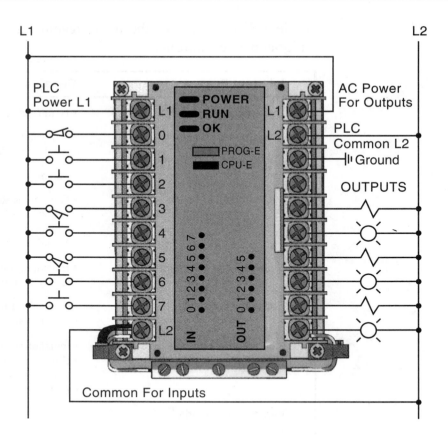

Figure 1-18. Programmable controller I/O connection diagram showing no physical connections between the inputs and outputs.

A typical example of the benefits of softwiring is a solenoid that is controlled by two limit switches connected in series (see Figure 1-19a). Changing the solenoid operation by placing the two limit switches in parallel (see Figure 1-19b) or by adding a third switch to the existing circuit (see Figure 1-19c) would take less than one minute in a PLC. In most cases, this simple program change can be made without shutting down the system. This same change to a hardwired system could take as much as thirty to sixty minutes of downtime, and even a half hour of downtime can mean a costly loss of production. A similar situation exists if there is a need to change a timer preset value or some other constant. A software timer in a PLC can be changed in as little as five seconds. A set of thumbwheel switches and a push button can be easily configured to input new preset values to any number of software timers. The time savings benefit of altering software timers, as opposed to altering several hardware timers, is obvious.

The hardware features of programmable controllers provide similar flexibility and cost savings. An intelligent CPU is capable of communicating with other intelligent devices. This capability allows the controller to be integrated into local or plantwide control schemes. With such a control configuration, a PLC can send useful English messages regarding the controlled system to an intelligent display. On the other hand, a PLC can receive supervisory information, such as production changes or scheduling information, from a host computer. A standard I/O system includes a variety of digital, analog, and special interface modules, which allow sophisticated control without the use of expensive, customized interface electronics.

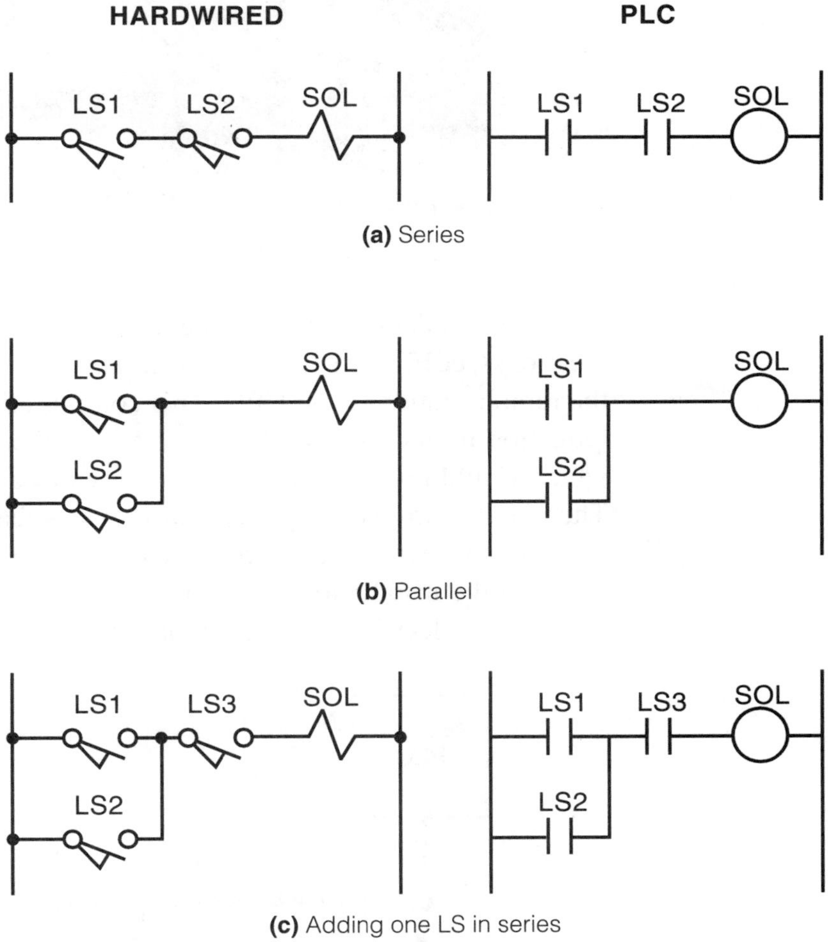

HARDWIRED   **PLC**

(a) Series

(b) Parallel

(c) Adding one LS in series

Figure 1-19. Example of hardwiring changes as opposed to softwiring changes.

EASE OF INSTALLATION

Several attributes make PLC installation an easy, cost-effective project. Its relatively small size allows a PLC to be conveniently located in less than half the space required by an equivalent relay control panel (see Figure 1-20). On a small-scale changeover from relays, a PLC's small, modular construction allows it to be mounted in the same enclosure where the relays were located. Actual changeover can be made quickly by simply connecting the input/output devices to the prewired terminal strips.

Figure 1-20. Space-efficient design of a PLC.

In large installations, remote input/output stations are placed at optimum locations (see Figure 1-21). A coaxial cable or a twisted pair of wires connects the remote station to the CPU. This configuration results in a considerable reduction in material and labor costs as compared to a hardwired system, which would involve running multiple wires and installing large conduits. The remote subsystem approach also means that various sections of a total system can be completely prewired by an OEM or PLC vendor prior to reaching the installation site. This approach considerably reduces the time spent by an electrician during an on-site installation.

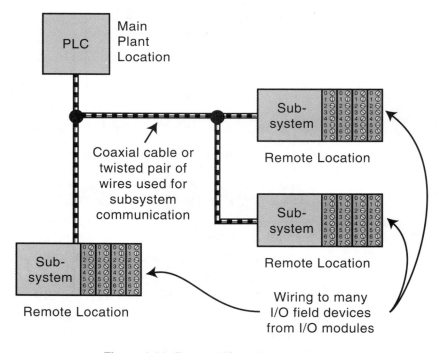

Figure 1-21. Remote I/O station installation.

EASE OF MAINTENANCE AND TROUBLESHOOTING

From the beginning, programmable controllers have been designed with ease of maintenance in mind. With virtually all components being solid-state, maintenance is reduced to the replacement of modular, plug-in components. Fault detection circuits and diagnostic indicators (see Figure 1-22), incorporated in each major component, signal whether the component is working properly or malfunctioning. In fact, most failures associated with a PLC-based system stem from failures directly related to the field input/output devices, rather than the PLC's CPU or I/O interface system (see Figure 1-23). However, the monitoring capability of a PLC system can easily detect and correct these field device failures.

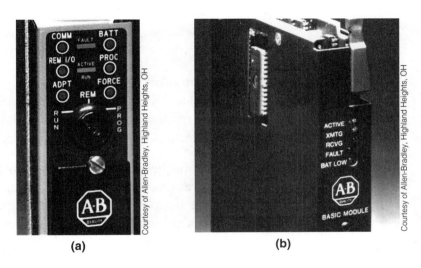

(a)   **(b)**

Figure 1-22. (a) A PLC processor and **(b)** an intelligent module containing several status indicators.

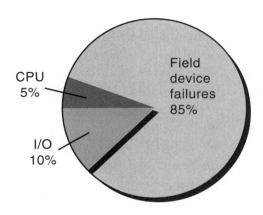

Figure 1-23. Failures in a PLC-based system.

With the aid of the programming device, any programmed logic can be viewed to see if inputs or outputs are ON or OFF (see Figure 1-24). Programmed instructions can also be written to enunciate certain failures.

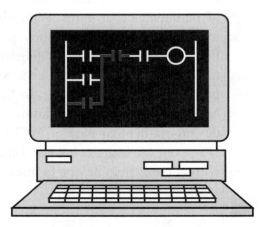

Figure 1-24. A programming device being used to monitor inputs and outputs, with highlighted contacts indicating an ON condition.

These and several other attributes of the PLC make it a valuable part of any control system. Once installed, its contribution will be quickly noticed and payback will be readily realized. The potential benefits of the PLC, like any intelligent device, will depend on the creativity with which it is applied.

It is obvious from the preceding discussion that the potential benefits of applying programmable controllers in an industrial application are substantial. The bottom line is that, through the use of programmable controllers, users will achieve high performance and reliability, resulting in higher quality at a reduced cost.

KEY TERMS

address
central processing unit (CPU)
execute
hardware
input/output system
interface
ladder diagram
programmable logic controller (PLC)
programming device
read
relay logic
scan
software
solid-state
write

CHAPTER TWO

NUMBER SYSTEMS AND CODES

I have often admired the mystical ways of Pythagoras and the secret magic of numbers.

—Sir Thomas Browne

 **CHAPTER HIGHLIGHTS**

In this chapter, we will explain the number systems and digital codes that are most often used in programmable controller applications. We will first introduce the four number systems most frequently used during input/output address assignment and programming: binary, octal, decimal, and hexadecimal. Then, we will discuss the binary coded decimal (BCD) and Gray codes, along with the ASCII character set and several PLC register formats. Since these codes and systems are the foundation of the logic behind PLCs, a basic knowledge of them will help you understand how PLCs work.

2-1 NUMBER SYSTEMS

A familiarity with number systems is quite useful when working with programmable controllers, since a basic function of these devices is to represent, store, and operate on numbers, even when performing the simplest of operations. In general, programmable controllers use binary numbers in one form or another to represent various codes and quantities. Although these number operations are transparent for the most part, there are occasions where a knowledge of number systems is helpful.

First, let's review some basics. The following statements apply to any number system:

- Every number system has a base or radix.

- Every system can be used for counting.

- Every system can be used to represent quantities or codes.

- Every system has a set of symbols.

The **base** of a number system determines the total number of unique symbols used by that system. The largest-valued symbol always has a value of one less than the base. Since the base defines the number of symbols, it is possible to have a number system of any base. However, number system bases are typically chosen for their convenience. The number systems usually encountered while using programmable controllers are base 2, base 8, base 10, and base 16. These systems are called binary, octal, decimal, and hexadecimal, respectively. To demonstrate the common characteristics of number systems, let's first turn to the familiar decimal system.

DECIMAL NUMBER SYSTEM

The **decimal number system**, which is the most common to us, was undoubtedly developed because humans have ten fingers and ten toes. Thus, the base of the decimal number system is 10. The symbols, or digits, used in this system are 0, 1, 2, 3, 4, 5, 6, 7, 8, and 9. As noted earlier, the total number of symbols (10) is the same as the base, with the largest-valued symbol being one less than the base (9 is one less than 10). Because the decimal system is so common, we rarely stop to think about how to express a number greater than 9, the largest-valued symbol. It is, however, important to note that the technique for representing a value greater than the largest symbol is the same for any number system.

In the decimal system, a *place value*, or *weight*, is assigned to each position that a number greater than 9 would hold, starting from right to left. The first position (see Figure 2-1), starting from the right-most position, is position 0, the second is position 1, and so on, up to the last position *n*. As shown in Figure 2-2, the **weighted value** of each position can be expressed as the base (10 in this case) raised to the power of *n* (the position). For the decimal system, then, the position weights from right to left are 1, 10, 100, 1000, etc. This method for computing the value of a number is known as the **sum-of-the-weights method**.

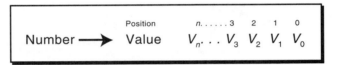

Figure 2-1. Place values.

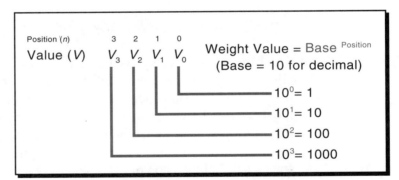

Figure 2-2. Weighted values.

The value of a decimal number is computed by multiplying each digit by the weighted value of its position and then summing the results. Let's take, for example, the number 9876. It can be expressed through the sum-of-the-weights method as:

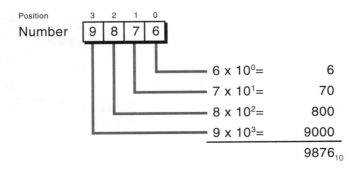

As you will see in other number systems, the decimal equivalent of any number can be computed by multiplying each digit by its base raised to the power of the digit's position. This is shown below:

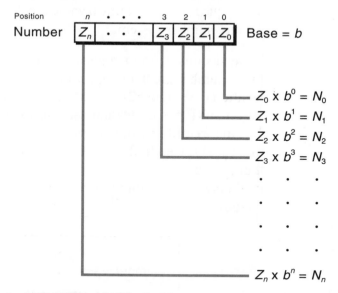

Therefore, the sum of N_0 through N_n will be the decimal equivalent of the number in base b.

BINARY NUMBER SYSTEM

The **binary number system** uses the number 2 as the base. Thus, the only allowable digits are 0 and 1; there are no 2s, 3s, etc. For devices such as programmable controllers and digital computers, the binary system is the most useful. It was adopted for convenience, since it is easier to design machines that distinguish between only two entities, or numbers (i.e., 0 and 1), rather than ten, as in decimal. Most physical elements have only two states: a light bulb is on or off, a valve is open or closed, a switch is on or off, and so on. In fact, you see this number system every time you use a computer—if you want to turn it on, you flip the switch to the 1 position; if you want to turn it off, you flip the switch to the 0 position (see Figure 2-3). Digital circuits can distinguish between two voltage levels (e.g., +5 V and 0 V), which makes the binary system very useful for digital applications.

Figure 2-3. The binary numbers, 1 and 0, on a computer's power switch represent ON and OFF, respectively.

As with the decimal system, expressing binary numbers greater than the largest-valued symbol (in this case 1) is accomplished by assigning a weighted value to each position from right to left. The weighted value (decimal equivalent) of a binary number is computed the same way as it is for a decimal number—only instead of being 10 raised to the power of the position, it is 2 raised to the power of the position. For binary, then, the weighted values from right to left are 1, 2, 4, 8, 16, 32, 64, etc., representing positions 0, 1, 2, 3, 4, 5, 6, etc. Let's calculate the decimal value that is equivalent to the value of the binary number 10110110:

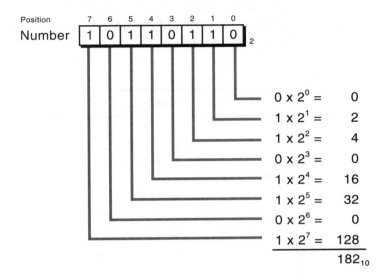

Position 7 6 5 4 3 2 1 0

Number $1\ 0\ 1\ 1\ 0\ 1\ 1\ 0$ ₂

$$0 \times 2^0 = 0$$
$$1 \times 2^1 = 2$$
$$1 \times 2^2 = 4$$
$$0 \times 2^3 = 0$$
$$1 \times 2^4 = 16$$
$$1 \times 2^5 = 32$$
$$0 \times 2^6 = 0$$
$$1 \times 2^7 = 128$$
$$182_{10}$$

Thus, the binary number 10110110 is equivalent to the number 182 in the decimal system. Each digit of a binary number is known as a **bit**; hence, this particular binary number, 10110110 (182 decimal), has 8 bits. A group of 4 bits is known as a **nibble**; a group of 8 bits is a **byte**; and a group of one or more bytes is a **word**. Figure 2-4 presents a binary number composed of 16 bits, with the **least significant bit (LSB)**, the lowest valued bit in the word, and the **most significant bit (MSB)**, the largest valued bit in the word, identified.

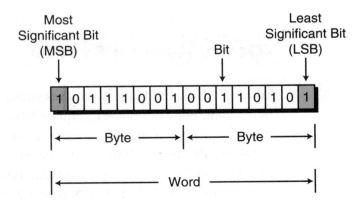

Figure 2-4. One word, two bytes, sixteen bits.

Counting in binary is a little more awkward than counting in decimal for the simple reason that we are not used to it. Because the binary number system uses only two digits, we can only count from 0 to 1—only one change in one digit location (OFF to ON) before a new digit position must be added. Conversely, in the decimal system, we can count from 0 to 9, equaling ten digit transitions, before a new digit position is added.

In binary, just like in decimal, we add another digit position once we run out of transitions. So, when we count in binary, the digit following 0 and 1 is 10 (one-zero, not ten), just like when we count 0, 1, 2...9 in decimal, another digit position is added and the next digit is 10 (ten). Table 2-1 shows a count in binary from 0_{10} to 15_{10}.

Decimal	Binary
0	0
1	1
2	10
3	11
4	100
5	101
6	110
7	111
8	1000
9	1001
10	1010
11	1011
12	1100
13	1101
14	1110
15	1111

Table 2-1. Decimal and binary counting.

OCTAL NUMBER SYSTEM

Writing a number in binary requires substantially more digits than writing it in decimal. For example, 91_{10} equals 1011011_2. Too many binary digits can be cumbersome to read and write, especially for humans. Therefore, the **octal numbering system** is often used to represent binary numbers using fewer digits. The octal number system uses the number 8 as its base, with its eight digits being 0, 1, 2, 3, 4, 5, 6, and 7. Table 2-2 shows both an octal and a binary count representation of the numbers 0 through 15 (decimal).

Decimal	Binary	Octal
0	0	0
1	1	1
2	10	2
3	11	3
4	100	4
5	101	5
6	110	6
7	111	7
8	1000	10
9	1001	11
10	1010	12
11	1011	13
12	1100	14
13	1101	15
14	1110	16
15	1111	17

Table 2-2. Decimal, binary, and octal counting.

Like all other number systems, each digit in an octal number has a weighted decimal value according to its position. For example, the octal number 1767 is equivalent to the decimal number 1015:

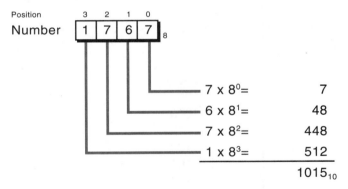

$$7 \times 8^0 = 7$$
$$6 \times 8^1 = 48$$
$$7 \times 8^2 = 448$$
$$1 \times 8^3 = 512$$
$$1015_{10}$$

As noted earlier, the octal numbering system is used as a convenient way of writing a binary number. The octal system has a base of 8 (2^3), making it possible to represent any binary number in octal by grouping binary bits in groups of three. In this manner, a very large binary number can be easily represented by an octal number with significantly fewer digits. For example:

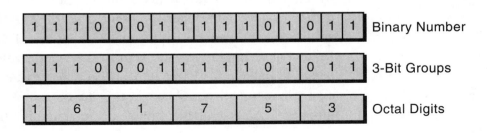

SECTION
1 | *Introductory Concepts*

So, a 16-bit binary number can be represented directly by six digits in octal. As you will see later, many programmable controllers use the octal number system for referencing input/output and memory addresses.

HEXADECIMAL NUMBER SYSTEM

The **hexadecimal (hex) number system** uses 16 as its base. It consists of 16 digits—the numbers 0 through 9 and the letters *A* through *F* (which represent the numbers 10 through 15, respectively). The hexadecimal system is used for the same reason as the octal system, to express binary numbers using fewer digits. The hexadecimal numbering system uses one digit to represent four binary digits (or bits), instead of three as in the octal system. Table 2-3 shows a hexadecimal count example of the numbers 0 through 15 with their decimal and binary equivalents.

Binary	Decimal	Hexadecimal
0	0	0
1	1	1
10	2	2
11	3	3
100	4	4
101	5	5
110	6	6
111	7	7
1000	8	8
1001	9	9
1010	10	A
1011	11	B
1100	12	C
1101	13	D
1110	14	E
1111	15	F

Table 2-3. Binary, decimal, and hexadecimal counting.

As with the other number systems, hexadecimal numbers can be represented by their decimal equivalents using the sum-of-the-weights method. The decimal values of the letter-represented hex digits A through F are used when computing the decimal equivalent (10 for A, 11 for B, and so on). The following example uses the sum-of-the-weights method to transform the hexadecimal number F1A6 into its decimal equivalent. The value of A in the example is 10 times 16^1, while F is 15 times 16^3. Thus, the hexadecimal number F1A6 is equivalent to the decimal number 61,862:

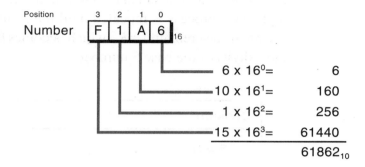

Like octal numbers, hexadecimal numbers can easily be converted to binary without any mathematical transformation. To convert a hexadecimal number to binary, simply write the 4-bit binary equivalent of the hex digit for each position. For example:

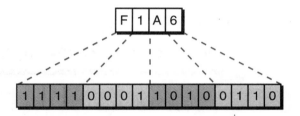

2-2 NUMBER CONVERSIONS

In the previous section, you saw how a number of any base can be converted to the familiar decimal system using the sum-of-the-weights method. In this section, we will show you how a decimal number can be converted to binary, octal, or any number system.

To convert a decimal number to its equivalent in any base, you must perform a series of divisions by the desired base. The conversion process starts by dividing the decimal number by the base. If there is a remainder, it is placed in the **least significant digit** (right-most) position of the new base number. If there is no remainder, a 0 is placed in least significant digit position. The result of the division is then brought down, and the process is repeated until the final result of the successive divisions is 0. This methodology may be a little cumbersome; however, it is the easiest conversion method to understand and employ.

As a generic example, let's find the base 5 equivalent of the number Z (see Figure 2-5). The first division ($Z \div 5$) gives an N_1 result and a remainder R_1. The remainder R_1 becomes the first digit of the base 5 number (the least significant digit). To obtain the next base 5 digit, the N_1 result is again divided by 5, giving an N_2 result and an R_2 remainder that becomes the second base 5 digit. This process is repeated until the result of the division ($N_n \div 5$) is 0, giving the last remainder R_n, which becomes the **most significant digit** (left-most digit) of the base 5 number.

Division	Remainder
$Z \div 5 = N_1$	R_1
$N_1 \div 5 = N_2$	R_2
$N_2 \div 5 = N_3$	R_3
$N_3 \div 5 = N_4$	R_4
$\bullet$	
$\bullet$	
$\bullet$	
$N_n \div 5 = 0$	R_n
New base 5 number is $(R_n ... R_4 R_3 R_2 R_1)_5$	

Figure 2-5. Method for converting a decimal number into any base.

Now, let's convert the decimal number 35_{10} to its binary (base 2) equivalent using this method:

Division	Remainder
$35 \div 2 = 17$	1
$17 \div 2 = 8$	1
$8 \div 2 = 4$	0
$4 \div 2 = 2$	0
$2 \div 2 = 1$	0
$1 \div 2 = 0$	1

Therefore, the base 2 (binary) equivalent of the decimal number 35 is 100011.

As another exercise, let's convert the number 1355_{10} to its hexadecimal (base 16) equivalent:

Division	Remainder
1355 ÷ 16 = 84	11
84 ÷ 16 = 5	4
5 ÷ 16 = 0	5

Thus, the hexadecimal equivalent of 1355_{10} is $54B_{hex}$ (remember that the hexadecimal system uses the letter *B* to represent the number 11).

There is another method, which is a little faster, for computing the binary equivalent of a decimal number. This method employs division by eight, instead of by two, to convert the number first to octal and then to binary from octal (three bits at a time).

For instance, let's take the number 145_{10}:

Division	Remainder
145 ÷ 8 = 18	1
18 ÷ 8 = 2	2
2 ÷ 8 = 0	2

The octal equivalent of 145_{10} is 221_8, so from Table 2-2, we can find that 221_8 equals 010010001 binary:

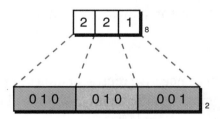

2-3 ONE'S AND TWO'S COMPLEMENT

The one's and two's complements of a binary number are operations used by programmable controllers, as well as computers, to perform internal mathematical calculations. To *complement* a binary number means to change it to a negative number. This allows the basic arithmetic operations of subtraction, multiplication, and division to be performed through successive addition. For example, to subtract the number 20 from the number 40, first complement 20 to obtain –20, and then perform an addition.

The intention of this section is to introduce the basic concepts of complementing, rather than to provide a thorough analysis of arithmetic operations. For more information on this subject, please use the references listed in the back of this book.

ONE'S COMPLEMENT

Let's assume that we have a 5-bit binary number that we wish to represent as a negative number. The number is decimal 23, or binary:

$$10111_2$$

There are two ways to represent this number as a negative number. The first method is to simply place a minus sign in front of the number, as we do with decimal numbers:

$$-(10111)_2$$

This method is suitable for us, but it is impossible for programmable controllers and computers to interpret, since the only symbols they use are binary 1s and 0s. To represent negative numbers, then, some digital computing devices use what is known as the **one's complement** method. First, the one's complement method places an extra bit (sign bit) in the most significant (left-most) position and lets this bit determine whether the number is positive or negative. The number is positive if the sign bit is 0 and negative if the sign bit is 1. Using the one's complement method, +23 decimal is represented in binary as shown here with the sign bit (0) indicated in bold:

$$\mathbf{0}\ 10111_2$$

The negative representation of binary 10111 is obtained by placing a 1 in the most significant bit position and inverting each bit in the number (changing 1s to 0s and 0s to 1s). So, the one's complement of binary 10111 is:

$$\mathbf{1}\ 01000_2$$

If a negative number is given in binary, its one's complement is obtained in the same fashion.

$$-15_{10} = \mathbf{1}\ 0000_2$$

$$+15_{10} = \mathbf{0}\ 1111_2$$

Two's Complement

The **two's complement** is similar to the one's complement in the sense that one extra digit is used to represent the sign. The two's complement computation, however, is slightly different. In the one's complement, all bits are inverted; but in the two's complement, each bit, from right to left, is inverted only after the first 1 is detected. Let's use the number +22 decimal as an example:

$$+22_{10} = \mathbf{0}\ 10110_2$$

Its two's complement would be:

$$-22_{10} = \mathbf{1}\ 01010_2$$

Note that in the negative representation of the number 22, starting from the right, the first digit is a 0, so it is not inverted; the second digit is a 1, so all digits after this one are inverted.

If a negative number is given in two's complement, its complement (a positive number) is found in the same fashion:

$$-14_{10} = \mathbf{1}\ 10010_2$$
$$+14_{10} = \mathbf{0}\ 01110_2$$

Again, all bits from right to left are inverted after the first 1 is detected. Other examples of the two's complement are shown here:

$$+17_{10} = \mathbf{0}\ 10001_2$$
$$-17_{10} = \mathbf{1}\ 01111_2$$

$$+7_{10} = \mathbf{0}\ 00111_2$$
$$-7_{10} = \mathbf{1}\ 11001_2$$

$$+1_{10} = \mathbf{0}\ 00001_2$$
$$-1_{10} = \mathbf{1}\ 11111_2$$

The two's complement of 0 does not really exist, since no first 1 is ever encountered in the number. The two's complement of 0, then, is 0.

The two's complement is the most common arithmetic method used in computers, as well as programmable controllers.

2-4 BINARY CODES

An important requirement of programmable controllers is communication with various external devices that either supply information to the controller or receive information from the controller. This input/output function involves the transmission, manipulation, and storage of binary data that, at some point, must be interpreted by humans. Although machines can easily handle this binary data, we require that the data be converted to a more interpretable form.

One way of satisfying this requirement is to assign a unique combination of 1s and 0s to each number, letter, or symbol that must be represented. This technique is called *binary coding*. In general, there are two categories of codes—those that represent numbers only and those that represent letters, symbols, and decimal numbers.

Several codes for representing numbers, symbols, and letters are standard throughout the industry. Among the most common are the following:

- ASCII

- BCD

- Gray

ASCII

Alphanumeric codes (which use a combination of letters, symbols, and decimal numbers) are used when information processing equipment, such as printers and cathode ray tubes (CRTs), must process the alphabet along with numbers and special symbols. These alphanumeric characters—26 letters (uppercase), 10 numerals (0-9), plus mathematical and punctuation symbols—can be represented using a 6-bit code (i.e., $2^6 = 64$ possible characters). The most common code for alphanumeric representation is **ASCII** (the American Standard Code for Information Interchange).

An ASCII (pronounced as-kee) code can be 6, 7, or 8 bits. Although a 6-bit code (64 possible characters) can accommodate the basic alphabet, numbers, and special symbols, standard ASCII character sets use a 7-bit code ($2^7 = 128$ possible characters), which provides room for lower case and control characters, in addition to the characters already mentioned. This 7-bit code provides all possible combinations of characters used when communicating with peripherals and interfaces.

An 8-bit ASCII code is used when parity check (see Chapter 4) is added to a standard 7-bit code for error-checking purposes (note that all eight bits can still fit in one byte). Figure 2-6a shows the binary ASCII code representation of the letter Z (132_8). This letter is generally sent and received in serial form between the PLC and other equipment.

Figure 2-6b illustrates a typical ASCII transmission, again using the character Z as an example. Note that extra bits have been added to the beginning and end of the character to signify the start and stop of the ASCII transmission. Appendix B shows a standard ASCII table, while Chapter 8 further explains serial communication.

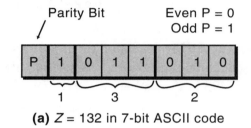

(a) Z = 132 in 7-bit ASCII code

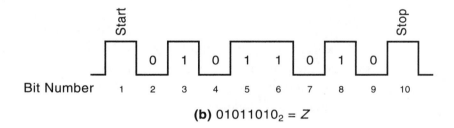

(b) $01011010_2 = Z$

Figure 2-6. (a) ASCII representation of the character Z and **(b)** the ASCII transmission of the character Z.

BCD

The **binary coded decimal (BCD)** system was introduced as a convenient way for humans to (1) handle numbers that must be input to digital machines and (2) interpret numbers that are output from machines. The best solution to this problem was to convert a code readily handled by man (decimal) to a code readily handled by processing equipment (binary). The result was BCD.

The decimal system uses the numbers 0 through 9 as its digits, whereas BCD represents each of these numbers as a 4-bit binary number. Table 2-4 illustrates the relationship between the BCD code and the binary and decimal number systems.

Decimal	Binary	BCD
0	0	0000
1	1	0001
2	10	0010
3	11	0011
4	100	0100
5	101	0101
6	110	0110
7	111	0111
8	1000	1000
9	1001	1001

Table 2-4. Decimal, binary, and BCD counting.

The BCD representation of a decimal number is obtained by replacing each decimal digit with its BCD equivalent. The BCD representation of decimal 7493 is shown here as an example:

$$\begin{array}{ccccc} \text{BCD} \rightarrow & 0111 & 0100 & 1001 & 0011 \\ \text{Decimal} \rightarrow & 7 & 4 & 9 & 3 \end{array}$$

Typical PLC applications of BCD codes include data entry (time, volume, weight, etc.) via thumbwheel switches (TWS), data display via seven-segment displays, input from absolute encoders, and analog input/output instructions. Figure 2-7 shows a thumbwheel switch and a seven-segment indicator field device.

(a)   (b)

Figure 2-7. (a) A seven-segment indicator field device and **(b)** a thumbwheel switch.

Nowadays, the circuitry necessary to convert from decimal to BCD and from BCD to seven-segment is already built into thumbwheel switches and seven-segment LED devices (see Figures 2-8a and 2-8b). This BCD data is converted internally by the PLC into the binary equivalent of the input data. Input and output of BCD data requires four lines of an input/output interface for each decimal digit.

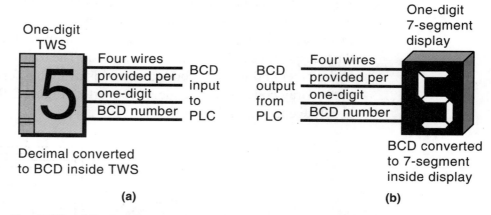

Figure 2-8. (a) Thumbwheel switch converts decimal numbers into BCD inputs for the PLC.
(b) The seven-segment display converts the BCD outputs from the PLC into a decimal number.

GRAY

The **Gray code** is one of a series of cyclic codes known as *reflected codes* and is suited primarily for position transducers. It is basically a binary code that has been modified in such a way that only one bit changes as the counting number increases. In standard binary, as many as four digits can change when counting with as few as four binary digits. This drastic change is seen in the transition from binary 7 to 8. Such a change allows a great chance for error, which is unsuitable for positioning applications. Thus, most encoders use the Gray code to determine angular position. Table 2-5 shows this code with its binary and decimal equivalents for comparison.

Gray Code	Binary	Decimal
0000	0	0
0001	1	1
0011	10	2
0010	11	3
0110	100	4
0111	101	5
0101	110	6
0100	111	7
1100	1000	8
1101	1001	9
1111	1010	10
1110	1011	11
1010	1100	12
1011	1101	13
1001	1110	14
1000	1111	15

Table 2-5. Gray code, binary, and decimal counting.

An example of a Gray code application is an optical absolute encoder. In this encoder, the rotor disk consists of opaque and transparent segments arranged in a Gray code pattern and illuminated by a light source that shines through the transparent sections of the rotating disk. The transmitted light is received at the other end in Gray code form and is available for input to the PLC in either Gray code or BCD code, if converted. Figure 2-9 illustrates a typical absolute encoder and its output.

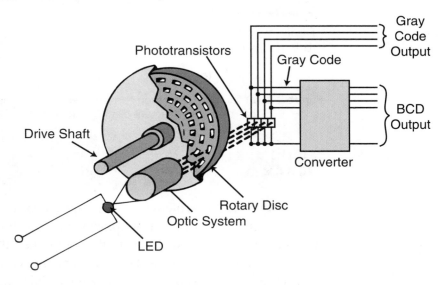

Figure 2-9. An absolute encoder with BCD and Gray outputs.

2-5 REGISTER WORD FORMATS

As previously mentioned, a programmable controller performs all of its internal operations in binary format using 1s and 0s. In addition, the status of I/O field devices is also read and written, in binary form, to and from the PLC's CPU. Generally, these operations are performed using a group of 16 bits that represent numbers and codes. Recall that the grouping of bits with which a particular machine operates is called a word. A PLC word is also called a **register** or *location*. Figure 2-10 illustrates a 16-bit register composed of a two-byte word.

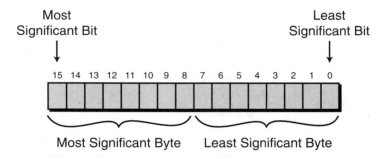

Figure 2-10. A 16-bit register/word.

Although the data stored in a register is represented by binary 1s and 0s, the format in which this binary data is stored may differ from one programmable controller to another. Generally, data is represented in either straight (noncoded) binary or binary coded decimal (BCD) format. Let's examine these two formats.

BINARY FORMAT

Data stored in binary format can be directly converted to its decimal equivalent without any special restrictions. In this format, a 16-bit register can represent a maximum value of 65535_{10}. Figure 2-11 shows the value 65535_{10} in binary format (all bits are 1). The binary format represents the status of a device as either 0 or 1, which is interpreted by the programmable controller as ON or OFF. All of these statuses are stored in registers or words.

Figure 2-11. A 16-bit register containing the binary equivalent of 65535_{10}.

If the most significant bit of the register in Figure 2-12 is used as a sign bit, then the maximum decimal value that the 16-bit register can store is $+32767_{10}$ or -32767_{10}.

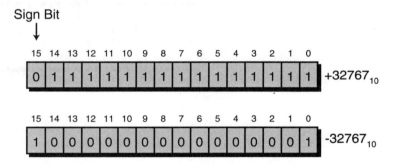

Figure 2-12. Two 16-bit registers with sign bits (MSB).

The decimal equivalents of these binary representations can be calculated using the sum-of-the-weights method. The negative representation of 32767_{10}, as shown in Figure 2-12, was derived using the two's complement method. As an exercise, practice computing these numbers (refer to Section 2-3 for help).

BCD FORMAT

The BCD format uses four bits to represent a single decimal digit. The only decimal numbers that these four bits can represent are 0 through 9. Some PLCs operate and store data in several of their software instructions, such as arithmetic and data manipulations, using the BCD format.

In BCD format, a 16-bit register can hold up to a 4-digit decimal value, with the decimal values that can be represented ranging from 0000–9999. Figure 2-13 shows a register containing the binary representation of BCD 9999.

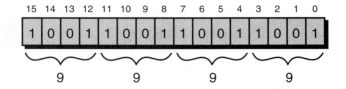

Figure 2-13. Register containing BCD 9999.

In a PLC, the BCD values stored in a register or word can be the result of BCD data input from a thumbwheel switch. A 4-digit thumbwheel switch will use a 16-bit register to store the BCD output data obtained during the read section of the scan (see Figure 2-14).

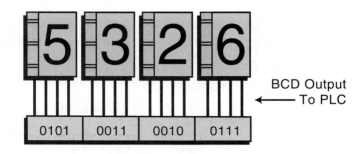

Figure 2-14. A 4-digit TWS using a 16-bit register to store BCD values.

EXAMPLE 2-1

Illustrate how a PLC's 16-bit register containing the BCD number 7815 would connect to a 4-digit, seven-segment display. Indicate the most significant digit and the least significant digit of the seven-segment display.

SOLUTION

Figure 2-15 illustrates the connection between a 16-bit register and a 4-digit, seven-segment display. The BCD output from the PLC register or word is sent to the seven-segment indicator through an output interface during the write, or update, section of the scan.

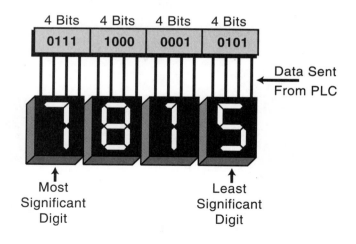

Figure 2-15. A 16-bit PLC register holding the BCD number 7815.

KEY TERMS

alphanumeric code
ASCII
base
binary coded decimal (BCD)
bit
byte
decimal number system
Gray code
hexadecimal number system
least significant bit (LSB)
least significant digit
most significant bit (MSB)
most significant digit
nibble
octal number system
one's complement
register
sum-of-the-weights method
two's complement
weighted value
word

CHAPTER THREE

LOGIC CONCEPTS

Science when well digested is nothing but good sense and reason.

—Leszinski Stanislas

CHAPTER HIGHLIGHTS   To understand programmable controllers and their applications, you must first understand the logic concepts behind them. In this chapter, we will discuss three basic logic functions—AND, OR, and NOT—and show you how, with just these three functions, you can make control decisions ranging from very simple to very complex. We will also introduce you to the fundamentals of Boolean algebra and its associated operators. Finally, we will explain the relationship between Boolean algebra and logic contact symbology, so that you will be ready to learn about PLC processors and their programming devices.

3-1   THE BINARY CONCEPT

The binary concept is not a new idea; in fact, it is a very old one. It simply refers to the idea that many things exist only in two predetermined states. For instance, a light can be on or off, a switch open or closed, or a motor running or stopped. In digital systems, these two-state conditions can be thought of as signals that are present or not present, activated or not activated, high or low, on or off, etc. This two-state concept can be the basis for making decisions; and since it is very adaptable to the binary number system, it is a fundamental building block for programmable controllers and digital computers.

Here, and throughout this book, binary 1 represents the presence of a signal (or the occurrence of some event), while binary 0 represents the absence of the signal (or the nonoccurrence of the event). In digital systems, these two states are actually represented by two distinct voltage levels, +V and 0V, as shown in Table 3-1. One voltage is more positive (or at a higher reference) than the other. Often, binary 1 (or logic 1) is referred to as TRUE, ON, or HIGH, while binary 0 (or logic 0) is referred to as FALSE, OFF, or LOW.

1 (+V)	0 (0V)	Example
Operating	Not operating	Limit switch
Ringing	Not ringing	Bell
On	Off	Light bulb
Blowing	Silent	Horn
Running	Stopped	Motor
Engaged	Disengaged	Clutch
Closed	Open	Valve

Table 3-1. Binary concept using positive logic.

Note that in Table 3-1, the more positive voltage (represented as logic 1) and the less positive voltage (represented as logic 0) were arbitrarily chosen. The use of binary logic to represent the more positive voltage level, meaning the occurrence of some event, as 1 is referred to as **positive logic**.

Negative logic, as illustrated in Table 3-2, uses 0 to represent the more positive voltage level, or the occurrence of the event. Consequently, 1 represents the nonoccurrence of the event, or the less positive voltage level. Although positive logic is the more conventional of the two, negative logic is sometimes more convenient in an application.

1 (+V)	0 (0V)	Example
Not operating	Operating	Limit switch
Not ringing	Ringing	Bell
Off	On	Light bulb
Silent	Blowing	Horn
Stopped	Running	Motor
Disengaged	Engaged	Clutch
Open	Closed	Valve

Table 3-2. Binary concept using negative logic.

3-2 LOGIC FUNCTIONS

The binary concept shows how physical quantities (binary variables) that can exist in one of two states can be represented as 1 or 0. Now, you will see how statements that combine two or more of these binary variables can result in either a TRUE or FALSE condition, represented by 1 and 0, respectively. Programmable controllers make decisions based on the results of these kinds of logical statements.

Operations performed by digital equipment, such as programmable controllers, are based on three fundamental logic functions—AND, OR, and NOT. These functions combine binary variables to form statements. Each function has a rule that determines the statement outcome (TRUE or FALSE) and a symbol that represents it. For the purpose of this discussion, the result of a statement is called an output (Y), and the conditions of the statement are called inputs (A and B). Both the inputs and outputs represent two-state variables, such as those discussed earlier in this section.

THE AND FUNCTION

Figure 3-1 shows a symbol called an AND **gate**, which is used to graphically represent the **AND** function. The AND output is TRUE (1) only if all inputs are TRUE (1).

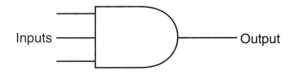

Figure 3-1. Symbol for the AND function.

An AND function can have an unlimited number of inputs, but it can have only one output. Figure 3-2 shows a two-input AND gate and its resulting output *Y*, based on all possible input combinations. The letters *A* and *B* represent inputs to the controller. This mapping of outputs according to predefined inputs is called a **truth table**. Example 3-1 shows an application of the AND function.

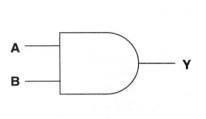

AND Truth Table		
Inputs		Output
A	B	Y
0	0	0
0	1	0
1	0	0
1	1	1

Figure 3-2. Two-input AND gate and its truth table.

EXAMPLE 3-1

Show the logic gate, truth table, and circuit representations for an alarm horn that will sound if its two inputs, push buttons PB1 and PB2, are 1 (ON or depressed) at the same time.

SOLUTION

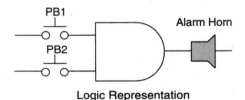

Logic Representation

PB1	PB2	Alarm Horn
Not pushed (0)	Not pushed (0)	Silent (0)
Not pushed (0)	Pushed (1)	Silent (0)
Pushed (1)	Not pushed (0)	Silent (0)
Pushed (1)	Pushed (1)	Sounding (1)

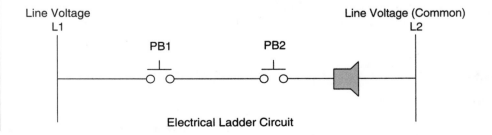

Electrical Ladder Circuit

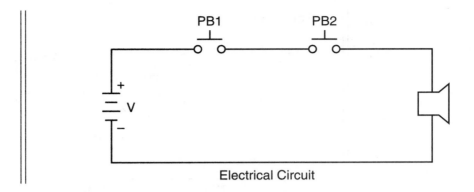

Electrical Circuit

THE OR FUNCTION

Figure 3-3 shows the OR gate symbol used to graphically represent the **OR** function. The OR output is TRUE (1) if one or more inputs are TRUE (1).

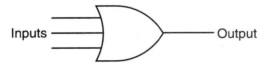

Figure 3-3. Symbol for the OR function.

As with the AND function, an OR gate function can have an unlimited number of inputs but only one output. Figure 3-4 shows an OR function truth table and the resulting output Y, based on all possible input combinations. Example 3-2 shows an application of the OR function.

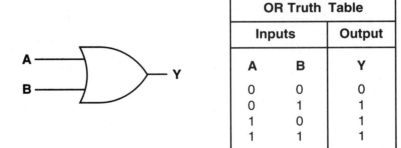

OR Truth Table		
Inputs		Output
A	B	Y
0	0	0
0	1	1
1	0	1
1	1	1

Figure 3-4. Two-input OR gate and its truth table.

EXAMPLE 3-2

Show the logic gate, truth table, and circuit representations for an alarm horn that will sound if either of its inputs, push button PB1 or PB2, is 1 (ON or depressed).

SOLUTION

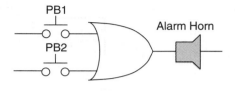

Logic Representation

PB1	PB2	Alarm Horn
Not pushed (0)	Not pushed (0)	Silent (0)
Not pushed (0)	Pushed (1)	Sounding (1)
Pushed (1)	Not pushed (0)	Sounding (1)
Pushed (1)	Pushed (1)	Sounding (1)

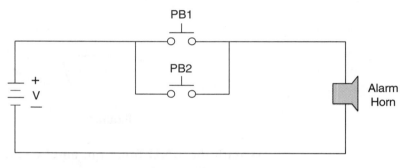

Electrical Circuit

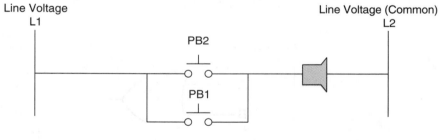

Electrical Ladder Circuit

THE NOT FUNCTION

Figure 3-5 illustrates the NOT symbol, which is used to graphically represent the **NOT** function. The NOT output is TRUE (1) if the input is FALSE (0). Conversely, if the output is FALSE (0), the input is TRUE (1). The result of the NOT operation is always the inverse of the input; therefore, it is sometimes called an *inverter*.

The NOT function, unlike the AND and OR functions, can have only one input. It is seldom used alone, but rather in conjunction with an AND or an OR gate. Figure 3-6 shows the NOT operation and its truth table. Note that an *A* with a bar on top represents NOT *A*.

Figure 3-5. Symbol for the NOT function.

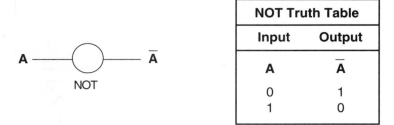

NOT Truth Table	
Input	Output
A	$\overline{A}$
0	1
1	0

Figure 3-6. NOT gate and its truth table.

At first glance, it is not as easy to visualize the application of the NOT function as it is the AND and OR functions. However, a closer examination of the NOT function shows it to be simple and quite useful. At this point, it is helpful to recall three points that we have discussed:

1. Assigning a 1 or 0 to a condition is arbitrary.

2. A 1 is normally associated with TRUE, HIGH, ON, etc.

3. A 0 is normally associated with FALSE, LOW, OFF, etc.

Examining statements 2 and 3 shows that logic 1 is normally expected to activate some device (e.g., if $Y = 1$, then motor runs), and logic 0 is normally expected to deactivate some device (e.g., if $Y = 0$, then motor stops). If these conventions were reversed, such that logic 0 was expected to activate some device (e.g., if $Y = 0$, then motor runs) and logic 1 was expected to deactivate some device (e.g., $Y = 1$, then motor stops), the NOT function would then have a useful application.

1. A NOT is used when a 0 (LOW condition) must activate some device.

2. A NOT is used when a 1 (HIGH condition) must deactivate some device.

The following two examples show applications of the NOT function. Although the NOT function is normally used in conjunction with the AND and OR functions, the first example shows the NOT function used alone.

EXAMPLE 3-3

Show the logic gate, truth table, and circuit representation for a solenoid valve (V1) that will be open (ON) if selector switch S1 is ON and if level switch L1 is NOT ON (liquid has not reached level).

SOLUTION

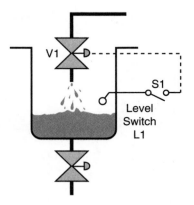

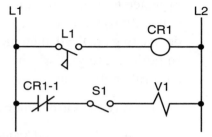

Logic Representation

S1	L1 ($\overline{L1}$)		V1
0	0	1	0
0	1	0	0
1	0	1	1
1	1	0	0

Truth Table

Electrical Ladder Circuit

Note: In this example, the level switch L1 is normally open, but it closes when the liquid level reaches L1. The ladder circuit requires an auxiliary control relay (CR1) to implement the not normally open L1 signal. When L1 closes (ON), CR1 is energized, thus opening the normally closed CR1-1 contacts and deactivating V1. S1 is ON when the system operation is enabled.

EXAMPLE 3-4

Show the logic gate, truth table, and circuit representation for an alarm horn that will sound if push button PB1 is 1 (ON or depressed) and PB2 is NOT 0 (not depressed).

SOLUTION

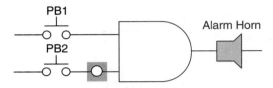

Logic Representation

PB1	PB2	Alarm Horn
Not pushed (0)	Not pushed (0)	Silent (0)
Not pushed (0)	Pushed (1)	Silent (0)
Pushed (1)	Not pushed (0)	Sounding (1)
Pushed (1)	Pushed (1)	Silent (0)

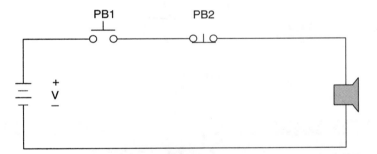

Electrical Circuit

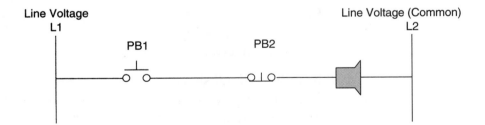

Electrical Ladder Circuit

Note: In this example, the physical representation of a field device element that signifies the NOT function is represented as a normally closed, or not normally open, switch (PB2). In the logical representation section of this example, the push button switch is represented as NOT open by the ─O─ symbol.

The two previous examples showed the NOT symbol placed at inputs to a gate. A NOT symbol placed at the output of an AND gate will negate, or invert, the normal output result. A negated AND gate is called a **NAND** gate. Figure 3-7 shows its logic symbol and truth table.

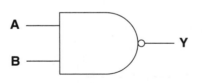

NAND Truth Table		
Inputs		Output
A	B	Y
0	0	1
0	1	1
1	0	1
1	1	0

Figure 3-7. Two-input NAND gate and its truth table.

The same principle applies if a NOT symbol is placed at the output of an OR gate. The normal output is negated, and the function is referred to as a **NOR** gate. Figure 3-8 shows its symbol and truth table.

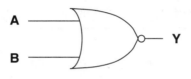

NOR Truth Table		
Inputs		Output
A	B	Y
0	0	1
0	1	0
1	0	0
1	1	0

Figure 3-8. Two-input NOR gate and its truth table.

3-3 PRINCIPLES OF BOOLEAN ALGEBRA AND LOGIC

An in-depth discussion of Boolean algebra is not required for the purposes of this book and is beyond the book's scope. However, an understanding of the Boolean techniques for writing shorthand expressions for complex logical statements can be useful when creating a control program of Boolean statements or conventional ladder diagrams.

In 1849, an Englishman named George Boole developed Boolean algebra. The purpose of this algebra was to aid in the logic of reasoning, an ancient form of philosophy. It provided a simple way of writing complicated combinations of "logical statements," defined as statements that can be either true or false.

When digital logic was developed in the 1960s, Boolean algebra proved to be a simple way to analyze and express digital logic statements, since all digital systems use a TRUE/FALSE, or two-valued, logic concept. Because of this relationship between digital logic and Boolean logic, you will occasionally hear logic gates referred to as Boolean gates, several interconnected gates called a Boolean network, or even a PLC language called a Boolean language.

Figure 3-9 summarizes the basic **Boolean operators** as they relate to the basic digital logic functions AND, OR, and NOT. These operators use capital letters to represent the wire label of an input signal, a multiplication sign (•) to represent the AND operation, and an addition sign (+) to represent the OR operation. A bar over a letter represents the NOT operation.

Logical Symbol	Logical Statement	Boolean Equation
A ──⊐ Y B ──	*Y* is 1 if *A* AND *B* are 1	$Y = A \cdot B$ or $Y = AB$
A ──⊐ Y B ──	*Y* is 1 if *A* OR *B* is 1	$Y = A + B$
A ──◯── Y	*Y* is 1 if *A* is 0 *Y* is 0 if *A* is 1	$Y = \overline{A}$

Figure 3-9. Boolean algebra as related to the AND, OR, and NOT functions.

In Figure 3-9, the AND gate has two input signals (*A* and *B*) and one output signal (*Y*). The output can be expressed by the logical statement:

Y is 1 if *A* AND *B* are 1.

The corresponding Boolean expression is:

$$Y = A \cdot B$$

which is read *Y equals A ANDed with B*. The Boolean symbol • for AND could be removed and the expression written as *Y = AB*. Similarly, if *Y* is the result of ORing *A* and *B*, the Boolean expression is:

$$Y = A + B$$

which is read *Y equals A ORed with B*. In the NOT operation, where *Y* is the inverse of *A*, the Boolean expression is:

$$Y = \overline{A}$$

which is read *Y equals NOT A*. Table 3-3 illustrates the basic Boolean operations of ANDing, ORing, and inversion. The table also illustrates how these functions can be combined to obtain any desired logic combination.

1. Basic Gates. Basic logic gates implement simple logic functions. Each logic function is expressed in terms of a truth table and its Boolean expression.

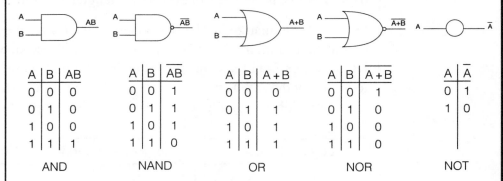

A	B	AB
0	0	0
0	1	0
1	0	0
1	1	1

AND

A	B	$\overline{AB}$
0	0	1
0	1	1
1	0	1
1	1	0

NAND

A	B	A+B
0	0	0
0	1	1
1	0	1
1	1	1

OR

A	B	$\overline{A+B}$
0	0	1
0	1	0
1	0	0
1	1	0

NOR

A	$\overline{A}$
0	1
1	0

NOT

2. Combined Gates. Any combination of control functions can be expressed in Boolean terms using three simple operators: (•), (+), and (⁻).

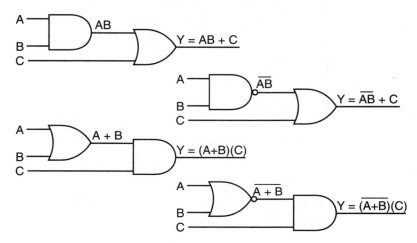

3. Boolean Algebra Rules. Control logic functions can vary from simple to very complex combinations of input variables. However simple or complex the functions may be, they satisfy the following rules. These rules are a result of a simple combination of basic truth tables and may be used to simplify logic circuits.

Commutative Laws

$A + B = B + A$

$AB = BA$

Associative Laws

$A + (B + C) = (A + B) + C$

$A(BC) = (AB)C$

De Morgan's Laws

$\overline{(A + B)} = \overline{A}\,\overline{B}$

$\overline{(AB)} = \overline{A} + \overline{B}$

$\overline{\overline{A}} = A, \ \overline{1} = 0, \ \overline{0} = 1$

$A + \overline{A}B = A + B$

$AB + AC + B\overline{C} = AC + B\overline{C}$

Distributive Laws

$A(B + C) = AB + AC$

$A + BC = (A + B)(A + C)$

Law of Absorption

$A(A + B) = A + AB = A$

Table 3-3. Logic operations using Boolean algebra.

4. Order of Operation and Grouping Signs. The order in which Boolean operations (AND, OR, NOT) are performed is important. This order will affect the resulting logic value of the expression. Consider the three input signals *A*, *B*, and *C*. Combining them in the expression $Y = A + B \cdot C$ can result in misoperation of the output device *Y*, depending on the order in which the operations are performed. Performing the OR operation prior to the AND operation is written $(A + B) \cdot C$, and performing the AND operation prior to the OR is written $A + (B \cdot C)$. The result of these two expressions is not the same.

The order of priority in Boolean expression is NOT (inversion) first, AND second, and OR last, unless otherwise indicated by grouping signs, such as parentheses, brackets, braces, or the vinculum. According to these rules, the previous expression $A + B \cdot C$, without any grouping signs, will always be evaluated only as $A + (B \cdot C)$. With the parentheses, it is obvious that *B* is ANDed with *C* prior to ORing the result with *A*. Knowing the order of evaluation, then, makes it possible to write the expression simply as $A + BC$, without fear of misoperation. As a matter of convention, the AND operator is usually omitted in Boolean expressions.

When working with Boolean logic expressions, misuse of grouping signs is a common occurrence. However, if the signs occur in pairs, they generally do not cause problems if they have been properly placed according to the desired logic. Enclosing two variables that are to be ANDed within parentheses is not necessary since the AND operator would normally be performed first. If two input signals are to be ORed prior to ANDing, they must be placed within parentheses.

To ensure proper order of evaluation of an expression, use parentheses as grouping signs. If additional signs are required brackets [ ], and then braces { } are used. An illustration of the use of grouping signs is shown below:

$$Y1 = Y2 + Y5\,[X1(X2 + X3)] + \{Y3[Y4(X5 + X6)]\}$$

5. Application of De Morgan's Laws. De Morgan's Laws are frequently used to simplify inverted logic expressions or to simply convert an expression into a usable form.

According to De Morgan's Laws:

$$\overline{AB} = \overline{A} + \overline{B}$$

$$\text{and } \overline{A + B} = \overline{A}\,\overline{B}$$

Table 3-3 continued.

3-4 PLC CIRCUITS AND LOGIC CONTACT SYMBOLOGY

Hardwired logic refers to logic control functions (timing, sequencing, and control) that are determined by the way devices are interconnected. In contrast to PLCs, in which logic functions are programmable and easily changed, hardwired logic is fixed and can be changed only by altering the way devices are physically connected or interwired. A prime function of a PLC is to replace existing hardwired control logic and to implement control functions for new systems. Figure 3-10a shows a typical hardwired relay logic circuit, and Figure 3-10b shows its PLC ladder diagram implementation. The important point about Figure 3-10 is not to understand the process of changing from one circuit to another, but to see the similarities in the representations. The ladder circuit connections of the hardwired relay circuit are implemented in the PLC via software instructions, thus all of the wiring can be thought of as being inside the CPU (*softwired* as opposed to hardwired).

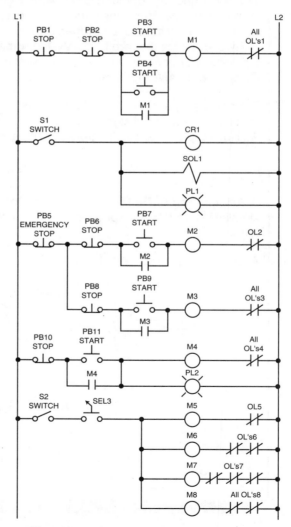

Figure 3-10a. Hardwired relay logic circuit.

The logic implemented in PLCs is based on the three basic logic functions (AND, OR, and NOT) that we discussed in the previous sections. These functions are used either alone or in combination to form instructions that will determine if a device is to be switched on or off. How these instructions are implemented to convey commands to the PLC is called the **language**. The most widely used languages for implementing on/off control and sequencing are ladder diagrams and Boolean mnemonics, among others. Chapter 9 discusses these languages at length.

The most conventional of the control languages is ladder diagram. Ladder diagrams are also called **contact symbology**, since their instructions are relay-equivalent contact symbols (i.e., normally open and normally closed contacts and coils).

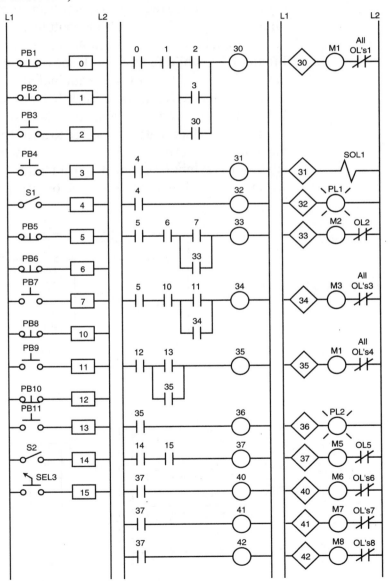

Figure 3-10b. PLC ladder diagram implementation of Figure 3-10a.

Contact symbology is a very simple way of expressing control logic in terms of symbols that are used on relay control schematics. If the controller language is ladder diagram, the translation from existing relay logic to programmed logic is a one-step translation to contact symbology. If the language is Boolean mnemonics, conversion to contact symbology is not required, yet is still useful and quite often done to provide an easily under-stood piece of documentation. Table 3-6a, shown later, provides examples of simple translations from hardwired logic to programmed logic. Chapter 11 thoroughly explains these translations.

The complete ladder circuit, in Figure 3-10, shown earlier, can be thought of as being formed by individual circuits, each circuit having one output. Each of these circuits is known as a **rung** (or network); therefore, a rung is the contact symbology required to control an output in the PLC. Some controllers allow a rung to have multiple outputs, but one output per rung is the convention. Figure 3-11a illustrates the top rung of the hardwired circuit from Figure 3-10, while Figure 3-11b shows the top rung of the equivalent PLC circuit. Note that the PLC diagram includes all of the field input and output devices connected to the interfaces that are used in the rung. A complete PLC ladder diagram program, then, consists of several rungs. Each rung controls an output interface that is connected to an **output device**, a piece of equipment that receives information from the PLC. Each rung is a combination of input conditions (symbols) connected from left to right between two vertical lines, with the symbol that represents the output at the far right.

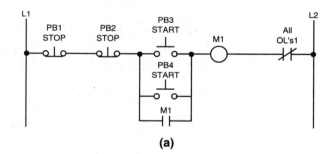

(a)

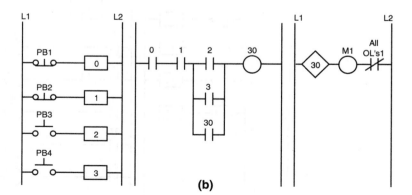

(b)

Figure 3-11. (a) Top rung of the hardwired circuit from Figure 3-10 and **(b)** its equivalent PLC circuit.

The symbols that represent the inputs are connected in series, parallel, or some combination to obtain the desired logic. These input symbols represent the **input devices** that are connected to the PLC's input interfaces. The input devices supply the PLC with field data. When completed, a ladder diagram control program consists of several rungs, with each rung controlling a specific output.

The programmed rung concept is a direct carryover from the hardwired relay ladder rung, in which input devices are connected in series and parallel to control various outputs. When activated, these input devices either allow current to flow through the circuit or cause a break in current flow, thereby switching the output devices ON or OFF. The input symbols on a ladder rung can represent signals generated by connected input devices, connected output devices, or outputs internal to the controller (see Table 3-4).

Input Devices	Output Devices
Push button	Pilot light
Selector switch	Solenoid valve
Limit switch	Horn
Proximity switch	Control relay
Timer contact	Timer

Table 3-4. ON/OFF input and output devices.

ADDRESSES USED IN PLCS

Each symbol on a rung will have a *reference number*, which is the address in memory where the current status (1 or 0) for the referenced input is stored. When a field signal is connected to an input or an output interface, its address will be related to the terminal where the signal wire is connected. The address for a given input/output can be used throughout the program as many times as required by the control logic. This PLC feature is an advantage when compared to relay-type hardware, where additional contacts often mean additional hardware. Sections 5-4 and 6-2 describe more about I/O interaction and its relationship with the PLC's memory and enclosure placement.

Figure 3-12 illustrates a simple electrical ladder circuit and its equivalent PLC implementation. Each "real" field device (e.g., push buttons PB1 and PB2, limit switch LS1, and pilot light PL1) is connected to the PLC's input and output modules (see Figure 3-13), which have a reference number—the address. Most controllers reference these devices using numeric addresses with octal (base 8) or decimal (base 10) numbering. Note that in the electrical ladder circuit, any complete electrical path (all contacts closed) from left to right will energize the output (pilot light PL1). To turn PL1 ON, then, one of the following two conditions must occur: (1) PB1 must be pressed and LS1 must be closed or (2) PB2 must be pressed and LS1 must be closed. Either of these two conditions will complete the electrical path and cause power to flow to the pilot light.

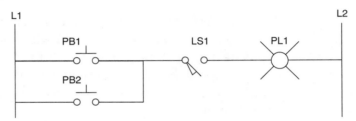

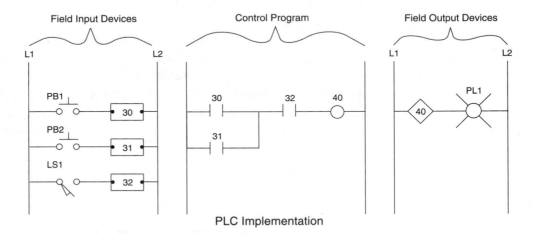

Figure 3-12. Electrical ladder circuit and its equivalent PLC implementation.

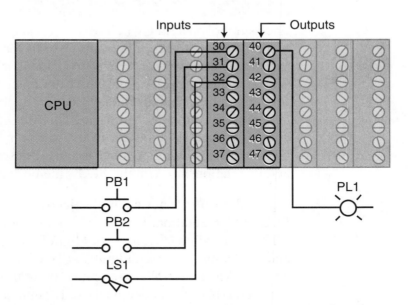

Figure 3-13. Field devices from Figure 3-12 connected to I/O module.

The same logic that applies to an electrical ladder circuit applies to a PLC circuit. In the PLC control program, power must flow through either addresses 30 (PB1) and 32 (LS1) or through addresses 31 (PB2) and 32 (LS1) to turn ON output 40. Output 40, in turn, energizes the light PL1 that is connected to the interface with address 40. In order to provide power to addresses 30, 31, or 32, the devices connected to the input interfaces addressed 30, 31, and 32 must be turned ON. That is, the push buttons must be pressed or the limit switch must close.

CONTACT SYMBOLS USED IN PLCS

Programmable controller contacts and electromechanical relay contacts operate in a very similar fashion. For example, let's take relay A (see Figure 3-14a) which has two sets of contacts, one **normally open** contact (A-1) and one **normally closed** contact (A-2). If relay coil A is not energized (i.e., it is OFF), contact A-1 will remain open and contact A-2 will remain closed (see Figure 3-14b). Conversely, if coil A is energized, or turned ON, contact A-1 will close and contact A-2 will open (see Figure 3-14c). The blue lines highlighting the coil and contacts denote an ON, or closed, condition.

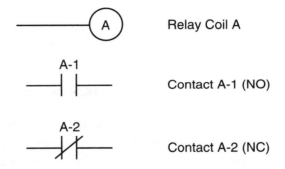

(a) Standard configuration for relay coil A with normally open contact A-1 and normally closed contact A-2.

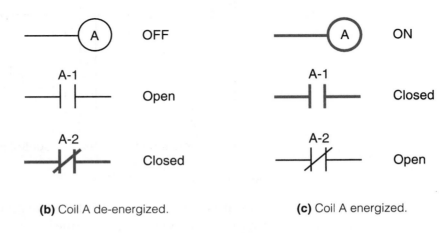

(b) Coil A de-energized.

(c) Coil A energized.

Figure 3-14. Relay and PLC contact symbols showing a relay coil and normally open and normally closed contacts.

Remember that when a set of contacts closes, it provides power flow, or continuity, in the circuit where it is used. Each set of available coils and its respective contacts in the PLC have a unique reference address by which they are identified. For instance, coil 10 will have normally open and normally closed contacts with the same address (10) as the coil (see Figure 3-15). Note that a PLC can have as many normally open and normally closed contacts as desired; whereas in an electromechanical relay, only a fixed number of contacts are available.

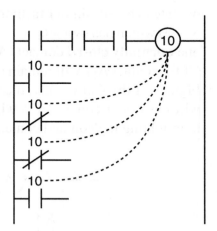

Figure 3-15. Multiple contacts from a PLC output coil.

A programmable controller also allows the multiple use of an input device reference. Figure 3-16 illustrates an example in which limit switch LS1 is connected to reference input module connection 20. Note that the PLC control program can have as many normally open and normally closed reference 20 contacts in as many rungs as needed.

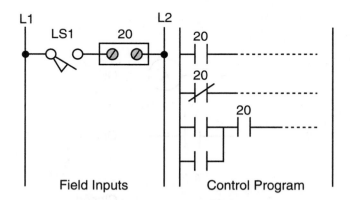

Figure 3-16. Input 20 has multiple contacts in the PLC control program.

The symbols in Table 3-5 are used to translate relay control logic to contact symbolic logic. These symbols are also the basic instruction set for the ladder diagram, excluding timer/counter instructions. Chapter 9 further explains these and more advanced instructions.

Symbol	Definition and Symbol Interpretation
⊣ ⊢	**Normally open contact.** Represents any input to the control logic. An input can be a connected switch closure or sensor, a contact from a connected output, or a contact from an internal output. When interpreted, the referenced input or output is examined for an ON condition. If its status is 1, the contact will close and allow current to flow through the contact. If the status of the referenced input/output is 0, the contact will remain open, prohibiting current from flowing through the contact.
⊣/⊢	**Normally closed contact.** Represents any input to the control logic. An input can be a connected switch closure or sensor, a contact from a connected output, or a contact from an internal output. When interpreted, the referenced input/output is examined for an OFF condition. If its status is 0, the contact will remain closed, thus allowing current to flow through the contact. If the status of the referenced input/output is 1, the contact will open, prohibiting current from flowing through the contact.
—◯—	**Output.** Represents any output that is driven by some combination of input logic. An output can be a connected device or an internal output. If any left-to-right path of input conditions is TRUE (all contacts closed), the referenced output is energized (turned ON).
—⊘—	**NOT output.** Represents any output that is driven by some combination of input logic. An output can be a connected device or an internal output. If any left-to-right path of input conditions is TRUE (all contacts closed), the referenced output is de-energized (turned OFF).

Table 3-5. Symbols used to translate relay control logic to contact symbolic logic.

The following seven points describe guidelines for translating from hardwired logic to programmed logic using PLC contact symbols:

- **Normally open contact.** When evaluated by the program, this symbol is examined for a 1 to close the contact; therefore, the signal referenced by the symbol must be ON, CLOSED, activated, etc.

- **Normally closed contact.** When evaluated by the program, this symbol is examined for a 0 to keep the contact closed; thus, the signal referenced by the symbol must be OFF, OPEN, deactivated, etc.

- **Output.** An output on a given rung will be energized if any left-to-right path has all contacts closed, with the exception of power flow going in reverse before continuing to the right. An output can control either a connected device (if the reference address is also a termination point) or an **internal output** used exclusively within the program. An internal output does not control a field device. Rather, it provides interlocking functions within the PLC.

- **Input.** This contact symbol can represent input signals sent from connected inputs, contacts from internal outputs, or contacts from connected outputs.

- **Contact addresses.** Each program symbol is referenced by an address. If the symbol references a connected input/output device, then the address is determined by the point where the device is connected.

- **Repeated use of contacts.** A given input, output, or internal output can be used throughout the program as many times as required.

- **Logic format.** Contacts can be programmed in series or in parallel, depending on the output control logic required. The number of series contacts or parallel branches allowed in a rung depends on the PLC.

Table 3-6a show how simple hardwired series and parallel circuits can be translated into programmed logic. A **series circuit** is equivalent to the Boolean AND operation; therefore, all inputs must be ON to activate the output. A **parallel circuit** is equivalent to the Boolean OR operation; therefore, any one of the inputs must be ON to activate the output. The STR and OUT Boolean statements stand for START (of a new rung) and OUTPUT (of a rung), respectively. Table 3-6b further explains Table 3-6a.

**KEY
TERMS**

**AND
Boolean operators
contact symbology
gate
input device
internal output
language
NAND
negative logic
NOR
normally closed
normally open
NOT
OR
output device
parallel circuit
positive logic
rung
series circuit
truth table**

(b)

(a) Series Circuit. In this circuit, if both switches LS1 and LS2 are closed, the solenoid SOL1 will energize.

(b) Parallel Circuit. In this circuit, if either of the two switches LS3 or LS4 closes, the solenoid SOL2 will energize.

(c) & (d) Series/Parallel Circuits. In a series/parallel circuit, the result of ORing two or more inputs is ANDed with one or more series or parallel inputs. In both of these examples, all of the relay circuit elements are normally open and must be closed to activate the pilot lights. Normally open contacts are used in the program.

(e) & (f) Parallel/Series Circuits. In a parallel/series circuit, the result of ANDing two or more inputs is ORed with one or more series inputs. In both of these examples, all of the relay circuit elements are normally open and must be closed to activate the output device. Normally open contacts are used in the program.

(f) Internal Outputs. Circuit (f) controls an electromechanical control relay. Control relays do not normally drive output devices, but rather drive other relays. They are used to provide additional contacts for interlocking logic. The internal output provides the same function in software; however, the number of contacts are unlimited and can be either normally open or normally closed.

(g) Normally Open Contacts. In the series circuit (g), the solenoid will energize if LS14 closes and CR1-1 is energized. CR1-1 is a contact from the control relay CR1 in circuit (f) and closes whenever CR1 is energized. In the program, CR1 was replaced by the internal output C1; therefore, the program uses a normally open contact from the internal output C1. SOL3 will energize when LS14 closes and C1 is energized.

(h) Normally Closed Contacts. In circuit (h), the solenoid will energize if LS14 closes and CR1-1 is not energized. The program uses a normally closed contact from the internal output C1. SOL3 will stay energized as long as the limit switch is closed and C1 is not energized.

Table 3-6. (a) Hardwired relay logic translated into PLC logic using contact symbols and **(b)** an explanation of the translation.

(a)

Relay Ladder Diagram	Contact or Ladder Diagram	Boolean Equation	Boolean Statements
(a) Series Circuit LS1 LS2 SOL1	X1 X2 Y1	$Y1 = X1 \cdot X2$	STR X1 AND X2 OUT Y1
(b) Parallel Circuit LS3 SOL2 / LS4	X3 X4 Y2	$Y2 = X3 + X4$	STR X3 OR X4 OUT Y2
(c) Series/Parallel Circuit LS5 CR1 PL1 (G) / LS6	X5 X6 C1 Y3	$Y3 = (X5 + X6) \cdot C1$	STR X5 OR X6 AND C1 OUT Y3
(d) Series/Parallel Circuit LS7 CR2 PL1 (G) / LS10 CR3	X7 X10 C2 C3 Y4	$Y4 = (X7 + X10) \cdot (C2 + C3)$	STR X7 OR X10 AND C2 OR C3 OUT Y4
(e) LS11 LS12 AL1 / LS13	X11 X12 X13 Y5	$Y5 = (X11 \cdot X12) + X13$	STR X11 AND X12 OR X13 OUT Y5
(f) Parallel/Series Circuit LS14 LS15 CR1 / LS16 LS17	X14 X15 X16 X17 Y6	$Y6 = (X14 \cdot X15) + (X16 \cdot X17)$	STR X14 AND X15 OR X16 AND X17 OUT Y6
(g) Series Circuit LS14 CR1-1 SOL3	X14 C1 Y7	$Y7 = X14 \cdot C1$	STR X14 AND C1 OUT Y7
(h) Series Circuit LS14 CR1-1 SOL4	X14 C1 Y10	$Y10 = X14 \cdot \overline{C1}$	STR X14 AND NOT C1 OUT Y10

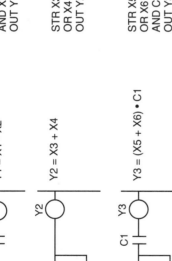
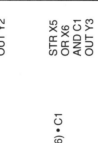

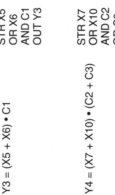

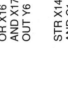

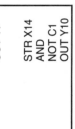

SECTION TWO

COMPONENTS AND SYSTEMS

- Processors, the Power Supply, and Programming Devices
- The Memory System and I/O Interaction
- The Discrete Input/Output System
- The Analog Input/Output System
- Special Function I/O and Serial Communication Interfacing

PROCESSORS, THE POWER SUPPLY, AND PROGRAMMING DEVICES

Unity makes strength, and since we must be strong, we must also be one.

—Grand Duke Friedrich von Baden

CHAPTER HIGHLIGHTS

The processor and the power supply are important parts of the central processing unit. In this chapter, we will take a look at these CPU components, concentrating on their roles and requirements in PLC applications. In addition, we will discuss the importance of CPU subsystem communications, error detection and correction, and power supply loading. Finally, we will present some of the most common programming devices for entering and editing the control program. The next chapter will discuss the other major component of the CPU—the memory system—and will explore the relationship between input/output field devices, memory, and the PLC.

4-1 INTRODUCTION

As mentioned in the first chapter, the central processing unit, or CPU, is the most important element of a PLC. The CPU forms what can be considered to be the "brain" of the system. The three components of the CPU are:

- the processor

- the memory system

- the power supply

Figure 4-1 illustrates a simplified block diagram of a CPU. CPU architecture may differ from one manufacturer to another, but in general, most CPUs follow this typical three-component organization. Although this diagram shows the power supply inside the CPU block enclosure, the power supply may be a separate unit that is mounted next to the block enclosure containing the processor and memory. Figure 4-2 shows a CPU with a built-in power supply. The programming device, not regarded as part of the CPU per se, completes the total central architecture as the medium of communication between the programmer and the CPU.

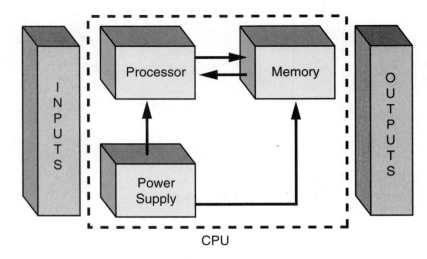

Figure 4-1. CPU block diagram.

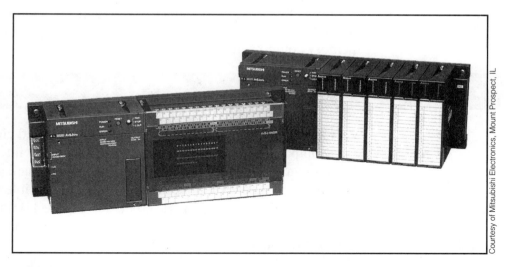

Figure 4-2. Two PLC CPUs with built-in power supplies (left with fixed I/O blocks and right with configurable I/O).

The term *CPU* is often used interchangeably with the word *processor*; however, the CPU encompasses all of the necessary elements that form the intelligence of the system—the processor plus the memory system and power supply. Integral relationships exist between the components of the CPU, resulting in constant interaction among them. Figure 4-3 illustrates the functional interaction between a PLC's basic components. In general, the

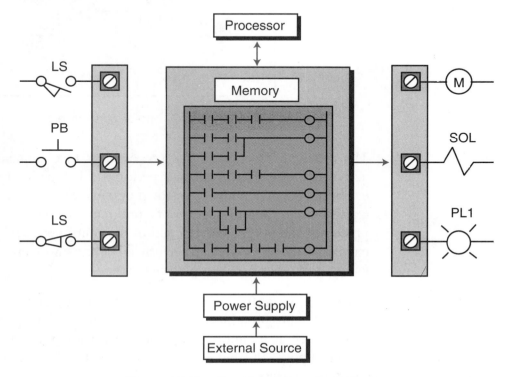

Figure 4-3. Functional interaction of a PLC system.

processor executes the control program stored in the memory system in the form of ladder diagrams, while the system power supply provides all of the necessary voltage levels to ensure proper operation of the processor and memory components.

4-2 PROCESSORS

Very small **microprocessors** (or micros)—integrated circuits with tremendous computing and control capability—provide the intelligence of today's programmable controllers. They perform mathematical operations, data handling, and diagnostic routines that were not possible with relays or their predecessor, the hardwired logic processor. Figure 4-4 illustrates a processor module that contains a microprocessor, its supporting circuitry, and a memory system.

Courtesy of Allen-Bradley, Highland Heights, OH

Figure 4-4. Allen Bradley's PLC processors—models 5/12, 5/15, and 5/25.

The principal function of the processor is to command and govern the activities of the entire system. It performs this function by interpreting and executing a collection of system programs known as the executive. The executive, a group of supervisory programs, is permanently stored in the processor and is considered a part of the controller itself. By executing the executive, the processor can perform all of its control, processing, communication, and other housekeeping functions.

The executive performs the communication between the PLC system and the user via the programming device. It also supports other peripheral communication, such as monitoring field devices; reading diagnostic data from the power supply, I/O modules, and memory; and communicating with an operator interface.

The CPU of a PLC system may contain more than one processor (or micro) to execute the system's duties and/or communications, because extra processors increase the speed of these operations. This approach of using several microprocessors to divide control and communication tasks is known as **multiprocessing**. Figure 4-5 illustrates a multiprocessor configuration.

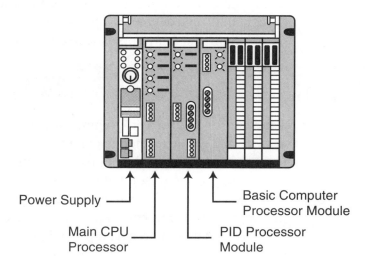

Power Supply — Basic Computer Processor Module

Main CPU Processor — PID Processor Module

Figure 4-5. A multiprocessor configuration.

Another multiprocessor arrangement takes the microprocessor intelligence away from the CPU, moving it to an intelligent module. This technique uses intelligent I/O interfaces, which contain a microprocessor, built-in memory, and a mini-executive that performs independent control tasks. Typical intelligent modules are proportional-integral-derivative (PID) control modules, which perform closed-loop control independent of the CPU, and some stepper and servo motor control interfaces. Figure 4-6 shows some intelligent I/O modules.

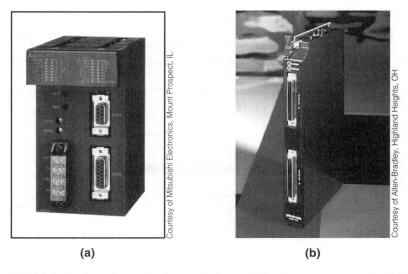

(a)   (b)

Figure 4-6. (a) A single-axis positioning module and **(b)** a temperature control interface.

The microprocessors used in PLCs are categorized according to their word size, or the number of bits that they use simultaneously to perform operations. Standard word lengths are 8, 16, and 32 bits. This word length affects the speed at which the processor performs most operations. For example, a 32-bit microprocessor can manipulate data faster than a 16-bit micro, since it manipulates twice as much data in one operation. Word length correlates with the capability and degree of sophistication of the controller (i.e., the larger the word length, the more sophisticated the controller).

4-3 PROCESSOR SCAN

The basic function of a programmable controller is to read all of the field input devices and then execute the control program, which according to the logic programmed, will turn the field output devices ON or OFF. In reality, this last process of turning the output devices ON or OFF occurs in two steps. First, as the processor executes the internal programmed logic, it will turn each of its programmed internal output coils ON or OFF. The energizing or de-energizing of these internal outputs will not, however, turn the output devices ON or OFF. Next, when the processor has finished evaluating all of the control logic program that turns the internal coils ON or OFF, it will perform an update to the output interface modules, thereby turning the field devices connected to each interface terminal ON or OFF. This process of reading the inputs, executing the program, and updating the outputs is known as the *scan*.

Figure 4-7 shows a graphic representation of the scan. The scanning process is repeated over and over in the same fashion, making the operation sequential from top to bottom. Sometimes, for the sake of simplicity, PLC manufacturers

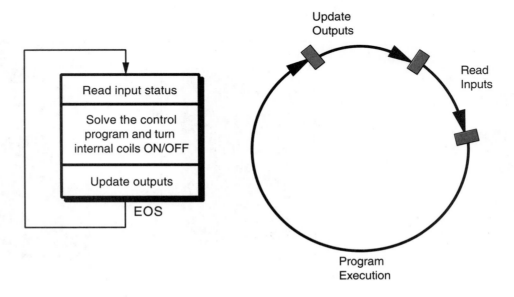

Figure 4-7. PLC total scan representation.

call the solving of the control program the **program scan** and the reading of inputs and updating of outputs the **I/O update scan**. Nevertheless, the total system scan includes both. The internal processor signal, which indicates that the program scan has ended, is called the *end-of-scan* (EOS) signal.

The time it takes to implement a scan is called the **scan time**. The scan time is the total time the PLC takes to complete the program and I/O update scans. The program scan time generally depends on two factors: (1) the amount of memory taken by the control program and (2) the type of instructions used in the program (which affects the time needed to execute the instructions). The time required to make a single scan can vary from a few tenths of a millisecond to 50 milliseconds.

PLC manufacturers specify the scan time based only on the amount of application memory used (e.g., 1 msec/1K of programmed memory). However, other factors also affect the scan time. The use of remote I/O subsystems can increase the scan time, since the PLC must transmit and receive the I/O update from remote systems. Monitoring control programs also adds time to the scan, because the microprocessor must send data about the status of the coils and contacts to a monitoring device (e.g., a PC).

The scan is normally a continuous, sequential process of reading the status of the inputs, evaluating the control logic, and updating the outputs. A processor is able to read an input as long as the input signal *is not* faster than the scan time (i.e., the input signal does not change state—ON to OFF to ON or vice versa—twice during the processor's scan time). For instance, if a controller has a total scan time of 10 msec (see Figure 4-8) and must monitor an input

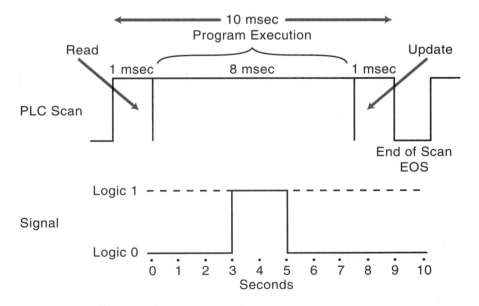

Figure 4-8. Illustration of a signal that will not be detected by a PLC during a normal scan.

signal that changes states twice during an 8 msec period (less than the scan), the programmable controller will not be able to "see" the signal, resulting in a possible machine or process malfunction. This scan characteristic must always be considered when reading discrete input signals and ASCII characters (see the ASCII section in Chapter 8). A programmable controller's scan specification indicates how fast it can react to inputs and still correctly solve the control logic. Chapter 9 provides more information about scan evaluation.

EXAMPLE 4-1

What occurs during the scanning operation of a programmable controller if the signal(s) from an input field device behave as shown in Figures 4-9a and 4-9b?

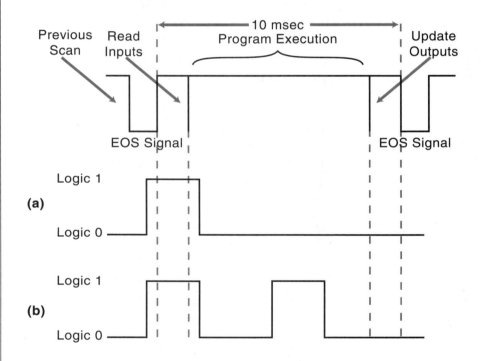

Figure 4-9. (a) Single-pulse and **(b)** double-pulse signals.

SOLUTION

In Figure 4-9a, the PLC will recognize the signal, even though it is shorter than the scan, because it was ON during the read section of the scan. In Figure 4-9b, the PLC will recognize the first signal, but it will not be able to detect the second pulse because this second ON-OFF-ON transition occurred in the middle of the scan. Thus, the PLC can not read it.

Note that although the signal in Figure 4-9a is shorter than the scan, the PLC recognizes it. However, the user should take precautions against signals that behave like this, because if the same signal occurs in the middle of the scan, the PLC will not detect it.

Also note that the behavior of the signal in Figure 4-9b will cause a misreading of the pulse. For instance, if the pulses are being counted, a counting malfunction will occur. These problems, however, can be corrected, as you will see later.

The common scan method of monitoring the inputs at the end of each scan may be inadequate for reading certain extremely fast inputs. Some PLCs provide software instructions that allow the interruption of the continuous program scan to receive an input or to update an output immediately. Figure 4-10 illustrates how immediate instructions operate during a normal program scan. These immediate instructions are very useful when the PLC must react instantaneously to a critical input or output.

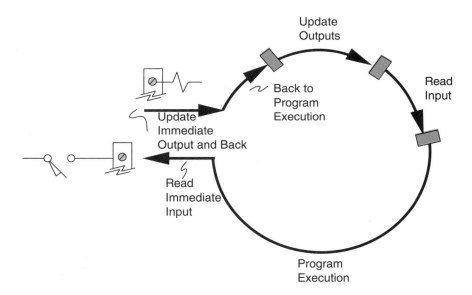

Figure 4-10. PLC scan with immediate I/O update.

Another method for reading extremely fast inputs involves using a *pulse stretcher*, or fast-response module (see Figure 4-11). This module stretches the signal so that it will last for at least one complete scan. With this type of interface, the user must ensure that the signal does not occur more than once per two scans; otherwise, some pulses will be lost. A pulse stretcher is ideal for applications with very fast input signals (e.g., 50 microseconds), perhaps from an instrumentation field device, that do not change state more than once per two scans. If a large number of pulses must be read in a shorter time than the scan time, a high-speed pulse counter input module can be used to read all the pulses and then send the information to the CPU.

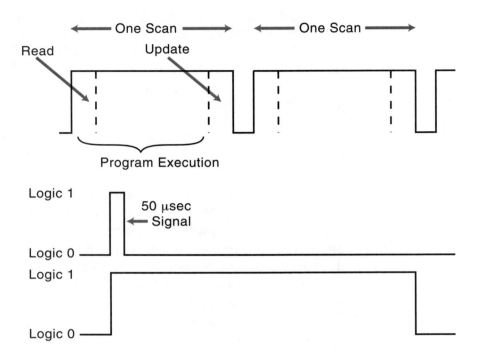

EXAMPLE 4-2

Referencing Figure 4-12, illustrate how, in one scan, **(a)** an immediate instruction will respond to an interrupt input and **(b)** the same input instruction can update an immediate output field device, like a solenoid.

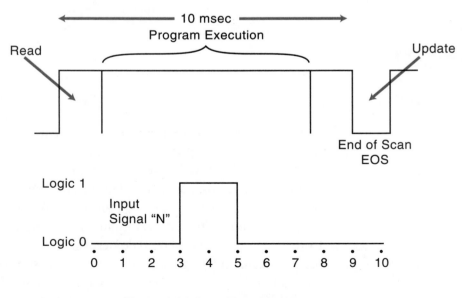

Figure 4-12. Example scan and signal.

SOLUTION

(a) As shown in Figure 4-13, the immediate instruction will interrupt the control program to read the input signal. It will then evaluate the signal and return to the control program, where it will resume program execution and update outputs.

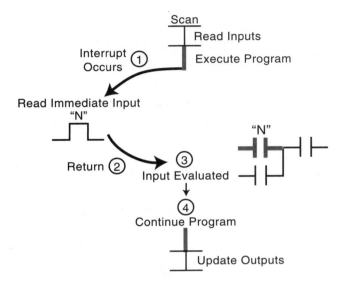

Figure 4-13. Immediate response to an interrupt input.

(b) Figure 4-14 depicts the immediate update of an output. As in part (a), the immediate instruction interrupts the control program to read and evaluate the input signal. However, the output is updated before normal program execution resumes.

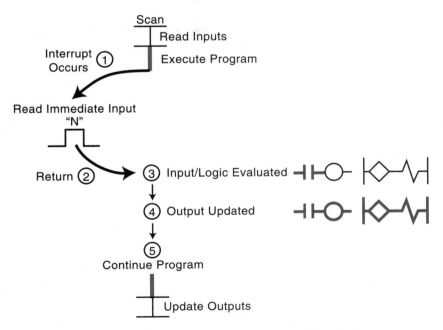

Figure 4-14. Immediate update of an output field device.

4-4 ERROR CHECKING AND DIAGNOSTICS

The PLC's processor constantly communicates with local and remote sub-systems (see Chapter 6), or *racks* as they may also be called. I/O interfaces connect these subsystems to field devices located either close to the main CPU or at remote locations. Subsystem communication involves data transfer exchange at the end of each program scan, when the processor sends the latest status of outputs to the I/O subsystem and receives the current status of inputs and outputs. An I/O subsystem adapter module, located in the CPU, and a remote I/O processor module, located in the subsystem chassis or rack, perform the actual communication between the processor and the subsystem. Figure 4-15 illustrates a typical PLC subsystem configuration.

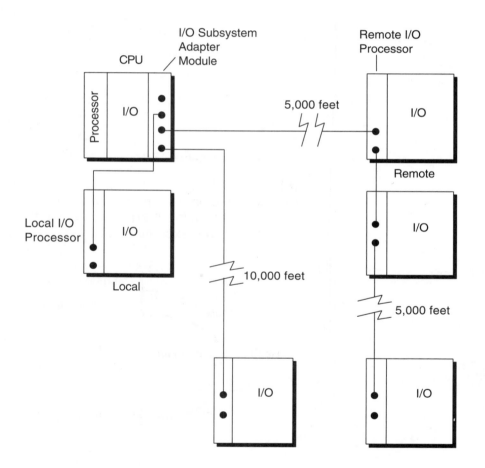

Figure 4-15. Typical PLC subsystem configuration.

The distance between the CPU and a subsystem can vary, depending on the controller, and usually ranges between 1,000 and 15,000 feet. The communication medium generally used is either twisted-pair, twinaxial, coaxial, or fiber-optic cable, depending on the PLC and the distance.

The controller transmits data to subsystems at very high speeds, but the actual speed varies depending on the controller. The data format also varies, but it is normally a serial binary format composed of a fixed number of data bits (I/O status), start and stop bits, and error detection codes.

Error-checking techniques are also incorporated in the continuous communication between the processor and its subsystems. These techniques confirm the validity of the data transmitted and received. The level of sophistication of error checking varies from one manufacturer to another, as does the type of errors reported and the resulting protective or corrective action.

ERROR CHECKING

The processor uses error-checking techniques to monitor the functional status of both the memory and the communication links between subsystems and peripherals, as well as its own operation. Common error-checking techniques include parity and checksum.

Parity. Parity is perhaps the most common error detection technique. It is used primarily in communication link applications to detect mistakes in long, error-prone data transmission lines. The communication between the CPU and subsystems is a prime example of the useful application of parity error checking. Parity check is often called **vertical redundancy check (VRC).**

Parity uses the number of 1s in a binary word to check the validity of data transmission. There are two types of parity checks: *even parity*, which checks for an even number of 1s, and *odd parity*, which checks for an odd number of 1s. When data is transmitted through a PLC, it is sent in binary format, using 1s and 0s. The number of 1s can be either odd or even, depending on the character or data being transmitted (see Figure 4-16a). In parity data transmission, an extra bit is added to the binary word, generally in the most significant or least significant bit position (see Figure 4-16b). This extra bit, called the **parity bit** (P), is used to make each byte or word have an odd or even number of 1s, depending on the type of parity being used.

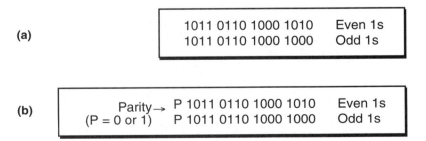

Figure 4-16. (a) A 16-bit data transmission of 1s and 0s and (b) the same transmission with a parity bit (P) in the most significant bit position.

Let's suppose that a processor transmits the 7-bit ASCII character C (1000011) to a peripheral device and odd parity is required. The total number of 1s is three, or odd. If the parity bit (P) is the most significant bit, the transmitted data will be P1000011. To achieve odd parity, P is set to 0 to obtain an odd number of 1s. The receiving end detects an error if the data does not contain an odd number of 1s. If even parity had been the error-checking method, P would have been set to 1 to obtain an even number of 1s.

Parity error checking is a single-error detection method. If one bit of data in a word changes, an error will be detected due to the change in the bit pattern. However, if two bits change value, the number of 1s will be changed back, and an error will not be detected even though there is a mistransmission.

In PLCs, when data is transmitted to a subsystem, the controller defines the type of parity (odd or even) that will be used. However, if the data transmission is from the programmable controller to a peripheral, the parity method must be prespecified and must be the same for both devices.

Some processors do not use parity when transmitting information, although their peripherals may require it. In this case, parity generation can be accomplished through application software. The parity bit can be set for odd or even parity with a short routine using functional blocks or a high-level language. If a nonparity-oriented processor receives data that contains parity, a software routine can also be used to mask out, or strip, the parity bit.

Checksum. The extra bit of data added to each word when using parity error detection is often too wasteful to be desirable. In data storage, for example, error detection is desirable, but storing one extra bit for every eight bits means a 12.5% loss of data storage capacity. For this reason, a data block error-checking method known as **checksum** is used.

Checksum error detection spots errors in blocks of many words, instead of in individual words as parity does. Checksum analyzes all of the words in a data block and then adds to the end of the block one word that reflects a characteristic of the block. Figure 4-17 shows this last word, known as the **block check character (BCC)**. This type of error checking is appropriate for memory checks and is usually done at power-up.

There are several methods of checksum computation, with the three most common being:

- cyclic redundancy check
- longitudinal redundancy check
- cyclic exclusive-OR checksum

Cyclic Redundancy Check. **Cyclic redundancy check (CRC)** is a technique that performs an addition of all the words in the data block and then stores the resulting sum in the last location, the block check character (BCC). This

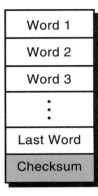

Figure 4-17. Block check character at the end of the data block.

summation process can rapidly reach an overflow condition, so one variation of CRC allows the sum to overflow, storing only the remainder bits in the BCC word. Typically, the resulting word is complemented and written in the BCC location. During the error check, all words in the block are added together, with the addition of the final BCC word turning the result to 0. A zero sum indicates a valid block. Another type of CRC generates the BCC using the remainder of dividing the sum by a preset binary number.

<u>Longitudinal Redundancy Check.</u> **Longitudinal redundancy check (LRC)** is an error-checking technique based on the accumulation of the result of performing an **exclusive-OR (XOR)** on each of the words in the data block. The exclusive-OR operation is similar to the standard OR logic operation (see Chapter 3) except that, with two inputs, only one can be ON (1) for the output to be 1. If both logic inputs are 1, then the output will be 0. The exclusive-OR operation is represented by the $\oplus$ symbol. Figure 4-18 illustrates the truth table for the exclusive-OR operation. Thus, the LRC operation is simply the logical exclusive-OR of the first word with the second word, the result with the third word, and so on. The final exclusive-OR operation is stored at the end of the block as the BCC.

Exclusive-OR Truth Table		
Inputs		Output
A	B	Y
0	0	0
0	1	1
1	0	1
1	1	0

Figure 4-18. Truth table for the exclusive-OR operation.

<u>Cyclic Exclusive-OR Checksum.</u> **Cyclic exclusive-OR checksum (CX-ORC)** is similar to LRC with some slight variations. The operation starts with a checksum word containing 0s, which is XORed with the first word of the block. This is followed by a left rotation of the bits in the checksum word. The next word in the data block is XORed with the checksum word and then

rotated left (see Figure 4-19). This procedure is repeated until the last word of the block has been logically operated on. The checksum word is then appended to the block to become the BCC.

A software routine in the executive program performs most checksum error-detecting methods. Typically, the processor performs the checksum computation on memory at power-up and also during the transmission of data. Some controllers perform the checksum on memory during the execution of the control program. This continuous on-line error checking lessens the possibility of the processor using invalid data.

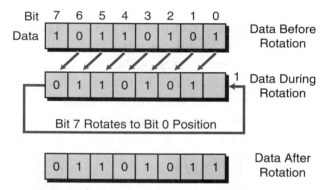

Figure 4-19. Cyclic exclusive-OR checksum operation.

EXAMPLE 4-3

Implement a checksum utilizing **(a)** LRC and **(b)** CX-ORC techniques for the four, 6-bit words shown. Place the BCC at the end of the data block.

word 1	1 1 0 0 1 1
word 2	1 0 1 1 0 1
word 3	1 0 1 1 1 0
word 4	1 0 0 1 1 1

SOLUTION

(a) Longitudinal redundancy check:

word 1	1 1 0 0 1 1
$\oplus$	$\oplus$
word 2	1 0 1 1 0 1
result	0 1 1 1 1 0
$\oplus$	$\oplus$
word 3	1 0 1 1 1 0
result	1 1 0 0 0 0
$\oplus$	$\oplus$
word 4	1 0 0 1 1 1
result	0 1 0 1 1 1

LRC data block:

word 1	1 1 0 0 1 1
word 2	1 0 1 1 0 1
word 3	1 0 1 1 1 0
word 4	1 0 0 1 1 1
BCC	0 1 0 1 1 1

(b) Cyclic exclusive-OR check:

Start with checksum word 000000.

CS start	0 0 0 0 0 0
⊕	⊕
word 1	1 1 0 0 1 1
result	1 1 0 0 1 1
left rotate	1 0 0 1 1 1
⊕	⊕
word 2	1 0 1 1 0 1
result	0 0 1 0 1 0
left rotate	0 1 0 1 0 0
⊕	⊕
word 3	1 0 1 1 1 0
result	1 1 1 0 1 0
left rotate	1 1 0 1 0 1
⊕	⊕
word 4	1 0 0 1 1 1
result	0 1 0 0 1 0
left rotate	1 0 0 1 0 0 (final checksum)

CX-ORC data block:

word 1	1 1 0 0 1 1
word 2	1 0 1 1 0 1
word 3	1 0 1 1 1 0
word 4	1 0 0 1 1 1
BCC	1 0 0 1 0 0

Error Detection and Correction. More sophisticated programmable controllers may have an error detection and correction scheme that provides greater reliability than conventional error detection. The key to this type of error correction is the multiple representation of the same value.

The most common error-detecting and error-correcting code is the **Hamming code**. This code relies on parity bits interspersed with data bits in a data word. By combining the parity and data bits according to a strict set of parity equations, a small byte is generated that contains a value that identifies the erroneous bit. An error can be detected and corrected if any bit is changed by any value. The hardware used to generate and check Hamming codes is quite complex and essentially implements a set of error-correcting equations.

Error-correcting codes offer the advantage of being able to detect two or more bit errors; however, they can only correct one-bit errors. They also present a disadvantage because they are bit wasteful. Nevertheless, this scheme will continue to be used with data communication in hierarchical systems that are unmanned, sophisticated, and automatic.

CPU DIAGNOSTICS

The processor is responsible for detecting communication failures, as well as other failures, that may occur during system operation. It must alert the operator or system in case of a malfunction. To do this, the processor performs **diagnostics**, or error checks, during its operation and sends status information to indicators that are normally located on the front of the CPU.

Typical diagnostics include *memory OK*, *processor OK*, *battery OK*, and *power supply OK*. Some controllers possess a set of fault relay contacts that can be used in an alarm circuit to signal a failure. The processor controls the fault relay and activates it when one or more specific fault conditions occur.

The relay contacts that are usually provided with a controller operate in a *watchdog timer* fashion; that is, the processor sends a pulse at the end of each scan indicating a correct system operation. If a failure occurs, the processor does not send a pulse, the timer times out, and the fault relay activates.

In some controllers, CPU diagnostics are available to the user during the execution of the control program. These diagnostics use internal outputs that are controlled by the processor but can be used by the user program (e.g., loss of scan, backup battery low, etc.).

4-5 THE SYSTEM POWER SUPPLY

The system power supply plays a major role in the total system operation. In fact, it can be considered the "first-line manager" of system reliability and integrity. Its responsibility is not only to provide internal DC voltages to the system components (i.e., processor, memory, and input/output interfaces), but also to monitor and regulate the supplied voltages and warn the CPU if something is wrong. The power supply, then, has the function of supplying well-regulated power and protection for other system components.

THE INPUT VOLTAGE

Usually, PLC power supplies require input from an AC power source; however, some PLCs will accept a DC power source. Those that will accept a DC source are quite appealing for applications such as offshore drilling operations, where DC sources are commonly used. Most PLCs, however, require a 120 VAC or 220 VAC power source, while a few controllers will accept 24 VDC.

Since industrial facilities normally experience fluctuations in line voltage and frequency, a PLC power supply must be able to tolerate a 10 to 15% variation in line voltage conditions. For example, when connected to a 120 VAC source, a power supply with a line voltage tolerance of ±10% will continue to function properly as long as the voltage remains between 108 and 132 VAC. A 220 VAC power supply with ±10% line tolerance will function properly as long as the voltage remains between 198 and 242 VAC. When the line voltage exceeds the upper or lower tolerance limits for a specified duration (usually one to three AC cycles), most power supplies will issue a shutdown command to the processor. Line voltage variations in some plants can eventually become disruptive and may result in frequent loss of production. Normally, in such a case, a constant voltage transformer is installed to stabilize line conditions.

Constant Voltage Transformers. Good power supplies tolerate normal fluctuations in line conditions, but even the best-designed power supply cannot compensate for the especially unstable line voltage conditions found in some industrial environments. Conditions that cause line voltage to drop below proper levels vary depending on application and plant location. Some possible conditions are:

- start-up/shutdown of nearby heavy equipment, such as large motors, pumps, welders, compressors, and air-conditioning units

- natural line losses that vary with distance from utility substations

- intraplant line losses caused by poorly made connections

- brownout situations in which line voltage is intentionally reduced by the utility company

A **constant voltage transformer** compensates for voltage changes at its input (the primary) to maintain a steady voltage to its output (the secondary). When operated at less than the rated load, the transformer can be expected to maintain approximately ±1% output voltage regulation with an input voltage variation of as much as 15%. The percentage of regulation changes as a function of the operated load (PLC power supply and input devices)—the higher the load, the more fluctuation. Therefore, a constant voltage transformer must be properly rated to provide ample power to the load. The rating of the constant voltage transformer, in units of volt-amperes (VA), should be selected based on the worst-case power requirements of the load. The recommended rating for a constant voltage transformer can be obtained from the PLC manufacturer. Figure 4-20 illustrates a simplified connection of a constant voltage transformer and a programmable controller.

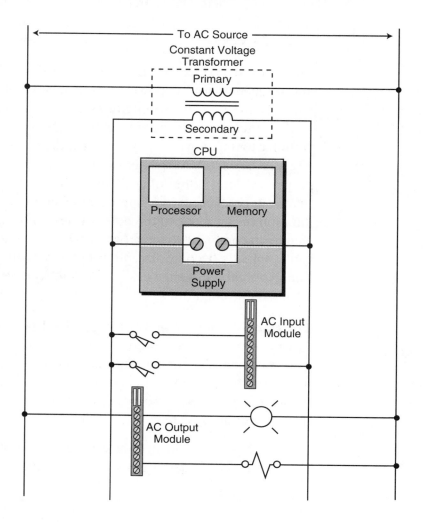

Figure 4-20. A constant voltage transformer connected to a PLC system (CPU and modules).

The Sola CVS standard sinusoidal transformer, or an equivalent constant voltage transformer, is suitable for programmable controller applications. This type of transformer uses line filters to remove high-harmonic content and provide a clean sinusoidal output. Constant voltage transformers that do not filter high harmonics are not recommended for programmable controller applications. Figure 4-21 illustrates the relationship between the output voltage and input voltage for a typical Sola CVS transformer operated at different loads.

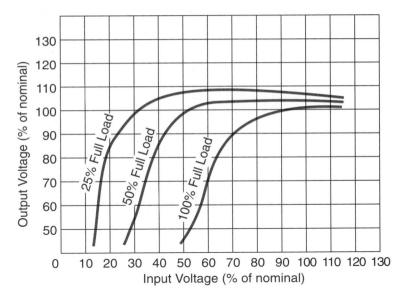

Figure 4-21. Relationship of input versus output voltages for a Sola unit.

Isolation Transformers. Often, a programmable controller will be installed in an area where the AC line is stable; however, surrounding equipment may generate considerable amounts of electromagnetic interference (EMI). Such an installation can result in intermittent misoperation of the controller, especially if the controller is not electrically isolated (on a separate AC power source) from the equipment generating the EMI. Placing the controller on a separate **isolation transformer** from the potential EMI generators will increase system reliability. The isolation transformer need not be a constant voltage transformer; but it should be located between the controller and the AC power source.

LOADING CONSIDERATIONS

The system power supply provides the DC power required by the logic circuits of the CPU and the I/O circuits. The power supply has a maximum amount of current that it can provide at a given voltage level (e.g., 10 amps at 5 volts), depending on the type of power supply. The amount of current that a given power supply can provide is not always sufficient to satisfy the requirements of a mix of I/O modules. In such a case, undercurrent conditions can cause unpredictable operation of the I/O system.

In most circumstances, an undercurrent situation is unusual, since most power supplies are designed to accommodate a mix of the most commonly used I/O modules. However, an undercurrent condition sometimes arises in applications where an excessive number of special purpose I/O modules are used (e.g., power contact outputs and analog inputs/outputs). These special purpose modules usually have higher current requirements than most commonly used digital I/O modules.

Power supply overloading can be an especially annoying condition, since the problem is not always easily detected. An overload condition is often a function of a combination of outputs that are ON at a given time, which means that overload conditions can appear intermittently. When power supply loading limits have been exceeded and overload occurs, the normal remedy is to either add an auxiliary power supply or to obtain a supply with a larger current capability. To be aware of system loading requirements ahead of time, users can obtain vendor specifications for I/O module current requirements. This information should include per point (single input or output) requirements and current requirements for both ON and OFF states. If the total current requirement for a particular I/O configuration is greater than the total current supplied by the power supply, then a second power supply will be required. An early consideration of line conditions and power requirements will help to avoid problems during installation and start-up.

Power Supply Loading Example. Undoubtedly, the best solution to a problem is anticipation of the problem. When selecting power supplies, current loading requirements, which can indicate potential loading problems, are often overlooked. For this reason, let's go over a load estimation example.

Consider an application where a PLC will control 50 discrete inputs and 25 discrete outputs. Each discrete input module can connect up to 16 field devices, while each output module can connect up to 8 field devices. In addition to this discrete configuration, the application requires a special servo motor interface module and five power contact outputs. The system also uses three analog inputs and three analog outputs.

Figure 4-22 illustrates the configuration of this PLC application. The first plug-in module is the power supply, then the processor module, and then the I/O modules.

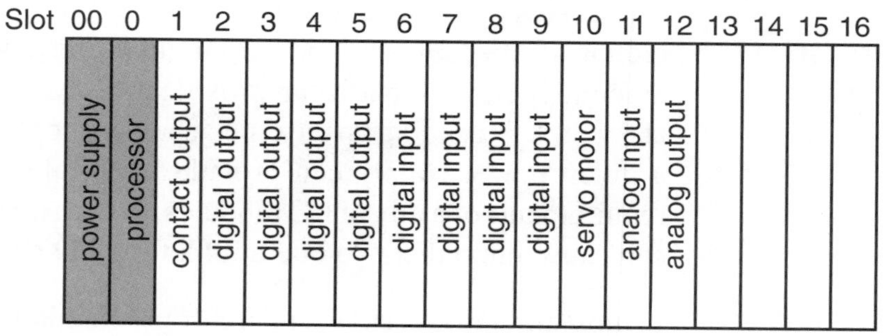

Application Note

Power supply requires one slot (slot 00).
Processor requires one slot (slot 0).
Twelve I/O slots are used, four are spare.
Auxiliary power supplies, if required, must be placed in slot 8.

Figure 4-22. Configuration of an example PLC.

The first step in estimating the load is to determine how many modules are required and then compute the total current requirement of these modules. Table 4-1 lists the module types, current requirements for all inputs and outputs ON at the same time, and the available power supplies for our programmable controller example.

Module Type	I/O Devices Connected	Connections per Module	# of Modules Required	Module Current @ On State	Total Current Required
Discrete in	50	16	4	250 mA	1000 mA
Discrete out	25	8	4	220 mA	880 mA
Contact	5	4	1	575 mA	575 mA
Analog in	3	4	1	600 mA	600 mA
Analog out	3	4	1	1200 mA	1200 mA
Servo motor	1	1	1	400 mA	400 mA
				TOTAL	4655 mA

Processor's current:	1.2 amps	
Power supplies available:	Type A	3 amps
	Type B	5 amps
	Type C	6 amps
Auxiliary power supply: (placement in slot 8)	Type AA	3 amps
	Type BB	5 amps

Table 4-1. Listing of modules and their current requirements.

The total power supply current required by this input/output system is 4655 mA, or 4.655 amps. Adding this current to the 1.2 amps required by the processor results in a total of 5.855 amps, the minimum current the power supply must provide to ensure the proper operation of the system. This total current indicates a worst-case condition, since it assumes that all I/Os are operating in the ON condition (which requires more current than the OFF condition).

For this example, there are several power supply options. These options include using a 6 amp power supply or using a combination of a smaller supply with an auxiliary source. If no expansion is expected, the 6 amp power source will suffice. Conversely, if there is a slight possibility for more I/O requirements, then an auxiliary supply will most likely be needed. The addition of an auxiliary supply can be done either at setup or when required; however, for the controller configuration in Figure 4-22, the auxiliary source must be placed in the eighth slot, resulting in I/O address changes if the auxiliary supply is added after setup. Therefore, the reference addresses in the program will have to be reprogrammed to reflect this change. Also, remember that the larger the power supply, the higher the price in most cases. You must keep all these factors in mind when configuring a PLC system and assigning I/O addresses to field devices.

4-6 PROGRAMMING DEVICES

Although the way to enter the control program into the PLC has changed since the first PLCs came onto the market, PLC manufacturers have always maintained an easy human interface for program entry. This means that users do not have to spend much time learning how to enter a program, but rather they can spend their time programming and solving the control problem.

Most PLCs are programmed using very similar instructions. The only difference may be the mechanics associated with entering the program into the PLC, which may vary from manufacturer to manufacturer. This involves both the type of instruction used by each particular PLC and the methodology for entering the instruction using a programming device. The two basic types of programming devices are:

- miniprogrammers
- personal computers

MINIPROGRAMMERS

Miniprogrammers, also known as *handheld* or *manual programmers*, are an inexpensive and portable way to program small PLCs (up to 128 I/O). Physically, these devices resemble handheld calculators, but they have a larger display and a somewhat different keyboard. The type of display is usually LED (light-emitting diode) or dot matrix LCD (liquid crystal display), and the keyboard consists of numeric keys, programming instruction keys, and special function keys. Instead of handheld units, some controllers have built-in miniprogrammers. In some instances, these built-in programmers are detachable from the PLC. Even though they are used mainly for editing and inputting control programs, miniprogrammers can also be useful tools for starting up, changing, and monitoring the control logic. Figure 4-23 shows a typical miniprogrammer along with a small PLC, in which miniprogrammers are generally used.

Most miniprogrammers are designed so that they are compatible with two or more controllers in a product family. The miniprogrammer is most often used with the smallest member of the PLC family or, in some cases, with the next larger member, which is normally programmed using a personal computer with special PLC programming software (discussed in the next section). With this programming option, small changes or monitoring required by the larger controller can be accomplished without carrying a personal computer to the PLC location.

Miniprogrammers can be intelligent or nonintelligent. Nonintelligent handheld programmers can be used to enter and edit the PLC program with limited on-line monitoring and editing capabilities. These capabilities are

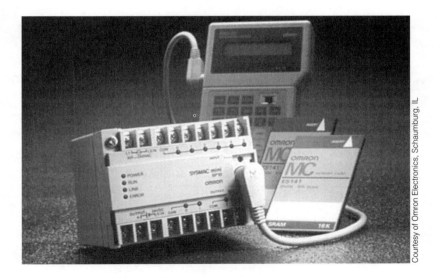

Figure 4-23. A typical miniprogrammer and a small PLC.

limited by memory and display size. Intelligent miniprogrammers are microprocessor-based and provide the user with many of the features offered by
personal computers during off-line programming (disconnected from the
PLC). These intelligent devices can perform system diagnostic routines
(memory, communication, display, etc.) and even serve as an operator
interface device that can display English messages about the controlled
machine or process.

Some miniprogrammers offer removable memory cards or modules, which
store a complete program that can be reloaded at any time into any member
of the PLC family (see Figure 4-24). This type of storage is useful in
applications where the control program of one machine needs to be duplicated
and easily transferred to other machines (e.g., OEM applications).

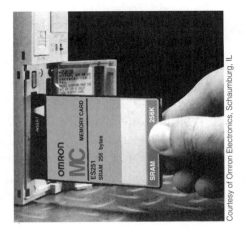

Figure 4-24. A removable memory card for a miniprogrammer.

PERSONAL COMPUTERS

Common usage of the personal computer (PC) in our daily lives has led to the practical elimination of dedicated PLC programming devices. Due to the personal computer's general-purpose architecture and standard operating system, most PLC manufacturers and other independent suppliers provide the necessary PC software to implement ladder program entry, editing, documentation, and real-time monitoring of the PLC's control program. The large screens of PCs can show one or more ladder rungs of the control program during programming or monitoring operation (see Figure 4-25).

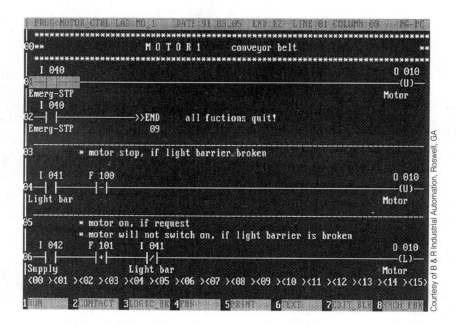

Figure 4-25. A PLC ladder diagram displayed on a personal computer.

Personal computers are the programming devices of choice not so much because of their PLC programming capabilities, but because PCs are usually already present at the location where the user is performing the programming. The different types of desktop, laptop, and portable PCs give the programmer flexibility—they can be used as programming devices, but they can also be used in applications other than PLC programming. For instance, a personal computer can be used to program a PLC, but it may also be connected to the PLC's local area network (see Figure 4-26) to gather and store, on a hard disk, process information that could be vital for future product enhancements. A PC can also communicate with a programmable controller through the RS-232C serial port, thus serving either as the data handler and supervisor of the PLC control or as the bridge between the PLC network and a larger computer system (see Figure 4-27).

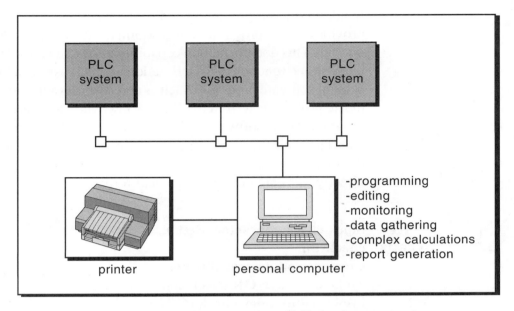

Figure 4-26. A PC connected to a PLC's local area network.

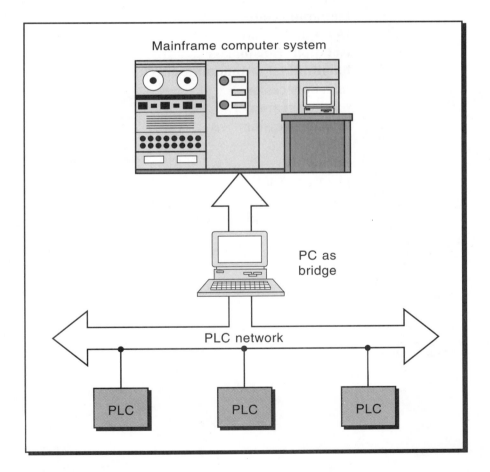

Figure 4-27. A PC acting as a bridge between a PLC network and a mainframe computer system.

In addition to programming and data collection activities, PC software that provides ladder programming capability often includes PLC documentation options. This documentation capability allows the programmer to define the purpose and function of each I/O address that is used in a PLC program. Also, general software programs, such as spreadsheets and databases, can communicate process data from the PLC to a PC via a software bridge or translator program. These software options make the PC almost invaluable when using it as a man/machine interface, providing a window to the inner workings of the PLC-controlled machine or process and generating reports that can be directly translated into management forms.

**KEY
TERMS**

block check character (BCC)
checksum
constant voltage transformer
cyclic exclusive-OR checksum (CX-ORC)
cyclic redundancy check (CRC)
diagnostics
exclusive-OR (XOR)
Hamming code
I/O update scan
isolation transformer
longitudinal redundancy check (LRC)
microprocessor
miniprogrammer
multiprocessing
parity
parity bit
program scan
scan time
vertical redundancy check (VRC)

CHAPTER FIVE

THE MEMORY SYSTEM AND I/O INTERACTION

The two offices of memory are collection and distribution.

—Samuel Johnson

CHAPTER HIGHLIGHTS Now that you've learned about the first three major components of the programmable controller, it's time to learn about the last—the memory system. Understanding the PLC's memory system will help you understand why it operates as it does, as well as how it interacts with I/O interfaces. In this chapter, we will discuss the different types of memory, including memory structure and capabilities. Then, we will explore the relationship between memory organization and I/O interaction. Finally, we will explain how to configure the PLC memory for I/O addressing.

5-1 MEMORY OVERVIEW

The most important characteristic of a programmable controller is the user's ability to change the control program quickly and easily. The PLC's architecture makes this programmability feature possible. The **memory** system is the area in the PLC's CPU where all of the sequences of instructions, or *programs*, are stored and executed by the processor to provide the desired control of field devices. The memory sections that contain the control programs can be changed, or reprogrammed, to adapt to manufacturing line procedure changes or new system start-up requirements.

MEMORY SECTIONS

The total memory system in a PLC is actually composed of two different memories (see Figure 5-1):

- the executive memory
- the application memory

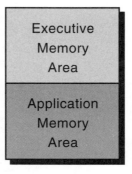

Figure 5-1. Simplified block diagram of the total PLC memory system.

The **executive memory** is a collection of permanently stored programs that are considered part of the PLC itself. These supervisory programs direct all system activities, such as execution of the control program and communication with peripheral devices. The executive section is the part of the PLC's memory where the system's available instruction software is stored (i.e., relay instructions, block transfer functions, math instructions, etc.). This area of memory is not accessible to the user.

The **application memory** provides a storage area for the user-programmed instructions that form the application program. The application memory area is composed of several areas, each having a specific function and usage. Section 5-4 covers the executive and application memory areas in detail.

5-2 MEMORY TYPES

The storage and retrieval requirements for the executive and application memory sections are not the same; therefore, they are not always stored in the same type of memory. For example, the executive requires a memory that permanently stores its contents and cannot be erased or altered either by loss of electrical power or by the user. This type of memory is often unsuitable for the application program.

Memory can be separated into two categories: volatile and nonvolatile. **Volatile memory** loses its programmed contents if all operating power is lost or removed, whether it is normal power or some form of backup power. Volatile memory is easily altered and quite suitable for most applications when supported by battery backup and possibly a disk copy of the program. **Nonvolatile memory** retains its programmed contents, even during a complete loss of operating power, without requiring a backup source. Nonvolatile memory generally is unalterable, yet there are special nonvolatile memory types that are alterable. Today's PLCs include those that use nonvolatile memory, those that use volatile memory with battery backup, as well as those that offer both.

There are two major concerns regarding the type of memory where the application program is stored. Since this memory is responsible for retaining the control program that will run each day, volatility should be the prime concern. Without the application program, production may be delayed or forfeited, and the outcome is usually unpleasant. A second concern should be the ease with which the program stored in memory can be altered. Ease in altering the application memory is important, since this memory is ultimately involved in any interaction between the user and the controller. This interaction begins with program entry and continues with program changes made during program generation and system start-up, along with on-line changes, such as changing timer or counter preset values.

The following discussion describes six types of memory and how their characteristics affect the manner in which programmed instructions are retained or altered within a programmable controller.

READ-ONLY MEMORY

Read-only memory (ROM) is designed to permanently store a fixed program that is not alterable under ordinary circumstances. It gets its name from the fact that its contents can be examined, or *read*, but not altered once information has been stored. This contrasts with memory types that can be read from and written to (discussed in the next section). By nature, ROMs are generally immune to alteration due to electrical noise or loss of power. Executive programs are often stored in ROM.

Programmable controllers rarely use read-only memory for their application memory. However, in applications that require fixed data, read-only memory offers advantages when speed, cost, and reliability are factors. Generally, the manufacturer creates ROM-based PLC programs at the factory. Once the manufacturer programs the original set of instructions, the user can never alter it. This typical approach to the programming of ROM-based controllers assumes that the program has already been debugged and will never be changed. This debugging is accomplished using a random-access memory–based PLC or possibly a computer. The final program is then entered into ROM. ROM application memory is typically found only in very small, dedicated PLCs.

RANDOM-ACCESS MEMORY

Random-access memory (RAM), often referred to as *read/write memory (R/W)*, is designed so that information can be written into or read from the memory storage area. Random-access memory does not retain its contents if power is lost; therefore, it is a volatile type of memory. Random-access memory normally uses a battery backup to sustain its contents in the event of a power outage.

For the most part, today's programmable controllers use RAM with battery support for application memory. Random-access memory provides an excellent means for easily creating and altering a program, as well as allowing data entry. In comparison to other memory types, RAM is a relatively fast memory. The only noticeable disadvantage of battery-supported RAM is that the battery may eventually fail, although the processor constantly monitors the status of the battery. Battery-supported RAM has proven to be sufficient for most programmable controller applications. If a battery backup is not feasible, a controller with a nonvolatile memory option (e.g., EPROM) can be used in combination with the RAM. This type of memory arrangement provides the advantages of both volatile and nonvolatile memory. Figure 5-2 shows a RAM chip.

Figure 5-2. A 4K words by 8 bits RAM memory chip.

PROGRAMMABLE READ-ONLY MEMORY

Programmable read-only memory (PROM) is a special type of ROM because it can be programmed. Very few of today's programmable controllers use PROM for application memory. When it is used, this type of memory is most likely a permanent storage backup for some type of RAM. Although a PROM is programmable and, like any other ROM, has the advantage of nonvolatility, it has the disadvantage of requiring special programming equipment. Also, once programmed, it cannot be easily erased or altered; any program change requires a new set of PROM chips. A PROM memory is suitable for storing a program that has been thoroughly checked while residing in RAM and will not require further changes or on-line data entry.

ERASABLE PROGRAMMABLE READ-ONLY MEMORY

Erasable programmable read-only memory (EPROM) is a specially designed PROM that can be reprogrammed after being entirely erased by an ultraviolet (UV) light source. Complete erasure of the contents of the chip requires that the window of the chip (see Figure 5-3) be exposed to a UV light source for approximately twenty minutes. EPROM can be considered a semipermanent storage device, because it permanently stores a program until it is ready to be altered.

EPROM provides an excellent storage medium for application programs that require nonvolatility, but that do not require program changes or on-line data entry. Many OEMs use controllers with EPROM-type memories to provide permanent storage of the machine program after it has been debugged and is fully operational. OEMs use EPROM because most of their machines will not require changes or data entry by the user.

Figure 5-3. A 4K by 8 bits EPROM memory chip.

An application memory composed of EPROM alone is unsuitable if on-line changes or data entry are required. However, many controllers offer EPROM application memory as an optional backup to battery-supported RAM. EPROM, with its permanent storage capability, combined with RAM, which is easily altered, makes a suitable memory system for many applications.

ELECTRICALLY ALTERABLE READ-ONLY MEMORY

Electrically alterable read-only memory (EAROM) is similar to EPROM, but instead of requiring an ultraviolet light source to erase it, an erasing voltage on the proper pin of an EAROM chip can wipe the chip clean. Very few controllers use EAROM as application memory, but like EPROM, it provides a nonvolatile means of program storage and can be used as a backup to RAM-type memories.

ELECTRICALLY ERASABLE PROGRAMMABLE READ-ONLY MEMORY

Electrically erasable programmable read-only memory (EEPROM) is an integrated circuit memory storage device that was developed in the mid-1970s. Like ROMs and EPROMs, it is a nonvolatile memory, yet it offers the same programming flexibility as RAM does.

Several of today's small and medium-sized controllers use EEPROM as the only memory within the system. It provides permanent storage for the program and can be easily changed with the use of a programming device (e.g., a PC) or a manual programming unit. These two features help to eliminate downtime and delays associated with programming changes. They also lessen the disadvantages of electrically erasable programmable read-only memory.

One of the disadvantages of EEPROM is that a byte of memory can be written to only after it has been erased, thus creating a delay. This delay period is noticeable when on-line program changes are being made. Another disadvantage of EEPROM is a limitation on the number of times that a single byte of memory can undergo the erase/write operation (approximately 10,000). These disadvantages are negligible, however, when compared to the remarkable advantages that EEPROM offers.

5-3 MEMORY STRUCTURE AND CAPACITY

BASIC STRUCTURAL UNITS

PLC memories can be thought of as large, two-dimensional arrays of single-unit storage cells, each storing a single piece of information in the form of 1 or 0 (i.e., the binary numbering format). Since each cell can store only one binary digit and *bit* is the acronym for "*bi*nary digi*t*," each cell is called a bit. A bit, then, is the smallest structural unit of memory. Although each bit stores information as either a 1 or a 0, the memory cells do not actually contain the numbers 1 and 0 per se. Rather, the cells use voltage charges to represent 1 and 0—the presence of a voltage charge represents a 1, the absence of a charge represents a 0. A bit is considered to be ON if the stored information is 1 (voltage present) and OFF if the stored information is 0 (voltage absent). The ON/OFF information stored in a single bit is referred to as the *bit status*.

Sometimes, a processor must handle more than a single bit of data at a time. For example, it is more efficient for a processor to work with a group of bits when transferring data to and from memory. Also, storing numbers and codes requires a grouping of bits. A group of bits handled simultaneously is called a *byte*. More accurately, a byte is the smallest group of bits that can be handled by the processor at one time. Although byte size is normally eight bits, this size can vary depending on the specific controller.

The third and final structural information unit used within a PLC is a *word*. In general, a word is the unit that the processor uses when data is to be operated on or instructions are to be performed. Like a byte, a word is also a fixed group of bits that varies according to the controller; however, words are usually one byte or more in length. For example, a 16-bit word consists of two bytes. Typical word lengths used in PLCs are 8, 16, and 32 bits. Figure 5-4 illustrates the structural units of a typical programmable controller memory.

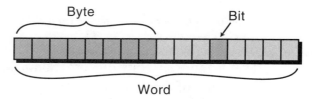

Figure 5-4. Units of PLC memory: bits, bytes, and words.

MEMORY CAPACITY AND UTILIZATION

Memory capacity is a vital concern when considering a PLC application. Specifying the right amount of memory can save the costs of hardware and time associated with adding additional memory capacity later. Knowing memory capacity requirements ahead of time also helps avoid the purchase of a controller that does not have adequate capacity or that is not expandable.

Memory capacity is nonexpandable in small controllers (less than 64 I/O capacity) and expandable in larger PLCs. Small PLCs have a fixed amount of memory because the available memory is usually more than enough to provide program storage for small applications. Larger controllers allow memory expandability, since the scope of their applications and the number of their I/O devices have less definition.

Application memory size is specified in terms of K units, where each K unit represents 1024 word locations. A 1K memory, then, contains 1024 storage locations, a 2K memory contains 2048 locations, a 4K memory contains 4096 locations, and so on. Figure 5-5 illustrates two memory arrays of 4K each; however, they have different configurations—the first configuration uses one-byte words (8 bits) and the other uses two-byte words (16 bits).

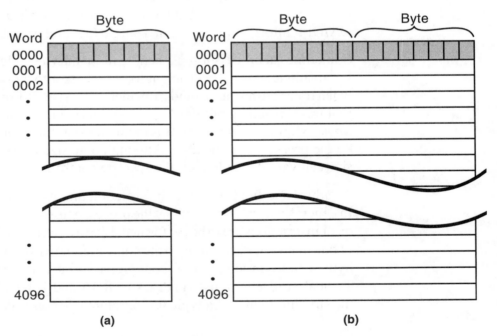

Figure 5-5. Block illustration of **(a)** a 4K by 8 bits storage location and **(b)** a 4K by 16 bits storage location.

The memory capacity of a programmable controller in units of K is only an indication of the total number of storage locations available. Knowing this maximum number alone is not enough to determine memory requirements.

Additional information concerning how program instructions are stored will help to make a better decision. The term *memory utilization* refers to the amount of data that can be stored in one location or, more specifically, to the number of memory locations required to store each type of instruction. The manufacturer can supply this data if the product literature does not provide it.

To illustrate memory capacity, let's refer to Figure 5-5. Suppose that each normally open and normally closed contact instruction requires 16 bits of storage area. With these memory requirements, the effective storage area of the memory system in Figure 5-5a is half that of Figure 5-5b. This means that, to store the same size control program, the system in Figure 5-5a would require 8K memory capacity instead of 4K, as in Figure 5-5b.

After becoming familiar with how memory is utilized in a particular controller, users can begin to determine the maximum memory requirements for an application. Although several rules of thumb have been used over the years, no one simple rule has emerged as being the most accurate. However, with a knowledge of the number of outputs, an idea of the number of program contacts needed to drive the logic of each output, and information concerning memory utilization, memory requirement approximation can be reduced to simple multiplication.

EXAMPLE 5-1

Determine the memory requirements for an application with the following specifications:

- 70 outputs, with each output driven by logic composed of 10 contact elements

- 11 timers and 3 counters, each having 8 and 5 elements, respectively

- 20 instructions that include addition, subtraction, and comparison, each driven by 5 contact elements

Table 5-1 provides information about the application's memory utilization requirements.

Instruction	Words of Memory Required
Examine ON or OFF (contacts)	1
Output coil	1
Add/subtract/compare	1
Timer/counter	3

Table 5-1. Memory utilization requirements.

SOLUTION

Using the given information, a preliminary estimation of memory is:

(a)   Control logic = 10 contact elements/output rung
Number of output rungs = 70

(b)   Control logic = 8 contact elements/timer
Number of timers = 11

(c)   Control logic = 5 contact elements/counter
Number of counters = 3

(d)   Control logic = 5 contact elements/math and compare
Number of math and compare = 20

Based on the memory utilization information from Table 5-1, the total number of words is:

(a)   Total contact elements   (70 x 10)   700
Total outputs   (70 x 1)   70
Total words   770

(b)   Total contact elements   (11 x 8)   88
Total timers   (11 x 3)   33
Total words   121

(c)   Total contact elements   (3 x 5)   15
Total counters   (3 x 3)   9
Total words   24

(d)   Total contact elements   (20 x 5)   100
Total math and compare   (20 x 1)   20
Total words   120

Thus, the total words of memory required for the storage of the instructions, outputs, timers, and counters is 1035 words (770 + 121 + 24 + 120), or just over 1K of memory.

The calculation performed in the previous example is actually an approximation because other factors, such as future expansion, must be considered before the final decision is made. After determining the minimum memory requirements for an application, it is wise to add an additional 25 to 50% more memory. This increase allows for changes, modifications, and future expansion. Keep in mind that the sophistication of the control program also affects memory requirements. If the application requires data manipulation and data storage, it will require additional memory. Normally, the enhanced instructions that perform mathematical and data manipulation operations will also have greater memory requirements. Depending on the PLC's manufacturer, the application memory may also include the data table and I/O table (discussed in the next section). If this is the case, then the amount of "real" user application memory available will be less than that specified. Exact memory usage can be determined by consulting the manufacturer's memory utilization specifications.

5-4 MEMORY ORGANIZATION AND I/O INTERACTION

The memory system, as mentioned before, is composed of two major sections—the system memory and the application memory—which in turn are composed of other areas. Figure 5-6 illustrates this memory organization, known as a **memory map**. Although the two main sections, system memory and application memory, are shown next to each other, they are not necessarily adjacent, either physically or by address. The memory map shows not only what is stored in memory, but also where data is stored, according to specific locations called *memory addresses*. An understanding of the memory map is very useful when creating a PLC control program and defining the data table.

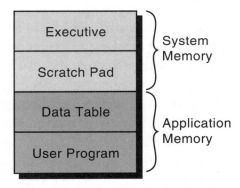

Figure 5-6. A simplified memory map.

Although two different programmable controllers rarely have identical memory maps, a generalized discussion of memory organization is still valid because all programmable controllers have similar storage requirements. In general, all PLCs must have memory allocated for four basic memory areas, which are as follows:

- **Executive Area.** The executive is a permanently stored collection of programs that are considered part of the system itself. These supervisory programs direct system activities, such as execution of the control program, communication with peripheral devices, and other system housekeeping activities.

- **Scratch Pad Area.** This is a temporary storage area used by the CPU to store a relatively small amount of data for interim calculations and control. The CPU stores data that is needed quickly in this memory area to avoid the longer access time involved with retrieving data from the main memory.

- **Data Table Area.** This area stores all data associated with the control program, such as timer/counter preset values and other stored constants and variables used by the control program or CPU. The data table also retains the status information of both the system inputs (once they have been read) and the system outputs (once they have been set by the control program).

- **User Program Area.** This area provides storage for programmed instructions entered by the user. The user program area also stores the control program.

The executive and scratch pad areas are hidden from the user and can be considered a single area of memory that, for our purpose, is called *system memory*. On the other hand, the data table and user program areas are accessible and are required by the user for control applications. They are called *application memory*.

The total memory specified for a controller may include system memory and application memory. For example, a controller with a maximum of 64K may have executive routines that use 32K and a system work area (scratch pad) of 1/4K. This arrangement leaves a total of 31 3/4K for application memory (data table and user memory). Although it is not always the case, the maximum memory specified for a given programmable controller normally includes only the total amount of application memory available. Other controllers may specify only the amount of user memory available for the control program, assuming a fixed data table area defined by the manufacturer. Now, let's take a closer look at the application memory and explore how it interacts with the user and the program.

APPLICATION MEMORY

The application memory stores programmed instructions and any data the processor will use to perform its control functions. Figure 5-7 shows a mapping of the typical elements in this area. Each programmable controller has a maximum amount of application memory, which varies depending on the size of the controller. The controller stores all data in the data table section of the application memory, while it stores programmed instructions in the user program section.

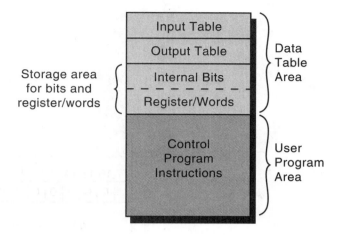

Figure 5-7. Application memory map.

Data Table Section. The data table section of a PLC's application memory is composed of several areas (see Figure 5-7). They are:

- the input table

- the output table

- the storage area

These areas contain information in binary form representing input/output status (ON or OFF), numbers, and codes. Remember that the memory structure contains cell areas, or bits, where this binary information is stored. Following is an explanation of each of the three data table areas.

<u>Input Table.</u> The **input table** is an array of bits that stores the status of digital inputs connected to the PLC's input interface. The maximum number of input table bits is equal to the maximum number of field inputs that can be connected to the PLC. For example, a controller with a maximum of 64 field inputs requires an input table of 64 bits. Thus, each connected input has an analogous bit in the input table, corresponding to the terminal to which the input is connected. The address of the input device is the bit and word location of its corresponding location in the input table. For example, the limit switch connected to the input interface in Figure 5-8 has an address of 13007_8 as its corresponding bit in the input table. This address comes from the word location 130_8 and the bit number 07_8, both of which are related to the module's rack position and the terminal connected to the field device (see Section 6-2). If the limit switch is OFF, the corresponding bit (13007_8) is 0 (see Figure 5-8a); if the limit switch is ON (see Figure 5-8b), the corresponding bit is 1.

During PLC operation, the processor will read the status of each input in the input module and place a value (1 or 0) in the corresponding address in the input table. The input table is constantly changing to reflect the changes of the input module and its connected field devices. These input table changes take place during the reading part of the I/O update.

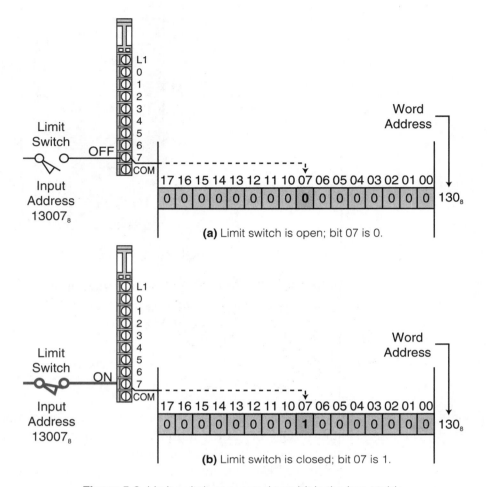

Figure 5-8. Limit switch connected to a bit in the input table.

Output Table. The **output table** is an array of bits that controls the status of digital output devices that are connected to the PLC's output interface. The maximum number of bits available in the output table equals the maximum number of output field devices that can interface with the PLC. For example, a PLC with a maximum of 128 outputs requires an output table of 128 bits.

Like the input table, each connected output has an analogous bit in the output table corresponding to the exact terminal to which the output is connected. The processor controls the bits in the output table as it interprets the control program logic during the program scan, turning the output modules ON and OFF accordingly during the output update scan. If a bit in the table is turned ON (1), then the connected output is switched ON (see Figure 5-9a); if a bit is cleared, or turned OFF (0), the output is switched OFF (see Figure 5-9b). Remember that the turning ON and OFF of field devices via the output module occurs during the update of outputs after the end of the scan.

Storage Area. The purpose of the **storage area** section of the data table is to store changeable data, whether it is one bit or a word (16 bits). The storage area consists of two parts: an *internal bit storage area* and a *register/word*

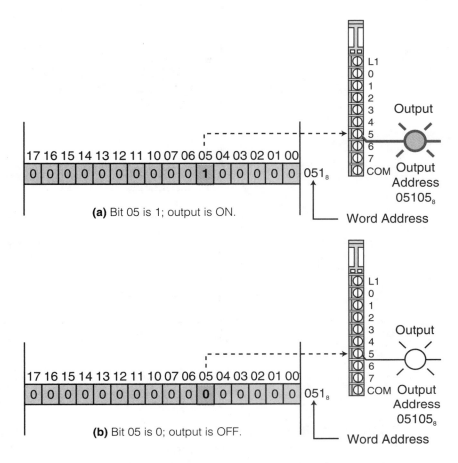

Figure 5-9. Field output connected to a bit in the output table.

storage area (see Figure 5-10). The internal bit storage area contains storage bits that are referred to as either *internal outputs, internal coils, internal (control) relays,* or *internals*. These internals provide an output, for interlocking purposes, of ladder sequences in the control program. Internal outputs do not directly control output devices because they are stored in addresses that do not map the output table and, therefore, any output devices.

When the processor evaluates the control program and an internal bit is energized (1), its referenced contact (the contact with this bit address) will change state—if it is normally open, it will close; if it is normally closed, it will open. Internal contacts are used in conjunction with either other internals or "real" input contacts to form interlocking sequences that drive an output device or another internal output.

The register/word storage area is used to store groups of bits (bytes and words). This information is stored in binary format and represents quantities or codes. If decimal quantities are stored, the binary pattern of the register represents an equivalent decimal number (see Chapter 2). If a code is stored, the binary pattern represents a BCD number or an ASCII code character (one character per byte).

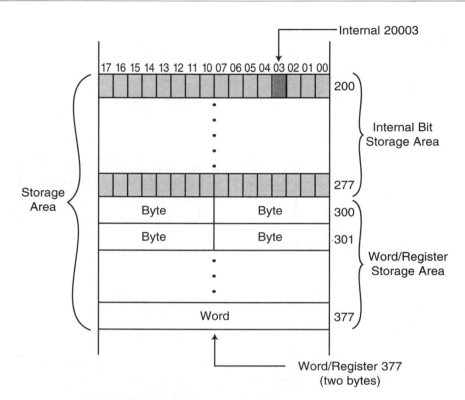

Figure 5-10. Storage area section of the data table.

Values placed in the register/word storage area represent input data from a variety of devices, such as thumbwheel switches, analog inputs, and other types of variables. In addition to input values, these registers can contain output values that are destined to go to output interface modules connected to field devices, such as analog meters, seven-segment LED indicators (BCD), control valves, and drive speed controllers. Storage registers are also used to hold fixed constants, such as preset timer/counter values, and changing values, such as arithmetic results and accumulated timer/counter values. Depending on their use, the registers in the register/word storage area may also be referred to as *input registers, output registers,* or *holding registers.* Table 5-2 shows typical constants and variables stored in these registers.

Constants	Variables
Timer preset values	Timer accumulated values
Counter preset values	Counter accumulated values
Loop control set points	Result values from math operations
Compare set points	Analog input values
Decimal tables (recipes)	Analog output values
ASCII characters	BCD inputs
ASCII messages	BCD outputs
Numerical tables	

Table 5-2. Constants and variables stored in register/word storage area registers.

EXAMPLE 5-2

Referencing Figure 5-11, what happens to internal 2301 (word 23, bit 01) when the limit switch connected to input terminal 10 closes?

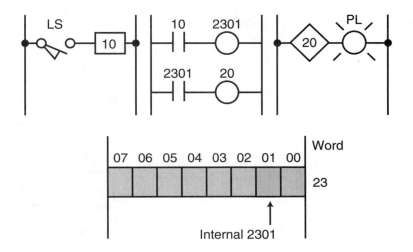

Figure 5-11. Open limit switch connected to an internal output.

SOLUTION

When LS closes (see Figure 5-12), contact 10 will close, turning internal output 2301 ON (a 1 in bit 01 of word 23). This will close contact 2301 ($\frac{2301}{\dashv\vdash}$) and turn real output 20 ON, causing the light PL to turn ON at the end of the scan.

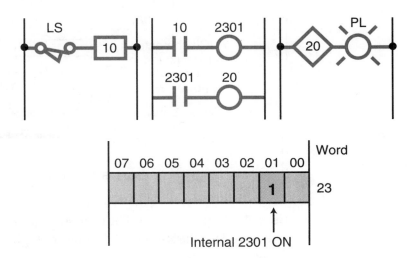

Figure 5-12. Closed limit switch connected to an internal output.

EXAMPLE 5-3

For the memory map shown in Figure 5-13, illustrate how to represent the following numbers in the storage area: **(a)** the BCD number 9876, **(b)** the ASCII character *A* (octal 101) in one byte (use lower byte), and **(c)** the analog value 2257 (1000 1101 0001 binary). Represent these values starting at register 400.

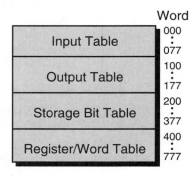

Figure 5-13. Memory map.

SOLUTION

Figure 5-14 shows the register data corresponding to the BCD number 9876, the ASCII character *A*, and the analog value 2257.

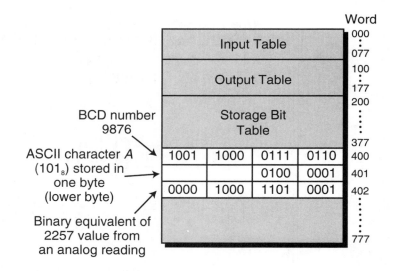

Figure 5-14. Solution for Example 5-3.

User Program Section. The user program section of the application memory is reserved for the storage of the control logic. All of the PLC instructions that control the machine or process are stored in this area. The processor's executive software language, which represents each of the PLC instructions, stores its instructions in the user program memory.

When a PLC executes its program, the processor interprets the information in the user program memory and controls the referenced bits in the data table that correspond to real or internal I/O. The processor's execution of the executive program accomplishes this interpretation of the user program.

The maximum amount of user program memory available is normally a function of the controller's size (i.e., I/O capacity). In medium and large controllers, the user program area is made flexible by altering the size of the data table so that it meets the minimum data storage requirements. In small controllers, however, the user program area is normally fixed. The amount of user program memory required is directly proportional to the number of instructions used in the control program. Estimation of user memory requirements is accomplished using the method described earlier in Section 5-3.

5-5 CONFIGURING THE PLC MEMORY—I/O ADDRESSING

Understanding memory organization, especially the interaction of the data table's I/O mapping and storage areas, helps in the comprehension of a PLC's functional operation. Although the memory map is often taken for granted by PLC users, a thorough understanding of it provides a better perception of how the control software program should be organized and developed.

DATA TABLE ORGANIZATION

The data table's organization, or *configuration* as it is sometimes called, is very important. The configuration defines not only the discrete device addresses, but also the registers that will be used for numerical and analog control, as well as basic PLC timing and counting operations. The intention of the following discussion of data table organization is not to go into detail about configuration, but to review what you have learned about the memory map, making sure that you understand how memory and I/O interact.

First, let's consider an example of an application memory map for a PLC. The controller has the following memory, I/O, and numbering system specifications:

- total application memory of 4K words with 16 bits

- capability of connecting 256 I/O devices (128 inputs and 128 outputs)

- 128 available internal outputs

- capability of up to 256 storage registers, selectable in groups of 8-word locations, with 8 being the minimum number of registers possible (32 groups of 8 registers each)

- octal (base 8) numbering system with 2-byte (16-bit) word length

To illustrate this memory map may seem unnecessary, but at this point, we do not know the starting address of the control program. This does not matter as far as the program is concerned; however, it does matter when determining the register address references to be used, since these register addresses are referred to in the control program (i.e., timer preset and accumulated values).

With this in mind, let's set the I/O table boundaries. Assuming the inputs are first in the I/O mapping, the input table will start at address 0000_8 and end at address 0007_8 (see Figure 5-15). The outputs will start at address 0010_8 and end at address 0017_8. Since each memory word has 16 bits, the 128 inputs require 8 input table words, and likewise for the outputs. The starting address for the internal output storage area is at memory location address 0020_8 and continues through address 0027_8 (8 words of 16 bits each totaling 128 internal output bits). Address 0030_8 indicates the beginning of the register/word storage area. This area must have a minimum of 8 registers, with a possibility of up to 256 registers added in 8-register increments. The first 8 required

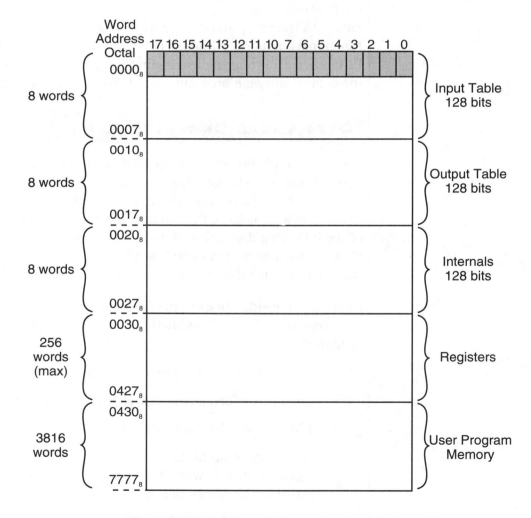

Figure 5-15. I/O table and user memory boundaries.

registers, then, will end at address 0037_8 (see Figure 5-16). Any other 8-register increments will start at 0040_8, with the last possible address being 0427_8, providing a total of 256 registers.

If all available storage registers are utilized, then the starting memory address for the control program will be 0430_8. This configuration will leave 3816 (decimal) locations to store the control software. Figure 5-15 showed this maximum configuration.

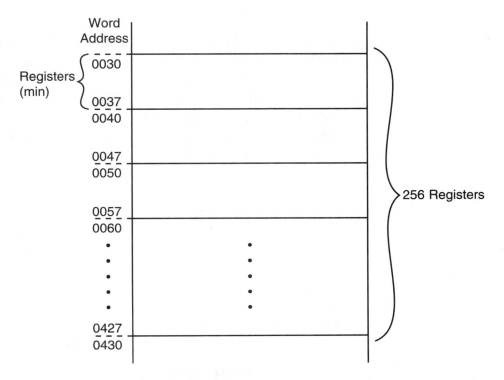

Figure 5-16. Breakdown, in groups of eight, of the register storage area at its maximum capacity.

Most controllers allow the user to change the range of register boundaries without any concern for starting memory addresses of the program. Nonetheless, the user should know beforehand the number of registers needed. This will be useful when assigning register addresses in the program.

I/O ADDRESSING

Throughout this text, we have mentioned that the programmable controller's operation simply consists of reading inputs, solving the ladder logic in the user program memory, and updating the outputs. As we get more into PLC programming and the application of I/O modules, we will review the relationship between the I/O address and the I/O table, as well as how I/O addressing is used in the program.

The input/output structure of a programmable controller is designed with one thing in mind—simplicity. Input/output field devices are connected to a PLC's I/O modules, which are located in the *rack* (the physical enclosure that houses a PLC's supplementary devices). The rack location of each I/O device is then mapped to the I/O table, where the I/O module placement defines the address of the devices connected to the module. Some PLCs use internal module switches to define the addresses used by the devices connected to the module. In the end, however, all of the input and output connections are mapped to the I/O table.

Assume that a simple relay circuit contains a limit switch driving a pilot light (see Figure 5-17). This circuit is to be connected to a PLC input module and output module, as shown in Figure 5-18. For the purpose of our discussion, let's assume that each module contains 8 possible input or output channels and that the PLC has a memory map similar to the one shown previously in Figure 5-15. The limit switch is connected to the number 5 (octal) terminal of the input module, while the light is connected to the number 6 (octal) terminal of the output module.

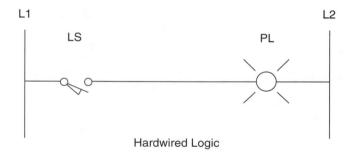

Figure 5-17. A relay circuit with a limit switch driving a pilot light.

Let's assume that, due to their placement inside the rack, the I/O modules' map addresses are word 0000 for the input module and word 0010 for the output module. Therefore, the processor will reference the limit switch as input 000005, and it will reference the light as output 001006 (i.e., the input is mapped to word 0000 bit 05, and the output is mapped to word 0010 bit 06). These addresses are mapped to the I/O table. Every time the processor reads the inputs, it will update the input table and turn ON those bits whose input devices are 1 (ON or closed). When the processor begins the execution of the ladder program, it will provide power (i.e., continuity) to the ladder element corresponding to the limit switch, because its reference address is 1 (see Figure 5-18). At this time, it will set output 001006 ON, and the pilot light will turn ON after *all* instructions have been evaluated and the end of scan (EOS)—where the output update to the module takes place—has been reached. This operation is repeated every scan, which can be as fast as every thousandth of a second (1 msec) or less.

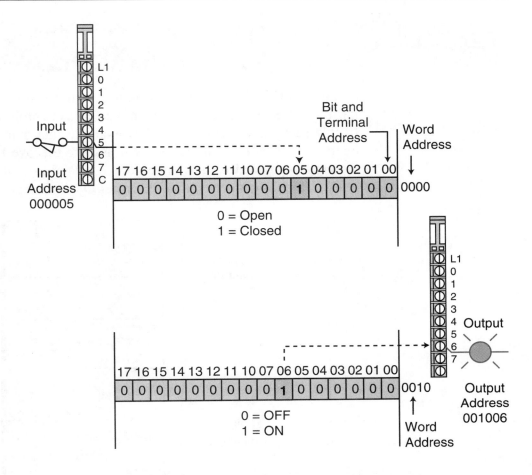

Figure 5-18. Input/output module connected to field devices.

Note that addresses 000005 and 001006 can be used as often as required in the control program. If we had programmed a contact at 001006 to drive internal output 002017 (see Figure 5-19), the controller would turn its internal output bit (002017) to 1 every time output 001006 turned ON. However, this output would not be directly connected to any output device. Note that internal storage bit 002017 is located in word 0020 bit 17.

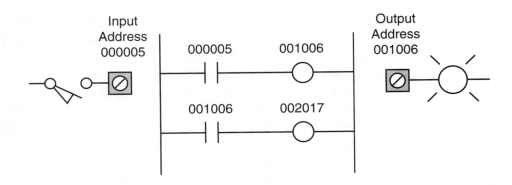

Figure 5-19. PLC ladder implementation of Figure 5-17 using an internal output bit.

5-6 SUMMARY OF MEMORY, SCANNING, AND I/O INTERACTION

So far, you have learned about scanning, memory system organization, and the interaction of input and output field devices in a programmable controller. In this section, we will present an example that summarizes these PLC operations. In this example, we will assume that we have a simple PLC memory, organized as shown in Figure 5-20, and a simple circuit (see Figure 5-21), which is connected to a PLC via I/O interfaces.

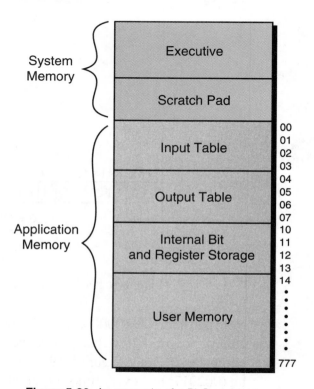

Figure 5-20. An example of a PLC memory map.

The instructions used to represent the simple control program, shown in Figure 5-22, are stored in the user memory section, where specific binary 1s and 0s represent the instructions (e.g., the $\overset{0010}{\dashv\vdash}$ instruction). During the PLC scan, the executive program reads the status of inputs and places this data into the input data table. Then, the programmable controller scans the user memory to interpret the instructions stored. As the logic is solved rung by rung according to the status of the I/O table, the results from the program evaluation are stored in the output table and the storage bit table (if the program uses internals). After the evaluation (program scan), the executive program updates the values stored in the output table and sends commands to the output modules to turn ON or OFF the field devices connected to their respective interfaces. Figure 5-23 on the page 26 shows the steps that will occur during the evaluation of the PLC circuit shown in Figure 5-22.

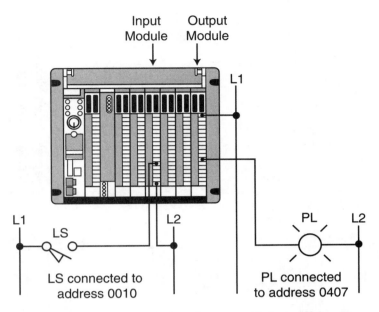

Figure 5-21. A simple circuit connected to a PLC via I/O interfaces.

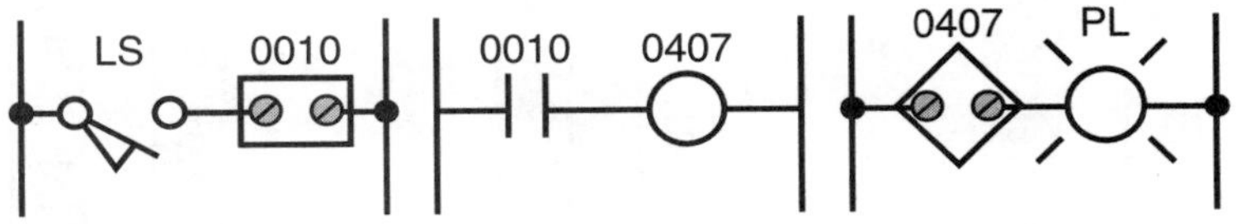

Figure 5-22. Instructions used to represent the control program.

5-7 MEMORY CONSIDERATIONS

The previous sections presented an analysis of programmable controller memory characteristics regarding memory type, storage capacity, organization, structure, and their relationship to I/O addressing. Particular emphasis was placed on the application memory, which stores the control program and data. Careful consideration must also be given to the type of memory, since certain applications require frequent changes, while others require permanent storage once the program is debugged. A RAM with battery support may be adequate in most cases, but in others, a RAM and an optional nonvolatile-type memory may be required.

It is important to remember that the total memory capacity for a particular controller may not be completely available for application programming. The specified memory capacity may include memory utilized by the executive routines or the scratch pad, as well as the user program area.

The application memory varies in size depending on the size of the controller. The total area available for the control program also varies according to the size of the data table. In small controllers, the data table is usually fixed, which

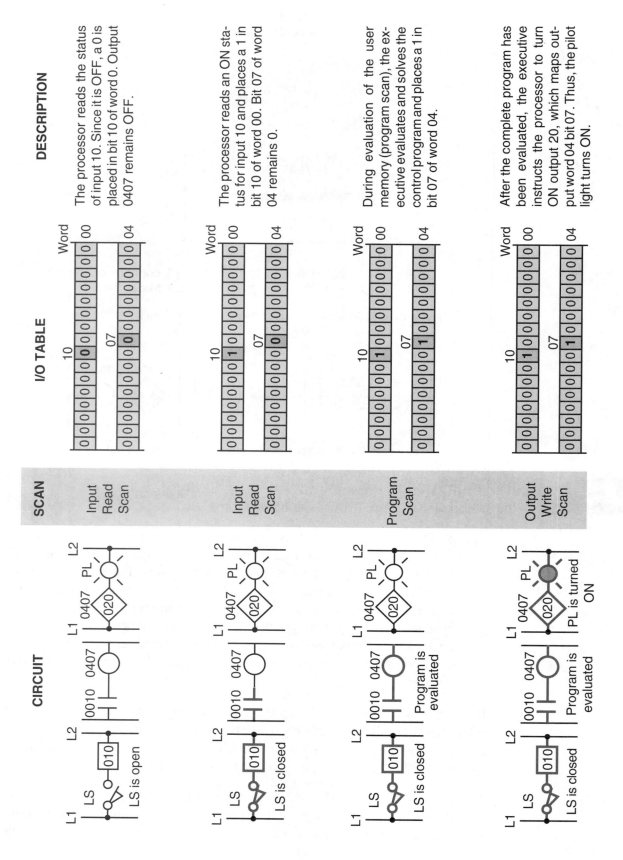

Figure 5-23. Steps in the evaluation of the PLC circuit shown in Figure 5-22.

SCAN	DESCRIPTION
Input Read Scan	The processor reads the status of input 10. Since it is OFF, a 0 is placed in bit 10 of word 0. Output 0407 remains OFF.
Input Read Scan	The processor reads an ON status for input 10 and places a 1 in bit 10 of word 00. Bit 07 of word 04 remains 0.
Program Scan	During evaluation of the user memory (program scan), the executive evaluates and solves the control program and places a 1 in bit 07 of word 04.
Output Write Scan	After the complete program has been evaluated, the executive instructs the processor to turn ON output 20, which maps output word 04 bit 07. Thus, the pilot light turns ON.

means that the user program area will be fixed. In larger controllers, however, the data table size is usually selectable, according to the data storage requirements of the application. This flexibility allows the program area to be adjusted to meet the application's requirements.

When selecting a controller, the user should consider any limitations that may be placed on the use of the available application memory. One controller, for example, may have a maximum of 256 internal outputs with no restrictions on the number of timers, counters, and various types of internal outputs used. Another controller, however, may have 256 available internal outputs that are restricted to 50 timers, 50 counters, and 156 of any combination of various types of internal outputs. A similar type of restriction may also be placed on data storage registers.

One way to ensure that memory requirements are satisfied is to first understand the application requirements for programming and data storage, as well as the flexibility required for program changes and on-line data entry. Creating the program on paper first will help when evaluating these capacity requirements. With the use of a memory map, users can learn how much memory is available for the application and, then, how the application memory should be configured for their use. It is also good to know ahead of time if the application memory is expandable. This knowledge will allow the user to make sound decisions about memory type and requirements.

KEY TERMS

application memory
data table
electrically alterable read-only memory (EAROM)
electrically erasable programmable read-only memory (EEPROM)
erasable programmable read-only memory (EPROM)
executive memory
input table
memory
memory map
nonvolatile memory
output table
programmable read-only memory (PROM)
random-access memory (RAM)
read-only memory (ROM)
scratch pad memory
storage area
user program memory
volatile memory

CHAPTER SIX

THE DISCRETE INPUT/OUTPUT SYSTEM

All science is concerned with the relationship of cause and effect.

—Laurence J. Peter

CHAPTER
HIGHLIGHTS
Input/output (I/O) systems put the "control" in programmable controllers. These systems allow PLCs to work with field devices to perform programmed applications. This chapter introduces the most common type of I/O system—the discrete interface—and explains its physical, electrical, and functional characteristics. You will learn how discrete I/O systems provide the connection between PLCs and the outside world. In the following two chapters, you will further explore the operation and installation of input/output systems, learning about analog and special function I/O interfaces.

6-1 INTRODUCTION TO DISCRETE I/O SYSTEMS

The discrete input/output (I/O) system provides the physical connection between the central processing unit and field devices that transmit and accept digital signals (see Figure 6-1). **Digital signals** are noncontinuous signals that have only two states—ON and OFF. Through various interface circuits and field devices (limit switches, transducers, etc.), the controller senses and measures physical quantities (e.g., proximity, position, motion, level, temperature, pressure, current, and voltage) associated with a machine or process. Based on the status of the devices sensed or the process values measured, the CPU issues commands that control the field devices. In short, input/output interfaces are the sensory and motor skills that exercise control over a machine or process.

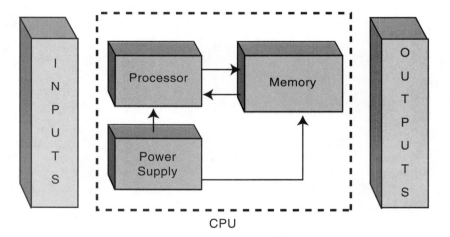

Figure 6-1. Block diagram of a PLC's CPU and I/O system.

The predecessors of today's PLCs were limited to just discrete input/output interfaces, which allowed interfacing with only ON/OFF-type devices. This limitation gave the PLC only partial control over many processes, because many process applications required analog measurements and manipulation of numerical values to control analog and instrumentation devices. Today's controllers, however, have a complete range of discrete and analog interfaces, which allow PLCs to be applied to almost any type of control. Figure 6-2 shows a typical discrete I/O system.

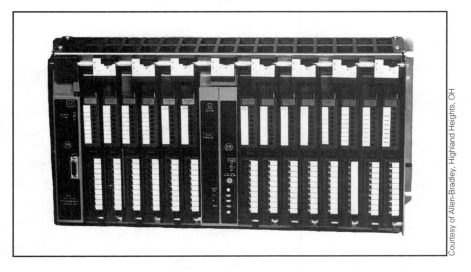

Figure 6-2. Typical discrete input/output system.

6-2 I/O RACK ENCLOSURES AND TABLE MAPPING

An **I/O module** is a plug-in–type assembly containing circuitry that communicates between a PLC and field devices. All I/O modules must be placed or inserted into a **rack enclosure**, usually referred to as a *rack*, within the PLC (see Figure 6-3). The rack holds and organizes the programmable controller's I/O modules, with a module's rack location defining the **I/O address** of its connected device. The I/O address is a unique number that identifies the input/output device during control program setup and execution. Several PLC manufacturers allow the user to select or set the addresses (to be mapped to the I/O table) for each module by setting internal switches (see Figure 6-4).

Figure 6-3. Example of an I/O rack enclosure.

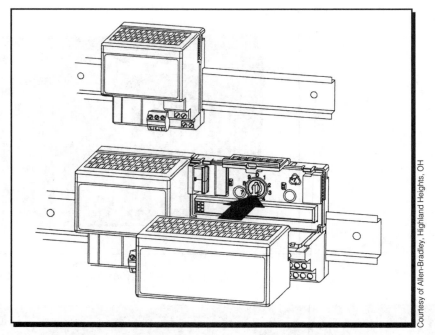

Figure 6-4. Internal switches used to set I/O addresses.

A rack, in general, recognizes the type of module connected to it (input or output) and the class of interface (discrete, analog, numerical, etc.). This module recognition is decoded on the back plane (i.e., the printed circuit board containing the data bus, power bus, and mating connectors) of the rack.

The controller's rack configuration is an important detail to keep in mind throughout system configuration. Remember that each of the connected I/O devices is referenced in the control program; therefore, a misunderstanding of the I/O location or addresses will create confusion during and after the programming stages.

Generally speaking, there are three categories of rack enclosures:

- master racks
- local racks
- remote racks

The term **master rack** (see Figure 6-5) refers to the rack enclosure containing the CPU or processor module. This rack may or may not have slots available for the insertion of I/O modules. The larger the programmable controller system, in terms of I/O, the less likely the master rack will have I/O housing capability.

A **local rack** (see Figure 6-6) is an enclosure, which is placed in the same area as the master rack, that contains I/O modules. If a master rack contains I/O modules, the master rack can also be considered a local rack. In general, a local rack (if not a master) contains a local I/O processor that sends data to and

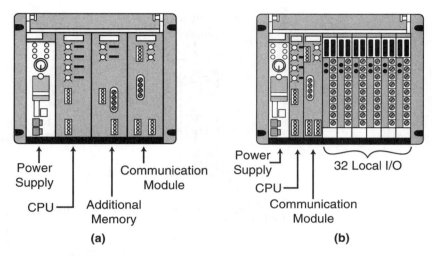

Figure 6-5. Master racks **(a)** without I/O modules and **(b)** with I/O modules.

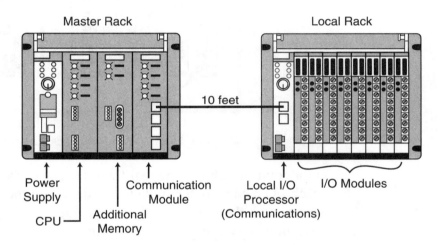

Figure 6-6. Local rack configuration.

from the CPU. This bidirectional information consists of diagnostic data, communication error checks, input status, and output updates. The I/O image table maps the local rack's I/O addresses.

As the name implies, **remote racks** (see Figure 6-7) are enclosures, containing I/O modules, located far away from the CPU. Remote racks contain an I/O processor (referred to as a remote I/O processor) that communicates input and output information and diagnostic status just like a local rack. The I/O addresses in this rack are also mapped to the I/O table.

The rack concept emphasizes the physical location of the enclosure and the type of processor (local, remote, or main CPU) that will be used in each particular rack. Every one of the I/O modules in a rack, whether discrete, analog, or special, has an address by which it is referenced. Therefore, each terminal point connected to a module has a particular address. This connec-

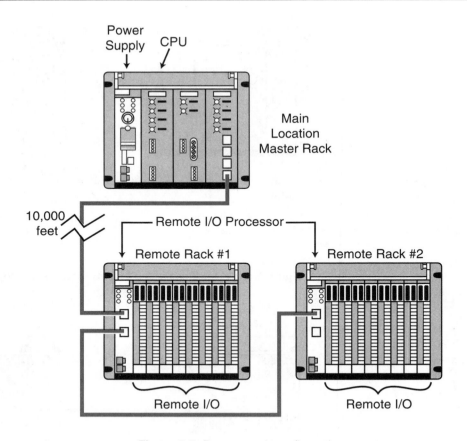

Power
Supply   CPU

Main
Location
Master Rack

10,000
feet

Remote I/O Processor

Remote Rack #1   Remote Rack #2

Remote I/O   Remote I/O

Figure 6-7. Remote rack configuration.

tion point, which ties the real field devices to their I/O modules, identifies each I/O device by the module's address and the terminal point where it is connected. This is the address that identifies the programmed input or output device in the control program.

I/O RACK AND TABLE MAPPING EXAMPLE

PLC manufacturers set specifications for placing I/O modules in rack enclosures. For example, some modules accommodate 2 to 16 field connections, while other modules require the user to follow certain I/O addressing regulations. It is not our intention in this section to review all of the different manufacturers' rules, but rather to explain how the I/O typically maps each rack and to illustrate some possible restrictions through a generic example.

As our example, let's use the PLC I/O placement specifications shown in Table 6-1. As Figure 6-8 illustrates, several factors determine the address location of each module. The type of module, input or output, determines the first address location from left to right (0 for outputs, 1 for inputs). The rack number and slot location of the module determine the next two address numbers. The terminal connected to the I/O module (0 through 7) represents the last address digit.

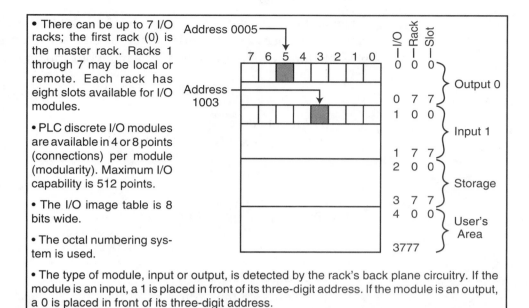

- There can be up to 7 I/O racks; the first rack (0) is the master rack. Racks 1 through 7 may be local or remote. Each rack has eight slots available for I/O modules.

- PLC discrete I/O modules are available in 4 or 8 points (connections) per module (modularity). Maximum I/O capability is 512 points.

- The I/O image table is 8 bits wide.

- The octal numbering system is used.

- The type of module, input or output, is detected by the rack's back plane circuitry. If the module is an input, a 1 is placed in front of its three-digit address. If the module is an output, a 0 is placed in front of its three-digit address.

Table 6-1. Specifications for the I/O rack enclosure example.

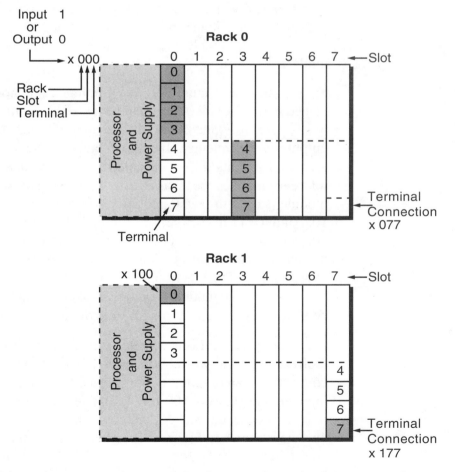

Figure 6-8. Illustration of the example I/O rack enclosure (x = 1 for inputs, 0 for outputs).

The maximum capacity of this system is 512 inputs or 512 outputs, or a total combination of 512 inputs and outputs that do not overlap addresses. The 512 possible inputs come from the following word addresses:

512 input addresses
(64 words × 8 bits/word)

1000_8 (word 100, bit 0)
•
•
to
•
•
1777_8 (word 177, bit 7)

While the 512 possible outputs come from word addresses:

512 output addresses
(64 words × 8 bits/word)

0000_8 (word 000, bit 0)
•
•
to
•
•
0777_8 (word 077, bit 7)

Again, note that the capacity is a total of 512 inputs and outputs together, not 512 each. If one input module takes a slot in the input table, the mirror image slot in the output table is taken by those inputs. The same applies for output modules.

For instance (see Figure 6-9), if a 4-point output module (see Figure 6-9b) is placed in rack 0, slot 0 (terminal addresses 0–3), the output table word 000_8, bits 0–3, represented by the shaded area in Figure 6-9c, will be mapped for outputs. Consequently, the input table image corresponding to the slot location 100_8, bits 0–3 (represented by the word *taken*) will not have a mapped reference input, since it has already been taken by outputs. If an 8-point input module is used in rack 0, slot 2 (see Figure 6-9a), indicating word location 102_8 (input = 1), the whole eight bits of that location in the input table (location 102_8 bits 0–7) would be taken by the mapping; the corresponding address in the output table (word location 002_8, bits 0–7 in Figure 6-9c) would not be able to be mapped. The bits from the output table that do not have a mapping due to the use of input modules could be used as internal outputs, since they cannot be physically connected output field devices (e.g., bits 4–7 of word 000).

For example, in Figure 6-9c, output addresses 0004 through 0007 (corresponding to word 000, bits 4–7 in the I/O table) cannot be physically connected to an output module because their map locations are taken by an input module (at word 100, bits 4–7). Therefore, these reference addresses can only be used as internal coil outputs. The use of these output bits as internal outputs is shown in Figure 6-10, where output 0004 (now used as an internal coil) will be turned ON if its logic is TRUE and contacts from this output can be used in other output rungs.

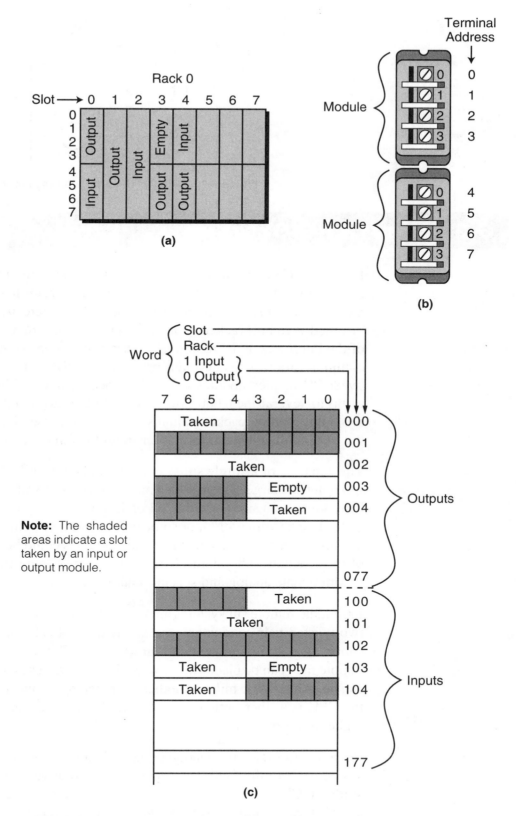

Note: The shaded areas indicate a slot taken by an input or output module.

Figure 6-9. Diagrams of **(a)** an I/O table, **(b)** two 4-point I/O modules in one slot, and **(c)** an I/O table mapping.

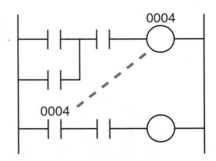

Figure 6-10. Output 0004 used as an internal coil.

6-3 REMOTE I/O SYSTEMS

In large PLC systems (upwards of 512 I/O), input/output subsystems can be located away from the central processing unit. A **remote I/O subsystem** is a rack-type enclosure, separate from the CPU, where I/O modules can be installed. A remote rack includes a power supply that drives the logic circuitry of the interfaces and a remote I/O adapter or processor module that allows communication with the main processor (CPU). The communication between I/O adapter modules and the CPU occurs in serial binary form at speeds of up to several megabaud (millions of bits transmitted per second). This serial information packet contains 1s and 0s, representing both the status of the I/O and diagnostic information about the remote rack.

The capacity of a single subsystem (rack) is normally 32, 64, 128, or 256 I/O points. A large system with a maximum capacity of 1024 I/O points may have subsystem sizes of either 64 or 128 points—eight racks with 128 I/O, sixteen racks with 64 I/O, or some combination of both sizes equal to 1024 I/O. In the past, only discrete interface modules could be placed in the racks of most remote subsystems. Today, however, remote I/O subsystems also accommodate analog and special function interfaces.

Individual remote subsystems are normally connected to the CPU via one or two twisted-pair conductors or a single coaxial cable, using either a *daisy chain, star*, or *multidrop* configuration (see Figure 6-11). The distance a remote rack can be placed away from the CPU varies among products, but it can be as far as two miles. Another approach for connecting remote racks to the CPU is a fiber-optic data link, which allows greater distances and has higher noise resistance.

Remote I/O offers tremendous materials and labor cost savings on large systems where the field devices are clustered at various, distant locations. With the CPU in a main control room or some other central area, only the communication link must be wired between the remote rack and the processor, replacing hundreds of field wires. Another advantage of remote I/O is that subsystems may be installed and started up independently, allowing maintenance of individual subsystems while others continue to operate. Also, troubleshooting and connection checks become much easier, since hundreds of wires do not need to be checked all the way back to the master rack.

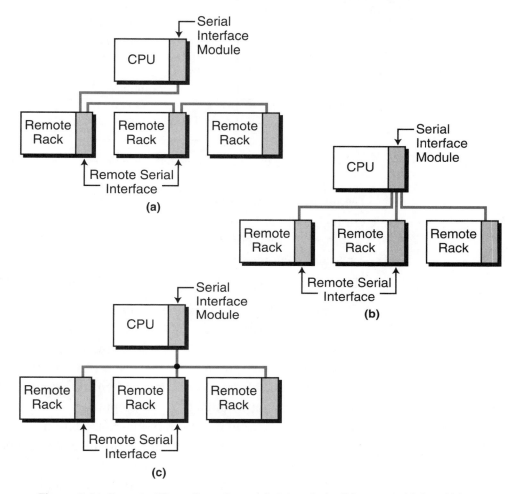

Figure 6-11. Remote I/O configurations: **(a)** daisy chain, **(b)** star, and **(c)** multidrop.

6-4 PLC INSTRUCTIONS FOR DISCRETE INPUTS

The most common class of input interfaces is digital (or discrete). **Discrete input interfaces** connect digital field input devices (those that send noncontinuous, fixed-variable signals) to input modules and, consequently, to the programmable controller. The discrete, noncontinuous characteristic of digital input interfaces limits them to sensing signals that have only two states (i.e., ON/OFF, OPEN/CLOSED, TRUE/FALSE, etc.). To an input interface circuit, discrete input devices are essentially switches that are either open or closed, signifying either 1 (ON) or 0 (OFF). Table 6-2 shows several examples of discrete input field devices.

Field Input Devices
Circuit breakers
Level switches
Limit switches
Motor starter contacts
Photoelectric eyes
Proximity switches
Push buttons
Relay contacts
Selector switches
Thumbwheel switches (TWS)

Table 6-2. Discrete input devices.

Many instructions are designed to manipulate discrete inputs. These instructions handle either *single bits*, which control one field input connection, or *multibits,* which control many input connections. Regardless of whether the instruction controls one discrete input or multiple inputs, the information provided by the field device is the same—either ON or OFF.

During our discussion of input modules, keep in mind the relationship between interface signals (ON/OFF), rack and module locations (where the input device is inserted), and I/O table mapping and addressing (used in the control program). Remember that each PLC manufacturer determines the addressing and mapping scheme used with its systems. Manufacturers may use a 1 for an input and a 0 for an output, or they may simply assign an I/O address for the input or output module inserted in a particular slot of a rack. Figure 6-12 illustrates a simplified 8-bit image table where limit switch LS1 is connected to a discrete input module in rack 0, which can connect 8 field inputs (0–7). Note that LS1 is known as input 014, which stands for rack 0, slot 1, connection 4.

When an input signal is energized (ON), the input interface senses the field device's supplied voltage and converts it to a logic-level signal (either 1 or 0), which indicates the status of that device. A logic 1 in the input table indicates an ON or CLOSED condition, and a logic 0 indicates an OFF or OPEN condition. PLC symbolic instructions, which include the normally open (⊣⊢) and normally closed (⊣/⊢) instructions, transfer this field status information into the input table.

For multibit modules that receive multiple inputs, such as thumbwheel switches used in register (BCD) interfaces, block transfer or get data instructions place input values into the data table (see Figure 6-13). Chapter 9 explains single-bit and multibit instructions in more detail.

Rack 0

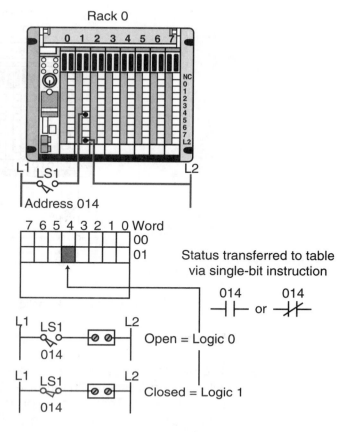

Figure 6-12. An 8-bit input image table.

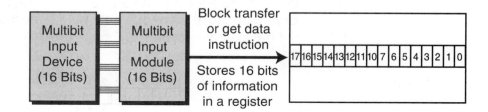

Figure 6-13. Block transfer and get data instructions transferring multibit input values into the data table.

Example 6-1

For the rack configuration shown in Figure 6-14, determine the address for each field device wired to each input connection in the 8-bit discrete input module. Assume that the first four slots of this 64 I/O micro-PLC are filled with outputs and that the second four slots are filled with inputs. Also, assume that the addresses follow a rack-slot-connection scheme and start at I/O address 000. Note that the number system is octal.

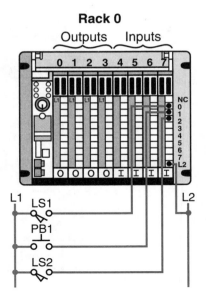

Figure 6-14. Rack configuration for Example 6-1.

SOLUTION

The discrete input module (where the input devices are connected) will have addresses 070 through 077, because it is located in rack 0, slot number 7. Therefore, each of the field input devices will have addresses as shown in Figure 6-15; LS1 will be known as input 070, PB1 as input 071, and LS2 as input 072. The control program will reference the field devices by these addresses. If LS1 is rewired to another connection in another discrete input, its address reference will change. Consequently, the address must be changed in the control program because there can only be one address per discrete field input device connection.

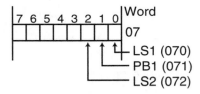

Figure 6-15. Field device addresses for the rack configuration in Example 6-1.

6-5 TYPES OF DISCRETE INPUTS

As mentioned earlier, discrete input interfaces sense noncontinuous signals from field devices—that is, signals that have only two states. Discrete input interfaces receive the voltage and current required for this operation from the back plane of the rack enclosure where they are inserted (see Chapter 4 for loading considerations). The signal that these discrete interfaces receive from input field devices can be of different types and/or magnitudes (e.g., 120 VAC, 12 VDC). For this reason, discrete input interface circuits are available in different AC and DC voltage ratings. Table 6-3 lists the standard ratings for discrete inputs.

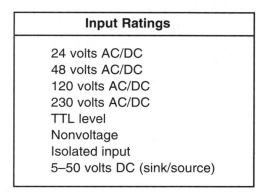

Input Ratings
24 volts AC/DC
48 volts AC/DC
120 volts AC/DC
230 volts AC/DC
TTL level
Nonvoltage
Isolated input
5–50 volts DC (sink/source)

Table 6-3. Standard ratings for discrete input interfaces.

To properly apply input interfaces, you should have an understanding of how they operate and an awareness of certain operating specifications. Section 6-9 discusses these specifications, while Chapter 20 describes start-up and maintenance procedures for I/O systems. Now, let's look at the different types of discrete input interfaces, along with their operation and connections.

AC/DC INPUTS

Figure 6-16 shows a block diagram of a typical **AC/DC input interface** circuit. Input circuits vary widely among PLC manufacturers, but in general, AC/DC interfaces operate similarly to the circuit in the diagram. An AC/DC input circuit has two primary parts:

- the power section

- the logic section

These sections are normally, but not always, coupled through a circuit that electrically separates them, providing isolation.

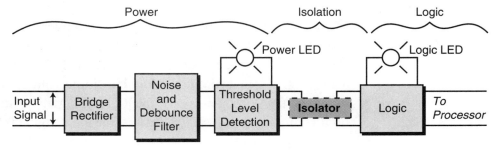

Figure 6-16. Block diagram of an AC/DC input circuit.

The power section of an AC/DC input interface converts the incoming AC voltage from an input-sensing device, such as those described in Table 6-2, to a DC, logic-level signal that the processor can use during the read input section of its scan. During this process, the bridge rectifier circuit of the interface's power section converts the incoming AC signal to a DC-level signal. It then passes the signal through a filter circuit, which protects the signal against bouncing and electrical noise on the input power line. This filter causes a signal delay of typically 9–25 msec. The power section's threshold circuit detects whether the signal has reached the proper voltage level for the specified input rating. If the input signal exceeds and remains above the threshold voltage for a duration equal to the filter delay, the signal is recognized as a valid input.

Figure 6-17 shows a typical AC/DC input circuit. After the interface detects a valid signal, it passes the signal through an isolation circuit, which completes the electrically isolated transition from an AC signal to a DC, logic-level signal. The logic circuit then makes the DC signal available to the processor through the rack's back plane data bus, a pathway along which data moves. The signal is electrically isolated so that there is no electrical connection between the field device (power) and the controller (logic). This electrical separation helps prevent large voltage spikes from damaging either the logic side of the interface or the PLC. An optical coupler or a pulse transformer provides the coupling between the power and logic sections.

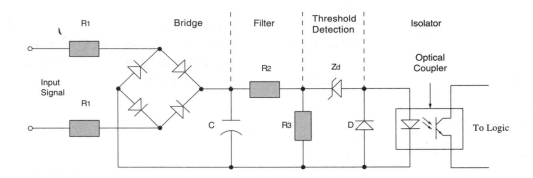

Figure 6-17. Typical AC/DC input circuit.

Most AC/DC input circuits have an LED (power) indicator to signal that the proper input voltage level is present (refer to Figure 6-16). In addition to the power indicator, the circuit may also have an LED to indicate the presence of a logic 1 signal in the logic section. If an input voltage is present and the logic circuit is functioning properly, the logic LED will be lit. When the circuit has both voltage and logic indicators and the input signal is ON, both LEDs must be lit to indicate that the power and logic sections of the module are operating correctly. Figure 6-18 shows AC/DC device connection diagrams.

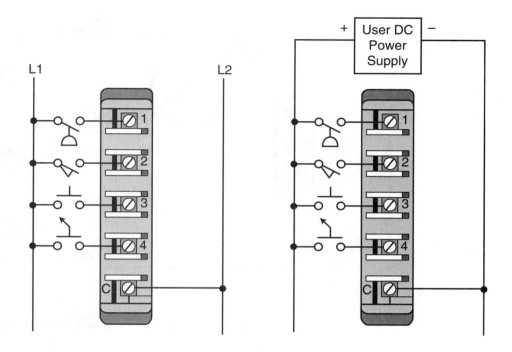

Figure 6-18. Device connections for **(a)** an AC input module and **(b)** a DC input module with common wire connection "C" used to complete the path from hot.

DC INPUTS (SINK/SOURCE)

A **DC input module** interfaces with field input devices that provide a DC output voltage. The difference between a DC input interface and an AC/DC input interface is that the DC input does not contain a bridge circuit, since it does not convert an AC signal to a DC signal. The input voltage range of a DC input module varies between 5 and 30 VDC. The module recognizes an input signal as being ON if the input voltage level is at 40% (or another manufacturer-specified percentage) of the supplied reference voltage. The module detects an OFF condition when the input voltage falls under 20% (or another manufacturer-specified percentage) of the reference DC voltage.

A DC input module can interface with field devices in both **sinking** and **sourcing** operations, a capability that AC/DC input modules do not have. Sinking and sourcing operations refer to the electrical configuration of the circuits in the module and field input devices. If a device *provides* current when it is ON, it is said to be sourcing current. Conversely, if a device *receives* current when it is ON, it is said to be sinking current. There are both sinking and sourcing field devices, as well as sinking and sourcing input modules. The most common, however, are sourcing field input devices and sinking input modules. Rocker switches inside a DC input module may be used to select sink or source capability. Figure 6-19 depicts sinking and sourcing operations and current direction.

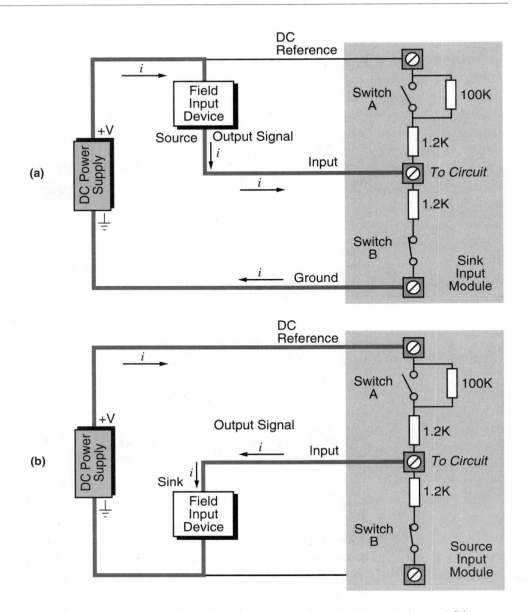

Figure 6-19. Current for **(a)** a sinking input module/sourcing input device and **(b)** a sourcing input module/sinking input device.

During interfacing, the user must keep in mind the minimum and maximum specified currents that the input devices and module are capable of sinking or sourcing. Also, if the module allows selection of a sink or source operation via selector switches, the user must assign them properly. A potential interface problem could arise, for instance, if an 8-input module was set for a sink operation and all input devices except one were operating in a source configuration. The source input devices would be ON, but the module would not properly detect the ON signal, even though a voltmeter would detect a voltage across the module's terminals. Figure 6-20 illustrates three field device connections to a DC input module with both sinking and sourcing input device capabilities.

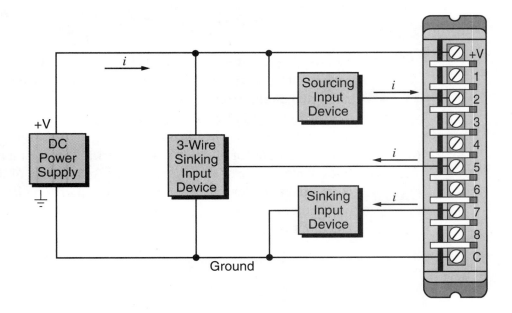

Figure 6-20. Field device connections for a sink/source DC input module.

The majority of DC proximity sensors used as PLC inputs provide a sinking sensor output, thereby requiring a sinking input module. However, if an application requires only one sinking output and the controller already has several sourcing inputs connected to a sourcing input module, the user may use the inexpensive circuit shown in Figure 6-21 to interface the sinking output with the sourcing input module. The sourcing current provided by this input is approximately 50 mA. Note that if the supply voltage (V_S) is increased, the current I_{out} will be greater than 50 mA.

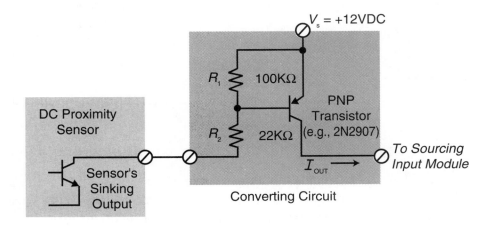

Figure 6-21. Conversion circuit interfacing a sinking output with a sourcing input module.

ISOLATED AC/DC INPUTS

Isolated input interfaces operate like standard AC/DC modules except that each input has a separate return, or *common*, line. Depending on the manufacturer, standard AC/DC input interfaces may have one return line per 4, 8, or 16 points. Although a single return line, provided in standard multipoint input modules, may be ideal for 95% of AC/DC input applications, it may not be suitable for applications requiring individual or isolated common lines. An example of this type of application is a set of input devices that are connected to different phase circuits coming from different power distribution centers. Figure 6-22 illustrates a sample device connection for an AC/DC input isolation interface capable of connecting five input devices.

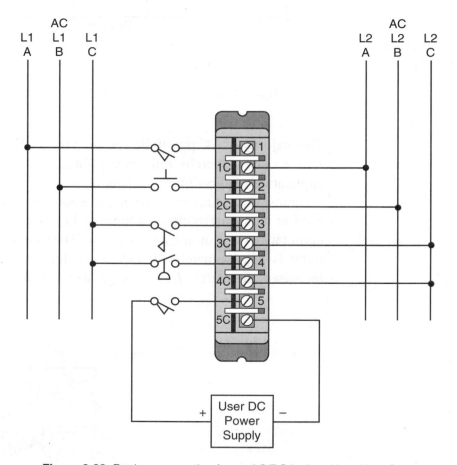

Figure 6-22. Device connection for an AC/DC isolated input interface.

Isolated input interfaces provide fewer points per module than their standard counterparts. This decreased modularity exists because isolated inputs require extra terminal connections to connect each of the return lines.

If isolation modules are not available for an application requiring singular return lines, standard interfaces may be used. However, the standard interfaces will lose inputs, because to keep isolation among inputs, they can have only one input line per return line. For example, a 16-point standard module with one common line per four points can accommodate four distinct isolated field input devices (each from a different source). However, as a result, it will lose 12 points. Figure 6-23 illustrates an 8-point module with different commons for every four inputs, thus allowing two possible isolated inputs.

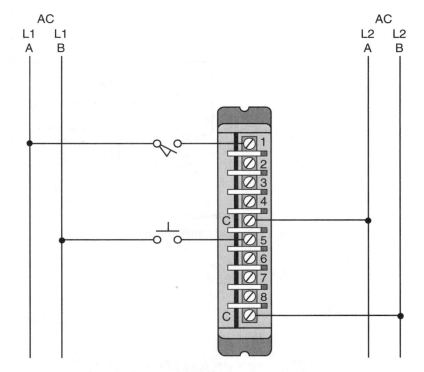

Figure 6-23. An 8-point standard input module used as an isolated module.

TTL INPUTS

Transistor-transistor logic (TTL) input interfaces allow controllers to accept signals from TTL-compatible devices, such as solid-state controls and sensing instruments. TTL inputs also interface with some 5 VDC–level control devices and several types of photoelectric sensors. The configuration of a TTL interface is similar to an AC/DC interface, but the input delay time caused by filtering is much shorter. Most TTL input modules receive their power from within the rack enclosure; however, some interfaces require an external 5-VDC power supply (rack or panel mounted).

Transistor-transistor logic modules may also be used in applications that use BCD thumbwheel switches (TWS) operating at TTL levels. These interfaces provide up to eight inputs per module and may have as many as sixteen inputs (high-density input modules). A TTL input module can also interface with thumbwheel switches if these input devices are TTL compatible. Figure 6-24 illustrates a typical TTL input module connection diagram with an external power supply.

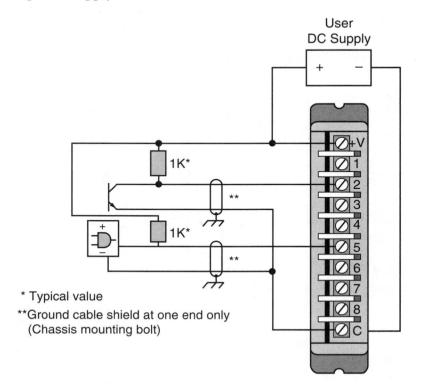

Figure 6-24. TTL input connection diagram.

REGISTER/BCD INPUTS

Multibit **register/BCD input modules** enhance input interfacing methods with the programmable controller through the use of standard thumbwheel switches. This register, or BCD, configuration allows groups of bits to be input as a unit to accommodate devices requiring that bits be in parallel form.

Register/BCD interfaces are used to input control program parameters to specific register or word locations in memory (see Figure 6-25). Typical input parameters include timer and counter presets and set-point values. The operation of register input modules is almost identical to that of TTL and DC input modules; however, unlike TTL input modules, register/BCD interfaces accept voltages ranging from 5 VDC (TTL) to 24 VDC. They are also grouped in modules containing 16 or 32 inputs, corresponding to one or two I/O registers (mapped in the I/O table), respectively. Data manipulation instructions, such as get or block transfer in, are used to access the data from the register input interface. Figure 6-26 illustrates a typical device connection for a register input.

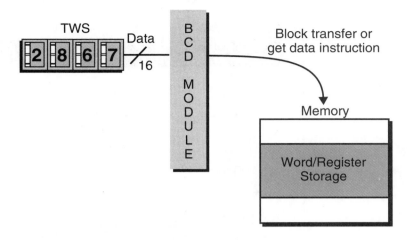

Figure 6-25. BCD interface inputting parameters into register/word locations in memory.

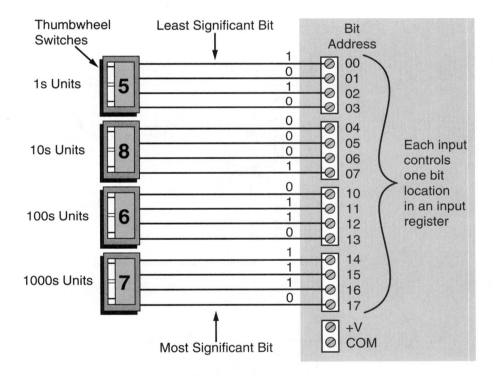

Figure 6-26. Register or BCD input module connection diagram.

Some manufacturers provide **multiplexing** capabilities that allow more than one input line to be connected to each terminal in a register module (see Figure 6-27). This kind of multiplexed register input requires thumbwheel switches that have an enable line (see Figure 6-28). When this line is selected, the TWS provides a BCD output at its terminals; when it is not selected, the TWS does not provide an output. If the TWS set provides four digits with one enable line (see Figure 6-29), then the enable line will make all of the outputs available

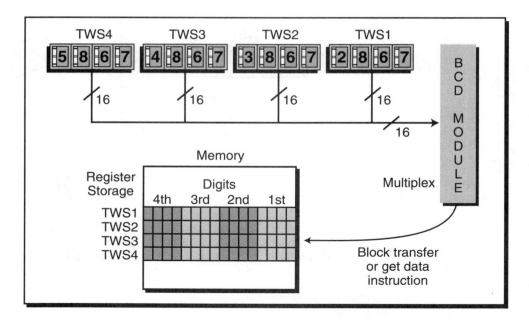

Figure 6-27. Multiplexing input module connection diagram.

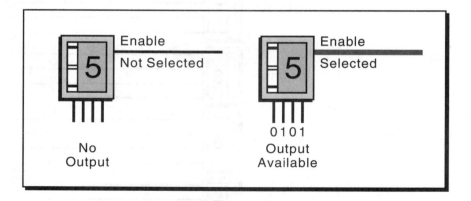

Figure 6-28. Single-digit TWS with enable line.

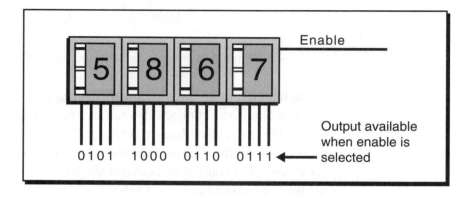

Figure 6-29. A 4-digit TWS with one common enable line.

when it is selected. This multiplexing technique minimizes the number of input modules required to read several sets of four-digit TWS. For instance, a 16-bit input module capable of multiplexing 6 input devices (6 × 16 = 96 total inputs) could receive information from six 4-digit thumbwheel switches. The user would not need to decode each of the six sets of 16 input groups, since the multiplexed module enables each group of 16 inputs to be read one scan at a time. However, the user may have to specify the register or word addresses where the 16-bit data will be stored through an instruction that specifies the storage location, along with the length or number of registers to be stored. Figure 6-30 illustrates a block diagram connection for a module capable of multiplexing four 4-digit TWS (four 16-bit input lines).

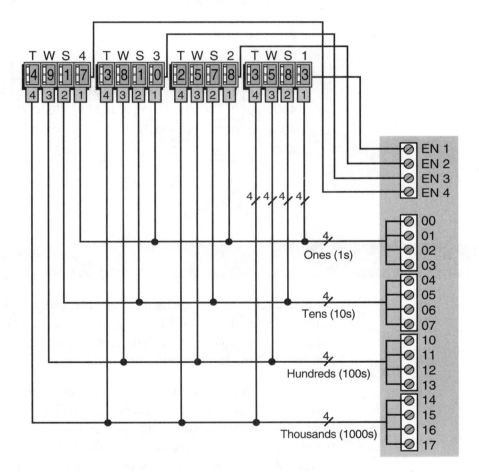

Figure 6-30. Block diagram of a multiplexed input module connected to four 4-digit TWS.

EXAMPLE 6-2

Referencing Figure 6-30, determine the values of the registers (in BCD) after an input transfer is made (in this case via a block transfer input of data). The input has a starting destination register of 4000 and a length of 4 registers (i.e., from registers 4000 to 4003). Assume that TWS set 1 is read first, TWS set 2 is read second, etc.

SOLUTION

The contents of register 4000 in BCD will be the BCD code equivalent of the first set of thumbwheel switches connected to the PLC register input module, and likewise for registers 4001, 4002, and 4003. Figure 6-31 shows the register contents. Note that the contents of each register does not represent the decimal equivalent of the binary pattern stored in that location, but rather the BCD equivalent. To change this number to decimal, you must convert the BCD pattern to its decimal equivalent using other instructions. For instance, the decimal equivalent of the binary (BCD) pattern stored in register 4000 is 13,699, not 3,583, as the TWS (BCD number) indicates.

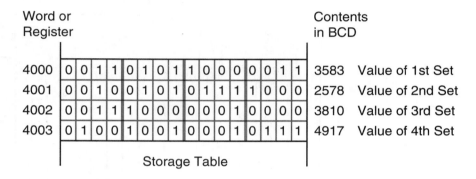

Word or Register	Bits	Contents in BCD	
4000	0 0 1 1 0 1 0 1 1 0 0 0 0 0 1 1	3583	Value of 1st Set
4001	0 0 1 0 0 1 0 1 0 1 1 1 1 0 0 0	2578	Value of 2nd Set
4002	0 0 1 1 1 0 0 0 0 0 0 1 0 0 0 0	3810	Value of 3rd Set
4003	0 1 0 0 1 0 0 1 0 0 0 1 0 1 1 1	4917	Value of 4th Set

Storage Table

Figure 6-31. Register contents for Example 6-2.

6-6 PLC INSTRUCTIONS FOR DISCRETE OUTPUTS

Like discrete input interfaces, **discrete output interfaces** are the most commonly used type of PLC output modules. These outputs connect the programmable controller with discrete output field devices. Many single-bit and multibit instructions are designed to manipulate discrete outputs.

During this discussion of output modules, keep in mind the relationship between output interface signals (ON/OFF), rack and module locations (where the output modules are inserted), and I/O table maps and addresses (used in the control program). Figure 6-32 illustrates a simplified 8-bit output image table. The coil of the motor starter (M1) is connected to a discrete output module (slot 7) in rack 0, which can connect 8 field inputs (0–7). Note that the starter will be known as output 077, which stands for rack 0, slot 7, terminal connection 7.

Output interface circuitry switches the supplied voltage from the PLC ON or OFF according to the status of the corresponding bit in the output image table. This status (1 or 0) is set during the execution of the control program and is sent to the output module at the end of scan (output update). If the signal from the processor is 1, the output module will switch the supplied voltage (e.g., 120 VAC) to the output field device, turning the output ON. If the signal received from the processor is 0, the module will deactivate the field device by switching to 0 volts, thus turning it OFF. Typically, an output coil (—O) instruction, like the one shown in Figure 6-32, activates the output interface when the reference address is logic 1 (ON).

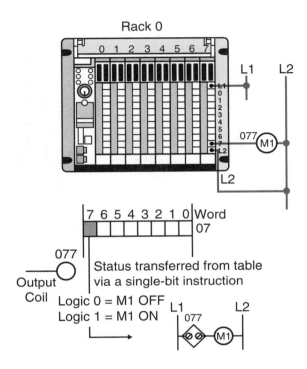

Figure 6-32. An 8-bit output image table with the module's L2 connection completing the path from L1 to L2.

Multibit outputs, such as BCD register outputs, use functional block instructions (e.g., block transfer out) to output a word or register to the module (see Figure 6-33). These instructions, in conjunction with input instructions, are heavily utilized during the programming and control of discrete I/O signals. Chapter 9 provides more information about the use and operation of functional block instructions.

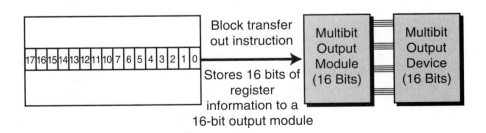

Figure 6-33. Functional block instruction transferring the output register contents to the module.

EXAMPLE 6-3

For the rack configuration shown in Figure 6-34, determine the addresses for each of the output field devices wired to the output connections in the 8-bit discrete input module. Assume that the first four slots of this 64 I/O micro-PLC are filled with outputs and that the second four are filled with inputs. The addressing scheme follows a rack-slot-connection convention (like Example 6-1), which starts at I/O address 000. Note that the number system is octal.

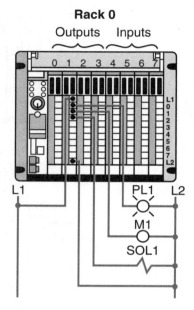

Rack 0

Outputs   Inputs

Figure 6-34. Rack configuration for Example 6-3.

SOLUTION

The field devices in this discrete output module will have addresses 010 through 017 because the module is located in rack 0, slot number 1 and the 8 field devices are connected to bits 0 through 7. Therefore, each of the field output devices will have the addresses shown in Figure 6-35—PL1 will be known as output 010, M1 as 011, and SOL1 as 012. Every time a bit address becomes 1, the field device with the corresponding address will be turned ON.

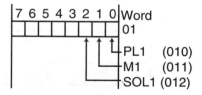

Figure 6-35. Field device addresses for the outputs in Example 6-3.

If M1 is rewired to another connection in another discrete output, the address that turns it ON and OFF will change. Consequently, the control program must be changed, since there can be only one reference address per discrete field output device connection.

6-7 DISCRETE OUTPUTS

Discrete output modules receive their necessary voltage and current from their enclosure's back plane (see Chapter 4 for loading considerations). The field devices with which discrete output modules interface may differ in their voltage requirements; therefore, several types and magnitudes of voltage are provided to control them (e.g., 120 VAC, 12 VDC). Table 6-4 illustrates some typical output field devices, while Table 6-5 lists the standard output ratings found in discrete output applications.

Output Devices
Alarms
Control relays
Fans
Horns
Lights
Motor starters
Solenoids
Valves

Table 6-4. Output field devices.

Output Ratings
12–48 volts AC/DC
120 volts AC/DC
230 volts AC/DC
Contact (relay)
Isolated output
TTL level
5–50 volts DC (sink/source)

Table 6-5. Standard output ratings.

AC OUTPUTS

AC output circuits, like input circuits, vary widely among PLC manufacturers, but the block diagram shown in Figure 6-36 depicts their general configuration. This block configuration shows the main sections of an AC output module, along with how it operates. The circuit consists primarily of the logic and power sections, coupled by an isolation circuit. An output interface can be thought of as a simple switch (see Figure 6-37) through which power can be provided to control an output device.

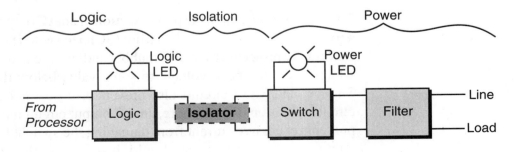

Figure 6-36. AC output circuit block diagram.

Output
Module

L1

"Switch"
controlled by
processor

Load

Output
Field
Device

L2

Logic 1– ON ("Switch" Closed)
Logic 0– OFF ("Switch" Open)

Figure 6-37. "Switch" function of an output interface.

During normal operation, the processor sends an output's status, according to the logic program, to the module's logic circuit. If the output is to be energized (reflecting the presence of a 1 in the output table), the logic section of the module will latch, or maintain, a 1. This sends an ON signal through the isolation circuit, which in turn, switches the voltage to the field device through the power section of the module. This condition will remain ON as long as the output table's corresponding image bit remains a 1. When the signal turns OFF, the 1 that was latched in the logic section unlatches, and the OFF signal passed through the isolation circuit provides no voltage to the power section, thus de-energizing the output device. Figure 6-38 illustrates a typical AC output circuit.

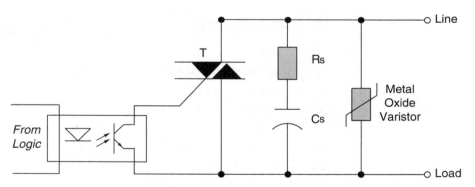

Figure 6-38. Typical AC output circuit.

The switching circuit in the power section of an AC output module uses either a triac or a silicon controlled rectifier (SCR) to switch power. The AC switch is normally protected by an RC snubber and/or a metal oxide varistor (MOV), which limits the peak voltage to some value below the maximum rating. Snubber and MOV circuits also prevent electrical noise from affecting the circuit operation. Furthermore, an AC output circuit may contain a fuse that prevents excessive current from damaging the switch. If the circuit does not contain a fuse, the user should install one that complies with the manufacturer's specifications.

As with input circuits, AC output interfaces may have LEDs to indicate operating logic signals and power circuit voltages. If the output circuit contains a fuse, it may also have a fuse status indicator. Figure 6-39 illustrates an AC output connection diagram. Note that power from the field (L1) supplies the voltage that the module uses to turn ON the output devices. Chapter 20 discusses other considerations for connecting AC outputs.

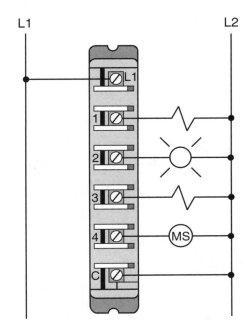

Figure 6-39. AC output module connection diagram.

DC OUTPUTS (SINK/SOURCE)

DC output interfaces control discrete DC loads by switching them ON and OFF. The functional operation of a DC output is similar to that of an AC output; however, the DC output's power circuit employs a power transistor to switch the load. Like triacs, transistors are also susceptible to excessive applied voltages and large surge currents, which can cause overdissipation and short-circuit conditions. To prevent these conditions, a power transistor is usually protected by a freewheeling diode placed across the load (field output device). DC outputs may also incorporate a fuse to protect the transistor during moderate overloads. These fuses are capable of opening, or breaking continuity, quickly before excessive heat due to overcurrents occurs.

As in DC inputs, DC output modules may have either sinking or sourcing configurations. If a module has a sinking configuration, current flows *from the load* into the module's terminal, switching the negative (return or common) voltage to the load. The positive current flows from the load to the common via the module's power transistor.

In a sourcing module configuration, current flows *from the module* into the load, switching the positive voltage to the load. Figure 6-40 illustrates a typical sourcing DC output circuit, and Figure 6-41 shows device connections for both sourcing and sinking configurations. Note that in sinking output devices, current flows into the device's terminal from the module (the module provides, or sources, the current). Conversely, the current in sourcing output devices flows out of the device's terminal into the module (the module receives, or sinks, the current).

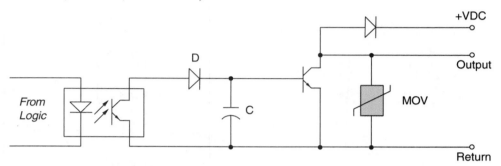

Figure 6-40. Typical sourcing DC output circuit.

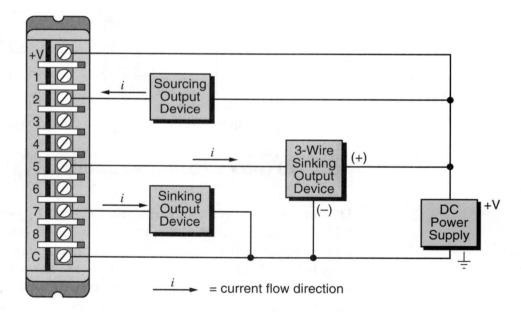

Figure 6-41. Field device connections for a sinking/sourcing DC output module.

ISOLATED AC AND DC OUTPUTS

Isolated AC and DC outputs operate in the same manner as standard AC and DC output interfaces. The only difference is that each output has its own return line circuit (common), which is isolated from the other outputs. This configuration allows the interface to control output devices powered by different sources, which may also be at different ground (common) levels.

A standard, nonisolated output module has one return connection for all of its outputs; however, some modules provide one return line per four outputs if the interface has eight or more outputs. Isolated interfaces provide less modularity (i.e., fewer points per module) than their standard counterparts, because extra terminal connections are necessary for the independent return lines. Figure 6-42 illustrates connections to an isolated AC output interface.

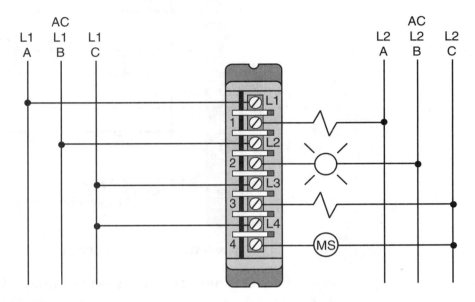

Figure 6-42. Connection diagram for an isolated AC output interface.

TTL OUTPUTS

TTL output interfaces allow a PLC to drive output devices that are TTL compatible, such as seven-segment LED displays, integrated circuits, and 5-VDC devices. Most of these modules require an external 5-VDC power supply with specific current requirements, but some provide the 5-VDC source voltage internally from the back plane of the rack. TTL modules usually have eight available output terminals; however, high-density TTL modules may be connected to as many as sixteen devices at a time. Typical output devices that use high-density TTL modules are 5-volt seven-segment indicators. Figure 6-43 illustrates typical output connections to a TTL output module. A TTL output interface requires an external power supply.

REGISTER/BCD OUTPUTS

Multibit **register/BCD output interfaces** provide parallel communication between the processor and an output device, such as a seven-segment LED display or a BCD alphanumeric display. Register output interfaces may also drive small DC loads with low current requirements (0.5 amps). Register

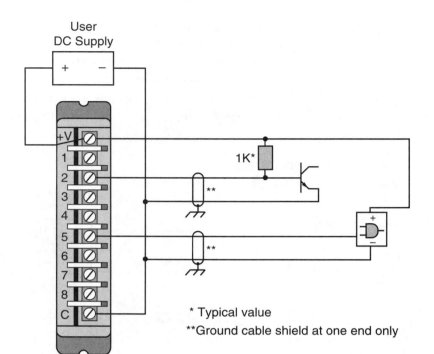

Figure 6-43. Connection diagram for a TTL output module.

output interfaces provide voltages ranging from 5 VDC (TTL level) to 30 VDC and have 16 or 32 output lines (one or two I/O registers). Figure 6-44 illustrates a typical device interface connection for a register output module.

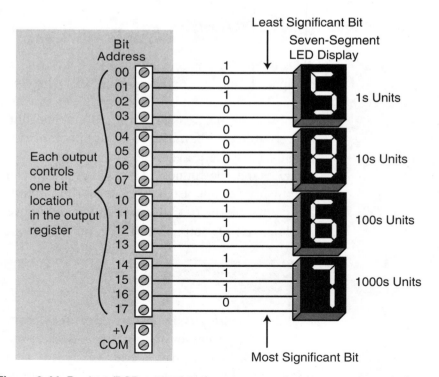

Figure 6-44. Register/BCD output interface connected to seven-segment indicators.

In a register output module, the information sent to the module originates in the register storage data table (see Figure 6-45). A 16-bit word or register is sent from this table to the module address specified by the data transfer or I/O register instruction (e.g., block transfer out). Once the data arrives at the module, it is latched and made available at the output circuits.

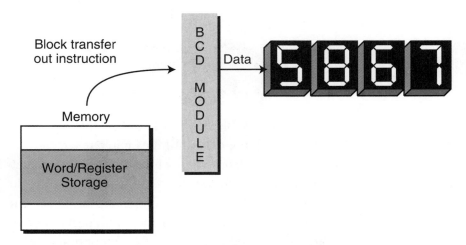

Figure 6-45. Output data table sending a 16-bit word to a register output module.

Register output modules may also have multiplexing capabilities (see Figure 6-46). As is the case with multiplexed inputs, multiplexed output devices (e.g., BCD display digits) require enable line capability to select the BCD display group that will receive the parallel, 16-bit data from the module (see Figure 6-47). A single-digit seven-segment display will be able to receive data if the enable is selected. Conversely, if the enable is not selected, the

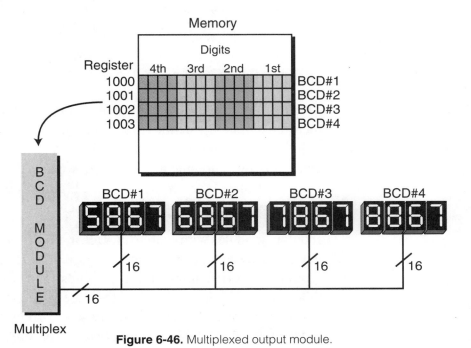

Figure 6-46. Multiplexed output module.

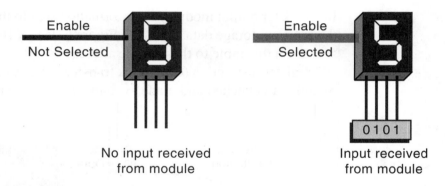

Figure 6-47. Single-digit seven-segment BCD display with enable line.

display will be blank or will contain the last data that was latched, because it may latch the data until the enable reselected and new data is available. If the BCD display contains four digits and one enable line (see Figure 6-48), the operation will be the same, except that the enable will control all four displays. With this option, one interface can control several groups of 16 or 32 outputs, depending on the modularity. For example, if a multiplexed output can handle four sets of 16-bit outputs, then it can drive up to four sets of 4-digit seven-segment indicators. Register data from the output table is sent to the module once a scan, updating each multiplexed set of output devices.

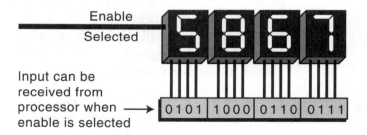

Figure 6-48. A 4-digit seven-segment BCD display with one common enable line.

The use of multiplexed outputs does not require special programming, since there are output instructions that specify the multiplexing operation. The only requirement is that the output devices (e.g., LED displays) must possess enable circuits allowing the module to connect the enable lines to each set of loads controlled by each set of 16 bits. Figure 6-49 shows a block diagram of a multiplexed output module with four sets of seven-segment LED indicators.

If output modules with enable lines are multiplexed, only passive-type output devices (i.e., seven-segment indicators, displays, etc.), as opposed to control-type elements (i.e., low-current solenoids), can be controlled. The reason for this is that while multiplexed outputs are very useful, their output data does not remain static for one channel, or set, of 16 bits or 32 bits; it changes for each circuit that is being multiplexed. The only way to use multiplexed

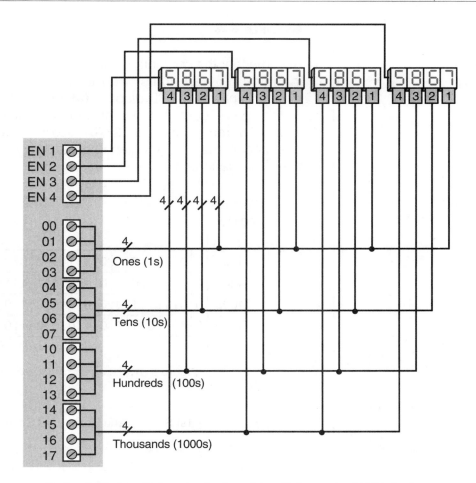

Figure 6-49. A multiplexed output module with four sets of LED displays.

modules and still have correctly operating output devices is to incorporate additional latching/enabling circuits into the output devices' hardware (see Figure 6-50). Such a situation may be encountered in the transmission of parallel data to instrumentation or computing devices that have enable and latching lines for incoming data.

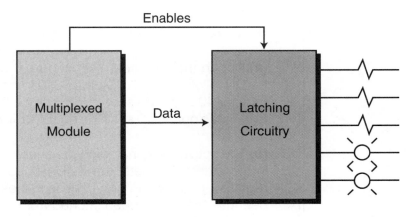

Figure 6-50. Latching/enabling circuit.

EXAMPLE 6-4

Assume that the contents of the registers in the storage table shown in Figure 6-51 are transferred to a BCD multiplexed module and, subsequently, to a BCD display. **(a)** What will be the value displayed on the seven-segment indicators during the third scan as shown in the timing diagram in Figure 6-52? **(b)** Also indicate, using Figure 6-49 as a reference, the lines (e.g., enable bits 0–17) that will be active during the third-scan transfer.

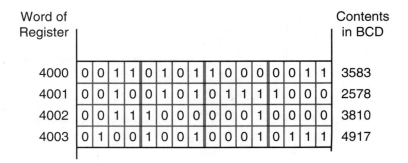

Word of Register																		Contents in BCD
4000	0	0	1	1	0	1	0	1	1	0	0	0	0	0	1	1	3583
4001	0	0	1	0	0	1	0	1	0	1	1	1	1	0	0	0	2578
4002	0	0	1	1	1	0	0	0	0	0	0	1	0	0	0	0	3810
4003	0	1	0	0	1	0	0	1	0	0	0	1	0	1	1	1	4917

Figure 6-51. Storage table for Example 6-4.

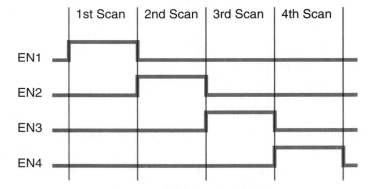

Figure 6-52. Timing diagram of enable signals from a BCD multiplexed module.

SOLUTION

(a) During the third scan (see Figure 6-53), the enable line EN3 will be ON, allowing the BCD data 3810 to go to BCD set #3. The value of register 4002 will be sent to the module through the wires connected to it. Since only BCD set #3 is enabled, it will accept all of the signals.

(b) The active lines, including the enable, are shown in blue. Note that in the other BCD sets, the BCD values from each set's respective register are shown in gray. These values may remain on the display because they have been latched from previous scans. They are not shown in blue because they are not active.

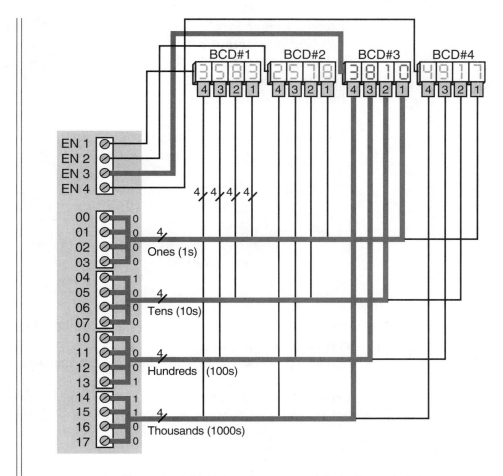

Figure 6-53. Multiplexed output module for Example 6-4.

CONTACT OUTPUTS

Contact output interfaces allow output devices to be switched by normally open or normally closed relay contacts. Contact interfaces provide electrical isolation between the power output signal and the logic signal through separation between contacts and between the coil and contacts. These outputs also include filtering, suppression, and fuses.

The basic operation of contact output modules is the same as that of standard AC or DC output modules. When the processor sends status data (1 or 0) to the module during the output update, the state of the contacts changes. If the processor sends a 1 to the module, normally open contacts close and normally closed contacts open. If the processor sends a 0, no change occurs to the normal state of the contacts.

Contact outputs can be used to switch either AC or DC loads, but they are normally used in applications such as multiplexing analog signals, switching small currents at low voltages, and interfacing with DC drives to control different voltage levels. High-power contact outputs are also available for applications that require the switching of high currents. Figure 6-54 shows a contact output circuit. The device connection for this output module is similar to an AC output module. In this circuit, one side (1A) goes to L1, while the other (1B) goes to the load.

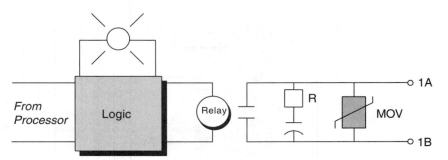

Figure 6-54. Contact output circuit.

Figure 6-55 illustrates an interfacing example where four analog voltage references are connected to a contact output module. These references represent preset speed values, which if connected to a speed drive controller, can be used to switch different motor velocities (e.g., two forward, two reverse). Note that each contact in this interface must be mutually exclusive—that is, only one contact can be closed at a time. Interlocking logic in the control program is necessary to prevent two or more output coils from being energized at the same time.

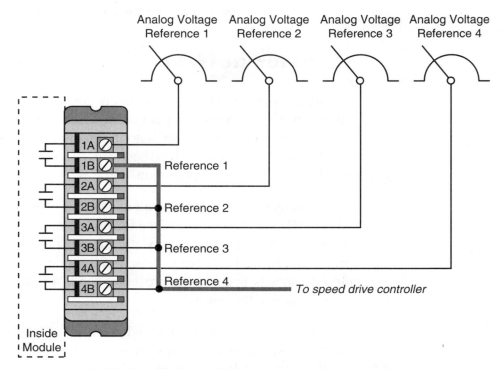

Figure 6-55. Example of a contact interface connection.

6-8 DISCRETE BYPASS/CONTROL STATIONS

Bypass/control stations are manual backup devices that are used in PLC systems to allow flexibility during start-up and output failure. By incorporating a selector switch that allows a field output device to be switched ON regardless of the state of its output module, these devices can override a PLC's output signal. Bypass devices can also be configured to place field outputs under PLC output control or to change them to an OFF condition.

Figure 6-56 shows a diagram of a typical bypass device. Bypass units provide 8 to 16 isolated points, each protected by a circuit breaker or fuse, for use with any PLC's discrete output modules. Bypass devices are placed between the PLC's output interface and the digitally controlled element (see Figure 6-57). Indicators, which are incorporated into the control system, show the ON/OFF state of the field device. Bypass units provide a way to control field devices without the PLC. These devices are very useful during maintenance situations, system start-up, and emergency disconnect of particular field devices.

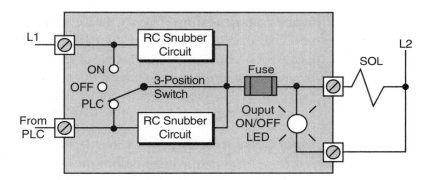

Figure 6-56. Typical bypass device.

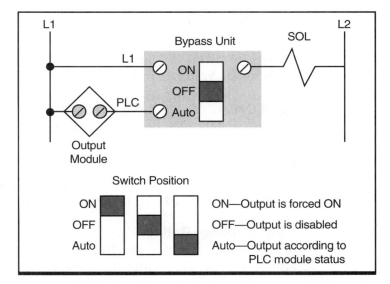

Figure 6-57. Bypass unit placed between the PLC and a field device.

6-9 INTERPRETING I/O SPECIFICATIONS

Perhaps with the exception of standard I/O current and voltage ratings, specifications for I/O circuits are all too often treated as a meaningless listing of numbers. Nevertheless, manufacturers' specifications provide valuable information about the correct and safe application of interfaces. These specifications place certain limitations on the module and also on the field equipment that it can operate. Failure to adhere to specifications can result in a misapplication of the hardware, leading to faulty operation or equipment damage. Table 6-6 provides an overview of the electrical, mechanical, and environmental specifications that should be evaluated for each PLC application. Following is a more detailed explanation of each specification. These specifications should also be evaluated for the interfaces covered in the next two chapters (analog and special function).

ELECTRICAL

Input Voltage Rating. This AC or DC value defines the magnitude and type of signal that will be accepted by the circuit. The circuit will usually accept a deviation from this nominal value of ±10–15%. This specification may also be called the *input voltage range*. For a 120 VAC–rated input circuit with a range of ±10%, the minimum and maximum acceptable input voltages for continuous operation will be 108 VAC and 132 VAC, respectively.

Input Current Rating. This value defines the minimum input current at the rated voltage that the input device must be capable of driving to operate the input circuit. This specification may also appear indirectly as the *minimum power requirement*.

Input Threshold Voltage. This value specifies the voltage at which the input signal is recognized as being absolutely ON. This specification is also called the *ON threshold voltage*. Some manufacturers also specify an OFF voltage, defining the voltage level at which the input circuit is absolutely OFF.

Input Delay. The input delay defines the duration for which the input signal must exceed the ON threshold before being recognized as a valid input. This specification is given as a minimum or maximum value. This delay is a result of filtering circuitry provided to protect against contact bounce and voltage transients. The input delay is typically 9–25 msec for standard AC/DC inputs and 1–3 msec for TTL or electronic inputs.

Output Voltage Rating. This AC or DC value defines the magnitude and type of voltage source that the I/O module can control. Deviation from this nominal value is typically ±10–15%. For some output interfaces, the output voltage is also the maximum continuous voltage. The output voltage

ELECTRICAL

Input Voltage Rating. An AC or DC value that specifies the magnitude and type of signal a circuit will accept.

Input Current Rating. The minimum current at the rated voltage an input device must be capable of driving.

Input Threshold Voltage. The voltage at which an input signal is recognized as being ON.

Input Delay. The duration for which an input signal must be ON to be recognized as a valid input.

Output Voltage Rating. An AC or DC value that specifies the magnitude and type of voltage that an I/O module can control.

Output Current Rating. The maximum current that a single output circuit can safely carry under load.

Output Power Rating. The maximum power an output module can dissipate with all circuits energized.

Current Requirements. The current demand that an I/O module places on the system power supply.

Surge Current (Max). The maximum current and duration for which an output circuit can exceed its maximum ON-state current rating.

OFF-State Leakage Current. The maximum leakage current that flows through the triac/transistor during its OFF state.

Output ON-Delay. The response time for an output to turn from OFF to ON after it receives an ON command.

Output OFF-Delay. The response time for an output to turn from ON to OFF after it receives an OFF command.

Electrical Isolation. A maximum value in volts defining the isolation between the I/O circuit and the controller logic.

Output Voltage/Current Ranges. The value of the voltage/current swing of the digital-to-analog converter.

Input Voltage/Current Ranges. The value of the voltage/current swing of the analog-to-digital converter.

Digital Resolution. A measure of how closely the converted analog I/O current or voltage signal approximates the actual analog value.

Output Fuse Rating. The type and rating of fuses that should be used in the interface.

MECHANICAL

Points Per Module. The number of input or output circuits that are on a single module.

Wire Size. The number of conductors and the largest gauge wire the I/O termination points will accept.

ENVIRONMENTAL

Ambient Temperature Rating. The maximum air temperature surrounding the I/O system for ideal operating conditions.

Humidity. The maximum air humidity surrounding the I/O system.

Table 6-6. Summary of I/O specifications.

specification may also be stated as the *output voltage range*, in which case both the minimum and maximum operating voltages are given. An output circuit rated at 48 VDC, for example, can have an absolute working range of 42 to 56 VDC.

Output Current Rating. This specification is also known as the *ON-state continuous current rating*, a value that defines the maximum current that a single output circuit can safely carry under load. The output current rating is a function of the electrical and heat dissipation characteristics of the component. This rating is generally specified at an ambient temperature (typically 0–60°C). As the ambient temperature increases, the output current decreases. Exceeding the output current rating or oversizing the manufacturer's fuse rating can result in a permanent short-circuit failure or other damage.

Output Power Rating. This maximum value defines the total power that an output module can dissipate with all circuits energized. The output power rating for a single energized output is the product of the output voltage rating and the output current rating expressed in volt-amperes or watts (e.g., 120 V × 2 A = 240 VA). This value for a given I/O module may or may not be the same if all outputs on the module are energized simultaneously. The rating for an individual output when all other outputs are energized should be verified with the manufacturer.

Current Requirements. The current requirement specification defines the current demand that a particular I/O module's logic circuitry places on the system power supply. To determine whether the power supply is adequate, add the current requirements of all the installed modules that the power supply supports, and compare the total with the maximum current the power supply can provide. The current requirement specification will provide a typical rating and a maximum rating (all I/O activated). An insufficient power supply current can result in an undercurrent condition, causing intermittent operation of field input and output interfaces.

Surge Current (Max). The surge current, also called the *inrush current*, defines the maximum current and duration (e.g., 20 amps for 0.1 sec) for which an output circuit can exceed its maximum ON-state continuous current rating. Heavy surge currents are usually a result of either transients on the output load or power supply line or the switching of inductive loads. Freewheeling diodes, Zener diodes, or RC networks across the load terminals normally provide output circuits with internal protection. If not, protection should be provided externally.

OFF-State Leakage Current. Typically, this is a maximum value that measures the small leakage current that flows through the triac/transistor during its OFF state. This value normally ranges from a few microamperes to a few milliamperes and presents little problem. It can present problems when switching very low currents or can give false indications when using a sensitive instrument, such as a volt-ohm meter, to check contact continuity.

Output-ON Delay. This specification defines the response time for the output to go from OFF to ON once the logic circuitry has received the command to turn ON. The ON response time of the output circuit affects the total time required to activate an output device. The worst-case time required to turn an output device ON after the control logic goes TRUE is the total of the two program scan times plus the I/O update, output-ON delay, and device-ON response times.

Output-OFF Delay. The output-OFF delay specification defines the response time for the output to go from ON to OFF once the logic circuitry has received the command to turn OFF. The OFF response time of the output circuit affects the total time required to deactivate an output device. The worst-case time required to turn an output device OFF after the control logic goes FALSE is the total of the two program scan times plus the I/O update, output-OFF delay, and device-OFF response times.

Electrical Isolation. This maximum value in volts defines the isolation between the I/O circuit and the controller logic circuitry. Although this isolation protects the logic side of the module from excessive input/output voltages or currents, the power circuitry of the module can still be damaged.

Output Voltage/Current Ranges. This specification is a nominal expression of the voltage/current swing of the D/A converter in analog outputs. This output will always be a proportional current or voltage within the output range. A given analog output module may have several hardware- or software-selectable, unipolar or bipolar ranges (e.g., 0 to 10 V, −10 to +10 V, 4 to 20 mA).

Input Voltage/Current Ranges. This specification defines the voltage/current swing of the A/D converter in analog inputs. This specification will always be a proportional current or voltage within the input range. A given analog input module may have several hardware- or software-selectable, unipolar or bipolar ranges (e.g., 0 to 10 V, −10 to +10 V, 4 to 20 mA).

Digital Resolution. This specification defines how closely the converted analog input/output current or voltage signal approximates the actual analog value within a specified voltage or current range. Resolution is a function of the number of bits used by the A/D or D/A converter. An 8-bit converter has a resolution of 1 part in 2^8 or 1 part in 256. If the range is 0 to 10 V, then the resolution is 10 divided by 256, or 40 mV/bit.

Output Fuse Rating. Fuses are often supplied as a part of the output circuit, but only to protect the semiconductor output device (triac or transistor). The manufacturer carefully selects the fuse that is employed or recommended for the interface based on the fusing current rating of the output switching device. Fuse rating incorporates a fuse opening time along with a current overload rating, which allows opening within a time frame that will avoid damage to the triac or transistor. The recommended specifications should be followed when replacing fuses or when adding fuses to the interface.

MECHANICAL

Points Per Module. This specification defines the number of input/output circuits that are on a single module (encasement). Typically, a module will have 1, 2, 4, 8, or 16 points per module. The number of points per module has two implications that may be of importance to the user. First, the less dense (fewer the number of points) a module is, the greater the space requirements are; second, the higher the density, the lower the likelihood that the I/O count requirements can be closely matched with the hardware. For example, if a module contains 16 points and the user requires 17 points, two modules must be purchased. Thus, the user must purchase 15 extra inputs or outputs.

Wire Size. This specification defines the number of conductors and the largest gauge wire that the I/O termination points will accept (e.g., two #14 AWG). The manufacturer does not always provide wire size specifications, but the user should still verify it.

ENVIRONMENTAL

Ambient Temperature Rating. This value is the maximum temperature of the air surrounding the input/output system for best operating conditions. This specification considers the heat dissipation characteristics of the circuit components, which are considerably higher than the ambient temperature rating itself. The ambient temperature rating is much less than the heat dissipation factors so that the surrounding air does not contribute to the heat already generated by internal power dissipation. The ambient temperature rating should never be exceeded.

Humidity Rating. The humidity rating for PLCs is typically 0–95% noncondensing. Special consideration should be given to ensure that the humidity is properly controlled in the area where the input/output system is installed. Humidity is a major atmospheric contaminant that can cause circuit failure if moisture is allowed to condense on printed circuit boards.

Proper observance of the specifications provided on the manufacturer's data sheets will help to ensure correct, safe operation of control equipment. Chapter 20 discusses other considerations for properly installing and maintaining input/output systems.

6-10 SUMMARY OF DISCRETE I/O

For the most part, all PLC system applications require the types of discrete I/O interfaces covered in this chapter. In addition to discrete interfaces, some PLC applications require analog and special I/O modules (covered in the next two chapters) to implement the required control.

All I/O interfaces accept input status data for the input table and accept processed data from the output table. This information is placed in or written from the I/O table (in word locations) according to the location or address of the modules. This address depends on the module's placement in the I/O rack enclosure; therefore, the placement of I/O interfaces is an important detail to keep in mind.

The software instructions that are generally used with discrete-type interfaces are basic relay instructions (ladder type), although multibit modules use functional block instructions as well as some advanced ladder functions. Chapter 9 explains these software instructions. Figure 6-58 shows several programmable controller input and output modules and enclosures.

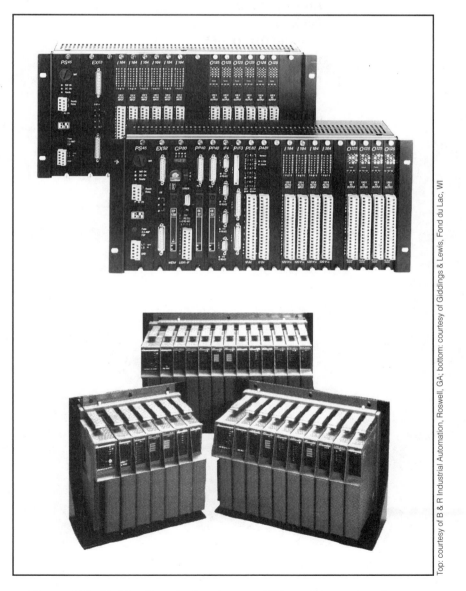

Top: courtesy of B & R Industrial Automation, Roswell, GA; bottom: courtesy of Giddings & Lewis, Fond du Lac, WI

Figure 6-58. PLC families sharing the use of I/O modules and enclosures.

KEY
TERMS
AC/DC I/O interface
bypass/control station
contact output interface
DC I/O interface
digital signal
discrete input interface
discrete output interface
I/O address
I/O module
isolated I/O interface
local rack
master rack
multiplexing
rack enclosure
register/BCD I/O interface
remote I/O subsystem
remote rack
sinking configuration
sourcing configuration
TTL I/O interface

CHAPTER SEVEN

THE ANALOG
INPUT/OUTPUT SYSTEM

*One line alone has no meaning; a second one
is needed to give it expression.*

—Eugène Delacroix

**CHAPTER
HIGHLIGHTS**
Although discrete I/O systems are invaluable tools for PLC controls, they cannot meet all the demands of new technological and application advances. Because they can interpret continuous signals, analog I/O interfaces are used in applications, such as batching and temperature control, where the simple two-state capabilities of discrete I/O systems are insufficient. This chapter explains the function and application of analog I/O interfaces, including a discussion of analog connections and instructions. In the next chapter, you will learn about another type of I/O system—special function interfaces— which are used to accomplish specific control tasks.

7-1 OVERVIEW OF ANALOG INPUT SIGNALS

Analog input modules, like the ones shown in Figure 7-1, are used in applications where the field device's signal is continuous (see Figure 7-2). Unlike discrete signals, which possess only two states (ON and OFF), **analog signals** have an infinite number of states. Temperature, for example, is an analog signal because it continuously changes by infinitesimal amounts. Consequently, a change from 70°F to 71°F is not just one change of 1°F, but rather an infinite number of smaller changes of a fraction of a degree.

Courtesy of Allen-Bradley, Highland Heights, OH

Figure 7-1. Analog input modules.

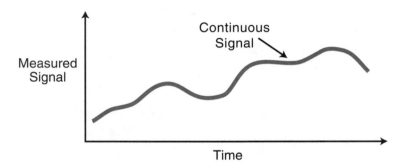

Figure 7-2. Representation of a continuous analog signal.

PLCs, like other digital computers, are discrete systems that only understand 1s and 0s. Therefore, they cannot interpret analog signals in their continuous form. **Analog input interfaces** translate continuous analog signals into discrete values that can be interpreted by PLC processors. These discrete values are subsequently used in the control program. Table 7-1 lists some devices that are typically interfaced with analog input modules.

Analog Inputs
Flow transducers
Humidity transducers
Load cell transducers
Potentiometers
Pressure transducers
Vibration transducers
Temperature transducers

Table 7-1. Devices used with analog input interfaces.

7-2 INSTRUCTIONS FOR ANALOG INPUT MODULES

Analog input modules digitize analog input signals, thereby bringing analog information into the PLC (see Figure 7-3). The modules store this multibit information in register locations inside the PLC. The analog instructions used with analog input modules are similar to, if not the same as, the instructions used with multibit discrete inputs. The only difference between them is that analog multibit instructions are the result of a digital transformation of the analog signal, while discrete multibit instructions are the result of many multibit devices (or separate signals) connected to the same number of discrete input connections.

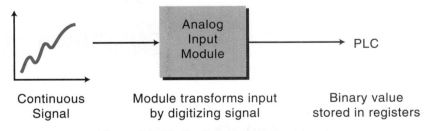

Continuous Signal   Module transforms input by digitizing signal   Binary value stored in registers

Figure 7-3. Digitization of an analog signal.

Figure 7-4 illustrates the sequence of events that occurs while reading an analog input signal. The module transforms the analog signal, through an analog-to-digital converter (A/D), into 12 bits of digital information that will be stored in register 1000 after the instruction is executed. After the PLC reads this information, the control program can reference the register address for comparisons, arithmetic calculations, etc. The analog value stored in the register will be in either BCD or binary format.

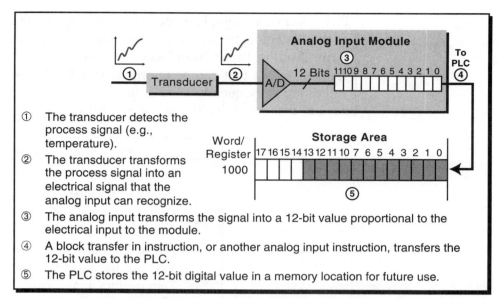

① The transducer detects the process signal (e.g., temperature).

② The transducer transforms the process signal into an electrical signal that the analog input can recognize.

③ The analog input transforms the signal into a 12-bit value proportional to the electrical input to the module.

④ A block transfer in instruction, or another analog input instruction, transfers the 12-bit value to the PLC.

⑤ The PLC stores the 12-bit digital value in a memory location for future use.

Figure 7-4. Steps in converting an analog signal to binary format.

EXAMPLE 7-1

What will the contents of register 1000 be after the multibit instruction shown in Figure 7-5 is executed? Note that the digitized value corresponding to the analog transformation shown in the figure is represented by 12 bits in binary format.

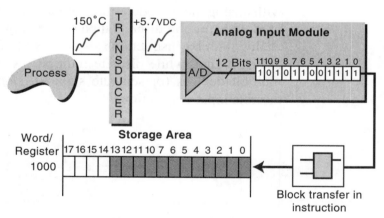

Figure 7-5. Multibit instruction.

SOLUTION

After the instruction is executed, the contents of register 1000 will be:

0000 1010 1100 1111

This number corresponds to the digitized value generated by the module. Since the value is represented in 12 bits, the preceding bits are filled with 0s. Note that the value stored in register 1000 is in binary. Its decimal equivalent, for computational purposes, is 2767.

7-3 ANALOG INPUT DATA REPRESENTATION

Field devices that provide an analog output as their signal (analog sensors or transducers) are usually connected to transmitters, which in turn, send the analog signal to the module. A **transducer** converts a field device's variable (i.e., pressure, temperature, etc.) into a very low-level electrical signal (current or voltage) that can be amplified by a **transmitter** and then input into the analog interface (see Figure 7-6).

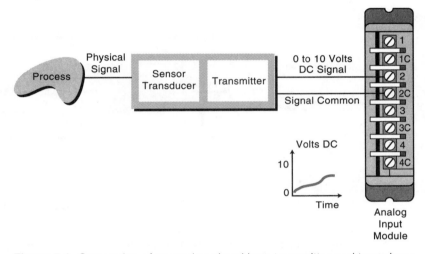

Figure 7-6. Conversion of an analog signal by a transmitter and transducer.

Due to the many types of transducers available, analog input modules have several standard electrical input ratings. Table 7-2 lists the standard current and voltage ratings for analog interfaces. Note that analog interfaces can be either *unipolar* (positive voltage only—i.e., 0 to +5 VDC) or *bipolar* (negative and positive voltages—i.e., –5 to +5 VDC).

Input Interfaces
4–20 mA
0 to +1 volts DC
0 to +5 volts DC
0 to +10 volts DC
1 to +5 volts DC
± 5 volts DC
± 10 volts DC

Table 7-2. Typical analog input interface ratings.

As mentioned earlier, an analog input module transforms an analog input signal via a sensor/transmitter unit into a discrete value that is readily understandable by man and machine (see Figure 7-7). This transformed value is the digital equivalent of the variable analog signal (e.g., pressure in psi) measured by the field device. The field sensing device sends a very low-level current or voltage analog input to the transmitter. The transmitter (sometimes incorporated in the same unit as the sensor) sends this information to the input module as an amplified current or voltage proportional to the signal being measured. Next, the analog input interface digitizes the current or voltage by converting it into an equivalent binary number. The interface then sends the digitized signal to the controller. Thus, the binary value that the PLC receives is the digital equivalent of the incoming analog signal.

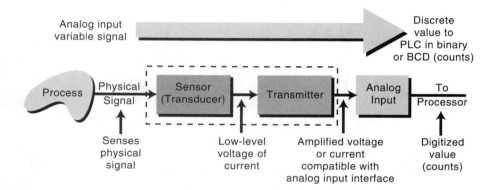

Figure 7-7. Transformation of an analog signal into a binary or BCD value.

An **analog-to-digital converter (A/D or ADC)** performs the signal conversion in an analog input module. The converter divides, or digitizes, the input signal into many digital counts, which represent the magnitude of the current or voltage. This division of the input signal is called **resolution**. The resolution of the module indicates how many parts the module's A/D will divide the input signal into; it is given as a function of how many bits the A/D uses during conversion. For example, if an A/D breaks down an input signal using 12 bits or 4096 parts (i.e., $2^{12} = 4096$) as shown in Figure 7-8, it has a 12-bit resolution (i.e., a 12-bit binary number with a value ranging from 0000 to 4095 decimal will represent the signal). In this case, the manufacturer could then use the remaining bits (bits 14–17) as status monitoring bits, representing module conditions such as *active*, *OK*, *channel operating*, etc.

An A/D converter transfers its digital-equivalent values to the processor, which in turn, makes them available for use in register or word locations. The format of these values varies according to the format used by the PLC; however, the most common formats are binary and BCD. In BCD format, the module or processor must perform an extra linearity computation to provide a valid BCD number.

Some PLCs also offer direct scale conversion of the input signal to equivalent engineering units (0 to 9999). Table 7-3 illustrates the conversion of psi values into engineering units and their decimal equivalents. The module

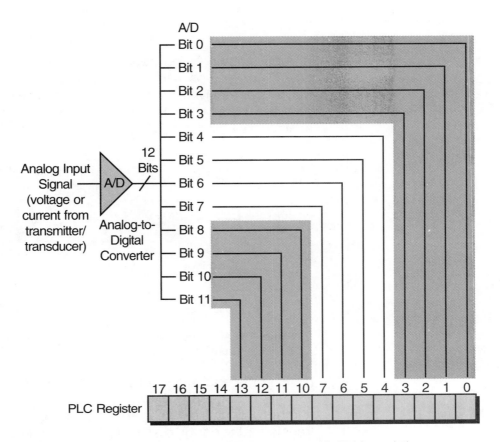

Figure 7-8. An analog-to-digital converter with 12-bit resolution.

Pressure psi	Analog Voltage Input	Digital Representation Engineering Units 0000-9999	Digital Representation Decimal Scale 0-4095
0	0V	0000	0
50	1V	1000	410
100	2V	2000	819
150	3V	3000	1229
200	4V	4000	1638
250	5V	5000	2047
300	6V	6000	2457
350	7V	7000	2866
400	8V	8000	3276
450	9V	9000	3685
500	10V	9999	4095

Table 7-3. Psi values translated into decimal equivalents and engineering units.

interprets the incoming 0 to 500 psi signal variable as a voltage ranging from 0 to 10 VDC. It then converts this voltage into an equivalent decimal value. A decimal value of 0 corresponds to 0 psi, while a decimal value of 4095 corresponds to 500 psi. The following examples illustrate how an A/D computes equivalent analog counts for an analog field signal.

EXAMPLE 7-2

An input module, which is connected to a temperature transducer, has an A/D with a 12-bit resolution (see Figure 7-9). When the temperature transducer receives a valid signal from the process (100 to 600°C), it provides, via a transmitter, a +1 to +5 VDC signal compatible with the analog input module.

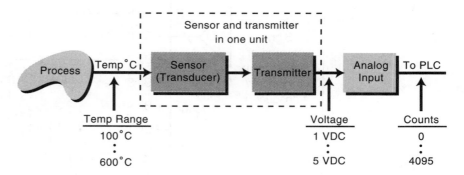

Figure 7-9. An A/D and an analog input module connected to a temperature-sensing device.

(a) Find the equivalent voltage change for each count change (the voltage change per degree Celsius change) and the equivalent number of counts per degree Celsius, assuming that the input module transforms the data into a linear 0 to 4095 counts, and **(b)** find the same values for a module with a 10-bit resolution.

SOLUTION

(a) The relationship between temperature, voltage signal, and module counts is:

Temperature	Voltage Signal	Input Counts
100°C	1 VDC	0
•	•	•
•	•	•
•	•	•
600°C	5 VDC	4095

The changes (Δ) in temperature, voltage, and input counts are 500°C, 4 VDC, and 4095 counts. Therefore, the voltage change for a 1°C temperature change is:

$$\Delta 500\,°C = \Delta 4 \text{ VDC}$$

$$1°C = \frac{4 \text{ VDC}}{500} = 8.0 \text{ mVDC}$$

The change in voltage for each input count is:

$$\Delta 4095 \text{ counts} = \Delta 4 \text{ VDC}$$

$$1 \text{ count} = \frac{4 \text{ VDC}}{4095} = 0.9768 \text{ mVDC}$$

Therefore, the corresponding number of counts per degree Celsius is:

$$\Delta 500°C = \Delta 4095 \text{ counts}$$

$$1°C = \frac{4095 \text{ counts}}{500} = 8.19 \text{ counts}$$

(b) A 10-bit resolution A/D will digitize the unipolar input signal into 1024 counts (i.e., $2^{10} = 1024$ counts, ranging from 0000 to 1023). The relationship between temperature, voltage signal, and counts is:

Temperature	Voltage Signal	Input Counts
100°C	1 VDC	0
•	•	•
•	•	•
•	•	•
500°C	4 VDC	1024

The changes in temperature, voltage, and counts are 500°C, 4 VDC, and 1023 counts. The voltage change per degree will be the same as in part (a) and is:

$$\Delta 500°C = \Delta 4 \text{ VDC}$$

$$1°C = \frac{4 \text{ VDC}}{500} = 8.0 \text{ mVDC}$$

The change in voltage per input count is:

$$\Delta 1023 \text{ counts} = \Delta 4 \text{ VDC}$$

$$1 \text{ count} = \frac{4 \text{ VDC}}{1023} = 3.91 \text{ mVDC}$$

Thus, the corresponding number of counts per degree Celsius is:

$$\Delta 500°C = \Delta 1023 \text{ counts}$$

$$1°C = \frac{1023 \text{ counts}}{500} = 2.046 \text{ counts}$$

EXAMPLE 7-3

A temperature transducer/transmitter (see Figure 7-10) provides a 0–10 VDC voltage signal that is proportional to the temperature variable being measured. The temperature measurement ranges between 0 and 1000°C. The analog input module accepts a 0–10 VDC unipolar signal range and converts it to a range of 0–4095 counts. The process application where this signal is being used detects low and high alarms at 100°C and 500°C, respectively.

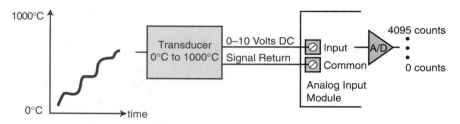

Figure 7-10. Temperature transducer/transmitter connected to an input module.

Find **(a)** the relationship (i.e., equation of the line) between the input variable signal (temperature) and the counts being measured by the PLC module and **(b)** the equivalent number of counts for each of the alarm temperatures specified.

SOLUTION

(a) Figure 7-11 shows the relationship between counts and the input signal in volts and degrees Celsius. Line Y describes the numerical relationship between the input signal and the number of counts (assuming a linear relationship).

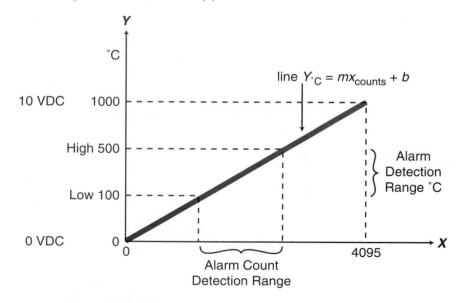

Figure 7-11. Relationship between counts and input signal.

To find the relationship between temperature and counts, find the numerical representation of the equation for line Y. This equation takes the form $Y = mX + b$ (see Appendix E), where m is the slope of the line and is described by:

$$m = \frac{Y_2 - Y_1}{X_2 - X_1} = \frac{°C_2 - °C_1}{\text{count 2} - \text{count 1}} = \frac{1000 - 0}{4095 - 0} = \frac{1000}{4095}$$

and Y_2, Y_1, X_2, and X_1 are known points. The value b is the value of Y, or °C, when X, or counts, equals 0. This value can be computed as:

$$b = Y_{°C} - mX_{\text{counts}}$$

where Y and X are values at known points (i.e., at 0°C and 0 counts). When X is at 0 counts, Y is at 0°C; therefore:

$$b = 0 - \left(\frac{1000}{4095}\right)0$$
$$= 0$$

Substituting the derived values for m and b into the equation $Y = mX + b$ produces the equation of line Y:

$$Y = mX + b$$
$$Y_{°C} = \frac{1000}{4095}X_{\text{counts}} + 0$$
$$= \frac{1000}{4095}X_{\text{counts}}$$

Using 4095 counts and 1000°C as the X and Y values when computing b would have derived the same equation (try it as an exercise).

(b) Based on the equation of line Y, the number of counts for each alarm range is:

$$Y_{°C} = \frac{1000}{4095}X_{\text{counts}}$$
$$X_{\text{counts}} = \frac{4095(Y_{°C})}{1000}$$

So, for the $Y_{°C}$ values of 100°C and 500°C, the X values are:

$$X_{\text{counts at }100°C} = \frac{4095(100)}{1000} = 409.5 \text{ counts}$$
$$X_{\text{counts at }500°C} = \frac{4095(500)}{1000} = 2047.5 \text{ counts}$$

Thus, the count value for 100°C is 409.5 counts and for 500°C is 2047.5 counts. Since count values must be whole numbers, rounding these values off yields 410 and 2048 counts, respectively. Therefore, at a count of 410, the low-level temperature alarm would be enabled; and at a count of 2048, the high-level temperature alarm would be enabled.

Another method for solving this problem is to determine the number of counts that are equivalent to 1°C. A change of 1000°C per 4095 counts can be expressed as:

$$\frac{\Delta \text{counts}}{\Delta \text{degrees}} = \frac{\text{max counts} - \text{min counts}}{\text{max degrees} - \text{min degrees}} = \frac{4095 - 0}{1000 - 0} = 4.095$$

Therefore, each degree is equivalent to 4.095 counts. The count value for 500°C would be (500)(4.095) = 2047.5 and for 100°C would be (100)(4.095) = 409.5. Rounding off these values yields 2048 and 410 counts, respectively—the same values we computed before. If the counts had not started at 0, an offset count addition would have been necessary for computing the number of counts per degree.

7-4 ANALOG INPUT DATA HANDLING

The previous section showed how an analog input module transforms an analog field signal into a discrete signal. Once the module digitizes the signal into binary counts, the processor can read the value and use the information. During the input reading section of the scan, the processor reads the value from the module and transfers the information to a location specified by the user. This location is usually a word or register storage area or an input register. The processor enters the count value into memory using instructions that differ from those used by standard discrete input modules, yet are similar to those used by multibit discrete input interfaces (see Figure 7-12).

Most analog modules provide more than one **channel**, or input, per interface. Therefore, they can connect to several input signals, as long as the signals are compatible with the module. The analog instructions used in PLCs take advantage of this multiple channel capability, inputting several values at a time into registers or words. Examples of these instructions are analog in, block transfer in, block in, and location in instructions (see Chapter 9). Some programmable controller manufacturers use other instructions, such as arithmetic instructions, to obtain count values from the analog module's address.

When a processor executes the instruction to read an analog input, it obtains the module's data during the next I/O scan and places the data in the destination register specified in the instruction. If multiple channels are to be

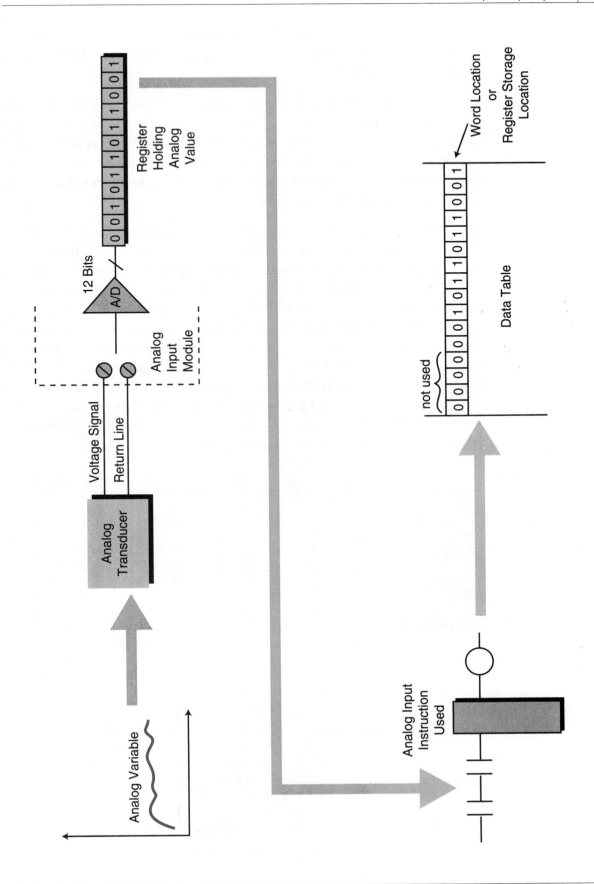

Figure 7-12. Process for inputting analog data to a word location.

read, the processor reads and stores one channel every scan. This does not cause a delay in signal processing, since the scan is very fast and the signals are rather slow in nature.

A processor can determine whether or not the module inserted in the enclosure is analog. If the module is analog, the processor will read the available data in groups of 16 bits, with 12 bits (depending on the resolution) displaying the analog value in binary or BCD. Some controllers may provide diagnostic information about the module and its channels by reading an extra word or register after all channels are input.

The physical location of a module within the rack or enclosure (see Chapter 5 for I/O enclosures) defines its address location. Figure 7-13 illustrates an example of an address for an analog module location. A typical instruction will reference a module's address location by specifying the module's rack and slot numbers, the number of channels or analog inputs used, and the starting register destination address. If a module uses eight channels and the destination storage register starts at address 200_8, the last storage register will be at address 207_8 (see Figure 7-14). The module may also send a status register; in which case, the bits in this register will indicate the status of each channel. The processor assigns the register range automatically according to the number of channels; however, the programmer must remember not to overlap the usage of already assigned registers.

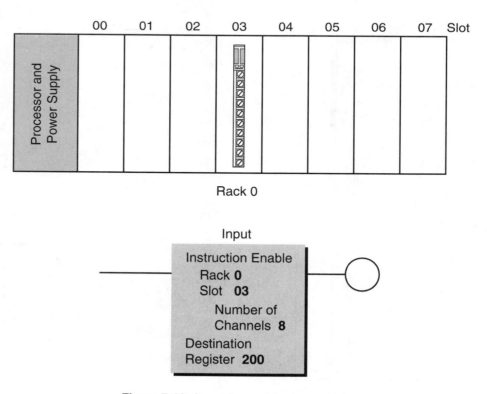

Figure 7-13. An addressed analog module.

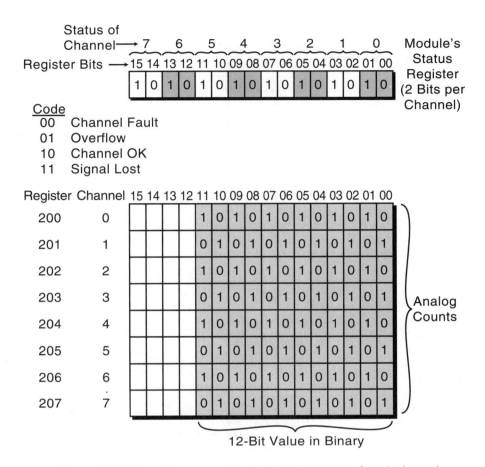

Figure 7-14. Bits within a register indicating the status of each channel.

7-5 ANALOG INPUT CONNECTIONS

Analog input modules usually provide a high input impedance (in the megaohm range) for voltage-type input signals. This allows the module to interface with high source-resistance outputs from input-sensing devices (e.g., transmitters or transducers). Current-type input modules provide low input impedance (between 250 and 500 ohms), which is necessary to properly interface with their compatible field sensing devices.

Analog input interfaces can receive either **single-ended** or **differential** inputs. The commons in single-ended inputs are electrically tied together, whereas differential inputs have individual return or common lines for each channel. Single-ended modules offer more points per module than their differential counterparts. Depending on the manufacturer, a module may be set to either single-ended or differential mode during software setup using rocker switches. Figure 7-15 illustrates typical analog connections for single-ended and differential inputs.

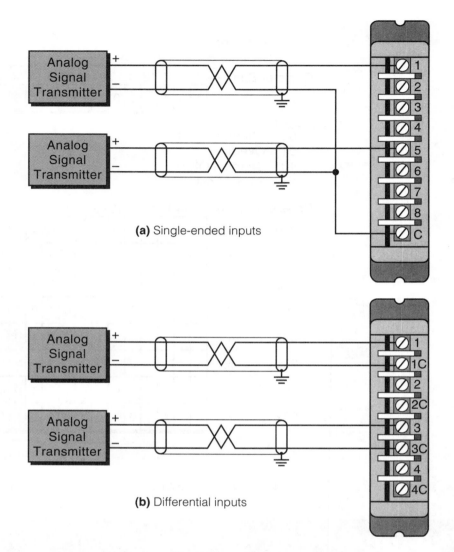

(a) Single-ended inputs

(b) Differential inputs

Figure 7-15. Connection diagrams for **(a)** single-ended and **(b)** differential analog input modules.

Each channel in an analog interface provides signal filtering and isolation circuits to protect the module from field noise. In addition to the noise precautions resident in the module, the user should consider protection from other electrical noise during the installation of the module (see Chapter 20). Shielded conductor cables should be used to connect both the input module and the transducer. These cables lower line impedance imbalances and maintain a good common mode rejection ratio of noise levels, such as power line frequencies.

Analog input interfaces seldom require external power supply sources because they receive their required power from the back plane of the rack or enclosure. These interfaces, however, draw more current than their discrete counterparts; therefore, loading considerations should be kept in mind during PLC system configuration and power supply selection.

7-6 OVERVIEW OF ANALOG OUTPUT SIGNALS

Analog output interfaces are used in applications requiring the control of field devices that respond to continuous voltage or current levels. An example of this type of field device is a volume adjust valve (see Figure 7-16). This type of valve, which is used in hydraulic-based punch presses, requires a 0–10 VDC signal to vary the volume of oil being pumped to the press cylinders, thereby changing the speed of the ram or platen. Table 7-4 lists some other common analog output devices.

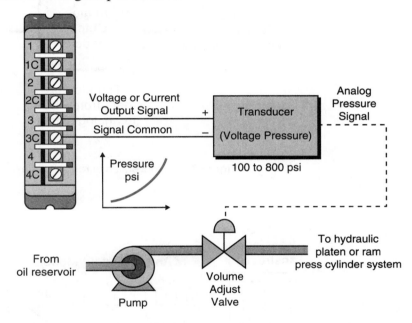

Figure 7-16. Representation of a volume adjust valve.

Analog Outputs
Analog valves
Actuators
Chart recorders
Electric motor drives
Analog meters
Pressure transducers

Table 7-4. Typical analog output field devices.

7-7 INSTRUCTIONS FOR ANALOG OUTPUT MODULES

Multibit analog output instructions, which are similar to those used with multibit discrete outputs, are used to send analog information to field devices. The controller transfers the contents of a register, generally specified by 12 bits, to the output module upon execution of the instruction (see Figure 7-17).

The module then transforms this value, whether BCD or binary, from digital to analog and passes it to the field control device. Figure 7-18 illustrates a multibit instruction transferring 12 bits of data from register 2000 to an analog output module that is connected to a control valve. These 12 bits of information, which are transferred to the field device for control, may be the result of other computations in the PLC program. Chapter 9 explains PLC instructions in more detail.

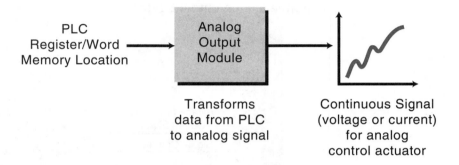

Figure 7-17. Conversion of register data to an analog signal.

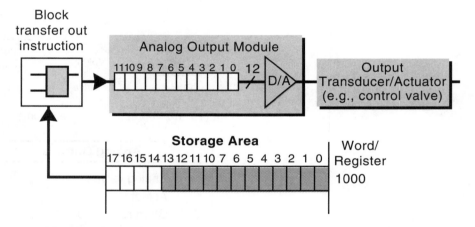

Figure 7-18. Steps in converting a binary value into an analog signal.

EXAMPLE 7-4

Figure 7-19 illustrates the binary transfer of information to an analog output module via a multibit instruction. Assume that the module converts a digital signal equal to the binary value 0000 0000 0000 (0 decimal) to an analog value that makes the control valve be completely closed, while it converts a value of 1111 1111 1111 (4095 decimal) to an analog value that makes the valve be fully open. What will the state of the valve be according to the contents of register 2000?

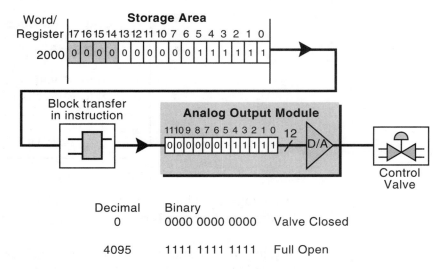

Figure 7-19. Block transfer of register contents to an analog output module.

SOLUTION

The value stored in register 2000 is 0000 0011 1111, which is equivalent to decimal 63. Thus, the valve is open approximately 1/65th, or 1.53%, of its fully open position (63 ÷ 4095 = 1.53%). Note that the position of the valve is determined by the decimal equivalent of the binary value, not the number of 1s and 0s—a binary number with half 1s and half 0s does not indicate that the valve is half open. If the value in the register had been in BCD, the output module would have converted the value to decimal to determine the valve position.

7-8 ANALOG OUTPUT DATA REPRESENTATION

Like analog inputs, **analog output interfaces** are usually connected to controlling devices through transducers (see Figure 7-20). These transducers amplify, reduce, or change the discrete voltage signal into an analog signal, which in turn, controls the output device. Since there are many types of controlling devices, transducers are available in several standard voltage and current ratings. Table 7-5 lists some of the standard ratings used in programmable controllers with analog output capabilities.

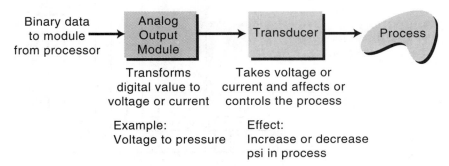

Figure 7-20. Analog output device connected to a transducer.

Output Interfaces
4–20 mA
10–50 mA
0 to +5 volts DC
0 to +10 volts DC
± 2.5 volts DC
± 5 volts DC
± 10 volts DC

Table 7-5. Analog ouput ratings.

An analog output interface operates much like an analog input module, except that the data direction is reversed. As mentioned earlier, a PLC processor can

only interpret digital binary numbers, so it assumes that all other devices operate in the same manner. An analog output module's responsibility, then, is to change the PLC's data from a binary value to an analog real-world signal that can be understood by field devices.

The data transformation that occurs in an output interface is exactly opposite of the transformation in an analog input interface (see Figure 7-21). A **digital-to-analog converter (D/A or DAC)** transforms the numerical data (BCD or binary) sent from the processor into an analog signal. This analog output value is proportional to the digital numerical value received by the module. Thus, the D/A converter creates a continuous analog signal with a magnitude proportional to the minimum and maximum capable analog voltages or currents of the field device (e.g., 0 to 10 VDC).

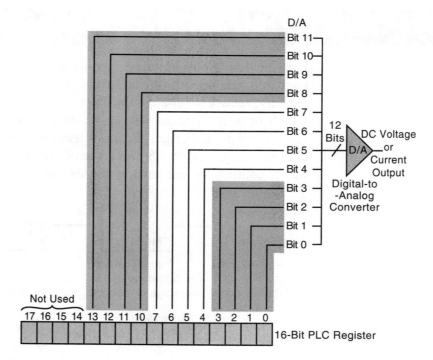

Figure 7-21. Digital-to-analog conversion of numerical data in a PLC register.

The resolution of a digital-to-analog converter is defined by the number of bits that it uses for the analog conversion. For example, a D/A with a 12-bit resolution creates an analog signal ranging from 0 to 4095 counts (4096 total values), which is proportional to a 12-bit digital signal ($2^{12} = 4096$). Therefore, the analog value 2047 in a 12-bit resolution is equal to half of the full range. For an analog field device with a range of 0 VDC (closed) to 10 VDC (fully open), a 2047 analog value would be equal to a 5 VDC signal. Table 7-6 shows the current, voltage, and psi output values from a D/A with a 12-bit resolution.

PLC Register		Output		Pressure
Decimal	**Binary**	**0–10 VDC**	**4–20 mA**	**(psi)**
0	0000 0000 0000 0000	0 VDC	4 mA	0 psi
2047	0000 0111 1111 1111	5 VDC	12 mA	1000 psi
4095	0000 1111 1111 1111	10 VDC	20 mA	2000 psi

Table 7-6. Output values for a 12-bit analog output module.

An analog output module ensures that the value provided by the processor is proportional to the signal or variable that is being controlled by the field device. For instance, if an output device provides pressure control ranging from 100 to 800 psi, the values from the processor, in counts, will be proportional to this range. Output modules can have both unipolar and bipolar configurations, which provide control voltages with either all positive values or negative and positive values, respectively.

EXAMPLE 7-5

A transducer connects an analog output module with a flow control valve capable of opening from 0 to 100% of total flow. The percentage of opening is proportional to a –10 to +10 VDC signal at the transducer's input. Tabulate the relationship between percentage opening, output voltage, and counts for the output module in increments of 10% (i.e., 10%, 20%, etc.). The bipolar output module has a 12-bit D/A (binary) with an additional sign bit that provides polarity to the output swing.

SOLUTION

Since the analog output module has a sign bit, it receives counts ranging from –4095 to +4095, which are proportional to the –10 to +10 VDC signal required by the transducer. Figure 7-22 graphically illustrates the relationship between the module's counts, the output voltage, and the percentage opening.

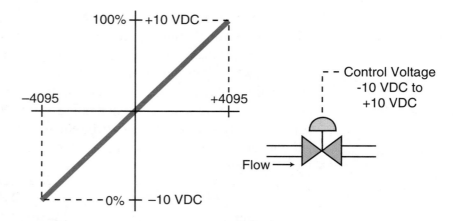

Figure 7-22. Relationship between counts, voltage, and percentage.

To formulate the desired table, first determine the equivalent values for each variable. Since the solution should be expressed in increments as a function of percentage, the percentage changes are calculated as follows:

ΔPercentage	ΔVoltage (−10 to +10)	ΔCounts (−4095 to +4095)
100	20	8190

$$1\% \text{ change as function of voltage} = \frac{20 \text{ VDC}}{100} = 0.2 \text{ VDC}$$

$$1\% \text{ change as function of counts} = \frac{8190}{100} = 81.90 \text{ counts}$$

Note that these computations are magnitude changes. To implement the table, the offset values for the voltage and counts must be added, taking into consideration the bipolar effect of the module and the negative-to-positive changes in counts. Therefore, to obtain the voltage and count equivalents per percentage change, add the offset voltage and count values when the percentage is at 0%. Thus:

$$\text{Percentage as function of voltage} = (0.2 \times P) - 10 \text{ VDC}$$

$$\text{Percentage as function of counts} = (81.9 \times P) - 4095 \text{ counts}$$

where P is the percentage to be used in the table. Therefore, to calculate the required table, multiply each voltage and count relationship by the desired percentage of opening (see Table 7-7).

The PLC's software program calculates output counts according to a predetermined algorithm. Sometimes, the output computations are expressed in engineering units that indicate a 0000 to 9999 (binary value or BCD) change in output value. These values must be ultimately converted to counts—in this case, −4095 to +4095 counts.

Percentage Opening	Output Voltage	Counts
0%	–10 VDC	–4095
10	–8	–3276
20	–6	–2457
30	–4	–1638
40	–2	–819
50	0	0
60	+2	+819
70	+4	+1638
80	+6	+2457
90	+8	+3276
100	+10	+4095

Table 7-7. Equivalent counts, voltages, and percentages.

7-9 ANALOG OUTPUT DATA HANDLING

In the previous section, we explained how a module transfers a signal to the transducer, which sends it to the controlling output device. Now, we will discuss how the processor handles this data, along with some common methods of linearizing output data to reflect engineering units.

The storage or I/O table section of a PLC's data table area holds the data to be sent to an analog output module (see Figure 7-23). This data comes from program computations that, when sent to the module, will control an analog output device. During the execution of the output update, the processor sends the register/word contents to the analog module specified by the address in the instruction. The module transforms the register/word's binary or BCD value into an analog output voltage or current. Since the program calculates the register/word value, the user should take precautions during programming to avoid computing or sending nonvalid ranges to the module. For example, if a word location containing a binary value of +5173 is sent to a 12-bit resolution module without checking for range validity, the module will be unable to interpret the data, thus emitting an incorrect analog output signal (5173 in binary uses more than 12 bits).

Like their input counterparts, analog output modules can handle more than one channel at a time, so one module can control several devices. The instructions that are used with these output interfaces provide the capability of transferring several words or register locations to the module. These instructions are called block transfer out, analog out, block out, or location out instructions (see Chapter 9). It is possible, however, to find PLCs that use arithmetic or other instructions to send data to the analog module address, using the destination register of the instruction.

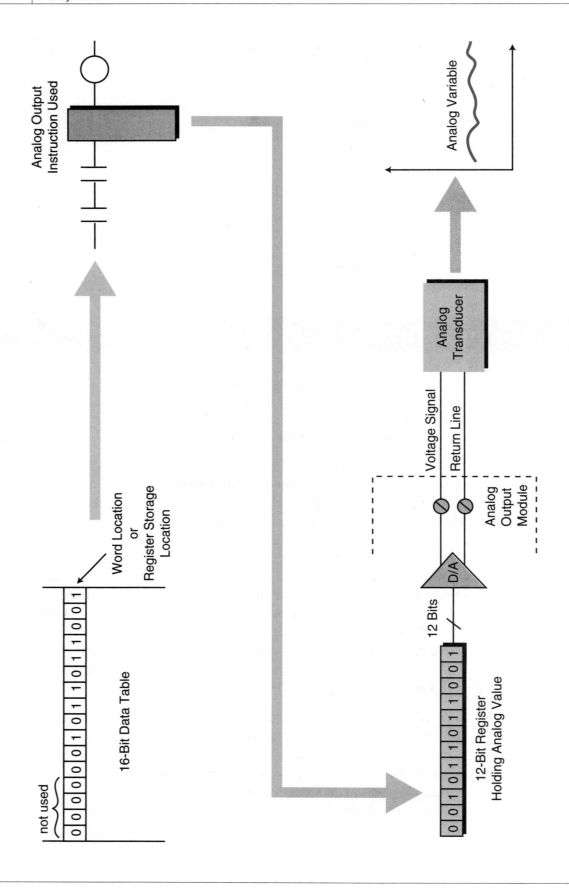

Figure 7-23. Transformation of binary storage table data into an analog signal.

Some PLC manufacturers offer software instructions that scale data within the module or during the execution of the analog output instruction. **Scaling** takes a value and sends it to the module as a linearized count value. For example, let's say an output module receives a BCD value of 5000, relating to an engineering unit (e.g., gallons per minute) halfway between 0000 and 9999 BCD. The software scaling instruction will change this value into the linearized, 12-bit, binary value 0111 1111 1111, or 2047 counts, which represents the halfway mark of the 0 to 4095 range.

Data transfers to analog modules with multiple output channels are updated one channel per scan. As with analog inputs, this update method does not create a noticeable delay, since the devices that respond to analog signals are slow in nature. The physical location of the module within the enclosure defines its address location (see Chapter 6 for I/O enclosures).

Figure 7-24 illustrates an example of an analog output module in an enclosure, along with its corresponding address location. A typical output instruction references a module by its slot and rack locations and the number of channels available or in use. A register called the *source register* stores the data to be transferred. The instruction specifies the starting source register address, and the starting source register transmits the specified number of channels. For example, if the starting register is 300_8 and the number of channels is four, the processor will send the data contained in registers 300_8 through 303_8 (see Figure 7-25).

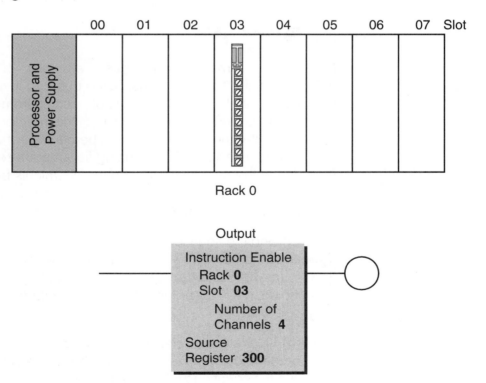

Figure 7-24. An addressed analog output module.

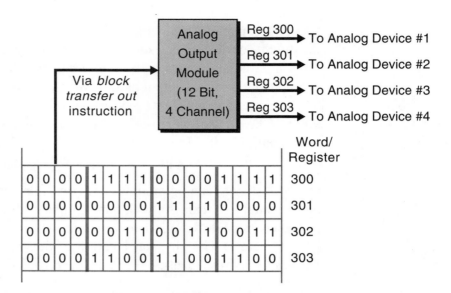

Figure 7-25. Transfer of data from a source register.

Remember that the analog output signal from the module depends on the register or word value it receives from the processor. In some situations, the value computed for a control action is based on a 0000 to 9999 range (engineering units). This value must be converted (if the output instruction does not provide scaling) to the output module count range (i.e., 0 to 4095 counts or −2048 to +2048 counts) before it can be transferred to the module. Example 7-6 addresses this type of conversion.

EXAMPLE 7-6

A programmable controller uses a bipolar −10 to +10 VDC signal to control the flow of material being pumped into a reactor vessel. The flow control valve has a range of opening from 0 to 100% to allow the chemical ingredient to flow into the reactor tank. The processor computes the required flow (the percentage of valve opening) through a predefined algorithm. Analog flow meters send feedback information to the processor about other chemicals being mixed. A register stores the computed value for percentage opening, ranging from 0000 to 9999 BCD (0 to 99.99%).

(a) Find the equation of the line defining the relationship between the analog output signal (in counts) and the analog output transformation from −4095 to +4095 counts. The module has a 12-bit resolution and includes a sign bit as a function of voltage output and percentage opening.

(b) Illustrate the relationship of outputs in counts to the computed percentage opening as stored in the PLC register (0000 to 9999). Also, find the equation that describes the relationship between the required counts and the available calculated value stored in the register.

SOLUTION

(a) Figure 7-26 shows line Y, which represents the number of counts as a function of voltage and percentage opening. The line has the form $Y = mX + b$, where m is the slope of the line and b is the value of Y when X is 0.

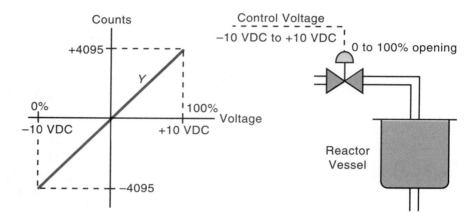

Figure 7-26. Representation of percentage opening and analog output counts.

The X-axis represents either the output voltage or the percentage opening, depending upon which equation is derived. The Y-axis represents the number of counts output by the module for each X value (% or VDC). The following equation expresses the number of counts as a function of voltage:

$$Y = mX + b$$

$$m = \frac{\Delta Y}{\Delta X} = \frac{4095 - (-4095)}{10\ \text{VDC} - (-10\ \text{VDC})} = \frac{8190}{20} \frac{\text{counts}}{\text{VDC}}$$

$$Y = \frac{8190}{20} X + b$$

To calculate b, replace Y with its value when X is 0 counts. When X is 0, Y is also 0; thus:

$$b = Y - \frac{8190}{20} X$$

$$b = 0 - \frac{8190}{20} (0)$$

$$b = 0$$

Therefore:

$$Y = \frac{8190}{20} X + 0$$

$$Y = \frac{8190}{20} X$$

This equation gives the value of Y in counts for any voltage X. The equation of line Y as a function of percentage can be computed in a similar manner:

$$Y = mX + b$$

$$m = \frac{\Delta Y}{\Delta X} = \frac{8190 \ \text{counts}}{100\%}$$

$$Y = \frac{8190}{100} X + b$$

To compute b, replace the count value Y when X is equal to 0%; this value is -4095 (refer to Figure 7-26). Therefore:

$$b = Y - \frac{8190}{100} X$$

$$b = -4095 - \frac{8190}{100}(0)$$

$$b = -4095$$

$$Y = \frac{8190}{100} X - 4095$$

This equation for Y gives the number of output counts for any percentage value X.

(b) Figure 7-27 shows the relationship between the output in counts and the value stored in the register, expressed as 0000 to 9999. This graph is very similar to the previous one; however, the output equation is expressed as a function of the register value used.

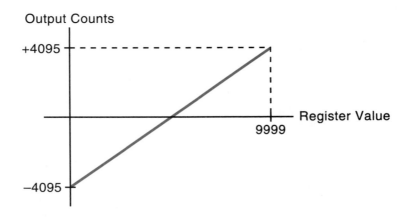

Figure 7-27. Output counts versus register values (0000–9999).

The equation for line Y showing the relationship between output counts and register value is:

$$Y = mX + b$$

$$m = \frac{\Delta \text{counts}}{\Delta \text{register value}} = \frac{8190}{9999}$$

$$Y = \frac{8190}{9999} X + b$$

The value of Y when X equals 0 is –4095, so:

$$b = Y - \frac{8190}{9999} X$$

$$b = -4095 - \frac{8190}{9999} (0)$$

$$b = -4095$$

Therefore:
$$Y = \frac{8190}{9999} X - 4095$$

The value of Y will be the output count for any value X (percentage) ranging from 0000 to 9999. If this type of equation is implemented in the PLC using standard decimal arithmetic instructions and a 0000 to 9999 register value encoded in BCD, the PLC's software must convert the values from BCD to decimal.

7-10 ANALOG OUTPUT CONNECTIONS

Analog output interfaces are available in configurations ranging from 2 to 8 outputs per module, but on average, most modules have 4 to 8 analog output channels. These channels can be configured as either single-ended or differential outputs. Differential is the most common configuration when individually isolated outputs are required.

Each analog output is electrically isolated from other channels and from the PLC itself. This isolation protects the system from damage due to overvoltage at the module's outputs. These interfaces may require external, panel-mounted power supplies; however, most analog modules receive their power from the PLC's power supply system. Current requirements for analog modules are higher than for discrete outputs and must be considered during the computation of current loading (see Chapter 4 for loading considerations). Figure 7-28 illustrates typical connections for both single-ended and differential analog output modules.

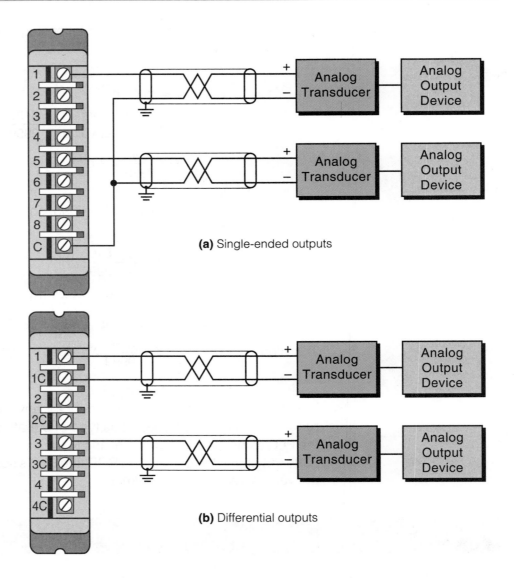

(a) Single-ended outputs

(b) Differential outputs

Figure 7-28. Connection diagrams for **(a)** single-ended and **(b)** differential analog output modules.

7-11 ANALOG OUTPUT BYPASS/CONTROL STATIONS

A PLC system may require the addition of a bypass/control station (see Figure 7-29). Bypass/control stations, which are placed between the PLC's analog interface and the controlled element, ensure continued production or control in a variety of abnormal process situations. A bypass/control station is very useful during start-up, override of analog outputs, and backup of analog outputs in case of failures.

During start-up, the operator can use a bypass/control station to manually position the final control elements through manipulation of initial control parameters, such as valve position, speed control, hydraulic servos, and

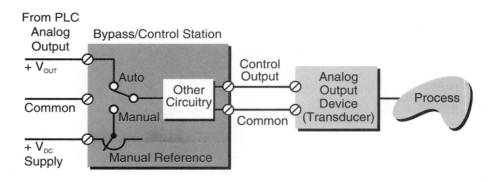

Figure 7-29. Block diagram of bypass/control backup unit.

pneumatic converters. This can be done without the PLC or prior to its checkout. When the final elements are working properly, the user can then perform a final check of the PLC and switch the bypass/control station to automatic mode for direct PLC control of the process.

KEY **analog input interface**
TERMS **analog output interface**
analog signal
analog-to-digital converter
channel
differential input/output
digital-to-analog converter
resolution
scaling
single-ended input/output
transducer
transmitter

CHAPTER EIGHT

SPECIAL FUNCTION I/O AND SERIAL COMMUNICATION INTERFACING

No rule is so general, which admits not some exception.

—Robert Burton

**CHAPTER
HIGHLIGHTS**
In previous chapters, we discussed analog and digital I/O interfaces. Although these types of interfaces allow control implementation in most types of applications, some processes require special types of signals. In this chapter, we will introduce special function I/O interfaces, which uniquely process analog and digital signals. We will also take a look at intelligent positioning, data-processing, and communication modules that expand the capabilities of PLCs. We will conclude with a discussion of peripheral interfacing and communication standards. When you finish this chapter, you will have learned about all the major components of programmable controllers—from processors to intelligent interfaces—and you will be ready to explore PLC programming.

8-1 INTRODUCTION TO SPECIAL I/O MODULES

Special function I/O interfaces provide the link between programmable controllers and devices that require special types of signals. These special signals, which differ from standard analog and digital signals, are not very common, occurring in only 5–10% of PLC applications. However, without special interfaces, processors would not be able to interpret these signals and implement control programs.

Special I/O interfaces can be divided into two categories:

- direct action interfaces
- intelligent interfaces

Direct action I/O interfaces are modules that connect directly to input and output field devices. These modules preprocess input and output signals and provide this preprocessed information directly to the PLC's processor (see Figure 8-1). All of the discrete and analog I/O modules discussed in Chapters 6 and 7, along with many special I/O interfaces, fall into this category. Special direct action I/O interfaces include modules that preprocess low-level and fast-input signals, which standard I/O modules can not read.

Special function **intelligent I/O interfaces** incorporate on-board microprocessors to add intelligence to the interface. These intelligent modules can perform complete processing tasks independent of the PLC's processor and program scan. They can also have digital, as well as analog, control inputs and outputs. Figure 8-2 illustrates an application of intelligent I/O interfaces. The method of allocating various control tasks to intelligent I/O interfaces is known as **distributed I/O processing**.

Special input/output modules are available along the whole spectrum of programmable controller sizes, from small controllers to very large PLCs. In general, special I/O modules are compatible throughout a family of PLCs.

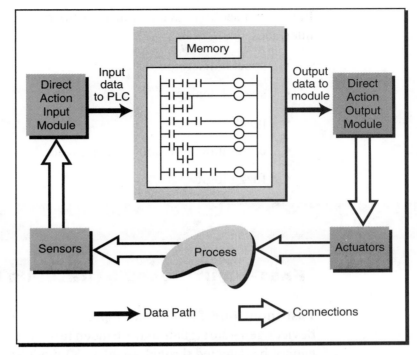

Figure 8-1. Direct action I/O interface application.

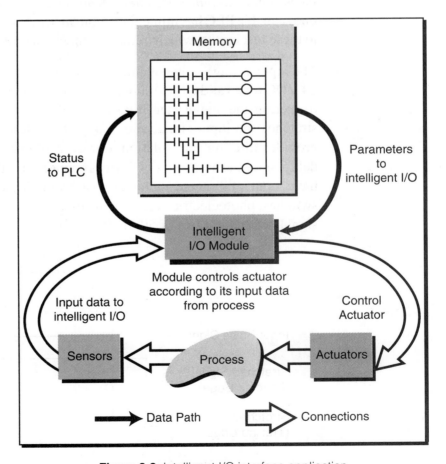

Figure 8-2. Intelligent I/O interface application.

In the next sections, we will discuss the most commonly found special I/O interfaces:

- special discrete

- special analog

- positioning

- communication/computer/network

- fuzzy logic

8-2 SPECIAL DISCRETE INTERFACES

FAST-INPUT/PULSE STRETCHER MODULES

Fast-input interfaces detect input pulses of very short duration. Certain devices generate signals that are much faster than the PLC scan time and thus cannot be detected through regular I/O modules. Fast-response input interfaces operate as *pulse stretchers,* enabling the input signal to remain valid for one scan. If a PLC has immediate input instruction capabilities, it can respond to these fast inputs, which initiate an interrupt routine in the control program.

The input voltage range of a fast-input interface is normally between 10 and 24 VDC for a valid ON (1) signal, with the leading or trailing edge of the input triggering the signal (see Figure 8-3). When the interface is triggered, it stretches the input signal and makes it available to the processor. It also provides filtering and isolation; however, filtering causes a very short input delay, since the normal input devices connected to this type of interface do not have contact bounce. Typical fast-input devices, including proximity switches, photoelectric cells, and instrumentation equipment, provide pulse signals with durations of 50 to 100 microseconds.

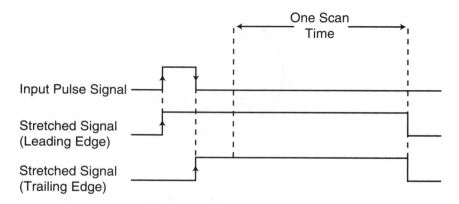

Figure 8-3. Pulse stretching in a fast-input module.

Connections to fast-input modules are the same as for standard DC input modules. Depending on the module, the field device must meet the sourcing or sinking requirements of the interface for proper operation. Usually, field devices must source a required amount of current to the fast-input module at the rated DC voltage.

WIRE INPUT FAULT MODULES

Wire input fault modules are special input interfaces designed to detect short-circuit or open-circuit connections between the module and input devices. Wire input fault modules operate like standard DC input modules in that they detect a signal and pass it to the processor for storage in the input table. These modules, however, are specially designed to detect any malfunction associated with the connections. Figure 8-4 illustrates a simplified block diagram of Allen-Bradley's wire input fault module. Typical applications of this module include critical input connections that must be monitored for correct wiring and field device operation.

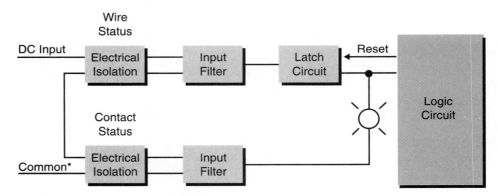

*All commons are tied together inside module

Figure 8-4. Wire input fault module diagram.

Wire input fault interfaces detect a short-circuit or open-circuit wire by sensing a change in the current. When the input is OFF (0), the interface sends a 6 mA current through a shunt resistor (placed across the input device) for each input; when the input is ON (1), the interface sends a 20 mA current. An opened or shorted input will disrupt this monitoring current, causing the module to detect a wire fault. The module signals this fault by flashing the corresponding status LED. The control program can also detect the fault and initiate the appropriate preprogrammed action.

Figure 8-5 illustrates a typical connection diagram for a wire input fault interface. Note that shunt resistors must be connected to the interface even though an input device is not wired to the module. The rating of the shunt resistor depends on the DC power supply voltage level used. This supply may range from 15 to 30 VDC. Although it is unlikely to occur, the total wire resistance of the connecting wire must not exceed the specified ohm rating for the DC supply voltage level. This wire resistance value is computed by multiplying the per foot ohm value by the total length of the wire connection. For example, a size 14 wire that has a resistance of 0.002525 ohms per foot should have a total wire resistance of less than 25 ohms when connected to a 15 VDC power supply. This implies that the wire should not exceed 9,900 feet in length (25 ÷ 0.002525 = 9,900).

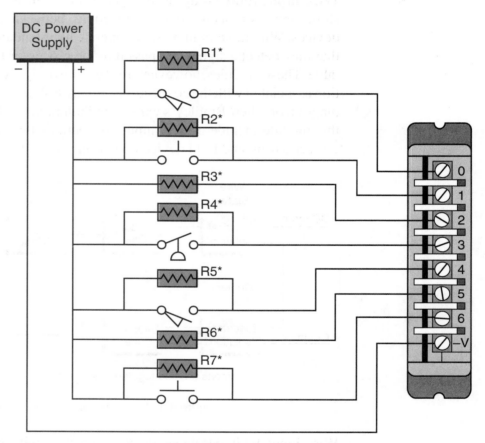

*Shunt resistors with 1/2 watt rating. Value depends on power supply voltage.

Figure 8-5. Wire input fault module connection diagram.

FAST-RESPONSE INTERFACES

Fast-response interfaces are extensions of fast-input modules. These interfaces detect fast inputs and respond with an output. The speed at which this occurs can be as short as 1 msec from the sensing of the input to the output response. The output response time is independent of the PLC processor and the scan time.

Fast-response modules have advantages that include the ability to respond to very fast input events, which require an almost immediate output response. For example, the detection of a feeder jam in a high-speed assembling or transporting line may require the module to send a fast disengage signal to the product feed, thus reducing the amount of product jammed or lost.

During operation, a fast-response module receives an enable signal from the processor (through the control program), which readies it for "catching" the fast input. Once the active module receives the signal, it sends an output and remains ON until the processor (via the ladder program) disables it, thereby resetting the output. Figure 8-6 illustrates a block diagram of this interface's operation, along with its logic and timing. Figure 8-7 illustrates how a fast-response interface functions. Furthermore, Figure 8-8 shows Allen-Bradley's

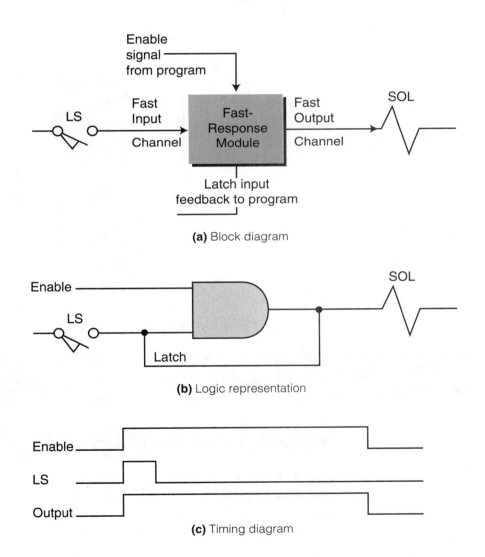

(a) Block diagram

(b) Logic representation

(c) Timing diagram

Figure 8-6. (a) Block diagram, **(b)** logic representation, and **(c)** timing diagram of a fast-response interface.

version of a fast-response module, called a High-Speed Logic Controller Module (1771-DR), which offers 8 inputs and 4 outputs that switch ON less than 1 msec after the detection of the fast input.

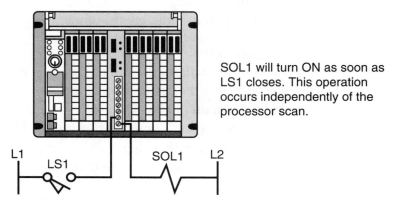

SOL1 will turn ON as soon as LS1 closes. This operation occurs independently of the processor scan.

Figure 8-7. Fast-response interface.

Courtesy of Allen-Bradley, Highland Heights, OH

Figure 8-8. Allen-Bradley's fast-response module (1771-DR).

8-3 SPECIAL ANALOG, TEMPERATURE, AND PID INTERFACES

WEIGHT INPUT MODULES

Weight input modules are special types of analog interfaces designed to read data from load cells, which are standard on storage tanks, reactor vessels, and other devices used in blending and batching operations. Figure 8-9 illustrates the configuration of a weight input application, while Figure 8-10 shows Allen-Bradley's weight input interface called the Weigh Scale Module (1771-WS). These weight modules support the industry standard of 2 or 3 millivolts per volt (mV/V) load cells.

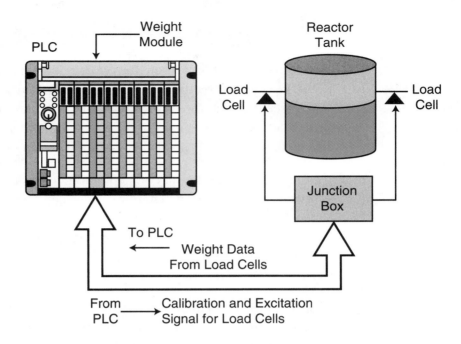

Figure 8-9. Weight input application configuration.

Figure 8-10. Allen-Bradley's Weigh Scale Module (1771-WS).

A weight input module provides the excitation voltage for load cells, as well as the necessary software for calibrating load cell circuits. A weight module sends an excitation voltage to a load cell and reads the signal created by the weight force exerted on the cell (see Chapter 13). The module's A/D converter then processes this information and passes it to the processor as a weight value. This eliminates the need for the PLC to convert the load cell's analog signal. Additionally, a weight module incorporates a calibration feature that avoids problems with calibration of the load cell system.

THERMOCOUPLE INPUT MODULES

In addition to standard analog voltage/current input interfaces that can receive signals directly from transmitters, special analog input interfaces can also accept signals directly from sensing field devices. **Thermocouple input modules**, which accept millivolt signals from thermocouple transducers, are an example of this type of special preprocessing interface.

Different types of thermocouple input modules are available, depending on the thermocouple used. These modules can interface with several types of thermocouples by selecting jumpers or rocker switches in the module. For example, an input module may be capable of interfacing with thermocouples of (ISA standard) type E, J, and K. Chapter 13 lists some of the ranges, types, and applications for the most commonly used thermocouples.

The operation of a thermocouple module is very similar to that of a standard analog input interface. The module amplifies, digitizes, and converts the input signal (in millivolts) into a digital signal. Depending on the manufacturer, the converted number will represent, in binary or BCD, the degrees Celsius or Fahrenheit being measured by the selected thermocouple.

Thermocouple modules do not provide a range of counts proportional to the measured temperature because thermocouples exhibit nonlinearities along their range. These nonlinearities usually occur between 0°C and the thermocouple's upper temperature limit. To determine the digital value of the incoming signal, the thermocouple input module's on-board microprocessor calculates the temperature (in °C or °F) that corresponds to the voltage reading. The microprocessor does this by referencing a thermocouple table (millivolts versus °C or °F) and performing a linear interpolation (see thermocouples in Chapter 13).

Thermocouple interfaces usually provide **cold junction compensation** for thermocouple (device) readings. This compensation allows the thermocouple to operate as though there were an ice-point reference (0°C), since all of the thermocouple's tables depicting the generation of electromotive force (emf) are referenced at this point.

In addition to cold junction compensation, thermocouple modules provide **lead resistance compensation** for a determined resistance value. Lead resistance deals with the loss of signal due to resistance in the wires. Thermocouple manufacturers can provide resistance values for given wire size lengths at known temperatures. Depending on the PLC manufacturer, thermocouple interfaces may provide different lead resistance compensations. One manufacturer may provide 200 ohms of compensation, while another may provide 100 ohms. If the lead resistance is greater than the available compensation, a calculation in the control program can add degrees Celsius to compensate for the resistance.

When possible, it is a good practice to use the same type of material for the lead wire as is used in the thermocouple. Smaller gauge wire provides a slightly faster response, but heavier gauge wire tends to last longer and resist contamination and deterioration at high temperatures. Figure 8-11 shows a typical thermocouple interface connection. Chapter 13 presents more information about thermal transducers.

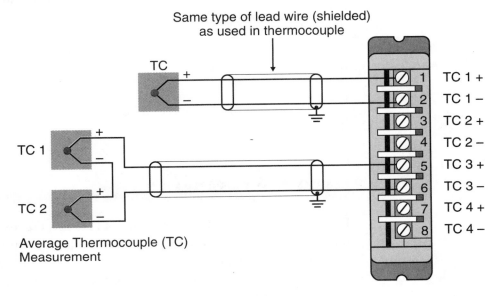

Figure 8-11. Thermocouple interface connection diagram.

The following example illustrates a case where a thermocouple performs compensation. Some typical uses of compensation are applications where very long lead wires are employed or where several thermocouples are connected in parallel.

EXAMPLE 8-1

A type J thermocouple is connected to a thermocouple module located in an I/O rack located 500 feet away. This thermocouple is connected to a heat trace circuit, which measures temperature ranges throughout the length of a process pipe. The thermocouple has 18 AWG lead wires that have a resistance of 0.222 ohms for each foot of double wire (positive and negative wire conductors) at 25°C. The thermocouple module has a lead resistance compensation of 50 ohms, and the manufacturer has a 0.05°C per ohm compensation error factor. Find the total lead resistance and the necessary compensation in degrees Celsius to be added to the value measured.

SOLUTION

The total lead resistance is computed as:

Lead resistance = (Thermocouple lead resistance)(Lead wire length)
= (0.222)(500)
= 111 ohms

The compensation requirement will be the difference between the total resistance and the module's compensation multiplied by the compensation error factor:

Compensation in °C = (111 − 50 ohms)(0.05°C/ohm)
= 3.05°C

Thus, a compensation of 3.05°C must be added to the thermocouple reading.

RTD INPUT MODULE

Resistance temperature detector (RTD) interfaces receive temperature information from RTD devices. RTDs are temperature sensors that have a wire-wound element whose resistance changes with temperature in a known and repeatable manner. An RTD in its most common form consists of a small coil of platinum, nickel, or copper protected by a sheath of stainless steel. These devices are frequently used for temperature sensing because of their accuracy, repeatability, and long-term stability.

The operation of RTD modules is similar to that of other analog input interfaces. These modules send a small (mA) current through the RTD and read the resistance to the current flow. In this manner, the module can measure changes in temperature, since the RTD changes resistance with changes in temperature.

An RTD module converts changes in resistance into temperature values, available to the processor in either °C or °F. Some interfaces are able to provide the processor with the resistance value in ohms in addition to temperature measurements. Depending on the manufacturer, the module may also be able to sense more than one type of RTD. Table 8-1 lists some of the most common RTD devices and their resistance ratings.

RTD devices are available in 2-, 3-, and 4-wire connections. Devices with a 2-wire scheme do not compensate for lead resistance; however, 3- and 4-wire RTDs do allow for lead resistance compensation. The most commonly used

RTD Type	Resistance Rating (ohms)	Temperature Range	
Platinum	100	−200 to 850°C	−328 to 1562°F
Nickel	120	−80 to 300°C	−112 to 572°F
Copper	10	−200 to 260°C	−328 to 500°F

Table 8-1. Common RTD types and their specifications.

RTD device is the 3-wire RTD. This type of device is used in applications requiring long lead wires, where wire resistance is significant in comparison to the ohms/°C sensitivity of the RTD element. It is a good practice to try to match the resistance of the lead wires by using quality cabling and heavy gauge wires (16–18 gauge). Figure 8-12 illustrates typical connections for an RTD module with 2-, 3-, and 4-wire RTDs. Chapter 13 explains more about resistance temperature detectors.

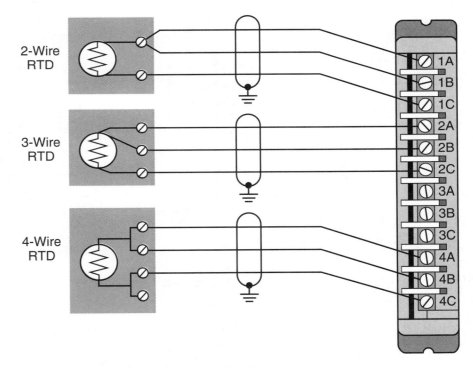

Figure 8-12. RTD connection diagram.

PID MODULES

Proportional-integral-derivative (PID) interfaces are used in process applications that require continuous closed-loop control employing the PID algorithm. These modules provide proportional, integral, and derivative control actions according to sensed parameters, such as pressure and temperature, which are the input variables to the system. PID control is often referred to as three-mode, closed-loop feedback control. Figures 8-13 and 8-14 illustrate PID control in block diagram form and process form, respectively.

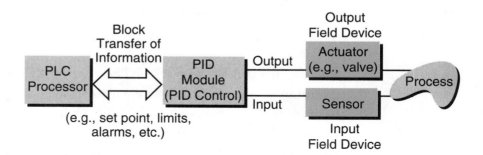

Figure 8-13. Block diagram of PID control.

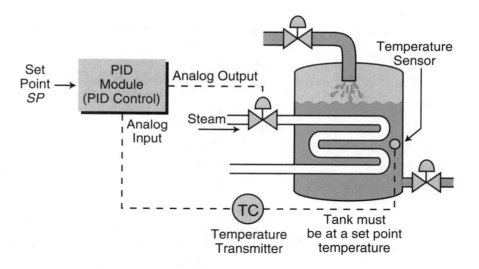

Figure 8-14. Illustration of a PID control process.

The basic function of closed-loop process control is to maintain certain process characteristics at desired set points. Process characteristics often deviate from their desired set point references as a result of load material changes, disturbances, and interactions with other processes (see Figure 8-15). During control, the actual process characteristics (liquid level, flow rate, temperature, etc.) are measured as the process variable (PV) and compared with the target set point (SP). If the process variable (actual value) deviates from the set point (desired value) an error (E) occurs ($E = SP - PV$). Once the module detects an error, the control loop modifies the control variable (CV) output to force the error to zero.

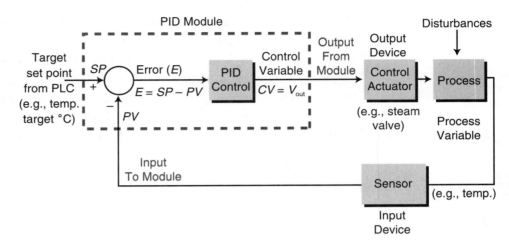

Figure 8-15. Closed-loop process control.

The following equation defines one of the control algorithms implemented by a PID module:

$$V_{out} = K_P E + K_I \int E dt + K_D \frac{dE}{dt}$$

where:

K_P = the proportional gain

$K_I = \frac{K_P}{T_I}$, which is integral gain (T_I = reset time)

$K_D = K_P T_D$, which is derivative gain (T_D = rate time)

$E = SP - PV$, which is error

V_{out} = the control variable output

The PID module receives the process variable in analog form and computes the error difference between the actual value and the set point value. It then uses this error difference in the algorithm computation to initiate a three-step, simultaneous, corrective action through a control variable output. First, the module formulates a *proportional* control action based on an output control variable that is proportional to the instantaneous error value ($K_P E$). Then, it initiates an *integral* control action (reset action) to provide additional compensation to the output control variable. This causes a change in the process variable in proportion to the value of the error over a period of time (K_I or K_P/T_I). Finally, the module initiates a *derivative* control action (rate action) adding even more compensation to the control output ($K_D = K_P T_D$). This action causes a change in the output control variable proportional to the rate of change of error. These three steps provide the desired control action in proportional (P), proportional-integral (PI), and proportional-integral-derivative (PID) control fashion, respectively.

A PID module receives primarily control parameter and set point information from the main processor. The module can also receive other parameters, such as maximum error and maximum/minimum control variable outputs for high and low alarms, if these signals are provided. During operation, the PID interface maintains status communication with the main CPU, exchanging module and process information. Figure 8-16 illustrates a block diagram of the PID algorithm and a typical PID module connection arrangement.

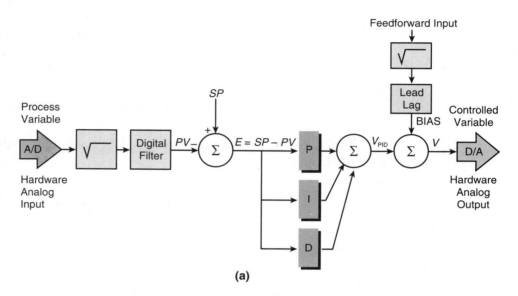

(a)

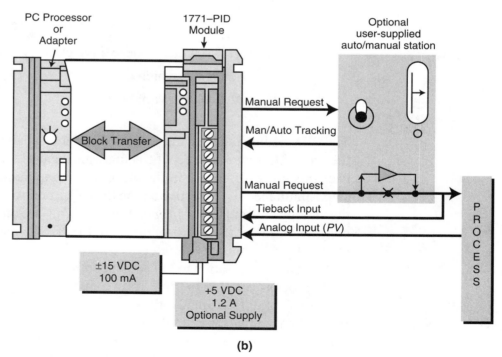

(b)

Figure 8-16. (a) Block diagram of the PID algorithm and **(b)** a connection diagram for Allen-Bradley's 1771-PID module.

Depending on the module used, PID interfaces can also receive data about the update time and the error deadband. The *update time* is the rate or period in which the output variable is updated. The *error deadband* is the quantity that is compared to the error signal (see Figure 8-17); if the error deadband is less than or equal to the signal error, no update takes place. Moreover, some modules also provide square root calculations of the process variable. To provide this calculation, the module performs a square root extraction of the process variable to obtain a linearized scaled output, which is then used by the PID loop. The control of flow by a PID is an example of an application using a square root extractor. Chapter 15, which describes process controller responses, explains more about PID.

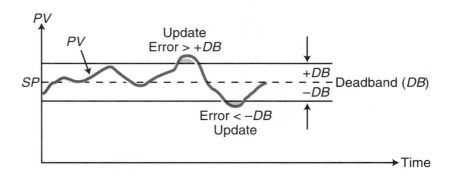

Figure 8-17. Error deadband.

8-4 POSITIONING INTERFACES

Positioning interfaces are intelligent modules that provide position-related feedback and control output information in machine axis control applications. This section covers the basic aspects of positioning motion control as it relates to PLC applications.

The motion control capabilities of positioning modules allow some programmable controllers to perform functions, using servo mechanisms (e.g., point-to-point control and axis positioning), that once required computer numerical control (CNC) machines.

POSITIONING INTERFACE INSTRUCTIONS

Positioning interfaces use PLC instructions that transfer blocks of data at a time (see Figure 8-18). This data includes initialization parameters, distances and limits, and velocities. Instructions, such as block transfer in/out and move data in/out, are typically used to implement this transfer of information.

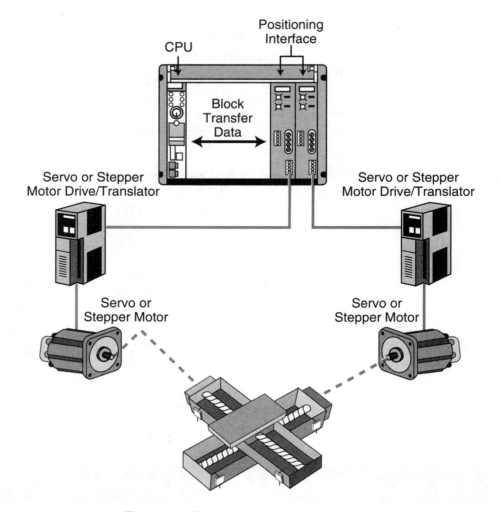

Figure 8-18. Positioning interface configuration.

ENCODER/COUNTER INTERFACES

Encoder/counter modules interface encoders and high-speed counter devices with programmable controllers. This type of module operates independently of the processor and I/O scan. An encoder/counter module is an integral part of a programmable controller system when it is used in applications requiring position information. Such applications include closed-loop positioning of machine tool axes, hoists, and conveyors, as well as cycle monitoring of high-speed machines, such as can-making equipment, stackers, and forming equipment.

There are two types of encoder/counter interfaces: absolute and incremental. *Absolute encoders* provide an angular measurement of the shaft. They provide this angular position (expressed in BCD, binary, or Gray code) in parallel to the encoder interface module. *Incremental encoders* measure shaft rotation over distance by outputting a fixed number of pulses per shaft rotation. The module provides two pulse signals that have a 90° phase difference (quadrature); it then determines the direction of rotation by sensing which of the two pulse channels is the leading waveform. Incremental encoders provide a *marker,* or *index,* channel that sends a pulse for every shaft revolution. This marker, which is an input to the module, can be used in conjunction with the module's limit switch channel input to establish a home position along the encoder's measurements. When the encoder interface is used in a counter configuration, however, only one input channel can be connected to a device that provides a pulse count.

During operation, an encoder/counter module (in incremental encoder mode) receives two pulse channel inputs that are counted and compared with a user-specified preset value. The interface may have one or two output lines available, which are energized once the incoming pulses are equal to, greater than, or less than the preset values. The input channels and output lines available are generally rated for TTL or for 12–48 VDC. The maximum input pulse frequency that an encoder/counter interface can properly count ranges between 50 and 60 kHz.

The communication between an encoder/counter interface and the processor is bidirectional. The module accepts the preset value and other control data from the processor and transmits values and status data to the PLC memory. The interface also lets the PLC know when the marker and limit switch are both energized, indicating a home position. On the other hand, the processor's control program, which tells the module to operate the outputs according to the count value received, enables the output controls. The control program also enables and resets the counter operation.

Typically, the length between the module and the encoder should not exceed 50 feet, and shielded cables should be used. Since encoder/counter modules have both inputs and outputs, they have isolation between the input and output circuits, as well as between the control logic and both I/O circuits. The use of separate power supplies, which must be provided by the user, enhances this isolation. Figure 8-19 shows the typical connections for an incremental encoder configuration.

STEPPER MOTOR INTERFACES

Stepper motor interfaces, as their name implies, are used in applications requiring control of stepper motors. Stepper motors are permanent-type magnet motors that translate incoming pulses, through a stepper translator, into mechanical motion. *Stepper* is a generic term that describes this type of brushless motor capable of making fixed angular motions in response to a step input.

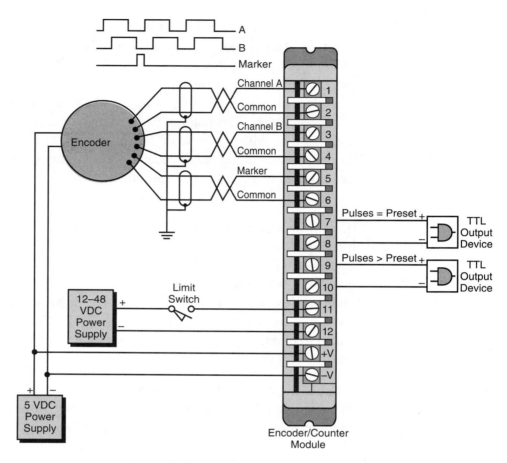

Figure 8-19. Encoder/counter interface connection diagram.

The motion of a stepper can be accelerated, decelerated, or maintained constantly by controlling the pulse rate output from a stepper module. The ability to respond to an input voltage (in the form of DC pulses) makes stepper motors well suited for incremental motor programmable control systems. Under controlled conditions, a stepper motor's motion follows the number of input pulses. This ability to respond to a fixed input enables the system to operate in an open-loop mode, leading to cost savings in the total system. However, in high-response applications, closed-loop operation is generally required (using encoder feedback). Figure 8-20 illustrates a simplified block diagram of a stepper motor system.

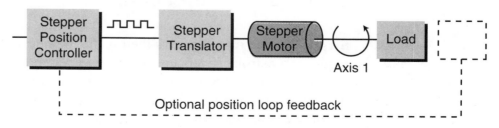

Figure 8-20. Block diagram of a stepper motor system.

A stepper interface generates a pulse train that is compatible with the stepper translator, indicating distance, rate, and direction commands to the motor. The motion induced can be rotational or linear, such as the forward or backward movement of a linear slide using leadscrews. Figure 8-21 shows a typical linear slide using a stepper motor that makes one revolution per 200 steps (resolution), thus yielding a 1.8° step angle (360/200 or 1/200th of a revolution). The stepper system shown in the figure provides a linear movement of 0.00125 inches per step because of the 4 threads per inch leadscrew. Example 8-2 illustrates how to calculate linear movement and step angle values.

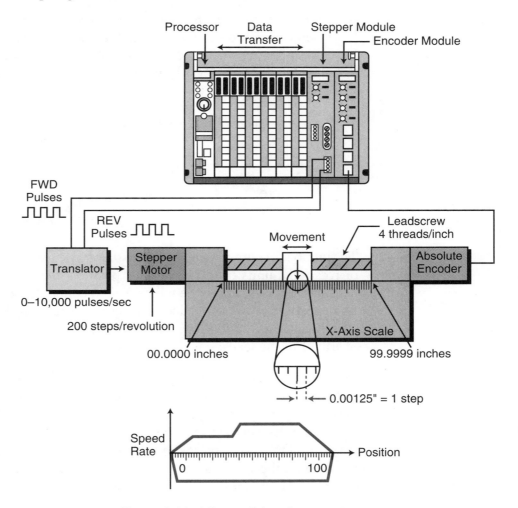

Figure 8-21. A linear slide using a stepper motor.

EXAMPLE 8-2

Referencing Figure 8-21, suppose that the 200-step motor is operating at half-stepping conditions (400 steps per revolution) and that the leadscrew has 5 threads per inch. What are the step angle and linear displacement per step used in the system?

SOLUTION

To compute the step angle, divide the number of degrees in one revolution (360°) by the number of steps required to turn the motor. Therefore, the step angle is:

$$\text{Step angle} = \frac{360°}{400}$$
$$= 0.9°$$

with a resolution of 1/400th of a revolution. Linear displacement is the number of inches moved in one step. To calculate this, multiply the number of threads it takes to move one inch by the number of steps in a revolution, since each thread requires one revolution (rotational-to-linear displacement). In this case, the leadscrew requires 5 revolutions to move one inch, and each revolution requires 400 steps.

$$1" \text{ travel} = (5 \text{ rev})(400 \text{ steps/rev})$$
$$= 2000 \text{ steps}$$

Therefore:
$$1 \text{ step} = \frac{1}{2000}$$
$$= 0.0005 \text{ inches}$$

The number of outburst pulses sent to the stepper, which translates into linear or rotational units of travel, defines position displacement. Therefore, the number of pulses sent to the motor from the module determines the motor's final position. The actual location also depends on the resolution of the stepper and the application, which defines the number of threads per inch of travel in the leadscrew.

The stepper's movement includes both the acceleration and deceleration of the motor. The acceleration part of the move is the time required to achieve the continuous speed rate of the motor (in pulses/sec). The continuous rate is the final pulse/sec rate sent to the motor (frequency). This frequency may vary from 1 to 20 kHz (pulses/sec). Conversely, the deceleration part is the time required for the speed rate to decrease to zero (pulses/sec). Acceleration and deceleration, also known as *ramps*, are specified as a function of time (seconds).

Stepper motor interfaces operate in two modes: *single-step profile mode* and *continuous profile mode*. In single-step mode, a PLC processor sends individual move sequences to the interface. These sequences include the acceleration and deceleration rates of the move, along with the final or continuous speed rate (see Figure 8-22). Once this move sequence is terminated, the processor may start another one by transferring the next move's profile information and commands. The processor can store several single-step mode profiles and send them to the module through the PLC program control.

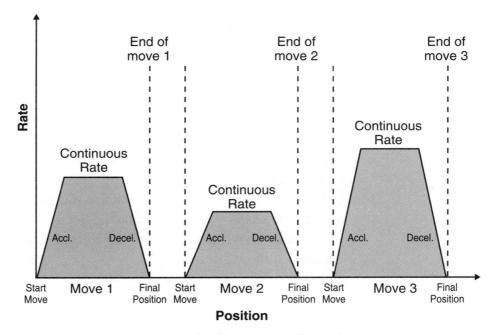

Figure 8-22. Single-step profile mode.

In continuous mode, the motion profile is cycled through various accelerations, decelerations, and continuous speed rates to form a blended motion profile (see Figure 8-23). Rather than requiring additional commands for motion speed changes, an interface in continuous mode receives the whole move profile in a single block of instructions. The interface then performs the step motor control duty until the motion is completed and the processor sends the next profile. As in the single-step mode, the processor can store several continuous mode profiles in its memory and send them to the interface during the program execution.

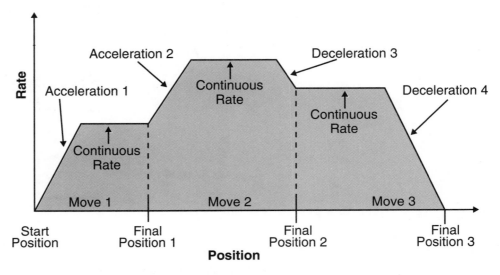

Figure 8-23. Continuous profile mode.

Each stepper interface used to control a stepper motor controls an axis, since the motion generated causes a movement about either the X-, Y-, or Z-axis (see Figure 8-24). Depending on the PLC manufacturer, more than one axis may be controlled using several stepper module interfaces. When multiple-axis motions are required, the axes can be controlled either independently or synchronously (see Figures 8-25a and 8-25b, respectively). When controlled independently, each axis is independent of the other, executing its own single-step or continuous profile mode. The beginning and end of each axis motion may be different. When controlled synchronously, the beginning and end of the motion commands for each axis occur at the same time. A profile of one of the axes may start later or end before the other axes (see Figure 8-25b), but the move that follows will not occur until all axes have started and ended their motions.

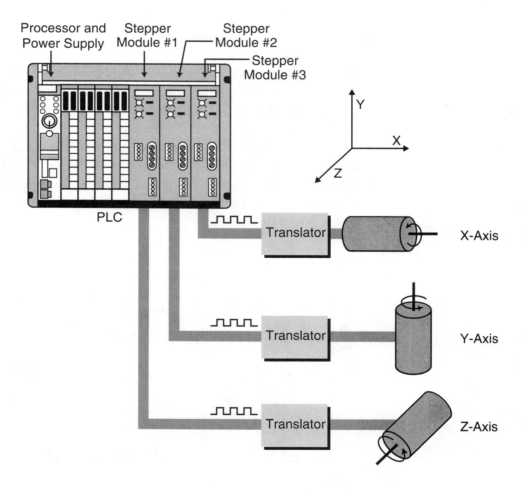

Figure 8-24. PLC system using stepper modules to control three axes.

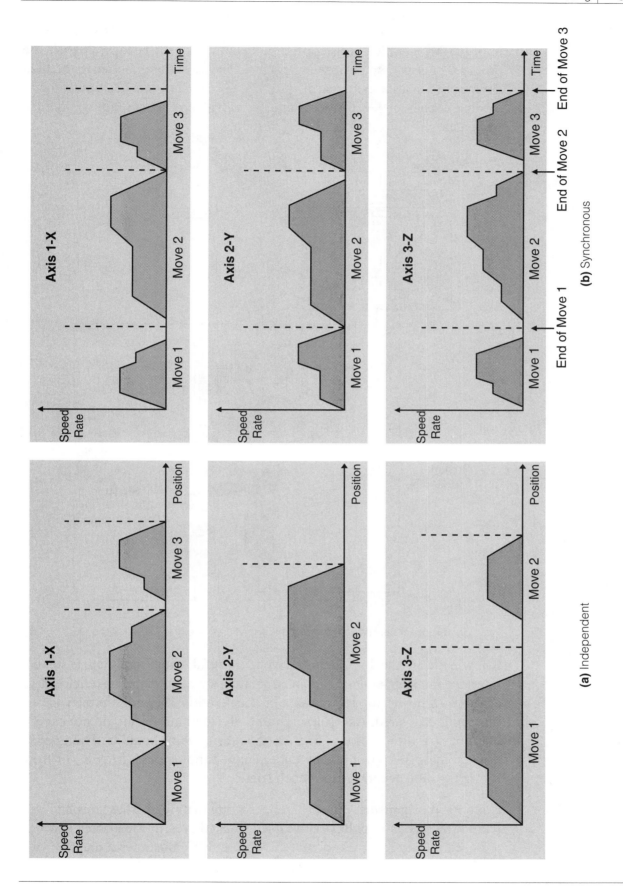

Figure 8-25. (a) Independent axis control and **(b)** synchronous axis control.

The use of a position/velocity feedback scheme (see Figure 8-26) can greatly improve the operation of a stepper motor control system, because this scheme provides closed-loop positioning control. The most common feedback field device used in a stepper control system is the encoder. In a position/velocity feedback scheme, the encoder is interfaced with an encoder input module to form a closed-loop stepper control system.

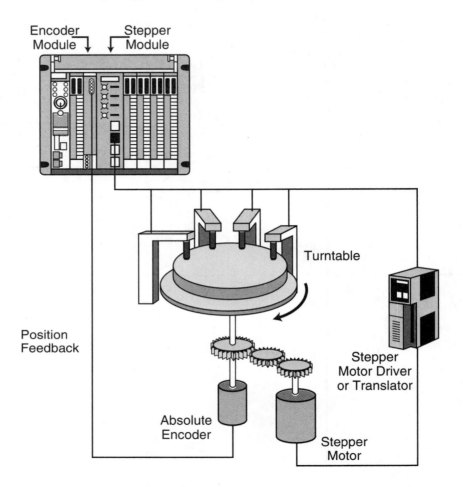

Figure 8-26. Stepper motor with a position/velocity feedback scheme.

Knowledge of the load being driven is useful when applying a stepper interface in a stepper motor application. Loads with high inertia require large amounts of power for acceleration or deceleration; therefore, proper inertia matching is desired. As a rule of thumb, the load inertia should not exceed ten times the rotor inertia. The friction of the system should be examined to prevent the system from being underdamped (not enough friction) or from losing position accuracy (too much friction).

Coupling mechanisms connect a stepper motor to its load. These mechanisms include metal bands, pulleys and cables, direct drives, and leadscrews, which are used mostly for linear actuation. Figure 8-27 illustrates a diagram of a typical stepper motor interface connection with jog forward and jog reverse capabilities. During jog forward, the operator pushes the jog forward push button, which turns the motor ON for as long as the button is pushed. This allows for the load to be moved forward slightly, perhaps to place it in a specific position. The jog reverse push button performs the same task but in the opposite direction.

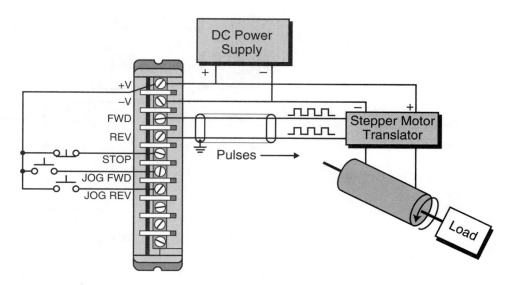

Figure 8-27. Stepper motor interface with jog forward and jog reverse capabilities.

SERVO MOTOR INTERFACES

Servo motor interfaces are used in applications requiring control of servo motors via servo drive controllers. A servo motor is a specially designed motor that contains a permanent magnet. The speed of a servo motor can be easily varied by changing the input voltage to the motor. A servo module provides the drive controller with a ±10 VDC signal, which defines the forward and reverse speeds of the servo motor. These modules are generally used when axis motion control, either linear or rotational, is required. A common linear motion example is a leadscrew assembly, which translates rotational movements from a servo motor into linear displacement (see Figure 8-28).

Applications that once employed clutch-gear systems or other mechanical arrangements to perform motion control now use servo interfaces. The advantages of servo control are shorter positioning time, higher accuracy, better reliability, and improved repeatability in the coordination of axis motion. Typical applications of servo positioning include grinders, metal-forming machines, transfer lines, material-handling machines, and the precise control of servo driver valves in continuous process applications.

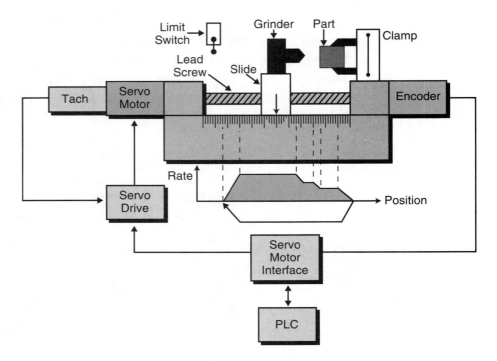

Figure 8-28. Servo motor interface application.

Servo positioning controls operate in a closed-loop system, requiring feedback information in the form of velocity or position. Servo control interfaces may receive velocity feedback in the form of a tachometer input, or positioning feedback in the form of an encoder input, or both. The feedback signal provides the module with information about the actual speed of the motor and the position of the axis. This information is then compared with the desired velocity and the desired position of the axis. If the module detects a difference between the desired and actual values, it will correct its output until the error between the feedback data and the set point velocity and position values is zero.

Figure 8-29 shows a servo control configuration block diagram. PLCs that have positioning control capabilities require two modules—one to implement the servo control task and one to receive feedback and close the loop. Some manufacturers, however, offer complete servo control for one axis in a single module.

Servo control, like stepper motor control, can occur in either single-step or continuous positioning mode (see Figure 8-30). Depending on the manufacturer, multiaxis control can also be synchronized in either single-step or continuous mode.

The PLC processor sends all of the move and position information, including acceleration, deceleration, and the final and feed velocities, to the servo module. In axis positioning applications, including those performed by servo

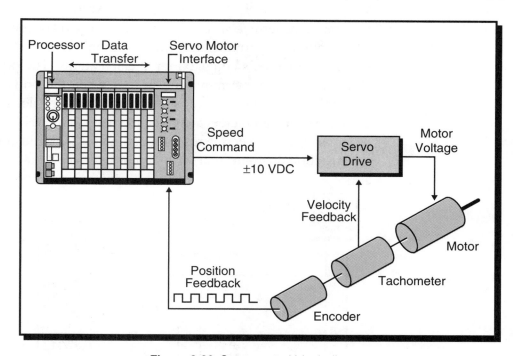

Figure 8-29. Servo control block diagram.

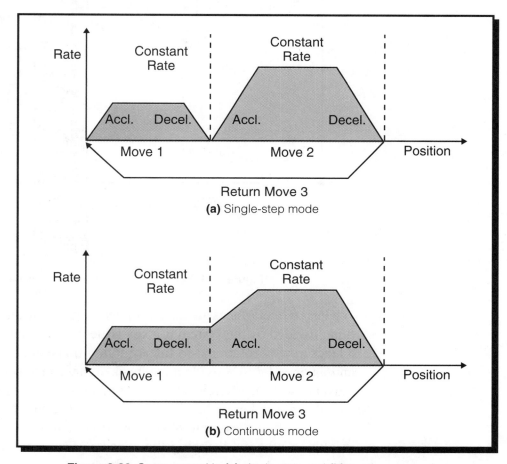

Figure 8-30. Servo control in **(a)** single-step and **(b)** continuous modes.

systems, the term *feed velocity* indicates a period of constant velocity. When the module is operating, the processor monitors its status without interfering with the module's complex, rapid calculations. The processor updates the module with a new move for an axis when the previous move has been completed and the module is ready for a new profile. The acceleration and deceleration parameters are given as speed in inches per minute per second (ipm/sec) at a specific resolution. Figure 8-31 illustrates a typical field connection diagram for a servo motor interface.

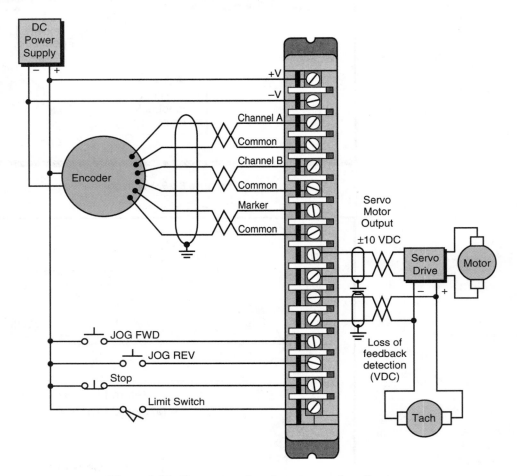

Figure 8-31. Servo motor interface connection diagram.

When servo interfaces are used for positioning control, the feedback resolution provided by the system is a key issue. For example, if an interface uses a leadscrew (a rotational-to-linear motion translator) for axis displacement and an encoder to provide a feedback signal to the servo module, the user must know the leadscrew pitch, the number of encoder pulses per revolution, and the multiplier value in the encoder section of the interface. Some interfaces allow the user to select a multiplier, thus providing better feedback resolution without changing the encoder. The example at the end of this section will show you how some of these parameters are used.

The feedback resolution of a servo positioning (linear) interface can be defined as:

$$\text{Feedback resolution} = \frac{\text{Pitch of motion translator}}{(\text{Encoder pulses per revolution})(\text{Feedback multiplier})}$$

Each servo interface has a predefined resolution, which varies from 0.001 to 0.0001 inches. A trade-off exists between axis velocity and feedback resolution, since axis speed is directly proportional to feedback resolution. Typical axis positioning speeds range from 500 to 1000 inches per minute (ipm) and encoder feedback input frequencies range up to 250 kHz. Remember that resolution, or accuracy, diminishes as the speed increases (e.g., a resolution of 0.0001 inches at 450 ipm will be 0.001 inches at 900 ipm).

EXAMPLE 8-3

A PLC system uses a servo interface to perform a one-axis positioning of a metal part. This part will be machined at a defined profile, which will be stored in the processor's memory. A leadscrew, which allows travel of 1/8th inch (0.125) per revolution, moves the part along an X-axis. A quadrature incremental encoder, which has a 200 kHz pulse frequency that provides 250 pulses per revolution, supplies position feedback information. The encoder is connected to an encoder feedback terminal in the servo interface that provides a software programmable multiplier of ×1, ×2, and ×4 increments per pulse (× = times).

(a) Find the feedback resolution and the number of pulses that will be received if the part travels 12.5 inches. **(b)** Also, describe a way to double the feedback resolution without changing the encoder.

SOLUTION

(a) Feedback resolution is a function of the leadscrew pitch and the product of the number of pulses per revolution generated by the encoder and the feedback multiplier. The leadscrew's pitch is 1/8th inch, which means that the part will travel 0.125 inches for every rotation (see Figure 8-32).

The feedback resolution is therefore:

$$\frac{0.125 \text{ inch/rev}}{250 \text{ pulses/rev} \times 1} = 0.0005 \text{ inches/pulse}$$

Thus, a metal part moving 12.5 inches will generate a position feedback of:

$$\frac{12.5 \text{ inches}}{0.0005 \text{ inches/pulse}} = 25,000 \text{ pulses}$$

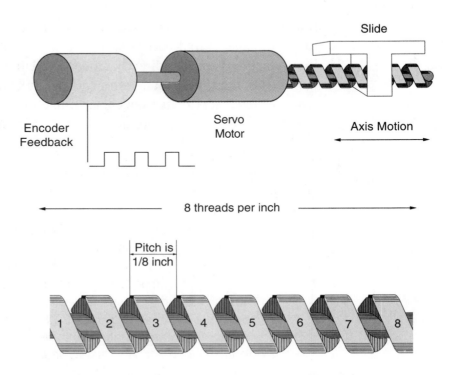

Figure 8-32. Leadscrew (linear) displacement system.

(b) Using a multiplier of ×2 would improve the 0.0005-inch resolution (movement per encoder pulse) to 0.00025 inches (0.0005 ÷ 2 = 0.00025). This ×2 multiplier option allows both of the quadrature pulses (A and B) to be counted, yielding twice as many pulses in one rotation.

8-5 ASCII, COMPUTER, AND NETWORK INTERFACES

Some special I/O modules aid in the communication of information to the real world. These intelligent modules accept data from and transmit data to field devices, including computers and other PLCs. This data is transmitted in one of the following forms:

- ASCII characters

- a computer language, such as BASIC or C

- a proprietary media, as in the case of a network

Local and remote I/O processors fall into the proprietary category of communication interfaces, since they communicate information through a network to the PLC's subsystems. However, they were discussed in the remote I/O section of Chapter 6, since these modules also fall under the discrete I/O category.

ASCII

ASCII input/output interfaces send and receive alphanumeric data between peripheral equipment and the controller. Typical peripheral devices include printers, video monitors, and displays. These special I/O interfaces are available with either basic communications circuitry only or with complete communication interface circuitry, including an on-board RAM buffer and a dedicated microprocessor (intelligent ASCII interface). The information exchange in either type of interface generally takes place via an RS-232C, RS-422, RS-485, or a 20 mA current loop standard communications link (see Section 8-7 for peripheral interfacing). An ASCII interface receives power from the back plane of the rack enclosure to which it is connected. Figure 8-33 shows an RS-232 ASCII interface.

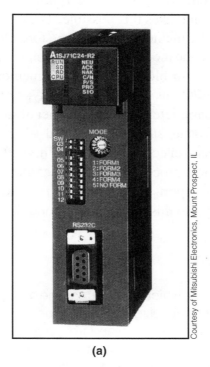
(a)

(b)

Figure 8-33. RS-232 ASCII interfaces from **(a)** Mitsubishi and **(b)** Allen-Bradley.

If an ASCII interface does not use a microprocessor, the main PLC processor handles all of the communications interfacing. This significantly slows down the communication process and the program scan, since the processor must handle each character or string of characters that is transmitted to or received from the module on a character-by-character (interrupt) basis. That is, the module interrupts the main CPU every time it receives a character from the peripheral, and the CPU accesses the module every time it needs to send a message to the peripheral. This communication speed is generally very slow, so for a character to be read, the scan time must be faster than the time required to accept one character. For example, if the scan time is 20 msec and the baud rate (i.e., the number of binary bits transmitted per second) is 300 (30 characters per second—1 ASCII character = 10 bits), a character will be received every 33.3 msec (1 second ÷ 30 characters = 1 character every 33.3 msec). Conversely, if the baud rate is 1200 (120 characters per second), more than one character will be transmitted from the peripheral per scan (one character every 8.33 msec). In this case, several characters will be lost since the PLC processor scans only once every 20 msec. This type of nonintelligent module, which does not have a microprocessor, is used in applications that require the communication of just a few characters, which are output at a relatively slow speed.

In an intelligent, or smart, ASCII interface, transmission between the peripheral and the module still occurs on an interrupt basis but at a faster transmission speed. An on-board microprocessor dedicated to performing I/O communication makes this possible. The on-board microprocessor contains its own RAM memory, which can store blocks of data that are to be transmitted. When the module receives the input data from the peripheral, the module transfers it in blocks to the PLC memory through a data transfer instruction at the I/O bus speed. With this type of interface, all of the initial communication parameters, such as number of stop bits, parity (even or odd) or nonparity, and baud rate, can be selected using either hardware (i.e., rocker switches or jumpers) or control software. This method significantly speeds up the communication process and increases data throughput. Applications requiring lengthy reports or fast information exchange with alphanumeric devices generally use this type of smart module.

EXAMPLE 8-4

A PLC system, which has a scan time of approximately 15 msec, uses a standard nonintelligent ASCII module. This ASCII interface reads and writes information to and from a remote alphanumeric keyboard/ display user interface. What is the maximum baud rate (bits per second) that can be used for proper transmission?

SOLUTION

A scan time of 15 msec implies that, for proper transmission, only one character can be received every 15 msec. Each ASCII character has

10 bits (7 for the code plus start, stop, and parity bits) that are used during each character transmission.

The inverse of the scan time provides the minimum time required by the processor to read an incoming character of 10 bits. Therefore, the time for one character (10 bits) is:

$$\frac{1}{scan} = \frac{1}{0.015} = 66.67 \text{ characters/scan}$$

The baud rate is:

$$(66.67)(10) = 666.7$$

Thus, the maximum baud rate would be 666.7 (or 667), which transmits 66 characters per second. However, since this is not a standard baud rate, the user would have to use a more standard one, perhaps a 600 baud rate.

COMPUTER MODULES—BASIC

BASIC modules, also referred to as *data-processing modules*, are intelligent I/O interfaces capable of performing computational tasks without burdening the PLC processor's computing time. In contrast to other intelligent I/O interfaces, such as servo controls, a BASIC module does not actually command or control specific field devices. Rather, it complements the performance of the PLC system.

In reality, a data-processing module is a personal computer packaged in an industrial I/O module, which inputs and runs user-written BASIC programs independently of the PLC's processor. The BASIC language instructions used in this type of interface are the same as those used in a regular personal computer; however, PLC manufacturers incorporate additional instructions in BASIC modules that allow them to access the PLC's memory (i.e., I/O data table). These added instructions are very useful when the module requires process information to perform BASIC-run calculations.

Some data-processing modules are able to run languages other than BASIC, such as PASCAL, C, or other high-level languages. These modules also contain added instructions that allow direct internal communication (data transfers) between the module, the PLC processor, and the memory. This communication generally occurs through move instructions, which transfer blocks of data to and from the module. Some typical move instructions are move block read and move block write. The user can directly initiate BASIC communication in three ways—through the module's programming port (using the terminal), upon recognition of a user-defined data decoding signal transferred from the PLC, or after the power-up initialization of the PLC system.

The programming port of a BASIC interface is generally compatible with the RS-232C, RS-422, and RS-485 communication standards (see Section 8-7), which are intended to support ASCII terminals or the manufacturer's PLC programming terminal. BASIC interfaces also have at least one serial peripheral port to provide interfacing with printers, asynchronous modems, and other serial peripherals. Under BASIC program control, the serial port is used to generate reports for operator interfaces or for local area networks of other personal computers that gather process data for storage purposes.

Other applications of computer modules are the implementation of artificial intelligence (AI) computations and number-crunching calculations. In AI applications, the computer interface accesses information from the PLC and processes it according to AI algorithms. Chapter 16 explains artificial intelligence. In number-crunching calculations, BASIC modules perform computations that would require awkward PLC programming.

With their vast data-handling capabilities, only the user's innovation limits the uses and applications of computer modules. The utilization of these interfaces in a PLC system is convincing proof of the successful integration of the personal computer's computing power with the PLC's powerful I/O handling and control capability.

NETWORK INTERFACE MODULES

Network interface modules (see Figure 8-34) allow a number of PLCs and other intelligent devices to communicate and pass PLC data over a high-speed local area communication network (see Chapter 18). Any device may interface with the network, because the network is not restricted to only products designed by the network's manufacturer.

Nowadays, many third-party suppliers manufacture products that are compatible with different PLC network environments. Among the most popular networks are:

- device-level bus networks (e.g., CANbus, Seriplex, etc.), which are used by discrete devices
- process field networks (e.g., Fieldbus and Profibus), which are used by analog devices
- Ethernet/IEEE 802.3 networks, used by PLC CPUs and computers
- proprietary networks, which are widely used by large PLC manufacturers

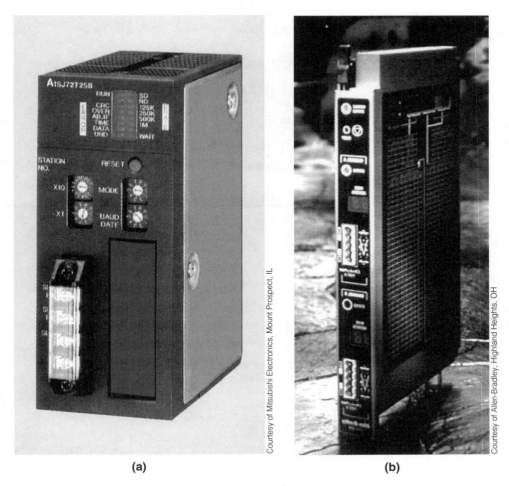

Figure 8-34. (a) Mitsubishi's MELSECNET/B interface and **(b)** Allen-Bradley's CANbus network interface.

A network interface module implements all of the necessary communication connections and protocols to ensure that a message is accurately passed along the network. In general, when a processor or other network device sends a message, its network interface transmits the message over the network at the network's baud rate speed. The receiving network interface accepts the transmission, passes the information to the CPU, and if necessary, sends a command to the intended field device. As you will see in Chapter 18, the speed and protocol for the communication link varies depending on the network.

Depending on the network type and configuration, a network module can be connected, at a distance of up to 10,000 feet, with 100 to 1000 devices (nodes). The communication media—twinaxial, coaxial, or twisted-pair—varies depending on the type of network. The different types of networks also utilize specific network interfaces. For example, a device-level CANbus network uses a CANbus-type interface. Chapter 19 provides more information on I/O bus networks. Figure 8-35 illustrates a typical configuration of a PLC network using the different types of network interface modules.

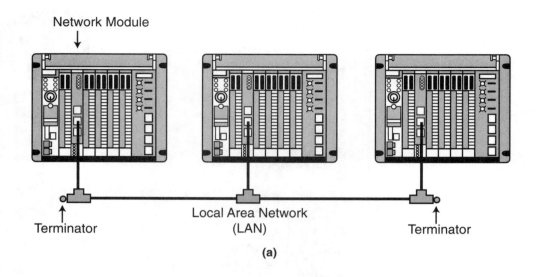

Network Module

Terminator

Local Area Network
(LAN)

Terminator

(a)

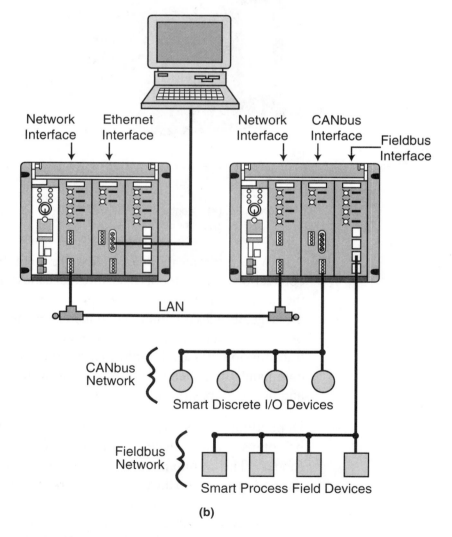

Network
Interface

Ethernet
Interface

Network
Interface

CANbus
Interface

Fieldbus
Interface

LAN

CANbus
Network

Smart Discrete I/O Devices

Fieldbus
Network

Smart Process Field Devices

(b)

Figure 8-35. (a) A standard PLC local area network and **(b)** a PLC local area network
with CANbus (device bus) and Fieldbus (process bus) subnetworks.

8-6 FUZZY LOGIC INTERFACES

Fuzzy logic interfaces, which are offered by a few PLC manufacturers, provide a way of implementing fuzzy logic algorithms in PLCs. Fuzzy logic algorithms analyze input data to provide control of a process. As shown in Figure 8-36, fuzzy logic modules do not function as actual input and output interfaces per se. Rather, they work with other input and output interfaces, providing an intelligent link between the two.

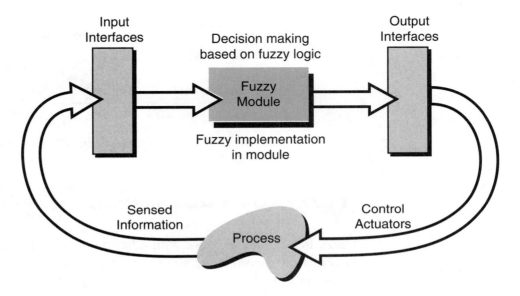

Figure 8-36. Fuzzy logic interface application.

Fuzzy logic modules are an integral part of the advanced capabilities of today's programmable controllers. They help to bridge the gap between the discrete and analog decision-making functions of a PLC. In essence, fuzzy logic modules allow PLCs to "reason," letting them interpret data in an analog-type form instead of just as ON or OFF. For example, a typical PLC connected to a temperature-sensing device can only sense whether a temperature is acceptable or unacceptable (see Figure 8-37a). That is, the temperatures between 60°F and 80°F are acceptable (logic 1); all other temperatures are unacceptable (logic 0). A PLC with fuzzy logic capabilities, however, can discern between the ranges of acceptable and unacceptable temperatures, judging a temperature to be either more acceptable or less acceptable (see Figure 8-37b). Thus, a fuzzy logic module can determine that 62°F is an acceptable temperature, but that it is not as acceptable as 70°F.

The "reasoning" capabilities of fuzzy modules allow them to provide fine-tuned control of analog processes, as well as nonlinear and time-variant processes, like tension and position control. These types of hard-to-control systems usually provide gross input deviations or insufficient input resolution, which often require human intuition and judgment. Fuzzy logic modules can provide this type of human-like judgment.

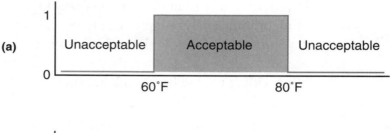

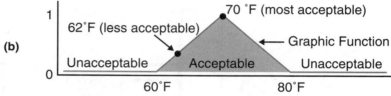

Figure 8-37. Temperature sensing in **(a)** a normal PLC and **(b)** a PLC with fuzzy logic capabilities.

FUZZY LOGIC ALGORITHMS

Fuzzy logic modules work with other modules to input and output process information according to fuzzy control algorithms. These algorithms are based on user-programmed *rules*, which are formed by *IF conditions* and *THEN actions*. A fuzzy module analyzes its inputs according to the IF conditions and then outputs control data according to the corresponding THEN action. For example, the temperature-sensing fuzzy logic algorithm shown in Figure 8-38 might have a rule stating that IF the input temperature is 75°F, THEN its level of acceptability is 0.5, so turn the output's controlling element (e.g., a servo valve) a little clockwise (perhaps 10 degrees to the right). The fuzzy algorithm determines how much the "little" amount is when the output is generated.

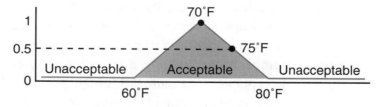

IF the temperature equals 75°F
THEN turn the output's controlling element a little clockwise

Figure 8-38. Example of a fuzzy logic algorithm.

Fuzzy logic control is even more practical when multiple rules exist. For example, a fuzzy I/O module may receive data from a field device measuring the input process temperature, as well as from a field device measuring the outside environmental temperature. In this case, the module could combine two rules to determine a more precise acceptability level, resulting in a more precise output action. For example, IF the input temperature is 75°F and IF the outside environmental temperature is 70°F, THEN the acceptability level is 0.63, so turn the control element a little less (perhaps 8 degrees) clockwise.

To provide reasoned control of a field device, a fuzzy logic module analyzes its rules according to its graphic function and then assigns each rule a *grade* to form what are known as *membership functions*. Membership functions classify input data and group the data into sets of values called *fuzzy sets*. A rule's grade indicates how well it fits into the membership function. The number of membership functions depends on the complexity of the control task and the number of inputs to the module.

Each membership function has labels associated with it. For instance, the membership function shown in Figure 8-39 has three labels: cool, nice, and hot. Thus, the rule "IF the temperature equals 65°F" has a grade of 0.5 cool and 0.5 nice, indicating that it is not totally nice but that it is not totally cool either. The same applies to the temperature 75°F, except that it is half nice and half hot. These grades are part of the control algorithm's fuzzy set, which is used to determine the control output. As we will explain in Chapter 17, a fuzzy set composed of several membership function may use up to seven labels to implement its rules.

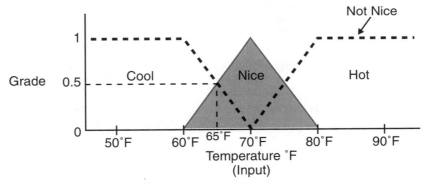

A reading of 65°F will have a grade of 0.5 nice temperature (50%) and 0.5 cool temperature (50%).

Figure 8-39. Membership functions used to create a grade.

Fuzzy logic interfaces allow the user to program the criteria for membership functions and fuzzy sets inside the module according to the control task requirements. A fuzzy module can be programmed through its serial port RS-232C serial port via a personal computer with specialized, manufacturer-provided fuzzy logic programming software.

FUZZY LOGIC AND I/O INTERACTION

Figure 8-40 shows Omron Electronics's Fuzzy Logic Unit (FLU), a fuzzy logic interface that can read process data from up to 8 input devices and write data to up to 4 output devices. This interface can perform up to 128 rules, each with a maximum of eight IF conditions and two THEN actions. The FLU, which works independently of the processor, can implement all of its fuzzy logic computations in 6 msec or less, thus providing fast implementation of fuzzy logic control.

Figure 8-40. Omron Electronics's Fuzzy Logic Unit (FLU) in a C200H PLC system.

As shown in Table 8-2, Omron's Fuzzy Logic Unit uses 10 words or registers of the programmable controller's data table to store its control parameters. The rack position of the FLU module determines the registers' addresses. Assuming that the placement of the module takes addresses 110 through 119, the module will use the addresses as follows:

- The first four bits (0–3) of the first word (word 110) contain, in BCD, the number of inputs that will be used with the FLU module. Bit 15 of this word turns on the fuzzy processing.

- The second word (word 111) specifies where the input data to be analyzed is stored in the PLC's memory. It indicates the starting register address, with the length of the data block being the BCD number from word 110.

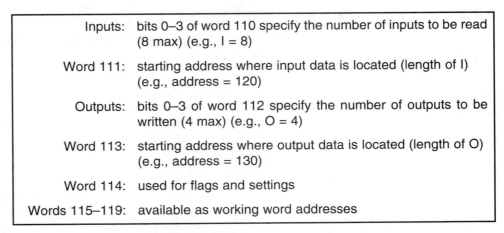

Inputs:	bits 0–3 of word 110 specify the number of inputs to be read (8 max) (e.g., I = 8)
Word 111:	starting address where input data is located (length of I) (e.g., address = 120)
Outputs:	bits 0–3 of word 112 specify the number of outputs to be written (4 max) (e.g., O = 4)
Word 113:	starting address where output data is located (length of O) (e.g., address = 130)
Word 114:	used for flags and settings
Words 115–119:	available as working word addresses

Table 8-2. Omron's FLU space requirements.

- Like the first word, the first four bits (0–3) of the third word (word 112) contain the number of outputs in BCD.

- The fourth word (word 113) contains the starting address for the storage of the output data, which is the result of the fuzzy logic computations. The length of the data block is the BCD number from word 112.

Because fuzzy logic modules work through other I/O interfaces, their input/output data must be transferred from/to the word address locations of the I/O modules working with them. Figure 8-41 illustrates the memory addresses (words) used by the Omron FLU in the previous example, along with the register locations of the corresponding I/O devices' input and output data.

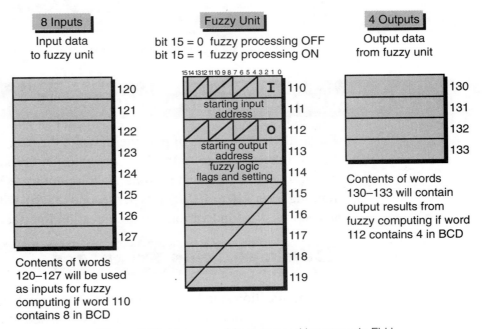

Figure 8-41. Memory addresses used by example FLU.

Block transfer instructions can be used to transfer data between the I/O modules and the fuzzy module (see Figure 8-42). Chapter 17 explains more about fuzzy logic control.

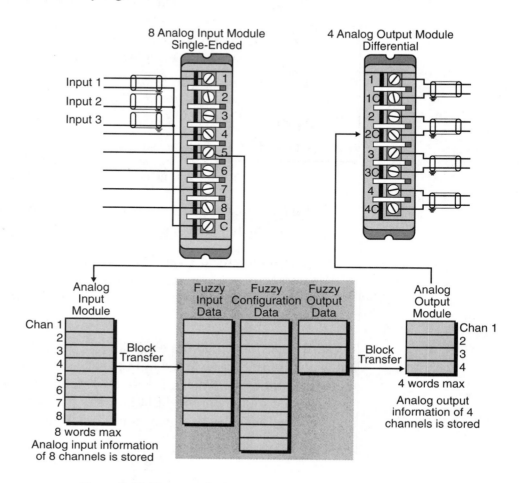

Figure 8-42. Data transfer between I/O modules and fuzzy module.

8-7 PERIPHERAL INTERFACING

Regardless of the type of peripheral used, the user must properly connect the peripheral device to the PLC or intelligent module to achieve correct communication. Typical peripherals communicate in serial form at speeds ranging from 110 to 19,200 bits per second (baud), with parity and nonparity, asynchronicity, and various communication interface standards.

COMMUNICATION STANDARDS

Communication standards fall into two categories: proclaimed and de facto. *Proclaimed standards* are officially established standards set by various electronics organizations, such as the Institute of Electrical and Electronics Engineers (IEEE) and the Electronic Industries Association (EIA). These institutions define public specifications through which manufacturers can establish communication schemes that allow compatibility among different manufacturers' products. Proclaimed standards, such as the IEEE 488 instrument bus, the EIA RS-232C, the EIA RS-422, and the EIA RS-485, are examples of well-defined proclaimed standards.

De facto standards are interface methods that have gained popularity through widespread use. Although these popular standards have been adopted throughout the industry, they have no official definition. Because they are not properly defined, some de facto standards cause interface problems; however, other standards, such as the 20 mA current loop, are good, well-defined de facto standards.

SERIAL COMMUNICATION

Serial communication, as the name implies, occurs in serial form through simple, twisted-pair cables. Serial data transmission is used for most peripheral communication devices, since these devices are slow in nature and require long cable connections. Serial communication allows peripheral equipment, such as terminals, modems, operator interface panels, and line printers, to receive ASCII information.

Two of the most popular standards for serial communication are the RS-232C and the 20 mA current loop. Other PLC standards are the RS-422 and RS-485, which improve performance and give greater flexibility in data communication interfaces.

The data communication links used with peripheral equipment can be unidirectional or bidirectional. If a peripheral is strictly either an input or an output device, then data transmission occurs in only one direction. In this case, a *unidirectional* serial signal line is all that is required to complete the link. Devices that serve as both input and output devices (e.g., video terminals) require bidirectional links. There are two ways to achieve this bidirectional communication. First, a single data line can be used as a shared communication line. The data can be sent in either direction, but only in one direction at a time. This operation is known as *half duplex*. If simultaneous bidirectional communication is required, two lines can connect the PLC to the peripheral. One line would be assigned permanently as an input, while the other would be a permanent output. This mode is known as *full duplex*. Figure 8-43 illustrates the unidirectional, half-duplex, and full-duplex communication methods.

(a)

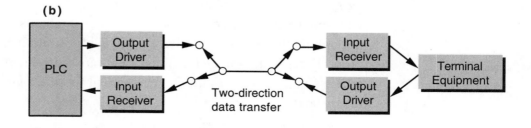

(b)

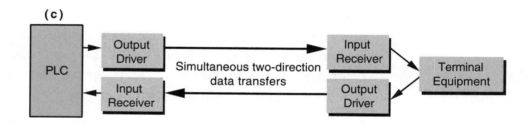

(c)

Figure 8-43. **(a)** Unidirectional, **(b)** half-duplex, and **(c)** full-duplex data communication formats.

EIA RS-232C. The EIA RS-232C is a proclaimed standard that defines the interfacing between data equipment and communication equipment that employs serial binary data interchange. This standard defines both the electrical signals and the mechanical details of the interface. A complete RS-232C interface consists of 25 data lines, which encompass all of the possible signals for simple and complex communication interfaces. Although several of these lines are specialized and a few are undefined, most peripherals require only three to five lines to operate properly. Table 8-3 describes the 25 data lines as specified by the EIA.

Figure 8-44a illustrates an RS-232C data communication system using a telephone modem, while Figure 8-44b shows the RS-232C wiring connections from a computer to a smart EIA PLC interface module. Figure 8-44c illustrates a typical RS-232C interface to a printer. Note that the communication between a computer and a PLC has few lines swapped if no modem or other data communication equipment is used. This configuration is

Pin Number	Description
1	Protective ground
2	Transmitted data
3	Received data
4	Request to send
5	Clear to send
6	Data set ready
7	Signal ground (common return)
8	Received line signal detector
9	(Reserved for data set testing)
10	(Reserved for data set testing)
11	Unassigned
12	Secondary received line signal detector
13	Secondary clear to send
14	Secondary transmitted data
15	Transmission signal element timing (DCE)
16	Secondary received data
17	Receiver signal element timing (DCE)
18	Unassigned
19	Secondary request to send
20	Data terminal ready
21	Signal quality detector
22	Ring indicator
23	Data signal rate selector (DTE/DCE)
24	Transmit signal element timing (DTE)
25	Unassigned

Table 8-3. EIA RS-232C data line descriptions.

called a *null modem* cable. The connection between a PLC and an RS-232C peripheral (printer, etc.) usually requires four wires; however, the user should refer to the connection specifications for both devices for specific details.

The RS-232C standard calls for certain electrical characteristics. Some of these specifications are as follow:

- The signal voltages at the interface point should be a minimum of +5 V and a maximum of +15 V for logic 0; for logic 1, the minimum is –15 V and the maximum is –5 V.

- The maximum recommended cable distance is 50 feet, or 15 meters; however, longer distances are permissible provided that the resulting load capacitance, measured at the interface point and including the signal terminator, does not exceed 2500 picofarads.

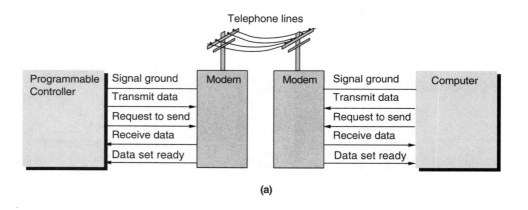

(a)

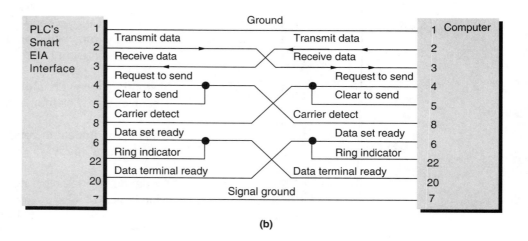

(b)

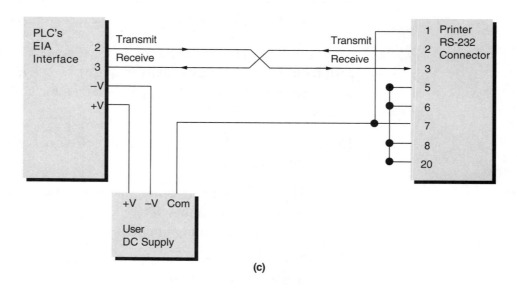

(c)

Figure 8-44. RS-232C communication connections for **(a)** a PLC to a modem, **(b)** a PLC to a computer, and **(c)** a PLC to a printer.

- The drivers used must be able to withstand open or short circuits between pins in the interface.

- The load impedance at the terminator side must be between 3000 and 7000 ohms, with no more than 2500 picofarads capacitance.

- Voltages under –3 V (logic 1) are called *mark* potentials (signal conditions); voltages above +3 V (logic 0) are called *space* voltages. The area between –3 V and +3 V is not defined.

Figure 8-45 illustrates a typical RS-232C serial ASCII pulse train. The transmission begins with a START bit (0) and ends with either one or two STOP bits (1). The transmission also includes parity, which can be even or odd (see Chapter 4 for parity).

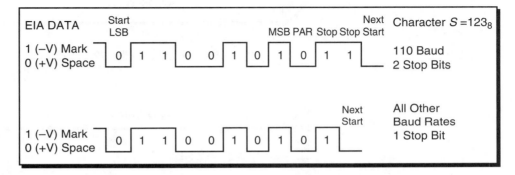

Figure 8-45. RS-232C serial ASCII pulse train.

EIA RS-422. The RS-422 standard overcomes some of the RS-232C shortcomings, including an upper data rate of 20K baud, a maximum cable distance of 50 feet, and an insufficient capacity to control additional loop-test functions for fault isolation. Like the RS-232C, the RS-422 standard still deals with the traditional serial/binary switch signals of two voltage levels across the interface. The RS-449 standard, which meets new operational requirements, defines the physical and mechanical specifications for the RS-422 electrical interface standard.

The RS-232C is an unbalanced link communication method, meaning that it specifies a primary station that is always in control (master/slave relationship). This primary station is responsible for setting logical states and operational modes of each secondary station, thereby controlling the entire data communication process. The RS-422, however, is a balanced link in which either station can configure itself and initiate transmission when both stations have identical data transfer and link control capabilities. The RS-422 specifies electrically balanced receivers and generators that tolerate and produce less noise. These provide superior performance up to 10 megabaud (10,000 K baud).

A balanced circuit in an RS-422 configuration employs differential signaling over a pair of wires for each circuit, while an unbalanced configuration signal (RS-232C) uses one wire for each circuit and a common return circuit. Figure 8-46 illustrates configurations for both RS-422 and RS-232C circuits.

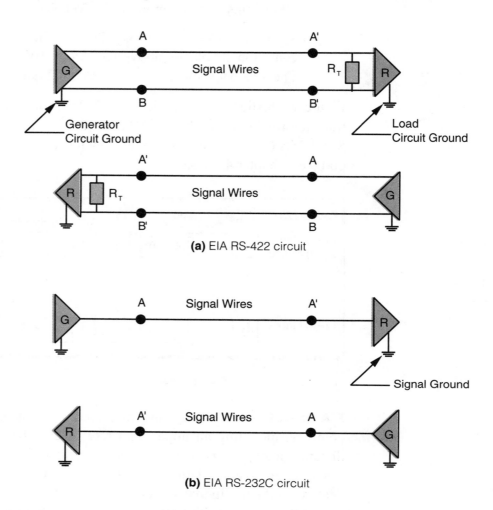

Figure 8-46. Circuit configurations for **(a)** RS-422 and **(b)** RS-232C connections (G = generator; R = receiver; R_T = optional cable termination; A, B, A', B' = interface points).

The RS-422 standard may be required when interconnecting cables are too long for effective unbalanced operation and noise in excess of 1 V can be measured across the signal conductors. The driver circuits for an RS-422 configuration are capable of furnishing the DC signal necessary to drive up to 10 parallel, connected RS-422 receivers. However, this capability involves considerations such as stub line lengths, data rate, grounding, fail-safe networks, etc. The standard does not specify cable characteristics, but to ensure proper operation, paired cables with metallic conductors should be employed and, if necessary, shielded.

The maximum allowable cable distance for the RS-422 standard is a function of the data transmission rate. Figure 8-47 illustrates the relationship between distance and data rate. The graph describes empirical measures using a 24 AWG copper conductor and a twisted-pair cable with a shunt capacitance of 52.5 pF/meter (16 pF/foot) terminated in a 100 ohm resistive load. The balanced electrical characteristics of RS-422 perform even better with an optimal cable termination of approximately 120 ohms in the receiver load.

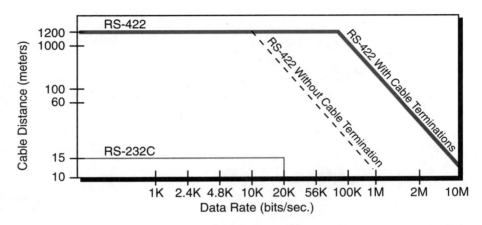

Figure 8-47. Cable distance versus data rate relationship for the RS-422 and RS-232C communications standards.

In reality, the curves in Figure 8-47 are conservative for RS-422 balanced operation. A cable can perform effectively, at lower data rates, at a distance of several miles with good engineering practice. However, if longer distances are required, the user should perform an analysis of the absolute loop resistance and the capacitance of the cable. In general, longer distances are possible when using 19 AWG cable, but the type and length of cable used must be capable of maintaining the necessary signal quality for the particular application.

The RS-449 mechanical standard, which supports the RS-422 electrical standard, offers several extra circuits (signals) that provide greater flexibility to the interface and accommodate new common return circuits. These additional functions and wires were beyond the capacity of an RS-232C 25-pin connector; therefore, the EIA selected a 37-pin connector for the RS-422 standard, because it satisfies interface channel requirements. If secondary channel operation is to be used as a low-speed TTY or acknowledgments channel, a separate 9-pin connector is also needed.

EIA RS-485. The RS-485 standard, like the RS-422, has dual transmitting and receiving lines (differential signals). This type of interface is best suited for industrial applications, because it provides better electrical isolation from the PLC or host than the RS-422 standard. It is also capable of being used in a network (e.g., multiple transmitters and receivers operated on a common media, such as twisted-pair cable). Distances of up to 4000 feet (1200 meters) can be attained with this standard.

20 mA Current Loop. The 20 mA current loop de facto standard consists of four basic wires: transmit plus, transmit minus, receive plus, and receive minus. Figure 8-48 illustrates the four lines used to form the 20 mA current loop. This de facto standard is also referred to as a TTY serial interface.

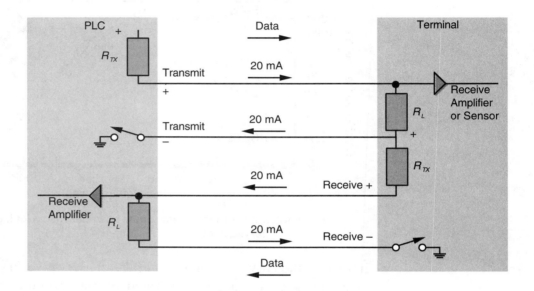

Figure 8-48. 20 mA current loop operation diagram.

In the 20 mA current loop standard, the opening and closing of current loops signifies 0s and 1s, respectively. When the current loop standard was first used in teletypewriters, rotating switch contacts in the sending teletypewriter connected and broke the loop; the corresponding 20 mA signal drove a print magnet in the receiving teletypewriter. Today, most 20 mA current loops electronically operate the opening switch and printer magnet arrangement.

To generate a current, the voltage in a 20 mA current loop is applied to a current limiting resistor at the data-sending end. This voltage is dropped across both the current limiting resistor (R_{TX}) and across the load resistor (R_L). The R values and the positive voltage applied to them must generate a flow current of 20 mA. Typically, a high voltage and high resistance (R_{TX}) are chosen, even though a low voltage and low resistance can be used. Current loop communications provide an advantage over other methods, since the wire resistance has no effect on the constant current loop. Voltage does not drop across the wire in current loop communications as it does in an RS-232C voltage-oriented interface, thus allowing the current loop interface to drive signals longer distances. To avoid this voltage drop, a current loop uses a constant source to generate the 20 mA current.

Converting a 20 mA current loop to an RS-232C interface can be done simply by employing an RS-232C-level receiver. The receiver drives a switching transistor on the transmission end, and an optical isolator and load resistor drive the RS-232C driver on the receiving end.

INTERFACE USES AND APPLICATIONS

Communications standards are used extensively in applications with a host PLC or with a computer in a network where one or more interfaces are used. Sometimes a PLC with an RS-232C or RS-422 communication interface must communicate with an RS-485 device. In this case, an RS-232C–to–RS-485 converter (or an RS-422–to–RS-485 converter) can provide this communication (see Figure 8-49). These converters provide electrical isolation, in addition to longer distance. Figure 8-50a illustrates one of B&R Industrial Automation's interface converters, which can be used for communicating between two PLCs (see Figure 8-50b) over a long distance (maximum of 500 m, or 16,500 ft). Each PLC starts its interfacing via RS-232 (or RS-422) and transfers to RS-485 to achieve the required distance.

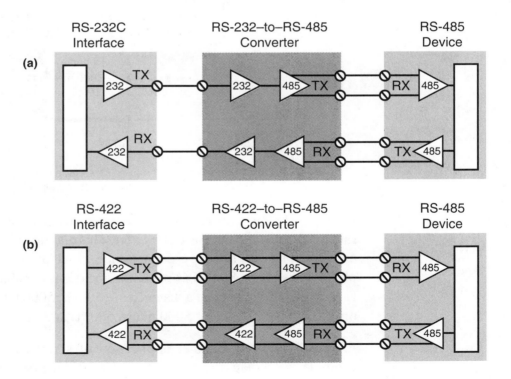

Figure 8-49. (a) RS-232C–to–RS-485 and **(b)** RS-422–to–RS-485 converters.

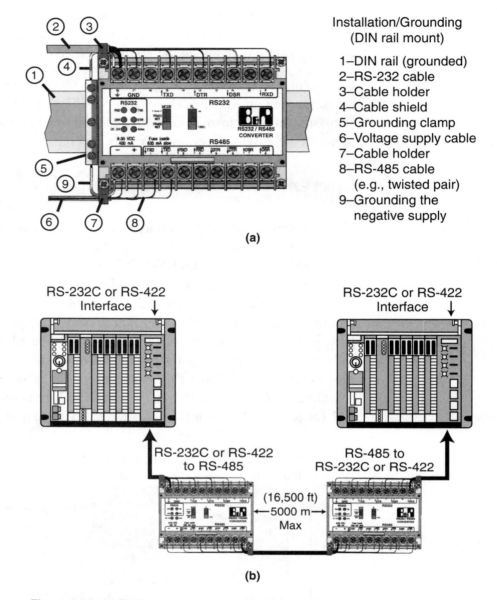

Figure 8-50. (a) B&R Industrial Automation's interface converter and **(b)** an example of the convertor communicating between two PLCs.

Figure 8-51 shows the relationship between transmission distance and data rate for the RS-485 interface converter. This diagram is based on a cable with an impedance of 110 Ω, a capacitance of 41 picofarads/m, and a cable ohmic resistance of 0.094 Ω/m. The converter is capable of driving a signal at rates of 115.2K baud at a distance of 1500 m (5000 ft). It is also capable of operating at a distance of 5000 m (16,500 ft) at a rate of 9.6K baud.

Figure 8-52 shows another application of serial communication. In this example, an isolated link coupler (1747-AIC) interface connects several Allen-Bradley SLC-500 PLC processors to a DH-485 network (RS-485–based). This link coupler provides a connection for each of the SLC-500

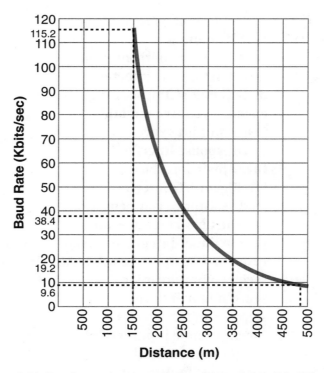

Figure 8-51. Baud rates for transmission distances in a RS-485 converter.

CPUs in the DH-485 network. The DH-485 network also interfaces with a personal computer through an RS-232–to–DH-485 communication interface. The maximum length of the main trunk of the DH-485 network is 4000 feet at a rate of 19.2K baud. This type of subnetwork is very useful for remote programming and data acquisition links of up to 32 devices.

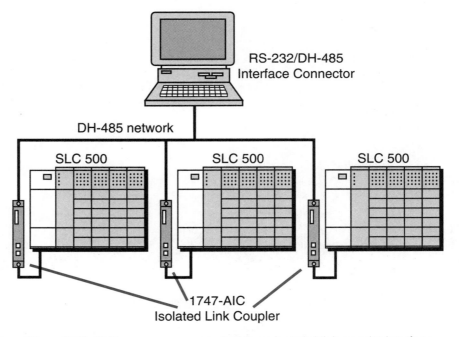

Figure 8-52. PLC processors connected to an isolated link coupler interface.

**KEY
TERMS**
ASCII I/O interfaces
BASIC module
cold junction compensation
direct action I/O interface
distributed I/O processing
encoder/counter module
fast-input interface
fast-response interface
fuzzy logic interface
intelligent I/O interface
lead resistance compensation
network interface module
proportional-integral-derivative (PID) interface
resistance temperature detector (RTD) interface
serial communication
servo motor interface
stepper motor interface
thermocouple input module
weight input module
wire input fault module

SECTION THREE

PLC PROGRAMMING

- Programming Languages
- The IEC 1131 Standard and Programming Language
- System Programming and Implementation
- PLC System Documentation

CHAPTER NINE

PROGRAMMING LANGUAGES

Language is only the instrument of science, and words are but the signs of ideas.

—Samuel Johnson

CHAPTER HIGHLIGHTS The programming languages used in programmable controllers have been evolving since the inception of the PLC in the late 1960s. In this chapter, we will introduce the three types of languages used in PLCs today—ladder, Boolean, and Grafcet. During our discussion of these languages, we will explain some of the versatile, powerful instructions associated with them. These instructions expand programming possibilities in areas such as data manipulation, network communication, data transfer, and program/flow controls, just to name a very few. After you gain a knowledge of these languages and instructions, you will be ready to explore the IEC 1131-3 standard for PLC programming languages, which includes ladder diagrams and the implementation of Boolean programming in an IEC 1131 environment. This programming language standard holds powerful capabilities for the future of PLC programming.

9-1 INTRODUCTION TO PROGRAMMING LANGUAGES

As PLCs have developed and expanded, programming languages have developed with them. Programming languages allow the user to enter a control program into a PLC using an established syntax. Today's advanced languages have new, more versatile instructions, which initiate control program actions. These new instructions provide more computing power for single operations performed by the instruction itself. For instance, PLCs can now transfer blocks of data from one memory location to another while, at the same time, performing a logic or arithmetic operation on another block. As a result of these new, expanded instructions, control programs can now handle data more easily.

In addition to new programming instructions, the development of powerful I/O modules has also changed existing instructions. These changes include the ability to send data to and obtain data from modules by addressing the modules' locations. For example, PLCs can now read and write data to and from analog modules. All of these advances, in conjunction with projected industry needs, have created a demand for more powerful instructions that allow easier, more compact, function-oriented PLC programs.

9-2 TYPES OF PLC LANGUAGES

The three types of programming languages used in PLCs are:

- ladder
- Boolean
- Grafcet

The ladder and Boolean languages essentially implement operations in the same way, but they differ in the way their instructions are represented and how they are entered into the PLC. The Grafcet language implements control instructions in a different manner, based on steps and actions in a graphic-oriented program.

LADDER LANGUAGE

The programmable controller was developed for ease of programming using existing relay ladder symbols and expressions to represent the program logic needed to control the machine or process. The resulting programming language, which used these original basic relay ladder symbols, was given the name **ladder language**. Figure 9-1 illustrates a relay ladder logic circuit and the PLC ladder language representation of the same circuit.

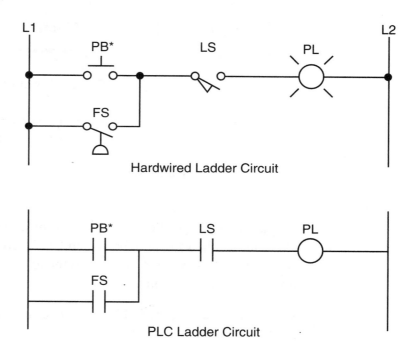

*Note: The PLC will know the elements PB, LS, FS, and PL by their addresses once the address assignment has been performed.

Figure 9-1. Hardwired logic circuit and its PLC ladder language implementation.

The evolution of the original ladder language has turned ladder programming into a more powerful instruction set. New functions have been added to the basic relay, timing, and counting operations. The term *function* is used to describe instructions that, as the name implies, perform a function on data—that is, handle and transfer data within the programmable controller. These instructions are still based on the simple principles of basic relay logic, although they allow complex operations to be implemented and performed.

New additions to the basic ladder logic also include function blocks, which use a set of instructions to operate on a block of data. The use of function blocks increases the power of the basic ladder language, forming what is known as **enhanced ladder language**. Figure 9-2 shows enhanced functions driven by basic relay ladder instructions. As shown in the figure, a block or a functional instruction between two contact symbols represents an enhanced functional block.

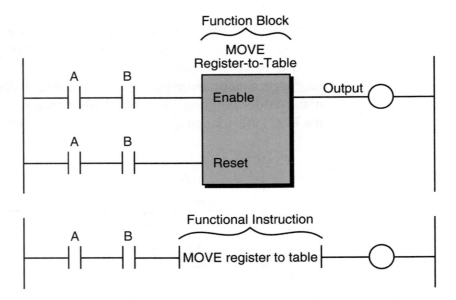

Figure 9-2. Enhanced functional block format.

The format representation of an enhanced ladder function depends on the programmable controller manufacturer; however, regardless of their format, all similar enhanced and basic ladder functions operate the same way. Throughout this chapter, we will refer to enhanced ladder instructions as block format instructions.

As indicated earlier, the ladder languages available in PLCs can be divided into two groups:

- basic ladder language

- enhanced ladder language

Each of these groups consists of many PLC instructions that form the language. The classification of which instructions fall into which categories differs among manufacturers and users, since a definite classification does not exist. However, a de facto standard has been created throughout the years that sorts the instructions into either the basic or enhanced ladder language. Table 9-1 shows a typical classification of basic and enhanced instructions. Sometimes, basic ladder instructions are referred to as *low-level language*, while enhanced ladder functions are referred to as *high-level language*. The

Basic	Enhanced
Relay contact	Double-precision arithmetic
Relay output	Square root
Timer	Sort
Counter	Move register
Latch	Move register to table
Jump to/go to	First in–first out
Master control relay	Shift register
End	Rotate register
Addition	Diagnostic block
Subtraction	Block transfer (in/out)
Multiplication	Sequencer
Division	PID
Compare (=, >, <)	Network
Go to subroutine	Logic matrix

Table 9-1. PLC instruction set classifications.

line that defines the grouping of PLC ladder instructions, however, is usually drawn between functional instruction categories. These instruction categories include:

- ladder relay

- timing

- counting

- program/flow control

- arithmetic

- data manipulation

- data transfer

- special function (sequencers)

- network communication

Although these categories are straightforward, the classification of them is subjective. For example, some people believe that basic ladder instructions include ladder relay, timing, counting, program/flow control, arithmetic, and some data manipulation. Others believe that only ladder relay, timing, and counting categories should be considered basic ladder instructions.

Regardless of classification, the effects of instruction categories are simple—the more instruction categories a PLC has, the more powerful its control capability becomes. Usually, small PLCs have only basic instructions with, perhaps, some enhanced instructions. Larger PLCs usually have more advanced instruction sets. However, recent advances in software development and I/O hardware have increased the computational power of small PLCs through advanced instructions. This new trend has made small PLCs very desirable in single, as well as distributed control, applications.

BOOLEAN

Some PLC manufacturers use **Boolean language**, also called *Boolean mnemonics*, to program a controller. The Boolean language uses Boolean algebra syntax (see Chapter 3) to enter and explain the control logic. That is, it uses the AND, OR, and NOT logic functions to implement the control circuits in the control program. Figure 9-3 shows a basic Boolean program.

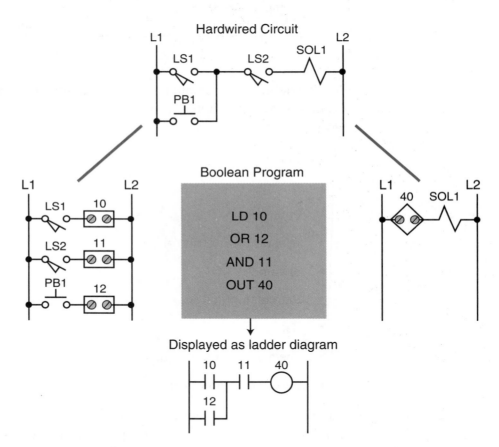

Figure 9-3. Hardwired logic circuit and its Boolean representation.

The Boolean language is primarily just a way of entering the control program into a PLC, rather than an actual instruction-oriented language. When displayed on the programming monitor, the Boolean language is usually viewed as a ladder circuit instead of as the Boolean commands that define the instruction. We will discuss Boolean programming, along with its instruction set, at the end of this chapter.

GRAFCET

Grafcet (Graphe Fonctionnel de Commande Étape Transition) is a symbolic, graphic language, which originated in France, that represents the control program as steps or stages in the machine or process. In fact, the English translation of Grafcet means "step transition function charts." As we will discuss in Chapter 10, Grafcet is the foundation for the IEC 1131 standard's sequential function charts (SFCs), which allow several PLC languages to be used in one control program.

Figure 9-4 illustrates a simple circuit represented in Grafcet. Note that Grafcet charts provide a flowchart-like representation of the events that take place in each stage of the control program. These charts use three components—steps, transitions, and actions—to represent events. The IEC 1131 standard's SFCs also use these components; however, the instructions inside the actions can be programmed using one or more possible languages, including ladder diagrams.

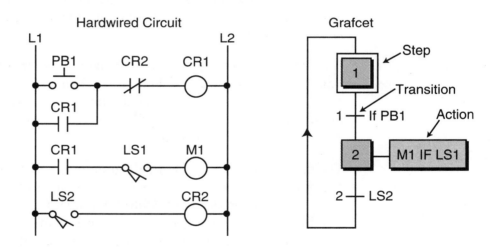

Figure 9-4. Hardwired logic circuit and its Grafcet representation.

Few programmable controllers may be directly programmed using Grafcet. However, several Grafcet software manufacturers provide off-line Grafcet programming using a personal computer. Once programmed in the PC, the Grafcet instructions can be transferred to a PLC via a translator or driver that translates the Grafcet program into a ladder diagram or Boolean language program. Using this method, a Grafcet software manufacturer can provide different PLCs that use the same "language." Figure 9-5 illustrates a typical translation that occurs when using Grafcet. Chapter 10 provides more detail about the versatility of this type of structural programming.

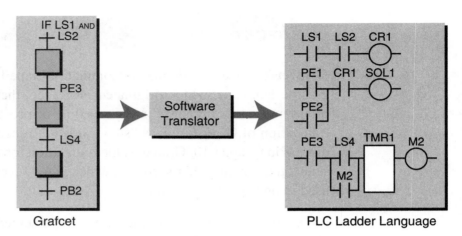

Figure 9-5. Grafcet translation.

9-3 LADDER DIAGRAM FORMAT

The ladder diagram language is a symbolic instruction set that is used to create PLC programs. The ladder instruction symbols can be formatted to obtain the desired control logic, which is then entered into memory. Since this type of instruction set consists of contact symbols, it is also referred to as *contact symbology*.

A thorough understanding of ladder diagram programming, including functional blocks, is extremely beneficial, even when using a PLC with IEC 1131 programming language capabilities. Because ladder diagrams are easy to use and implement, they provide a powerful programming tool when used in the IEC 1131 environment.

The main functions of a ladder diagram program are to control outputs and perform functional operations based on input conditions. Ladder diagrams use rungs to accomplish this control. Figure 9-6 shows the basic structure of a ladder rung. In general, a rung consists of a set of input conditions (represented by contact instructions) and an output instruction at the end of the rung (represented by a coil symbol). The contact instructions for a rung may be referred to as *input conditions*, *rung conditions*, or the *control logic*.

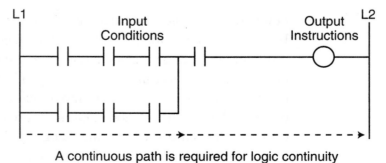

A continuous path is required for logic continuity

Figure 9-6. Ladder rung structure.

A ladder rung is TRUE (i.e., energizing an output or functional instruction block) when it has logic continuity. Logic continuity exists when power flows through the rung from left to right. The execution of logic events that enable the output provide this continuity. In a ladder rung, the left-most side (left power line) simulates the L1 line of a relay ladder diagram, while the right-most side (right power line) simulates the L2 line of the electromechanical representation. Continuity occurs when a path between these two lines contains contact elements in a closed condition, allowing power to flow from left to right. These contact elements either close or remain closed according to the status of their reference inputs. Figure 9-7 illustrates several continuous paths that provide continuity and energize the output of the rung. Power continuity is normally represented on a PLC's monitoring device (e.g., a PC) by bold or emphasized lines, as shown in Figure 9-8a. Figure 9-8b illustrates power continuity through only one energized contact element; note that the output is not ON. We will explain how these contact symbols are interpreted to be ON or OFF in the next section.

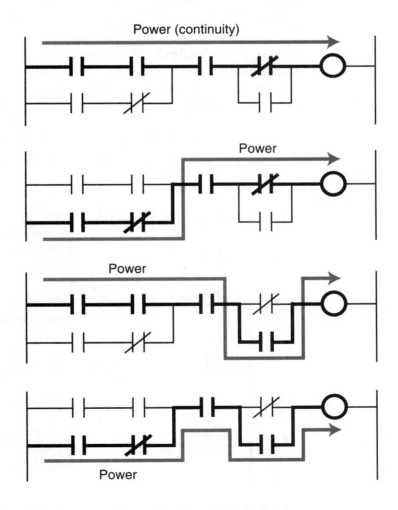

Figure 9-7. Illustration of several different continuity paths in a ladder rung.

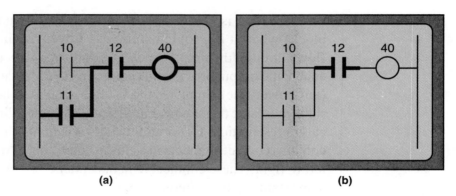

Figure 9-8. Monitoring device showing **(a)** power continuity through the rung—inputs 11 and 12 are ON, turning output 40 ON—and **(b)** power continuity through only input 12, thus output 40 is not ON.

When a ladder diagram contains a functional block, contact instructions are used to represent the input conditions that drive (or *enable*) the block's logic. A functional block can have one or more enable inputs that control its operation. In addition, it can have one or more output coils, which signify the status of the function being performed. For example, the block shown in Figure 9-9a has an enable block line, which when energized (i.e., continuity exists), will activate the block to perform the instruction. Thus, this instruction says: IF the enable is ON because the desired logic has continuity, THEN execute the block instruction. Depending on the instruction, other enable lines (see Figure 9-9b) may drive the block using reset or other control functions.

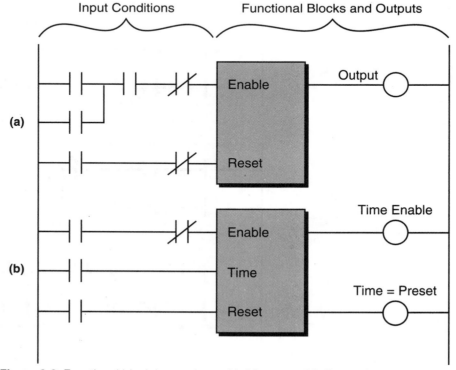

Figure 9-9. Functional block instructions with **(a)** one enable line and one output and **(b)** one enable line, a start timing command, and two outputs.

To make a block active at all times without any driving logic, the user can omit all contact logic and place a continuity line in the block during programming (see Figure 9-10).

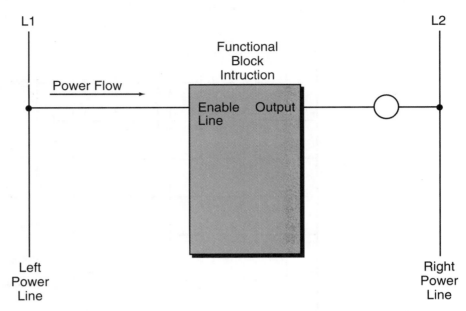

Figure 9-10. A functional block instruction that is always enabled.

The **ladder rung matrix** determines the maximum number of ladder contact elements that can be used to program a rung (see Figure 9-11). The size of this matrix differs among both PLC manufacturers and the programming devices used (CRT screens versus miniprogrammers). For functional block operations, a ladder matrix may have less available ladder contact elements because the functional block instruction display takes up room in

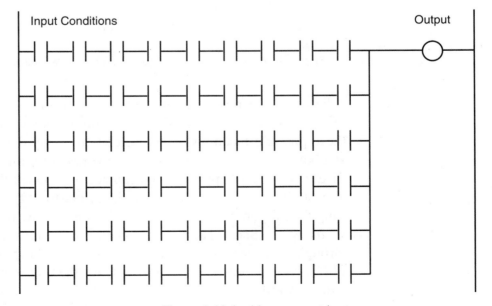

Figure 9-11. Ladder rung matrix.

the matrix (see Figure 9-12a). In PLCs with enhanced ladder format functional instructions instead of block-type instructions, the ladder matrix may use one or more contact symbol spaces to represent the instruction in the programming device (see Figure 9-12b).

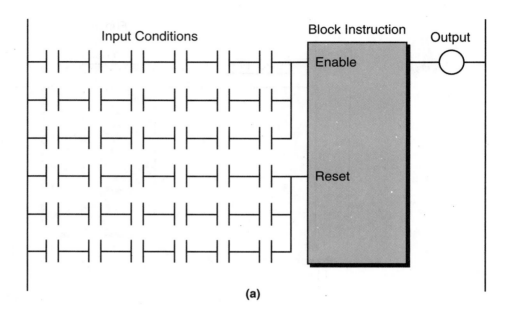

Figure 9-12. Ladder matrix with **(a)** functional block instructions and **(b)** enhanced ladder format functional instructions.

A ladder matrix represents all the possible locations where a contact symbol instruction can be placed. The programming device usually displays all of these possible locations on the screen, allowing the user to place contact symbols in the desired locations. However, according to the maker of the PLC, certain rules apply to contact placement. One rule, which is present in almost all PLCs, prevents reverse (i.e., right-to-left) power flow in a ladder rung (see Figure 9-13). PLC logic does not allow reverse power to avoid *sneak paths*. Sneak paths occur when power flows in a reverse direction through an undesired field device, thus completing a continuity path. If a PLC's logic requires reverse power flow, the user must reprogram the rung with forward power flow to all contact elements. The next example illustrates the solution to the reverse power flow rung in Figure 9-13.

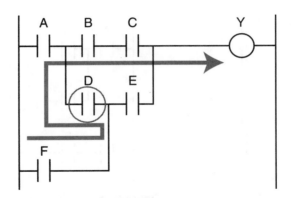

Figure 9-13. Reverse power flow at contact D.

EXAMPLE 9-1

Solve the logic rung shown in Figure 9-13 so that no reverse power flow condition exists. The reverse condition is not part of the required logic for the output to be energized.

SOLUTION

The forward power flow of the logic determines output *Y*. Let's implement it using logic concepts. The output *Y* is defined, using forward paths only, as:

$$Y = \overbrace{(A \bullet B \bullet C)}^{\text{1st line}} + \overbrace{(A \bullet D \bullet E)}^{\text{2nd line}} + \overbrace{(F \bullet E)}^{\text{3rd line}}$$

which can be minimized, using Boolean algebra's distributed rule, to (see Chapter 3):

$$Y = A \bullet (B \bullet C + D \bullet E) + (F \bullet E)$$

Figure 9-14 shows the implementation of this logic gate, while Figure 9-15 gives the ladder-equivalent solution.

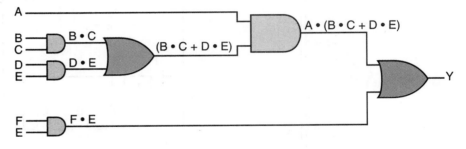

Figure 9-14. Logic solution for Example 9-1.

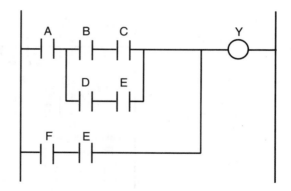

Figure 9-15. Ladder diagram implementation for Example 9-1.

EXAMPLE 9-2

Solve the ladder logic shown in Figure 9-13 so that no reverse power flow exists. Assume that the reverse path logic through contact D and then forward through contacts B and C is required in the PLC logic solution to energize the output.

SOLUTION

Following the same procedure as in Example 9-1, we can obtain the desired logic for output *Y* using Boolean logic expressions. Therefore, output *Y*, including the reverse power flow logic, is represented by:

$$Y = \overbrace{(A \bullet B \bullet C)}^{\text{1st line}} + \overbrace{(A \bullet D \bullet E)}^{\text{2nd line}} + \overbrace{(F \bullet E)}^{\text{3rd line}} + \overbrace{(F \bullet D \bullet B \bullet C)}^{\text{Reverse path}}$$
$$= A \bullet (B \bullet C + D \bullet E) + F(E + D \bullet B \bullet C)$$

The term $F \bullet D \bullet B \bullet C$ implements the reverse power flow sequence that output *Y* requires. Figure 9-16 shows the ladder diagram of this solution.

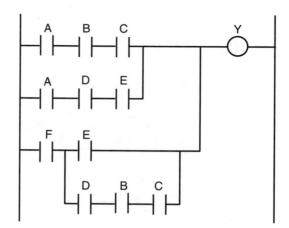

Figure 9-16. Ladder diagram implementation for Example 9-2.

9-4 LADDER RELAY INSTRUCTIONS

Ladder relay instructions are the most basic instructions in the ladder diagram instruction set. These instructions represent the ON/OFF status of connected inputs and outputs. Ladder relay instructions use two types of symbols: contacts and coils. **Contacts** represent the input conditions that must be evaluated in a given rung to determine the control of the output. **Coils** represent a rung's outputs. Table 9-2 lists common ladder relay instructions.

In a program, each contact and coil has a referenced address number, which identifies what is being evaluated and what is being controlled. The address number references the I/O table location of the connected input/output or the internal or storage bit output. A contact, regardless of whether it represents an input/output connection or an internal output, can be used throughout the control program whenever the condition it represents must be evaluated.

The format of the rung contacts in a PLC program depends on the desired control logic. Contacts may be placed in whatever series, parallel, or series/ parallel configuration is required to control a given output. When logic continuity exists in at least one left-to-right contact path, the rung condition is TRUE; that is, the rung controls the given output. The rung condition is FALSE if no path has continuity.

Ladder Relay Instructions		
(Purpose: To provide hardwired relay capabilities in a PLC)		
Instruction	**Symbol**	**Function**
Examine-ON/Normally Open	—| |—	Tests for an ON condition in a reference address
Examine-OFF/Normally Closed	—|/|—	Tests for an OFF condition in a reference address
Output Coil	—( )—	Turns real or internal outputs ON when logic is 1
NOT Output Coil	—(∅)—	Turns real or internal outputs OFF when logic is 1
Latch Output Coil	—(L)—	Keeps an output ON once it is energized
Unlatch Output Coil	—(U)—	Resets a latched output
One-Shot Output	—(OS)—	Energizes an output for one scan or less
Transitional Contact	—|↑|—	Closes for one scan when its trigger contact makes a positive transition

Table 9-2. Ladder relay instructions.

The relay-type instructions covered in this section are the most basic programmable controller instructions. They provide the same capabilities as hardwired relay logic, but with greater flexibility. These instructions provide the ability to examine the ON/OFF status of specific bit addresses in memory and control the state of internal and external outputs.

EXAMINE-ON/NORMALLY OPEN ─┤ ├─

An *examine-ON* instruction, referred to as a *normally open* (NO) contact instruction, tests for an ON condition in a reference address. This reference address can be an input table bit corresponding to an input device, an output bit in the internal bit storage section of the data table, or an output table bit corresponding to an output device (see Chapter 5 for I/O addressing).

During the execution of an examine-ON instruction in the control program, the processor examines the reference address of the instruction for an ON condition. If the reference address is logic 0 (OFF), the processor will not change the state of the normally open contact; thus, it does not provide continuity to the rung (see Figure 9-17a). However, if the reference address is logic 1 (ON), the processor will close the normally open condition to provide power flow in the rung (see Figure 9-17b).

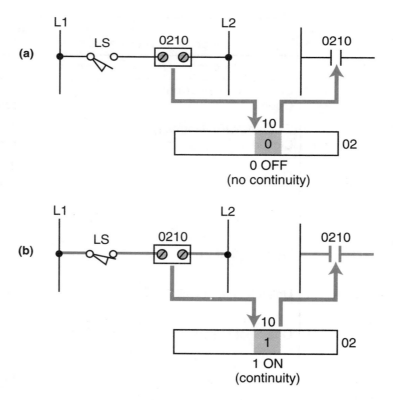

Figure 9-17. (a) An examine-ON instruction with a logic 0 reference address and **(b)** an examine-ON instruction with a logic 1 reference address.

EXAMINE-OFF/NORMALLY CLOSED

An *examine-OFF* instruction, also called a *normally closed* (NC) contact instruction, tests for an OFF condition in the reference address. Like an examine-ON instruction, the address can reference the input table, the output table, or the internal bit storage section of the output table.

During the execution of an examine-OFF instruction, the processor examines the reference address for an OFF condition. If the reference address has a logic 0 status (OFF), the instruction will continue to provide power (continuity) through the normally closed contacts (see Figure 9-18a). If the reference address has a logic 1 status (ON), the instruction will open the normally closed contact, thus breaking continuity to the rung (see Figure 9-18b). An examine-OFF instruction can be associated with a logic NOT function, so that if the reference address is NOT ON, logic continuity will be provided.

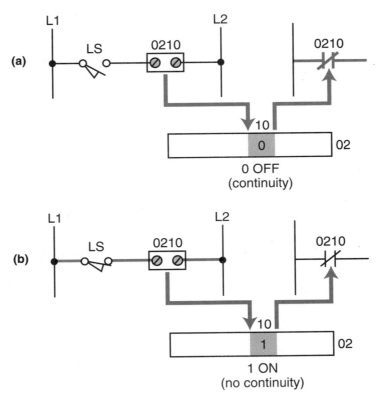

Figure 9-18. (a) An examine-OFF instruction with a logic 0 reference address and **(b)** an examine-OFF instruction with a logic 1 reference address.

OUTPUT COIL

An *output coil* instruction controls either a real output (connected to the PLC via output interfaces) or an internal output (control relay). This instruction uses an output coil address bit in the internal storage area as its reference address. The —( )— symbol may also represent an output coil instruction.

During the execution of an output coil instruction, the processor evaluates all the input conditions in the ladder rung. If no continuity exists, the processor places a 0 in the output coil address bit, indicating an OFF condition to the output coil instruction (see Figure 9-19a). However, if the processor detects continuity in any path, the processor places a logic 1 in the output coil address bit referenced by the instruction (see Figure 9-19b). This logic 1 status indicates an ON condition to the output coil instruction. Therefore, if the output coil address references an output bit in the output table, the processor will turn ON the corresponding output. This will turn ON the field device connected to the terminal referenced by the output coil address. Remember that the processor turns ON the device only after it has completely solved (scanned) the ladder program and updated the output at the end of the scan.

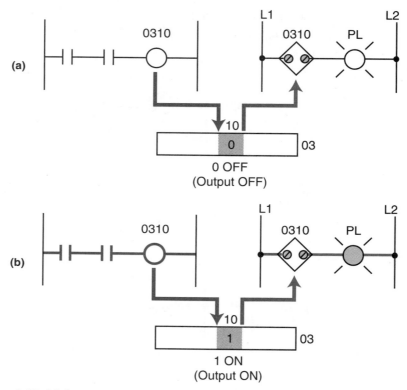

Figure 9-19. (a) An output coil instruction with a logic 0 reference address and **(b)** an output coil instruction with a logic 1 reference address.

When an output coil is used as an *internal output*, its coil address maps an internal bit storage address, rather than an output table bit that maps a real field device. In this case, when the output coil is turned ON, the corresponding bit in the internal bit storage area becomes logic 1. These internal outputs are used when a program requires interlocking sequences or when a real output is not necessary.

Normally open and normally closed reference contacts for an output coil open and close according to the status of the output coil. Figure 9-20 illustrates an example of a simple ladder diagram with normally open and normally closed contacts driving an output rung. For output 20 to turn ON, two things must happen: (1) PB1 must be pushed to turn ON reference input 10 and (2) limit switch LS1 must not be activated to keep reference input 11 OFF. In this case, the processor examines input 10 for an ON condition and input 11 for an OFF condition; if both logic conditions are met, it energizes output 20. With output 20 ON, the normally open contact 20 will close, turning internal output 100 ON. Also, the normally closed contact 20 will open because the test for an OFF condition at output 20 is not true (reference 20 is ON); therefore, it will turn internal output 101 OFF. At the EOS, the pilot light (PL1) will be lit because the processor will send a 1 to the output module, which will latch the logic 1 signal until continuity in the rung (output 20) is disrupted. Note that outputs 100 and 101 do not control real output devices because they reference internal bits that are not mapped to the I/O table.

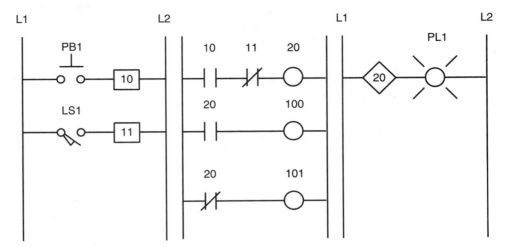

Figure 9-20. Normally open and normally closed contacts driving real and internal output coils.

NOT OUTPUT COIL

A *NOT output coil* instruction (recall the NOT logic function) is essentially the opposite of an output coil instruction. If continuity is not present in the rung, the instruction turns the referenced output bit ON. If continuity is present, it turns the output OFF. Also, when a NOT output coil is ON, its reference contacts change state (normally open contacts close, normally closed ones open). If a NOT output coil is OFF, then the opposite occurs—the normally open reference contacts stay open and the normally closed ones remain closed. The —(/)—symbol represents the NOT output coil in some programmable controllers.

A NOT output coil instruction can be tricky to implement. Therefore, it is often easier to obtain a NOT output coil ladder rung by applying Boolean logic rules to the logic expression of the output rung. An example of this rung configuration follows.

EXAMPLE 9-3

(a) Implement the equivalent ladder rung logic shown in Figure 9-21 using a NOT output coil instruction, and **(b)** implement the NOT Y logic without using a NOT coil.

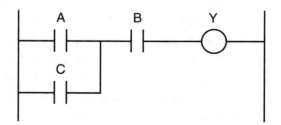

Figure 9-21. Ladder rung for Example 9-3.

SOLUTION

(a) The ladder logic expression representing output Y is:

$$Y = (A + C) \bullet B$$

Using De Morgan's Law (see Chapter 3), the NOT Y function can be expressed as:

$$\overline{Y} = \overline{(A + C) \bullet B}$$
$$= \overline{(A + C)} + \overline{B}$$
$$= (\overline{A} \bullet \overline{C}) + \overline{B}$$

Figure 9-22 shows the implementation of this logic using a NOT output coil. Output Y will be ON if A and B are ON or if C and B are ON (note that A, B, and C are examine-OFF instructions). Remember that the NOT output is ON if continuity does not exist and OFF if continuity is present. The circuit shown in Figure 9-22 is logically identical to the one in Figure 9-21.

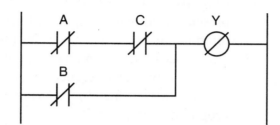

Figure 9-22. Implementation of Figure 9-21 using a NOT coil.

(b) The easiest way to implement a logic NOT function in the rung in Figure 9-21 would be to use the same rung, except that the output Y would be a NOT coil. If we cannot use a NOT coil, then we can implement the NOT by adding another rung as shown in Figure 9-23. Here, output *Z* is essentially the implementation of the NOT output *Y* coil.

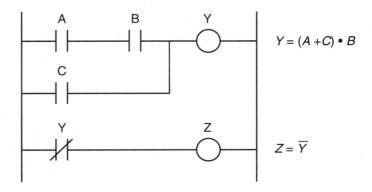

$$Y = (A + C) \bullet B$$

$$Z = \overline{Y}$$

Figure 9-23. Implementation of the NOT *Y* logic without a NOT coil.

LATCH OUTPUT COIL

A *latch coil* instruction causes an output to remain energized even if the status of the contacts that caused the output to energize changes. If any rung path has logic continuity, this instruction turns the output ON and keeps it ON, even if logic continuity or system power is lost. The latched output will remain ON until it is unlatched by an unlatch output instruction. An unlatch instruction is the only automatic (programmed) way to reset a latched output. Although most PLCs allow latching of internal and external outputs, some controllers will latch internal outputs only. A latch output coil instruction may also be referred to as a *set coil* instruction, which can be unlatched by a *reset coil* instruction.

UNLATCH OUTPUT COIL

An *unlatch coil* instruction resets a latched output with the same reference address. When any rung path has logic continuity, this instruction turns OFF the latched reference address coil, or rather unlatches it to an OFF condition. Figure 9-24 illustrates the use of latch and unlatch coils.

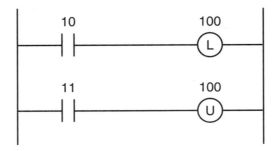

Figure 9-24. Latch and unlatch coil instructions.

Latch and unlatch instructions may occur in block form as shown in Figure 9-25. The only difference between the ladder and block forms is that, in block form, latching and unlatching are performed in the same instruction. If the unlatch input is ON (continuity), the output coil will remain OFF. Note that the latch and unlatch outputs in Figure 9-24 can have ladder logic rungs in between them, while the ones shown in Figure 9-25 cannot. A latch/unlatch block may also be called a *set/reset* block.

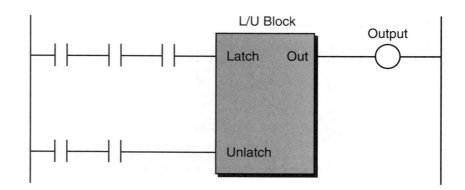

Figure 9-25. Latch/unlatch functional block instruction.

ONE-SHOT OUTPUT   ─(OS)─

A *one-shot output* instruction operates in a manner similar to an output coil instruction—if the ladder rung has continuity, the one-shot output will be energized (ON). However, the length of time that a one-shot output is ON is one scan or less, depending on where it is located in the program.

One-shot outputs are used to reset conditions in one scan. Note that when using a one-shot output to reset other output rungs or functional blocks, the logic to be reset must be programmed after the one-shot rung is programmed. Figure 9-26 illustrates a one-shot output and its timing diagram.

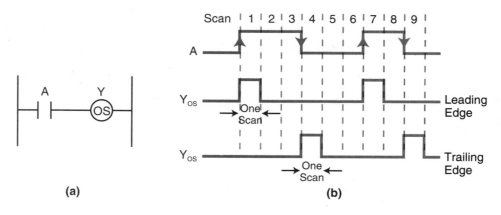

Figure 9-26. (a) A one-shot output instruction and **(b)** its timing diagram.

Depending on the controller used, a one-shot output may trigger a leading-edge or a trailing-edge signal. A leading-edge trigger turns the one-shot output ON for one scan after the OFF-to-ON transition of the input. A trailing-edge trigger turns the output ON for one scan after the ON-to-OFF transition of the input.

TRANSITIONAL CONTACT

A *transitional contact* instruction provides a one-shot pulse when its referenced trigger signal makes either an OFF-to-ON (leading-edge) transition or an ON-to-OFF (trailing-edge) transition. In a leading-edge transitional instruction, the contact will close for exactly one program scan whenever the trigger signal goes from OFF to ON. The contact will allow logic continuity for that one scan and then open again, even though the triggering signal may stay ON. The triggering signal must turn OFF and ON again for the transitional contact to reclose. Conversely, in a trailing-edge transitional instruction, an OFF-to-ON transition of the trigger signal turns the contact ON for one scan. The contact address (trigger) may be an external input/output or an internal output.

Programmable controllers that do not provide one-shot output instructions generally provide transitional contact instructions. Like a one-shot output, a transitional contact is used to reset conditions in one scan, for example, to reset a latched coil (i.e., unlatch it). Figure 9-27 shows circuit applications for both leading-edge and trailing-edge transitional contacts, along with their respective timing diagram.

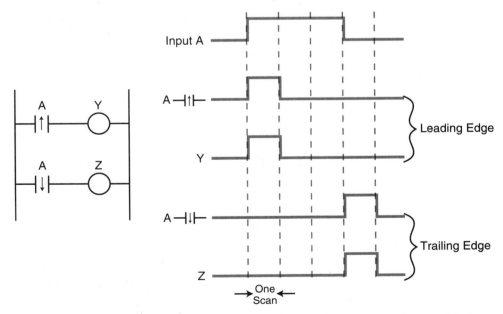

Figure 9-27. Leading- and trailing-edge transitional contact instructions and their timing diagrams.

9-5 LADDER RELAY PROGRAMMING

LADDER SCAN EVALUATION

Scan evaluation is an important concept, since it defines the order in which the processor executes a ladder diagram. The processor starts solving a ladder program after it has read the status of all inputs and stored this information in the input table. The solution starts at the top of the ladder program, beginning with the first rung and proceeding one rung at a time. As the processor solves the control program, it examines the reference address of each programmed instruction, so that it can assess logic continuity for the rung being solved. Even if the output conditions in the rung being solved affect previous rungs, the processor will not return to the previous rung to resolve it.

To make this clearer, let's examine the diagram in Figure 9-28, which illustrates four simple rungs. The normally open contact 10, which we will assume corresponds to a push button, activates the first rung. If contact 10 turns ON, it will turn output 100 ON. In the next rungs, contact 100 will turn output 101 ON, contact 101 will turn output 102 ON, and contact 102 will turn output 103 ON. Even though they are connected to different rungs, all of these outputs turn ON in the same scan, because the processor updates the real output devices connected to the modules when it finishes the program scan. In this case, if outputs 100, 101, 102, and 103 were connected to pilot lights, they would all turn ON at the same time.

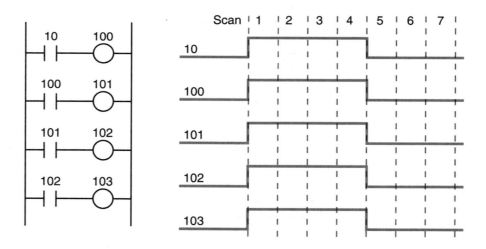

Figure 9-28. Ladder rung where all outputs turn ON in the same scan.

Figure 9-29 illustrates the same ladder logic as in Figure 9-28 but with the placement of rungs reversed. Assuming that input 10 is pushed in the first scan, the processor must make four scans before it energizes output 103. The logic the processor uses in the first scan is as follows: (1) When input 10 is pushed, the processor examines reference 102 and finds it OFF (logic 0);

therefore, output 103 stays OFF. (2) In the second rung, contact 101 is OFF; therefore, output 102 remains OFF. (3) In the third rung, contact 100 is OFF, so output 101 remains OFF. (4) In the fourth rung, contact 10 is ON because the push button is pushed, so output 100 turns ON. In the next scan (second), if the push button remains ON, output 101 will turn ON because, at the end of the first scan, the reference address 100 was set to logic 1. This logic will continue until the fourth scan, when all four outputs will be ON. The outputs will turn OFF in the same way once the push button is released.

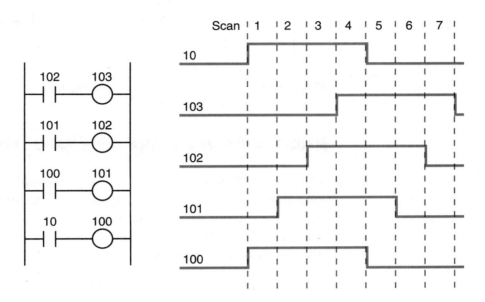

Figure 9-29. Ladder rung where the outputs turn ON in different scans.

The physical operation of a circuit like the one in Figure 9-29 is almost impossible to observe while a PLC is running the control program because a PLC completes its scan in milliseconds. All the pilot lights would seem to come ON at the same time, even if they actually came on in different scans. The only way to observe the ladder outputs would be to use single-scan PLC operation. With single-scan operation, the processor reads the inputs, executes the logic, updates the outputs, and stops until another single scan is executed. Single-scan operation is generally used during the testing of a control program.

The important thing to remember about a ladder program is that for an output to have an effect on another rung in the same scan, it must be programmed before that rung. If it is not, order of execution problems can arise, especially when using transitional contacts and one-shot outputs to reset and unlatch other rungs. Figure 9-30 illustrates this type of programming order problem, where the output unlatch instruction will never occur. Once contact 10 closes, latching output 100, only the closing of contact 11 will unlatch the output. When contact 12 closes, it triggers the one-shot output 11 (or the transitional contact 12) for one scan. However, at the end of the scan, the one-shot output turns OFF, so it is not able to unlatch coil 100 in the next scan.

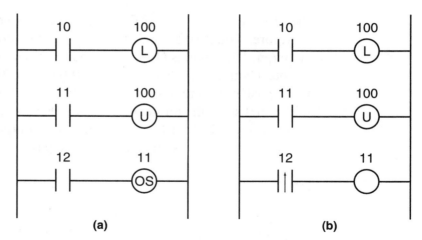

(a) **(b)**

Figure 9-30. (a) The one-shot output and **(b)** the transitional contact will never unlatch coil 100.

PROGRAMMING NORMALLY CLOSED INPUTS

So far in our discussion, we have tried to avoid presenting input device connections that are in the normally closed condition. The reason for this is simple—we did not want to confuse you. Understanding how to program a normally closed input device is a difficult concept to comprehend at first. Once you learn it, try explaining it to someone else and watch their reaction.

To explain how to program normally closed inputs, let's look at the following example. Suppose we want to implement logic identical to the simple hardwired circuit shown in Figure 9-31. Implementing the same logic means that the pilot light PL1 in the PLC should behave in the same manner as the one in the hardwired circuit—if PB1 is not pushed, PL1 will be ON; if PB1 is pushed, PL1 will be OFF. Figures 9-32 and 9-33 show two possible methods for programming PB1 and implementing the logic. At first glance, you may think that the solution in Figure 9-32 is the answer, but that is not true; Figure 9-33 is the correct implementation.

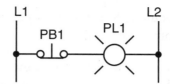

Figure 9-31. Hardwired logic.

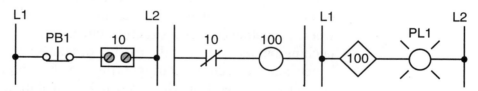

Figure 9-32. Logic implementation with PB1 programmed as a normally closed contact.

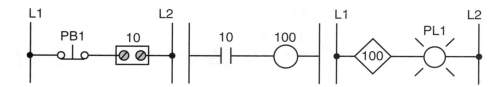

Figure 9-33. Logic implementation with PB1 programmed as a normally open contact.

In Figure 9-32, the reference address of PB1 (input 10) is programmed as a normally closed contact (examine OFF) that drives output coil 100, which is connected to pilot light PL1. When the PLC starts, it reads the status of the input device connected to input 10 and stores this data in the input table. If PB1 is not pushed (see Figure 9-34a), the processor reads input 10 as logic 1 (power flowing to the module). During the execution of the ladder logic, the PLC will evaluate the examine-OFF instruction, and since the reference (input 10) is ON, it will open the normally closed contact, disrupting continuity. Thus, output 100 will be OFF, and PL1 will not turn ON. Conversely, if PB1 is pushed (see Figure 9-34b), the input module at location 10 will be logic 0 (power not flowing to the module). The processor's examination for an OFF condition at reference 10 will then be TRUE; therefore, the instruction will provide continuity to the rung and turn output 100 and PL1 ON.

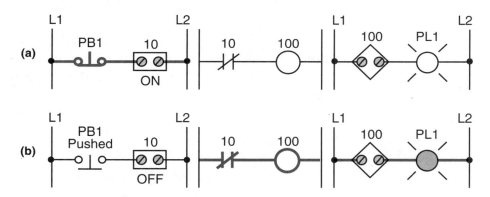

Figure 9-34. Power flow through the circuit shown in Figure 9-32 with **(a)** PB1 not pushed and **(b)** PB1 pushed.

In Figure 9-33, the normally closed input condition has been programmed as an examine-ON instruction. During operation (see Figure 9-35a), if PB1 is not pushed, the input module 10 will read an ON status. When the processor evaluates the ladder rung, its examination for an ON condition at reference 10 will be TRUE. Therefore, contact 10 will close to provide power to the rung, turning output 100 and PL1 ON. On the other hand, if PB1 is pushed (see Figure 9-35b), the input will have an OFF status and the processor will store a logic 0 in the input table. During the evaluation of the rung, the processor will find its examination for an ON condition at reference 10 to be FALSE (input 10 is OFF), and continuity will not occur because the contacts will remain open. Thus, output 100 and PL1 will be OFF.

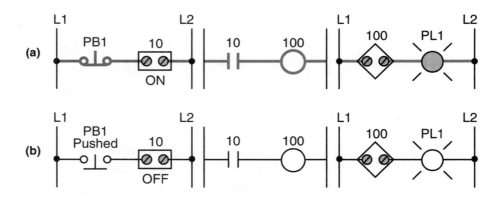

Figure 9-35. Power flow through the circuit shown in Figure 9-33 with **(a)** PB1 not pushed and **(b)** PB1 pushed.

The programming solution for a normally closed input connection, as shown in Figure 9-33, exemplifies the following: for a normally closed wired input device to behave as a normally closed device when connected, it must be programmed as an examine-ON, or normally open, contact instruction. Discrete inputs to a PLC can be made to act as normally open or normally closed contacts, regardless of their original configuration. This ability to examine a single device for either an open or closed state is the key to the flexibility of PLCs—no matter how a device is wired (normally open or normally closed), the controller can be programmed to perform the desired action without changing the wiring. Remember that the programming state of an input depends not only on how it is wired, but also on the desired control action. The following example shows a case in which the PLC programming of one push button with two contacts differs depending on which contact is wired to the module.

EXAMPLE 9-4

Show the PLC implementation of the hardwired logic shown in Figure 9-36 for the following scenarios using only one push button connection: **(a)** with the normally open contact connected to the input module and **(b)** with the normally closed contact connected to the input module. Describe the operation of each implementation as well.

Use input address 10 for the push button and addresses 30 and 31 for pilot lights PL1 and PL2, respectively. Indicate the lights in the ON condition (without PB1 being pushed) using a shaded PL indicator.

SOLUTION

Examining the circuit in Figure 9-36 shows that, if PB1 is not pushed, PL1 should be OFF. PL2 should be ON because the other contact of PB1 (the normally closed one) provides power to PL2. We can wire any

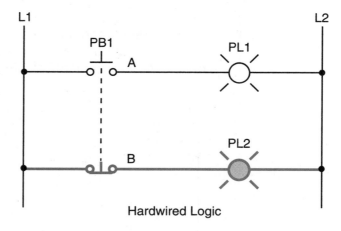

Figure 9-36. Hardwired logic for Example 9-4.

of PB1's two connections (A or B) to the input module to satisfy the required logic. Remember that we can make any contact act as we desire in the PLC program (i.e., as a normally open or normally closed contact).

(a) Figure 9-37 shows the solution for the normally open contact connection. An examine-ON instruction drives PL1, and an examine-OFF instruction drives PL2. When PB1 contact A is not pushed, PL1 is OFF and PL2 is ON. The first rung implements a push button wired as normally open to act as a normally open push button, while the second rung implements a push button wired as normally open to act as a normally closed push button.

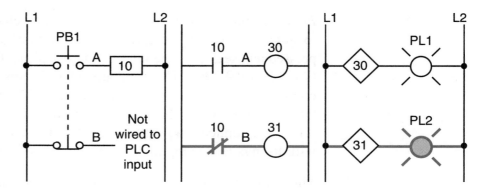

Figure 9-37. Normally open implementation of Figure 9-36.

(b) Figure 9-38 shows the circuit solution for the normally closed contact connection. In this solution, an examine-OFF instruction drives PL1. During operation, PB1 contact B provides power to the module if it is not pushed; therefore, the reference address (10) is logic 1. The normally closed contact with address 10 will be open as long as PB1 is not depressed, keeping PL1 (output 30) OFF. In the second rung, an examine-ON instruction drives the output for PL2 (31), which is closed as long as PB1 is not pushed. The first rung implements a push button wired as normally closed to act as a normally open push button, while the second rung implements a push button wired as normally closed to act as a normally closed push button.

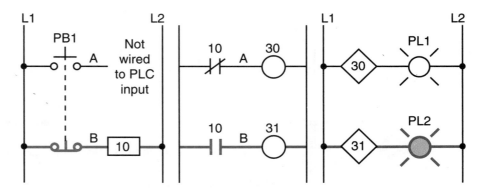

Figure 9-38. Normally closed implementation of Figure 9-36.

As illustrated in the previous example, a normally open input can be programmed in a PLC to behave like a normally closed device and vice versa. However, for fail-safe reasons, normally closed input devices should be wired to the input module as normally closed devices and then programmed as examine-ON instructions, so that they behave like normally closed devices. A wired normally open device *must not* be programmed to act as a normally closed device, especially if it is being used to interrupt continuity when a device is pushed or closed.

Figure 9-39a shows an example of a normally closed stop push button used to stop the power to a motor. During operation, when the start PB has been pressed and sealed by the internal motor contact (100), the motor turns ON (see Figure 9-39b). The normally closed stop PB interrupts the power continuity to the motor output coil contact. The pressing of this stop push button is the only way the motor can be stopped (see Figure 9-39c). However, if the wire connection for the stop PB is accidentally cut, the motor circuit will disengage (see Figure 9-39d).

This same logic operation can also be achieved using a normally open stop PB instead of a normally closed one and implementing it as a normally closed circuit in the PLC program (see Figure 9-40a). When the start button is pushed, the motor turns ON (see Figure 9-40b); if the stop PB is pressed, the motor turns OFF (see Figure 9-40c). However, there is no way to stop the motor from running if the normally open stop PB wire is cut (see Figure 9-40d). The programmed examine-OFF instruction corresponding to the stop PB will never disrupt continuity in this situation. The only way to stop the motor is to shut down power to the whole PLC system. This type of PLC system configuration is dangerous and should be avoided at all times.

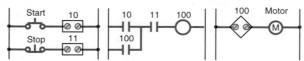

(a) The normally closed stop push button is programmed as normally open. Contact 100 is used as an interlock with the start push button after the start is pushed. When the start push button is pressed, the motor turns ON.

(b) After the start push button is pressed and released, the motor remains ON.

(c) If the stop push button is pressed when the motor is ON, the motor will turn OFF.

(d) If the stop push button connection breaks when the motor is ON, the motor will turn OFF.

Figure 9-39. Normally closed stop push button programmed as normally open.

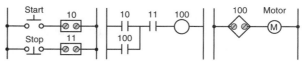

(a) The normally open stop push button is programmed as normally closed. When the start push button is pressed, the motor turns ON.

(b) After the start push button is pressed and released, the motor remains ON.

(c) If the stop push button is pressed when the motor is ON, the motor will turn OFF.

(d) If the stop push button connection breaks when the motor is ON, pressing the stop push button will not turn the motor OFF. This is a dangerous situation.

Figure 9-40. Normally open stop push button programmed as normally closed.

9-6 TIMERS AND COUNTERS

PLC timers and counters are internal instructions that provide the same functions as hardware timers and counters. They activate or deactivate a device after a time interval has expired or a count has reached a preset value. Timer and counter instructions are generally considered internal outputs. Like relay-type instructions, timer and counter instructions are fundamental to the ladder diagram instruction set.

Timer instructions may have one or more *time bases* (TB) which they use to time an event. The time base is the resolution, or accuracy, of the timer. For instance, if a timer must time a 10 second event, the user must choose the number of times the time base must be counted to get to 10 seconds. Therefore, if the timer has a time base of 1 second, then the timer must count ten times before it activates its output. This number of counts is referred to as *ticks*. The most common time bases are 0.01 sec, 0.1 sec, and 1 sec. Table 9-3 shows the number of ticks required for a 10 second count, based on different time bases.

Required Time	Number of Ticks	Time Base (secs)
10 sec	10	1.00
10 sec	100	0.10
10 sec	1000	0.01
Note: Required time = (# of ticks)(Time base)		

Table 9-3. Time bases.

Timers are used in applications to add a specific amount of delay to an output in the program. Applications of PLC timers are innumerable, since they have completely replaced hardware timers in automated control systems. As an example, timers may be used to introduce a 0.01 second delay in a control program. The program may require such a delay because the PLC turns ON its outputs very quickly as compared to the hardwired relay system it is replacing. This small delay will slow down the response of other components so that proper operation occurs.

Counter instructions are used to count events, such as parts passing on a conveyor, the number of times a solenoid is turned ON, etc. Counters, along with timers, must have two values, a *preset value* and an *accumulated value*. These values are stored in register or word locations in the data table. The preset value is the target number of ticks or counting numbers that must be achieved before the timer or counter turns its output ON. The accumulated value is the current number of ticks (timer) or counts (counter) that have elapsed during the timer or counter operation. The preset value is stored in a preset register, while the accumulated value is kept in an accumulated register. Both of these registers are defined during the programming of the instruction. Either the basic ladder format or the block instruction format can be used to implement timers and counters.

EXAMPLE 9-5

During a machine modernization project, it is found that part of a relay ladder circuit (see Figure 9-41), when translated into a PLC circuit, does not work correctly. This malfunction is due to the fact that in the hardwired circuit, relay CR5, which is driven by device LS4, had enough delay time to synchronize with the rest of the circuit so that the solenoid actuation was correct. Now that it has been implemented in the PLC, CR5 no longer has this delay. The delay needed is estimated at 3 AC cycles (60 Hz) and the time bases available in the PLC are 0.01, 0.1, and 1 sec. Which time base should be used to create the delay and how many ticks must the delay last?

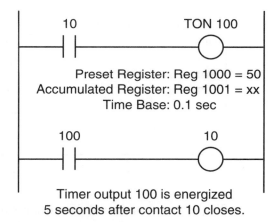

Timer output 100 is energized
5 seconds after contact 10 closes.

Figure 9-41. Example relay ladder circuit.

SOLUTION

The estimated delay of 3 AC cycles translates into 60 Hz (i.e., 60 cycles/sec). So:

$$1 \text{ cycle} = \frac{1}{60} = 16.66 \text{ msec}$$

$$3 \text{ cycles} = \frac{3}{60} = 50.00 \text{ msec}$$

Thus, the required delay is 50 msec. Therefore, the only time base small enough to use is 0.01 sec. Using this time base, the timer must count 5 ticks to create the delay.

9-7 TIMER INSTRUCTIONS

PLCs provide several types of timer instructions. However, PLC manufacturers may provide different definitions for each type of timer function offered. Table 9-4 presents a list of typical timer instructions.

| Timer Instructions |||
| *(Purpose: To provide hardware timer capabilities in a PLC)* |||
Instruction	Symbol	Function
ON-Delay Energize Timer	TON ◯	Energizes an output after a set time period when logic 1 exists
ON-Delay De-energize Timer	TON ◯	De-energizes an output after a set time period when logic 1 exists
OFF-Delay Energize Timer	TOF ◯	Energizes an ouput after a set time period when logic 0 exists
OFF-Delay De-energize Timer	TOF ◯	De-energizes an output after a set time period when logic 0 exists
Retentive ON-Delay Timer	RTO ◯	Energizes an output after a set time period when logic 1 exists and then retains the accumulated value
Retentive Timer Reset	RTR ◯	Resets the accumulated value of a retentive timer

Table 9-4. Timer instructions.

The function of the various timer instructions is essentially the same, differing only in the type of output provided. Figure 9-42 illustrates the two formats used for timers. A block format timer has one or two inputs, depending on the programmable controller. These inputs are called the *control line* and the

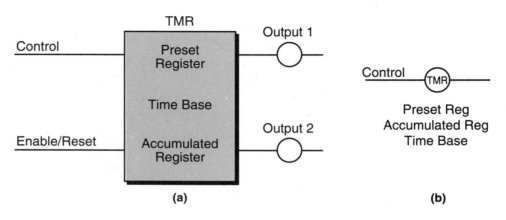

Figure 9-42. (a) Block format and **(b)** ladder format timer instructions.

enable/reset line. If the control line is TRUE (i.e., it has continuity) and the enable line is also TRUE, the block function will start timing. A ladder format timer generally has only one input, which is the control line. If the control line is ON, the timer will start timing.

Common to both timer formats is the use of a preset register to hold the preset value and an accumulated register to store the accumulated value. Some PLCs allow the user to enter a constant value directly into the timer to set the preset value. This particular value, however, must be entered into a predefined register for that specific timer address.

A timer's time base is selectable depending on the PLC used (e.g., 0.01 sec, 0.1 sec, 1.0 sec, etc.). When the accumulated tick count equals the preset count, the timer executes its timing function and sets the output condition, which depends on the type of timer used (e.g., ON-delay energize, etc.).

It is important to note that when PLC timers replace hardwired timers, they replace the time-delay contacts associated with the timers, but *not* the instantaneous contacts that may be available from a hardwired timer. Figure 9-43 illustrates an example showing both time-delay and instantaneous

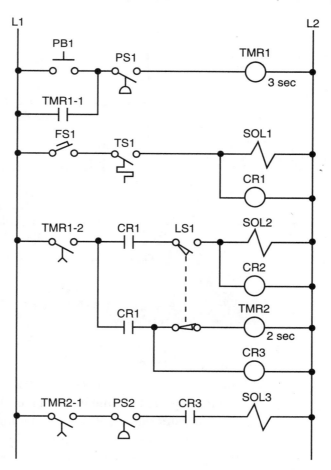

Figure 9-43. Hardwired circuit with time-delay and instantaneous contacts.

hardwired timer contacts. Timer TMR1 in line 1 has an instantaneous contact in line 2 (TMR1-1), which is used to seal PB1, and a time-delay contact (TMR1-2) in line 5. For this type of ladder logic translation into a PLC program, the user must "trap" the timer through interlocking, so that the instantaneous timer seal can be accomplished. Chapter 11 presents this type of programming example.

ON-DELAY ENERGIZE TIMER

An *ON-delay energize timer* (TON) output instruction either provides time-delayed action or measures the duration for which some event occurs. Once the rung has continuity, the timer begins counting time-based intervals (ticks) and counts down until the accumulated time equals the preset time. When these two values are equal, the timer energizes the output and closes the timed-out contact associated with the output (see Figure 9-44). The timed contact can be used throughout the program as either a normally open or normally closed contact. If logic continuity is lost before the timer times out, the timer resets the accumulated register to zero.

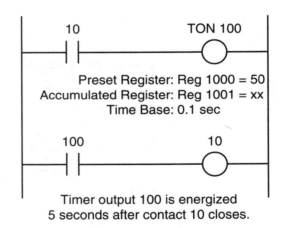

Preset Register: Reg 1000 = 50
Accumulated Register: Reg 1001 = xx
Time Base: 0.1 sec

Timer output 100 is energized
5 seconds after contact 10 closes.

Figure 9-44. ON-delay energize timer instruction.

ON-DELAY DE-ENERGIZE TIMER

An *ON-delay de-energize timer* (TON) instruction operates in a manner similar to an ON-delay energize timer instruction, except that the timer's output is already ON. This instruction de-energizes the output once the rung has continuity and the time interval has elapsed (accumulated register value = preset register value). PLC manufacturers provide either ON-delay energize or ON-delay de-energize timers, since it is easy to program one from the other. Figure 9-45 illustrates a timing diagram for both types of ON-delay timer instructions.

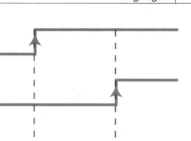

Figure 9-45. Timing diagram for (a) an ON-delay energize timer and (b) an ON-delay de-energize timer.

OFF-DELAY ENERGIZE TIMER

An *OFF-delay energize timer* (TOF) output instruction provides time-delayed action. If the control line rung does not have continuity, the timer begins counting time-based intervals until the accumulated time value equals the programmed preset value. When these values are equal, the timer energizes the output and closes the timed-out contact associated with the output (see Figure 9-46). The timed contact can be used throughout the program as either a normally open or normally closed contact. If logic continuity occurs before the timer times out, the accumulated value resets to zero.

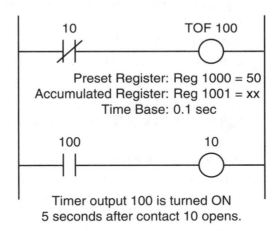

Preset Register: Reg 1000 = 50
Accumulated Register: Reg 1001 = xx
Time Base: 0.1 sec

Timer output 100 is turned ON
5 seconds after contact 10 opens.

Figure 9-46. OFF-delay energize timer instruction.

OFF-DELAY DE-ENERGIZE TIMER

An *OFF-delay de-energize timer* (TOF) instruction is similar to its OFF-delay energize counterpart; however, this timer's output is ON and will be de-energized once the rung loses continuity and the time interval has elapsed (accumulated register value = preset register value). Like ON-delay timers, PLC manufacturers usually provide either OFF-delay energize or de-energize timers. Figure 9-47 shows timing diagrams for both types of OFF-delay timers.

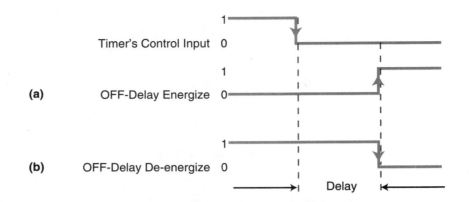

Figure 9-47. Timing diagram for **(a)** an OFF-delay energize timer and **(b)** an OFF-delay de-energize timer.

RETENTIVE ON-DELAY TIMER

A *retentive ON-delay timer* (RTO) output instruction is used if the timer's accumulated value must be retained even if logic continuity or system power is lost. If any rung path has logic continuity, the timer begins counting time-based intervals until the accumulated time equals the preset value. The accumulated register retains this accumulated value, even if power or logic continuity is lost before the timer has timed out. When the accumulated time equals the preset time, the timer energizes the output and turns ON (closes) the timed-out contact associated with the output. Again, these timer contacts can be used throughout the program as normally open or normally closed contacts. A retentive timer reset instruction resets a retentive timer's accumulated value.

RETENTIVE TIMER RESET

A *retentive timer reset* (RTR) output instruction is the only way to automatically reset the accumulated value of a retentive timer. If any rung path has logic continuity, then this instruction resets the accumulated value of its referenced retentive timer to zero. Note that the retentive timer reset address will be the same as the retentive timer output instruction it is resetting.

9-8 COUNTER INSTRUCTIONS

There are two basic types of counters: those that can count up and those that can count down. Depending on the controller, the format of these counters may vary. Some PLCs use the ladder format (output coil), while others use functional block format. Figure 9-48 illustrates these two formats, while Table 9-5 presents common counter instructions.

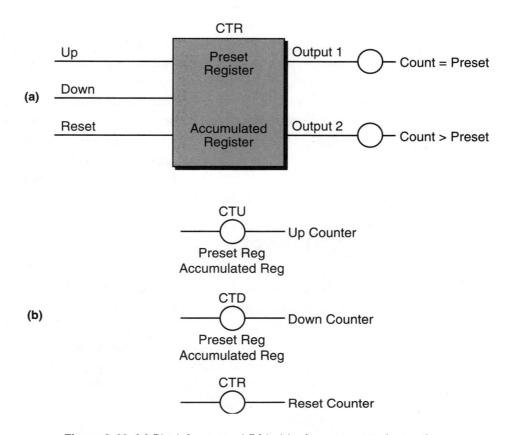

Figure 9-48. (a) Block format and **(b)** ladder format counter instructions.

Counter Instructions		
(Purpose: To provide hardware counter capabilities in a PLC)		
Instruction	**Symbol**	**Function**
Up Counter	CTU —○—	Increases the accumulated register value every time a referenced event occurs
Down Counter	CTD —○—	Decreases the accumulated register value every time a referenced event occurs
Counter Reset	CTR —○—	Resets the accumulated value of an up or down counter

Table 9-5. Counter instructions.

UP COUNTER

An *up counter* (CTU) output instruction adds a count, in increments of one, every time its referenced event occurs. In a control application, this counter turns a device ON or OFF after reaching a certain count (i.e., the preset value in the preset register). Also, this counter can keep track of the number of parts (e.g., filled bottles, machined parts, etc.) that pass a certain point. An up counter increases its accumulated value (the count value in its accumulated register) each time the up-count event makes an OFF-to-ON transition. When the accumulated value reaches the preset value, the counter turns ON the output, finishes the count, and closes the contact associated with the referenced output. After the counter reaches the preset value, it either resets its accumulated register to zero or continues its count for each OFF-to-ON transition, depending on the controller. In the latter case, a reset instruction is used to clear the accumulated value.

DOWN COUNTER

A *down counter* (CTD) output instruction decreases the count value in its accumulated register by one every time a certain event occurs. In practical use, a down counter is used in conjunction with an up counter to form an *up/down counter*, given that both counters have the same reference registers. In an up/down counter, the down counter provides a way to correct data that is input by the up counter. For example, while an up counter counts the number of filled bottles that pass a certain point, a down counter with the same reference address can subtract one from the accumulated count value every time it senses an empty or improperly filled bottle. Depending on the programmable controller, the down counter will either stop counting down at zero or at a specified maximum negative value. In a block format instruction, a down count occurs every time the down input of the counter transitions from OFF to ON.

COUNTER RESET

A *counter reset* (CTR) output instruction resets up counter and down counter accumulated values to zero. When programmed, a counter reset coil has the same reference address as the corresponding up/down counter coils. If the counter reset rung condition is TRUE, the reset instruction will clear the referenced address. The reset line in a block format counter instruction sets the accumulated count to zero (accumulated register = 0). Figure 9-49 illustrates a typical block-formatted counter rung with up, down, and reset counter instructions. The counter will count up when contact 10 closes, count down when contact 11 closes, and reset register 1003 to 0 when contact 12 closes. If the count is equal to 15 as a result of either an up or down count, output 100 will be ON. If the contents of register 1003 are greater than 15, output 101 will be ON. Output 102 will be ON if the accumulated count value is less than 15.

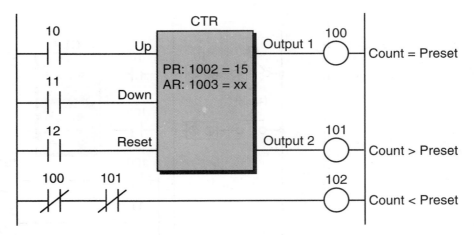

Figure 9-49. Counter function block with up, down, and reset counter instructions.

EXAMPLE 9-6

Figure 9-50 illustrates a block counter instruction being used to count parts as detected by a photoelectric eye (PE) input. The preset value of counts is 500. Modify this circuit so that it will automatically reset every time the counter reaches 500. Also, add the instructions necessary to implement an output coil that indicates that the count has reached 500.

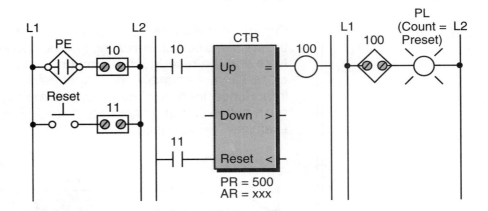

Figure 9-50. Functional block counter instruction.

SOLUTION

Figure 9-51 illustrates a circuit that will automatically reset the counter. When the preset and accumulated counts are equal, the counter output 100 turns ON, latching output 101 to indicate a reached count. This same counter output resets the counter. Remember that the PLC has already evaluated all inputs, so the counter is reset in the following scan. The previous input 11 is used to manually unlatch output 101.

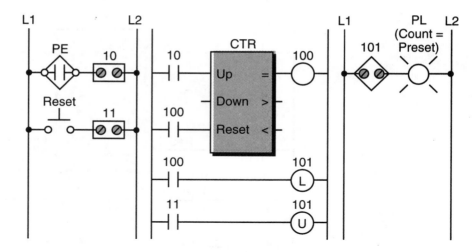

Figure 9-51. Automatically resetting counter.

EXAMPLE 9-7

Referencing the solution to Example 9-6 (i.e., see Figure 9-51), implement the count detection circuit using interlocking standard outputs and contacts instead of latch/unlatch coils.

SOLUTION

Figure 9-52 illustrates an interlocking circuit that latches, or traps, the counter's output, indicating that the count value has been reached. Note that the reset push button (input 11) is programmed normally closed from a normally open input device. If this input was of safety importance, then the circuit would have incorporated a normally closed push button (wired as normally closed) that was programmed as an examine-OFF instruction.

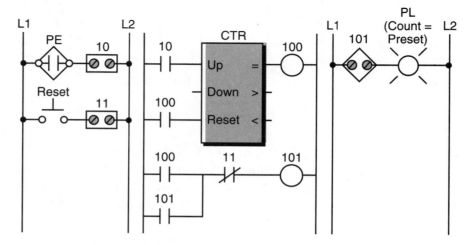

Figure 9-52. Solution to Example 9-7.

9-9 PROGRAM/FLOW CONTROL INSTRUCTIONS

Program/flow control instructions direct the flow of operations, as well as the execution of instructions, within a ladder program. They perform these functions using branching and return instructions, which are executed when certain already programmed control logic conditions occur. Typically, program/flow control instructions form a "fence" within a program. This fence contains groups of other ladder instructions that are used to implement the desired function. Figure 9-53 illustrates a fence created using program/flow control instructions.

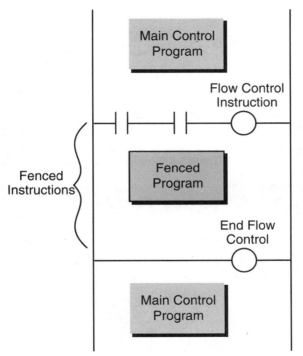

Figure 9-53. A fence created using a program/flow control instruction.

Some programmable controllers, depending on their capabilities and the scope of their application, use several types of program/flow control instructions. These instructions allow the controller to efficiently perform special user-programmed routines that are executed only when required. This reduces the scan time, thereby optimizing total system response.

Table 9-6 shows some of the most commonly used program/flow control instructions. These instructions are generally used in pairs. When paired, the first instruction starts the flow control change, sending the PLC to a special routine of instructions in another section of the control program. The other instruction returns the PLC to the program it was running when the flow control change occurred.

Program/Flow Control Instructions

(Purpose: To direct the evaluation/execution of instructions in a ladder program)

Instruction	Symbol	Function
Master Control Relay	MCR —◯—	Activates/deactivates the execution of a group of ladder rungs
Zone Control Last State	ZCL —◯—	Determines whether or not a group of ladder rungs will be evaluated
End	END —◯—	Identifies the last rung of an MCR or ZCL instruction
Jump To	JMP —◯—	Jumps to a specified rung in the program if certain conditions exists
Go To Subroutine	GOSUB —◯—	Goes to a specified subroutine in the program if certain conditions exist
Label	—┤LBL├—	Identifies the target rung of a JMP or GOSUB instruction
Return	RET —◯—	Terminates a ladder subroutine

Table 9-6. Program/flow control instructions.

MASTER CONTROL RELAY

A *master control relay* (MCR) output instruction activates or deactivates the execution of a group or zone of ladder rungs (see Figure 9-54). An MCR rung is used in conjunction with an END rung (discussed later) to fence a group of

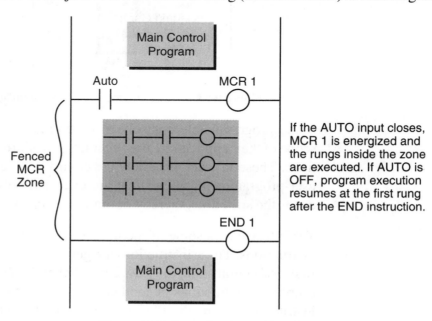

If the AUTO input closes, MCR 1 is energized and the rungs inside the zone are executed. If AUTO is OFF, program execution resumes at the first rung after the END instruction.

Figure 9-54. Example of an MCR instruction.

rungs. The fence consists of an MCR rung with conditional inputs at the beginning of the zone and an END rung with no conditional inputs at the end of the zone. When the MCR rung condition is TRUE, it activates the referenced output, allowing all rung outputs within the zone to be controlled by their respective rung input conditions. When the MCR output is turned OFF, it de-energizes all nonretentive (nonlatched) outputs within the zone.

ZONE CONTROL LAST STATE

A *zone control last state* (ZCL) instruction is similar to an MCR instruction—it determines whether or not a group of ladder rungs will be evaluated. In this instruction, a ZCL output with conditional inputs occurs at the start of the fenced zone, while an END ZCL output with no conditional inputs occurs at the end of the zone. When the referenced ZCL output is activated, the outputs within the zone are controlled by their respective input conditions. When the ZCL output is turned OFF, the outputs within the zone stay in their last state.

END

An *end* (END) instruction signifies the last rung of a master control relay or zone control last state instruction. This instruction is usually unconditional (i.e., programmed without any conditions to energize). An end instruction reference address may or may not reference a MCR or ZCL. If a reference is included, the END instruction will end that particular MCR or ZCL. If the instruction does not include a reference address, it will terminate the latest MCR or ZCL instruction.

JUMP TO

A *jump to* (JMP) instruction allows the control program sequence to be altered if certain conditions exist. If the rung condition is TRUE, the jump to coil reference address tells the processor to jump forward and execute the target rung. The jump to address label specifies the target rung to jump to. Using this instruction, a PLC can alter the order of execution of the control program to execute a rung that needs immediate attention. Figure 9-55 illustrates a jump to instruction. This instruction may also be called a *go to* instruction. Note that care should be exercised when jumping over timers and counters. Jumping over timers and counters will cause the timing and counting instructions not to be executed.

GO TO SUBROUTINE

Like a jump to instruction, a *go to subroutine* (GOSUB) output instruction also allows normal program execution to be altered if certain conditions exist. In this instruction, if the rung condition is TRUE, the GOSUB coil reference

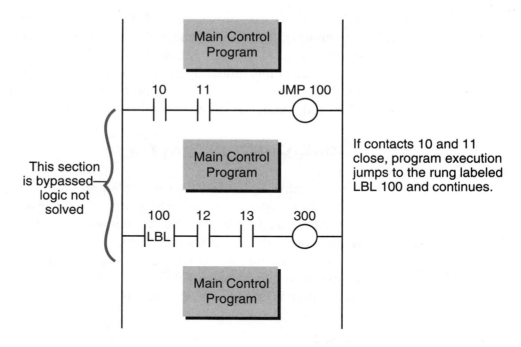

Figure 9-55. Example of a jump to instruction.

address tells the processor to jump to the ladder rung with a label (LBL) instruction having the same reference number. The processor then continues the program execution until it encounters a return coil. Each subroutine in the program must begin with a labeled rung and end with an unconditional return instruction. A go to subroutine instruction may also be called a *jump to subroutine* (JSB) instruction.

A GOSUB instruction is very useful whenever a subroutine in the program is either referenced by several sections of the main control program or is referenced on a timely basis (i.e., look up analog interpretation table every 10 seconds). Subroutines are generally located at the end of the control program and are sometimes located in an area specified by the PLC maker (see Figure 9-56). If a PLC does not have a reserved subroutine area, the user can create one by programming a dummy rung with direct control to another dummy rung at the end of the programmed subroutines (see Figure 9-57). For proper programming documentation order, the subroutine area should be located at the end of the control program.

LABEL

A *label* (LBL) instruction identifies the ladder rung that is the target destination of a jump to or GOSUB instruction. The label instruction reference number must match that of the jump to or GOSUB instruction with which it is used. A label instruction does not contribute to logic continuity

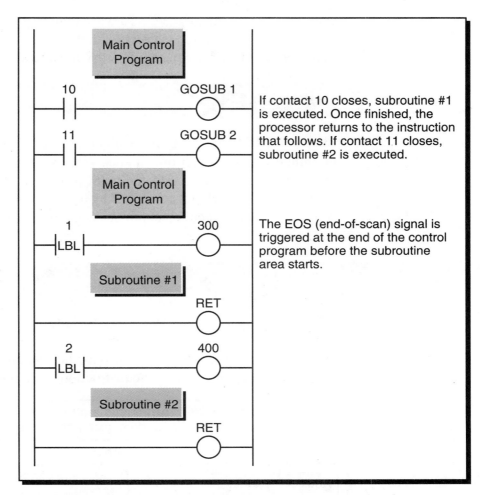

If contact 10 closes, subroutine #1 is executed. Once finished, the processor returns to the instruction that follows. If contact 11 closes, subroutine #2 is executed.

The EOS (end-of-scan) signal is triggered at the end of the control program before the subroutine area starts.

Figure 9-56. PLC with assigned subroutines at the end of the program.

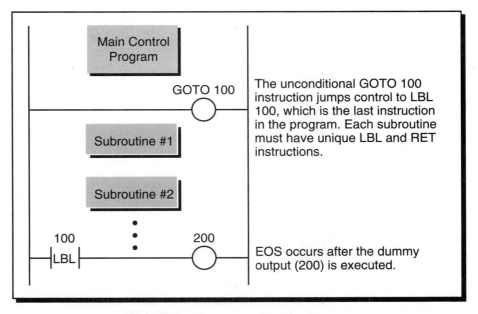

The unconditional GOTO 100 instruction jumps control to LBL 100, which is the last instruction in the program. Each subroutine must have unique LBL and RET instructions.

EOS occurs after the dummy output (200) is executed.

Figure 9-57. User-created subroutine area.

and, for all practical purposes, is always logically TRUE. This instruction is always the first condition instruction in the referenced rung. A label instruction referenced by a unique address can only be defined once in a program.

RETURN

A *return* (RET) instruction terminates a ladder subroutine and is programmed with no conditional inputs. When the control program encounters this instruction, it returns to the main program, going to the ladder rung immediately following the GOSUB instruction that initiated the subroutine. Normal program execution continues from that point. Each subroutine must have a return instruction.

9-10 ARITHMETIC INSTRUCTIONS

Arithmetic instructions in a PLC include the basic four operations of addition, subtraction, multiplication, and division. In addition to these four math functions, large PLCs may also include square root operations. Table 9-7 lists these typical arithmetic instructions and their symbols.

Arithmetic Instructions		
(Purpose: To allow PLCs to perform mathematical functions with register data)		
Instruction	**Symbol**	**Function**
Addition—Ladder	ADD (+)	Adds the values stored in two registers
Addition—Block	ADD	Adds the values stored in two registers
Subtraction—Ladder	SUB (−)	Subtracts the values stored in two registers
Subtraction—Block	SUB	Subtracts the values stored in two registers
Multiplication—Ladder	MUL (×)	Multiplies the values stored in two registers
Multiplication—Block	MUL	Multiplies the values stored in two registers
Division—Ladder	DIV (÷)	Finds the quotient of the values in two registers
Division—Block	DIV	Finds the quotient of the values in two registers
Square Root—Block	SQR	Calculates the square root of a register value

Table 9-7. Arithmetic instructions.

Like other instructions, arithmetic instructions may be in either the basic ladder format or the functional block format; however, operation in either format is essentially the same. Figure 9-58 illustrates these formats. Most arithmetic instructions require three reference registers, which define the two operand registers and the destination register of the operation. Some instructions, such as multiplication and division, may use four registers. Most arithmetic operations in a PLC require only **single-precision arithmetic**, meaning that the values of the operands and the result can be held in one register each. If operations dealing with larger numbers are required, a PLC may offer **double-precision arithmetic** instructions. Double precision means that the system uses double the number of registers to hold the operands and result, because it must store larger numbers. For example, a double-precision addition instruction would use a total of six registers, two for each operand and two for the result.

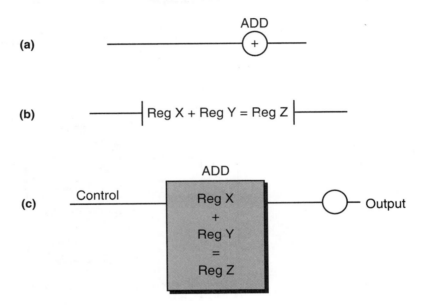

Figure 9-58. (a) Coil, **(b)** contact, and **(c)** block format arithmetic instructions.

As discussed earlier, a register can hold a maximum value of 65,535 in 16 bits (all 1s) if there is no sign bit. If the most significant bit is used as the sign bit, then a register may hold a maximum value of +32,767 and a minimum value of –37,767. If the result value of the operation is larger than the value a register can hold, an overflow condition will exist, and the instruction will turn ON an overflow bit or output. The numerical format used in math operations will vary depending on the PLC but is usually three, four, or five digits (BCD or binary). Note that in single-precision BCD, the maximum register value is 9999 (unsigned) or ±999 (signed).

In the following discussion, we will present arithmetic instructions in both ladder and block formats to familiarize you with the differences between them. Note that the ladder format may require other ladder data transfer instructions to obtain the arithmetic operands. In functional block format, some manufacturers offer the ability to "cascade" block functions (see Figure 9-59). Cascading is very useful when dealing with multiple arithmetic operations, since one instruction will activate the next one when finished. Other manufacturers allow arithmetic operations to be performed in block form (see Figure 9-60); that is, using blocks of several contiguous registers as the operands and storing the results in another block of registers.

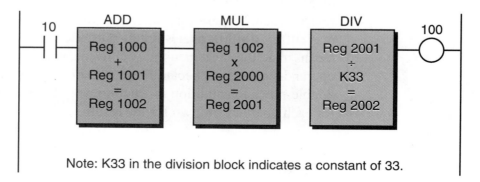

Note: K33 in the division block indicates a constant of 33.

Figure 9-59. "Cascading" allows several functional block arithmetic operations to be performed sequentially.

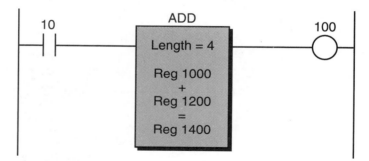

Note: The contents of registers 1000, 1001, 1002, and 1003 will be added to registers 1200, 1201, 1202, and 1203. The results will be stored in registers 1400, 1401, 1402, and 1403.

Figure 9-60. Arithmetic operations performed in block form.

ADDITION—LADDER

The *addition* (ADD) ladder instruction adds the values stored in two referenced memory locations. Different controllers access these values differently. Some instruction sets use a *get* (GET) data transfer instruction to access the two operand register values (see Figure 9-61), while others simply reference the two registers using contact symbols (see Figure 9-58b). The processor stores the sum of the values in the register referenced by the ADD coil. If the addition operation is enabled only when certain rung conditions are TRUE, then the input conditions should be programmed before the addition rung. One bit in the addition result register usually signals an overflow condition.

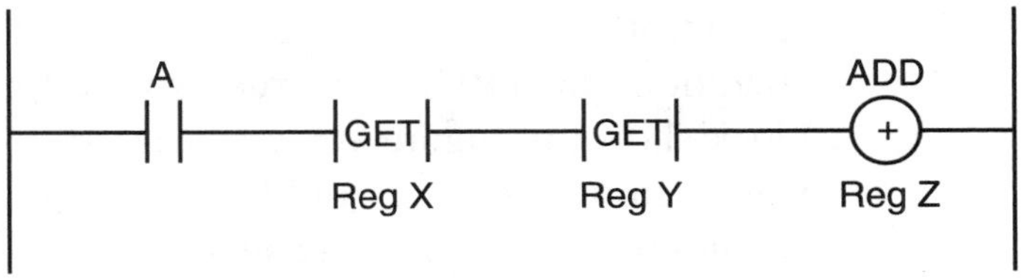

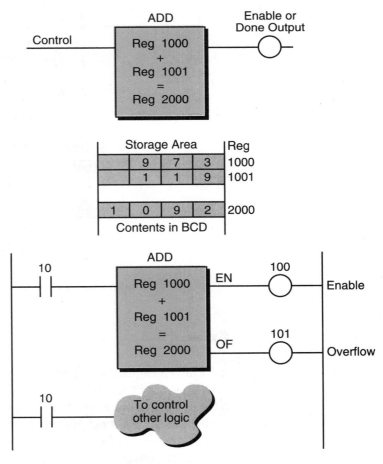

If A closes, the contents of register X and register Y are added and stored in register Z. If A does not close, no addition is performed. If contact A was omitted, the addition would be performed in every scan.

Figure 9-61. Ladder format addition.

ADDITION—BLOCK

An *addition* (ADD) functional block adds two values stored within the controller and places the sum in a specified register. The operand values can be fixed constants, values contained in I/O or holding registers, or variable numbers stored in any memory location. Figure 9-62 illustrates a typical addition functional block.

Figure 9-62. Addition functional block.

A control line enables the operation of an addition block. When the rung conditions are TRUE, the processor performs the addition function. In the block shown in Figure 9-62, register 1000 and register 1001 can be preset values, storage registers, or I/O registers. Each time an OFF-to-ON transition enables the control line, the instruction adds the values in these two registers and places the result in register 2000. The done, or enable, output coil indicates that the operation has been completed. This output remains ON as long as the control line is TRUE. An overflow of the addition operation energizes the overflow output of the block. If the operation overflows, some PLCs will clamp, or store, the results at the maximum value that the register can hold. Others will store the difference between the maximum count value and the actual overflow value.

Some controllers use double-precision addition when working in block format (see Figure 9-63). This operation is identical to simple ladder addition, but the PLC uses two registers each to hold the operands and two registers to store the result.

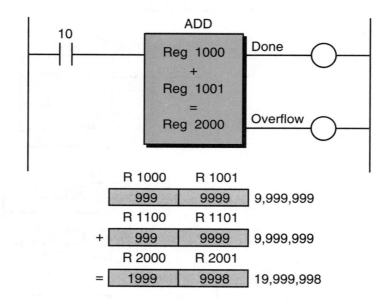

Figure 9-63. Double-precision addition block.

EXAMPLE 9-8

In Figure 9-64, two ingredients are added to a reactor tank for mixing. Analog input modules, which provide 12-bit information in BCD, send data about the two ingredients' flows to the PLC. The values are stored in registers 1000 and 1001. Implement instructions to keep track of the total amount of the combined ingredients, so that this information can be displayed on a monitor for the operator.

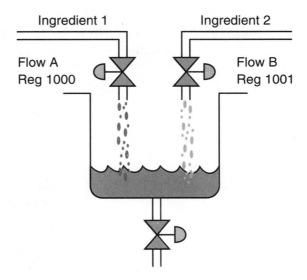

Figure 9-64. Flow of two ingredients into a reactor tank.

SOLUTION

One register can hold the total of both ingredients after the addition of the two ingredients' flows. Figure 9-65 shows the use of an ADD instruction to store the BCD result in register 2000. Note that this ADD instruction is always active.

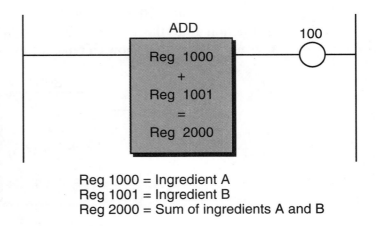

Reg 1000 = Ingredient A
Reg 1001 = Ingredient B
Reg 2000 = Sum of ingredients A and B

Figure 9-65. Solution to Example 9-8.

SUBTRACTION—LADDER

The *subtraction* (SUB) ladder instruction subtracts the values stored in two registers. As in an addition instruction, if the rung is enabled, the subtraction operation occurs. A GET data transfer instruction usually accesses the two registers used by a SUB instruction. The subtraction result register will usually have an underflow bit to represent a negative result. Figure 9-66 shows a rung with a SUB instruction.

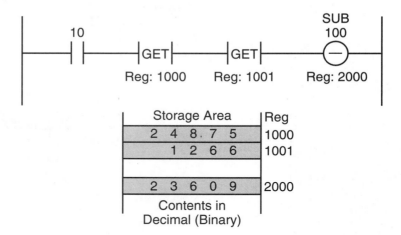

Storage Area	Reg
2 4 8 . 7 5	1000
1 2 6 6	1001
2 3 6 0 9	2000

Contents in
Decimal (Binary)

If contact 10 closes, the value in register 1001 is subtracted from the value in register 1000 (Reg 1000 – Reg 1001) and the result is stored in register 2000. If contact 10 does not close, no subtraction is performed.

Figure 9-66. Ladder format subtraction instruction.

SUBTRACTION—BLOCK

The *subtraction* (SUB) functional block, as in the ladder format subtraction instruction, finds the difference between two values and stores the result in a register. Figure 9-67 shows a typical subtraction functional block.

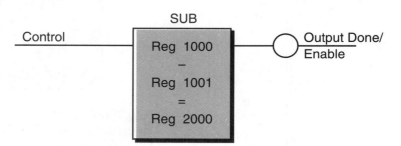

Figure 9-67. Subtraction functional block.

The control input in a subtraction block operates the same way as in an addition block. When the rung condition is logic 1, the controller performs the block operation. Three registers hold the data during the operation. The values that these registers can hold vary in format and may or may not include a sign bit. For example, referring to Figure 9-67, register 1000 could contain 9009 decimal and register 1001 could hold –10,020. The result of this operation would be +19,029 [9009 – (–10,020)], which would be stored in register 2000. Since the formats for subtraction vary, sometimes the result register may not include a sign bit. In this case, the controller will provide three outputs (see Figure 9-68): a positive result output (register 1000 greater than register 1001), an equal result output (register 1000 equal to register 1001), and a negative result output (register 1000 less than register 1001). The block will energize the output that corresponds to the result value. A three-output subtraction block essentially performs a comparison function.

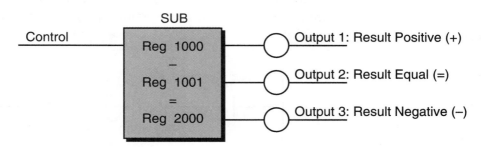

Figure 9-68. Subtraction block with sign outputs.

Some controllers allow a constant to be added to another register, through the block function, by placing an indicator, such as the letter K, in front of the number (e.g., K1035 = constant 1035). Controllers that do not provide I/O transfer instructions may use subtraction blocks to transfer analog or multibit I/O values to and from the I/O table. They do this by subtracting a constant of 0 from the input/output data and then storing the result in the target register.

Figure 9-69 illustrates an example of a SUB block instruction used to read an analog input and write an analog output. If contact 10 closes, the SUB operation is executed. Register 100 specifies the reference address of the input or output module (analog or multibit). During the reading of an input, a constant of 0 (register 1001) is subtracted from the input module's input value (register 100) and the result is stored in register 1000 for use by the control program. During the writing of an output, a constant of 0 (register 1001) is subtracted from the value in register 1000 and the result is sent to the output module (register 100).

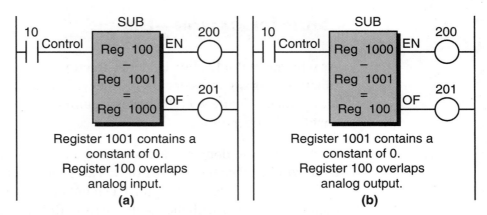

Figure 9-69. Subtraction block used to **(a)** read an analog input and **(b)** write an analog output.

MULTIPLICATION—LADDER

A *multiplication* (MUL) ladder instruction multiplies the values from two operand registers. It then uses two other registers to hold the result of the multiplication (see Figure 9-70). The reason why the result is held in two registers is that, normally, the product of two 4-digit numbers is an 8-digit number. Some controllers provide two adjacent registers in which to store the result.

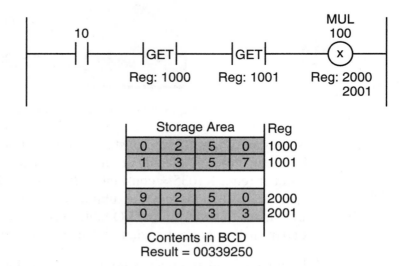

Figure 9-70. Ladder format multiplication instruction.

One or two output coils reference the two result registers in a multiplication instruction, depending on the PLC. GET instructions access the operand registers. If a condition must be present to enable the operation, it should be programmed before the multiplication rung accesses the two operands. In Figure 9-70, if contact 10 closes, the contents of registers 1000 and 1001 will be multiplied and stored in registers 2000 and 2001.

MULTIPLICATION—BLOCK

As with the multiplication ladder instruction, a *multiplication* (MUL) block function uses two registers to store the result and one register to hold each of the operands. Figure 9-71 illustrates a multiplication block, with a control line enabling its operation.

A PLC may use double-precision for a multiplication block, meaning that there will be twice the number of registers for the operands and result. This allows, for example, an 8-digit BCD number to be multiplied by another 8-digit BCD number with the result (up to 16 digits BCD) stored in four result registers. Other controllers use *scaling*, in which the result of the multiplica-

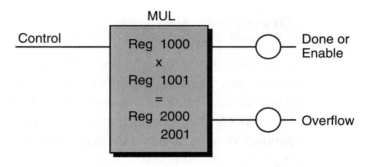

Figure 9-71. Multiplication functional block.

tion is held temporarily in two registers and then multiplied by a scale value (see Figure 9-72). For example, assume that a PLC has a 4-digit BCD format and that registers 1000 and 1001 contain the values 9001 and 8172, respectively, with a scaling value of –5 (or 10^{-5}). If the controller uses scaling, it will hold the result (73,556,172) in two result registers temporarily (as 7355 and 6172), and then multiply it by the scaling value (10^{-5}), resulting in 735.56172. Thus, the result register will contain the value 736 (rounded off). Knowing that the result has been scaled, the user can compute the actual result, which is 736×10^5 (73,600,000).

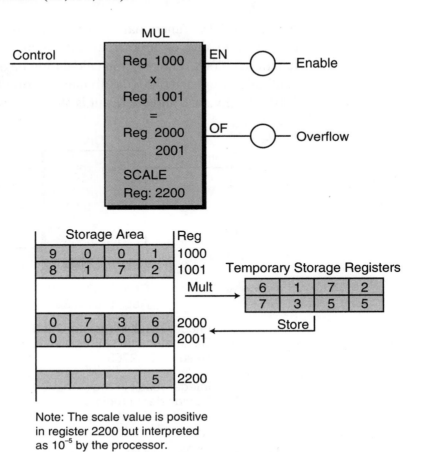

Note: The scale value is positive in register 2200 but interpreted as 10^{-5} by the processor.

Figure 9-72. Multiplication function block with scaling.

DIVISION—LADDER

A *division* (DIV) instruction finds the quotient of two numbers. This quotient is held in two result registers and referenced by the output coil. The first result register generally holds the integer part of the quotient, while the second result register holds the decimal fraction part. Both operands used in a division operation may be obtained through GET instructions. Figure 9-73 shows a rung using a division instruction.

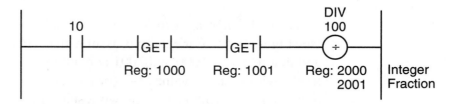

Figure 9-73. Ladder format division instruction.

DIVISION—BLOCK

A *division* (DIV) functional block finds the quotient of two numbers, storing the result in one or more registers. Figure 9-74 illustrates this type of functional block. The division calculation begins after the control rung has continuity. Register 1000 (the dividend) is divided by the contents of register 1001 (the divisor), and the result is stored in two contiguous destination

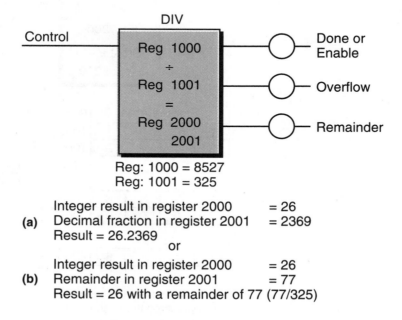

Reg: 1000 = 8527
Reg: 1001 = 325

(a) Integer result in register 2000 = 26
Decimal fraction in register 2001 = 2369
Result = 26.2369
or

(b) Integer result in register 2000 = 26
Remainder in register 2001 = 77
Result = 26 with a remainder of 77 (77/325)

Figure 9-74. Division functional block with the second result register storing **(a)** the decimal fraction and **(b)** the remainder.

registers. In this case, the destination registers are register 2000, which holds the integer part of the result, and register 2001, which holds the decimal fraction part. Depending on the controller, the second result register may hold the remainder instead of a decimal fraction. Some controllers also allow a scaling factor to be specified in a division block. This scaling factor permits fractional results, which would otherwise be lost, to be scaled and stored in a register.

Depending on the PLC used, a division block can have three possible outputs. When energized, the top output generally represents a successful division, the middle output represents an overflow or error (divide by zero), and the lower output indicates whether or not the result has a remainder.

SQUARE ROOT—BLOCK

A *square root* (SQR) block instruction generally has two or three registers—one that holds the value to be operated on and one or two other registers that hold the result of the square root operation. One of the result registers may hold the integer part of the result while the other holds the fractional part. The processor may also provide scaling. Once the control rung has continuity, the square root operation takes place. Of the possible block outputs, the first one represents a successful or valid operation, nd the second one indicates an overflow condition. Figure 9-75 illustrates a square root block instruction.

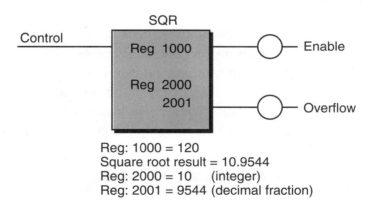

Reg: 1000 = 120
Square root result = 10.9544
Reg: 2000 = 10   (integer)
Reg: 2001 = 9544 (decimal fraction)

Figure 9-75. Square root functional block.

The square root instruction is useful in applications like the calculation of flow rate from a differential pressure (DP) orifice flow meter (see Figure 9-76). In this application, the flow rate (Q) is equal to a constant (K) times the square root of the differential pressure ($\Delta P_A = P_{out} - P_{in}$). The analog input value from the DP flow meter must have its square root value extracted and then multiplied by the constant. The resulting value will give the volume per unit time (ft³/min) of the flow. Chapter 13 further discusses differential pressure transducers and their measurements.

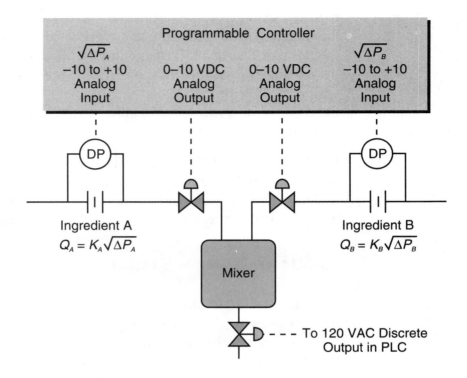

Figure 9-76. Square root instruction application in a DP flow meter.

9-11 DATA MANIPULATION INSTRUCTIONS

Data manipulation instructions are enhancements of the basic ladder diagram instruction set. Whereas relay-type instructions are limited to the control of internal and external outputs based on the status of specific bit addresses, data manipulation instructions allow multibit operations. Data manipulation instructions handle operations that take place within one, two, or more registers. Table 9-8 presents some data manipulation instructions.

DATA COMPARISONS

Data comparison (CMP) instructions, as the name implies, compare the values stored in two registers. These instructions are useful when checking for a proper range of values in the control or data entry section of the application program. In some controllers, data compariso instructions are expressed in the basic ladder format, while in other controllers, they are block instructions. In both formats, they provide three basic data comparisons: *compare equal to, compare greater than,* and *compare less than.* Based on the results of these comparisons, the processor can turn outputs ON or OFF and perform other operations.

Data Manipulation Instructions		
(Purpose: To provide multibit, multiregister operations in a PLC)		
Instruction	**Symbol**	**Function**
Data Comparison	CMP/LIM	Compares the values stored in two registers
Logic Matrix	AND/OR/NAND NOR/NOT/XOR	Performs logic operations on two or more registers
Data Conversion	ABS/COMPL INV/BIN-BCD	Changes the value stored in a register to another format
Set Constant Parameters	SET	Loads a register with a fixed value
Increment	INCR	Increases the contents of a register by one
Shift	SHIFT	Moves the bits in a register to the right or left
Rotate	ROT	Shifts register bits right/left and moves the shifted-out bit to the other end of the register
Examine Bit	XBON/XBOFF	Examines the status of a single bit in a memory location

Table 9-8. Data manipulation instructions.

Comparison instructions that use the basic ladder format operate in a manner similar to arithmetic instructions (see Figure 9-77). If the rung has continuity, the instruction performs a comparison; if the comparison is TRUE, the instruction passes continuity to the output coil. Typical comparison instruc-

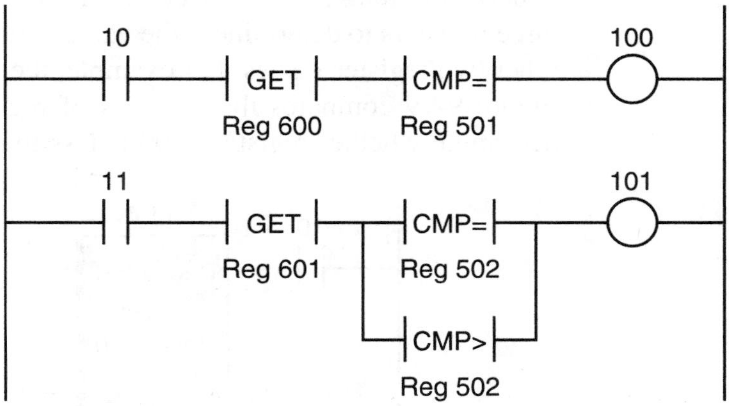

If contact 10 closes, the contents of register 600 are compared to the contents of register 501; if they are equal, coil 100 is turned ON. If contact 11 closes, the contents of register 601 are compared to the contents of register 502; if they are greater than or equal to register 502, output 101 is turned ON.

Figure 9-77. Ladder format comparisons.

335

tions are *greater than* (>), *less than* (<), and *equal to* (=), along with combinations of these such as *less than or equal to* (≤), *greater than or equal to* (≥), and *not equal to* (≠). A GET instruction accesses the first register to be compared to the comparison (CMP) register. Note that all ladder conditions are programmed before the GET and CMP instructions.

The compare functional block, shown in Figure 9-78a, compares the contents of two registers, register 2000 and register 2001, for a specific comparison, in this case, equal to. The block instruction energizes output coil 100 when the comparison occurs, and it energizes output coil 101 if the comparison has been satisfied. As shown in Figure 9-78b, some PLCs may also have one comparison block, which has several outputs, that performs multiple compare functions at the same time. This type of comparison block compares the data in the registers and then turns ON the output corresponding to the outcome of the comparison (i.e., less than, greater than, equal to).

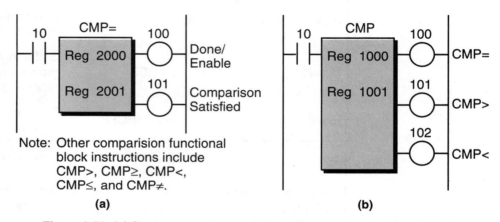

Figure 9-78. (a) Single-comparison and **(b)** multicomparison functional blocks.

Some controllers offer another comparison option that uses another register to perform a *limit* (LIM) instruction. This instruction compares the values in three registers to determine if the value in the middle register is between the other two register values. For example, the limit functional block shown in Figure 9-79 compares the contents of registers 1100, 1200, and 1300 to determine whether register 1200 is less than or equal to register 1100 and

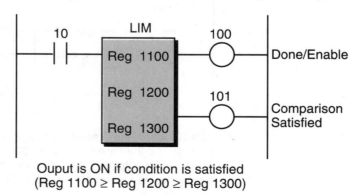

Ouput is ON if condition is satisfied
(Reg 1100 ≥ Reg 1200 ≥ Reg 1300)

Figure 9-79. Comparison block using a limit instruction.

whether register 1200 is greater than or equal to register 1300 (i.e., R1100 ≥ R1200 ≥ R1300). If the comparison is TRUE, the limit instruction energizes the comparison-satisfied output. The done/enable output is ON whenever the instruction is enabled.

Some controllers that do not have compare block capabilities can perform a comparison of two registers using a subtraction block (see Section 9-10). In this case, three output coils signal whether the result of the subtraction is positive (greater than), equal (equal to), or negative (less than).

EXAMPLE 9-9

Figure 9-80 shows a section of the program from Example 9-8 in which an ADD instruction was used to keep track of the two ingredients being poured into a reactor tank. The first two ladder rungs open the valves for ingredients A and B, allowing them to be poured into the tank once the Start Adding Ingredients command is ON (input 10). Implement an instruction block that ensures that the valves close when ingredient A reaches 500 gallons and ingredient B reaches 750 gallons.

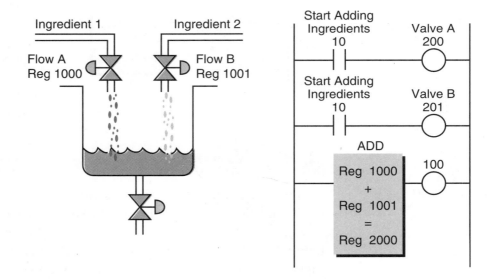

Figure 9-80. Mixing application and its corresponding ladder program.

SOLUTION

Figure 9-81 illustrates the use of two compare instructions that detect the target ingredient amounts. The outputs of these compare instructions are used to interlock and break continuity to each of the valve's circuits. Note that the values of the ingredient flows (the contents of registers 1000 and 1001) are compared with two constants (K). Also, note that the comparison made is greater than or equal to (≥) to avoid missing the compare (equal) because of a minuscule movement in the analog input reading.

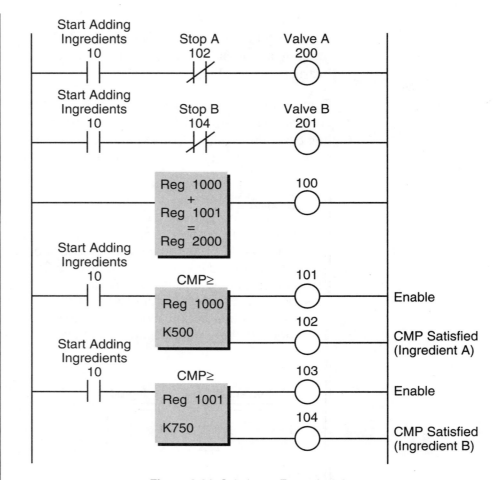

Figure 9-81. Solution to Example 9-9.

LOGIC MATRIX

A *logic matrix* functional block performs AND, OR, exclusive-OR, NAND, NOR, and NOT logic operations on two or more registers (see Chapter 3 for logic functions). A logic function performed between two registers can be thought of as a matrix operation of length one, since each operand has one register. Figure 9-82 shows a typical logic matrix function block. A logic matrix operation between two registers may be used to mask out certain bits of the source or original register and then pass only the status of those bits used in the mask to the result register (see Figure 9-83).

An enabled control input triggers the performance of a logic matrix function block. The block specifies the type of logic function to be performed, while the user specifies the registers inside the block. These are generally holding or storage registers. Referring to Figure 9-82, registers 1000 and 1100 hold the operand values, while register 2000 holds the result of the operation. The length of the operation indicates the number of words or registers adjacent to each of the register operands, providing data in matrix form.

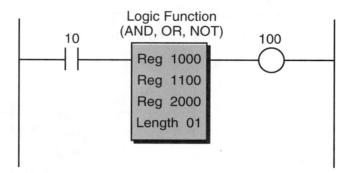

Figure 9-82. Logic matrix functional block.

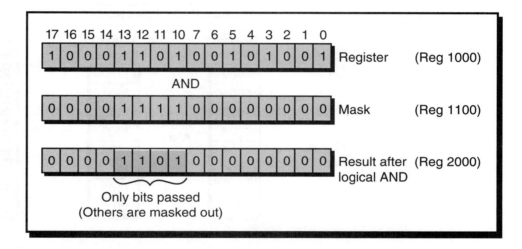

Figure 9-83. Logic matrix function block used to mask out bits.

During its operation, a logic matrix function block has three possible outputs. It energizes the top output when the control line is enabled, it energizes the middle output once the operation is done, and it energizes the lower output if an error occurs. As an example, let's examine the logic function block shown in Figure 9-84, which has a length of 8 and an AND logic function. When the control input enables the block, the logic function will AND the contents of registers 1000 through 1007 with the contents of registers 1100 through 1107, placing the result of the operation in registers 2000 through 2007. Each register typically holds 16 bits of data. So, in this case, the function block will AND the 128 bits in registers 1000–1007 with the 128 bits in registers 1100–1107 and store the result (128 bits) in another matrix (registers 2000 through 2007).

Some controllers have only two operand registers (e.g., R1000 and R1001). When these controllers perform a logic operation, they store the result in one of the operand registers, erasing the operand data previously stored in that register. A data transfer of that register's contents to another register(s) prior to executing the logic matrix block can prevent loss of operand data when using this type of block function.

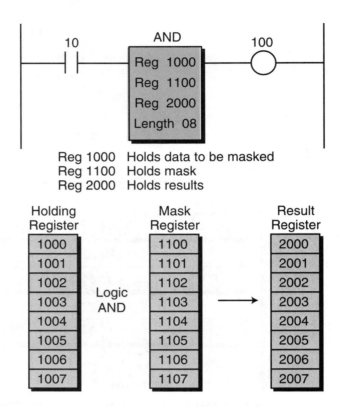

Reg 1000 Holds data to be masked
Reg 1100 Holds mask
Reg 2000 Holds results

Figure 9-84. Logic matrix function block example.

DATA CONVERSIONS

Data conversion instructions change the contents of a given register from one format to another. Typical data conversion instructions include BCD-to-binary, binary-to-BCD, absolute, complement, and inversion.

A *BCD-to-binary* (BCD-BIN) data conversion instruction (see Figure 9-85) converts BCD input data from field devices, such as thumbwheel switches, into binary format. This conversion allows the input data to be used in math operations. Conversely, a *binary-to-BCD* (BIN-BCD) instruction converts data from the PLC into BCD format, so that field devices that operate in BCD (e.g., seven-segment LED indicators) can use it (see Figure 9-86).

The operation of a data conversion block is basically the same regardless of whether it is performing a BCD-BIN or a BIN-BCD conversion. When the control input is enabled, the block converts the contents of the first register (either BCD or BIN) into binary or BCD, depending on the conversion instruction. It then places the result of the conversion in the second register and energizes the block output when the instruction is finished. Some PLCs allow multiple registers to be converted at the same time by specifying a length in the instruction (see Figure 9-87).

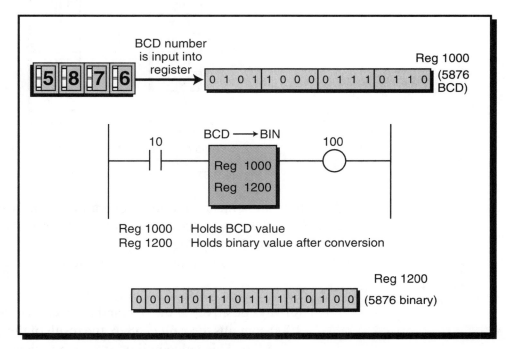

Figure 9-85. BCD-to-binary data conversion.

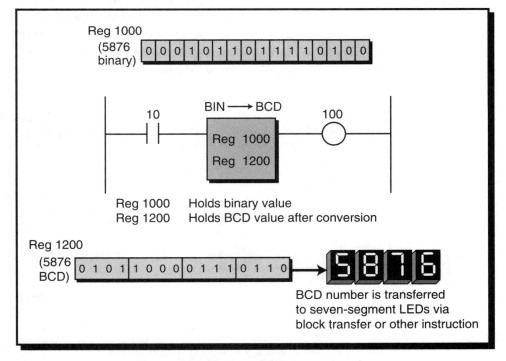

Figure 9-86. Binary-to-BCD data conversion.

Length = 8

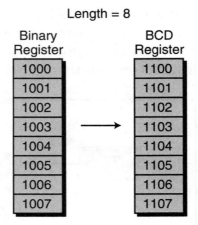

Binary Register → BCD Register

8 registers are converted after execution

Figure 9-87. Multiple-register binary-to-BCD conversion.

Absolute, complement, and invert operations usually occur within a single register. In other words, the operation stores the result in the register location that the operand occupied. Figure 9-88 shows a typical absolute/complement/invert block, which operates as follows:

- An *absolute* (ABS) functional block computes the absolute value (always positive) of the operand register's contents. Thus, if register 1000 contains the value –5876, the result of the block instruction will be +5876. This value will be stored in register 1000.

- A *complement* (COMPL) functional block changes the sign of the operand register's contents. For example, if register 1000 contains the value +5876, the result of the complement instruction will be –5876. Similarly, if register 1000 held the value –7654, the result of the complement would be +7654.

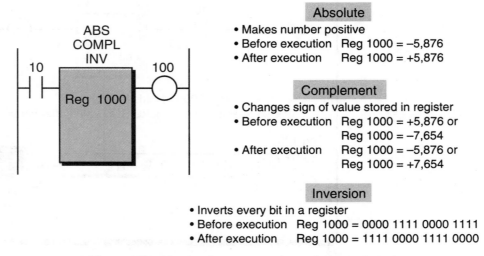

ABS
COMPL
INV

10   100

Reg 1000

Absolute
- Makes number positive
- Before execution   Reg 1000 = –5,876
- After execution   Reg 1000 = +5,876

Complement
- Changes sign of value stored in register
- Before execution   Reg 1000 = +5,876 or
                      Reg 1000 = –7,654
- After execution   Reg 1000 = –5,876 or
                      Reg 1000 = +7,654

Inversion
- Inverts every bit in a register
- Before execution   Reg 1000 = 0000 1111 0000 1111
- After execution   Reg 1000 = 1111 0000 1111 0000

Figure 9-88. Absolute/complement/invert functional block.

- An *invert* (INV) functional block inverts all of the bits in the operand register. If the binary number in register 1000 is 0000 1111 0000 1111, the number will be 1111 0000 1111 0000 after the instruction, and the block output will be ON when the instruction is finished.

SET CONSTANT PARAMETERS

Sometimes a constant value, which will be used later in the program for comparisons or set points must be stored in a register. For this reason, some PLCs provide a *set constant parameters* (SET) block instruction, which allows a fixed value to be assigned to a register. When a set constant parameters block is enabled (see Figure 9-89), it sets the referenced register (register 1000) equal to the value specified (in BCD, binary, etc.) and turns ON the output when the operation is completed. This instruction is very useful when resetting storage or I/O registers to zero during their initialization.

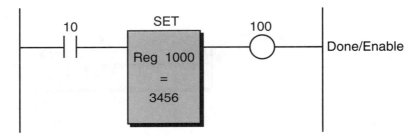

After Execution Reg 1000 = 3,456

Figure 9-89. Set constant parameters functional block.

INCREMENT

An *increment* (INCR) instruction (see Figure 9-90) increases the contents of a register by one. This instruction is useful, for example, when keeping track of events or the number of executions of a routine. An increment block may also be used with a counter that has a large preset count to keep track of how many times the maximum count has taken place.

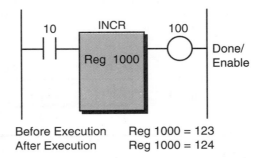

Before Execution   Reg 1000 = 123
After Execution   Reg 1000 = 124

Figure 9-90. Increment functional block.

SHIFT AND ROTATE

A *shift* (SHIFT) instruction moves the bits in a register to the right or to the left. Figure 9-91 illustrates the execution of a right-shift instruction. A left-shift instruction is identical, except that the bit is moved in the opposite direction (shifting out the most significant bit). Shift blocks use *bit-in* and *bit-out* variables to specify the location of the bit whose value will be shifted—a bit-in variable is the value to be added to a register, while a bit-out variable is the value to be deleted from a register. These bits can be real I/O locations that can be used to input or output data through the shift operation.

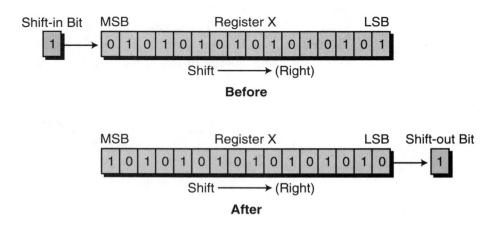

Figure 9-91. Right-shift execution.

A *rotate* (ROT) instruction, like a shift instruction, shifts data to the right or left; but instead of losing the shift-out bit, this bit becomes the shift-in bit at the other end of the register (rotated bit). Figure 9-92 illustrates the operation of a right-rotate instruction.

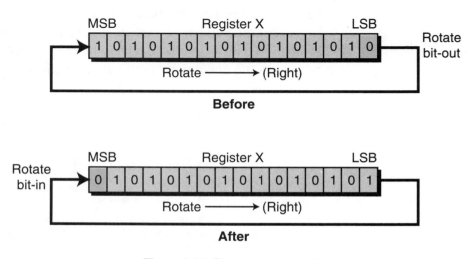

Figure 9-92. Right-rotate execution.

Figure 9-93 illustrates functional blocks for both the shift and rotate functions. The control input enables the blocks' operation. Some block instructions have right and left lines to determine the shift or rotate direction; others may indicate shift-right (SHFR) or shift-left (SHFL) and rotate-right (ROTR) or rotate-left (ROTL) in the instruction. Also, a shift/rotate block may have several variables available inside it, depending on the PLC model. In general,

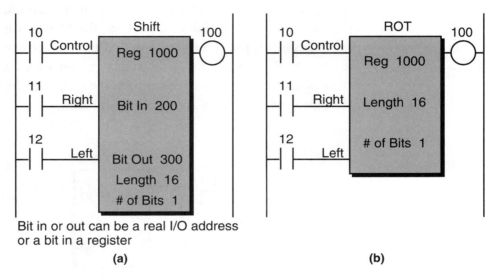

Bit in or out can be a real I/O address
or a bit in a register

(a)   **(b)**

Figure 9-93. (a) Shift and **(b)** rotate functional blocks.

the first register stores the data to be shifted or rotated. If a length is specified, the first register is the starting location. For example (see Figure 9-94), if the length is 3 in a right-shift instruction, then the block operation will encompass 48 bit shifts (16 bits/register×3 registers = 48 bits), starting from register 1000 and ending at register 1002. The number of bits indicates the number of bit shifts or bit rotates that take place simultaneously when the control input goes from OFF to ON. Shift and rotate instructions are very useful in applications where a PLC must track the status of inputs along a path of travel (e.g., overhead conveyors in a parts-painting process).

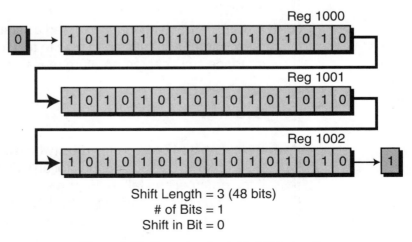

Shift Length = 3 (48 bits)
# of Bits = 1
Shift in Bit = 0

Figure 9-94. Example of a right-shift instruction.

EXAMINE BIT

An *examine bit* (XB) functional block examines the status (ON or OFF) of a single point, or bit, in a memory location. This type of instruction is used when "flags" are set during a PLC program and then later tested and compared. A flag is a bit that is specially marked for later examination. In an *examine bit ON* (XBON) instruction, the block examines the bit position specified in the register for an ON condition. It then energizes the output if the status of the bit is ON. Conversely, in an *examine bit OFF* (XBOFF) instruction, the instruction energizes the output if the specified bit is OFF. Figure 9-95 illustrates an examine bit block.

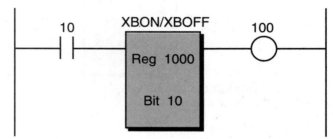

Examines bit 10 of register 1000 for an
ON (XBON) or an OFF (XBOFF) status.

Figure 9-95. Examine bit functional block.

EXAMPLE 9-10

A PLC application controls a batching process where the reading of a temperature input (Batch Temp) is critical to the process. The process's temperature transducer is connected to a four-channel, 0–10 VDC analog input module with a 12-bit resolution. The remaining four bits of each channel are used as status indicators for the module (see Figure 9-96). Illustrate how to test for a fault in this analog input interface's critical temperature measurement.

SOLUTION

By testing bit 17 of register 1000 (which is the destination of the critical temperature reading channel) for an OFF condition, we can determine if the channel has failed. Figure 9-97 shows how an XBOFF instruction accomplishes this test. If bit 17 is OFF, a fault has occurred; if it is ON, the channel is OK. The instruction that drives the logic of this

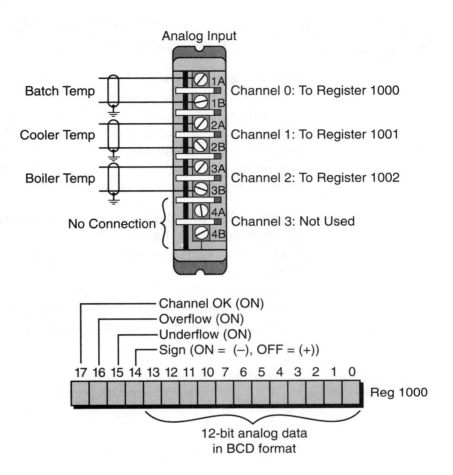

Figure 9-96. Analog input interface for a batching application.

instruction is a contact that is closed by the program when the reading of the analog signal takes place. For the instruction to be operational at all times, even when no reading is taking place, the instruction block enable line must be programmed directly to the left power rail without any contact instructions.

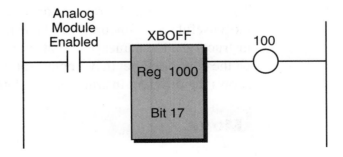

Figure 9-97. A fault has occurred if register 1000 bit 17 is 0 (OFF).

9-12 DATA TRANSFER INSTRUCTIONS

Data transfer instructions move, or transfer, numerical data within a PLC, either in single register units or in blocks (a group of registers). These instructions can move data to or from any location in the memory data table, with the exception of user-restricted areas. Typical uses of data transfer instructions are the movement of constant and/or preset values to counters and timers, the reading of analog inputs and multibit input modules, and the transferring of data to output modules.

As with other instructions, data transfer instructions may be in either ladder or functional block format, although block format is most common. The ladder format functions used to transfer data are the *get* (GET) and *put* (PUT) instructions (see Figure 9-98), which are generally used with PLCs that provide basic ladder format implementation of arithmetic and data comparison instructions. A GET data transfer instruction accesses data from a certain register, whereas a PUT instruction stores data in a specified register.

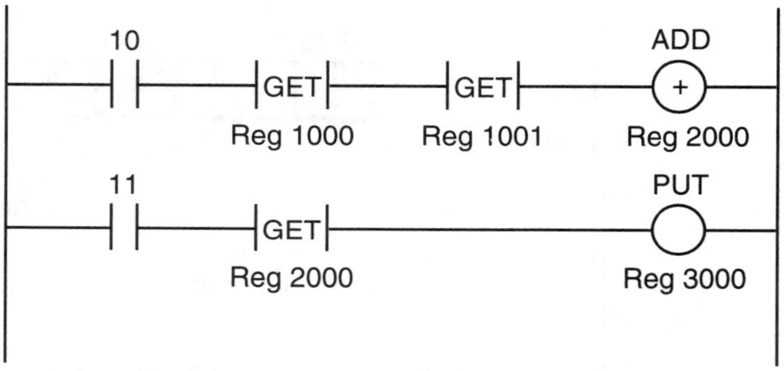

If contact 10 closes, the contents of registers 1000 and 1001 are added and stored in register 2000. If contact 11 closes, the contents of register 2000 are transferred (stored) into register 3000. The contents of register 2000 are not altered.

Figure 9-98. GET and PUT instructions used in the ladder format.

The functional block group of data transfer instructions forms perhaps the most useful set of functions available in enhanced PLCs, after the basic relay instructions. The names of the data transfer instructions may differ depending on the controller, yet they implement the same transfer functions. Table 9-9 shows the different instructions available with data transfer operations.

MOVE

A *move* (MOV) functional block instruction transfers information from one location to another, with the destination location being a single bit or register. Figure 9-99 shows *move bit* (MOVB) and *move register* (MOVR) functional blocks. Some PLCs also offer *move byte* instructions.

Data Transfer Instructions		
(Purpose: To move numerical data within a PLC)		
Instruction	**Symbol**	**Function**
Move	MOV/MOVB/ MOVR/MOVM	Transfers information from one location to another
Move Block	MOVBK	Moves data from a group of register locations to another location
Table Move	REG-TABLE/ TABLE-REG	Transfers data from a block or table to a register
Block Transfer— In/Out	BKXFER	Stores a block of data in specified memory or register locations
ASCII Transfer	ASCII XFER	Transmits ASCII data between a peripheral device and a PLC
First In–First Out Transfer	FIFO	Constructs a table or queue for storing data
Sort	SORT	Sorts the data in a block of registers in ascending/descending order

Table 9-9. Data transfer instructions.

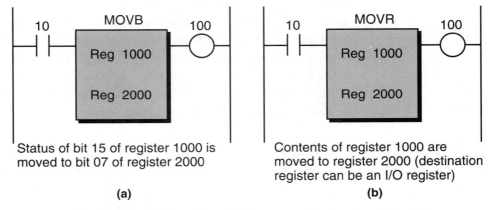

Status of bit 15 of register 1000 is moved to bit 07 of register 2000

Contents of register 1000 are moved to register 2000 (destination register can be an I/O register)

(a)   (b)

Figure 9-99. (a) Move bit and **(b)** move register functional blocks.

Some PLCs perform a move function to special word table locations. In this case, the PLC automatically coverts the copied data to the proper numerical format for the destination location. For example, a register or word might contain a BCD value that, when transferred to another register or word, is stored as a binary value, thus executing a BCD-to-binary conversion within the move instruction.

Another type of move instruction, a *move mask* (MOVM) instruction, masks certain bits within the register. Figure 9-100 illustrates this type of move block. The move mask block transfers the data in register 1000 to register 1100, with the exception of the bits specified by a 0 in the mask register 2000.

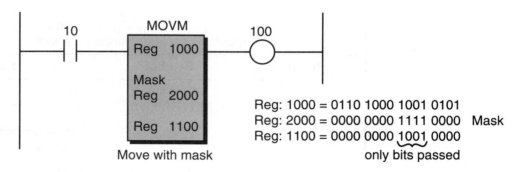

Figure 9-100. Move mask functional block.

Yet another move instruction found in some controllers is the *move status* instruction. This block function transfers system or I/O module status data to a storage/result register. This information can then be masked, compared, or examined to determine the status of major or minor faults in the system or an I/O module. With this information, the controller can take corrective action through the control program, if necessary.

MOVE BLOCK

A *move block* (MOVBK) instruction copies a group of register or word locations from one place to another. The length of the block is generally user-specified. Figure 9-101 illustrates a move block instruction. When energized, the control input triggers the execution of this block. The block function then transfers data from locations 1000 through 1023 (length = 20) to locations 2000 through 2023, respectively. The data in registers 1000 through 1023 is left unchanged. In some PLCs, the user can specify how many locations can be transferred during one scan (rate per scan).

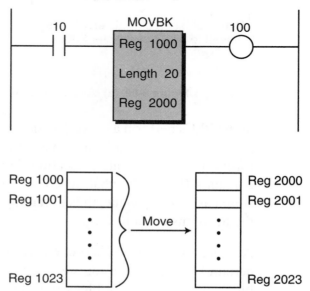

Figure 9-101. Move block functional block.

TABLE MOVE

A *table move* instruction transfers data from a block or table to a register or word in memory. There are two types of table move instructions: *table-to-register* (TABLE-REG) and *register-to-table* (REG-TABLE). The main characteristic of a table move block is the manipulation of a pointer register, which specifies the particular table location in which the register or word value will be stored. Figure 9-102 shows a table move block.

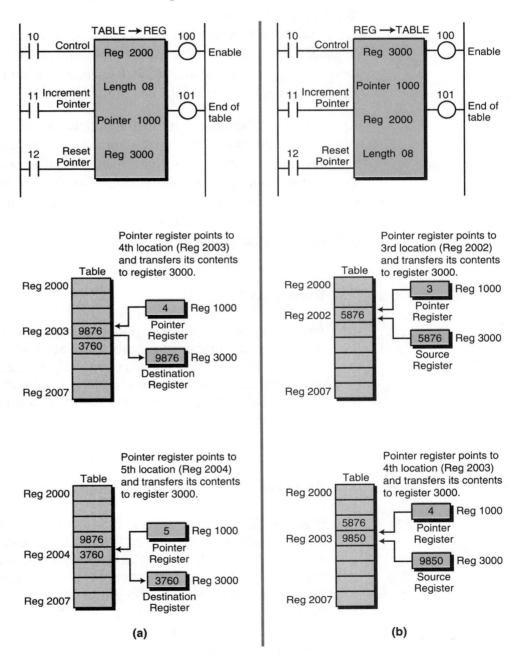

Figure 9-102. (a) Table-to-register and **(b)** register-to-table functional block.

The transition of the control input from OFF to ON enables a table move instruction, which then increments the contents of the pointer register every time the middle input, the increment (INCR) pointer, transitions from OFF to ON. The bottom input of the table move block resets the pointer to zero (initialize to top of table). If data must be stored to or retrieved from a specific table location, the pointer register can be loaded with the appropriate value, which points to the specified location. A set parameter or move register instruction loads this information prior to the table move.

Referencing Figure 9-102, the length specifies the number of word locations in the table to be moved (8 in this example), beginning at the starting location (register 2000). After the table move block transfers the data from these eight locations, it energizes the top output. It energizes the middle output when the pointer register has reached the end of the table.

Applications of the table move instruction include the loading of new data into a table, the storage of input information (e.g., analog) from special modules, and the input of error information from a controlled process. It is also useful when changing preset parameters in timers and counters and when simultaneously driving a group of 16 outputs through I/O registers. A table move instruction is also used when looking up values in a table for comparison, linear interpolation, etc.

EXAMPLE 9-11

A batching system operates during an eight-hour shift, where several batch sizes are processed at the rate of approximately one batch per hour. Implement instructions to store the batch information, including the batch size in gallons and the time of day when the batch was finished. Register 1000 holds the value of the total batch, while register 1500 holds the time of day (in hours and minutes) in BCD format (HHMM).

SOLUTION

Figure 9-103 illustrates a register-to-table instruction that will transfer the outputs of registers 1000 and 1500, using the same pointer register to store the information to two tables simultaneously. This ensures that the pointer points to a batch amount that corresponds to the time of the batch (see Figure 9-104). The Batch Done signal, perhaps coming from the opening of the discharge valve, triggers the register-to-table instruction. Once the storing of the register into the table has taken place, the instruction's enable/done output increments the pointer. The pointer is incremented in only one of the blocks to avoid a double increment. The increment occurs after both register-to-table instructions have been executed to ensure that the data is stored by the same pointer counter. Note that the Batch Done signal is a transitional contact, so the register-to-table instruction only transfers the register data once to its appropriate table location.

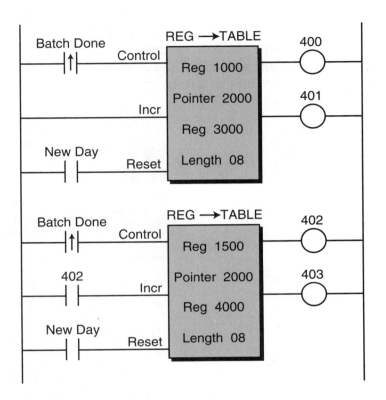

Figure 9-103. Register-to-table instruction used for storing batch information.

Table 8-Hour Shift			Table Time of Day		
Table 3000			Table 4000		
1	323	R3000	1	0714	R4000
2	401	R3001	2	0823	R4001
3	378	R3002	3	0914	R4002
4	303	R3003	4	1017	R4003
5	350	R3004	5	1130	R4004
6	400	R3005	6	1224	R4005
7	320	R3006	7	1330	R4006
8	318	R3007	8	1422	R4007
Batch (in gallons)			Time (in hr: min)		

Figure 9-104. Table 3000 stores batch sizes and table 4000 stores the time of day the batches were completed.

BLOCK TRANSFER—IN/OUT

Some PLCs provide *block transfer* (BXFER) instructions, which are primarily used with special I/O modules to transfer blocks of data. The two basic types of block transfers are *block transfer in* and *block transfer out*. Figure 9-105 shows a block transfer in/out instruction. The module address location of the transfer data may be explicitly marked as the rack and slot location of the interface. For example, the block transfer input in Figure 9-106 shows that the contents to be read from the intelligent module (rack 01, slot 03, 8 channels) will be stored in registers 1000 through 1007.

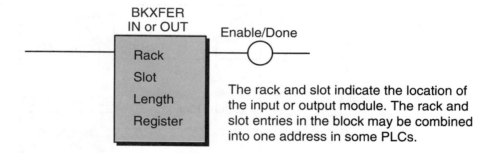

Figure 9-105. Block transfer in/out functional block.

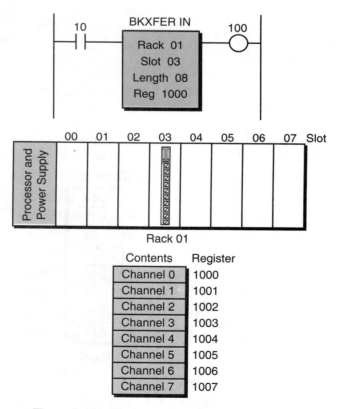

Figure 9-106. Block transfer in instruction.

The control input, when enabled, executes a block transfer instruction. During a block transfer in function, the instruction stores data about the I/O module in memory locations or registers starting at the specified register location. The block length specifies how many locations are needed to store the I/O module data. For example, the data from an analog input module with four input channels can be read all at once, if the length is specified as four. A block transfer out instruction operates in a similar manner, with the address of the output module determining the destination of the data transfer. The top output of the block transfer instruction, when energized, signals the completion of the transfer operation.

ASCII TRANSFERS

An *ASCII transfer* (ASCII XFER) instruction transmits ASCII-formatted data between a PLC and a peripheral device. This functional block, which is under program control, operates in conjunction with an ASCII communications module. The communication of ASCII data usually occurs in two ways: reading data from a peripheral or writing data to a peripheral. This functional block is widely used in applications that require report generation. Figure 9-107 illustrates a typical read/write ASCII functional block.

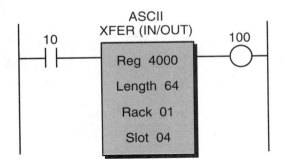

A total of 64 ASCII characters are sent (XFER OUT) or received (XFER IN). Each register location holds 2 characters (one character per byte).

Figure 9-107. ASCII transfer in/out functional block.

The control input activates an ASCII transfer (in or out) instruction. When reading data, the instruction allows the special I/O module to perform a read function. The processor then reads the data from the module and stores it in special memory locations (from the first register to the last, as specified by the length). The I/O address in the block indicates the location of the module. When writing data, the processor transfers information from the location where it is stored to the address where the module is located.

Some ASCII transfer instructions use a pointer register to access specific characters in the table (e.g., to decode a specific input character from the data table). Other ASCII instructions allow the user to specify how many bytes or characters are transmitted during a scan. The speed of transmission (baud rate) is a function of the scan time, which depends on the number of ASCII devices that are active at one time. An ASCII transfer instruction assumes that proper baud rates, start/stop bits, and parity have been established in the I/O module hardware.

FIFO STACK TRANSFERS

A *first-in–first-out* (FIFO) instruction constructs a table or queue where data is stored. The basic function of this operation is similar to a shift register instruction, in which one word (16 bits) is shifted within the stack each time the instruction is executed. The data is always shifted in the order in which it was received—the first word shifted in will be the first word shifted out. FIFO is, in essence, a first-come–first-served format. Figure 9-108 shows a typical FIFO block instruction.

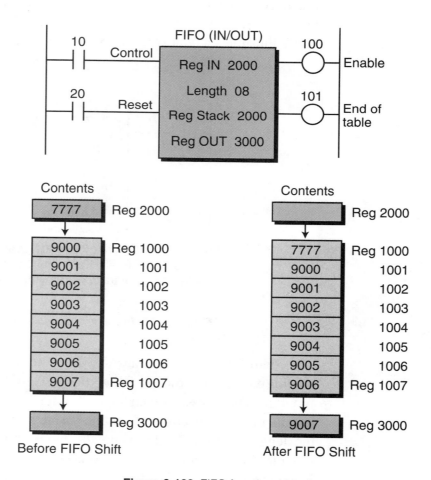

Figure 9-108. FIFO functional block.

A FIFO operation consists of two parts: a *FIFO input* (FIFO IN) instruction and a *FIFO output* (FIFO OUT) instruction. The FIFO IN instruction loads the queue, while the FIFO OUT instruction unloads it. FIFO instructions are useful for storing, and later retrieving, large groups of temporary data as it is received. A typical application of a FIFO instruction is the storage and retrieval of data that is synchronized to the external movement of parts on a conveyor or transfer machine.

An OFF-to-ON transition of the control input logic initiates a FIFO block. Some blocks may have a reset signal to reset the FIFO stack (clear stack). For a FIFO instruction, the *register in* holds the data that will be transferred to the queue. This data is placed in the FIFO stack when the control input is ON. The data in the last position of the stack is output through the *register out*. The FIFO length specifies the length of the stack.

The FIFO instruction is very useful when trying to keep the values obtained from a process in a "moving window." For example, Figure 9-109 illustrates a temperature profile as a function of time. If the desired window is from t_0 to t_1, the values can be kept in a FIFO stack. Thus, the stack will always contain the last t_0–t_1 values.

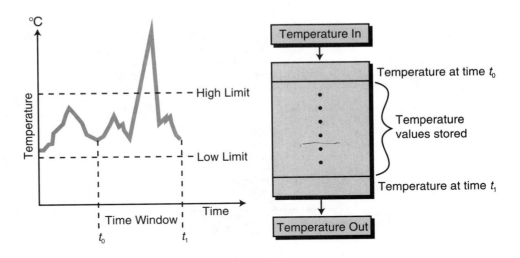

Figure 9-109. Temperature profile.

SORT

The *sort* (SORT) block function sorts a block of registers, in ascending or descending order, according to their contents. Figure 9-110 shows a sort block in which the closing of contact 10 enables the instruction. This block sorts registers 1000 through 1017 in ascending order and then stores the results in registers 2000 through 2017. This type of function is very useful when computing the median of sample readings—an operation that requires the sample to be in numerical order. PLCs provide either ascending, descending, or both types of sorting.

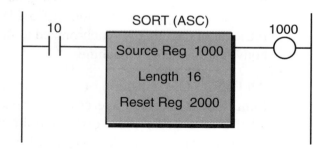

Figure 9-110. Ascending sort functional block.

9-13 SPECIAL FUNCTION INSTRUCTIONS

As the name implies, **special function instructions** perform operations that do not fall under any other PLC instruction categories. These functions are usually available in medium to large controllers. Table 9-10 lists the common special function instructions.

Special Function Instructions		
(Purpose: To allow specialized operations in a PLC)		
Instruction	**Symbol**	**Function**
Sequencer	SEQ	Outputs data in a time-driven or event-driven manner
Diagnostic	DIAG	Compares actual input data with reference data
Proportional-Integral-Derivative	PID	Provides closed-loop control of a process

Table 9-10. Special function instructions.

SEQUENCERS

A *sequencer* (SEQ) block is a powerful instruction that simulates a drum timer. A sequencer is analogous to a music box mechanism, in which each peg produces a tone as the cylinder rotates and strikes the resonators. In a sequencer, each peg (bit) can be interpreted as a logic 1 and no peg as a logic 0.

A sequencer table, which is similar to a spread-out music box cylinder, provides sequencer information. Figure 9-111 illustrates a cylinder and sequencer table comparison. The number of bits in a sequencer can vary from

8 to 64 or more. The width of the table may also vary, as may the size of a cylinder. Through I/O registers, which map real output points, each step in a sequencer table can become an output representing one of the pegs.

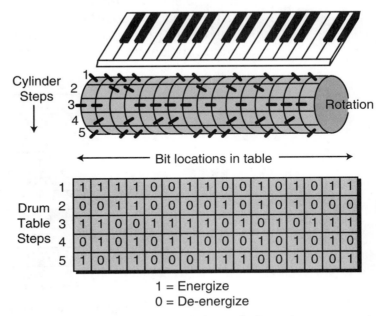

Figure 9-111. Comparison of a music box cylinder and a sequencer table.

Figure 9-112 shows a typical sequencer functional block. An OFF-to-ON transition of the control input initiates this block, causing the contents of the sequencer table to be output in a sequential manner. The pointer register points to each step being output (i.e., the table register location). Every time the control input is energized, the pointer register is automatically incremented, thus pointing to the next table location. Depending on the PLC, either an event or time may drive the control input line; therefore, sequencers may be either event driven or time driven. A reset pointer input can reset the pointer register to zero (point to step 1), if needed. The sequence length and width specify how many steps and bits are in the table, respectively. Whenever the sequencer instruction is enabled, it energizes the block's first output. The second output indicates the end of the sequencer table.

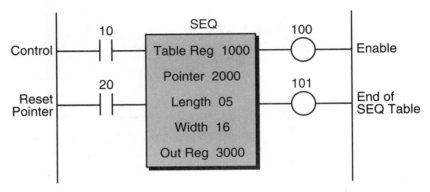

Figure 9-112. Sequencer functional block.

EXAMPLE 9-12

A PLC application calls for the implementation of ten different steps that take place in a sequential manner. For the purpose of detecting a fault in a troubleshooting condition, the process step code, as shown in Figure 9-113, should be revealed in a seven-segment display to the operator. Implement an instruction block that will satisfy this application.

Step	Code
1	1023
2	4576
3	4588
4	5101
5	5130
6	5417
7	5418
8	7809
9	7810
10	7900

Figure 9-113. Process step code.

SOLUTION

Figure 9-114 shows a way to display the code number using a 16-bit register output connected to a four-digit, seven-segment display. A sequencer instruction transfers the codes from the sequencer table to the output register. The output register matches (i.e., is mapped to) the location of the 16-bit output interface (i.e., rack 0, slot 7, corresponding to word 07). Every time the Start of Process Step signal goes from OFF to ON, the sequencer will output the process code to the indicator.

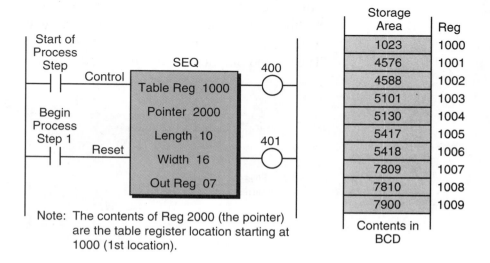

Note: The contents of Reg 2000 (the pointer) are the table register location starting at 1000 (1st location).

Figure 9-114a. Sequencer instruction block.

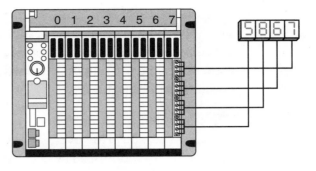

Figure 9-114b. Seven-segment display.

DIAGNOSTICS

A *diagnostics* (DIAG) block instruction compares two memory blocks, one containing actual input conditions and the other containing a reference condition. The instruction compares these blocks bit by bit to determine if they are identical. If a miscomparison occurs, the instruction stores the bit number and the state of the bit in a holding register. Diagnostic instructions are useful for signaling machine malfunctions.

Figure 9-115 illustrates a diagnostic function block. An energized control input initiates the block function. The block then compares the contents of the first register locations (1000 through 1007) with the contents of the second

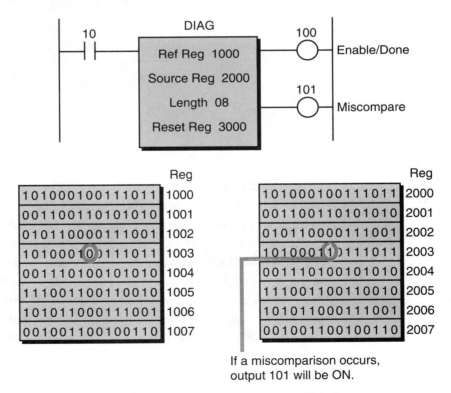

Figure 9-115. Diagnostics functional block.

reference register locations (2000 through 2007). If it finds a difference, it stores this information in the result register (register 3000) without altering the contents of the other register locations. When the instruction is finished, it energizes the top output. The instruction energizes the second output if a miscomparison occurs.

The controlled machine generally determines the reference conditions for inputs and outputs. However, some controllers allow reference conditions to be "taught" to the PLC. These controllers gather reference teaching conditions using sequencer input, block transfer in, and other instructions, depending on the model used.

PID

PLCs that are capable of performing analog control using the PID algorithm use *proportional-integral-derivative* (PID) functional blocks. The user specifies certain parameters associated with the algorithm to control the process correctly. Figure 9-116 illustrates a typical PID block.

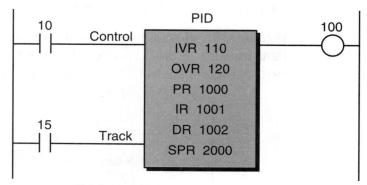

Register 110 maps analog input module
Register 120 maps analog output module

Figure 9-116. PID functional block.

An energized control input enables a PID block's automatic operation. The bottom input track, when energized, determines whether the PID variables are being tracked but not output. If the block is not enabled (i.e., in manual mode), the controller can still track the variables when the track line is enabled. The user specifies the input variable register (IVR) and the output variable register (OVR), which are associated with the locations of the analog modules (input and output). The proportional register (PR), integral register (IR), and derivative register (DR) hold the gain values that must be specified for the three parts of the control process. The set point register (SPR) holds the target value for the process set point. Depending on the controller, the user can specify other block variables, such as dead times, high and low limits, and rate of update. The top output of the PID block indicates an active loop control, while the middle and bottom outputs indicate low- and high-limit alarms, respectively. Some PLC manufacturers provide a fill-in-the-blanks screen (see Figure 9-117) during the programming of a PID instruction, so that the user can input the different parameters.

Figure 9-117. Fill-in-the-blanks screen.

Some controllers provide PID capabilities without the PID block instruction. In this case, the controller uses a special PID module that contains all of the input/output parameters. An output instruction, such as block transfer out or move data, transfers the set point and gain parameters to the module during initialization of the program. The control program can alter this module data if any parameter changes are required. Chapter 15, which explains process controllers and loop tuning, provides more information about PID control.

9-14 NETWORK COMMUNICATION INSTRUCTIONS

Local area networks (LANs) provide communication channels between independent computers (referred to as *nodes*) located in a small radius. Because they connect different computers, LANs have created a need for instructions that communicate and exchange information between the PLCs in a network. Therefore, PLC manufacturers now offer **network communication instructions**, which transfer information like contact status, output coil status, and register status between PLCs. These network instructions are often specific to the manufacturer's family of PLCs.

Table 9-11 describes typical instructions used in a PLC network environment. These instructions are very easy to implement; however, the programmer must enforce compliance with the PLC network's rules. Also, the programmer should assign registers and organize the program to avoid confusion on the network.

Network Communication Instructions		
(Purpose: To allow communication through a local area network)		
Instruction	**Symbol**	**Function**
Network Output	—○— NET	Passes one-bit status information from a PLC to a network
Network Contact	—| |— NET	Captures status information from a network output
Network Send	NET SEND	Sends register information to a network
Network Receive	NET RCV	Captures available register data in a network
Send Node	SEND NODE	Sends register data to a specific node in a network
Get Node	GET NODE	Retrieves register data from a specific node in a network

Table 9-11. Network communication instructions.

Once a PLC executes a network communication instruction and updates it at the EOS, the processor passes the information to the network hardware (modules or internal boards) for processing and transmission. The format of the instruction may differ, depending on the controller—some controllers use data transfer instructions to access the network, while others use specific instructions. Therefore, the instructions presented here are guidelines to illustrate implementation.

The organization of a network depends on how it is configured. In some controllers, the network interface is built into the main CPU, while in others, it is in an interface module. Regardless of format, both network interfaces perform the same function—network communications. If a PLC's network interface is installed in the I/O racks, the manufacturer may provide one of a number of ways to set up that particular PLC for the network. Some PLCs may configure the network during the configuration stage, when the network module slot location is specified. Other controllers may automatically recognize where the network interface is located. Yet in other PLCs, a network software instruction, similar to a block transfer in or block transfer out instruction, specifies the network module's slot location.

The output coils and contacts in a network may be referred to as *network outputs* and *network contacts,* while the registers in a network may be called *network registers*. Network outputs are internal outputs that are often located in a special area of the data table, along with the network registers. These network elements may be part of an internal storage area with additional LAN capabilities. For example (see Figure 9-118), if a PLC has 512 possible internal outputs, 64 of them may be used as network outputs; likewise, if it has 128 storage registers, 32 of them may be used as network registers. These network-mapped addresses, if used, will be sent automatically if the network is active. Chapter 18 explains local area network operation and configuration more extensively.

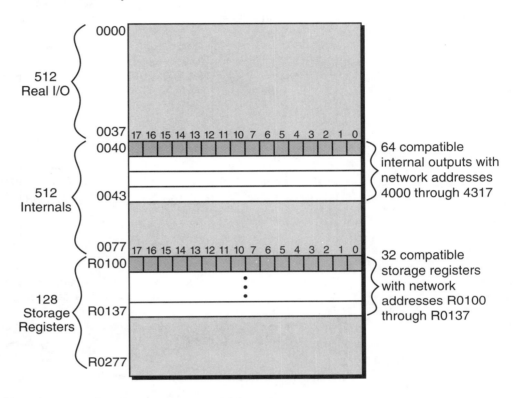

Figure 9-118. Mapping of network-compatible addresses with all numbers in octal.

Now, let's explore the operational function of some network instructions. In this discussion, we will assume that the programmable controller specifies the slot location of the network interface during the total system configuration. If this was not the case, then the PLC would require a slot entry specification for each instruction.

NETWORK OUTPUT

A *network output* instruction, shown in Figure 9-119, is used in conjunction with a network contact to pass one-bit status information from a PLC to the network. If continuity exists in the logic path of the network output, the network output instruction will turn ON its corresponding reference address. It will then send the information about the status of the reference address to the network interface for LAN transmission. Depending on the controller, the reference address must be a valid network coil. After transmission, the status of the output is available to all network stations or nodes (PLCs).

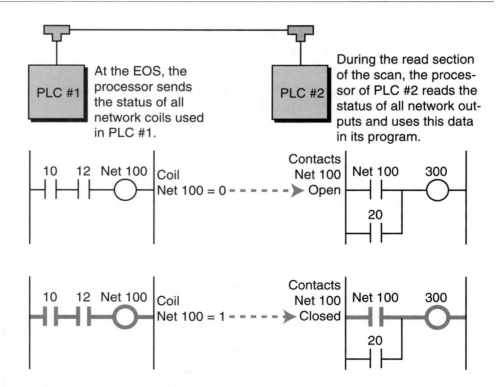

Figure 9-119. Operation of a network output coil and a network contact instructions. Note that contact 20 in PLC #2 is a local contact.

NETWORK CONTACT

A *network contact* instruction captures the status information from a network output. The reference address of the network contact must be the same as that of an active network output; otherwise, the contact (examine ON or examine OFF) will never be evaluated. The reference must also be a valid reference address, which may differ among PLC manufacturers.

Figure 9-119 illustrated the operation of a network contact instruction used in conjunction with a network output instruction. In this instruction, the processor obtains information from the network as it reads the inputs, during which it reads the status buffer of the network module as though it were a small data table. If the referenced network output address is logic 1, the controller will perform an evaluation and open or close the referenced contacts to provide or remove continuity. This evaluation depends on how the network contact is programmed (normally open or normally closed).

NETWORK SEND

A *network send* (NET SEND) instruction sends register information to a local area network. The activation of this functional block is the same as for other blocks—if the rung is TRUE, then the instruction is performed, sending the contents of the register to the network line. The instruction may provide two outputs to indicate that the operation has been performed and that no error was detected (output 1 and output 2, respectively).

Figure 9-120 illustrates a typical network send instruction block. If the specified length is more than one, the network may receive several transmitted registers; the registers to be transmitted will start at the first register and end at last register (first + length). A network send instruction generally operates in conjunction with a network receive instruction.

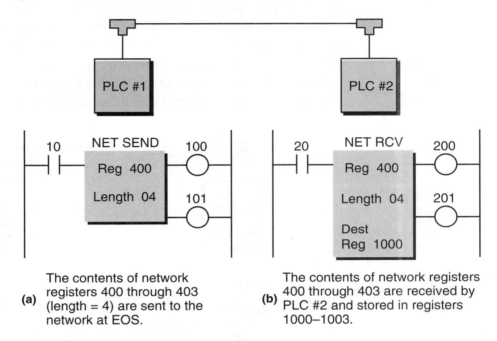

(a) The contents of network registers 400 through 403 (length = 4) are sent to the network at EOS.

(b) The contents of network registers 400 through 403 are received by PLC #2 and stored in registers 1000–1003.

Figure 9-120. (a) Network send and **(b)** network receive instructions.

NETWORK RECEIVE

A *network receive* (NET RCV) block function captures the available registers in the network's lines and stores their information in the receiving PLC's data table (register area). The user must make sure that the register information requested (i.e., register address numbers) matches the addresses used by the NET SEND instructions. For instance, if a NET SEND instruction uses network registers 400 to 403 (length of 4), the PLC that will retrieve those network registers must reference the same network registers in its NET RCV instructions.

Figure 9-120 illustrated the use of a network receive instruction. Once a network instruction captures the register information, it stores the data in the destination register(s), as specified by the length of the block. Of the two outputs available, the first one represents the completion of the operation, while the second one indicates if an error has occurred.

SEND NODE

A *send node* (SEND NODE) instruction operates in a more direct way than a network send function. This instruction transmits register information to specific PLCs (nodes) connected to the network. Essentially, a send node function implements a *copy to* function, where several registers from the sending node are written to another node.

Figure 9-121 illustrates a send node instruction. Continuity in the instruction's control line enables the block, which sends the contents of the starting register through the last register to the specified node. The block stores the information from the starting register through the last one in destination registers. The completion of the instruction turns ON the first output, while a network error condition energizes the second output.

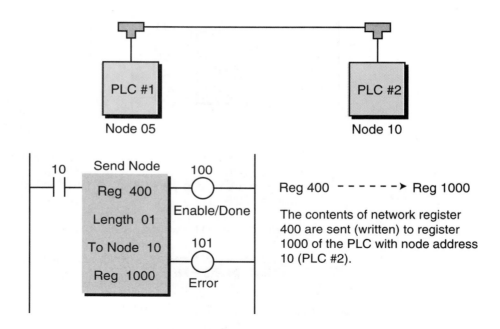

Figure 9-121. Send node functional block operation.

GET NODE

A *get node* (GET NODE) instruction retrieves register information from another PLC node. This instruction essentially copies the register from the requested node to the requesting node.

Figure 9-122 illustrates the use of a get node function. When the block is enabled, it requests the contents of the specified registers in the target node and stores the data in the destination registers of the PLC executing the get node instruction. The first output is energized when the instruction is completed; the second output is energized if a communication error occurs during network transmission.

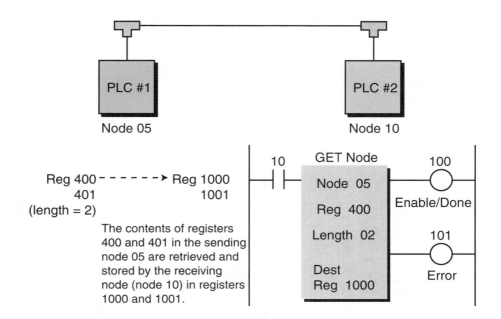

Figure 9-122. Get node functional block operation.

9-15 BOOLEAN MNEMONICS

As discussed in Section 9-2, Boolean mnemonics is a PLC language based primarily on the Boolean operators AND, OR, and NOT. A complete Boolean instruction set consists of the Boolean operators and other mnemonic instructions, which implement all of the functions of the basic ladder diagram instruction set. A mnemonic instruction is written in an abbreviated form, using three or four letters that imply the operation of the instruction. Table 9-12 lists a typical set of Boolean instructions and their equivalent ladder diagram symbols. The Boolean language is used to enter logic into a PLC's memory. However, a PLC may display the entered Boolean information as a ladder diagram on the programming terminal.

Enhanced Boolean output operators, which perform additional control functions, are a result of further enhancements to the Boolean instruction set. Figure 9-123 shows a short Boolean program and its equivalent ladder diagram representation. Chapter 3 discusses the principles of Boolean algebra, which are applied in the Boolean language. The next chapter illustrates other forms of Boolean programming utilizing the IEC 1131-3 instruction list language.

Mnemonic	Function	Description	Ladder Equivalent
LD/STR	Load/start	Starts a logic sequence with a NO contact	
LR/STR NOT	Load/start not	Starts a logic sequence with a NC contact	
AND	And point	Makes a NO contact series connection	
AND NOT	And not point	Makes a NC contact series connection	
OR	Or point	Makes a NO contact parallel connection	
OR NOT	Or not point	Makes a NC contact parallel connection	
OUT	Energize coil	Terminates a sequence with an output coil	
OUT NOT	De-energize coil	Terminates a sequence with a NOT output coil	
OUT CR	Energize internal coil	Terminates a sequence with an internal output	
OUT L	Latch output coil	Terminates a sequence with a latch output	
OUT U	Unlatch output coil	Terminates a sequence with an unlatch output	
TMR	Timer	Terminates a sequence with a timer	
CTU	Up counter	Terminates a sequence with an up counter	
ADD	Addition	Terminates a sequence with an addition function	
SUB	Subtraction	Terminates a sequence with a subtraction function	
MUL	Multiplication	Terminates a sequence with a multiplication function	
DIV	Division	Terminates a sequence with a division function	
CMP	Compare (=, <, >)	Terminates a sequence with a compare function	
JMP	Jump	Terminates a sequence with a jump function	
MCR	Master control relay	Terminates a sequence with an MCR output	
END	End MCR, jump, or program	Terminates a sequence with an end of control flow function	
ENT	Enter value for register	Used to enter preset values of registers	Not required

Table 9-12. Boolean instructions and their ladder diagram equivalents.

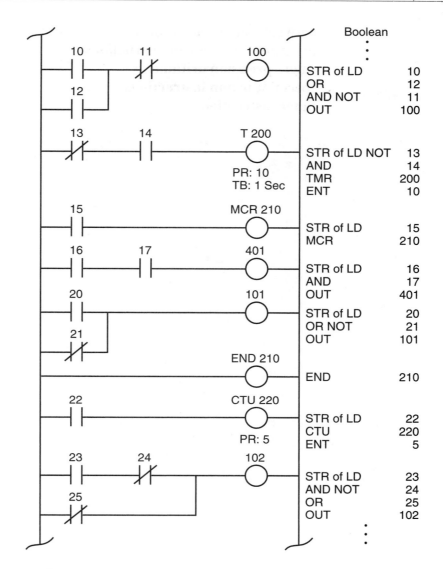

Figure 9-123. Boolean program and its ladder diagram representation.

Key Terms

arithmetic instructions
Boolean language
coil
contact
counter instructions
data manipulation instructions
data transfer instructions
double-precision arithmetic
enhanced ladder language
Grafcet
ladder language
ladder relay instructions
ladder rung matrix

network communications instructions
program/flow control instructions
single-precision arithmetic
special function instructions
timer instructions

CHAPTER TEN

THE IEC 1131 STANDARD AND PROGRAMMING LANGUAGE

Thought is the blossom; language the opening bud; action the fruit behind it.

—Henry Ward Beecher

 **CHAPTER HIGHLIGHTS**

As we have discussed in the previous chapters, programming a PLC can be a difficult task due to increased interlocking requirements in the control program as it becomes larger and more complicated. Additionally, each PLC manufacturer offers a different set of instructions within its PLC family. Many of these instruction sets are not applicable to other PLCs, and there is no easy way to translate an already written PLC program to another brand of PLC's programming format.

In this chapter, we will introduce you to the IEC 1131 standard, which attempts to simplify and standardize PLC programming. We will explain the languages used with the IEC standard, as well as discuss how these languages are implemented in the control program using sequential function charts to ease interlocking. Moreover, we will explain how some manufacturers use the IEC standard to implement a PLC-like environment without a programmable controller.

10-1 INTRODUCTION TO THE IEC 1131

The International Electrotechnical Commission (IEC) SC65B-WG7 committee developed the **IEC 1131 standard** in an effort to standardize programmable controllers. One of the committee's objectives was to create a common set of PLC instructions that could be used in all PLCs. Although the IEC 1131 standard reached the status of international standard in August 1992, the effort to create a global PLC standard has been a very difficult task to accomplish due to the diversity of PLC manufacturers and the problem of program incompatibility among PLC brands. However, the inroads that have been made so far have had a tremendous impact on the way PLCs will be programmed in the future.

The IEC 1131 standard for programmable controllers consists of five parts:

- general information
- equipment and test requirements
- programming languages
- user guidelines
- messaging services (communications)

Although there are five parts in the IEC 1131 standard, the third part—programming languages—provides all of the information about instructions and programming standards. The other four sections describe the different guidelines to be used for the testing and communication of language instructions, as well as the methodology that must be employed by the programmable controller user.

The IEC 1131 programming language standard is referred to as the **IEC 1131-3 programming standard**, since part 3 of the standard deals with programming languages—hence the dash three (-3). In this chapter, we will refer to the actual programming language as the IEC 1131-3 standard and to the overall standard as the IEC 1131.

LANGUAGES AND INSTRUCTIONS

The IEC 1131-3 standard defines two graphical languages and two text-based languages for use in PLC programming. The graphical languages use symbols to program control instructions, while the text-based languages use character strings to program instructions.

Graphical languages

- ladder diagrams (LD)

- function block diagram (FBD)

Text-based languages

- instruction list (IL)

- structured text (ST)

Additionally, the IEC 1131-3 standard includes an object-oriented programming framework called **sequential function charts (SFCs)**. SFC is sometimes categorized as an IEC 1131-3 language, but it is actually an organizational structure that coordinates the standard's four true programming languages (i.e., LD, FBD, IL, and ST). The SFC structure is much like a flowchart-type of programming framework, utilizing different languages for different control tasks and also routing control program actions. The SFC structure has its roots in the early French standard of Grafcet (IEC 848).

The IEC 1131-3 standard is a graphic/object-oriented block programming method, which increases the programming and troubleshooting flexibility of its programmable controllers. It allows sections of a program to be individually grouped as tasks, which can then be easily interlocked with the rest of the program. Thus, a complete IEC 1131-3 program may be formed by many small task programs represented inside SFC graphic blocks. The combination of languages available in the IEC 1131-3 standard also enhances PLC programming and troubleshooting by providing not only a better programming language, but also a better method for implementing control solutions.

The IEC 1131-3 uses a wide variety of standard data functions and function blocks, which operate on a large number of data variable types. Table 10-1 shows some examples of these data types and functions, as well as some typical function blocks. *Data variable type* refers to the kind of data received by the controller (e.g., binary, real numbers, time data, etc.), while *data functions* are the operations performed on the data (e.g., comparison, invert, addition, etc.). *Function blocks* are sets of data function instructions that work on blocks of data. Moreover, *variable scope* refers to the extent that a variable can be used in an application. For example, global variables can be used by any program in an application, while local variables can only be used by one particular program. Note that, in addition to the standard types of variables, functions, and blocks, the IEC 1131-3 allows for other types of vendor- and user-defined PLC programming elements. Thus, the IEC 1131-3 does not specify a set number of programming features, but rather establishes the groundwork for standard and additional functions.

Variables	Description
Data variable types	• Bit-based strings (Boolean or bit, byte, word) • Integers (signed and unsigned) • Real • Time (time, date—e.g., Time_Of_Day) • ASCII character strings • Vendor- and user-defined (single and arrays)
Data functions	• Bit-based (Boolean: AND, OR, NOT, etc.) • Numerical/arithmetic (ADD, SUB, MUL, DIV, SQR, LOG, LN, SIN, COS, TAN, etc.) • Data function conversions • Select functions (LIMIT, MAX, MIN, etc.) • Comparisons (>, <, =, >=, =<, <>) • ASCII string functions (LENgth, LEFT, RIGHT, INSERT, REPLACE, DELETE, etc.) • Vendor- and user-defined functions
Function blocks	• Set/reset—bistable—latch/unlatch • Edge trigger detection ($\neq,\uparrow,\downarrow$) • Counters (up, down, up/down) • Timers (TON, TOF) • Vendor- and user-defined blocks
Variable scope	• Global • Local

Table 10-1. Data variable types, functions, and blocks.

The IEC 1131 standard's data type and function flexibility allows programmable controller manufacturers to provide instructions they consider necessary, but that are not defined within the standard. Such instructions may include specific application instructions, such as a servo positioning instruction used with a particular vendor's intelligent servo control module. While this instruction may fall within the programmability parameters of the standard, it may not be available in other PLCs that comply with the standard. Thus, the IEC 1131 standard lets vendors enhance their IEC 1131-3 instruction sets by adding more powerful, customized instructions. It also allows users to create their own instructions, in block form, to perform a specific task.

DECLARING VARIABLES

During the implementation of a control system, the user must name, or *declare*, the variables used. This variable declaration is nothing more than the mapping of I/O addresses, indicating which field devices are wired to which I/O modules (see Chapter 5). Figure 10-1a shows a limit switch (LS1) implemented in a standard programmable controller environment. In this configuration, the device is declared (or named) in the control program as its address—10. In an IEC 1131-3 environment, however, a device can be described by any alphanumeric name. This name can include underscores (_). Hence, the limit switch can be declared as a variable named Limit_Switch_1, Clamp_Limit_Switch, or another appropriate name (see Figure 10-1b). From the moment a variable is declared, it will be known by that name throughout the control program, regardless of the IEC 1131-3 programming language used. The name assigned to a variable is not case sensitive; that is, it can be declared in uppercase, lowercase, or a combination of the two. Therefore, the user may choose the appropriate name representations for the purposes of program appearance (e.g., the use of uppercase for a main variable name and lowercase for a secondary variable name).

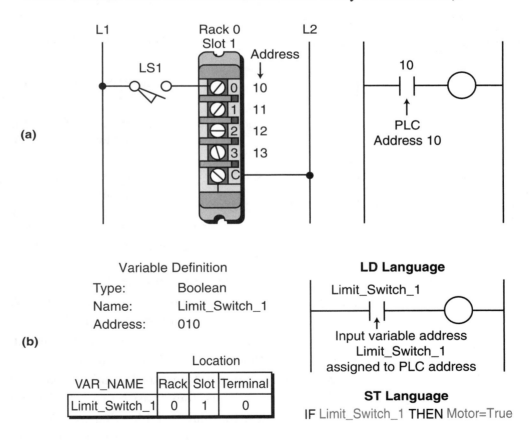

Figure 10-1. Limit switch addressed in **(a)** a standard PLC environment and **(b)** an IEC 1131-3 environment.

When declaring a variable, the user must specify the variable type in addition to the variable name. This allows the PLC to know what type of data the device corresponding to the variable transmits. The IEC 1131-3 supports many different local and global data variable types (see Table 10-2); however, the three most common are:

- Boolean
- integer
- real

Class	Variable Type	Description	Example
Discrete	Boolean	TRUE (1)/ FALSE (0)	Limit switch, motor, push button
Analog	Integer	−128 or +3764	Timer value, TWS input, integer calculation
	Real (floating point)	−34.573 or +1.35 × 10³	Analog I/O, general computation
ASCII string	Message strings	"Temperature":= Temp_Value_Var	Display information on a monitor or printer
System	Internal	Control relays (Bool)	Internal relay coils, timer outputs
		Integer and real variables	Internal arithmetic computations
	Input	Variables connected to input interfaces	Discrete/analog inputs, TWS inputs, analog interfaces
	Output	Variables connected to output interfaces	Discrete/analog outputs, outputs to LED displays, analog interfaces

Table 10-2. Data variable types.

Boolean variables are single-bit variables, meaning that the data transmitted and received is in the form of 1s and 0s. Discrete I/O variables fall under this category; therefore, they must be specified as "Bool" (short for Boolean) variables in the control program. Many nondiscrete variables, such as analog input signals that are read through an analog input card, are **integer variables**, because they transmit data in the form of whole numbers (e.g., 2042, −127, etc.). Thus, they must be specified in the control program as integer variables. Internal variables that transmit fractional and floating-point data (i.e., a number multiplied by an exponential expression—2.7×10^2) are **real variables** and, again, must be classified as such.

EXAMPLE 10-1

Implement the Boolean variable declaration (variable names and variable types) for the input devices shown in Figure 10-2a for use in a control program. Assume that the controller being used follows the rack-slot-terminal address configuration (e.g., rack 0, slot 0, terminal 3 is address 003). Figure 10-2b shows the wiring to the input module.

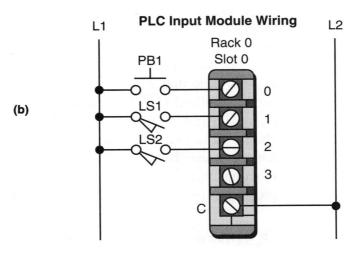

Hardwired Circuit

(a)

Description of Inputs

PB1: Used to manually start conveyor sequence
LS1: Detects parts in automatic start
LS2: Detects a no-jam condition

(b)

PLC Input Module Wiring

Figure 10-2. (a) A traditional hardwired circuit and **(b)** its wiring diagram.

SOLUTION

Figure 10-3 shows a sample variable declaration for this example. All of the input devices are discrete; therefore, they are specified as Boolean variables. PB1 is named MAN_START_PB, LS1 is named AUTO_PART_Detect, and LS2 is named NO_JAM_Detect.

Note that these variable names, which can be chosen by the user, describe the operational functions of the input devices.

Input Variable Name	Variable Type	Address Location		
		Rack	Slot	Terminal
MAN_START_PB	Bool	0	0	0
AUTO_PART_Detect	Bool	0	0	1
NO_JAM_Detect	Bool	0	0	2

Figure 10-3. Boolean variable declaration.

10-2 IEC 1131-3 PROGRAMMING LANGUAGES

While the IEC 1131-3 programming standard provides great new potential for programmable controller users, it is actually based on the relay ladder logic that has been inherent in PLCs since their inception. The IEC 1131-3 is based on the ladder logic used in PLC ladder diagrams (including functional blocks) because of its simplicity of use, representation, and to some extent, programmability. The IEC 1131-3, however, reduces the need for complex interlocking circuits within PLC ladder diagram circuits. It enhances the languages previously used in programmable controllers and incorporates them with a powerful framework—sequential function charts—making interlocking, interpretation of the control program, and implementation of the control system much easier for both the programmer and the final user of the system. With this in mind, let's briefly discuss the four languages that are used with the IEC 1131-3 standard—ladder diagrams, function block diagrams, instruction list, and structured text—along with sequential function charts. Note that, when programming in the IEC 1131-3, any of these languages may be used either alone or as a group, with or without sequential function charts. In Section 10-4, we will list all available IEC 1131-3 programming instructions.

LADDER DIAGRAMS (LD)

Ladder diagram language (LD) uses a standardized set of ladder programming symbols to implement control functions. This type of programming language is essentially the one that has always been available in PLCs (see Figure 10-4). Users familiar with current PLC ladder diagrams can use the same programming techniques and methods when using this language in an IEC 1131-3 environment. However, as we will explain later, interlocking ladder diagram programming is much easier to implement in the IEC 1131-3 format due to the use of sequential function charts.

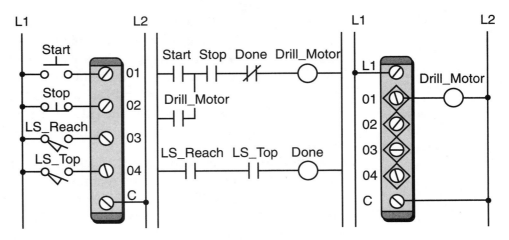

Figure 10-4. Ladder diagram representation of a PLC program.

FUNCTION BLOCK DIAGRAM (FBD)

Function block diagram (FBD) is a graphical language that allows the user to program elements (e.g., PLC function blocks) in such a way that they appear to be wired together like electrical circuits. Figure 10-5 illustrates this type of function block diagram configuration. Some IEC 1131-3 systems use logic symbols to represent the function blocks. Note that the output logic of the block in Figure 10-5 does not incorporate an output coil because the output is represented by the variable assigned to the output of the block. This

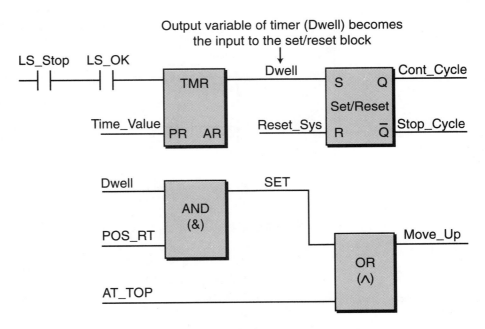

Section of a control program using a timer,
set/reset, AND, and OR function blocks

Figure 10-5. Function block diagram language.

variable can be used throughout the program in other instructions and as a control output through the address mapping performed during variable declaration. The user may still choose to use an output coil representation if desired; however, it will only be allowed in the last (right-most) block. The FBD language uses both standard and vendor-specified function blocks. The block functions typically used with the IEC 1131 standard include, for the most part, the block functions discussed in Chapter 9.

In addition to standard and vendor-specified functions, the IEC 1131-3 allows users to "build" their own function blocks according to control program requirements. This is referred to as *encapsulating* a block function. The advantage of creating user-defined blocks is that they can be built using other function blocks, instruction list, or structured text programming with or without ladder diagram instructions. This allows great flexibility in function block programming. Encapsulation also lets the user store a newly created block in a library and use it as many times as needed in the program, just like any other function block. Example 10-2 illustrates how ladder diagrams can be used to create a custom function block.

EXAMPLE 10-2

Illustrate how the hardwired start/stop circuit shown in Figure 10-6 can be implemented using ladder diagrams in a custom-built function block to turn ON motor M1 and pilot light PL1.

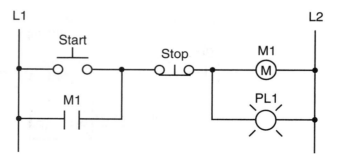

Figure 10-6. Start/stop circuit.

SOLUTION

Figure 10-7 illustrates the ladder diagram equivalent of the hardwired start/stop circuit. Note that there are two rungs for the two outputs and that both the input and output variables are specified with the same names that they had in the hardwired circuit.

To implement this simple ladder diagram as a function block, it must be programmed or stored in an encapsulated block (see Figure 10–8a). The final function block will look like the diagram shown in Figure

10-8b. Note that the inputs to the start/stop block will act according to the logic used to program the block. If the driving logic to the start input is ON, then the motor and light will turn ON. If the stop input is ON, then both the motor and light outputs will be OFF. The two input variables (the START and STOP commands), as well as the two output variables (the MOTOR and PILOT_LIGHT signals), are Boolean variables.

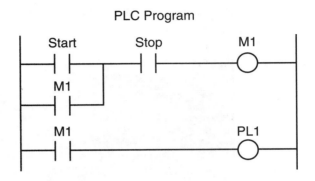

Figure 10-7. Ladder diagram equivalent of the circuit in Figure 10-6.

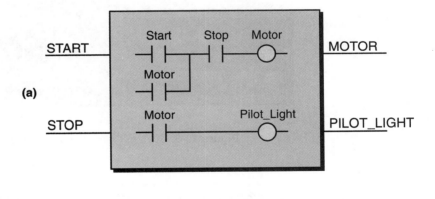

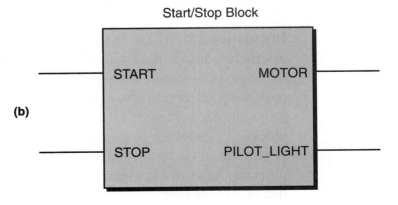

Figure 10-8. (a) Encapsulated ladder diagram and **(b)** start/stop block function.

The flexibility of custom block creation is enhanced by the fact that the user can build custom blocks using ladder diagrams or any of the other IEC 1131-3 languages (IL and ST). Also, custom blocks can be used in conjunction with other standard or vendor-specified function blocks. This allows the programmer to create very powerful function blocks that can be integrated into any ladder diagram or function block diagram. Figure 10-9 shows a custom block instruction that was created in a B&R Industrial Automation PLC using their instruction list language.

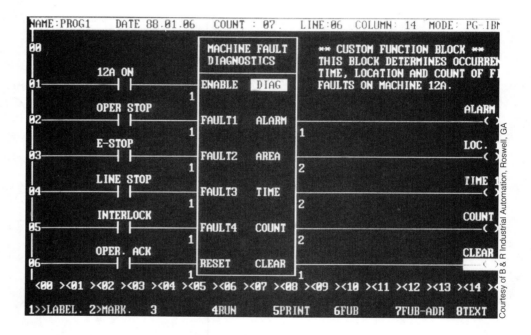

Figure 10-9. Custom function block from B & R Industrial Automation.

INSTRUCTION LIST (IL)

Instruction list (IL) is a low-level language similar to the machine or assembly language used with microprocessors (see Figure 10-10). This type of language is useful for small applications, as well as applications that require speed optimization of the program or a specific routine in the program. As mentioned earlier, IL can be used to create custom function blocks. A typical application of IL might involve the initialization to zero (i.e., reset) of the accumulated value registers for all the timers in a control program. As shown in Figure 10-11, a programmer could use IL to create a function block that would load the contents of all the timers' accumulated registers (AR) with a value of zero.

Instructions		Comments
LD	b1	(*current result:=TRUE*)
AND	b2	(*current result:=b1 AND b2*)
ANDN	b3	(*current result:=b1 AND b2 AND NOT b3*)
ST	b0	(*b0:=current result*)

Note: The current result is held in a result register. The last instruction stores the result register as the variable b0.

Figure 10-10. Example of the machine/assembly language used in microprocessors.

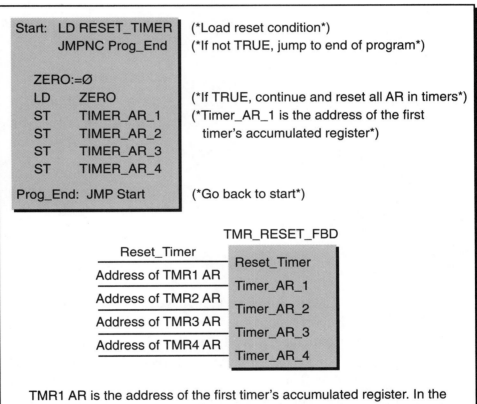

```
Start:  LD RESET_TIMER        (*Load reset condition*)
        JMPNC Prog_End        (*If not TRUE, jump to end of program*)

        ZERO:=Ø
        LD      ZERO          (*If TRUE, continue and reset all AR in timers*)
        ST      TIMER_AR_1    (*Timer_AR_1 is the address of the first
        ST      TIMER_AR_2       timer's accumulated register*)
        ST      TIMER_AR_3
        ST      TIMER_AR_4

Prog_End: JMP Start           (*Go back to start*)
```

TMR_RESET_FBD

Reset_Timer		Reset_Timer
Address of TMR1 AR		Timer_AR_1
Address of TMR2 AR		Timer_AR_2
Address of TMR3 AR		Timer_AR_3
Address of TMR4 AR		Timer_AR_4

TMR1 AR is the address of the first timer's accumulated register. In the FBD, this address is known as Timer_AR_1 so that the IL program can interpret it. The result of the IL program will be that the values in the specified accumulated registers will be reset to 0. The variable Reset_Timer will trigger the block and start the IL instruction. The IL routine will cycle back to start while the block is enabled by the Reset_Timer variable being ON. There are also ways to "pulse" just once through the program so that the instruction is executed only one time, if enabled.

Figure 10-11. Instruction list custom function block that resets timer values to zero.

STRUCTURED TEXT (ST)

Structured text (ST) is a high-level language that allows structured programming, meaning that many complex tasks can be broken down into smaller ones. ST resembles a BASIC- or PASCAL-type computer language (see Figure 10-12), which uses subroutines to perform different parts of the control function and passes parameters and values between the different sections of the program. Like LD, FBD, and IL, the structured text language utilizes variable definitions to identify input and output field devices and any other internally created variables that are used in the program. ST also supports iterations, such as WHILE...DO and REPEAT...UNTIL, as well as other conditional executions, such as IF...THEN...ELSE. Moreover, structured text language supports Boolean operations (AND, OR, etc.) and a variety of specific data, such as time of day information.

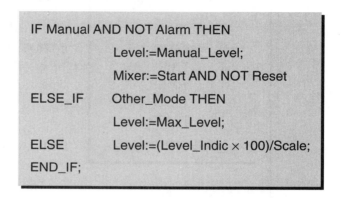

```
IF Manual AND NOT Alarm THEN
            Level:=Manual_Level;
            Mixer:=Start AND NOT Reset
ELSE_IF    Other_Mode THEN
            Level:=Max_Level;
ELSE       Level:=(Level_Indic × 100)/Scale;
END_IF;
```

Figure 10-12. Example of a BASIC-like computer program.

The structured text language is extremely useful for executing routines like report generation, where English-like instructions explain what is being done. Remember that ST can be used to encapsulate, or create, a function block that will perform a certain task when triggered by the control logic (see Figure 10-13). This function block routine can be used repeatedly throughout the control program.

Some PLC manufacturers enhance the standard features of ST by using it to integrate real-time force I/O and monitoring I/O (analog and digital) data in the same manner as a standard PLC would using ladder diagrams. For example, an ST instruction such as FORCE Variable_One would force Variable_One to be ON regardless of any other conditions, as long as Variable_One is Boolean. If the variable was analog, the instruction may be FORCE Variable_One = 5000; in which case, the value of the analog variable would be set to 5000 during the forcing.

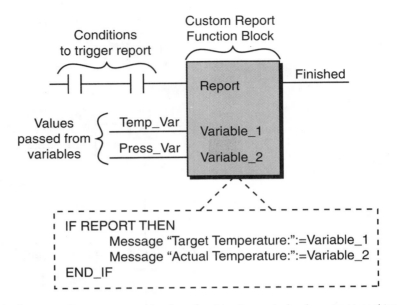

Figure 10-13. Report generation function block created using structured text.

Structured text programming is particularly suited to applications involving data handling, computational sorting, and intensive mathematical applications utilizing floating-point values. ST is also the best language for implementing artificial intelligence (AI) computations, fuzzy logic, and decision making.

SEQUENTIAL FUNCTION CHARTS (SFC)

Sequential functional chart, or SFC, is a graphical "language" that provides a diagrammatic representation of control sequences in a program. Basically, sequential function chart is a flowchart-like framework that can organize the subprograms or subroutines (programmed in LD, FBD, IL, and/or ST) that form the control program. SFC is particularly useful for sequential control operations, where a program flows from one step to another once a condition has been satisfied (TRUE or FALSE).

The SFC programming framework contains three main elements that organize the control program:

- steps
- transitions
- actions

A **step** is a stage in the control process. For example, the mixing application shown in Figure 10-14 has three steps—the initial step, the mixing step, and the emptying step. When the control program receives an input, it will execute each of these steps starting with step 1. Each step may or may not have an **action** associated with it. An action is a set of control instructions prompting

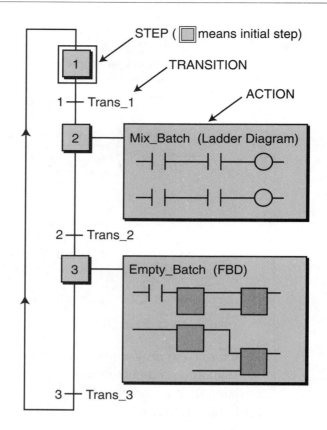

Figure 10-14. Sequential function chart of a mixing process.

the PLC to execute a certain control function during that step. An action may be programmed using any one of the four IEC 1131-3 languages. After the PLC executes a step/action, it must receive a **transition** before it will proceed to the next step. A transition can take the form of a variable input, a result of a previous action, or a conditional IF statement (e.g., IF Temp_1≥100). So, for the application shown in Figure 10-15, the PLC will execute action 2 only after step 1 receives a valid input and transition 1

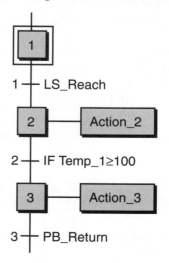

Figure 10-15. Transitions in a sequential function chart.

occurs (i.e., the limit switch LS_Reach triggers). After the PLC finishes action 2, it will wait for transition 2 (IF Temp_1≥100) to occur and then move to step 3.

As mentioned earlier, the sequential function chart language has its origin in the French standard Grafcet, a flowchart-like programming language. The Grafcet graphic language also uses steps, transitions, and actions, which operate in the same manner as in SFC. In Grafcet, when a step is active, the processor scans the I/O logic and program pertinent to the step's action, as well as the logic for the transition immediately after it (i.e., the transition that deactivates the step and action).

Like Grafcet, SFC is similar to a flowchart in the way control is passed from one step to another (see Figure 10-16). Also, like in Grafcet, SFC can be programmed to directly relate to timing or event diagrams. Figure 10-17 shows a comparison of a timing diagram and its related Grafcet and SFC programs. As shown in the timing diagram (see Figure 10-17a), if the condition Part_Present_LS is satisfied (the limit switch closes), the Advance_Solenoid output will turn ON. Once the Part_In_Position_LS variable is ON, the Clamp_Solenoid output will turn ON. Then, when the At_Depth_LS condition becomes TRUE, the Drill_Motor output will turn ON for 10 seconds. Note that the Clamp_Solenoid output is also activated during the Drill_Motor action. Once the time expires, the timing diagram indicates that the Clamp_Solenoid and Drill_Motor outputs will both turn OFF and stay OFF, while the Return_Solenoid output turns ON. No further

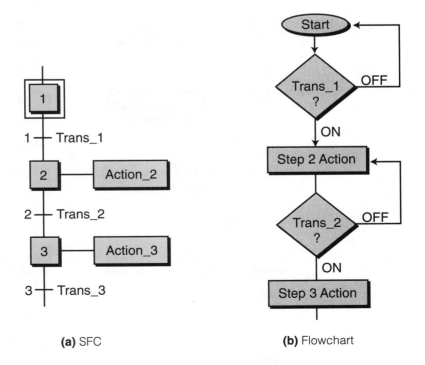

(a) SFC   **(b)** Flowchart

Figure 10-16. Comparison of **(a)** an SFC diagram and **(b)** a flowchart.

action will occur until the At_Top_LS command is satisfied, at which time, the process will stop and the Return_Solenoid output will reset for another sequence. Figures 10-17b and 10-17c illustrate the timing diagram as implemented in Grafcet and SFC, respectively. Both of these programming languages graphically represent the timing diagram implementation using

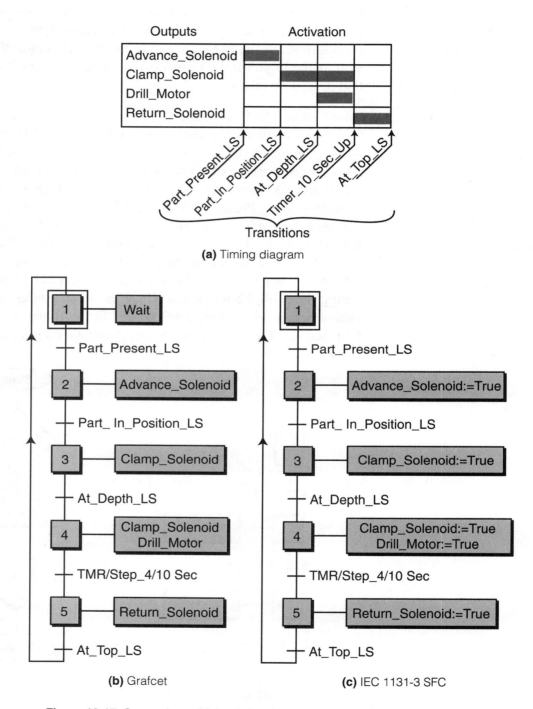

(a) Timing diagram

(b) Grafcet

(c) IEC 1131-3 SFC

Figure 10-17. Comparison of **(a)** a timing diagram with its associated **(b)** Grafcet and **(c)** SFC programs.

steps, actions, and transitions. The actions represent the activation of the solenoid and drill motor, while the transitions represent the limit switch inputs and timer status.

The major difference between Grafcet and SFC is that Grafcet employs only written action statements, such as Open_*Variable* (e.g., Open_Valve) to implement its action blocks and turn devices ON and OFF. SFC, on the other hand, implements actions in a number of ways using LD, IL, ST, and FBD or a combination of these languages, including custom function blocks. For example, in action 2 of the Grafcet program in Figure 10-17b, the statement Advance_Solenoid indicates the turning ON of the field device associated with the output variable assigned to Advance_Solenoid. In other words, if an output variable is stated in a Grafcet action, it will become TRUE or ON. In the SFC-equivalent program in Figure 10-17c, the step 2 instruction indicates that the Advance_Solenoid will be equal to TRUE (ON). Thus, SFC does not actually contain a statement of the output variable, but rather an instruction that turns the device ON or OFF (TRUE or FALSE) during that action.

Sequential function charts can be thought of as building-block objects used to create the "total" control program, or the big picture, while the other languages are used to implement detailed programming within the SFC. In fact, SFCs can have what are known in Grafcet terms as **macrosteps**, which allow one master sequential function chart to have other sequential function charts as its actions (see Figure 10-18). These smaller, embedded sequential function charts, which have their own steps, transitions, and actions, are similar to subroutines in a program.

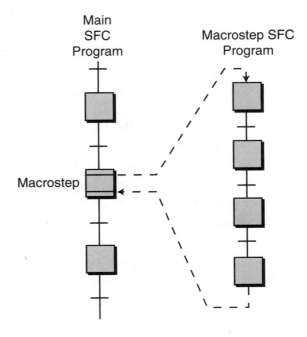

Figure 10-18. Macrostep within an SFC program.

One of the greatest advantages of sequential function charts is that they are easier to troubleshoot than standard ladder diagram programs. For example, in the sequential function chart shown earlier in Figure 10-17c, if the action Clamp_Solenoid (solenoid ON) at step 3 does not make the transition to step 4, it is easy to recognize that a problem occurred at the transition after step 3, which corresponds to the activation of the At_Depth_LS transition. Thus, an SFC pinpoints the step or transition where a fault occurs.

PROGRAMMING LANGUAGE NOTATION

As we have noted, sequential function charts can provide the infrastructure for a control program, which is then built using one or more of the four IEC 1131-3 programming languages. In the next section, we will further explain how SFCs can be used implement a control program. However, let's first review the similarities between programming notations in the ladder diagram (LD), function block diagram (FBD), structured text (ST), and instruction list (IL) languages.

Figure 10-19 shows a simple ladder diagram and its FBD, ST, and IL language equivalents. Note that the ST language (see Figure 10-19c) uses two operators, AND and &, to denote the AND function. The := symbol is used in an ST program to assign an output variable (e.g., Valve_3) to a logic expression. In instruction list (see Figure 10-19d), the first instruction (instruction LD) loads the status of variable Limit_S_1 to the accumulator register, which IL calls the *result register*. The second instruction (instruction AND) ANDs the status of Limit_S_1 with the variable Start_Cycle and stores the outcome back in the result register. The third instruction (instruction ST) stores the contents of the result register as the output variable, Valve_3. This process is similar to Boolean programming language.

As demonstrated, the instructions used to implement control sequences in each programming language are very similar in their construction, as well as their visual representation. Depending on the PLC application, an SFC may use one or more of these languages to program instructions inside its actions. To differentiate between languages, some software manufacturers include starting and ending commands that define the language being used. Other manufacturers allow the mixing of languages without any differentiation between them. Figure 10-20 illustrates a group of instructions that have been labeled with a differentiation mnemonic. The term #Language=*name* signals the beginning of a language, and #ENDlanguage*name* signals the end of it.

Bool_Var (Boolean variable) Inputs: Limit_S_1 for Limit Switch 1
                                        Start_Cycle for Start Cycle PB

Bool_Var (Boolean variable) Outputs: Valve_3 for Solenoid Valve #3

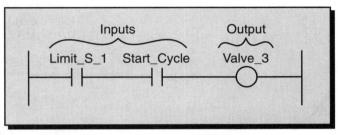

(a) Ladder diagram (LD)

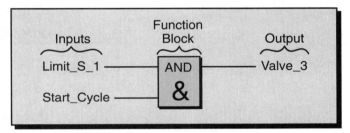

(b) Function block diagram (FBD)

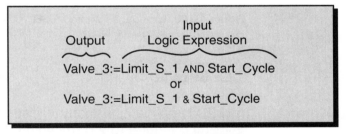

(c) Structured text (ST)

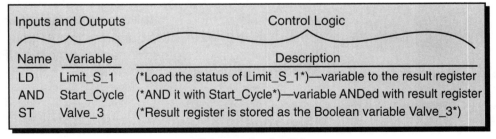

Inputs and Outputs		Control Logic
Name	Variable	Description
LD	Limit_S_1	(*Load the status of Limit_S_1*)—variable to the result register
AND	Start_Cycle	(*AND it with Start_Cycle*)—variable ANDed with result register
ST	Valve_3	(*Result register is stored as the Boolean variable Valve_3*)

(d) Instruction list (IL)

Figure 10-19. Implementation of a simple program in **(a)** ladder diagram, **(b)** function block diagram, **(c)** structured text, and **(d)** instruction list.

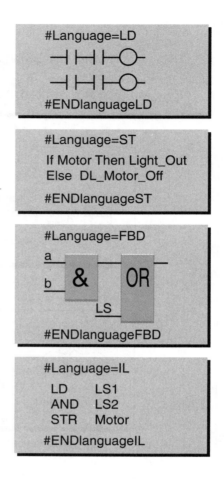

Figure 10-20. Languages within an SFC differentiated by beginning and ending language labels.

EXAMPLE 10-3

In PLC applications, many limit switches exhibit a "bouncing" behavior (see Figure 10-21), meaning that the switch opens and closes several times before finally turning ON or OFF. Develop an encapsulated custom function block (see Figure 10-22), which will provide 50 msec debouncing capabilities, that can be stored in a library and used to program all bouncing input limit switches. Note that debouncing must be performed for both the OFF-to-ON and the ON-to-OFF transitions.

SOLUTION

Figure 10-23 illustrates the timing diagram of the limit switch input. It shows that a 50 msec delay (shown in blue) should exist in the OFF-to-ON and ON-to-OFF transitions to filter any bouncing signals. Timers can be used to implement both delays.

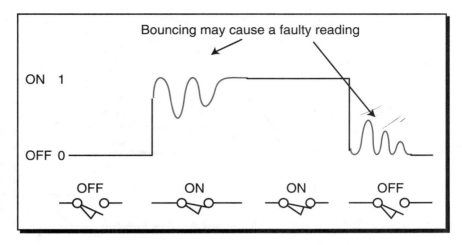

Figure 10-21. Bouncing behavior in a limit switch.

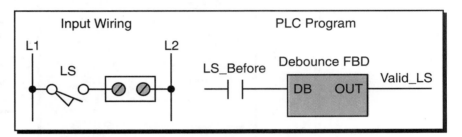

Figure 10-22. Rough diagram of an encapsulated debouncing function block.

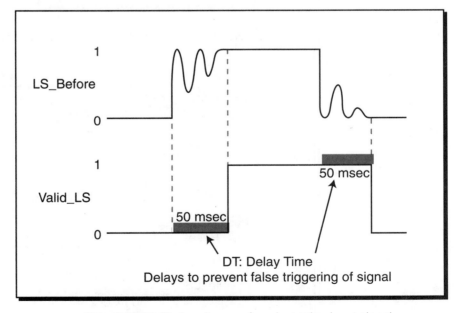

Figure 10-23. Timing diagram for a bouncing input signal.

Figure 10-24 illustrates the implementation of a debouncing circuit using ladder diagrams and an ON-delay energize timer. Figure 10-25 shows the corresponding timing diagram. Note that the output of the latch/unlatch output (102) is the actual input, in this case the limit

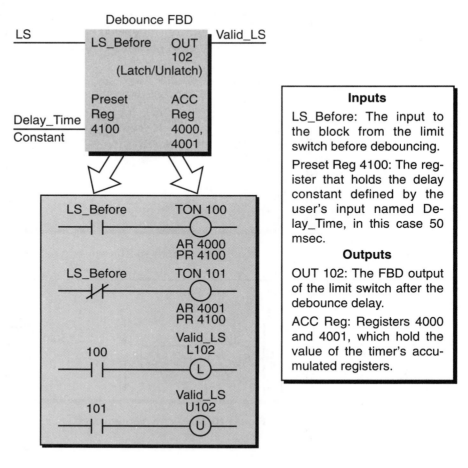

Figure 10-24. Debouncing function block programmed using ladder diagram.

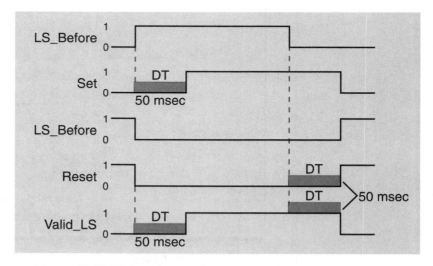

Figure 10-25. Timing diagram for the ladder circuit in Figure 10-24.

switch signal after passing through the debouncing circuit. Figure 10-26 illustrates the same type of debouncing filter implementation using FBD. Note that the output of the set/reset (S/R), or bistable, block will also be the debounced limit switch input (Valid_LS). The variable T_Delay will be an integer that is a preset time value of 50 msec. The input signals LS_Before (limit switch before debouncing) and Valid_LS (limit switch after debouncing) are both Bool (TRUE/FALSE) variables. Once created, the function block diagram can be encapsulated as a custom block as shown in Figure 10-27a. It can then be used with any input that requires a 50 msec debounce filter (see Figure 10-27b). The encapsulated block can satisfy any debounce requirement as long as the T_Delay variable is specified accordingly.

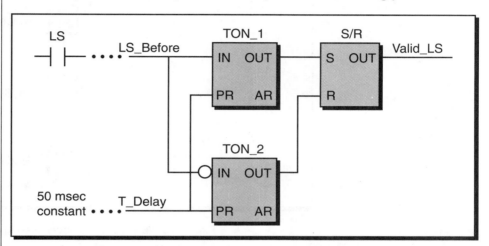

Figure 10-26. Debouncing circuit programmed using FBD.

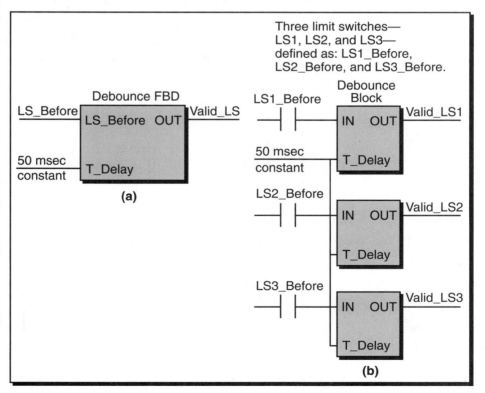

Figure 10-27. (a) FBD as an encapsulated custom block and **(b)** a custom block used to debounce three limit switch signals.

SFC FORMAT

Sequential function charts represent the order of events in a sequential process. An SFC divides a process into many steps, which are represented by various graphic components (see Figure 10-28). All of these components are used to form one or more charts that comprise the complete control program.

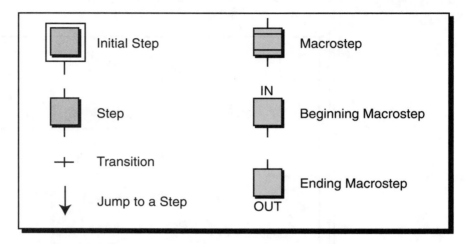

Figure 10-28. Graphic symbols used in SFCs.

Figure 10-29, for example, illustrates a small control program composed of three SFCs, each with its own independent initial step. By having independent steps, the control program starts scanning all of these charts when it first begins program execution, providing a parallel beginning. Chart 3

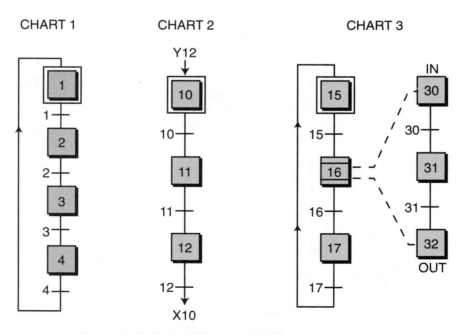

Figure 10-29. Three SFCs representing a control process.

also has a macrostep, which can be considered to be a subroutine or subprogram chart, but its initial step (IN step 30) is not independent. Chart 2 has a different link representation than charts 1 and 3 between its last step (12) and its first step (10), meaning that instead of using an arrow to link these steps, it uses jump instructions. The jump to instruction, programmed after the last step, uses an X followed by the step number to specify which step to go to—in this case, step 10. The jump from instruction, which is programmed before the initial step, uses a Y and the transition number (i.e., Y12) to indicate where the jump is from. This X*step number* and Y*transition number* notation is used throughout SFCs to distinguish between step and transition variables. Some 1131-3 systems use the letters *S* and *T* to denote steps and transitions, respectively, instead of the letters *X* and *Y*.

Sequential function charts are classified by *levels*, depending on how much detail they show. The SFC representations in Figure 10-29 are level 0 charts, because they do not specify any of the actions in their steps and do not define their transitions. Level 1 and level 2 charts (see Figure 10-30) show the actions associated with their steps. A level 1 chart represents its actions with names, comments, or descriptions of the control action executed in each step. It may also describe what occurs in each transition, or it may show the transition conditions in ST, along with the variables that will trigger them. A level 2 chart actually shows the instructions (in LD, FBD, ST, or IL) that implement the control action. In addition, it may specify an action description name like the ones used in level 1 charts; however, this name is shown in parentheses to avoid confusion with the instruction programming.

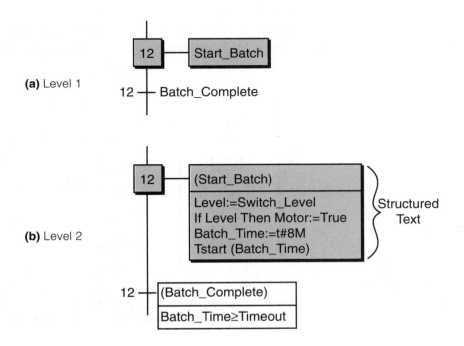

Figure 10-30. (a) Level 1 and **(b)** level 2 sequential function charts.

Each step and transition in an SFC has an ON status or condition if it is active and an OFF status if it is inactive. A dot, or *token*, indicates the ON/OFF status of a step or transition. As illustrated in Figure 10-31, the dot in the step 11 block indicates that the step is active, meaning that the status of X11 is ON. Some manufacturers refer to the ON/OFF status of a step or transition as its *Boolean activity* or *Boolean attribute* because of the TRUE/FALSE nature of the signal activity.

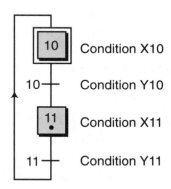

Figure 10-31. The dot in step 11 indicates that it is ON.

Figure 10-32a illustrates a step being activated by a transition, while Figure 10-32b shows a step being deactivated by a transition. As shown in the timing diagram in Figure 10-32a, Y9 and X10 are both FALSE during time *a1* because the Y9 transition has not occurred and, therefore, has not passed the token to step 10 (i.e., activated it). Once a condition or variable triggers transition Y9 (turns it ON), step 10 becomes active and the step condition X10 becomes TRUE. In Figure 10-32b, the timing diagram shows that step 12 is active (X12 is ON) during time *b1* and becomes deactivated the moment transition Y12 turns ON at time *b2*.

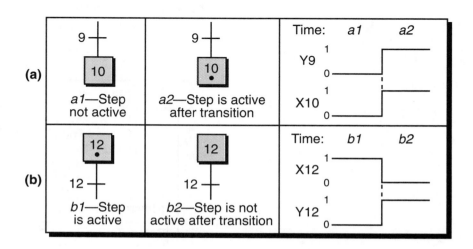

Figure 10-32. (a) An inactive step activated by a transition and **(b)** an active step deactivated by a transition.

EXAMPLE 10-4

Figure 10-33 shows an SFC in three different stages: (a) step 3 active, (b) step 4 active after being triggered by transition IN_1, and (c) step 4 turned OFF by the triggering of transition IN_2. Using a timing diagram, graphically illustrate the status of the steps (Xs) and the transitions (Ys) in each of these three phases.

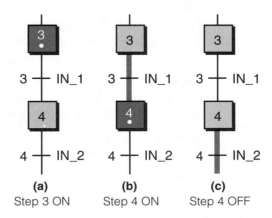

Figure 10-33. Control being passed through an SFC.

SOLUTION

Figure 10-34 shows the timing diagrams for each of the three stages in Figure 10-33. When step 3 is active (with token), X3 is ON and its action will be executed. Once the transition IN_1 occurs (Y3 goes from OFF to ON), the token passes to step 4 for execution of its action; thus, X4 becomes ON. Step 4 will remain active (ON) until transition IN_2 (Y4) becomes TRUE, at which time, the control token will pass to the next step. Note that a transition does not need to remain ON once the token is passed to the next step down the chart. For example, the transition Y3 signal turned OFF immediately after passing the token to step 4; the dotted line in the timing diagram indicates this.

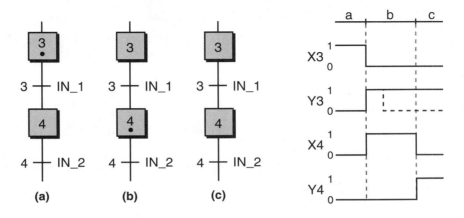

Figure 10-34. Timing diagram for the chart in Figure 10-33.

TRANSITIONS

As described in the previous example, the triggering condition of a transition can be a momentary pulse that quickly goes from OFF to ON to OFF. Figure 10-35 shows two pulse transitions, Y9 and Y10, which activate and deactivate step 10. These transitions can be programmed so that either the leading edge or the trailing edge of the pulse triggers the move to the next step. In Figure 10-36, transition 9 is programmed as a leading-edge transition using an AND condition. In this configuration, the turning ON of signal A will initiate the transition to step 10 as long as signal B is already ON. Transition 10 is also programmed using an AND condition; however, it is a trailing-edge transition. This means that, as long as signal D is active, the turning OFF of signal C will turn OFF step 10. This type of transition is similar to leading- and trailing-edge transitionals in ladder diagrams.

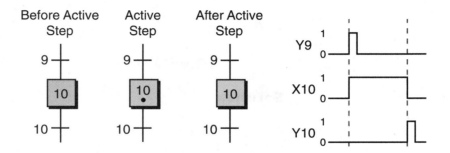

Figure 10-35. Example of momentary transition pulses.

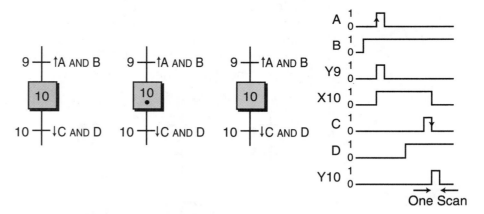

Figure 10-36. Leading- and trailing-edge transition pulses.

A timing element can be included in a transition to determine how long a step will be active. For instance, step 11 in Figure 10-37a will be active and its action executed for a period of 100 seconds because transition Y11 includes a timer set for 100 seconds. A timing transition instruction can also

be combined with Boolean logic combinations (AND, OR, NOT) and
IF...THEN instructions. For example, in Figure 10-37b, the control program
in the action in step 11 will lower a part down in a punch press and wait for
at least 10 seconds. However, it will also wait for the Down_Pos input to
be TRUE before deactivating the step 11 action and moving control to the
next step.

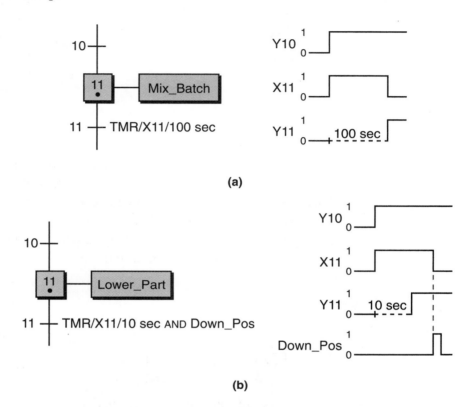

(a)

(b)

Figure 10-37. (a) A timed transition and **(b)** a timed transition combined with a
Boolean logic function.

10-3 SEQUENTIAL FUNCTION CHART PROGRAMMING

The signal that triggers a transition may be the result of an external variable
or a step's output. For example, in Figure 10-38, step 10's action instructions
(in this case, an LD, ST, and FBD control sequence) control the status of the
transition Time_Up, which will move control execution to the next step.
When step 10 becomes active, the Mix_Start action begins, and the processor
scans all the I/O in the action and executes the program as described by the
action's instructions. If Mix_Rdy is TRUE (in the LD part of the action),
then the motor will be turned on for 30 seconds as specified by the timer.
Once the 30 seconds have elapsed, the timer's Boolean output variable
Time_Up, which is defined as an internal Bool variable, will be TRUE,
initiating the transition to the next step.

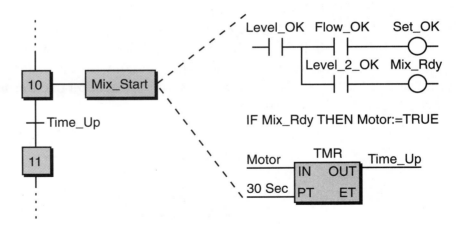

Figure 10-38. Action output as a trigger for a transition.

Transitions can also be logically combined with other instructions, most commonly with the structured text language. For instance, in Figure 10-39, the transition from step 12 to 13 will occur if the command Set_OK inside the action of step 12 (labeled as Action_1) is TRUE and the signal Level_Switch is TRUE. Set_OK is an internal output, while Level_Switch is a direct input signal connected to a PLC input module.

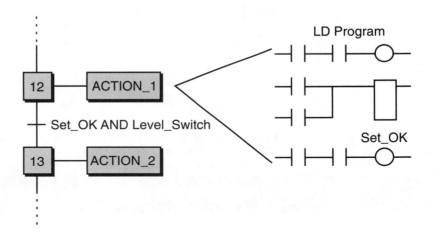

Figure 10-39. Combination of an internal output and an external variable as a transition.

PROGRAMMING NORMALLY CLOSED TRANSITIONS

As explained in the previous chapter, a normally closed input device should be programmed as normally open in a PLC for it to operate as a normally closed device. The reason for this is safety. When programmed as normally open, the device will lose continuity and turn OFF if its connection to the input module is cut. This provides fail-safe operation. This same criteria applies for a normally closed device in a PLC using IEC 1131-3 program-ming—all normally closed devices should be programmed as normally open, regardless of the language used.

Normally closed devices must also be programmed carefully when used as triggering variables in an SFC transition. If the normally closed device is not actuated (e.g., a normally closed limit switch is closed), the transition from one step to the next one will be in one scan. Let's take a closer look. Figure 10-40a illustrates a part of a simple chart in which the normally closed limit switch LS_1 is used to trigger the transition from step 10 to step 11. Note that the timing diagram, which represents the Boolean activity, indicates that LS_1 is ON when not activated. Thus, the transition from step 10 to 11 will occur as soon as step 10 is active (one scan). To trigger the transition from step 10 to step 11 upon the activation of LS_1 (normally closed LS_1 opening), the transition must be programmed as NOT LS_1. This way, if LS_1 opens, the NOT LS_1 instruction will trigger the transition. Note that in Figure 10-40b, the limit switch opened momentarily to trigger the transition to step 11. It is a good idea to study timing diagrams when programming a normally closed device to observe the required behavior of the transition.

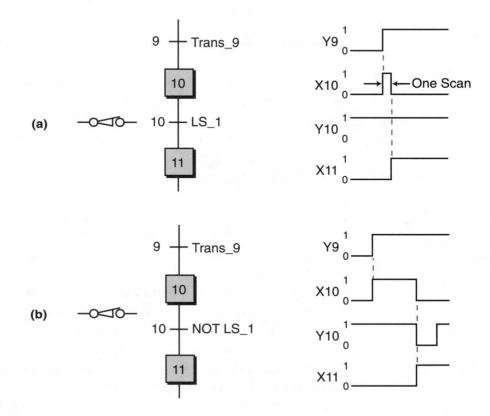

Figure 10-40. The transition from step 10 to step 11 will **(a)** occur in one scan unless **(b)** transition 10 is programmed as NOT LS_1.

Figure 10-41 illustrates a simple start/stop hardwired motor circuit and its timing diagram. When the momentary normally open start push button is pressed and the normally closed stop push button is not pressed, the motor will be ON and its motor contacts M1-1 will seal the start push button, meaning that the motor will remain ON until the stop PB is pressed. When the stop PB is pressed, the circuit will lose continuity and the motor will turn OFF. Logically speaking, as shown in the timing diagram in Figure 10-41, the motor will be ON if both the start PB (wired as normally open) and the stop PB (wired as normally closed) are ON (1), in other words, start is ON (Start=1) and stop is NOT OFF (Stop=1). Therefore, the logic expression that will turn M1 ON is M1=Start AND Stop.

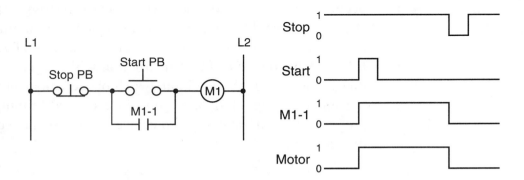

Figure 10-41. A hardwired start/stop motor circuit and its timing diagram.

This logic expression indicates that M1 will be ON if the start PB is pushed and the stop PB is not pushed (normally closed). However, the logic expression does not provide latching capabilities, meaning that if the start PB is pushed once and released, the motor M1 will not stay ON. As we will explain shortly, in the SFC implementation of this M1 logic expression, the latching or interlocking of the M1 logic expression is not required.

Figure 10-42 illustrates the SFC implementation of the hardwired circuit in Figure 10-41, along with its timing diagram. In the SFC, the logic expression that triggers transition 1 (Start_AND_Stop) is the same logic expression that turns motor M1 ON in the hardwired circuit, but without interlock. The program does not require interlocking between the push buttons because it does not need to remember that the start PB was pressed to keep the motor ON. Once the momentary start PB is pressed, step 1 (no action) transitions to step 2, where the action turns ON the motor and keeps it in that state. The program will turn the motor OFF as soon as transition Y2 is triggered, meaning that the NOT_Stop condition occurred. As soon as the stop push button is pressed (see the timing diagram in Figure 10-42), transition Y2 will be satisfied and the control token will be transferred from step X2 (motor ON) to step X1, turning off the action in X2 and, consequently, motor M1.

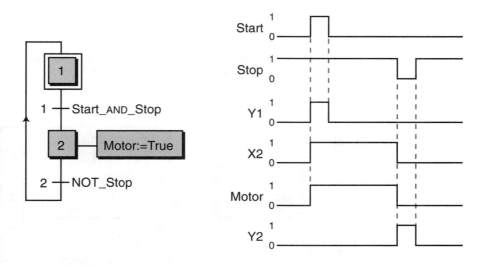

Figure 10-42. SFC implementation of the hardwired circuit in Figure 10-41.

EXAMPLE 10-5

Figure 10-43 illustrates a block diagram of PLC input devices used to control the ON/OFF state of two motors, Motor_1 and Motor_2. Assume that the pair of start/stop push buttons used with Motor_1 has a normally open start and a normally closed stop, while the start/stop push buttons used with Motor_2 are both normally open (for illustration purposes). Using SFCs, implement two independent programs in the PLC system that will control the start/stop sequence of the two motors.

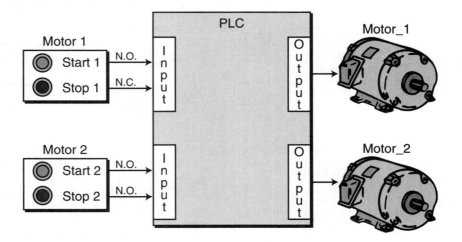

Figure 10-43. Block diagram of a program controlling two motors.

SOLUTION

Figure 10-44 shows the SFC charts for the two push button stations, while Figure 10-45 shows the corresponding timing diagrams. Note that the logic for the transitions that turn the motors ON is different. For

Motor_1, the logic takes into consideration that the normally closed stop push button is wired as normally closed. For Motor_2, the logic shows that the stop push button is a normally open push button wired as open to an input module.

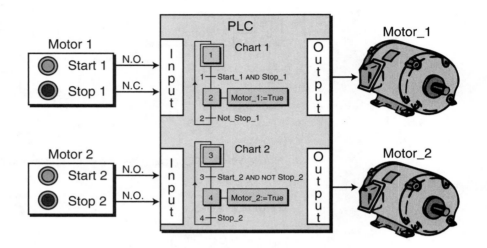

Figure 10-44. SFC charts for Motor_1 and Motor_2.

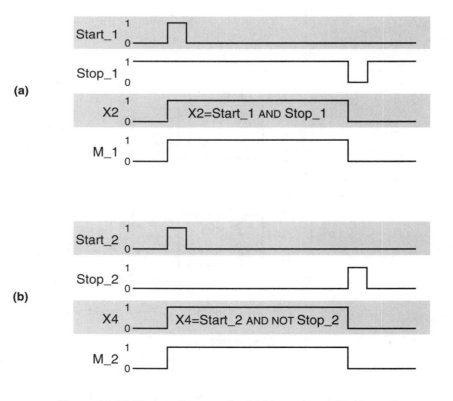

Figure 10-45. Timing diagrams for **(a)** Motor_1 and **(b)** Motor_2.

As illustrated in Example 10-5, the programming of an input field device depends on how it is wired to the input interface. A timing diagram can provide tremendous help in determining the appropriate logic for a required transition. Note that the same type of fail-safe circuit that is required in ladder diagrams must also be incorporated when programming SFCs. A fail-safe start/stop circuit can be implemented using ladder diagrams in an action, as illustrated in Figure 10-46.

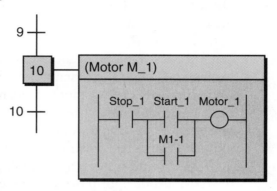

Figure 10-46. Fail-safe circuit implemented in an SFC using ladder diagrams.

DIVERGENCES AND CONVERGENCES

So far, we have only discussed sequential function charts that have one link between their steps and transitions. However, SFCs can have multiple links between these program elements (see Figure 10-47). These multiple links can be one of two types:

- divergences

- convergences

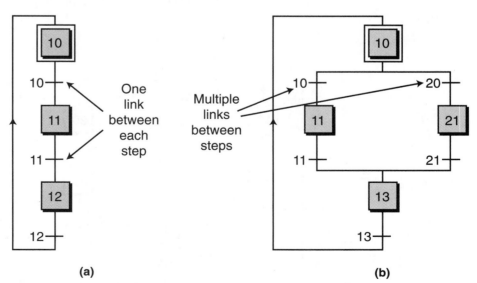

(a)   **(b)**

Figure 10-47. An SFC with **(a)** one link between the steps and transitions and **(b)** multiple links between steps and transitions.

A **divergence** is when an SFC element has many links going out of it, while a **convergence** is when an element has many links coming into it. Both divergences and convergences can have either OR or AND configurations, which relate to the Boolean logic operators of the same name.

OR Divergences and Convergences. Figure 10-48 shows an *OR divergence*, or *single divergence,* which connects one step to many transitions. An OR divergence allows an active step to pass its token to one of several steps via connecting transitions; thus, it "diverts" one step to several transitions. Although an OR divergence connects a step with several transitions, the step can only activate one of these transitions at a time. In other words, like an exclusive-OR (XOR) function, the transitions must be mutually exclusive, triggering only one transition. Depending on the IEC 1131-3 system, an OR divergence must have either mutually exclusive triggering signals (i.e., when one transition is ON, the others are OFF) or programmed logic that creates a mutually exclusive situation (i.e., only one divergence path can be triggered at a time). Some systems avoid multiple divergence paths by selecting either the left-most or right-most divergence if several triggering conditions occur at once. This prioritizes divergence path selection.

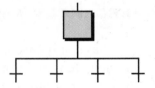

Figure 10-48. OR divergence.

Figure 10-49 shows an SFC with an OR divergence after step 1. Once step 1 is activated, either step 10 or 20 can be activated if either transition 1 or 2 is triggered. These two transitions have mutually exclusive triggering condi-

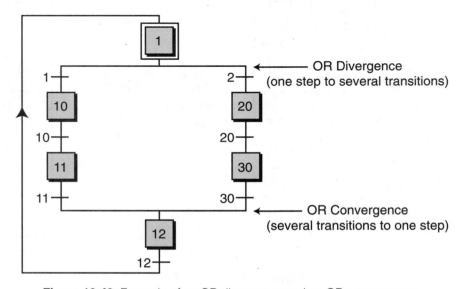

Figure 10-49. Example of an OR divergence and an OR convergence.

tions, so that the token advances in only one branch of the divergence. Therefore, if transition 1 is triggered, step 10 becomes active; if transition 2 is triggered, step 20 becomes active.

An *OR convergence*, also called a *single convergence*, is used to link several transitions to the same step (see Figure 10-50). An OR convergence is the opposite of an OR divergence; it "converges" several transitions to one step. Referring to Figure 10-49, this SFC illustrates an OR convergence in addition to an OR divergence. The OR convergence indicates that either of two links, one containing transition 11 and the other containing transition 30, can pass the control token to step 12. Because of the mutually exclusive requirement of the transition triggers, OR convergences and divergences are well suited for programming alarm circuit SFCs like the one shown in Figure 10-51. In this program, if the circuit is working properly after initialization, the program will begin the control sequence (transition 1 to step 20); whereas if an error occurs, the program will initiate an alarm action (transition 2 to step 30), which will sound an alarm until the alarm acknowledgment is triggered (transition 30). From step 1, the program can pass the token through only one path (either transition 1 or 2), but not both, because of the logical mutual exclusivity of the OR programming.

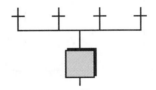

Figure 10-50. OR convergence.

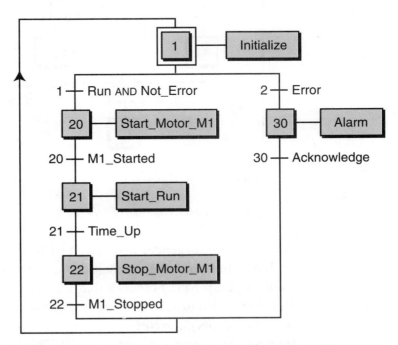

Figure 10-51. An alarm circuit with an OR divergence and an OR convergence.

AND Divergences and Convergences. An *AND divergence*, also called a *double divergence*, provides a link from one transition to many steps in parallel form (see Figure 10-52). Unlike an OR divergence, an AND divergence can pass the token through several branches at once. For example, if transition 1 in Figure 10-53 is triggered, the program will pass control to both step 40 and step 50 at the same time. The parallel lines that represent an AND convergence indicate that it passes control to all the steps below it in parallel.

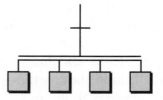

Figure 10-52. AND divergence.

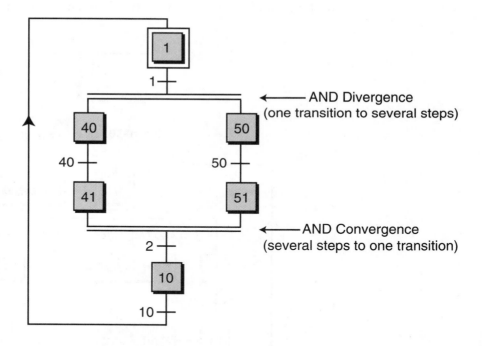

Figure 10-53. Example of an AND divergence and an AND convergence.

An *AND convergence*, also referred to as a *double convergence*, links multiple steps to a single transition (see Figure 10-54). It is most commonly used to group SFC branches that were separated by an AND divergence. Referring to Figure 10-53, once steps 41 and 51 both have the token (i.e., their actions are ON), the SFC program will wait for transition 2 to trigger and then pass the control token to step 10. This is an AND convergence function

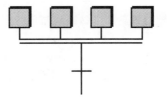

Figure 10-54. AND convergence.

because both steps 41 AND 51 will be deactivated by the transition. If more than two links converge at the transition, then all the steps immediately preceding the convergence must be active before the transition can occur. When it does occur, all the steps will converge to the step following the transition. For example, if only step 51 is active and transition 2 occurs, the SFC will not pass control to step 10. When both steps 41 and 51 are active and transition 2 is TRUE, then control will pass to step 10.

AND divergences and convergences are ideal for running control programs in a synchronized, parallel manner. For example, Figure 10-55 illustrates a sequential function chart depicting two processes that occur in parallel (at the same time). When transition 1 becomes active, it diverts activity to two program sections, each controlling one of the processes. Each program section, Process1 and Process2, must be completed (steps 21 and 31 active) before transition 2 can occur, transferring control back to step 1. Note that in an SFC transition like transition 2, which has the trigger variable True, the transition is always triggered. When used in an application, this type of AND convergence transition simply waits for both processes to finish before transferring control to the next step.

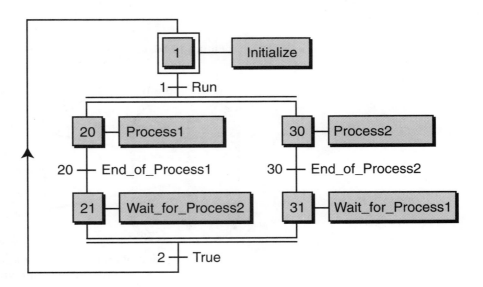

Figure 10-55. An SFC using AND convergences and divergences to run two processes in parallel.

SUBPROGRAMS

As illustrated in Figure 10-56, a process may have several main program charts executing different main control tasks within the PLC system. Depending on the IEC 1131-3 software system, these main programs may utilize one or more **subprograms** (smaller, independent programs) to implement specialized control sequences (see Figure 10-57). For example, ISaGRAF, a software system manufacturer who produces an IEC 1131– compatible program that runs in a "soft PLC" environment, provides the user

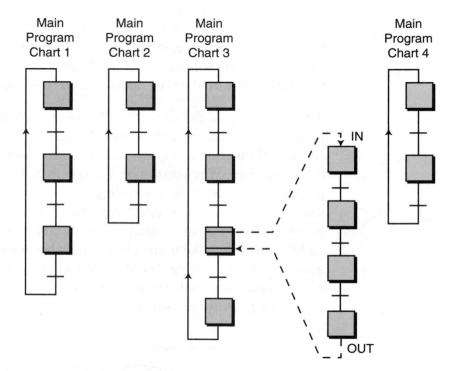

Figure 10-56. Process with several SFC programs.

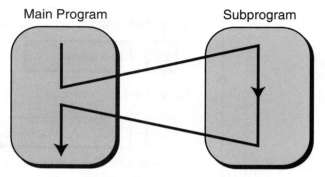

Execution of parent (main) program
is suspended until subprogram ends.

Figure 10-57. Execution of a subprogram within a main program—main program is suspended until subprogram ends.

with the ability to have a main program with one or more subprograms organized in a "father-child" relationship (see Figure 10-58). A father program can "call" (i.e., jump to) any of the child programs in a process, but a child program can only have one father program.

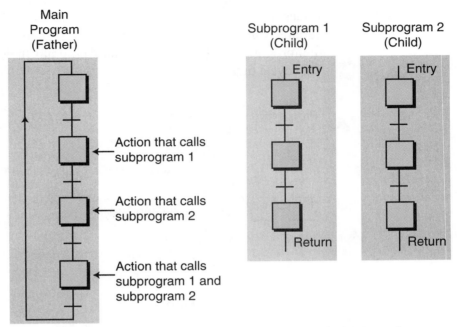

Father (main program) can call any child (subprogram)

Figure 10-58. Subprograms organized in a "father-child" relationship.

Subprograms are similar in operation to macrosteps, except that macrosteps can actually be considered an SFC type of subroutine. They are also similar to custom function blocks in the sense that they can be used over and over where needed to implement a control function. Subprograms can be written in any of the IEC 1131-3 languages and can be called directly from an SFC action using any of the four languages. In contrast, a macrostep routine can only be called from the macrostep action that contains it. Custom-built function blocks, on the other hand, can be called from any main program's action once they are in the SFC program library. These function blocks, however, cannot pass completed information to the main program like a subprogram can.

Subprograms differ from custom blocks and macrosteps because they can pass and receive variables and values in a manner similar to a computer program. For example, the statement (in ST):

Actual_Weight:=SP_Weighing (Max_Wt, Tare_Wt)

states that the variable Actual_Weight will be equal to the value computed by the subprogram SP_Weighing (SP denotes subprogram), which receives the data values of the variables Max_Wt and Tare_Wt from the main program.

When a program calls a subprogram, it asks for a value to be delivered when execution returns from the subprogram (see Figure 10-59). This value may be Boolean, real, or integer. In the previous Actual_Weight example, the main program obtains or computes the variable values Max_Wt and Tare_Wt, which are passed to the subprogram. The subprogram SP_Weighing uses these two values to compute a value that is passed to the variable Actual_Weight. This variable value can then be used in the main program. Because subprograms run miniprograms within the larger control program, they can dramatically affect the scan cycle time.

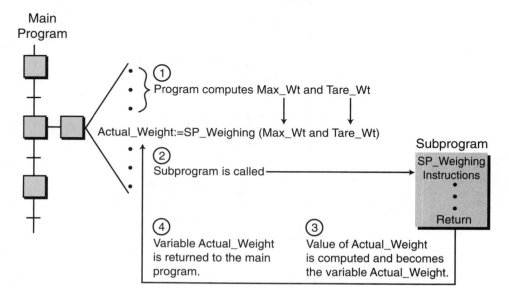

Figure 10-59. Interpretation of a subprogram call from a main program.

The syntax for calling subprograms may differ slightly from one software system to another. Nevertheless, all subprograms execute a small routine and then return a desired computed value to the main program. Figure 10-60 illustrates how an SFC program calls a subprogram from an instruction in one of its actions. In this example, step 11's action (Action_11) has several

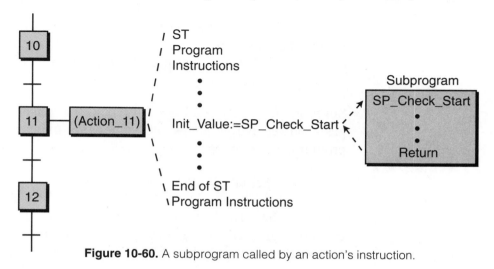

Figure 10-60. A subprogram called by an action's instruction.

ST instructions with the instruction Init_Value:=SP_Check_Start initiating a subprogram named SP_Check_Start. This subprogram calculates the value of the variable Init_Value and sends this data back to the main program, so that the main program can use the variable value in the control process.

Figure 10-61 illustrates several subprogram example calls using other languages. The SUBPROG_1 subprogram will be called and executed once it is found directly in the program (IL and ST) or once the conditions are satisfied (LD and FBD). Remember that the subprogram can be written in any of the languages, regardless of the calling language. The PLC's manufacturer can provide IEC 1131-3 software system specifications for properly passing and receiving subprogram parameters.

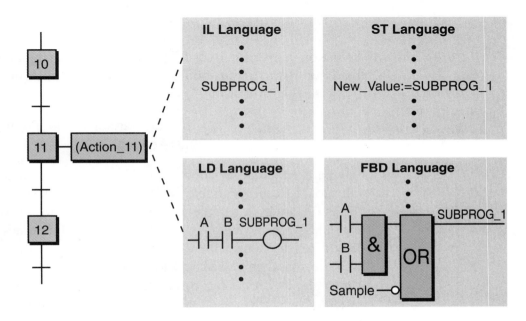

Figure 10-61. Subprogram calls in IL, ST, LD, and FBD languages. SUBPROG_1 is defined as a subprogram during the program structure definition.

An SFC transition can also call a subprogram, as shown in Figure 10-62. In this example, transitions 1 and 2 call for the subprograms ErrEval and EvalCond, which are mutually exclusive. These subprograms determine whether the process should be executed or whether an alarm condition should be set. The two subprogram calls follow the syntax:

$$\text{Subprogram\_}Name();$$

This syntax specifies the subprogram name and the return condition (), which is a Boolean result that triggers the transition. The value returned by the subprogram yields the following conditions:

Return value = 0   →   FALSE condition
Return value <> 0   →   TRUE condition

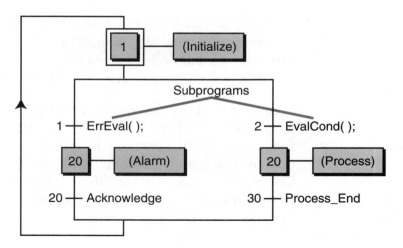

Figure 10-62. Subprogram call from a transition.

An action can also call a subprogram directly using an instruction syntax that is similar to a transition. For example, the following instruction:

Action:
    Result_Variable:=Sub_Program();
End_Action;

may be used to call a subprogram that will give a Boolean TRUE/FALSE value to the Result_Variable, which can be used to trigger a transition. Figure 10-63 illustrates a sample subprogram call from an SFC action. In this example, when the action in step 1 is activated, it initiates a subprogram that determines the value of the variable Init. The value of this variable (expressed in Boolean) is then passed back to the main program, where it is used to either trigger the start of a macrostep process program or sound an alarm. The variable labels Error and OK must have been declared as Boolean variables during programming (e.g., Error:=False, OK:=True) for the proper transition to occur. The (P) in the action name in step 1 indicates a pulse-type action (momentary), which we will discuss in the next section.

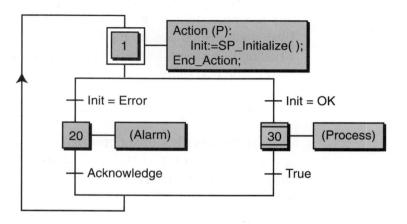

Figure 10-63. Subprogram call from an SFC action.

10-4 TYPES OF STEP ACTIONS

An action in a sequential function chart is executed when its corresponding step is active. When the step becomes active, the software control instructions contained in the action will be executed and scanned until the token is transitioned to the next step in the chart. A step action can take several forms, depending on the desired operation and result. These types of actions are:

- Boolean actions

- pulse actions

- normal actions

- SFC actions

BOOLEAN ACTIONS

A **Boolean action** assigns a Boolean value (i.e., TRUE/FALSE) to a variable during the step's action. A Boolean variable may be a real output or an internal output. The instruction simply reflects the state (ON/OFF) of the corresponding variable with respect to the state of its action. Let's take, for example, the action shown in Figure 10-64. Once step 20 is active (X20 is ON), the variable Bool_Var_1 will be turned ON as long as the step is active. The variable /Bool_Var_2—i.e., NOT Bool_Var_2 (/ = NOT)—is the NOT value of the active step X20 and, accordingly, of the variable Bool_Var_2. The variables Bool_Var_3 and Bool_Var_4, followed by (S) and (R) respectively, indicate set and reset instructions to the variable. The set (S) parameter becomes active when the step becomes active, setting the variable to TRUE. The set variable stays active until it is reset in the same step

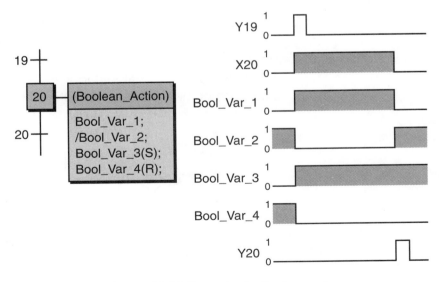

Figure 10-64. Example of a Boolean action.

or in another step; however, it keeps the variable as TRUE, even when the step is deactivated. Conversely, the reset (R) parameter resets the variable to FALSE when the step activity is TRUE. The reset action remains FALSE until the variable is set. Figure 10-65 shows a similar example with different variables. Note that the Solenoid_2(R) instruction resets the variable Solenoid_2, which was set to ON in a previous action.

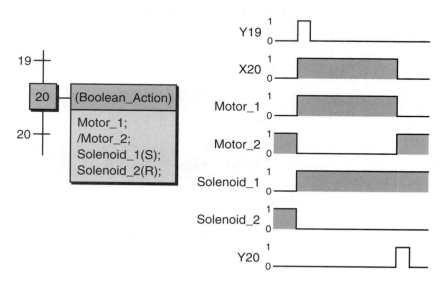

Figure 10-65. Example of a Boolean action controlling a motor and a solenoid.

EXAMPLE 10-6

Using SFC Boolean actions, implement a chart that will turn ON and OFF two pilot lights according to the timing diagram shown in Figure 10-66. In the timing diagram, PLight_1 is ON for one second while PLight_2 is OFF, then PLight_1 is OFF for one second while PLight_2 is ON. Assume that a normally open push button labeled as Start initiates the pilot light sequence and that a normally open push button labeled as Reset resets the whole operation, turning both pilot lights OFF. Include a light enable (Light_EN) pilot light indicator that is ON at the start of the operation and OFF when the operation is reset.

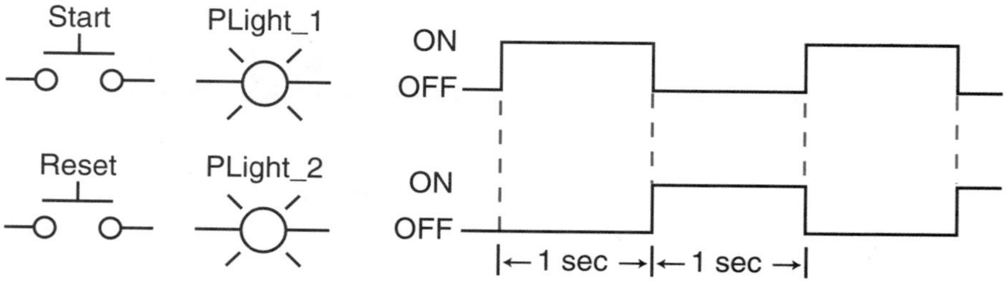

Figure 10-66. Timing diagrams for two pilot lights.

SOLUTION

Figure 10-67 illustrates the desired timing diagram of the two inputs (Start and Reset) and the three pilot lights (PLight_1, PLight_2, and Light_EN). Figure 10-68 depicts the SFC implementation of this timing diagram, where the initial step sets both PLight_1 and PLight_2 to an OFF (FALSE) state. Once the Start push button is pushed, the token passes to step 2, which has no action, and continues to the

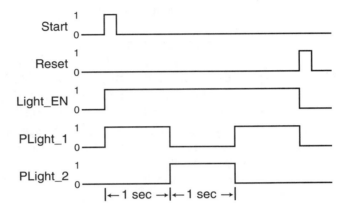

Figure 10-67. Timing diagram for the SFC implementation in Example 10-6.

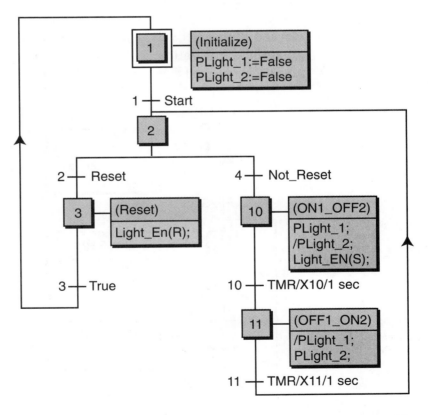

Figure 10-68. SFC implementation of the two pilot lights in Figure 10-66.

OR divergence. At the OR divergence, control goes to step 10 (ON1_OFF2) if the Reset push button is not pressed (Not_Reset), thereby turning ON PLight_1, keeping PLight_2 OFF (opposite state of the step activity), and turning ON Light_EN using a set parameter. The timer transition Y10 is triggered one second after step X10 is activated, passing control to step X11, which reverses the state of the pilot lights using Boolean actions. Like the Y10 transition, the Y11 transition also allows one second of activation before it turns OFF the step and passes the token to step 2, where the sequence is repeated.

Conversely, if the Reset push button is pressed (Reset), the program activates step 3, which resets the light enable output and transitions the sequence to step 1, where the program will wait until the Start push button is pressed. Note that this SFC program requires the operator to depress the Reset push button input at transition 2 for at least two seconds in order to reset the lights to OFF. The reason for this is that the program may be at the opposite OR divergence (transition 4), which will last for two seconds before the reset signal can be scanned at transition 2.

The implementation of the previous example could have been done many different ways using Boolean actions. For instance, instead of using the /PLight_2 and /PLight_1 instructions in steps 10 and 11, the program could have specified only the ON conditions of PLight_1 and PLight_2 in steps 10 and 11, respectively, letting the transition trigger turn OFF the variables. A **stand-alone action** could also have been programmed to detect the reset function and send the program back to step 1 in the main chart. Figure 10-69

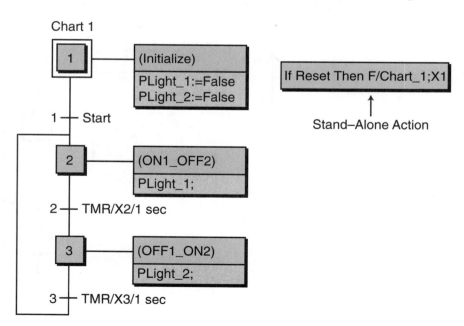

Figure 10-69. Implementation of the process in Example 10-6 using a stand-alone action.

shows this stand-alone configuration, along with an alternative set of Boolean actions for this program. Although a stand-alone action is not linked to the program, it will direct a transition move to a specified step if its logical conditions are satisfied. A stand-alone action basically acts as an interrupt jump to instruction, specifying the chart program and the step to go to. Note that a stand-alone action is active at all times, ready to force the program to the specified step. If the Reset push button in Figure 10-69 is pressed, the stand-alone action will force the program to go to step 1 of the Chart_1 program, regardless of where it is in the execution of the Chart_1 program. Also, in this configuration, the Reset push button may be pushed momentarily, so it does not require a two-second push like it did before.

PULSE ACTIONS

Pulse actions allow the execution of one or more instructions in a step's action only once after the activation of the step. That is, once the step is activated, a pulse action will occur only once before the step is deactivated. Depending on the IEC 1131 software system, the instructions in the action may be in one or more of the available languages. The typical syntax of an SFC pulse action looks like the block in Figure 10-70.

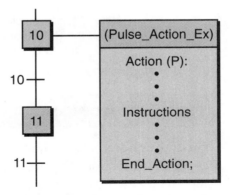

Figure 10-70. Syntax of a pulse action.

The notation (P) indicates a pulse action. A pulse action may be represented in a timing diagram as shown in Figure 10-71, where its execution is shown at the start of the step activity. Figure 10-72 illustrates a typical SFC with a

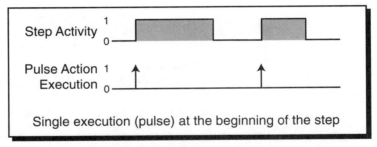

Figure 10-71. Execution of a pulse action.

pulse action implementing a count-up (add by one) instruction using ST instructions. Pulse actions are well suited for applications that require one-time execution of an action, for instance, the initialization of variables in a process. Pulse action instructions are similar in operation to the one-shot functions discussed in Chapter 9.

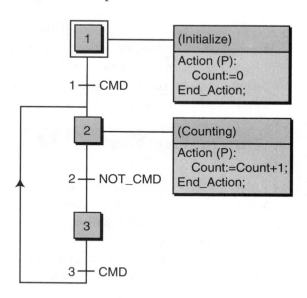

Note: Step 3 is included as a dummy step to wait for the CMD (command signal to count) to go from OFF to ON to count again.

Figure 10-72. A count-up instruction implemented as a pulse action.

NORMAL ACTIONS

Normal actions, also called *nonstored actions*, incorporate IEC 1131-3 language instructions that are executed continuously during the activity of a step. In other words, the instructions within a normal action will be executed and scanned over and over until the step is deactivated (see Figure 10-73). The basic syntax for a normal instruction is shown in Figure 10-74, where (N) indicates normal. Normal actions may also omit the (N) parameter in the instruction syntax.

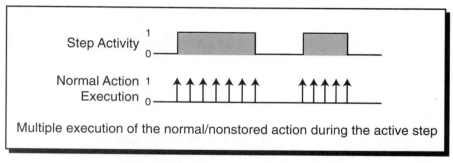

Figure 10-73. Execution of a normal action.

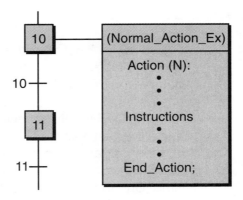

Figure 10-74. Syntax for a normal action.

Figure 10-75 shows an example of a counting program using a normal action in step 2. Note that step 1 uses a pulse action to set the value of the variable R_Count to zero. As the next example illustrates, the normal action in step 2 (programmed using ST language) performs a counting procedure on the rising edge of the signal Cmd (command) and stores the total count value as variable R_Count. This counting procedure is executed for as long as step 2 is active.

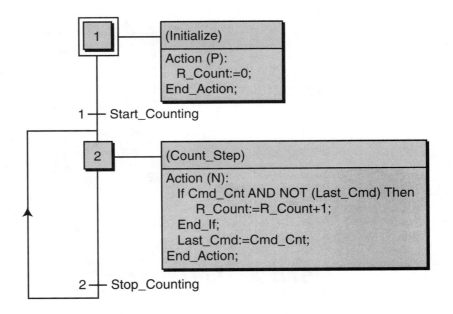

Figure 10-75. Example of a counting program using a normal action.

EXAMPLE 10-7

Referring to Figure 10-75, explain the operation of step 2. Also, draw a timing diagram of the signals indicating when the counter variable R_Count begins and ends during each count.

SOLUTION

Figure 10-76 illustrates the timing diagram of step 2. The variable R_Count increases its value by one every time the signal Cmd_Cnt goes from OFF to ON. The IF condition in step 2 of Figure 10-75 ensures that the signal is tested to make sure that it has gone OFF after the OFF-to-ON transition. The Last_Cmd:=Cmd_Cnt instruction traps the last value of Cmd_Cnt, so that the count does not get executed again without Cmd_Cnt going OFF first. When the action is deactivated by the Stop_Counting transition variable, the status of Cmd_Cnt and Last_Cmd are reset to OFF (not stored). Note that the R_Count value is reset to zero at step 1. However, the value of R_Count will be stored as a normal integer value in the program until it is changed, as in this example, in step 1.

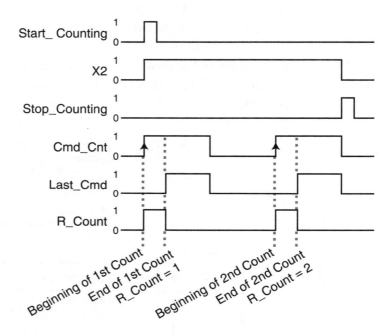

Figure 10-76. Timing diagram of step 2 in Figure 10-75.

SFC ACTIONS

An **SFC action** is a child-type SFC sequence program that can be activated (started) or deactivated (killed) when the step is active. Remember that a child program belongs to a father, or main, program. SFC actions may have normal, set, or reset parameters that influence the operation of the SFC action (see Table 10-3). Figure 10-77 illustrates a batching process SFC that uses SFC actions. The main SFC program has two child programs, Batch_Mix and Batch_Pump, which are activated by the main (father) program. The main SFC program uses normal, set, and reset operands.

Syntax	Description
Child_Prog_Name (N);	Normal: An SFC action with an (N) parameter is started when the step becomes active. The normal, or nonstored, child action is killed when the step is deactivated. The (N) parameter is optional in the syntax of this action.
Child_Prog_Name (S);	Set: An SFC action with an (S) parameter is started when the step becomes active. This set (S) action remains activated when the step is deactivated.
Child_Prog_Name (R);	Reset: An SFC action with an (R) parameter is killed when the step becomes active. This reset (R) action is used to turn off a set SFC action.

Table 10-3. Syntax for SFC action parameters.

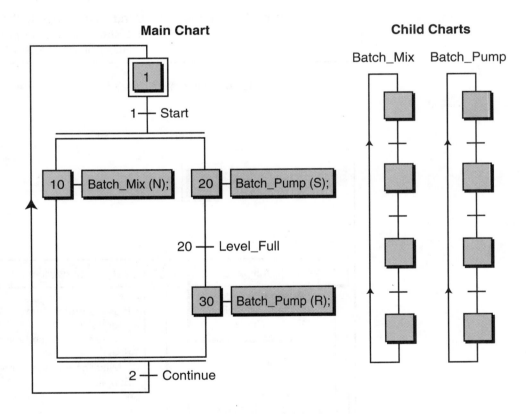

Figure 10-77. Batching process implemented using SFC actions.

Once Start is triggered, the SFC activates both of the child programs. The Batch_Mix program has a normal (nonstored) parameter, while the Batch_Pump program has set and reset parameters. The Batch_Pump program becomes active as soon as step 20 is activated. It remains active until the signal Level_Full is turned ON, activating step 30 and resetting, or killing, the Batch_Pump program.

SFC actions may be started or killed using any of the programming languages, depending on the IEC 1131-3 software system manufacturer. The syntax differs slightly from one system to another and may take the form shown in Table 10-4. The start and kill instructions have the same effects as the set (S) and reset (R) parameters, respectively. Figure 10-78 illustrates an SFC action using structured text. The starting and killing of the child program can be either nonstored or pulse actions, but in this example, the

Syntax	Description
START (Child_Prog_Name);	Starts, or activates, the SFC program Child_Prog_Name. The child program will be active until it is killed in another step.
KILL (Child_Prog_Name);	Kills, or deactivates, an SFC program started by a START instruction.
STATUS (Child_Prog_Name);	Sends a message to the operator indicating the status of a child program: either active (TRUE) or inactive (FALSE).

Table 10-4. Alternative syntax for SFC action parameters.

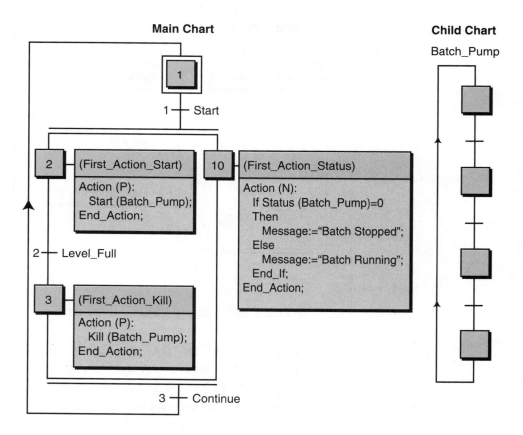

Figure 10-78. An SFC action programmed using ST and alternative SFC action syntax.

start and kill of the Batch_Pump program are both pulse actions. The status action (step 10) is a nonstored action used to send a message, perhaps to a display, to inform the operator of whether the batch is running or not running.

10-5 IEC 1131-3 SOFTWARE SYSTEMS

In addition to the implementation of the IEC 1131-3 in PLCs, many manufacturers of software systems provide the IEC 1131-3 standard in different hardware platforms and operating systems, such as Windows and Unix. These software systems emulate the operation of a programmable controller (i.e., they are software PLCs or "soft PLCs") in the hardware platform being used (e.g., a PC). They support either a third-party I/O system or one or more of a PLC manufacturer's I/O through the use of built-in drivers that communicate with an I/O rack (see Figure 10-79).

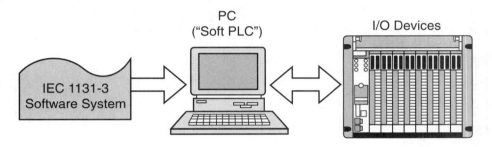

Figure 10-79. A software PLC interfaced with I/O devices.

The Paradym-31 software system from Wizdom Controls, Inc. provides an IEC 1131-3 graphical programming environment in a Windows-based software platform. This system allows the user to employ LD, FBD, or a custom-built function block language to program the actions in the SFC application. The user must program custom function blocks in C code. In fact, the Paradym-31 system compiles the entire IEC 1131 program in an ANSI C code and then downloads it to a hardware platform or to a third-party controller and its system.

Another software system, which offers a full implementation of all five IEC 1131-3 languages, is ISaGRAF from TranSys, Inc. and CJ International. This system provides a thorough set of instructions for all languages and several SFC-type actions. ISaGRAF also allows the user to test or simulate a PLC program in a personal computer, making it easier to debug an entire application or parts of it without actual hardware and I/O connections. ISaGRAF can run in a variety of operating systems, including OS-9, VRTX, VXWorks, ControlWare, DOS, and Windows NT. This software package can also transfer a control program to a programmable controller using a PortPack tool driver. Table 10-5 lists the ISaGRAF set of instructions for each of the IEC 1131-3 languages.

LADDER DIAGRAM SYMBOLS

Symbol	Description
─┤ ├─	Normally open contact (examine-ON)
─┤/├─	Normally closed contact (examine-OFF)
─( )─	Output coil
─(/)─	NOT, or inverted, output coil
─(SET)─	Set output coil
─(RST)─	Reset output coil
─(RET)─	Return—used as the conditional end of a program or to return from a ladder diagram subprogram
─(JMP)─	Jump to a label "LAB"

INSTRUCTION LIST OPERATORS

Operator	Modifier	Operand	Description
LD	N	Variable, constant	Loads operand
ST	N	Variable	Stores current result
S		Bool variable	Sets to TRUE
R		Bool variable	Resets to FALSE
AND (&)	N(	Bool variable	Boolean AND
OR	N(	Bool variable	Boolean OR
XOR	N(	Bool variable	Exclusive-OR
ADD	(	Variable, constant	Addition
SUB	(	Variable, constant	Subtraction
MUL	(	Variable, constant	Multiplication
DIV	(	Variable, constant	Division
GT	(	Variable, constant	Test:>
GE	(	Variable, constant	Test:>=
EQ	(	Variable, constant	Test:=
LE	(	Variable, constant	Test:<=
LT	(	Variable, constant	Test:<
JMP	C N	Label	Jump to a label
RET	C N		Returns from subprogram
)			Delayed operation/execution

Note: *N* is used for negation (NOT) of the operand.
(is used to indicate that the operation is to be delayed.
C is used to indicate a conditional operation.

STRUCTURED TEXT OPERATORS AND STATEMENTS

Type	Operators
Boolean	• NOT: Boolean negation • AND(&): logical AND • OR: logical OR • XOR: logical exclusive-OR • REDGE: rising-edge detection • FEDGE: falling-edge detection
Arithmetic	• Addition: + • Subtraction: − • Multiplication: × • Division: /
Comparison	• Less than: < • Greater than: > • Less than or equal to: <= • Greater than or equal to: >= • Equal to: = • Not equal to: <>
Basic statements	• Assignment (:+) • RETURN statement • IF-THEN-ELSE structure • CASE statement • WHILE iteration statement • REPEAT iteration statement • FOR iteration statement • EXIT statement
Extensions	• TSTART–TSTOP: timer control • SYSTEM function • OPERATE function
Controls the execution of SFC child programs	• GSTART: starts an SFC program • GKILL: kills an SFC program • GFREEZE: freezes an SFC program • GRST: restarts a frozen program • GSTATUS: gets current status of an SFC program
Acesses the status of an SFC step	• GSnnn.x: Boolean value that represents the activity of the step • GSnnn.t: time elapsed since last activation of the step (*nnn* is the reference number of the SFC step)

Table 10-5. ISaGRAF instruction set.

FUNCTION BLOCK DIAGRAM OPERATORS AND FUNCTIONS

Standard Operators	Functions	Standard Function Blocks	
Data manipulation	• Assignment • Analog negation	Boolean	• SR: set dominant bistable • RS: reset dominant bistable • R_Trig: rising-edge detection • F_Trig: falling-edge detection • SEMA: semaphore
Boolean operations	• Boolean AND • Boolean OR • Boolean exclusive-OR	Counting	• CTU: up counter • CTD: down counter • CTUD: up/down counter
Arithmetic operations	• Addition • Subtraction • Multiplication • Division	Timers	• TON: ON-delay timing • TOF: OFF-delay timing • TP: pulse timing
Logic operations	• Analog bit-to-bit AND mask • OR mask • Exclusive-OR mask	Integer analogs	• CMP: full comparison function block • StackInt: stack of integer analogs
Comparison tests	• Less than • Less than or equal to • Greater than • Greater than or equal to • Equal to • Not equal to	Real analogs	• AVERAGE: running average for *N* samples • HYSTER: Boolean hysteresis on difference of reals • LIM_ALRM: high/low limit alarm with hysteresis • INTEGRAL: integration over time • DERIVATE: differentiation according to time • PID: proportional-integral-derivative control
Data conversion	• Convert to Boolean • Convert to integer analog • Convert to real analog • Convert to timer • Convert to message	Signal generation	• BLINK: blinking Boolean signal • SIG_GEN: signal generator
Other message concatenations	• System access • Operate I/O channel		

Standard Functions

Math	*Trigonometric*	*Register control*	*Data manipulation*	*Data conversion*	*String management*
• Absolute value • Exponent • Power calculation • Logarithm • Square root • Truncate decimal part	• Arc cosine • Arc sine • Arc tangent • Cosine • Sine • Tangent	• Rotate left • Rotate right • Shift left • Shift right	• Minimum • Maximum • Limit • Modulo • Multiplexer (4 or 8 entries) • Binary selector • Odd parity • Random value	• Character→ASCII code • ASCII code→character *Array manipulation* • Create array of integer values • Read/write array element	• Get string length • Insert string • Delete substring • Find substring • Replace substring • Extract left, middle, or part • Get time of day

Table 10-5 continued.

PLC LANGUAGES SIMILAR TO THE IEC 1131-3

PLC manufacturers may adapt their programmable controller languages to embrace some of the qualities of the IEC 1131-3 standard. These qualities usually reflect the ease of programming found when using sequential function charts to encapsulate parts of a ladder program into an action. This added software versatility enhances a programmable controller system tremendously by speeding up program development, minimizing interlocking sequences, and reducing system troubleshooting time.

For instance, PLC Direct, a PLC manufacturer, offers programmable controllers with both standard ladder programming language instructions (RLL—relay ladder logic) and RLL Plus, which is their software language that incorporates some of the features of sequential function charts. In fact, the RLL Plus language closely follows the activation of a horizontal flowchart. As an example, let's examine a machine press application. The sequence chart in Figure 10-80 shows the sequential steps for implementing the pressing and stamping routine, which can be programmed using either standard ladder diagrams (see Figure 10-81a) or RLL Plus (see Figure 10-81b). The highlighted sections of the program in Figure 10-81a indicate the interlocking requirements for the operation shown in the flowchart. While both the ladder diagram and the RLL Plus programs implement the same control and use the same inputs and outputs, the RLL Plus program is much easier to understand and troubleshoot. For example, if the press system stops at SG S0003 (stage step 0003) and the coil output does not jump to SG S0004 (stage step 0004), then the fault must have occurred in either the Press Down output (Y1) or the Lower Limit input (X4). By investigating just this area of the PLC program, rather than the whole ladder diagram, the troubleshooting technician can find the fault more quickly.

The RLL Plus programming language, like sequential function charts, executes each stage's ladder diagram actions when that stage is active. When the control program starts, the initial stage (ISG) is activated. Jump instructions, driven by the ladder diagram contacts that form the transition logic, pass the token from stage to stage. The last rung in the active stage performs the transition logic. The RLL Plus software also supports divergences and convergences, along with the use of timers and counters in the implementation of transitions. Subroutine implementations are also available through the use of call instructions in the stage programming. Figure 10-82 presents the stage (SFC step) instructions typically used with the RLL Plus programming language.

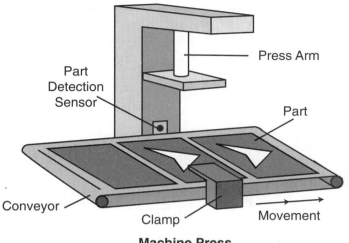

Machine Press

Operation

(0) The machine is inactive

(1) The operator presses the Start PB to start the machine.

(2) The machine checks for a part. If the part is present, the process continues. If it is not, the conveyor moves until a part is present.

(3) A clamp locks the part in place.

(4) The press stamps the part.

(5) The clamp is unlocked and the finished piece is moved out of the press.

(6) The process stops if the machine is in one-cycle mode or continues if it is in automatic mode.

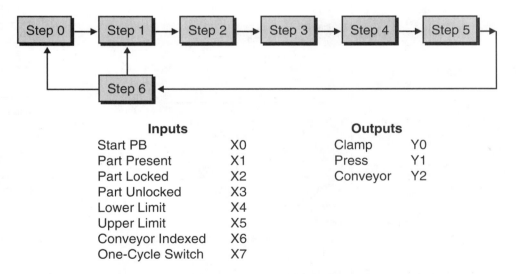

Inputs		Outputs	
Start PB	X0	Clamp	Y0
Part Present	X1	Press	Y1
Part Locked	X2	Conveyor	Y2
Part Unlocked	X3		
Lower Limit	X4		
Upper Limit	X5		
Conveyor Indexed	X6		
One-Cycle Switch	X7		

Note: For this PLC an X denotes an input and a Y denotes an output.

Figure 10-80. Pressing and stamping routine.

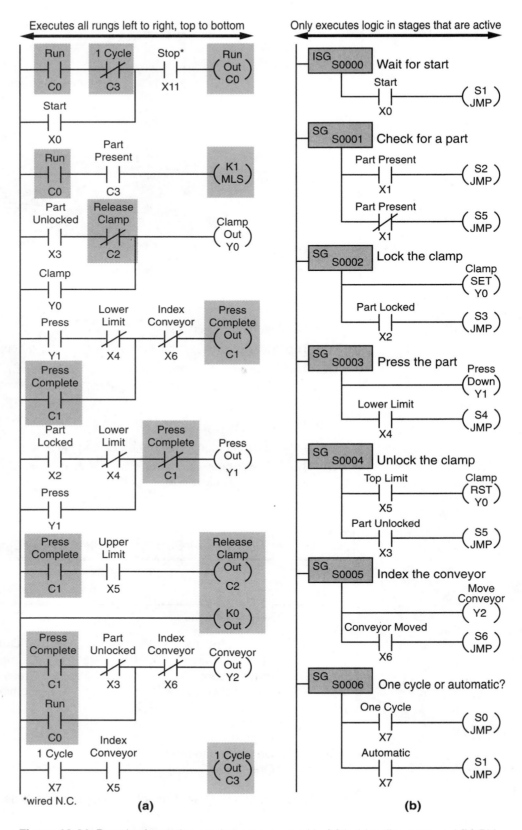

Figure 10-81. Pressing/stamping routine programmed in **(a)** ladder diagrams and **(b)** RLL Plus. SG denotes a stage step and ISG denotes an initial stage step.

Initial Stage (ISG) The initial stage instruction is used to signal the starting point of the user application program. The ISG instruction will be active on power up and PROGRAM to RUN transitions. (*aaa* = Stage memory location)	ISG S *aaa*
Stage (SG) Stage instructions are used to create structured programs. They are program segments that can be activated or deactivated with control logic. (*aaa* = Stage memory location)	SG S *aaa*
Jump (JMP) The JMP coil deactivates the active stage and activates a specified stage when there is power flow to the coil. (*aaa* = Stage memory location)	S *aaa* —(JMP)
Not Jump (NJMP) The NJMP coil deactivates the active stage and activates a specified stage when there is no power flow to the coil. (*aaa* = Stage memory location)	S *aaa* —(NJMP)
Converge Stages (CV) Converge stages is a group of stages that, when all stages are active, will activate another stage specified by the associated converge jump(s) (CVJMP). One scan after the CVJMP is executed, the converge stages will be deactivated. (*aaa* = Stage memory location)	CV S *aaa*
Converge Jump (CVJMP) The CVJMP coil deactivates the active CV stages and activates a specified stage when there is power flow to the coil. (*aaa* = Stage memory location)	S *aaa* —(CVJMP)
Block Call/Block/Block End (BCALL w/BLK and BEND) The BCALL coil activates a block of stages when there is power flow to the coil. BLK is the label that marks the beginning of a block of stages. BEND is the label used to mark the end of a block of stages. (*aaa* = C memory location)	C *aaa* —(BCALL) BLK C *aaa* —(BEND)

Figure 10-82. RLL Plus stage instructions.

Referring to Figure 10-81b, note that the program uses set and reset output instructions (SG2 and SG4, respectively) to turn ON and OFF the clamp (output Y0). Just like in an SFC, this is required because standard outputs in a stage are turned OFF once the control token has been passed to another stage. In this case, set and reset parameters were used because the clamp output solenoid needed to be ON from stage 2 through stage 4. Figure 10-83a shows the equivalent sequential function chart diagram of the program shown in Figure 10-81b. Figure 10-83b illustrates the flowchart of the process, which closely resembles the operation of the SFC program.

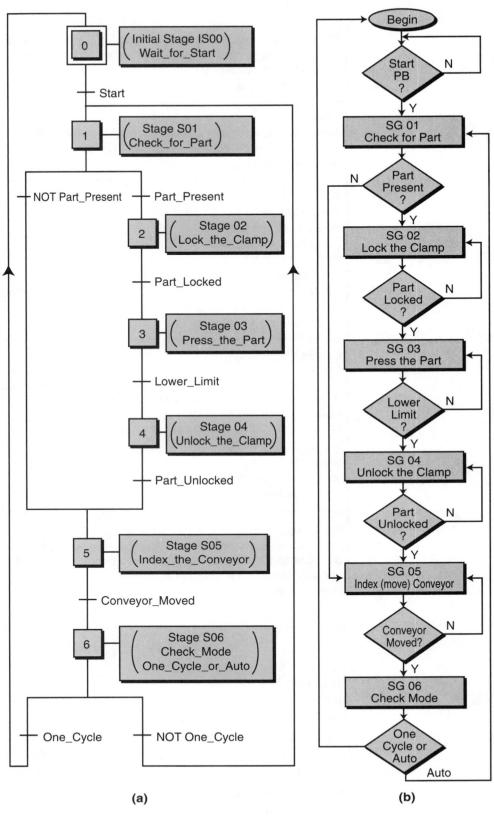

Figure 10-83. (a) An SFC program for the press/stamp control program and **(b)** its corresponding flowchart.

EXAMPLE 10-8

Referencing Figures 10-81b and 10-83a, implement an additional stage that monitors a normally closed stop push button and resets the completed pressing operation. This stage should be monitored at all times and, upon activation (i.e., after resetting all outputs), should return to the initial stage.

SOLUTION

The monitoring stage of the Stop PB must be activated as soon as the Start PB is pressed, which is when the program starts executing control. Figure 10-84 illustrates the Stop PB monitoring implementation. Note that stage S500 is ON (set) as soon as Start is pressed in the initial stage. As the PLC scans the control program during execution,

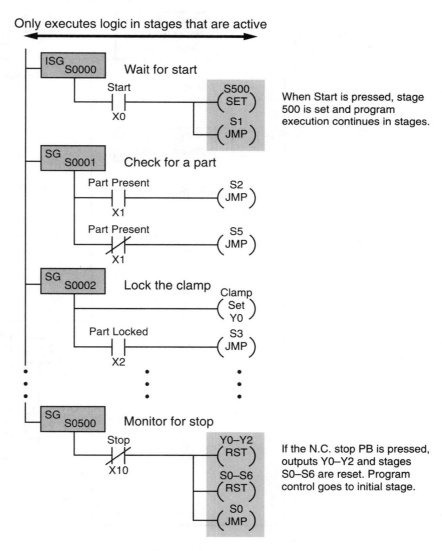

Figure 10-84. Implementation of a stop-monitoring block.

it also scans the Stop signal, which if pressed, resets all outputs (Y0 to Y2) and stages (S0 to S6) and then jumps to the initial stage (S0). Figure 10-85 shows the equivalent SFC level 1 implementation chart. Note that in the SFC, stage (step) 499 has been included so that a parallel AND divergence can be implemented and the Stop PB can be scanned (the actions in steps 4 and 5 do not execute any instructions). As the NOT Stop transition occurs (NOT Stop because of the normally closed wiring), the token passes to stage 500 for a one scan reset of all outputs, then, the token goes back to the initial stage.

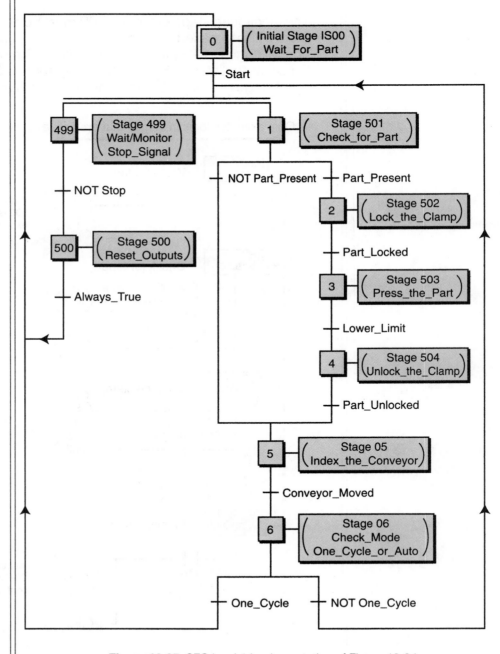

Figure 10-85. SFC level 1 implementation of Figure 10-84.

10-6 SUMMARY

The IEC 1131-3 standard provides PLC users with tremendous advantages in both the programming and troubleshooting of a control system. Although not all PLC manufacturers offer an IEC 1131-3 language for their products, the trend is leaning toward the use of an SFC-type of structured programming, including one or more of the programming languages, in most PLCs.

PLCs and software systems that support all or part of the IEC 1131 standard have better documented programs than other systems because of the structure required to implement the control program. Other IEC 1131-3 characteristics, such as the necessity to declare variables to the I/O system, provide immediate benefits to anyone who is troubleshooting the system. The same holds true for anyone else who must modify the program after installation.

Even though the IEC 1131-3 programming method reduces program design time, users must employ a few guidelines to obtain maximum benefits from the method. Table 10-6 lists some rules that will help to obtain the maximum benefits of IEC 1131-3 programming and troubleshooting. For PLC users and programmers, one of the most important advantages associated with the IEC 1131-3 is the option to choose the language for the programming and implementation of the control system.

PROGRAMMING GUIDELINES

- Be consistent in the definition of the control outputs and routines that will take place in actions.
- Define variables with proper, easy-to-reference names, especially the I/O variables.
- Be consistent in the programming of transitions. For instance, program transition conditions from inside the actions or from external inputs to avoid double usage of transition variables within steps.
- Interlocking should be done, when possible, in the transitions. Do not perform interlocking in one action for another action, since one action may be ON while the other one is OFF.
- Document the actions and transitions properly so that troubleshooting personnel understands how the machine or process is being controlled.

TROUBLESHOOTING GUIDELINES

- When there is a malfunction, locate the step that is active at that time.
- Find out the status of the transition elements that form the logic after the step where the operation halted. If it is an external input variable, check for hardware connections and interfacing; if it is an internal variable (coil, contact), check the step logic to see if the triggering signal is occurring.
- The active step and its following transition are generally the location in the program where a fault may occur and where the program stops.

Table 10-6. Rules for IEC 1131-3 programming and troubleshooting.

One of the greatest obstacles to achieving a programming standard common to all PLCs is that PLC manufacturers cautiously protect their proprietary ways of using ladder and function block instructions in order to maintain competitive advantages. This, however, does not mean that PLC manufacturers will not evolve their languages into IEC 1131-3–type languages that are transportable within their own family of PLCs. In the future, IEC 1131-3 "translators" (see Figure 10-86), which will be able to transport an IEC 1131-3 program from one PLC to another via PC software, may solve the transportability problem between different PLC brands. Regardless of potential and present obstacles, the IEC 1131 standard will surely set the pace for all PLC manufacturers wanting to continue their quest for improvement in control programming, troubleshooting, and system training.

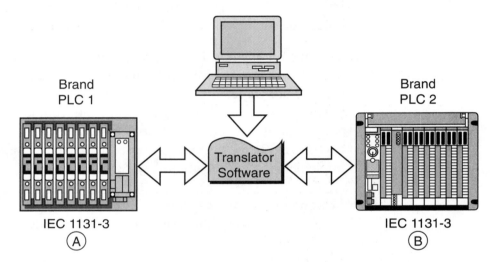

Figure 10-86. IEC 1131-3 translator.

KEY TERMS

action
Boolean action
Boolean variable
convergence
divergence
function block diagram (FBD)
IEC 1131 standard
IEC 1131-3 programming standard
instruction list (IL)
integer variable
ladder diagram language (LD)
macrostep
normal action
pulse action
real variable

sequential function charts (SFC)
SFC action
stand-alone action
step
structured text (ST)
subprogram
transition

SYSTEM PROGRAMMING AND IMPLEMENTATION

He that invents a machine augments the power of man and the well-being of mankind.

—Henry Ward Beecher

 **CHAPTER HIGHLIGHTS** The implementation of a control program requires complex organizational and analytical skills, which change depending on the application. Because they are so varied, we cannot explain how to solve every specific control task. Nevertheless, we can provide you with techniques and guidelines for completing this problem-solving process. In this chapter, we will introduce a strategy for implementing a control program, which includes program organization, system configuration, and I/O programming. These strategies also apply to PLCs with the IEC 1131-3 programming standard. Additionally, we will present both simple and complex PLC programming examples. After you finish this chapter, you will be ready to learn how to document the PLC system—the last step in implementing the control program.

11-1 CONTROL TASK DEFINITION

A user should begin the problem-solving process by defining the **control task**, that is, determining what needs to be done. This information provides the foundation for the control program. To help minimize errors, the control task should be defined by those who are familiar with the operation of the machine or process. Proper definition of the task is directly related to the success of the control program.

Control task definition occurs at many levels. All of the departments involved must work together to determine what inputs are required, so that everyone understands the purpose and scope of the project. For example, if a project involves the automation of a manufacturing plant in which materials will be retrieved from the warehouse and sent to the automatic packaging area, personnel from both the warehouse and packaging areas must collaborate with the engineering group during the system definition. Management should also be involved if the project requires data reporting.

If the control task is currently done manually or through relay logic, the user should review the steps of the manual procedure to determine what improvements, if any, can be made. Although relay logic can be directly implemented in a PLC, the procedure should be redesigned, when possible, to meet current project needs and to capitalize on the capabilities of programmable controllers.

11-2 CONTROL STRATEGY

After the control task has been defined, the planning of its solution can begin. This procedure commonly involves determining a **control strategy**, the sequence of steps that must occur within the program to produce the desired output control. This part of the program development is known as the development of an algorithm. The term *algorithm* may be new or strange to some readers, but it need not be. Each of us follows algorithms to accomplish certain tasks in our daily lives. The procedure that a person follows to go from home to either school or work is an algorithm—the person exits the house, gets into the car, starts the engine, and so on. In the last of a finite number of steps, he or she reaches the destination.

The PLC strategy implementation for a control task closely follows the development of an algorithm. The user must implement the control from a given set of basic instructions and produce the solution in a finite number of steps. If developing an algorithm to solve the problem becomes difficult, he or she may need to return to the control task definition to redefine the problem. For example, we cannot explain how to get from where we are to Bullfrog County, Nevada unless we know both where we are and where Bullfrog County is. As part of the problem definition, we need to know if a particular method of transportation is required. If there is a time constraint, we need to know that too. We cannot develop a control strategy until we have all of this problem definition information.

The fundamental rule for defining the program strategy is *think first, program later*. Consider alternative approaches to solving the problem and allow time to polish the solution algorithm before trying to program the control function. Adopting this philosophy will shorten programming time, reduce debugging time, accelerate start-up, and focus attention where it is needed—on design when designing and on programming when programming.

Strategy formulation challenges the system designer, regardless of whether it is a new application or the modernization of an existing process. In either case, the designer must review the sequence of events and optimize control through the addition or deletion of steps. This requires a knowledge of the PLC-controlled field devices, as well as input and output considerations.

11-3 IMPLEMENTATION GUIDELINES

A programmable controller is a powerful machine, but it can only do what it is told to do. It receives all of its directions from the control program, the set of instructions or solution algorithms created by the programmer. Therefore, the success of a PLC control program depends on how organized the user is. There are many ways to approach a problem; but if the application is approached in a systematic manner, the probability of mistakes is less.

The techniques used to implement the control program vary according to the programmer. Nevertheless, the programmer should follow certain guidelines. Table 11-1 lists programming guidelines for new applications and modernizations. New applications are new systems, while modernizations are upgraded existing control systems that have functioned previously without a PLC (i.e., through electromechanical control or individual, analog, loop controllers).

New Applications	Modernizations
• Understand the desired function of the system.	• Understand the actual process or machine function.
• Review possible control methods and optimize the process operation.	• Review machine logic of operation and optimize when possible.
• Flowchart the process operation.	• Assign real I/O and internal addresses to inputs and outputs.
• Implement the flowchart by using logic diagrams or relay logic symbology.	• Translate relay ladder diagram into PLC coding.
• Assign real I/O addresses and internal addresses to inputs and outputs.	
• Translate the logic implementation into PLC coding.	

Table 11-1. Programming guidelines.

As mentioned previously, understanding the process or machine operation is the first step in a systematic approach to solving the control problem. For new applications, the strategy should follow the problem definition. Reviewing strategies for new applications, as well as revising the actual method of control for a modernization project, will help detect errors that were introduced during the planning stages.

The programming stage reveals the difference between new and modernization projects. In a modernization project, the user already understands the operation of the machine or process, along with the control task. An existing relay ladder diagram, like the one shown in Figure 11-1, usually defines the sequence of events in the control program. This ladder diagram can be almost directly translated into PLC ladder diagrams.

New applications usually begin with specifications given to the person who will design and install the control system. The designer translates these specifications into a written description that explains the possible control strategies. The written explanation should be simple to avoid confusion. The designer then uses this explanation to develop the control program.

11-4 PROGRAM ORGANIZATION AND IMPLEMENTATION

Organization is a key word when programming and implementing a control solution. The larger the project, the more organization is needed, especially when a group of people is involved.

In addition to organization, a successful control solution also depends on the ability to implement it. The programmer must understand the PLC control task and controlled devices, choose the correct equipment for the job

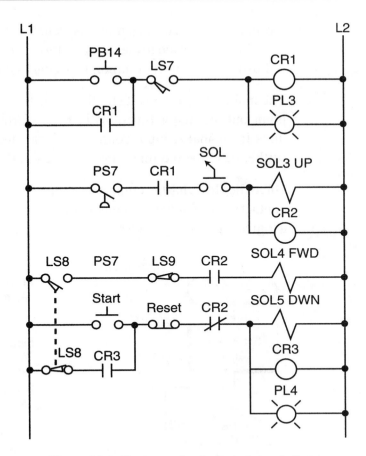

Figure 11-1. Electromechanical relay circuit diagram.

(hardware and software), and understand the PLC system. Once these preliminary details are understood, the programmer can begin sketching the control program solution. The work performed during this time forms an important part of the system or project documentation. Documenting a system once it is installed and working is difficult, especially if you do not remember how you got it to work in the first place. Therefore, documenting the system throughout its development will pay off in the end.

CREATING FLOWCHARTS AND OUTPUT SEQUENCES

Flowcharting is a technique often used when planning a program after a written description has been developed. A flowchart is a pictorial representation that records, analyzes, and communicates information, as well as describes the operational process in a sequential manner. Figure 11-2 illustrates a simple flowchart. Each step in the chart performs an operation, whether it is an input/output, decision, or data process.

In a flowchart, broad concepts and minor details, along with their relationship to each other, are readily apparent. Sequences and relationships that are hard to extract from general descriptions also become obvious when expressed through a flowchart. Even the flowchart symbols themselves have specific meanings, which aid in the interpretation of the solution algorithm. Figure 11-3 illustrates the most common flowchart symbols and their meanings.

The main flowchart itself should not be long and complex; instead, it should point out the major functions to be performed (e.g., compute engineering units from analog input counts). Several smaller flowcharts can be used to further describe the functions specified in the main flowchart.

Once the flowchart is completed, the user can employ either logic gates or contact symbology to implement the logic sequences. Logic gates implement a logical output sequence given specific real and/or internal input conditions,

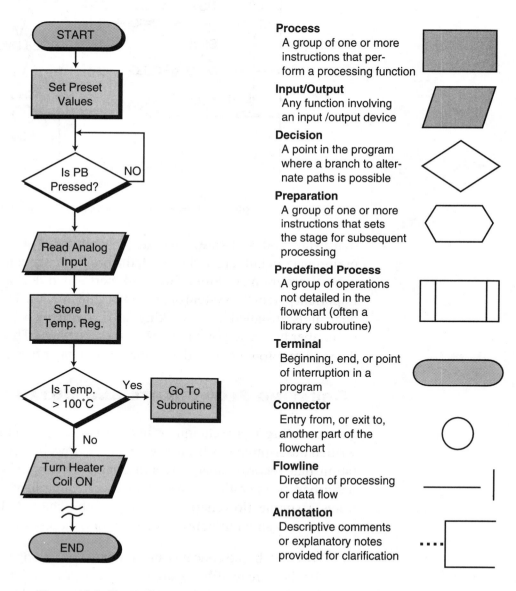

Process
A group of one or more instructions that per-form a processing function

Input/Output
Any function involving an input /output device

Decision
A point in the program where a branch to alter-nate paths is possible

Preparation
A group of one or more instructions that sets the stage for subsequent processing

Predefined Process
A group of operations not detailed in the flowchart (often a library subroutine)

Terminal
Beginning, end, or point of interruption in a program

Connector
Entry from, or exit to, another part of the flowchart

Flowline
Direction of processing or data flow

Annotation
Descriptive comments or explanatory notes provided for clarification

Figure 11-2. Simple flowchart.

Figure 11-3. Flowchart symbols.

while PLC contact symbology directly implements the logic necessary to program an output rung. Figure 11-4 illustrates both of these programming methods. Users should employ whichever method they feel most comfortable with or, perhaps, a combination of both (see Figure 11-5). Logic gate diagrams, however, may be more appropriate in controllers that use Boolean instruction sets.

Inputs and outputs marked with an X on a logic gate diagram, as in Figure 11-4b, represent real I/O in the system. If no mark is present, an I/O point is an internal. The labels used for actual input signals can be either the actual device names (e.g., LS1, PB10, AUTO, etc.) or symbolic letters and numbers that are associated with each of the field elements. During this stage, the user should prepare a short description of the logic sequence.

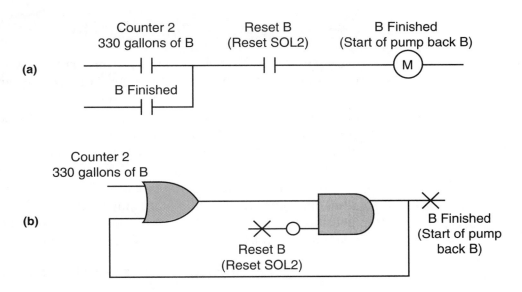

Figure 11-4. (a) PLC contact symbology and **(b)** logic gate representation of a logic sequence.

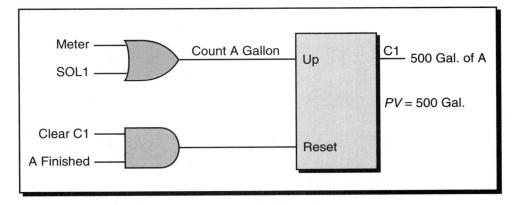

Figure 11-5. A combination of logic gates and contact symbology.

CONFIGURING THE PLC SYSTEM

PLC configuration should be considered during flowcharting and logic sequencing. The PLC's configuration defines which I/O modules will be used with which types of I/O signals, as well as where the modules will be located in the local or remote rack enclosures. The modules' locations determine the I/O addresses that will be used in the control program.

During system configuration, the user should consider the following: possible future expansions; special I/O modules, such as fast-response or wire fault inputs; and the placement of interfaces within a rack (all AC I/O together, all DC and low-level analog I/O together, etc.). Consideration of these details, along with system configuration documentation, will result in a better system design.

REAL AND INTERNAL I/O ASSIGNMENT

The assignment of inputs and outputs is one of the most important procedures that occurs during the programming organization and implementation stages. The I/O assignment table documents and organizes what has been done thus far. It indicates which PLC inputs are connected to which input devices and which PLC outputs drive which output devices. The assignment of internals, including timers, counters, and MCRs, also takes place here. These assignments are the actual contact and coil representations that are used in the ladder diagram program. In applications where electromechanical relay diagrams are available (e.g., modernization of a machine or process), identification of real I/O can be done by circling the devices and then assigning them I/O addresses (see Example 11-1).

Table 11-2 shows an I/O address assignment table for real inputs and outputs, while Table 11-3 shows an I/O address assignment table for internals. These assignments can be extracted from the logic gate diagrams or ladder symbols

Module Type	I/O Address			Description
	Rack	Group	Terminal	
Input	0	0	0	LS1—Position
	0	0	1	LS2—Detect
	0	0	2	Sel Switch—Select 1
	0	0	3	PB1—Start
Output	0	0	4	SOL1
	0	0	5	PL1
	0	0	6	PL2
	0	0	7	Motor M1
Output	0	1	0	SOL2
	0	1	1	PL3

Table 11-2. I/O address assignment table for real inputs and outputs.

Device	Internal	Description
CR7	1010	CR7 replacement
TDR10	T200	ON-delay timer 12 sec
CR10	1011	CR10 replacement
CR14	1012	CR14 replacement
—	1013	Setup interlock

Table 11-3. I/O address assignment table for internal outputs.

that were used to describe the logic sequences. They can also come from the circled elements on an electromechanical diagram. The numbers used for the I/O addresses depend on the PLC model used. These addresses can be represented in octal, decimal, or hexadecimal. The description section of the table specifies the field devices that correspond to each address.

The table of address assignments should closely follow the input/output connection diagram (see Figure 11-6). Although industry standards for I/O representations vary among users, inputs and outputs are typically represented by squares and diamonds, respectively. The I/O connection diagram forms part of the documentation package.

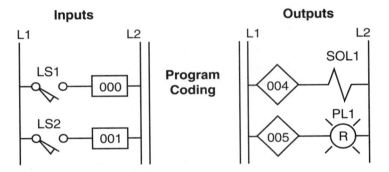

Figure 11-6. Partial connection diagram for the I/O address assignment in Table 11-2.

During the I/O assignment, the user should group associated inputs and outputs. This grouping will allow the monitoring and manipulation of a group of I/O simultaneously. For instance, if 16 motors will be started sequentially, they should be grouped together, so that monitoring the I/O registers associated with the 16 grouped I/O points will reveal the motors' starting sequence. Due to the modularity of an I/O system, all the inputs and all the outputs should be assigned at the same time. This practice will prevent the assignment of an input address to an output module and vice versa.

EXAMPLE 11-1

For the circuit shown in Figure 11-7, **(a)** identify the real inputs and outputs by circling each, **(b)** assign the I/O addresses, **(c)** assign the internal addresses (if required), and **(d)** draw the I/O connection diagram.

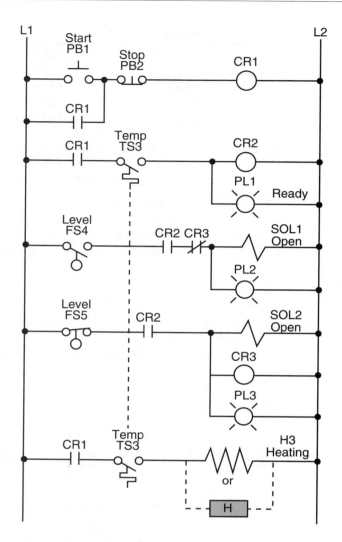

Figure 11-7. Electromechanical relay circuit.

Assume that the PLC used has a modularity of 8 points per module. Each rack has 8 module slots, and the master rack is number 0. Inputs and outputs can have any address as long as the correct module is used. The PLC determines whether an input or output module is connected in a slot. The number system is octal, and internals start at address 1000_8.

SOLUTION

(a) Figure 11-8 shows the circled real input and output connections. Note that temperature switch TS3 is circled *twice* even though it is only *one* device. In the address assignment, only one of them is referenced, and only one of them is wired to an input module.

(b) Table 11-4 illustrates the assignment of inputs and outputs. It assigns all inputs and all outputs, leaving spare I/O locations for future use.

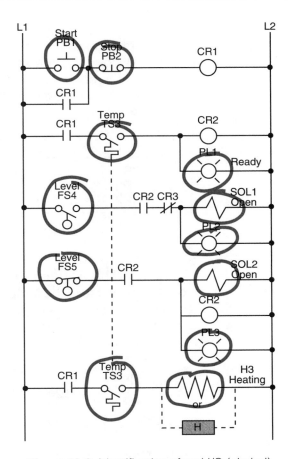

Figure 11-8. Identification of real I/O (circled).

Module Type	I/O Address			Description
	Rack	Group	Terminal	
Input	0	0	0	Start PB1
	0	0	1	Stop PB2
	0	0	2	Temp TS3
	0	0	3	Level FS4
	0	0	4	Level FS5
	0	0	5	—
	0	0	6	—
	0	0	7	—
Spare	0	1	0	Not used
	•	•	•	
	•	•	•	
	0	1	7	
Output	0	2	0	PL1 Ready
	0	2	1	SOL1 Open
	0	2	2	PL2
	0	2	3	SOL2 Open
	0	2	4	PL3
	0	2	5	H3 Heating
	0	2	6	—
	0	2	7	—

Table 11-4. I/O address assignment.

(c) Table 11-5 presents the output assignments, including a description of each internal. Note that control relay CR2 is not assigned as an internal since it is the same as the output rung corresponding to PL1. When the control program is implemented, every contact associated with CR2 will be replaced by contacts with address 020 (the address of PL1).

Device	Internal	Description
CR1	1000	Control relay CR1
CR2	—	Same as PL1 Ready
CR3	—	Same as SOL2 Open

Table 11-5. Internal output assignment.

(d) Figure 11-9 illustrates the I/O connection diagram for the circuit in Figure 11-7. This diagram is based on the I/O assignment from part (b). Note that only one of the temperature switches, the normally open TS3 switch, is a connected input. The logic programming of each switch should be based on a normally open condition (see Chapter 9 for more about input connections).

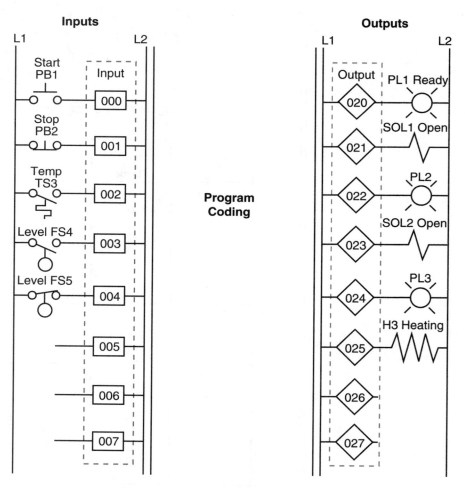

Figure 11-9. I/O connection diagram.

REGISTER ADDRESS ASSIGNMENT

The assignment of addresses to the registers used in the control program is another important aspect of PLC organization. The easiest way to assign registers is to list all of the available PLC registers. Then, as they are used, describe each register's contents, description, and function in a register assignment table. Table 11-6 shows a register assignment table for the first 15 registers in a PLC system, ranging from address 2000_8 to address 2016_8.

Register	Contents	Description
2000	Analog input	Temperature input temp 3 (inside)
2001	Analog input	Temperature input temp 4 (outside)
2002	Spare	–
2003	Spare	–
2004	TWS input	Set point (SP1) input from TWS panel 1
2005	TWS input	Set point volume (V1) from TWS panel 2
2006	Constant 2350	Timer constant of 23.5 sec (0.01 sec TB)
2007	Accumulated	Accumulated value for counter R2010
2010	Spare	–
2011	Spare	–
2012	Constant 1000	Beginning of look-up table (value #1)
2013	Constant 1010	Look-up value #2
2014	Constant 1023	Look-up value #3
2015	Constant 1089	Look-up value #4
2016	Constant 1100	Look-up value #5

Table 11-6. Register assignment table.

ELEMENTS TO LEAVE HARDWIRED

During the assignment of inputs and outputs, the user should decide which devices will not be wired to the controller. These elements will remain part of the electromechanical control logic. These elements usually include devices that are not frequently switched off after start, such as compressors and hydraulic pumps. Components like emergency stops and master start push buttons should also remain hardwired, principally for safety purposes. This way, if the controller is faulty and an emergency occurs, the user can shut down the system without PLC intervention.

Figure 11-10 provides an example of system components that are typically left hardwired. Note that the normally open PLC Fault Contact 1 (or watchdog timer contact) is wired in series with other emergency conditions. This contact stays closed when the controller is operating correctly, but opens when a fault occurs. The system designer can also use this contact if an emergency occurs to disable the PLC system's operation.

PLC fault contacts are safety contacts that are available to the user when implementing or enhancing a safety circuit. When a PLC is operating correctly, the normally open fault contact closes and the normally closed one opens when the PLC is first turned on. As shown in Figure 11-10, these contacts are connected in series with the hardwired circuit, so that if the PLC fails during standard operation, the normally open contacts will open. This will shut down the hardwired circuit at the point where the PLC becomes the controlling element. This circuit also uses a safety control relay (SCR) to control power to the rest of the control components. The normally closed fault contacts are used to indicate an alarm condition.

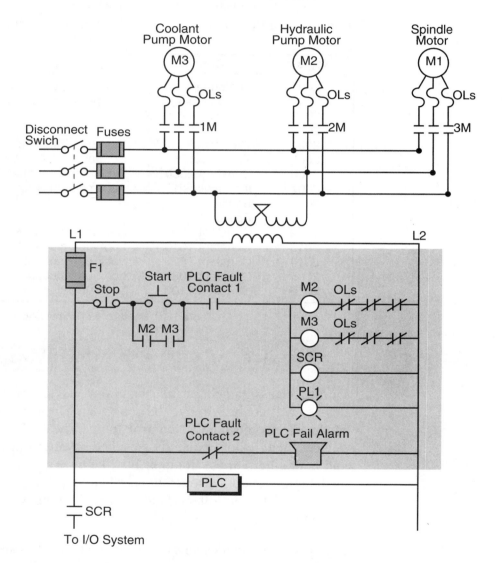

Figure 11-10. Hardwired components in a PLC system.

In the diagram shown in Figure 11-10, an emergency situation (including a PLC malfunction) will remove power (L1) to the I/O modules. The turning OFF of the safety control relay (SCR) will open the SCR contact, stopping the flow of power to the system. Furthermore, the normally closed PLC fault contact (PLC Fault Contact 2) in the hardwired section will alert personnel of a system failure due to a PLC malfunction. The designer should implement this type of alarm in the main PLC rack, as well as in each remote I/O rack location, since remote systems also have fault contacts incorporated into the remote controllers. This allows subsystem failures to be signaled promptly, so that the problem can be fixed without endangering personnel.

SPECIAL INPUT DEVICE PROGRAMMING

Some PLC circuits and input connections require special programming. One example, which we discussed in Chapter 9, is the programming of normally closed input devices. Remember that the programming of a device is closely related to how that device should behave in the control program.

Normally Closed Devices. An input device that is wired as a normally open input can be programmed to act as either a normally open or a normally closed device. The same rule applies for normally closed inputs. Generally, if a device is wired as a normally closed input and it must act as a normally closed input, its reference address is programmed as normally open. As the following example illustrates, however, a normally closed device in a hardwired circuit is programmed as normally closed when it is replaced in the PLC control program. Since it is not referenced as an input, the program does not evaluate the device as a real input.

EXAMPLE 11-2

For the circuit in Figure 11-11, draw the PLC ladder program and create an I/O address assignment table. For inputs, use addresses 10_8 through 47_8. Start outputs at address 50_8 and internals at address 100_8.

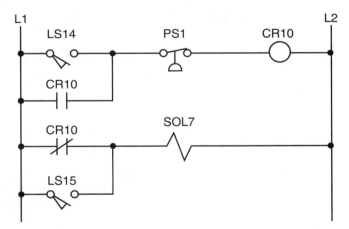

Figure 11-11. Electromechanical relay circuit.

SOLUTION

Figure 11-12 shows the equivalent PLC ladder diagram for the circuit in Figure 11-11. Table 11-7 shows the I/O address assignment table for this example. The normally closed contact (CR10) is programmed as normally closed because internal coil 100 references it and requires it to operate as a normally closed contact.

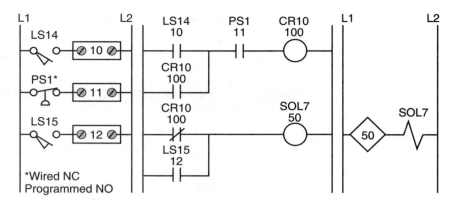

Figure 11-12. PLC ladder diagram of the circuit in Figure 11-11.

I/O Address	Device	Type
10	LS14	Input
11	PS1	Input
12	LS15	Input
50	SOL7	Output
100	CR10	Internal

Table 11-7. I/O address assignment table.

Master Control Relays. Another circuit the programmer should be aware of is a master control relay (MCR). In electromechanical circuit diagrams, an MCR coil controls several rungs in a circuit by switching ON or OFF the power to those rungs. In a hardwired circuit, there is no definite end to an MCR except when the circuit is followed all the way through. For example, in Figure 11-13, the MCR output in line 1 controls the power to the hardwired

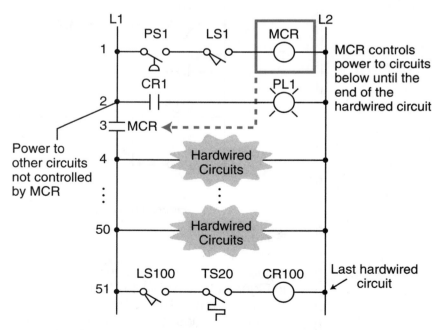

Figure 11-13. Electromechanical relay circuit with a master control relay.

elements from line 3, where the MCR contact is located, to the last element in line 51. If the master control relay is ON, power will flow to these rungs (lines 4 through 51). If the master control relay is OFF, power will not flow and these devices will not implement the control action. This configuration is equivalent to a hardwired subprogram or subroutine—if the MCR is ON, the rungs are executed; if it is OFF, the rungs are not executed. At line 2 in the circuit, power branches to other circuits that are not affected by the MCR's action. These circuits are the regular hardwired program.

During the translation from a hardwired ladder circuit to PLC symbology, the programmer must place an END MCR instruction after the last rung the MCR should control. Figure 11-14 illustrates the placement of the MCR instruction for the circuit in Figure 11-13. To provide proper fencing for the program's MCR control section, internal output coil 1000, labeled CR1 (line 1 of PLC program), was inserted so that PL1 would not be inside the fenced MCR area. This is the way the hardwired circuit operates. The END1

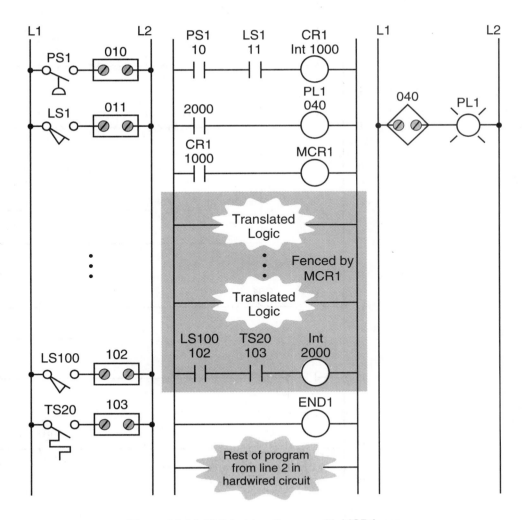

Figure 11-14. PLC ladder diagram with MCR fence.

instruction ends the MCR fence. The instructions corresponding to the hardwired circuits that branch from line 2 in the electromechanical diagram of Figure 11-13 are located after the END1 instruction. Figure 11-15 illustrates a partial ladder rung of a more elaborate circuit with this type of MCR condition. The corresponding PLC program should have an END MCR after the rung containing the PL3 output.

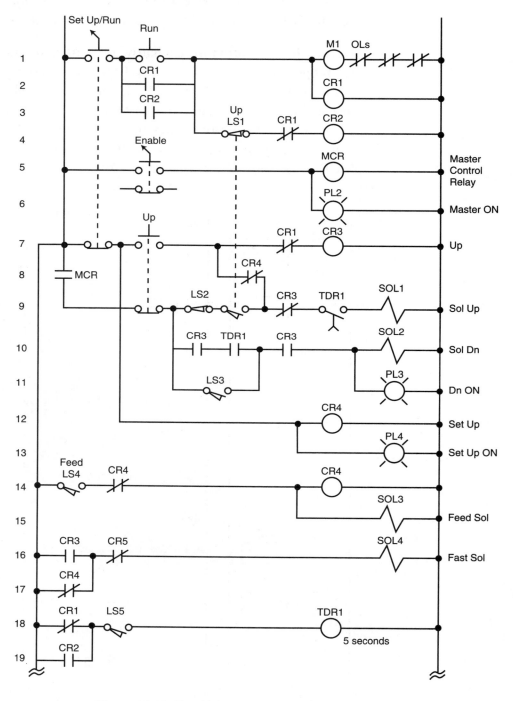

Figure 11-15. Electromechanical relay circuit with an MCR.

EXAMPLE 11-3

Highlight the sections of the circuit in Figure 11-15 that will be under the control of a PLC MCR. What additional measures must be taken to include or bypass other hardwired circuits within the MCR fence?

SOLUTION

Figure 11-16 highlights the circuits that must be fenced under the MCR instruction. Note that solenoid SOL1 and part of its driving logic are not included in the MCR fencing because SOL1, CR3, and TDR1 can also be turned ON by logic prior to the MCR fence (see Figure 11-17). For the MCR fence to be properly programmed, the PLC program

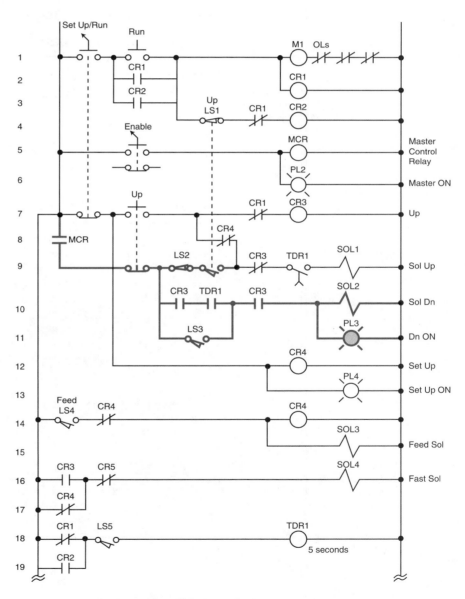

Figure 11-16. MCR-controlled program elements.

must include two internal control relays that take SOL1 out of the fence. Figure 11-18 illustrates the fenced circuit with the additional internals (CR1000 and CR1001). Note that the instructions in this diagram have the same names as in the hardwired circuit. The solenoid SOL1 will be outside of the MCR fence because it can be turned ON by either the outside logic (highlighted section in Figure 11-17) or the logic inside the MCR fence (highlighted section in Figure 11-18).

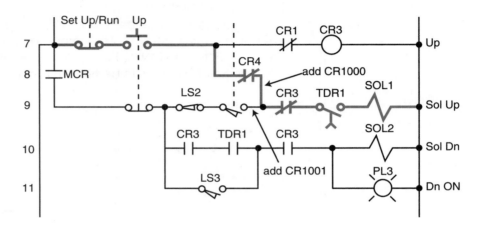

Figure 11-17. SOL1 activated by logic outside of the MCR fence.

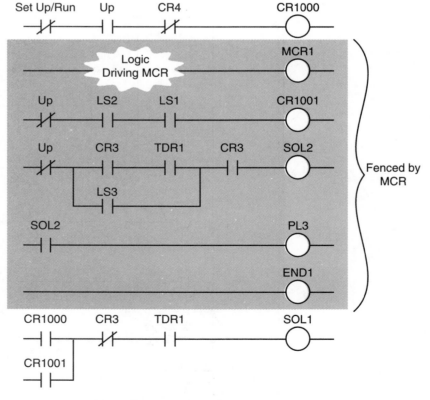

Figure 11-18. MCR fence.

Bidirectional Power Flow. The circuit in Figure 11-19 illustrates another condition that can cause programming problems: the possibility of bidirectional power flow through the normally closed CR4 contact in line 8. To solve the bidirectional flow problem, the programmer must know whether or not CR4 influences the two output rungs to which it is connected. These rungs are the CR3 control relay output and the solenoid SOL1 output (rungs 7 and 9, respectively). Figure 11-19 illustrates the two paths that can occur in the hardwired circuit. PLCs only allow forward paths; therefore, if a reverse path is necessary for this circuit's logic, the CR4 contact must be included in the logic driving the CR3 output (see Figure 11-19b). Chapter 9 provides more details about reverse and bidirectional power flow.

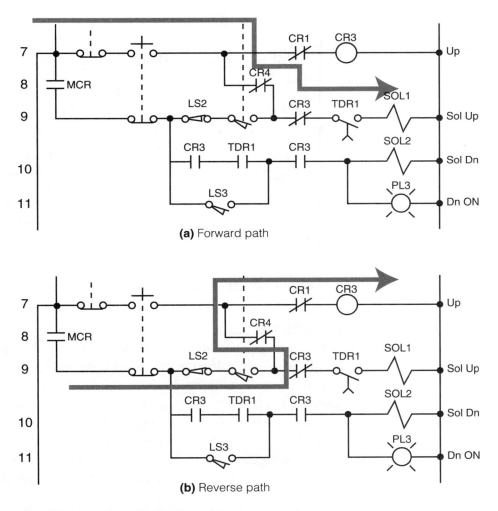

Figure 11-19. (a) Forward and **(b)** reverse power flow in a hardwired circuit.

Instantaneous Timer Contacts. The electromechanical circuit shown in Figure 11-15 specifies an instantaneous timer contact (the normally open TDR1 contact in line 10). This type of contact, however, is usually unavailable in PLCs. To implement an instantaneous timer contact (i.e., a contact that closes or opens once the timer is enabled), the programmer must use an internal output to trap the timer, then use the internal's contact as an instantaneous contact to drive the timer's logic.

In the electromechanical circuit in Figure 11-20a, if PB1 and LS1 both close, the timer will start timing and the instantaneous contact (TMR1-1) will close, thus sealing PB1. If PB1 is released (OFF), the timer will continue to time because the circuit is sealed. Figure 11-20b illustrates the technique for trapping a timer. In this PLC program, an internal output traps the instantaneous contact from the circuit's electromechanical timer. Thus, the contacts from this internal drive the timer. If a trap does not exist, the timer will start timing when PB1 and LS1 both close, but will stop timing as soon as PB1 is released.

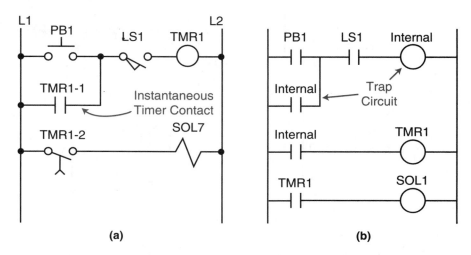

Figure 11-20. (a) An instantaneous timer contact in a hardwired circuit and **(b)** a trapped timer in a PLC circuit.

Complicated Logic Rungs. When a logic rung is very confusing, the best programming procedure is to isolate it from the other rungs. Then, reconstruct all of the possible logic paths from right to left, starting at the output and ending at the beginning of the rung. If a section of a rung, like the one discussed in Example 11-3, directly connects or interacts with another rung, it may be easier to create an internal output at the point where the two rungs cross. Then, use the internal output to drive the rest of the logic. For the circuit shown in Figure 11-15, this cross point is in line 9 at the normally closed contact CR4 between normally open LS1 and normally closed CR3.

PROGRAM CODING/TRANSLATION

Program coding is the process of translating a logic or relay diagram into PLC ladder program form. This ladder program, which is stored in the application memory, is the actual logic that will implement the control of the machine or process. Ease of program coding is directly related to how orderly the previous stages (control task definition, I/O assignment, etc.) have been done. Figure 11-21 shows a sample program code generated from logic gates and electromechanical relay diagrams (internal coil 1000 replaces the control relay). Note that the coding is a PLC representation of the logic, whether it is a new application or a modernization. The next sections examine this coding process closer and present several programming examples.

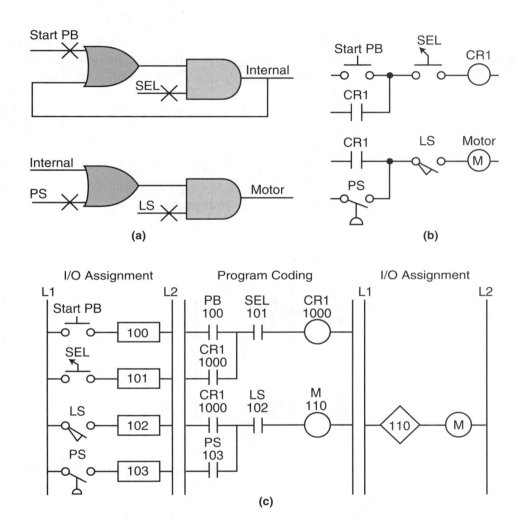

Figure 11-21. Translation from **(a)** logic gates and **(b)** an electromechanical relay diagram into **(c)** PLC program coding.

11-5 DISCRETE I/O CONTROL PROGRAMMING

In this section, we will present several programming examples that illustrate the modernization of relay systems. We will also present examples relating to new PLC control implementations. These examples will deal primarily with discrete controls. The next section will explain more about analog I/O interaction and programming.

CONTROL PROGRAMMING AND PLC DESCRIPTIONS

Modernization applications involve the transfer of a machine or process's control from conventional relay logic to a programmable controller. Conventional hardwired relay panels, which house the control logic, usually present maintenance problems, such as contact chatter, contact welding, and other electromechanical problems. Switching to a PLC can improve the performance of the machine, as well as optimize its control. The machine's "new" programmable controller program is actually based on the instructions and control requirements of the original hardwired system.

Throughout this section, we will use the example of a midsized PLC capable of handling up to 512 I/O points (000 to 777 octal) to explain how to implement and configure a PLC program. The I/O structure of the controller has 4 I/O points per module. The PLC has eight racks (0 through 7), each one with eight slots, or groups, where modules can be inserted. Figure 11-22 illustrates this configuration.

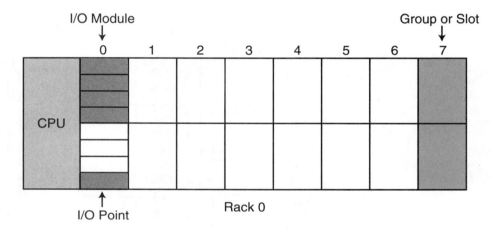

Figure 11-22. Example PLC configuration.

The PLC can accept four-channel analog input modules, which can be placed in any slot location. When analog I/O modules are used, discrete I/O cannot be used in the same slot. The PLC can also accept multiplexed register I/O. These multiplexed modules require two slot positions and provide the enable (select) lines for the I/O devices. The software instructions available in this PLC are similar to those presented in Chapter 9.

Addresses 000 through 777 octal represent input and output device connections mapped to the I/O table. The first digit of the address represents the rack number, the second digit represents the slot, and the third digit specifies the terminal connection in the slot. The PLC detects whether the slot holds an input or an output.

Point addresses 1000_8 to 2777_8 may be used for internal outputs, and register storage starts at register 3000_8 and ends at register 4777_8. Two types of timer and counter formats can be used—ladder format and block format—but all timers require an internal output to specify the ON-delay output. Ladder format timers place a "T" in front of the internal output address, while block format timers specify the internal output address in the block's output coil.

Throughout the examples presented in this section and the next, we will use addresses 000_8 through 027_8 for discrete inputs and addresses 030_8 through 047_8 for discrete outputs. Analog I/O will be placed in the last slot of the master rack (0) whenever possible. During the development of these examples, you will discover that sometimes the assignment of internals and registers is performed parallel to the programming stages.

SIMPLE RELAY REPLACEMENT

This relay replacement example involves the PLC implementation of the electromechanical circuit shown in Figure 11-23. The hardware timer TMR1 requires instantaneous contacts in the first rung, which are used to latch the

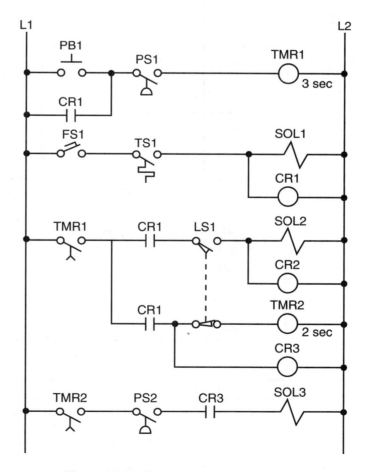

Figure 11-23. Electromechanical relay circuit.

rung. If the instantaneous TMR1 contacts are implemented using a PLC time-delay contact, then PB1 must be pushed for the timer's required time preset to latch the rung. This instantaneous contact will be implemented by trapping the timer with an internal output.

Tables 11-8 and 11-9 show the I/O address and internal output assignments for the electromechanical circuit's real I/O. Table 11-10 presents the register assignment table. Note that internals do not replace control relays CR1 and CR2 since the output addresses 030 and 031 corresponding to solenoids SOL1 and SOL2 are available. Therefore, addresses 030 and 031 can replace the CR1 and CR2 contacts, respectively, everywhere they occur in the program. The normally open contact LS1 connects limit switch LS1 to the PLC input interface; and the normally open LS1 reference, programmed with an examine-OFF instruction, implements the normally closed LS1 in the program. Figure 11-24 illustrates the PLC program coding solution.

| Module Type | I/O Address | | | Description |
	Rack	Group	Terminal	
Input	0	0	0	PB1
	0	0	1	PS1
	0	0	2	FS1
	0	0	3	TS1
Input	0	0	4	LS1
	0	0	5	PS2
	0	0	6	—
	0	0	7	—
Output	0	3	0	SOL1
	0	3	1	SOL2
	0	3	2	SOL3
	0	3	3	—

Table 11-8. I/O address assignment.

Device	Internal	Description
TMR1	1000	Used to trap TMR1
CR1	—	Same as SOL1 (030)
CR2	—	Same as SOL2 (031)
TMR1	1001	Timer TMR1
TMR2	1002	Timer TMR2
CR3	1003	Replace CR3

Table 11-9. Internal address assignment.

Register	Description
4000	Preset timer count for 3 sec
4001	Accumulated count timer 1001
4002	Preset timer count for 2 sec
4003	Accumulated count timer 1002

Table 11-10. Register assignment.

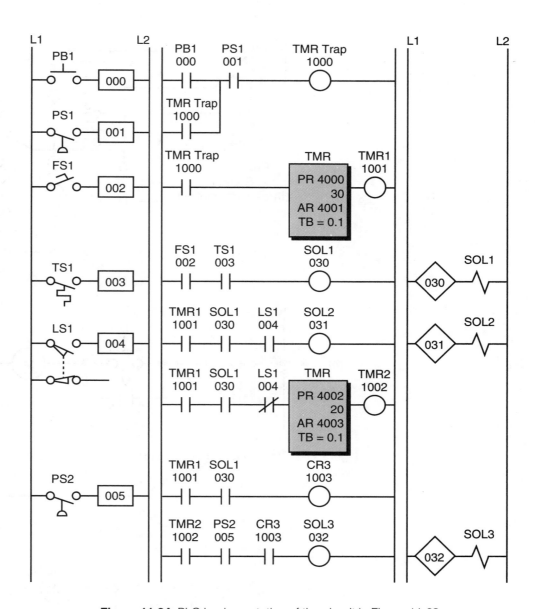

Figure 11-24. PLC implementation of the circuit in Figure 11-23.

SIMPLE START/STOP MOTOR CIRCUIT

Figure 11-25 shows the wiring diagram for a three-phase motor and its corresponding three-wire control circuit, where the auxiliary contacts of the starter seal the start push button. To convert this circuit into a PLC program, first determine which control devices will be part of the PLC I/O system; these are the circled items in Figure 11-26. In this circuit, the start and stop push buttons (inputs) and the starter coil (output) will be part of the PLC system. The starter coil's auxiliary contacts will not be part of the system because an internal will be used to seal the coil, resulting in less wiring and fewer

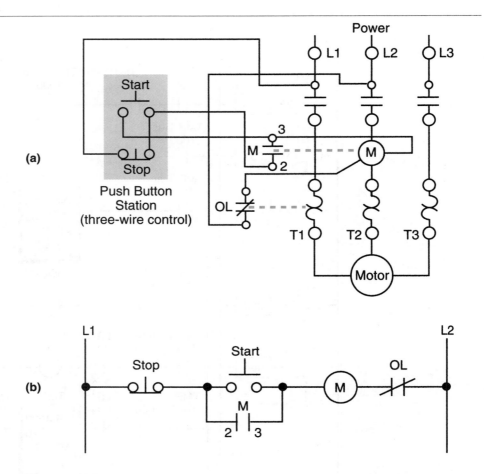

Figure 11-25. (a) Wiring diagram and **(b)** relay control circuit for a three-phase motor.

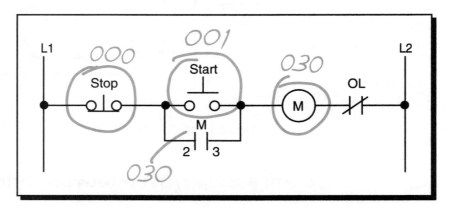

Figure 11-26. Real inputs and outputs to the PLC.

connections. Table 11-11 shows the I/O address assignment, which uses the same addressing scheme as the circuit diagram (i.e., inputs: addresses 000 and 001, output: address 030).

To program the PLC, the devices must be programmed in the same logic sequence as they are in the hardwired circuit (see Figure 11-27). Therefore, the stop push button will be programmed as an examine-ON instruction

| Module Type | I/O Address | | | Description |
	Rack	Group	Terminal	
Input	0	0	0	Stop PB (NC)
	0	0	1	Start PB
	0	0	2	—
	0	0	3	—
Output	0	3	0	Motor M1
	0	3	1	—
	0	3	2	—
	0	3	3	—

Table 11-11. I/O address assignment.

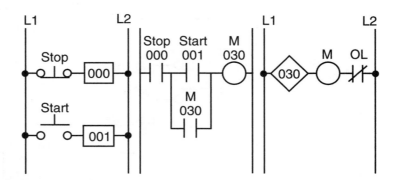

Figure 11-27. PLC implementation of the circuit in Figure 11-25.

(a normally open PLC contact) in series with the start push button, which is also programmed as an examine-ON instruction. This circuit will drive output 030, which controls the starter. If the start push button is pressed, output 030 will turn ON, sealing the start push button and turning the motor ON through the starter. If the stop push button is pressed, the motor will turn OFF. Note that the stop push button is wired as normally closed to the input module. Also, the starter coil's overloads are wired in series with the coil.

In a PLC wiring diagram, the PLC is connected to power lines L1 and L2 (see Figure 11-28). The field inputs are connected to L1 on one side and to the module on the other. The common, or return, connection from the input module goes to L2. The output module receives its power for switching the load from L1. Output terminal 030 is connected in series with the starter coil and its overloads, which go to L2. The output module also directly connects to L2 for proper operation. Note that, in the motor control circuit's wiring diagram (see Figure 11-29), the PLC output module is wired directly to the starter coil.

Although the three-phase motor has a three-wire control circuit, its corresponding PLC control circuit has only two wires. This two-wire configuration is similar to a three-wire configuration because it provides low-voltage release; however, it does not provide low-voltage protection. Referring to

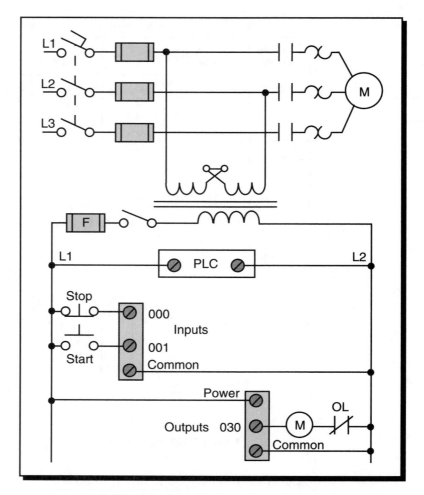

Figure 11-28. PLC wiring diagram of a three-phase motor.

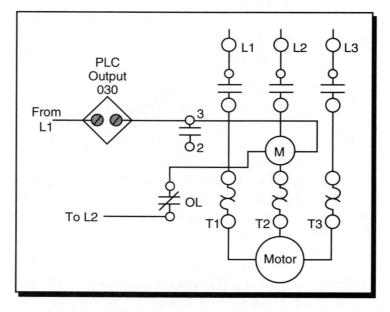

Figure 11-29. Motor control circuit's wiring diagram.

Figure 11-29, the starter's seal-in contacts (labeled as 3—| |—2) are not used and are shown as unconnected. If the motor is running and the overloads open, the motor will stop, but the circuit will still be ON. Once the overloads cool off and the overload contacts close, the motor will start again immediately. Depending on the application, this situation may not be desirable. For example, someone may be troubleshooting the motor stoppage and the motor may suddenly restart. Making the auxiliary contact an input and using its address to seal the start push button can avoid this situation by making the two-wire circuit act as a three-wire circuit (see Figure 11-30). In this configuration, if the overloads open while the motor is running, the coil will turn off and their auxiliary contacts will break the circuit in the PLC.

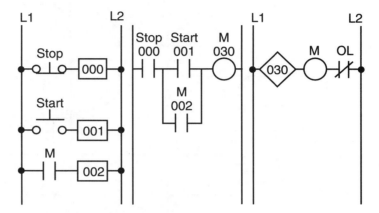

Figure 11-30. Two-wire circuit configured as a three-wire circuit.

FORWARD/REVERSE MOTOR INTERLOCKING

Figure 11-31 illustrates a hardwired forward/reverse motor circuit with electrical and push button interlockings. Figure 11-32 shows the simplified wiring diagram for this motor. The PLC implementation of this circuit

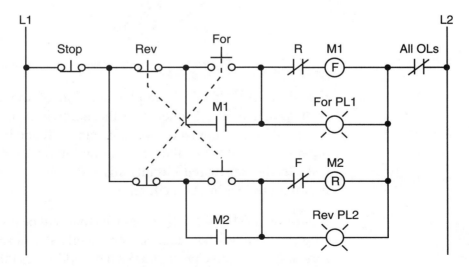

Figure 11-31. Hardwired forward/reverse motor circuit.

should include the use of the overload contacts to monitor the occurrence of an overload condition. The auxiliary starter contacts (M1 and M2) are not required in the PLC program because the sealing circuits can be programmed using the internal contacts from the motor outputs. Low-voltage protection can be implemented using the overload contact input so that, if an overload occurs, the motor circuit will turn off. However, after the overload condition passes, the operator must push the forward or reverse push button again to restart the motor.

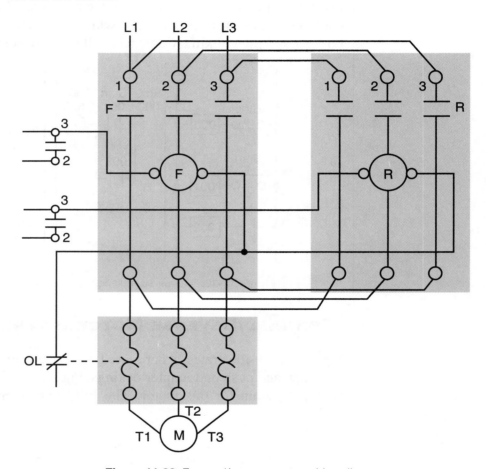

Figure 11-32. Forward/reverse motor wiring diagram.

For simplicity, the PLC implementation of the circuit in Figure 11-31 includes all of the elements in the hardwired diagram, even though the additional starter contacts (normally closed R and F in the hardwired circuit) are not required, since the push button interlocking accomplishes the same task. In the hardwired circuit, this redundant interlock is performed as a backup interlocking procedure.

Figure 11-33 shows the field devices that will be connected to the PLC. The stop push button has address 000, while the normally open sides of the forward and reverse push buttons have addresses 001 and 002, respectively. The overload contacts are connected to the input module at address 003. The

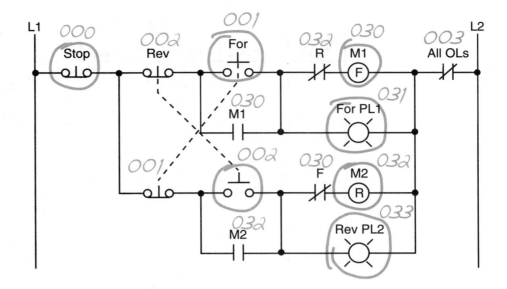

Figure 11-33. Real inputs and outputs to the PLC.

output devices—the forward and reverse starters and their respective interlocking auxiliary contacts—have addresses 030 and 032. The forward and reverse pilot light indicators have address 031 and 033, respectively. Additionally, the overload light indicators have addresses 034 and 035, indicating that the overload condition occurred during either forward or reverse motor operation. The addresses for the auxiliary contact interlocking using the R and F contacts are the output addresses of the forward and reverse starters (030 and 032). The ladder circuit that latches the overload condition (forward or reverse) must be programmed before the circuits that drive the forward and reverse starters as we will explain shortly. Otherwise, the PLC program will never recognize the overload signal because the starter will be turned off in the circuit during the same scan when the overload occurs. If the latching circuit is after the motor starter circuit, the latch will never occur because the starter contacts will be open and continuity will not exist.

Table 11-12 shows the real I/O address assignment for this circuit. Figure 11-34 shows the PLC implementation, which follows the same logic as the hardwired circuit and adds additional overload contact interlockings. Note that the motor circuit also uses the overload input, which will shut down the motor. The normally closed overload contacts are programmed as normally open in the logic driving the motor starter outputs. The forward and reverse motor commands will operate normally if no overload condition exists because the overload contacts will provide continuity. However, if an overload occurs, the contacts in the PLC program will open and the motor circuit will turn OFF. The overload indicator pilot lights (OL Fault Fwd and OL Fault Rev) use latch/unlatch instructions to latch whether the overload occurred in the forward or reverse operation. Again, the latching occurs before the forward and reverse motor starter circuits, which will turn off due

Module Type	I/O Address			Description
	Rack	Group	Terminal	
Input	0	0	0	Stop PB (wired NC)
	0	0	1	Forward PB (wired NO)
	0	0	2	Reverse PB (wired NO)
	0	0	3	Overload contacts
Input	0	0	4	Acknowledge OL/Reset PB
	•	•	•	
	•	•	•	
	•	•	•	
Output	0	3	0	Motor starter M1 (FWD)
	0	3	1	Forward PL1
	0	3	2	Motor starter M2 (REV)
	0	3	3	Reverse PL2
Output	0	3	4	Overload condition FWD
	0	3	5	Overload condition REV
	0	3	6	—
	0	3	7	—

Table 11-12. I/O address assignment.

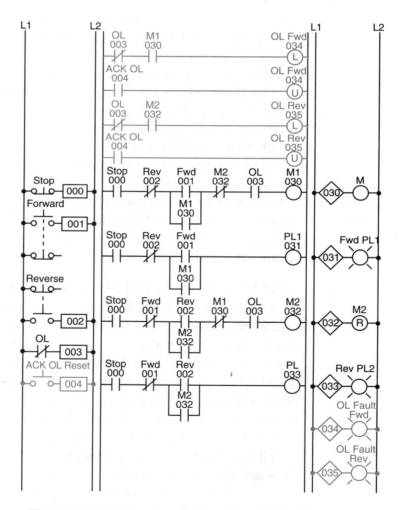

Figure 11-34. PLC implementation of the circuit in Figure 11-31.

to the overload. An additional normally open acknowledge overload reset push button, which is connected to the input module, allows the operator to reset the overload indicators. Thus, the overload indicators will remain latched, even if the physical overloads cool off and return to their normally closed states, until the operator acknowledges the condition and resets it.

Figure 11-35 illustrates the motor wiring diagram of the forward/reverse motor circuit and the output connections from the PLC. Note that the auxiliary contacts M1 and M2 are not connected. In this wiring diagram, both the forward and reverse coils have their returns connected to L2 and not to the overload contacts. The overload contacts are connected to L1 on one side and to the PLC's input module on the other (input 003). In the event of an overload, both motor starter output coils will be dropped from the circuit because the PLC's output to both starters will be OFF.

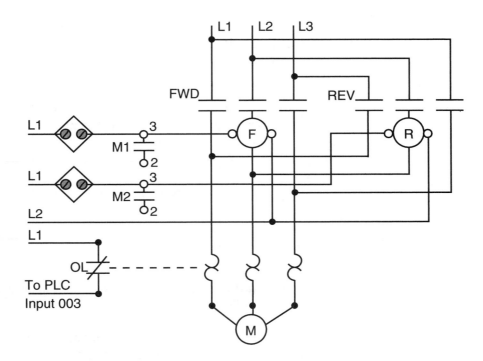

Figure 11-35. Forward/reverse motor wiring diagram.

REDUCED-VOLTAGE-START MOTOR CONTROL

Figure 11-36 illustrates the control circuit and wiring diagram of a 65% tapped, autotransformer, reduced-voltage-start motor control circuit. This reduced-voltage start minimizes the inrush current at the start of the motor (locked-rotor current) to 42% of that at full speed. In this example, the timer must be set to 5.3 seconds. Also, the instantaneous contacts from the timer in lines 2 and 3 must be trapped.

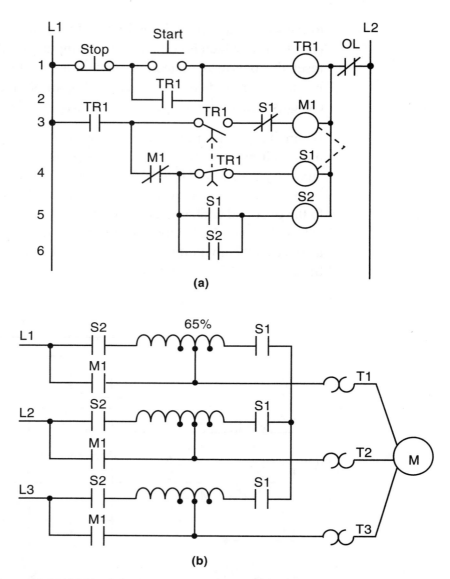

Figure 11-36. (a) Hardwired relay circuit and **(b)** wiring diagram of a reduced-voltage-start motor.

Figure 11-37 illustrates the hardwired circuit with the real inputs and outputs circled. The devices that are not circled are implemented inside the PLC through the programming of internal instructions. Tables 11-13, 11-14, and 11-15 show the I/O assignment, internal assignment, and register assignment, respectively. Figure 11-38 illustrates the PLC implementation of the reduced-voltage-start circuit. The first line of the PLC program traps the timer with internal output 1000. Contacts from this internal replace the instantaneous timer contacts specified in the hardwired control circuit. This PLC circuit implementation does not provide low-voltage protection, since the interlocking does not use the physical inputs of M1, S1, and S2. If low-voltage protection is required, then the starter's auxiliary contacts or the overload contacts can be programmed as described in the previous examples. If the auxiliary contacts or the overloads are used as inputs, they must be pro-

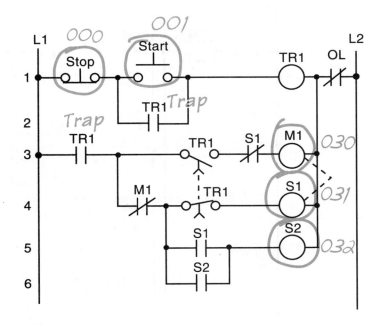

Figure 11-37. Real inputs and outputs to the PLC.

Module Type	I/O Address			Description
	Rack	**Group**	**Terminal**	
Input	0	0	0	Stop PB (NC)
	0	0	1	Start PB (NO)
Output	0	3	0	Motor Starter M1
	0	3	1	S1
	0	3	2	S2

Table 11-13. I/O address assignment.

Device	Internal	Description
—	1000	Trap timer circuit
Timer	1001	Timer

Table 11-14. Internal address assignment.

Register	Description
4000	Preset register value 53, time base 0.1 sec for 5.3 sec (timer output is 1001)
4001	Accumulated register for timer output 1001

Table 11-15. Register assignment.

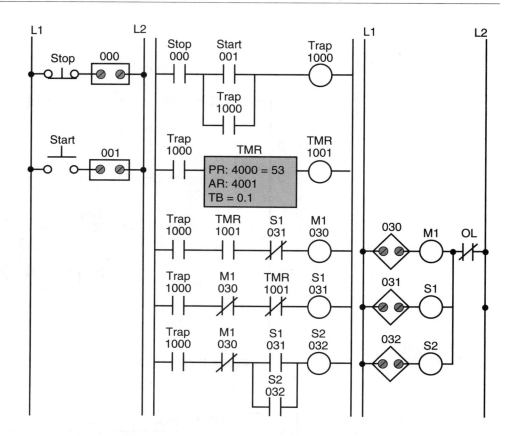

Figure 11-38. PLC implementation of the circuit in Figure 11-36.

grammed as normally open (closed when the overloads are closed and the motor is running) and placed in series with contact 1000 in line 3 of the PLC program. If the overloads open, the circuit will lose continuity and M1 will turn OFF.

AC MOTOR DRIVE INTERFACE

A common PLC application is the speed control of AC motors with variable speed (VS) drives. The diagram in Figure 11-39 shows an operator station used to manually control a VS drive. The programmable controller implementation of this station will provide automatic motor speed control through an analog interface by varying the analog output voltage (0 to 10 VDC) to the drive.

The operator station consists of a speed potentiometer (speed regulator), a forward/reverse direction selector, a run/jog switch, and start and stop push buttons. The PLC program will contain all of these inputs except the potentiometer, which will be replaced by an analog output. The required input field devices (i.e., start push button, stop push button, jog/run, and forward/reverse) will be added to the application and connected to input modules, rather than using the operator station's components. The PLC program will contain the logic to start, stop, and interlock the forward/reverse commands.

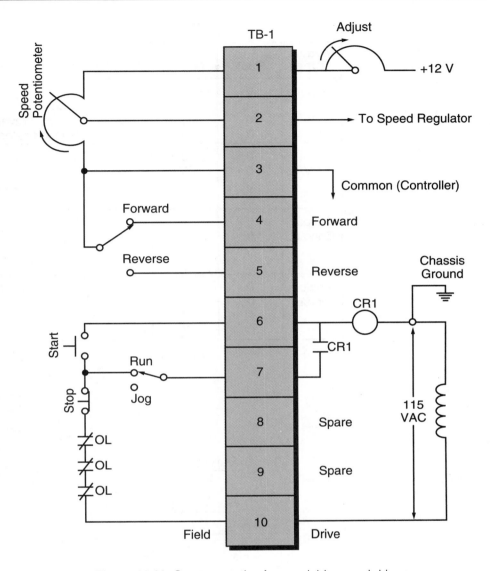

Figure 11-39. Operator station for a variable speed drive.

Table 11-16 shows the I/O address assignment table for this example, while Figure 11-40 illustrates the connection diagram from the PLC to the VS drive's terminal block (TB-1). The connection uses a contact output interface to switch the forward/reverse signal, since the common must be switched. To activate the drive, terminal TB-1-6 must receive 115 VAC to turn ON the internal relay CR1. The drive terminal block TB-1-8 supplies power to the PLC's L1 connection to turn the drive ON. The output of the module (CR1) is connected to terminal TB-1-6. The drive's 115 VAC signal is used to control the motor speed so that the signal is in the same circuit as the drive, avoiding the possibility of having different commons (L2) in the drive (the start/stop common is not the same as the controller's common). In this configuration, the motor's overload contacts are wired to terminals TB-1-9 and TB-1-10, which are the drive's power (L1) connection and the output interface's L1 connection. If an overload occurs, the drive will turn OFF

| Module Type | I/O Address | | | Description |
	Rack	Group	Terminal	
Input	0	0	0	Start
	0	0	1	Stop
	0	0	2	Forward/reverse selector
	0	0	3	Run/jog selector
Output 115 VAC	0	3	0	Drive enable (L1 from drive)
	0	3	1	
	0	3	2	
	0	3	3	
Output Contact	0	3	4	Forward
	0	3	5	Reverse
	0	3	6	
	0	3	7	
Analog Output	0	7	0	Analog speed reference 0–10 VDC
	0	7	1	
	0	7	2	
	0	7	3	

Table 11-16. I/O address assignment.

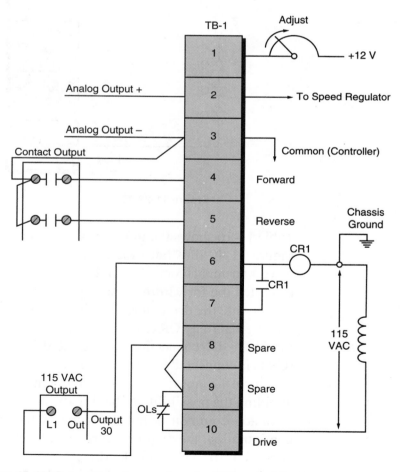

Figure 11-40. Connection diagram from the PLC to the VS drive's terminal block.

because the drive's CR1 contact will not receive power from the output module. This configuration, however, does not provide low-voltage protection, since the drive and motor will start immediately after the overloads cool off and reclose. To have low-voltage protection, the auxiliary contact from the drive, CR1 in terminal TB-1-7, must be used as an input in the PLC, so that it seals the start/stop circuit.

Figure 11-41 shows the PLC ladder program that will replace the manual operator station. The forward and reverse inputs are interlocked, so only one of them can be ON at any given time (i.e., they are mutually exclusive). If the jog setting is selected, the motor will run at the speed set by the analog output when the start push button is depressed. The analog output connection simply allows the output to be enabled when the drive starts. Register 4000 holds the value in counts for the analog output to the drive. Internal 1000, which is used in the block transfer, indicates the completion of the instruction.

Sometimes, a VS drive requires the ability to run under automatic or manual control (AUTO/MAN). Several additional hardwired connections must be made to implement this dual control. The simplest and least expensive way to do this is with a selector switch (e.g., a four-pole, single-throw, single-break selector switch). With this switch, the user can select either the automatic or manual option. Figure 11-42 illustrates this connection. Note

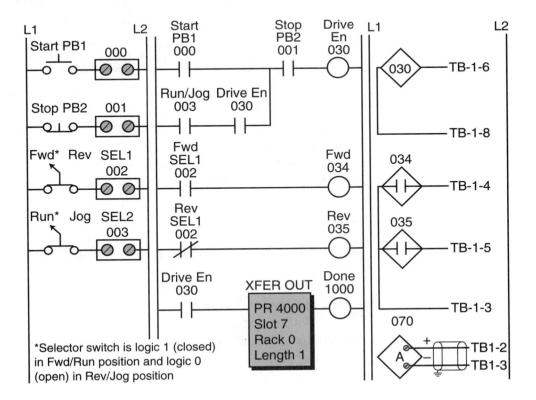

Figure 11-41. PLC implementation of the VS drive.

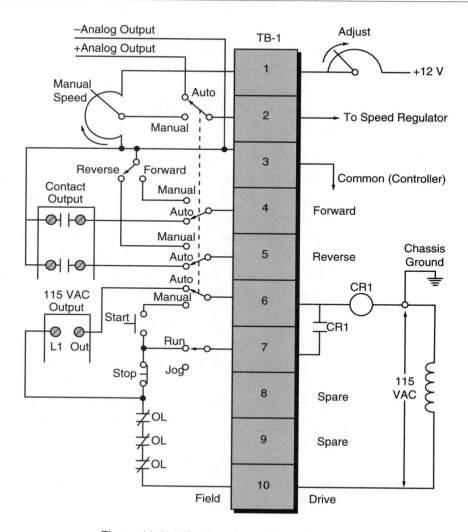

Figure 11-42. VS drive with AUTO/MAN capability.

that the start, stop, run/jog, potentiometer, and forward/reverse field devices shown are from the operator station. These devices are connected to the PLC interface under the same names that are used in the control program (refer to Figure 11-41). If the AUTO/MAN switch is set to automatic, the PLC will control the drive; if the switch is set to manual, the manual station will control the drive.

CONTINUOUS BOTTLE-FILLING CONTROL

In this example (see Figure 11-43), we will implement a control program that detects the position of a bottle via a limit switch, waits 0.5 seconds, and then fills the bottle until a photosensor detects a filled condition. After the bottle is filled, the control program will wait 0.7 seconds before moving to the next bottle. The program will include start and stop circuits for the outfeed motor and the start of the process. Table 11-17 shows the I/O address assignment, while Tables 11-18 and 11-19 present the internal and register assignments, respectively. These assignments include the start and stop process signals.

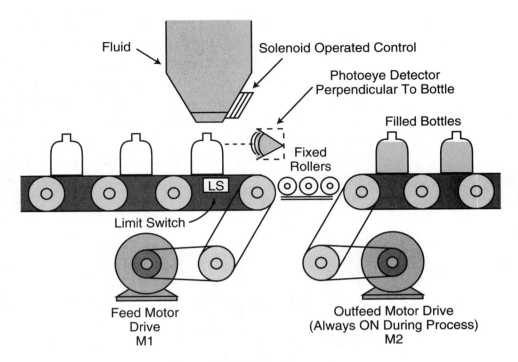

Figure 11-43. Bottle-filling system.

Module Type	I/O Address			Description
	Rack	Group	Terminal	
Input	0	0	0	Start process PB1
	0	0	1	Stop process PB2 (NC)
	0	0	2	Limit switch (position detect)
	0	0	3	Photoeye (level detect)
Output	0	3	0	Feed motor M1
	0	3	1	Outfeed motor M2 (system ON)
	0	3	2	Solenoid control
	0	3	3	—

Table 11-17. I/O address assignment.

Device	Internal	Description
Timer	1001	Timer for 0.5 sec delay after position detect
Timer	1002	Timer for 0.7 sec delay after level detect
—	1003	Bottle filled, timed out, feed motor M1

Table 11-18. Internal output assignment.

Register	Description
4000	Preset value 5, time base 0.1 sec (1001)
4001	Accumulated value for 1001
4002	Preset value 7, time base 0.1 sec (1002)
4003	Accumulated value for 1002

Table 11-19. Register assignment.

Figure 11-44 illustrates the PLC ladder implementation of the bottle-filling application. Once the start push button is pushed, the outfeed motor (output 031) will turn ON until the stop push button is pushed. The feed motor M1 will be energized once the system starts (M2 ON); it will stop when the limit switch detects a correct bottle position. When the bottle is in position and 0.5 seconds have elapsed, the solenoid (032) will open the filling valve and remain ON until the photoeye (PE) detects a proper level. The bottle will remain in position for 0.7 seconds, then the energized internal 1003 will start the feed motor. The feed motor will remain ON until the limit switch detects another bottle.

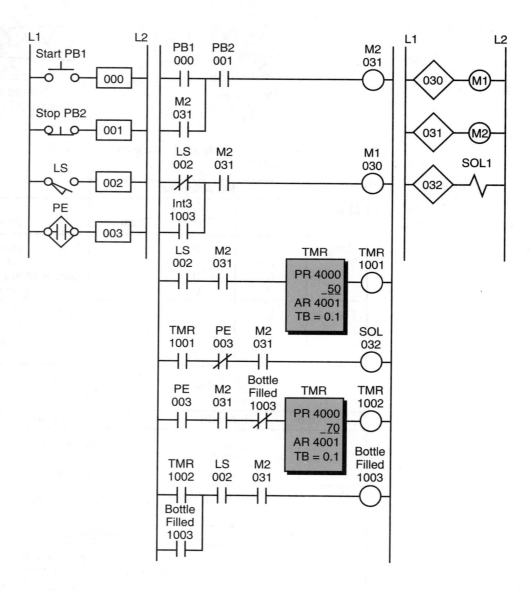

Figure 11-44. PLC implementation of the bottle-filling application.

LARGE RELAY SYSTEM MODERNIZATION

This example presents the modernization of a machine control system that will be changed from hardwired relay logic to PLC programmed logic. The field devices to be used will remain the same, with the exception of those that the controller can implement (e.g., timers, control relays, interlocks, etc.). The benefits of modernizing the control of this machine are:

- a more reliable control system

- less energy consumption

- less space required for the control panel

- a flexible system that can accommodate future expansion

Figure 11-45 illustrates the relay ladder diagram that presently controls the logic sequence for this particular machine. For the sake of simplicity, the diagram shows only part of the total relay ladder logic.

An initial review of the relay ladder diagram indicates that certain portions of the logic should be left hardwired—lines 1, 2, and 3. This will keep all emergency stop conditions independent of the controller. The hydraulic pump motor (M1), which is energized only when the master start push button is pushed (PB1), should also be left hardwired. Figure 11-46 illustrates these hardwired elements. Note that the safety control relay (SCR) will provide power to the rest of the system if M1 is operating properly and no emergency push button is depressed. Furthermore, the PLC fault contact can be placed in series with the emergency push buttons and also connected to a PLC failure alarm. During proper operation, the PLC will energize the fault coil, thus closing PLC Fault Contact 1 and opening PLC Fault Contact 2.

Continuing the example, we can now start assigning the real inputs and outputs to the I/O assignment document. We will assign internal output addresses to all control relays, as well as timers and interlocks from control relays. Tables 11-20 and 11-21 present the assignment and description of the inputs and outputs, as well as the internals. Note that inputs with multiple contacts, such as LS4 and SS3, have only one connection to the controller.

Figure 11-47 shows the PLC program coding (hardwired relay translation) for this example. This ladder program illustrates several special coding techniques that must be used to implement the PLC logic. Among these techniques are the software MCR function, instantaneous contacts from timers, OFF-delay timers, and the separation of rungs with multiple outputs.

An MCR internal output, specified through the program software, performs a function similar to a hardwired MCR. Referring to the relay logic diagram in Figure 11-45, if the MCR is energized, its contacts will close, allowing power to flow to the rest of the system. In the PLC software, the internal MCR

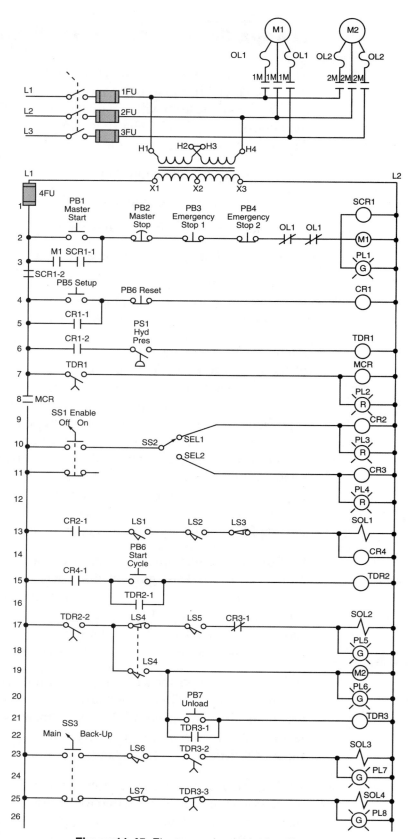

Figure 11-45. Electromechanical relay diagram.

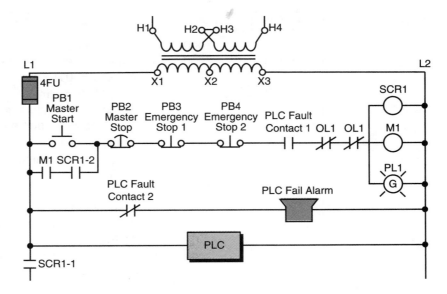

Figure 11-46. Elements of the moderization example system to be left hardwired.

Module Type	I/O Address			Description
	Rack	Group	Terminal	
Input	0	0	0	PB5—Setup PB
	0	0	1	PB6—Reset (wired NC)
	0	0	2	PS1—Hydraulic pressure switch
	0	0	3	SS1—Enable selector switch (NC contact left unconnected)
Input	0	0	4	SEL1—Select 1 position
	0	0	5	SEL2—Select 2 position
	0	0	6	LS1—Limit switch up (position 1)
	0	0	7	LS2—Limit switch up (position 2)
Input	0	1	0	LS3—Location set
	0	1	1	PB6—Start load cycle
	0	1	2	LS4—Trap (wired NC)
	0	1	3	LS5—Position switch
Input	0	1	4	PB7—Unload PB
	0	1	5	SS3—Main/backup (wired NO)
	0	1	6	LS6—Maximum length detect
	0	1	7	LS7—Minimum length backup
Output	0	3	0	PL2—Setup OK
	0	3	1	PL3—Select 1
	0	3	2	PL4—Select 2
	0	3	3	SOL1—Advance forward
Output	0	3	4	SOL2—Engage
	0	3	5	PL5—Engage ON
	0	3	6	M2—Run motor
	0	3	7	PL6—Motor run ON
Output	0	4	0	SOL3—Fast stop
	0	4	1	PL7—Fast stop ON
	0	4	2	SOL4—Unload with backup
	0	4	3	PL8—Backup ON

Table 11-20. I/O address assignment.

Device	Internal	Description
CR1	1000	CR1 (Setup Rdy)
TDR1	2000	Timer preset 10 sec register 3000 (accumulated register 3001)
MCR	MCR1700	First MCR address
CR2	—	Same as PL3 address
CR3	—	Same as PL4 address
CR4	—	Same as SOL1
—	1001	To set up internal for instantaneous contact of TDR2
TDR2	2001	Timer preset 5 sec register 4002 (accumulated register 4003)
—	1002	To set up internal for instantaneous contact of TDR3
TDR3	2002	Timer preset 12 sec register 4004 (accumulated register 4005)

Table 11-21. Internal address assignment.

1700 accomplishes this same function (for this example, MCR1700 is the first available address for MCRs). If the MCR coil is not energized, the PLC will not execute the ladder logic that is fenced between the MCR coil and the END MCR instruction.

An internal will not replace the control relay CR2 in line 9 since the PL3 contacts in line 10 can be used instead. This technique can be used whenever a control relay is in parallel with a real output device. Moreover, we do not need to separate the coils in lines 17 and 18 of the hardwired logic. This has already been done, since the PLC used here does not allow rungs with multiple outputs. Using separate rungs for each output is always a good practice.

The normally closed inputs that are connected to the input modules are programmed as normally open, as explained in the previous sections. The limit switch LS4 has two contacts—a normally open one and a normally closed one in lines 17 and 19, respectively, of Figure 11-45. However, only one set of contacts needs to be connected to the controller. In this example, we have selected the normally closed contact LS4. Although the normally open contact is not connected to the controller, its hardwired function can still be achieved by programming LS4 as a normally closed ladder contact.

Applications such as this one also require timers with instantaneous contacts, which are not available in most PLCs. An instantaneous contact is one that opens or closes when the timer is enabled. In most PLCs, an internal coil is used as a substitute for an instantaneous contact. Line 15 in the hardwired logic shows that, if PB6 is pressed and CR4 is closed, the timer TDR2 will start timing and contact TDR2-1 will seal PB6. This arrangement requires special PLC implementation. If we use software timer contacts, the

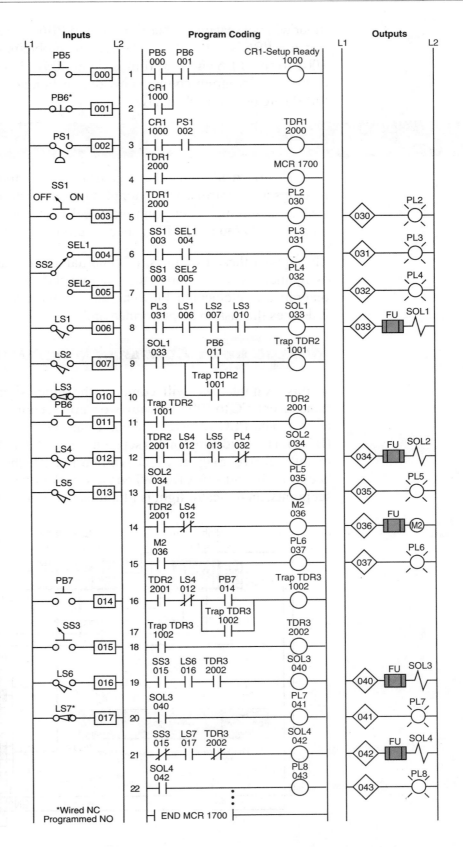

Figure 11-47. PLC implementation of the circuit in Figure 11-45.

timer will not seal until it has timed out. If PB6 is released, the timer will reset because PB6 is not sealed. To solve this problem, we can use internal coil 1001 to seal PB6 and start timing timer 2001 (TDR2). Lines 9, 10, and 11 of the PLC program coding show this technique. The time delay contacts (2001) are used for ON delays.

11-6 ANALOG I/O CONTROL PROGRAMMING

This section deals with the organization and implementation of analog readings and controls. The examples presented describe new additions to a system, assuming no existing electrical ladder logic is available. The examples are based on the PLC specifications described in Section 11-5.

Throughout these examples, the internal address assignments and register assignments will develop as the program coding from the flowchart or logic diagram is implemented. Flowcharting is important during this stage because it defines the task to be performed and the steps required to perform it.

ANALOG INPUT COMPARISON AND DATA LINEARIZATION

In this example, we will compare the input signal from a temperature transducer (0°C to 1000°C) with two alarm set points (a low alarm and a high alarm). The PLC receives set point data via two sets of 4-digit switches (BCD). The valid range of this set point data is 100 to 850°C. The analog input module receives a signal, which is proportional to the temperature, that ranges between −10 and +10 VDC. When the signal is over or under either of the two set points, an indicator light is illuminated. Figure 11-48 illustrates a simple

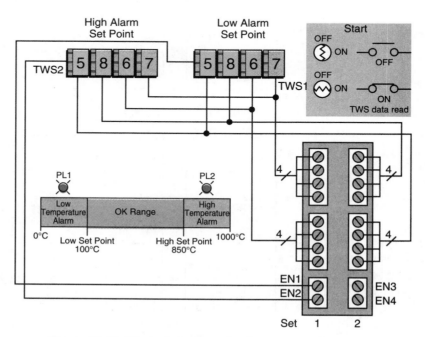

Figure 11-48. Elements in an analog input comparison system.

diagram of the elements used in this system. The thumbwheel switches (TWS) are connected to a register input module with multiplexing (MUX) capability. The TWS inputs are read only once when the spring-loaded, key-operated switch is turned ON. Figure 11-49 shows the analog input relationship between counts and degrees Celsuis. Figure 11-50 illustrates a flowchart of the required steps for this example, while Figure 11-51 shows a flowchart of the subroutines used in the program.

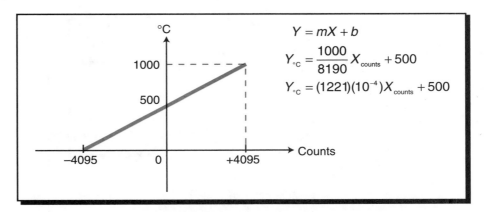

Figure 11-49. Relationship between counts and degrees Celsuis.

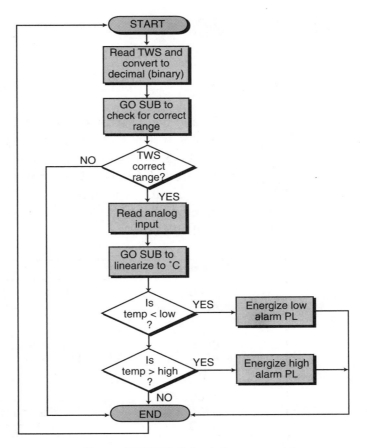

Figure 11-50. Flowchart of process steps.

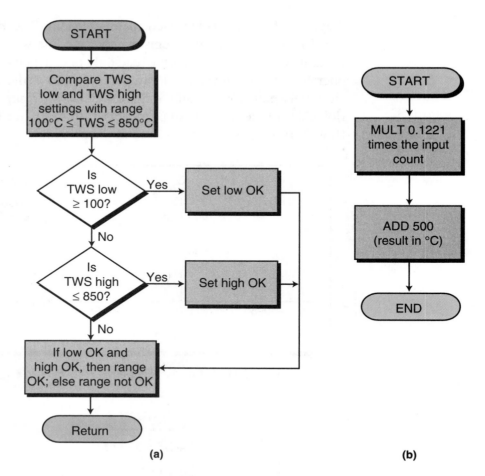

Figure 11-51. Subroutine flowcharts: **(a)** check for correct range and **(b)** linearize to °C.

Tables 11-22 and 11-23 present the register and internal output address assignment tables, respectively. Table 11-24 lists the I/O address assignment, while Figure 11-52 illustrates the final PLC circuit implementation. Moreover, Figures 11-53 and 11-54 show the subroutine circuit programs. A block transfer input instruction, used to read the TWS, selects slot location 1 and reads 8 bits (2 digits). It then automatically goes to the next slot (slot 2) to get the other 8 bits.

Register	Description
4000	Low limit alarm TWS (BCD)
4001	High limit alarm TWS (BCD)
⋮	
4010	Low limit alarm decimal °C
4011	High limit alarm decimal °C
⋮	
4100	Analog input counts (temperature)
4101	Used in multiplication temp result
4102	Result temperature in °C

Table 11-23. Register assignment.

Device	Internal	Description
—	1000	Xfer in TWS (MUX) enabled (TWS Read)
—	1001	BCD-to-binary conversion enabled (BIN Done)
—	1002	Xfer in analog input enabled (Temp Read)
—	1003	Compare temp with low alarm
—	1004	Low alarm condition (Temp < Low)
—	1005	Low temp alarm PL1
—	1006	Compare temp with high alarm
—	1007	High alarm condition (Temp > High)
—	1010	High temp alarm PL2
—	1100	Subroutine to check for valid ranges (CMP for Low)
—	1101	Low temp input more than 100°C
—	1102	Compare for high range (CMP for High)
—	1103	High temp input less than 850°C
—	1104	1 = Range OK, 0 = Range not OK
—	1200	Go to subroutine to linearize input counts to °C
—	1201	Addition of 500 enabled—reg 4102 has temp value

Table 11-23. Internal output assignment.

Module Type	I/O Address			Description
	Rack	Group	Terminal	
Input	0	0	0	Start analog reading and TWS input
	0	0	1	
	0	0	2	
	0	0	3	
Output	0	0	4	Not used
	0	0	5	
	0	0	6	
	0	0	7	
Register Input (low byte)	0	1	0	Least significant two digits of low alarm and high alarm set points; register multiplexes in both inputs
	0	1	1	
	0	1	2	
	0	1	3	
	0	1	4	
	0	1	5	
	0	1	6	
	0	1	7	
Register Input (high byte)	0	2	0	Most significant two digits of low alarm and high alarm set points; both are multiplexed in.
	0	2	1	
	0	2	2	
	0	2	3	
	0	2	4	
	0	2	5	
	0	2	6	
	0	2	7	
Output	0	3	0	Low alarm PL1 indicator
	0	3	1	High alarm PL2 indicator
	0	3	2	
	0	3	3	
Analog Input	0	7	0	Channel 1 analog input temp
	0	7	1	Channel 2 spare
	0	7	2	Channel 3 spare
	0	7	3	Channel 4 spare

Table 11-24. I/O address assignment.

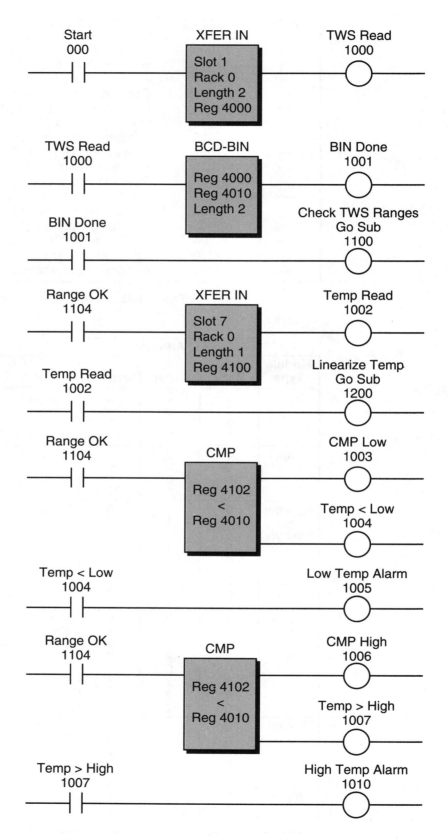

Figure 11-52. PLC implementation of the analog input comparison system.

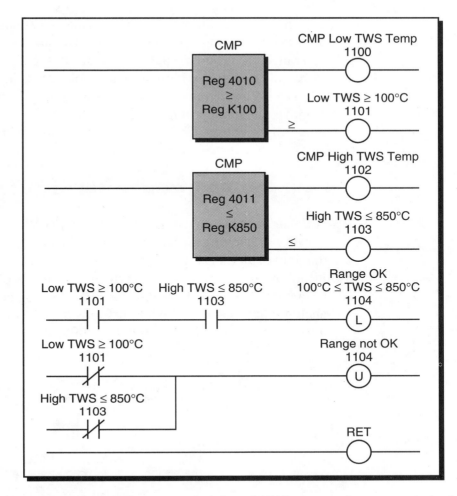

Figure 11-53. Subroutine 1100—check for valid TWS range and convert to decimal.

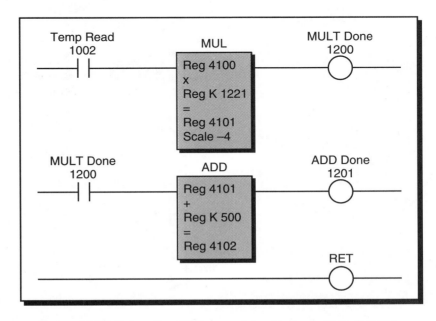

Figure 11-54. Subroutine 1200—convert analog counts to degrees.

The check range subroutine compares the values entered for the low and high temperature alarms with the constants 100 and 850. If the values are within that range, the program latches internal 1104, indicating that the range is OK. If the values are not within the acceptable range, internal 1104 remains OFF (0). The latch is required because the subroutine is not executed during every scan, yet the main program uses the OK or not OK signal during its regular program execution. The check range subroutine is only executed when the key switch is turned ON.

ANALOG POSITION READING FROM AN LVDT

A linear variable differential transformer (LVDT) provides position feedback for the moving mechanism of a machine. Figure 11-55 illustrates a block diagram of an LVDT application. The LVDT has a range of ±10 inches from its null position; therefore, the effective total range is 20 inches from a zero reference. The LVDT provides a ±10 VDC signal and is connected to an analog input module, which transforms the –10 to +10 VDC voltage swing into counts ranging from –4095 to +4095.

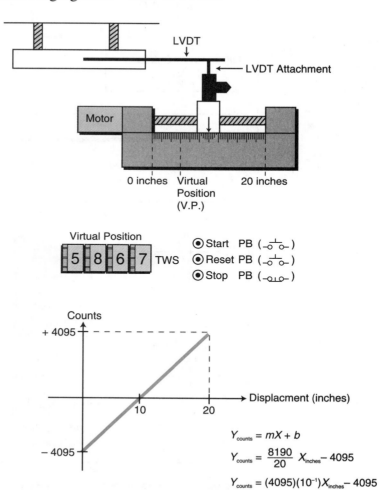

$$Y_{counts} = mX + b$$

$$Y_{counts} = \frac{8190}{20} X_{inches} - 4095$$

$$Y_{counts} = (4095)(10^{-1})X_{inches} - 4095$$

Figure 11-55. LVDT analog position reading system.

When the start push button starts the machine, the moving piece must move to the virtual starting position (V.P.) defined by the set of 4-digit TWS. The TWS settings range from 00.00 to 20.00; the decimal point will be implemented in the controller. When the machine finishes its cycle, the moving piece must return to the virtual position. The machine cycle may end at either side of the virtual starting position.

Figure 11-56 illustrates the flowchart for this system, while Tables 11-25, 11-26, and 11-27 show the I/O address assignment, register assignment, and internal assignment, respectively. Figure 11-57 presents the PLC program solution for this example.

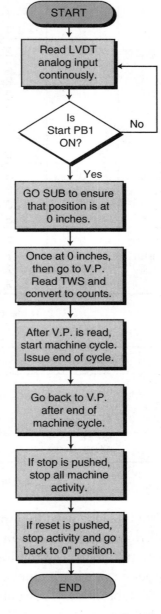

Figure 11-56. Flowchart of the LVDT reading and virtual position calculations.

Module Type	I/O Address			Description
	Rack	Group	Terminal	
Input	0	0	0	Start PB1—to virtual position
	0	0	1	Stop PB2—stop machine (NC)
	0	0	2	Reset PB3—reset to 0" position
	0	0	3	
Register Input (high byte)	0	1	0	Most significant two digits of TWS channel 1 (virtual position in decimal points)
	0	1	1	
	0	1	2	
	0	1	3	
	0	1	4	
	0	1	5	
	0	1	6	
	0	1	7	
Register Input (low byte)	0	2	0	Least significant two digits of TWS channel 1 (virtual position in decimal points)
	0	2	1	
	0	2	2	
	0	2	3	
	0	2	4	
	0	2	5	
	0	2	6	
	0	2	7	
Output	0	3	0	Forward command
	0	3	1	Reverse command
	0	3	2	
	0	3	3	
Analog Input	0	7	0	Channel 1 LVDT analog input
	0	7	1	Channel 2 spare
	0	7	2	Channel 3 spare
	0	7	3	Channel 4 spare

Table 11-25. I/O address assignment.

Register	Description
4000	TWS value in BCD; virtual position
4001	TWS value in binary after conversion
4002	Subtraction of 4095 (−4095)
4003	Virtual position in counts (equation)
4100	LVDT analog value in counts

Table 11-26. Register assignment.

Subroutines are used to implement the flowchart, to facilitate interlocking and programming. Latch instructions enable the subroutines, allowing the program to go to a subroutine until its operation has been performed. Once a subroutine finishes its function, it sends an unlatch signal signifying the end of the subroutine. This unlatch signal triggers the execution of the next subroutine.

Device	Internal	Description
—	1000	Start machine command
—	1001	LVDT analog input established (LVDT Read)
—	1100	Latch for enable to go to subroutine
—	1150	Compare LVDT position with 0 inches
—	1151	Position reached—0" position
—	1152	Energize reverse motor command from this sub
—	1153	One-shot position 0" found
—	1200	Latch to enable to go to subroutine (TWS Read)
—	1250	Read TWS block enable (TWS Read Sub)
—	1251	Convert output from BCD to binary (decimal)
—	1252	Multiply (according to equation) enable
—	1253	Subtract enabled
—	1254	Compare enabled
—	1255	V.P. found—1254 ON (Pos $\geq$ V.P.)
—	1256	Energize forward motor from this sub
—	1257	One-shot position V.P. found
—	1300	Latch to enable to go to subroutine to return to V.P.
—	1350	Compare LVDT with V.P. ($\geq$)—Return to V.P. sub (Pos > V.P.)
—	1351	Position $\geq$ V.P.—reverse motor (Ahead of V.P.)
—	1352	Compare LVDT with V.P. ($\leq$) (Pos < V.P.)
—	1353	Position $\leq$ V.P.—forward motor (Behind V.P.)
—	1354	Latch found V.P. from Pos $\geq$ V.P. (Reverse ahead of V.P.)
—	1355	One shot found V.P. from reverse
—	1356	Latch found V.P. from Pos $\leq$ V.P. (Forward behind V.P.)
—	1357	One shot found V.P. from forward
—	1360	Reverse motor from this sub
—	1361	Forward motor from this sub
—	1362	One shot found V.P. from $\leq$ or from $\geq$ after cycle
—	1400	Latch a reset to go to sub for 0" position after reset
—	1700	Latch to go to machine cycle
—	1750	Go sub machine cycle
—	1777	End of cycle signal (Cycle Done)

Table 11-27. Internal output assignment.

Figures 11-58, 11-59, and 11-60 present the subroutine codes. In Figure 11-58 (check for 0-inch position), the compare instruction checks for the LVDT count to be less than or equal to the compare constant –4090, rather than strictly equal to the value –4095. If the instruction checked for the value to be strictly equal to –4095, then fluctuations inherent in the LVDT's count output could cause the PLC to not latch this value. So, once the LVDT passes –4090 counts, it latches this value and assumes that the position is at 0 inches.

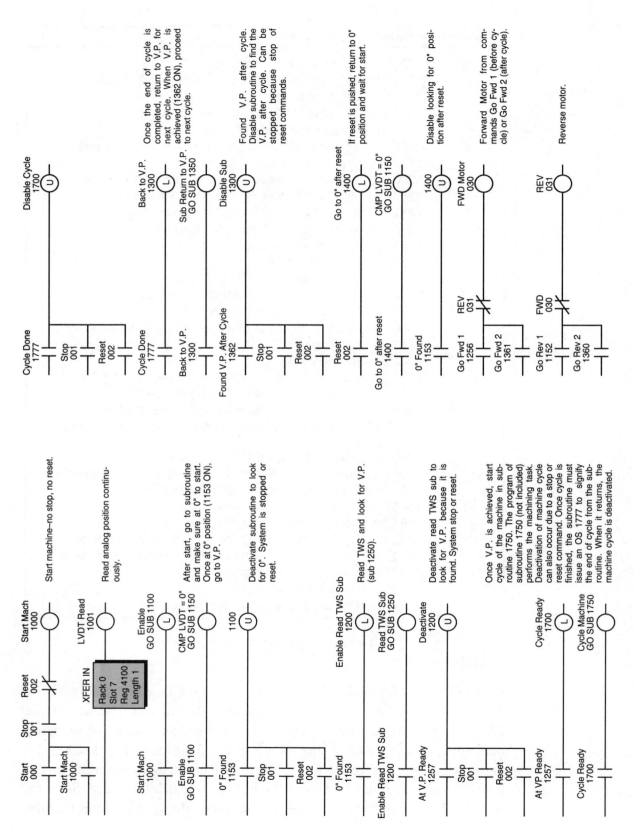

Figure 11-57. PLC implementation of the LVDT analog position reading example.

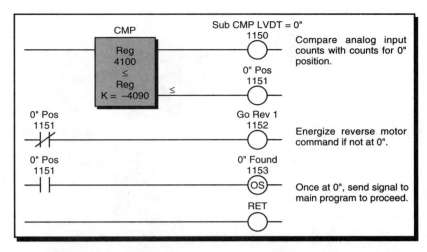

Figure 11-58. Subroutine 1150 brings moving part to 0" position.

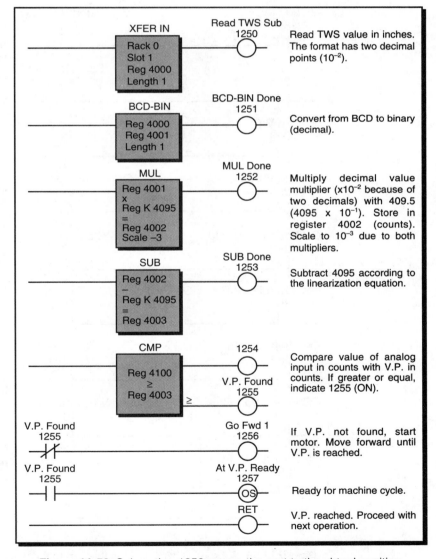

Figure 11-59. Subroutine 1250 moves the part to the virtual position.

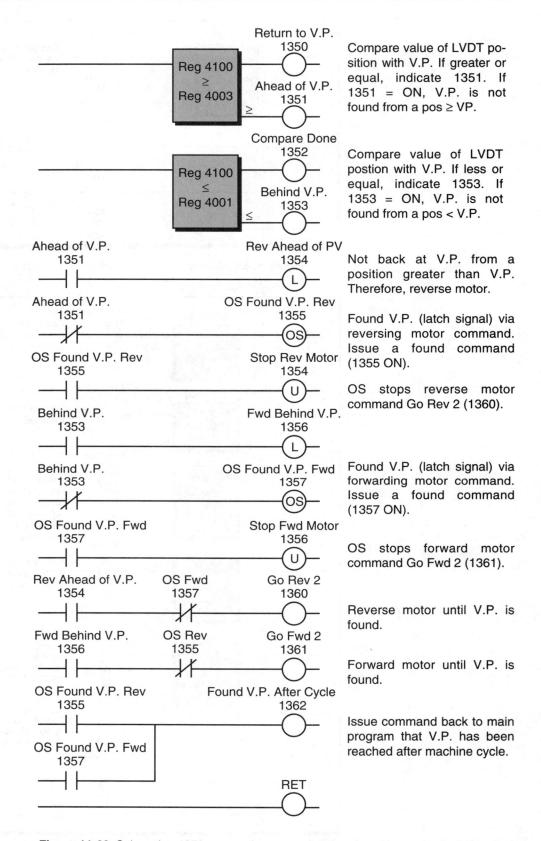

Figure 11-60. Subroutine 1350 returns the part to the virtual position at the end of cycle.

In Figure 11-59, scale multiplication allows the virtual position, which has two decimal points (10^{-2}) to be multiplied by the multiplication constant ($4095 \times 10^{-1} = 409.5$); thus, the final scale is 10^{-3}. This routine allows the motor to move the part to the virtual position as specified by the LVDT. Once the virtual position has been reached, the system is ready to start the machine cycle (one-shot output 1257). The machine cycle subroutine will return an end-of-cycle signal (output 1777) when finished, which disables the cycle subroutine (see Figure 11-57).

When the end of cycle has occurred, the PLC will tell the motor to move either forward or backward, depending on the moving part position at the end of cycle. The interlocking performed by output rungs 1354 and 1355 (refer to Figure 11-60) allow the motor to move in reverse if the part is farther than the virtual position (current position > V.P.). Rungs 1356 and 1357 perform the opposite function if the position of the part is closer than the virtual position (current position < V.P.).

The one-shot circuits used in the LVDT application prevent the system from moving the motor forward or backward until the part is at exactly the virtual position in counts. Analog count signals may jump one or two counts in either direction (up or down). This can result in instability, causing the forward and reverse signals to clash. The logic that is employed in this subroutine will detect, once the part crosses the virtual position (one-shot outputs 1355 and 1357 in Figure 11-60), whether the part is coming from a reverse motor or forward motor operation. Once the part is detected (i.e., when the one-shot is triggered), a minor jump in analog counts will not affect the operation, since the program has already determined that the part has just passed the virtual position. After the part stops at the virtual position, both the forward and reverse motor commands from the subroutine are inhibited.

LINEAR INTERPOLATION OF NONLINEAR INPUTS

In some PLC applications, the analog input signal received does not have a linear relationship with the signal being measured. That is, the ratio of change in the measurement variable is not the same throughout the measurement range. For example, a pressure transducer measuring hydraulic pressure (see Figure 11-61) may not provide a signal that is a linear representation of psi changes versus voltage changes (and therefore input counts). Sometimes the system that is being controlled creates these nonlinearities. The use of look-up tables and linear interpolation methods based on premeasured values can circumvent nonlinearity problems. In linear interpolation, the PLC stores known measured values in a table and then refers to this table during the reading of every measurement (analog counts) to determine the value of the variable (e.g., psi). It calculates this value by interpolating between the known measured values of the variable below and above the actual analog count reading. The more known values in the table, the more accurate the interpolated values will be.

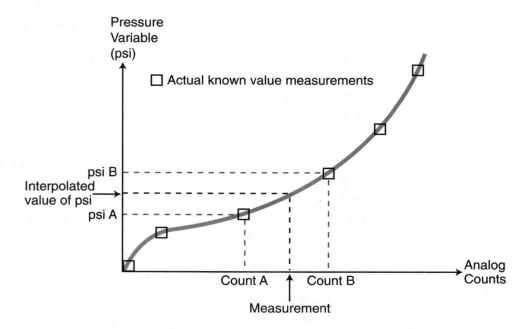

Figure 11-61. Nonlinear input signals from a pressure transducer.

To provide an example of this type of application, we will use a system with a pressure transducer that provides a 0 to +10 VDC output. The range of the pressure measurement is from 0 to 1000 psi. Measurement tests have shown that the relationship associated with the transducer's measurement is nonlinear. Figure 11-62 illustrates the difference between a linear curve and the actual nonlinear measurements. The analog input module transforms the 0 to 10 VDC signal into counts ranging from 0 to +4095. Table 11-28 shows the test counts for different psi pressure values.

Let's assume that the control algorithm requires the input measurement to be converted to engineering units (in this case, psi). Since we cannot perform this conversion based on a linear equation, we must obtain the psi values by estimating a pressure according to an input count reading. The PLC performs this linear interpolation by looking through tables (groups of contiguous registers) for a psi value equivalent to the counts. The two tables the PLC uses are the psi measurement table and its corresponding count value table. The psi table starts at register 3100, and the count table starts at register 3000. Table 11-29 shows these two tables, along with the corresponding pointer values. The pointer (register 4000) points at a register in the table according to a specified offset (table-to-register instruction). For instance, if the pointer value is 3 (reg 4000 = 3), then it points to psi register 3102 and count register 3002. The contents of the pointer register are in decimal, while the other registers are in octal. Figure 11-63 shows the flowchart for the look-up and interpolation procedures.

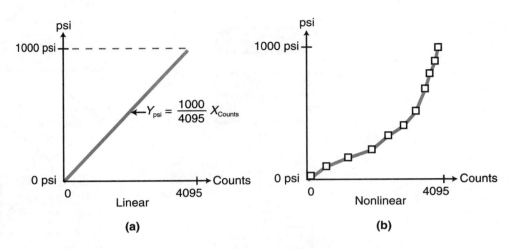

Figure 11-62. (a) Linear behavior and (b) actual nonlinear measurements.

psi Measurement	Analog Input Counts
0	0
50	600
100	1200
150	1500
200	1950
250	2280
300	2900
400	3300
600	3700
800	3820
1000	4090

Table 11-28. Sample psi measurements and corresponding counts.

psi Table		Counts Table		Pointer
Register	psi	Register	Counts	Reg 4000
3100	0	3000	0	1
3101	50	3001	600	2
3102	100	3002	1200	3
3103	150	3003	1500	4
3104	200	3004	1950	5
3105	250	3005	2280	6
3106	300	3006	2900	7
3107	400	3007	3300	8
3110	600	3010	3700	9
3111	800	3011	3820	10
3112	1000	3012	4090	11

Table 11-29. Look-up tables for psi and count values.

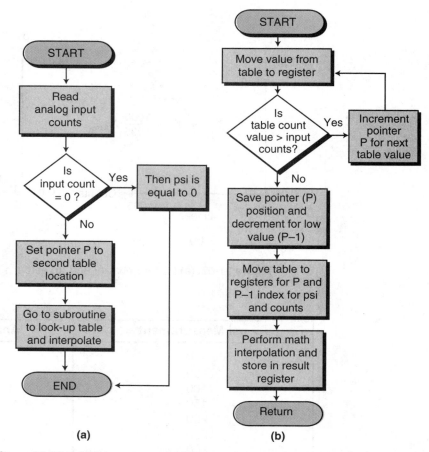

Figure 11-63. (a) Main program look-up procedure and **(b)** interpolation subroutine.

Table 11-30 shows the register assignment table for this example, and Table 11-31 shows the internal output assignment. The analog input module, with address 070, is the only real input.

Register	Description
3000	Look-up table storage (counts reg 3000–3012)
3100	Look up table storage (psi reg 3100–3112)
4000	Pointer P – 1
4050	Pointer P
4100	Analog input counts (pressure)
4150	psi result register (computed after interpolation)
4400	Low psi register $R_{psi\text{-}low}$
4450	High psi register $R_{psi\text{-}high}$
4500	Low count register $R_{counts\text{-}low}$
4550	High count register $R_{counts\text{-}high}$
4600	Temporary register (subtract)
4601	Temporary register (subtract)
4602	Temporary register (multiply)
4603	Temporary register (multiply)
4604	Temporary register (divide)

Table 11-30. Register assignment.

Internal	Description
1000	Analog transfer enabled
1001	Compare for input = 0 counts
1002	Input counts are 0
1003	Move constant 0 to computed psi register
1004	Move constant 2 as pointer (enabled)
⋮	
1100	Go sub interpolation, move table to register
1101	Compare enabled
1102	High counts ≥ input counts (High cts)
1103	Increment (ADD) pointer enabled
⋮	
1200	Go to calculate math, store pointer
1201	Subtract 1 from pointer enabled
1202	Move table to register enabled (Low cts)
1203	Move table to register enabled (Low psi)
1204	Move table to register enabled (High psi)
1205	Subtract enabled
1206	Subtract enabled
1207	Multiply enabled
1210	Subtract enabled
1211	Divide enabled
1212	Add enabled

Table 11-31. Internal output assignment.

Figure 11-64 shows the method used to interpolate psi values based on the contents of each register used. The PLC program shown in Figure 11-65 implements the linear interpolation by finding the high and low counts and psi values. Using the two pointers, the PLC obtains the high and low values for the counts and psi through table-to-register instructions. The PLC program compares its current analog counts with the count values in the table registers to find a table location with a greater value. This comparison starts with the lowest count values in the table. If the actual value (analog counts) is more than the value pointed to by the pointer, the PLC increments the pointer (adds 1) and tests a new table value. Once the program finds a register in the table that contains a value greater than the analog reading, it stores the pointer value associated with this register in register 4050, thus pointing to the register at pointer P in Figure 11-64. Then, the value contained in register 4050 is decreased by one and stored in register 4000 to point to the register at point P − 1 in Figure 11-64. These two registers (4000 and 4050) point to the low psi/low counts and high psi/high counts, respectively, which will be used to complete the interpolation.

In the software program presented, we have considered that the actual count value may be 0 counts. When the count value is 0, the equivalent psi is 0 and the program does not perform the subroutine shown in Figure 11-66. If it did perform the subroutine, the program would enter into a loop error.

| psi Table | | Counts Table | | Pointer Register |
Register	Contents	Register	Contents	
4400	Low psi	4500	Low counts	4000 @ P − 1
4150	Computed psi	4100	Actual counts	
4450	High psi	4550	High counts	4050 @ P

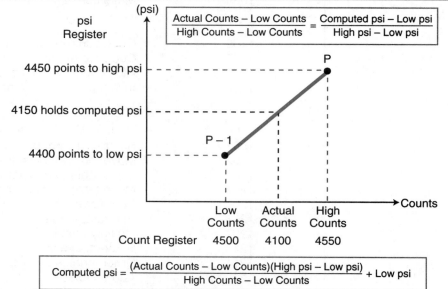

$$\frac{\text{Actual Counts} - \text{Low Counts}}{\text{High Counts} - \text{Low Counts}} = \frac{\text{Computed psi} - \text{Low psi}}{\text{High psi} - \text{Low psi}}$$

psi
Register

(psi)

4450 points to high psi

4150 holds computed psi

4400 points to low psi

P

P − 1

Low
Counts

Actual
Counts

High
Counts

Counts

Count Register   4500   4100   4550

$$\text{Computed psi} = \frac{(\text{Actual Counts} - \text{Low Counts})(\text{High psi} - \text{Low psi})}{\text{High Counts} - \text{Low Counts}} + \text{Low psi}$$

Figure 11-64. Interpolation method.

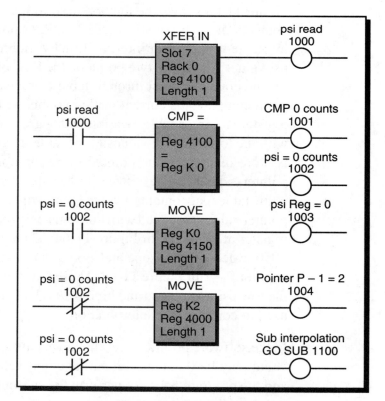

Figure 11-65. Main program.

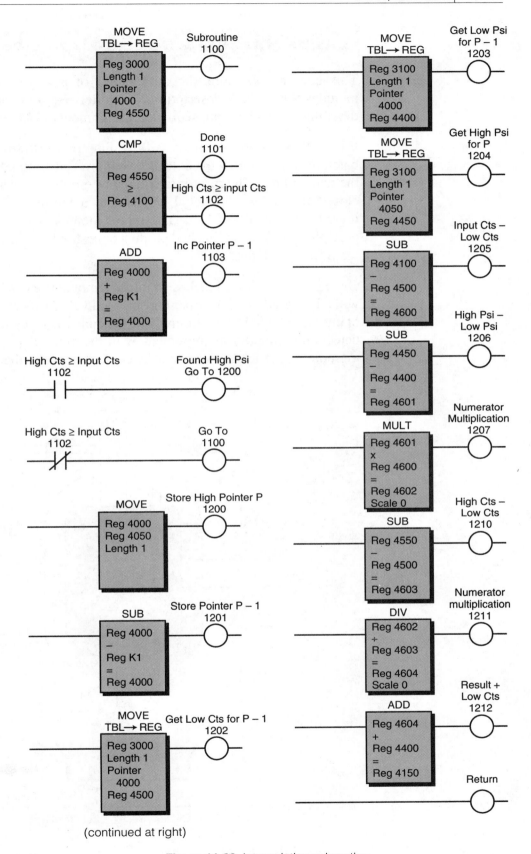

Figure 11-66. Interpolation subroutine.

(continued at right)

LARGE BATCHING CONTROL APPLICATION

This example explains the automation of a large batching process. It includes the process description, controller requirements, flowchart, logic diagrams for each output sequence, assignment of I/O, and program coding.

Figure 11-67 shows the process flow diagram illustrating the elements this batching application will control. Two ingredients, A and B, will be mixed in the reactor tank. The reactor tank must be empty (indicated by the normally closed liquid level switch LLS) and at a temperature of 100°C before ingredient A can be added. The mixer motor must be off to avoid liquid precipitation, and the finished product tank should be in a set position, which the limit switch detects.

Once the reactor tank reaches an initial temperature of 100°C, the controller will add ingredient A by opening solenoid valve 1 (SOL1) until 100 gallons of ingredient A have been poured into the tank. LLS1, which is normally open, detects the quantity of ingredient A in the tank. This switch closes when ingredient A reaches the proper level. At this point, the controller will add

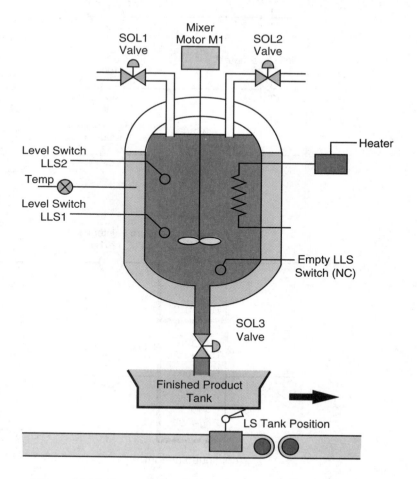

Figure 11-67. Process flow diagram of the batching application.

ingredient B by opening SOL2. LLS2 detects the quantity of ingredient B, which should be 400 gallons. The temperature should be at 100°C ± 10% during the Add Ingredients step. If the temperature drops, the PLC will turn ON the heater automatically while the process continues.

When the reactor tank contains both ingredients, the controller will raise the temperature to 300°C ± 10% and turn ON the mixer for 20 minutes. The PLC will control the temperature automatically at predefined set points during the process.

SOL3 will activate the drain valve when the mixing is completed. This operation will reset the process until another finished product tank is placed in position, and the cycle starts again. The system should incorporate pilot light indicators to alert the operator to the status of the batching process.

This application must be capable of reading analog signals from the process. In this case, the voltage comes from a temperature transducer (0–10 volts), which has a range of 0 to 500°C (50°C/volt). The heater coil's ON/OFF control switch controls the temperature. The application also requires standard 110 VAC input and output modules. Figure 11-68 shows a flowchart of this process. It illustrates what has been described in the control task definition and serves as a preparation for the logic diagrams.

Figure 11-69 shows the logic diagrams for this example. The logic diagrams map the initial implementation of the logic required to control each of the process sequences. These diagrams represent the conditions required for a rung to be energized. Real I/Os are marked with an X. The first logic diagram shows the initial requirements for starting the process. The start push button, when pressed, enables the Start Mix output if the following conditions are satisfied: the tank is in position, SOL3 is closed, and the stop push button is not pressed. Pilot lights PL1 and PL2 indicate that the tank is in position and that the system Start Mix signal is enabled.

Logic diagram 2 in Figure 11-69 sets the initial temperature (T1) at 100°C. The logic indicates that the mixer motor (M1) must be off, the Start Mix input enabled, and the reactor tank empty. The Ready to Mix input in diagram 2 is an interlock from logic diagram 6; it disables T1 when T2 is being set. Note that the OR function uses the Empty signal with the initial Set to T1 signal to ensure that, even when the tank is still adding ingredients, the temperature control will maintain the temperature at T1.

Logic diagram 3 controls the Ready to Add signal, which allows ingredient A to be added. Here, the output Temp OK1 (T1 = 100°C) indicates that the tank is at the proper temperature. As long as the Start Mix signal is still enabled, the process is ready to add the first ingredient.

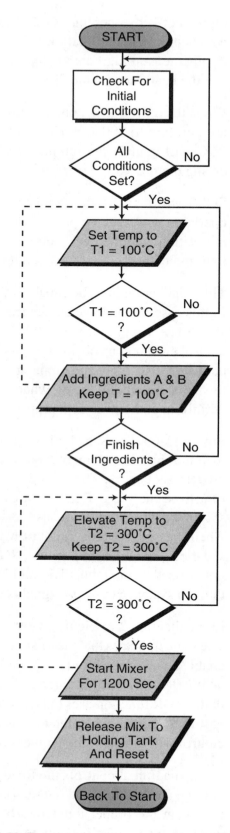

Figure 11-68. Flowchart for the large batching application process.

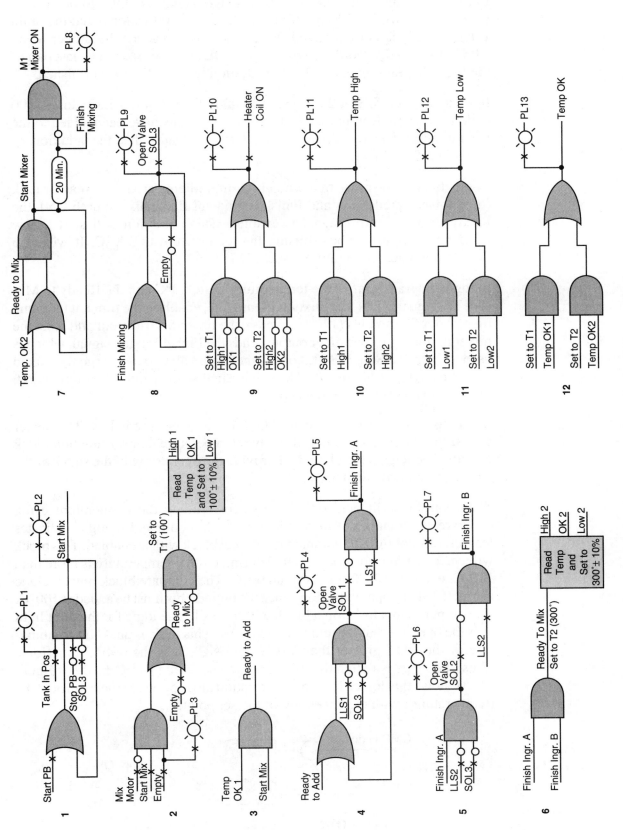

Figure 11-69. Logic diagrams for the batching application.

In logic diagram 4, the Ready to Add A signal enables SOL1 to open. This action occurs while LLS1 is still open (less than 100 gallons) and the drain valve (SOL3) is not energized. When the liquid reaches the proper level, LLS1 closes and, according to the logic, SOL1 de-energizes. The last part of the logic diagram indicates that the addition of ingredient A is finished.

In logic diagram 5, SOL2 opens to add ingredient B until LLS2 closes (500 gallon level), indicating that 400 additional gallons have been added to the reactor tank. The remainder of the logic indicates that the addition of ingredient B is finished.

Logic diagram 6 shows that when both ingredients are in the reactor tank (the Finish Ingredient A and Finish Ingredient B signals are both ON), the Ready to Mix control signal is enabled. This condition will start a new temperature control block, raising the temperature to 300°C. It will also disable the other temperature control (T1).

In logic diagram 7, after the temperature is at 300°C and the Ready to Mix (Set T2) signal is ON, the mixer will turn ON, enabling the timer at the same time. After 20 minutes (1200 seconds), the timer will time out and reset the mixer motor logic. The timer output sends the Finish Mixing signal, which is used to energize SOL3. SOL3 opens the drain valve to discharge the mixed ingredients (logic diagram 8). The valve remains open until the empty switch returns to its normal state (closed).

Logic diagram 9 turns the heater ON if the temperature is low. The heater can be turned on from either of the two temperature control function block outputs. Sequences 10, 11, and 12 provide the operator with the status of the temperature inside the tank.

The controller will perform the logic for reading the temperature using compare functional block instructions. Once the command, or logic, indicates temperature control, the compare functional block will be enabled. This block will perform three comparisons to determine if the temperature is more than 110°C, equal to 100°C, or less than 90°C. The compare block must include a limit (LIM) compare function, since the ingredients must be added at 100°C. The output of this functional block will be OK1. The logic for the pilot light tells the operator that the temperature is OK. This logic is an AND combination of the NOT greater than 110°C and NOT less than 90°C functions. Thus, the range is within the tolerances as specified (100°C ± 10%). Figure 11-70 shows this logic. The limit instruction also applies to the control of T2 (temperature), with the exception of the set point.

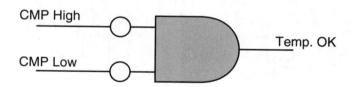

Figure 11-70. Logic diagram for Temp OK signal.

The assignment of real I/O begins by addressing the real inputs and outputs. Table 11-32 illustrates the assignment of I/O for this application example. Note that the modularity for the digital I/O is four points per module. The analog module contains two input channels, which occupy one half of a group (four locations).

| Module Type | I/O Address | | | Description |
	Rack	Group	Terminal	
Input	0	0	0	Start mix PB
	0	0	1	Stop PB
	0	0	2	Tank position LS
	0	0	3	Empty LLS (NC)
Input	0	0	4	LLS1 (100 gal)
	0	0	5	LLS2 (500 gal)
	0	0	6	Not used
	0	0	7	Not used
Output	0	1	0	PL tank in position
	0	1	1	Start system (Start Mix PL)
	0	1	2	Empty reactor tank
	0	1	3	Solenoid valve 1 (ingredient A)
Output	0	1	4	PL valve 1 open
	0	1	5	Finish A
	0	1	6	SOL valve 2
	0	1	7	PL valve 2
Output	0	2	0	Finish B
	0	2	1	Mixer (M1)
	0	2	2	PL mixer ON
	0	2	3	Solenoid valve 3 (drain)
Output	0	2	4	PL valve 3 open
	0	2	5	Heater coil
	0	2	6	PL heater ON
	0	2	7	PL temp high
Output	0	3	0	PL temp low
	0	3	1	PL temp OK
	0	3	2	Not used
	0	3	3	Not used
Analog Input	0	3	4	Input 34 connected to transmitter
	0	3	5	Corresponds to I/O register 3034
	0	3	6	Input 36 left as is
	0	3	7	Register 3036

Table 11-32. I/O address assignment.

Table 11-33 shows the assignment of internals. This table lists several internal coil addresses representing coil relay conditions related to the logic diagram. The coils associated with the compare functional blocks are internals, which are used to describe the temperature conditions, such as high and low. I/O register 3034 stores the analog value of the temperature. This register will be compared to the storage registers that hold the equivalent values of the temperature ranges.

Device	Internal	Description
Logic	1000	Set to T1
CMP block	1001	CMP High 1
LIM	1002	CMP for range OK 1
CMP block	1003	CMP Low 1
Logic	1004	Temp OK 1
Logic	1005	Ready to add
Logic	1006	Ready to mix/set to T2
CMP block	1007	CMP High 2
LIM	1010	CMP for range OK 2
CMP block	1011	CMP Low 2
Logic	1012	Temp OK 2
Logic	1013	Start mixer
Timer	2000	Timer preset 20 min (1200 sec) register 4000 (accumulated register 4001)

Table 11-33. Internal output assignment.

Translating the logic diagrams into PLC diagrams is the next step after tabulating the input and output assignments. The program coding follows the logic diagram sequences previously specified and uses the information from the I/O and internal tables as references for the addresses. Figure 11-71 shows the program coding for this example.

The ladder logic shown in the program coding is the implementation of each logic diagram. The internals are assigned as specified in the internal assignment table. Several storage registers, added in the compare blocks, hold preset values. These values correspond to the equivalent temperature set points used, including the tolerances (i.e., 110°C, 100°C, 300°C, 270°C, etc.).

The voltage received from the temperature transmitter ranges from 0 to 10 V, representing 0 to 500°C. Thus, each volt represents a change of 50°C. The controller used in this example receives the voltage signal and converts it to a count ranging from 0000 to 9999. These counts are proportional to the voltage and, therefore, to the temperature.

The first set point (register 4000) contains the count value 2200, which is proportional to 110°C (100°C ± 10%); register 4003 contains the value 1800, which is equivalent to 90°C (100°C ± 10%). Registers 4001 and 4002 contain the values of 2040 (102°C) and 1960 (98°C), respectively. The controller uses these values to detect a small range in which the temperature is very close to 100°C, thus starting the ingredient addition. These two registers are used in a LIM compare block that detects when the temperature is between 98 and 102°C. The PLC does not compare the value to 100°C because the value may never be exactly 100°C during the time it is being read. The preset values of the other compare blocks are specified in the same manner.

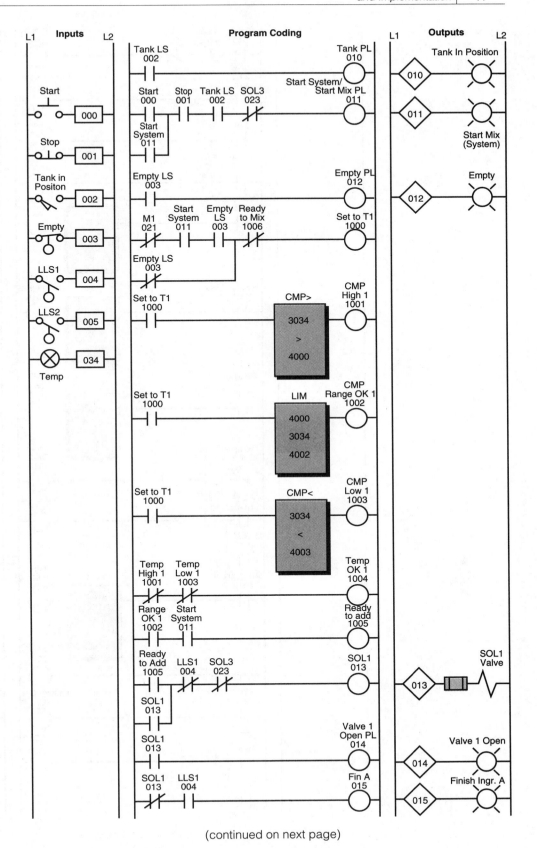

(continued on next page)

Figure 11-71. PLC implementation of the batching application.

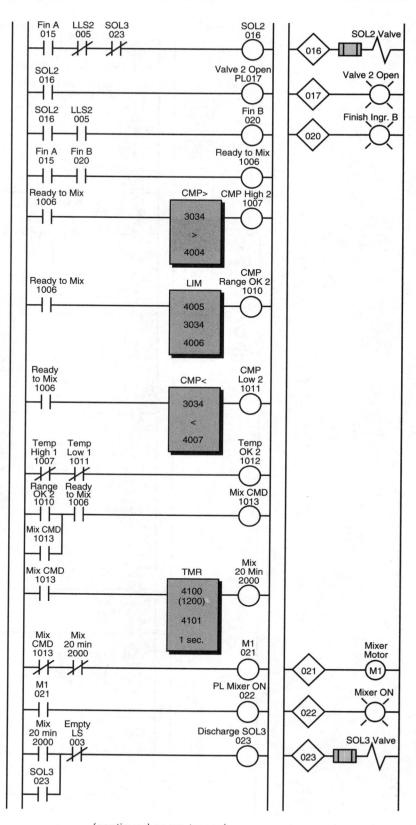

(continued on next page)

Figure 11-71 continued.

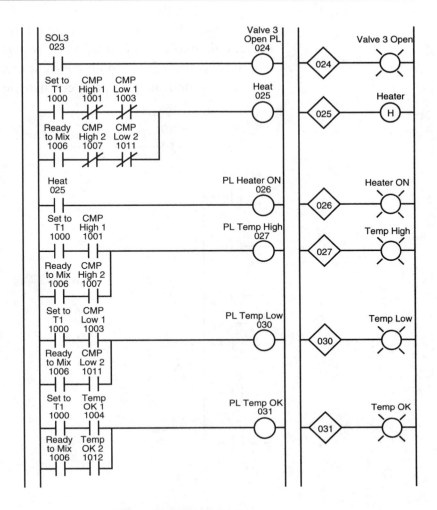

Figure 11-71 continued.

11-7 SHORT PROGRAMMING EXAMPLES

This section presents several examples of logic series that are often encountered when programming a controller. For convenience, the examples are implemented using the most basic ladder diagram instructions. Therefore, they may require more instructions than they would if they were programmed using a higher level instruction set.

EXAMPLE 1: INTERNAL STORAGE BITS

Most programming devices are limited in the number of series contacts or parallel branches that a rung can have. This limitation can be overcome through the use of internal storage bits. Figure 11-72a illustrates a PLC program that was translated directly from a hardwired relay diagram that requires seven parallel OR branches. If the programmable controller had only allowed five OR branches, an internal could have been used to break the circuit into two circuits, as shown in Figure 11-72b. The program's operation is the same in both configurations. This technique would also be valid if the contacts were arranged in series.

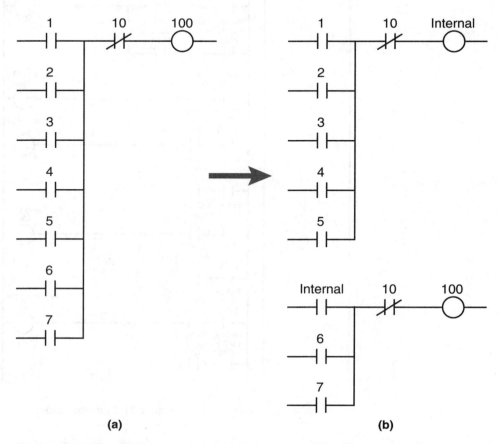

(a)   (b)

Figure 11-72. (a) A relay circuit using seven rungs that is **(b)** converted to five rungs using an internal.

EXAMPLE 2: START/STOP CIRCUIT

The start/stop circuit shown in Figure 11-73 can be used to start or stop a motor or process or to simply enable or disable some function. To start a motor, the ladder output only needs to reference the motor output address. If the intent of the circuit is to detect that some process is enabled, the output can be referenced with an internal address.

In Figure 11-73, the stop push button and emergency stop inputs are programmed as normally open. They are programmed this way because these types of inputs are usually wired normally closed. As long as the stop push button and the emergency stop push button are not pushed, the programmed contacts will allow logic continuity. Since the start push button (normally open) is a momentary device (i.e., it allows continuity only when pressed), a contact from the motor output is used to seal the circuit. Often, the seal-in contact is an input from the motor starter contacts.

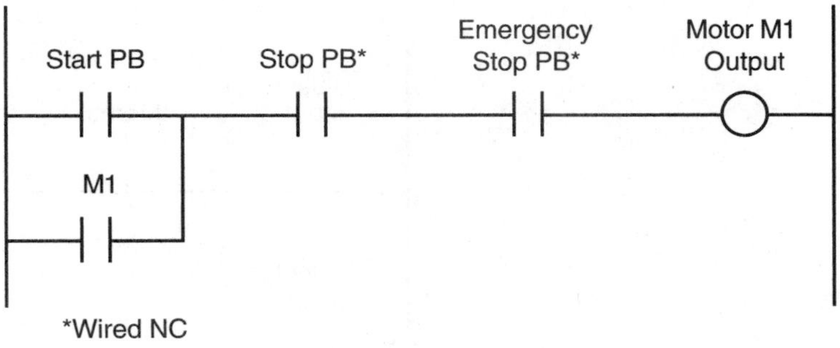

Figure 11-73. Start/stop circuit.

EXAMPLE 3: EXCLUSIVE-OR CIRCUIT

The exclusive-OR circuit in Figure 11-74 is used to prevent an output from energizing if two conditions, which can activate the output independently, occur simultaneously. Thus, if either input A or B is activated, the output will be energized. However; if both are activated, the output will not be energized.

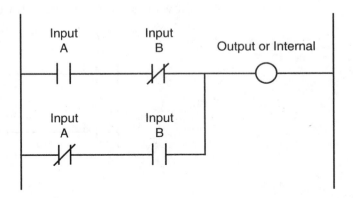

Figure 11-74. Exclusive-OR circuit.

EXAMPLE 4: ONE-SHOT SIGNAL

The one-shot (transitional output) signal in Figure 11-75 is a program-generated pulse output that, when triggered, is ON for the duration of one program scan and then turns OFF. A momentary signal (e.g., a push button) or an output that comes ON and stays ON for some time (e.g., a motor) can enable a one-shot signal. Whichever input signal is used, the leading-edge (OFF-to-ON) transition of the input signal triggers the one-shot signal, which stays ON for one scan and then goes OFF. The signal remains OFF until the trigger is activated, causing it to come ON again. Clear or reset signals are typically one-shot signals; the one-shot signal is perfect for this application, since it stays ON for only one scan.

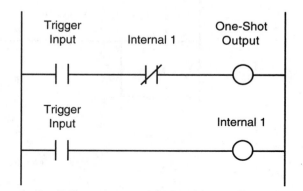

Figure 11-75. One-shot output circuit.

EXAMPLE 5: TRAILING-EDGE ONE-SHOT SIGNAL

A trailing-edge one-shot signal (see Figure 11-76) generates a pulse with a one-scan duration. This signal reacts like the one-shot signal in Example 4; however, the trigger for this pulse is the trailing edge of the trigger pulse. Figure 11-76 shows the timing diagram for each element's activation.

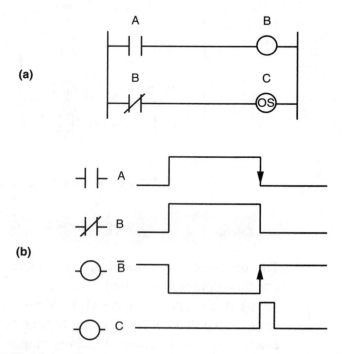

Figure 11-76. (a) A trailing-edge one-shot output circuit and **(b)** its corresponding timing diagram.

EXAMPLE 6: INITIALIZATION USING AN MCR

The logic circuit shown in Figure 11-77 can be used to set up parameters during an initialization period. These parameters include timer and counter preset values, high- and low-limit set point values, and any other preset or starting values. Typically, the initialization period occurs only once during the program, either when the system is first powered up or when power is restored after a power loss.

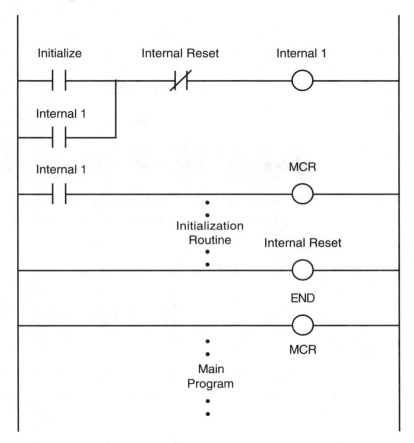

Figure 11-77. Initialization circuit using an MCR.

EXAMPLE 7: SYSTEM START-UP HORN

A start-up horn logic circuit (see Figure 11-78) is used to signal that moving equipment (e.g., conveyor motors) is about to start. The setup output signal in this example is similar to a start/stop circuit; but instead of starting the system, it enables the timer, allowing the horn to sound for 10 seconds. The horn sounds when the start input is closed and stops when the timer times out or the reset input opens. The system can start, if the setup signal remains ON, when the horn delay timer times out.

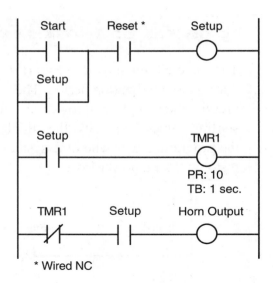

* Wired NC

Figure 11-78. Start-up horn circuit.

EXAMPLE 8: OSCILLATOR CIRCUIT

An oscillator logic circuit (see Figure 11-79) is a simple timing circuit that generates a periodic output pulse of any duration. The TMR1 output generates this pulse.

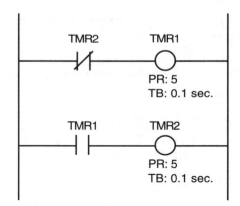

Figure 11-79. Oscillator circuit.

EXAMPLE 9: ANNUNCIATOR FLASHER CIRCUIT

A flasher circuit (see Figure 11-80) toggles an output ON and OFF continually. In this circuit, an oscillator circuit output (TMR1) is programmed in series with an alarm condition. As long as the alarm condition is TRUE, the annunciator output will flash. The output in this case is a pilot light; however, this same logic could be used in conjunction with a horn, which would pulse during the alarm condition. Any number of alarm conditions can be programmed using the same flasher circuit.

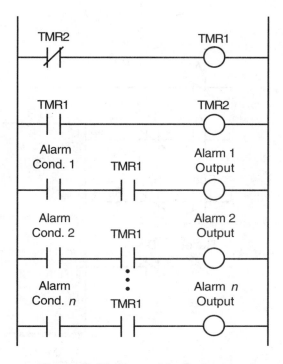

Figure 11-80. Annunciator flasher circuit.

EXAMPLE 10: SELF-RESETTING TIMER

The self-resetting timer shown in Figure 11-81 provides a one-scan pulse each time the timer is energized. The specified preset value of the timer determines the repetition of this pulse.

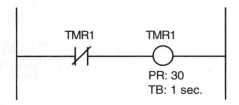

Figure 11-81. Self-resetting timer circuit.

EXAMPLE 11: SCAN COUNTER

The circuit shown in Figure 11-82 computes scan time. This short program counts the number of times two consecutive scans occur during a time interval, which is defined by the timer (e.g., 10 seconds). Once the time interval elapses, the program multiplies the number of two-scan counts by two. It then divides the time interval (10 seconds) by the number of total scans, thus computing the scan time, which is stored in a result register. The result register is scaled so that the scan time is expressed in milliseconds.

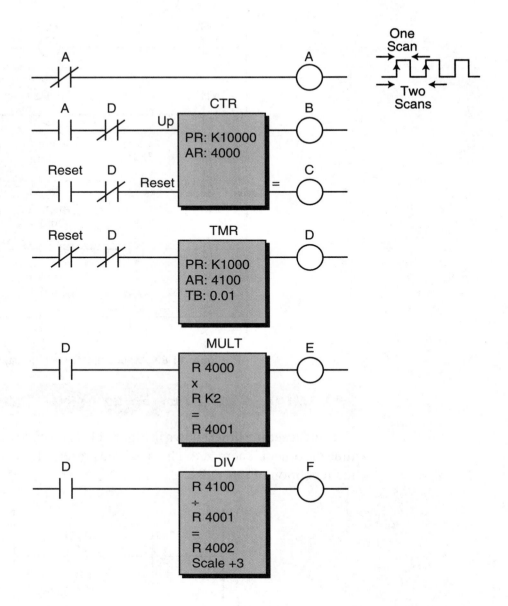

Figure 11-82. Scan counter circuit.

EXAMPLE 12: SEQUENTIAL MOTOR STARTING

This example (see Figure 11-83) illustrates how several motors or other devices can be started sequentially, as opposed to all at once. For simplicity, we used an ON-delay timer to delay the start of each motor. However, this approach is impractical for starting a large number of motors. If a large number of motors will be started, other techniques that do not require as many timers as motors (e.g., shift registers, self-resetting timers, oscillator circuits, etc.) should be used.

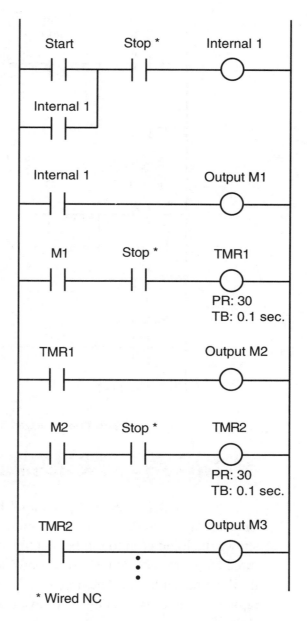

Figure 11-83. Sequential motor-starting circuit.

EXAMPLE 13: DELAYED DE-ENERGIZE DEVICE

This example (see Figure 11-84) illustrates the use of an OFF-delay timer to de-energize a motor or another device after a delay period. Note that the output of the OFF-delay timer before the Stop Motor 1 push button is pressed is TRUE, thus keeping the TMR1 contact in line 1 closed. When the Stop Motor 1 push button (wired as normally closed) is depressed while the motor is running, it energizes the internal output, enabling the OFF-delay timer. When the timer times out, the contacts open and the motor de-energizes.

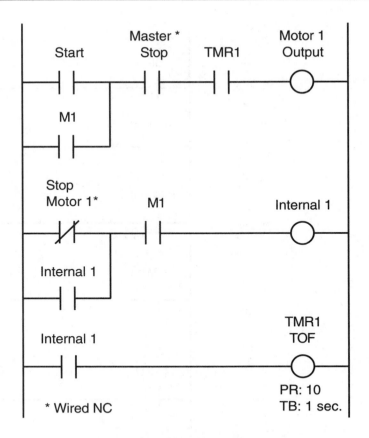

Figure 11-84. Delayed de-energize circuit.

EXAMPLE 14: 24-HOUR CLOCK

A 24-hour clock has many applications, but it is typically used to display the time of day or to determine the time a report is generated. Figure 11-85 shows the logic used to implement this type of clock. This logic consists of three counters: one counts 60 seconds, another counts 60 minutes, and the third counts 24 hours. The time is displayed by outputting the accumulated register value of each counter to seven-segment BCD displays.

EXAMPLE 15: ELIMINATION OF BIDIRECTIONAL POWER FLOW

Sometimes, when converting relay logic to program logic, you will find relay circuits that allow power to flow bidirectionally, as shown in Figure 11-86a. In this circuit, power can flow down through CR3 or up through CR3 to make a complete path. PLCs do not allow bidirectional power flow, so the circuit must be restructured to establish a circuit for each power flow direction. The result, shown in Figure 11-86b, is two separate circuits that allow unidirectional, left-to-right power flow.

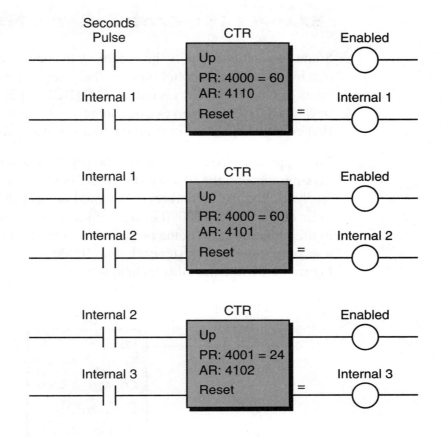

Figure 11-85. 24-hour clock circuit.

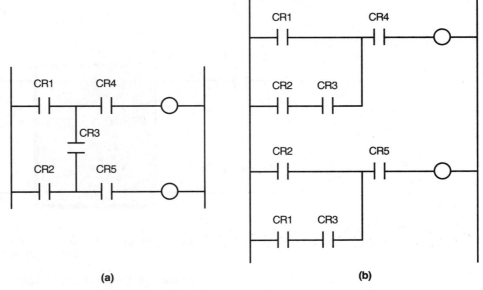

| (a) | (b) |

Figure 11-86. (a) Hardwired circuit allowing bidirectional power flow at CR3 and **(b)** reconfiguration to eliminate bidirectional power flow.

531

EXAMPLE 16: EXCEEDING THE MAXIMUM COUNT

Some applications require the capability to count events that will exceed the maximum allowable count number that a register can hold. The maximum count in most controllers is either 9999 (BCD) or 32767 (binary). Counting beyond BCD 9999 involves cascading two counters, where the output of the first counter is used to input an up-count signal to the second counter.

This approach does not work for binary format however. If 32767 is the maximum count, the first register would contain a value of 1 (after 32767 is reached), and the second register would contain 00000. This result would indicate a count of 100000 instead of the actual count of 32768. A solution to this situation is to set the preset value of the first counter to 9999 and use a second counter to register each time 10000 counts occur. The sequences in Figure 11-87 illustrate this technique.

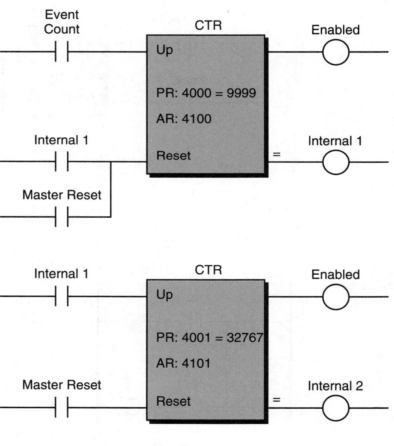

Preset Registers: 4000 = 9999
4001 = 32767

Maximum Count is 327679999
in Accumulated Registers 4100 and 4101

Figure 11-87. Binary counting program.

EXAMPLE 17: PUSH-TO-START/PUSH-TO-STOP CIRCUIT

Often, it is desirable for a single push button to perform both the start (enable) and stop (disable) functions. In this example (see Figure 11-88a), when the push button (PB1) is depressed for the first time, internal output 2 turns ON and remains ON. If the push button is depressed again, internal output 2 turns OFF. The second logic rung detects the first time the push button is pressed, while the first rung detects the second time the button is depressed. The simplified timing diagram in Figure 11-88b shows the operation of this circuit.

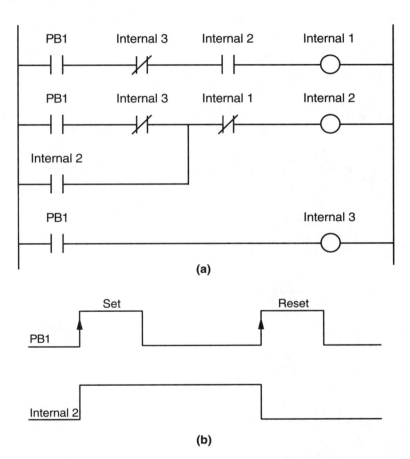

Figure 11-88. (a) A push-to-start/push-to-stop circuit and **(b)** its timing diagram.

KEY TERMS
control strategy
control task
flowcharting
program coding

PLC SYSTEM
DOCUMENTATION

*If you cannot—in the long run—tell everyone
what you have been doing, your doing has
been worthless.*

—Erwin Schroedinger

CHAPTER HIGHLIGHTS While proper PLC programming is important for a well-run application program, all of that work is lost without adequate system documentation. Without documentation, system activities, such as changes, installation, and maintenance, are difficult to accomplish. In this chapter, we will explain how to document all aspects of a PLC—from system configuration to register assignments. We will also explore some different documentation methods. After that, you will be ready to learn how to implement a PLC process control system.

12-1 INTRODUCTION TO DOCUMENTATION

Documentation is an orderly collection of recorded information about both the operation of a machine or process and the hardware and software components of its control system. These records are a valuable reference during system design, installation, start-up, debugging, and maintenance.

To the system designer, documentation should be a working tool that is used throughout the design phase. If the various documentation components are created and kept current during system design, they will provide the following benefits:

- They will provide an easy way to communicate accurate information to all those involved with the system.

- They will serve as a reference to the designer during and after the design phase.

- They will help the designer, or someone else, answer questions, diagnose possible problems, and modify the program if requirements change.

- They will serve as training material both for the operators who will interface with the system and for the maintenance personnel who will maintain it.

- They will allow the system to be reproduced or altered to serve other purposes.

Proper documentation comes from the compilation of hardware, as well as software, information. The engineering or electrical group that designs the system usually provides this information to the end user. Although documentation is often thought of as extraneous, it is actually a vital system component and a good engineering practice. In this chapter, we will explain the requirements of a good PLC documentation package, which will facilitate the understanding of the control system. Table 12-1 lists the components of a thorough documentation package.

Documentation Components
System abstract
System configuration
I/O wiring connection diagram
I/O address assignments
Internal storage address assignments
Storage register assignments
Variable declaration
Control program printout
Stored control program

Table 12-1. Components of a good PLC documentation package.

12-2 STEPS FOR DOCUMENTATION

SYSTEM ABSTRACT

A good system design starts with a thorough understanding of the problem and a good description of the process to be controlled. This assessment is followed by a systematic approach that will lead to the implementation of the control system. Once the system is finished, the personnel involved in the design should provide a global description, or abstract, of the scheme and procedure used to control the process.

A **system abstract** should provide the following:

- a clear statement of the control problem or task

- a description of the design strategy or philosophy used to implement the solution to the problem, which defines the functions of the major hardware and software components of the system, as well as why they were selected

- a statement of the objectives to be achieved

As an example of an abstract, let's examine a warehouse with a PLC that controls some conveyors. The statement of the task would specify that the PLC should control the conveyors so that the parts are sorted correctly (see Figure 12-1). The design philosophy would indicate that a single CPU, located in the warehouse's central area, controls two product conveyor lines. It would also indicate that remote subsystems, located in rooms 4 and 5, control the sorters for those areas. Moreover, it would specify that the programmable controller will gather data about total production from both lines and report it in printed form at the end of each shift. Finally, the statement of the objectives (i.e., 90% sorting accuracy) would allow the user to measure the success of the control implementation. This system abstract will convey general design information to the end user or anyone else who needs to understand the original control task.

ABC Warehouse Automation:

System Abstract

Task:

- PLC must sort parts correctly. Type A parts belong in area A; type B parts belong in area B.

Design Philosophy:

- Control room CPU operates conveyor A and conveyor B. Conveyor A is directed to sorter 4; conveyor B is directed to sorter 5.
- Subsystem 00, located in room 4, controls sorter 4.
- Subsystem 01, located in room 5, controls sorter 5.
- PLC will report total production data (parts sorted correctly) at the end of each shift. This data will be printed out on the control room printer.

Objective:

- The PLC system should achieve a 90% average of correctly sorted parts.

Figure 12-1. System abstract.

SYSTEM CONFIGURATION

As the name implies, the **system configuration** is a system arrangement diagram. In fact, it is a pictorial drawing of the hardware elements defined in the system abstract. It shows the location, simplified connections, and minimum details of the system's major hardware components (i.e., CPU, subsystems, peripherals, GUIs, etc.). Figure 12-2 illustrates a typical system configuration diagram.

The system configuration not only indicates the physical location of subsystems, but also the designation of the I/O rack address assignments. Referencing the rack address assignments allows for quick location of

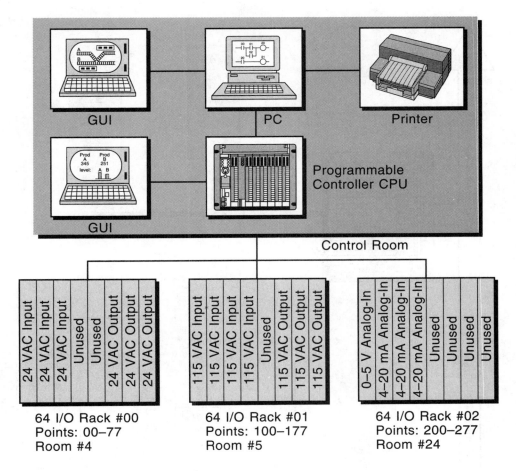

Figure 12-2. System arrangement diagram.

specific I/O devices. For example, during start-up the user can easily determine that I/O point 0200 (LS, PB, etc.), located in subsystem 02, is housed in room number 24.

If the programmable controller system involves a network framework with other components, the system configuration should show a general block diagram of the whole network (all nodes) and the major devices connected to it. For example, Figure 12-3 illustrates a PLC-based system in which a network interfaces with two other networks, a process bus network and a device bus network. This system configuration diagram immediately gives a broad picture of the total system.

I/O WIRING CONNECTION DIAGRAM

An **I/O wiring connection diagram** shows the actual connections of field input and output devices to the PLC module. This drawing normally includes power supplies and subsystem connections to the CPU. Figure

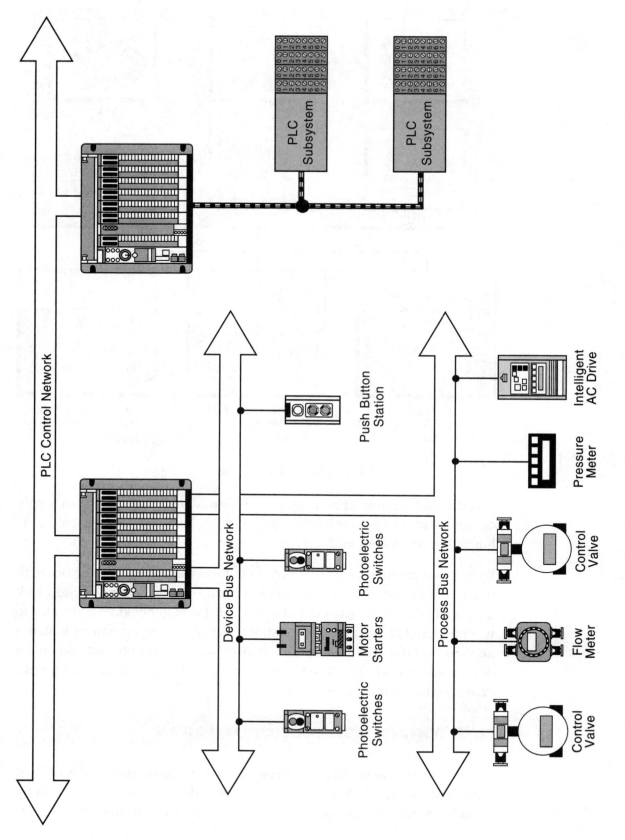

Figure 12-3. System configuration diagram of a PLC interfacing with process bus and device bus networks.

PLC Control Network

Device Bus Network

Process Bus Network

PLC Subsystem

PLC Subsystem

Push Button Station

Intelligent AC Drive

Pressure Meter

Photoelectric Switches

Control Valve

Motor Starters

Flow Meter

Photoelectric Switches

Control Valve

12-4 illustrates an example of an I/O wiring connection diagram. This diagram shows the rack, group, and module locations of each field device to illustrate the termination address of each I/O point. If the field devices are not wired directly to the I/O module, then the diagram should show terminal block numbers (see Figure 12-5). This way, anyone troubleshooting the PLC system will know which points to check in the terminal blocks. Good I/O wiring documentation is invaluable during installation, as well as for later reference.

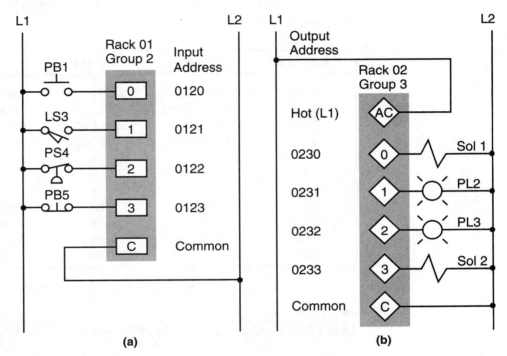

Figure 12-4. I/O wiring connection diagrams for **(a)** inputs and **(b)** outputs.

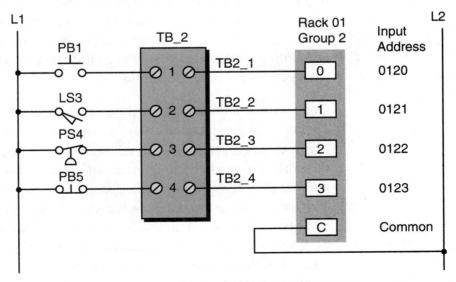

Figure 12-5. Input connection diagram indicating terminal block numbers.

I/O ADDRESS ASSIGNMENTS

An **I/O address assignment document** identifies each field device by address (rack, group, and terminal), the type of input or output module (e.g., 115 VAC, 24 VAC), the type of field device (e.g., limit switch, solenoid), and the function the device performs in the field. Table 12-2 shows a typical I/O address assignment document. This assignment document is similar to the I/O assignment table that will be completed prior to developing the control program.

Address	I/O type	Device	Function
0120	115 VAC in	PB	Start push button PB 1
0121	115 VAC in	LS	Up limit #2
0122	115 VAC in	PS (NC)	Hydraulic pressure OK
0123	115 VAC in	PB (NC)	Reset PB 2
•	•	•	•
•	•	•	•
•	•	•	•
0230	24 VAC out	Sol	Retract #1
0231	24 VAC out	PL	#2 in position
0232	24 VAC out	PL	Running
0233	24 VAC out	Sol	Fast up #3

Table 12-2. I/O address assignment.

INTERNAL STORAGE ADDRESS ASSIGNMENTS

An **internal storage address assignment document** is an important part of the total documentation package. Because internals are used for programming timers, counters, and control relay replacements and are not associated with any field devices, programmers tend to use them freely, without accounting for their usage. However, just as with real I/O, misuse of internals can result in system misoperation.

Good documentation of internals simplifies field modifications during start-up. For example, imagine a start-up situation involving the modification of one or more program rungs by adding extra interlocking. For this modification, the user must utilize internal coils that are not already assigned. If the internal I/O address assignment document is current and accurate, showing both used and unused addresses, the user can quickly locate available internal addresses. This saves time and avoids confusion. The internal address assignment document indicates the address, type, and function of each internal in the program. Table 12-3 illustrates a typical I/O address assignment document for internals.

Internal	Type	Description
1000	Coil	Used to latch position
1001	Coil	Set up instantaneous timer contact
1002	Compare	Used for CMP equal
1003	Add	Addition positive
•	•	•
•	•	•
•	•	•
T100	Timer	On-delay timer—motor 1
C400	Counter	Count pieces on conveyor #1
•	•	•
•	•	•
•	•	•

Table 12-3. Internal storage address assignment.

STORAGE REGISTER ASSIGNMENTS

Each available system register, whether a user storage register or an I/O register, should be properly identified. Most applications use registers to store or hold information for timers, counters, or comparisons. Keeping an accurate record of the use of and changes to these registers is very critical. Just as with I/O assignment documents, the **storage register assignment** table should show whether or not a register is being used. Table 12-4 shows a typical documentation form for register assignments.

Register	Contents	Description
3036	Temperature input T1	I/O register with analog module
3040	Temperature input T2	I/O register with analog module
4000	1200	20 sec preset of TDR3
4001	2000	Count preset for CMP=
4002	5000	Count preset for CMP>
•	•	•
•	•	•
•	•	•
4100	0	Not used
to	•	•
4200	•	•

Table 12-4. Storage register assignment table.

VARIABLE DECLARATION

In an IEC 1131-3 programming environment (discussed in Chapter 10), the documentation of the physical I/O addresses, internal storage addresses, and storage address assignments requires that the devices connected to the PLC via its I/O be declared, or defined, as variables. Table 12-5 illustrates a typical variable declaration. A proper variable declaration, which includes the name of the input, output, or internal, should be included in each of the assignment documents (e.g., I/O assignment, storage register assignment).

Address	Type	Description	Var_Name
0120	IN	Start push button PB1	START_PB1
0121	IN	Up limit switch LS2	Up_LS2
0122	IN	Hydraulic pressure OK	HYD_PRES_OK
0123	IN	Reset PB2	RESET_PB2
•	•	•	•
•	•	•	•
•	•	•	•
0230	OUT	Retract solenoid SOL1	Retract_SOL1
0231	OUT	Pilot light PL2 in position	In_Pos_PL2
0232	OUT	Pilot lights running	RUN_PL
0233	OUT	Fast up solenoid SOL3	Fast_Up_SOL3

Table 12-5. Variable declaration.

CONTROL PROGRAM PRINTOUT

The **control program printout** is a hard copy of the control logic program stored in the controller's memory. Whether stored in ladder form or some other language, the hard copy should be an exact replica of the controller's memory. Figure 12-6 shows a typical ladder printout in its basic format.

A basic hard copy printout shows each programmed instruction with the associated address of each input and output. This printout, however, does not readily provide information about each instruction's function or which field device is being evaluated or controlled. For this reason, the program coding alone, without the previously mentioned documentation, is not adequate for interpretation of the control system. Most manufacturers provide a documentation package that allows the programming device, generally a PC (personal computer), to enter labels or mnemonic nomenclature for the control program elements.

The extent of the control program printout and documentation varies from one PLC manufacturer to another. This documentation may or may not include information pertaining to the input/output connection diagram.

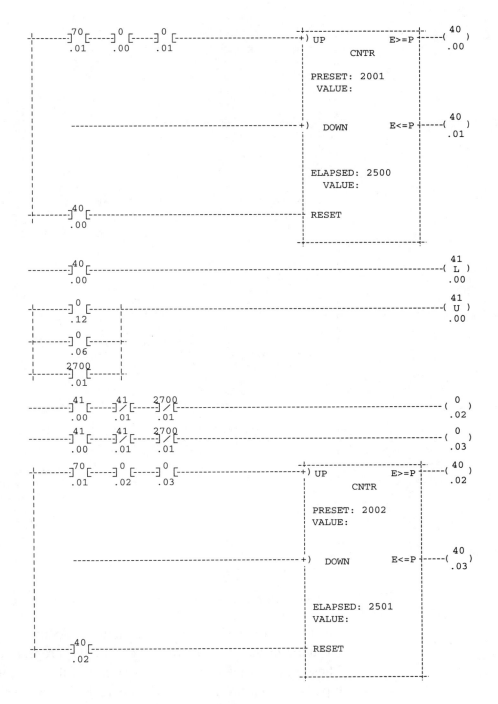

Figure 12-6. Ladder diagram printout.

Figure 12-7 illustrates a ladder control program with generic documented elements in the ladder rung. Sometimes, only the I/O address number represents these ladder diagram elements. Most PLC manufacturers' documentation allows the user to set global or generic mnemonic comments and then cross-reference the mnemonics with the inputs and outputs (real and internal) used in the system.

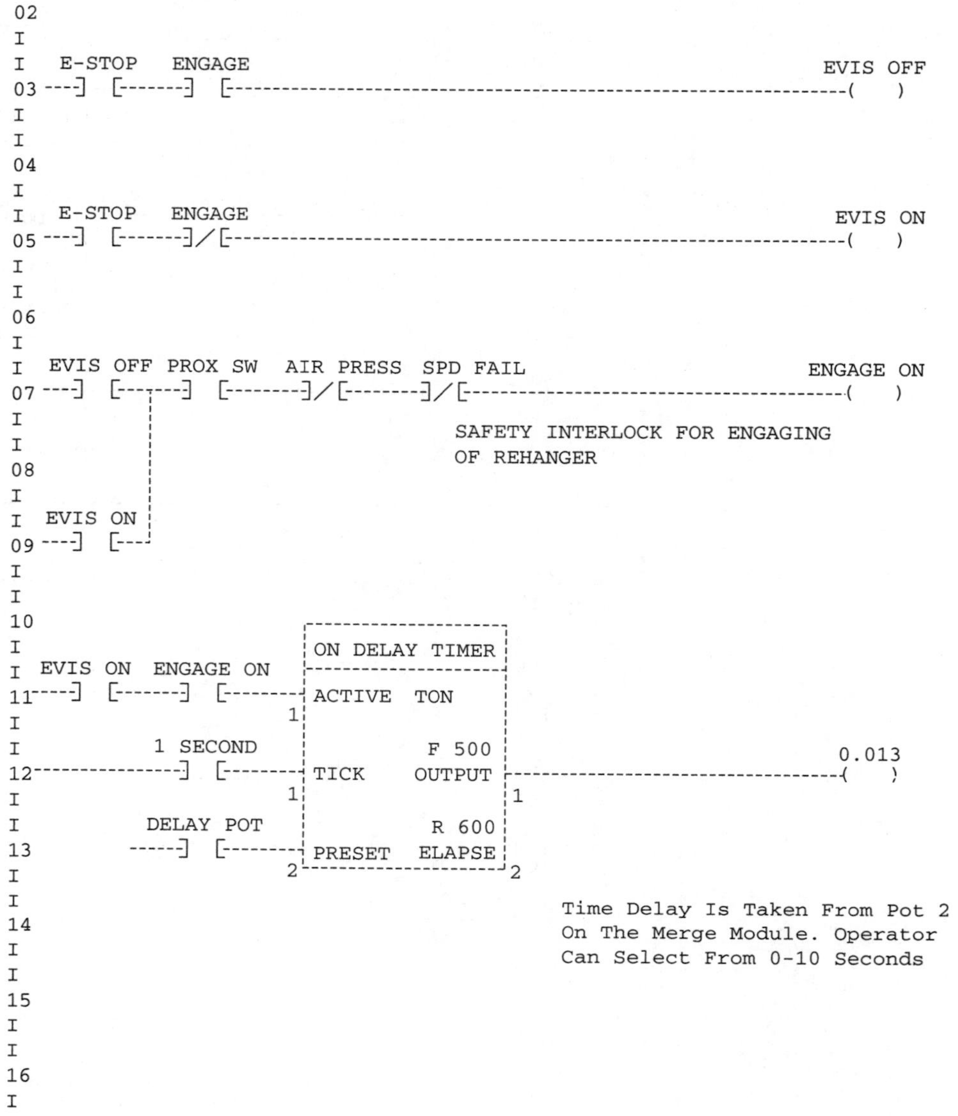

```
02
I
I   E-STOP     ENGAGE                                                      EVIS OFF
03 ---] [-------] [----------------------------------------------------------(   )
I
I
04
I
I   E-STOP     ENGAGE                                                      EVIS ON
05 ---] [-------]/[----------------------------------------------------------(   )
I
I
06
I
I   EVIS OFF PROX SW  AIR PRESS SPD FAIL                                ENGAGE ON
07 ---] [---┬---] [-------]/[-------]/[---------------------------------------(   )
I          ┊
I          ┊                          SAFETY INTERLOCK FOR ENGAGING
08         ┊                          OF REHANGER
I          ┊
I   EVIS ON┊
09 ---] [---┘
I
I
10
I                      ┌─────────────────────┐
I   EVIS ON  ENGAGE ON │  ON DELAY TIMER      │
11 ---] [-------] [------┤ ────────────────────
I                    1  │ ACTIVE   TON         │
I          1 SECOND     │            F 500     │                        0.013
12 ---------------] [----┤ TICK     OUTPUT ├------------------------------(   )
I                    1  │                  │  1
I          DELAY POT    │            R 600     │
13 -----] [----------┤ PRESET  ELAPSE │
I                    2  └─────────────────────┘ 2
I
I
14                                     Time Delay Is Taken From Pot 2
I                                      On The Merge Module. Operator
I                                      Can Select From 0-10 Seconds
15
I
I
16
I
```

Figure 12-7. Ladder control program printout with manufacturer's documentation.

Most IEC 1131 software systems include a documentation package that uses the defined variables as the labels for the programmed control elements. These systems also provide a summary of the variable declaration and the types of variables declared. Figure 12-8 shows a typical IEC 1131 level 1 chart printout.

The controller's memory always holds the latest software revision of the program; therefore, the user should have the most recent hard copy when examining the system. Changes are frequently made to the program during start-up, so these changes should be immediately documented, even though this is time consuming. Another good practice is to obtain the latest hard copy of the program after any field changes have taken place.

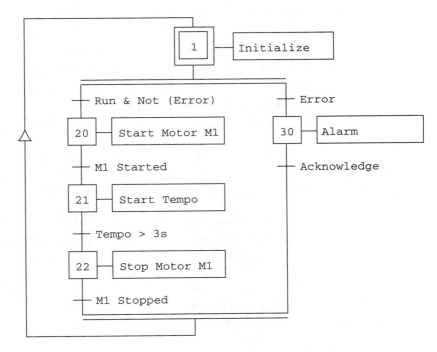

Figure 12-8. IEC 1131 level 1 chart printout.

CONTROL PROGRAM STORAGE

For the most part, PLC programming occurs at a location other than where the controller will finally be installed. For this reason, the user should save the control program on a storage medium, such as a cassette tape, a floppy disk, or an electronic memory module. This practice allows the user to send or carry the stored program to the installation site and reload it into the controller's memory quickly. This approach is usually employed when the system uses a volatile-type memory, but it is also used with nonvolatile memory for backup purposes.

The reproducible, stored control program, like any other form of documentation, should be kept accurate and current. A good practice is to always have two copies, in case one is damaged or misplaced. Also, make sure that the stored program agrees with the latest hard copy of the control logic.

12-3 PLC DOCUMENTATION SYSTEMS

Documentation plays a very important role in the design of any programmable controller–based system. This documentation can be a tedious, costly task requiring several skilled people to implement the many phases of documentation (e.g., drafting, table preparation, and I/O assignment). Therefore, as an alternative to creating this time-consuming documentation by hand, several manufacturers in the PLC support industry have developed sophisticated, yet simple, systems for documenting a total PLC system. These systems speed up the documentation procedure and reduce the manpower needed for the task. They increase the total program development productivity by reducing programming errors and increasing documentation throughput. In addition to the standard types of documentation previously discussed, documentation systems even provide several other useful documents.

Powerful and popular documentation systems, like the ones shown in Figure 12-9, offer numerous advantages and cost savings over manual documentation methods. Some of these advantages are as follows:

- Documentation systems provide electronic cut-and-paste capabilities, macros, copy functions, generic addressing capabilities, and address exchange functions.

- They provide a multiple-character, wide-field labeling capability for contacts and elements. In addition to ladder element labels, these systems also allow unlimited comments to be placed anywhere else on the ladder drawing.

- These systems provide the capability for a complete range of I/O elements and hardwired I/O drawings with integrated, automatic cross-referencing of the logic program.

- Documentation systems can upload and download programs for most PLC systems.

Courtesy of Mitsubishi Electronics, Mount Prospect, IL

Figure 12-9. Mitsubishi's PLC documentation system for A1S series PLCs.

Besides the ladder printout, documentation systems also provide input and output usage (assignment) reports. These reports list the controller's I/O addresses, illustrating how each point is used. These systems can also generate construction mnemonics documentation (e.g., contacts, limit switches, etc.), as well as a complete report of all instructions available for use in the PLC. Moreover, full cross-referenced reports provide direct information about all register contents and where each element is used in the program. An important advantage of the program listings produced by documentation systems is that they show, on a single document, all of the vital information about the control program.

Documentation systems are capable of uploading, verifying, and storing the PLC program directly from the controller or from a cassette, floppy disk, micro disk, or other storage media. Even though many third-party documentation and software programming systems exist, many PLC manufacturers now incorporate this capability into their own systems. In addition to documentation and programming capabilities, some of these systems also provide graphic user interfaces, or GUIs, which create user-friendly graphics that visually depict the control process.

12-4 CONCLUSION

Much can be said about documentation and its relevance to the total system package. We cannot overemphasize the importance of establishing—from the outset—complete, accurate documentation of the control problem and its solution. If the system documentation is created during the system design phase, as it should be, it will not become an unwelcome burden imposed on designers as the project nears completion.

Documentation may seem trivial to some or too much work to others. Whether designing their own control system or subcontracting the design, users should ensure that a good documentation package is delivered with the equipment. A well-designed system is one that is not just put to work during start-up, but can also be maintained, expanded, modified, and kept running without difficulty. Good documentation will definitely help both the designers and the end users in these tasks. Remember that, regardless of the application, a design is not good unless its documentation is also good.

Key Terms

control program printout
documentation
internal storage address assignment document
I/O address assignment document
I/O wiring connection diagram
storage register assignment document
system abstract
system configuration diagram

PLC PROCESS APPLICATIONS

- Data Measurements and Transducers
- Process Responses and Transfer Functions
- Process Controllers and Loop Tuning

CHAPTER THIRTEEN

DATA MEASUREMENTS AND TRANSDUCERS

When you measure what you are speaking about and express it in numbers, you know something about it; but when you cannot measure it, when you cannot express it in numbers, your knowledge is of a meager and unsatisfactory kind.

—William Thomson, Lord Kelvin

CHAPTER HIGHLIGHTS

As in any technical discipline, PLC users must know how to use and apply the measurement instruments and equipment that are connected to different field devices. In this chapter, we will explain how to interpret the data produced by measurement equipment, as well as how to deal with unexpected, erroneous measurements. We will also discuss transducers, a type of measurement device. This discussion, which will include displacement, pressure, flow, and vibration transducers, will explain how these devices turn measurement data into electrical signals. In the next chapter, you will apply this knowledge to process control applications.

13-1 BASIC MEASUREMENT CONCEPTS

DATA INTERPRETATION

Data interpretation and representation are very important when working with on-line process control operations. Measurement devices provide the control system with important information about the inner workings of the process. Therefore, every user must clearly understand what data is being collected by the measurement devices and how it should be interpreted. This will help the user to apply the control program correctly, so that the process will behave in a predictable manner.

To understand the data-gathering process, you must first understand how instrumentation and data-collecting devices interpret data readings. These devices can interpret data sampling readings four different ways, each with a different meaning. These methods for interpreting information include:

- mean
- median
- mode
- standard deviation

Mean. The **mean** is the average value of a set of readings. This value is useful in applications that require an estimation of future or expected readings. To illustrate the mean, let's use an instrument that emits a signal at set time intervals (every 10 seconds). This signal ranges from 2 to 20 mV and represents the mean value of the measurements taken during the 10-second time interval. That is, each signal's value is the average of the readings taken since the last signal. Let's suppose that the instrument last emitted a 13 mV signal and that it will send another signal in 10 seconds. Meanwhile, the instrument collects data every two seconds, resulting in values of 14 mV, 14.5 mV, 15 mV, 14.7 mV, and 14.8 mV. The mean of these readings, expressed as $\overline{X}$, is defined as:

$$\overline{X} = \frac{X_1 + X_2 + X_3 + \dots X_n}{n}$$

or

$$\overline{X} = \frac{\sum_{n=1}^{i} X_n}{n}$$

Therefore, at the new reporting time, the instrument will emit a 14.6 mV signal—the mean value of the five readings (i.e., $n = 5$):

$$\overline{X} = \frac{14 \text{ mV} + 14.5 \text{ mV} + 15 \text{ mV} + 14.7 \text{ mV} + 14.8 \text{ mV}}{5}$$

$$= 14.6 \text{ mV}$$

Median. The **median** is the middle value of a set of readings that are organized in ascending order. The following equations define the median:

$$M = X_{\left(\frac{m+1}{2}\right)} \qquad \text{for an odd number of samples}$$

$$M = \frac{X_{\left(\frac{m}{2}\right)} + X_{\left(\frac{m}{2}+1\right)}}{2} \qquad \text{for an even number of samples}$$

where:

$M =$ the median

$m =$ the total number of readings

$X =$ a reading value

The readings from the previous example placed in ascending order are 14 mV, 14.5 mV, 14.7 mV, 14.8 mV, and 15 mV. This is an odd number of values (i.e., $m = 5$). Therefore:

$$M = X_{\left(\frac{5+1}{2}\right)}$$

$$= X_3$$

The value X_3 is the third value in ascending order, so the median is 14.7 mV. Note that for an even number of samples the median is the mean of the two center values.

The median calculation provides statistical information about the data measurements taken and is more tolerant of errors than the mean calculations. Referencing the previous example, if the 14 mV reading had been erroneously read as 20 mV due to noise in the system, the mean would have increased to 15.8 mV, whereas the median would have remained 14.7 mV. Thus, the median value is not greatly affected by extreme deviations caused by measurement errors.

Mode. The **mode** is the most frequent value or values in a set of data. The mode value for the following set of instrumentation readings—14 mV, 14.5 mV, 14 mV, 14.5 mV and 14.5mV—is 14.5 mV, because it is the most frequent value. If six readings had been taken and the sixth one was 14 mV, two mode values would have existed, 14.5 mV and 14 mV (three occurrences of each).

Mode values occur mostly in discrete processes, where events are not broken down into infinitesimal readings, as in analog processes. PLC count readings from analog input modules rarely contain a significant mode value, since the continuous nature of the signal constantly introduces changes into the readings. Therefore, the mode is not as valuable as the mean and median in determining measurement errors.

Standard Deviation. Often, an application requires information not only about the mean value of a set of process readings, but also about how these readings are distributed in relation to the mean. The standard deviation provides valuable information about a group of data, thus aiding in the quantitative evaluation of the sample measurements.

To demonstrate standard deviation, let's examine a set of five instrument readings: 9 mV, 9.5 mV, 15 mV, 19.7 mV, and 19.8 mV. The mean of this sample is 14.6 mV, yet the readings are very dispersed. **Standard deviation** measures the spread of these values in relation to the mean and is expressed as:

$$\sigma = \sqrt{\frac{\sum_{n=1 \text{ to } i}\left(\overline{X} - X_n\right)^2}{n - 1}}$$

where:

$\sigma =$ the standard deviation

$\overline{X} =$ the calculated mean

$n =$ the number corresponding to each reading, starting at 1 and ending at the last reading, i

This formula computes the deviation of each sample from the mean. The larger the standard deviation value, the more spread out the values (samples) are from the mean. For our instrument readings, the standard deviation will have a value of $\sigma = 5.26$:

$$\sigma = \sqrt{\frac{31.36 + 26.01 + 0.16 + 26.01 + 27.04}{5 - 1}}$$

$$= \sqrt{\frac{110.58}{4}}$$

$$= 5.26$$

When the data values are evenly distributed around the mean in a bell form, they are said to have a *normal distribution* or *Gaussian distribution* (see Figure 13-1). The standard deviation in a Gaussian (normal) distribution measurement provides information that allows for a quantitative determination about how the data is spread. In a normal distribution, several conclusions can be obtained:

- 68% of all readings lie within $\pm 1\sigma$ (see Figure 13-2a)

- 95% of all readings lie within $\pm 2\sigma$ (see Figure 13-2b)

- 99% of all readings lie within $\pm 3\sigma$ (see Figure 13-2c)

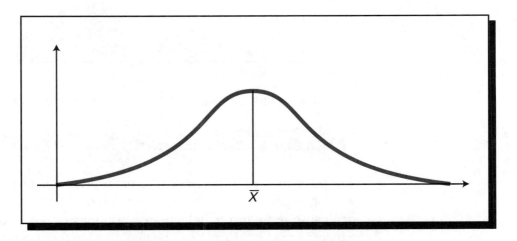

Figure 13-1. Normal distribution curve.

For example, if a set of reactor vessels in a continuous process has two temperature control loops, one that maintains a 358°C temperature with a standard deviation of 40°C and another that maintains a 358°C temperature with a standard deviation of 20°C, we know that the latter provides us with a more peaked graph about the mean. In fact, 68% of the temperature readings in the second loop lie between 338°C and 378°C, while in the first loop, 68% of the readings lie between 318°C and 398°C.

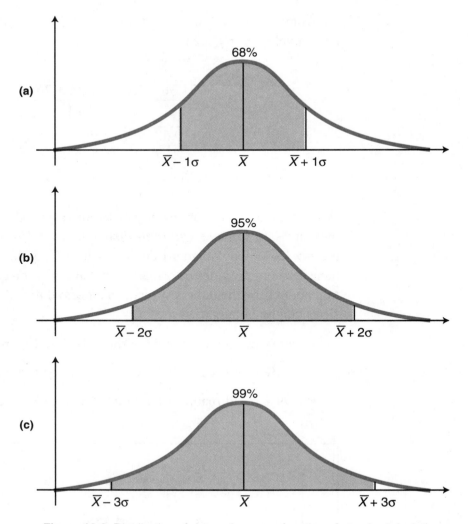

Figure 13-2. Distribution of data values as a function of standard deviation.

MEASUREMENT ERRORS

The possibility of measurement errors is present in any system that produces a finished good. Equipment malfunctions or misreadings of process variables can cause these errors, which are variations or deviations from the true or expected reading. Errors can be classified in three categories:

- gross errors

- system errors

- random errors

Gross errors are the result of human miscalculation, *system errors* are the result of the instrument itself or the environment, and *random errors* are the result of unexpected actions in the process line. Table 13-1 shows some examples of these types of errors, along with ways to predict and prevent them.

	Gross Errors	System Errors	Random Errors
Possible Causes	• Readings using an incorrect scale • False instrument readings • Improper setting of zero adjustment • Use of the wrong formula • Inexact recording of data • Incorrect instrument settings	• Instrument not calibrated properly • Unaccounted for nonlinearities • Worn-out parts • Loss of power due to improper application	• Changes in environment (e.g., pressure and temperature) • Variation of materials used in the process line • Vibration in conveyor line • Any other physical disturbance
Prediction	• No prediction can be drawn	• Compare readings to those of standard calculations • Determine the system error so that a cumulative error can be measured and expected	• Application of statistical analysis obtained by collecting data
Prevention or Reduction	• Pay more attention to correct displays • Have different persons take the same readings • Take multiple readings from an instrument • Be aware of instrument capabilities	• Calibrate instruments regularly • Emphasize regular maintenance of machinery and instruments • Monitor the consistency of the technique	• Physical adjustments should be made to instruments and the process line, so that they can withstand disturbances

Table 13-1. Measurement error types and their causes.

13-2 INTERPRETING ERRORS IN MEASUREMENTS

The discovery of errors is invaluable when controlling a machine or process because error information helps the user improve the system. Errors can be discovered in anticipation of the outcome (error prediction) or after a product is made (error detection). Error prediction is more useful than error detection, but it is harder to implement. Detection of an error after a product is made is fairly easy, since the final product can be checked against a reference model that matches all specifications. Although error prediction is much more useful, error detection is better than not discovering the error at all. For example, it is better to stop production of a machined piece because it has been found that the piece does not meet the customer's specifications than to ship a bad product to the customer.

Once an error is detected, it can be interpreted using statistical analysis. This type of statistical data analysis is, in fact, part of the foundation of artificial intelligence systems. These systems continuously collect data about a process and adjust production parameters accordingly. They then store their data measurements in a global database for use in later statistical analysis (see Chapter 16 for more about artificial intelligence systems).

In automated control systems, the controlling system and the process itself usually generate system errors. Several events, composed of a mix of several process errors, may combine to form a compounded system error. Likewise, guarantee errors, caused by errors in raw materials or supplies, may also generate system errors. Because their cause can be found, system errors can be predicted and corrected.

Unknown events that occur during the process create random errors. Therefore, unlike system errors, random errors can only be detected and corrected, not predicted. Most of the time, the user must employ statistical analysis to detect and remedy these errors.

INTERPRETING COMBINED ERRORS

Combined errors are errors caused by the interaction of two or more independent variables, each one causing a different problem. The system propagates the interaction of these variables; therefore, combined errors are also called **propagation errors**. By calculating statistical data about the sample before propagation and knowing the average and standard deviation requirements for the final product, the user can predict the outcome of the final product and make corrections for propagation errors throughout the process.

The value of an outcome formed by several variables (e.g., materials going into a batching process) is directly related to the average value of each variable. For instance, if a batching process uses two ingredients, A and B, and their average weights are $\overline{A}$ and $\overline{B}$, then the final weight of a mix containing both materials would be $\overline{A} + \overline{B}$. This outcome is the addition of both $\overline{A}$ and $\overline{B}$ because the operation to be performed is a blending, which implies that the quantities are added. Thus, the final outcome is directly related to the equation that governs the process being performed. In real life, the actual equation of a process is very hard to obtain; it is usually only approximated.

Standard deviation specifies how each sample value relates to the mean. Accordingly, the standard deviation of an outcome product can predict how the value of the final product will be spread out about its mean in relation to each of its component variables. This information forecasts the variance of the final product value. In the previous blending example, the average weight outcome (W) is represented by:

$$W = \overline{A} + \overline{B}$$

where $\overline{A}$ and $\overline{B}$ are the average weights of the ingredient products. If the distribution follows the normal (bell) curve, then:

- 68% of all samples lie within $W \pm 1\sigma_w$, or $(\overline{A} + \overline{B}) \pm 1\sigma_w$

- 95% of all samples lie within $W \pm 2\sigma_w$, or $(\overline{A} + \overline{B}) \pm 2\sigma_w$

- 99% or all samples lie within $W \pm 3\sigma_w$, or $(\overline{A} + \overline{B}) \pm 3\sigma_w$

where σ_w is the standard deviation of the final product.

However, to find the actual standard deviation, we must define the relationship between ingredients A and B and σ_w. To obtain an equation that allows two or more input variables, let's define the function K as the equation governing the final product and/or process. After numerous sample observations (n), the final product formula (K_n) will be a function of the amount of ingredients A and B added during the sample observations—A_n and B_n. That is:

$$K_n = K(A_n, B_n)$$

We can conclude that the most likely value for the function (the average value) is:

$$K_n = K(\overline{A}, \overline{B})$$

where the final outcome is a function of the two averages. We can define any deviation of a sample observation from the mean as ΔK_n, which is expressed as:

$$\Delta K_n = K_n - K(\overline{A}, \overline{B})$$

or

$$\Delta K_n = K(A_n, B_n) - K(\overline{A}, \overline{B})$$

If the deviation from the mean is 0 ($\Delta K_n = 0$), implying that the value of the nth observation is the same as the mean, then we would have:

$$K(A_n, B_n) = K(\overline{A}, \overline{B})$$

Based on differential calculus theory, we can transform the ΔK_n term into partial derivatives as:

$$\Delta K_n = \frac{\partial K}{\partial A} \Delta A_n + \frac{\partial K}{\partial B} \Delta B_n$$

By taking the average value of the sum of the squares and performing the square root of the right-hand term, we have:

$$\sigma_K = \sqrt{\left(\frac{\partial K}{\partial A}\right)^2 \sigma_A^2 + \left(\frac{\partial K}{\partial B}\right)^2 \sigma_B^2}$$

where σ_K is the standard deviation of the final outcome and σ_A and σ_B are the standard deviations of the independent variables A and B. The other terms are the partial derivatives of the function. This equation indicates that an approximate standard deviation of a function (product) can be predicted by knowing the standard deviations of the independent variables and the function of the process itself. The following example illustrates the use of this function.

EXAMPLE 13-1

A manufacturing plant produces sphere-shaped pellets. These pellets are heated for a period of time to make specific changes in the sphere size. After numerous observations, quality control has determined that the radius (r) has a mean of 1.0 inch and a standard deviation (σ_r) of 0.0008 inches. The pellet material weight (W) has a mean value of 0.15 lbs/in^3 and a standard deviation (σ_W) of 0.00082 lbs/in^3.

(a) Find the probable sphere weight of the final product and its standard deviation. (b) Make suggestions about how this information could be used.

SOLUTION

(a) The total weight (W_t) of the sphere can be calculated as volume times weight ($V \times W$):

$$V = \tfrac{4}{3}\pi r^3$$
$$W_t = VW$$
$$= \tfrac{4}{3}\pi r^3 W$$

Therefore, the final total weight of the product under normal process conditions is:

$$W_t = \tfrac{4}{3}(3.1416)(1.0)^3(0.15)$$
$$= 0.628 \text{ lbs}$$

The standard deviation is calculated using the formula:

$$\sigma_{W_t} = \sqrt{\left(\frac{\partial W_t}{\partial W}\right)^2 \sigma_w^2 + \left(\frac{\partial W_t}{\partial r}\right)^2 \sigma_r^2}$$
$$= \sqrt{\left(\tfrac{4}{3}\pi r^3\right)^2 \left(\sigma_w\right)^2 + \left(4W\pi r^2\right)^2 \left(\sigma_r\right)^2}$$
$$= \sqrt{(17.545)(6.7 \times 10^{-7}) + (3.553)(6.4 \times 10^{-7})}$$
$$= 0.003746 \text{ lbs}$$

(b) The previous calculations show that, based on the samples obtained for the average radius and average weight of the produced part, the standard deviation of the finished product can be estimated at 0.003746 lbs. If this value is within the range specified by quality control, the product will be acceptable. On the other hand, if the value for average final weight fluctuates greatly, producing an unacceptable standard deviation value, the process must be altered so that the part meets quality control specifications. These process alterations could include raising or lowering the heat to control the radius of the part (by expansion), thus shaping the sphere so that its weight is within the desired standard deviation range. This process adjustment would, however, require a definition of the amount of heat needed to alter the shape and size of the pellets. In order to make this kind of process adjustment, the system must be capable of measuring samples during the manufacturing process via transducers and other measuring equipment.

INTERPRETING GUARANTEE ERRORS

Guarantee errors are known values that state that a product or material's specifications will be within a specified arithmetic deviation from the mean. For example, if a supplier specifies that a metal part used in an assembly line has a length of 26 centimeters with a guarantee deviation (error) of less than 0.1%, then the length of its supplied parts is within a range of 26 cm ± 0.026 cm. Moreover, if the manufacturer specifies a ±3σ standard deviation, then 99% of the parts will be within ±0.026 cm of the mean.

To anticipate the possible value (outcome) of a process using guarantee limits, the arithmetic worst-case scenario must be calculated. The following example illustrates how two variables can be manipulated according to their guarantee values to obtain the process outcome's worst-case condition for error tolerance.

EXAMPLE 13-2

An electric heater with a current control system has a resistance value of 150 ohms. The resistor has a guarantee deviation of ±0.15% of total resistance. The current, which is controlled by a PLC's analog output, has a ±0.1% guarantee limit at 4.5 amps. Find the nominal power at the heater and the error (deviation from the true mean) as guaranteed by the limits.

SOLUTION

The equation that describes power dissipation is:

$$P = I^2R$$

where:

$P =$ the power dissipation
$I =$ the current
$R =$ the resistance

Therefore, the nominal power calculation for the heater is:

$$P = (4.5)^2(150)$$
$$= 3037.5 \text{ watts}$$

The guarantee limits of the component are:

$$I = 4.5 \ \text{amps} \pm 0.1\%$$
$$= 4.5 \pm 0.0045 \ \text{amps}$$

$$R = 150 \ \text{ohms} \pm 0.15\%$$
$$= 150 \pm 0.225 \ \text{ohms}$$

The variations in power, ΔP_I and ΔP_R (power variations due to current and resistance, respectively), caused by each of the guarantee error limits are:

$$\Delta P_I = \frac{\partial P}{\partial I} \Delta I = 2IR\Delta I$$
$$= 2(4.5)(150)(\pm 0.0045)$$
$$= \pm 6.075 \ \text{watts}$$

$$\Delta P_R = \frac{\partial P}{\partial R} \Delta R = I^2 \Delta R$$
$$= (4.5)^2 (\pm 0.225)$$
$$= \pm 4.556 \ \text{watts}$$

Adding these variation values to the nominal power value yields the expected worst-case value:

$$P = P_{\text{nominal}} \pm \Delta P_I \pm \Delta P_R$$
$$= 3037.5 \pm 6.075 \pm 4.556 \ \text{watts}$$
$$= 3037.5 \pm 10.631 \ \text{watts}$$
$$= 3037.5 \ \text{watts} \pm 0.35\%$$

Thus, the variation in outcome power based on guarantee variable errors is 0.35% of the total power.

13-3 TRANSDUCER MEASUREMENTS

This section deals primarily with two measuring techniques that are used to implement transducer circuits. These techniques involve the use of bridge circuits and linear variable differential transformer (LVDT) mechanisms. For example, to detect pressure and changes in pressure, you can use a strain gauge, which is based on the bridge circuit technique, or a Bourdon tube, which is based on the LVDT mechanism technique. A knowledge of how these transducer measurement circuits work will give you a better perspective of not only how they are used, but also where functional errors may occur when measurement problems arise.

BRIDGE CIRCUIT TECHNIQUES

Bridge circuits use resistive elements to sense measurement changes. Depending on how the circuit is configured, the bridge will change the voltage or current of its output in proportion to changes in its resistive measurement element. This resistance change generally creates a *bridge imbalance*. Under normal (balanced) operation of a bridge circuit, the current that passes through one section of a current-sensitive bridge is the same as that in the other section (or in a voltage-sensitive bridge, the voltage differential between the two sections is zero). An imbalance occurs when the resistance of one element changes, thus creating a current or voltage offset that is proportional to the resistance change. The bridge circuit utilizes this offset measurement to determine the value of the measured variable. Figure 13-3 shows a bridge circuit.

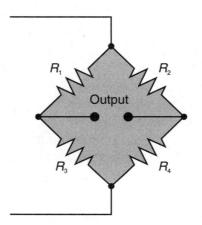

Figure 13-3. Simple bridge circuit.

Voltage-Sensitive Bridge. A *voltage-sensitive bridge* senses a voltage differential at the output of the bridge that is proportional to the resistance change in the bridge. Figure 13-4 illustrates a voltage-sensitive bridge, where D is the detector device and R_D is its resistance. The value of R_D for a voltage-sensitive bridge is very high. This amount of resistance could be provided by the input impedance of an amplifier module of a PLC. The following example illustrates the relationship between the resistors in a voltage-sensitive bridge circuit. Note that a change in the resistance of R_4 (the measuring element) creates the bridge imbalance; the other resistors have fixed, known values.

EXAMPLE 13-3

For the voltage-sensitive circuit shown in Figure 13-4, find **(a)** the equation that describes the voltage differential measurement between point A and point B and **(b)** the bridge resistance ratio when the voltage differential is 0 (balanced state).

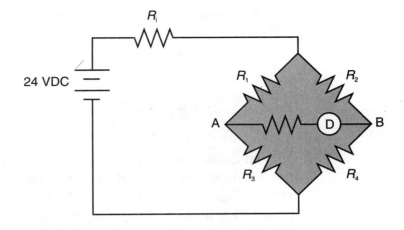

Figure 13-4. Voltage-sensitive bridge circuit.

SOLUTION

(a) Assuming that $R_D = \infty$ (i.e., it is very large) and the excitation voltage impedance (R_i) equals 0, the voltages at points V_A and V_B are:

$$V_A = \left(\frac{R_3}{R_1 + R_3}\right)V$$

$$V_B = \left(\frac{R_4}{R_2 + R_4}\right)V$$

The voltage differential between points A and B is:

$$\Delta V = V_A - V_B$$

$$= V\left[\left(\frac{R_3}{R_1 + R_3}\right) - \left(\frac{R_4}{R_2 + R_4}\right)\right]$$

$$= V\left[\frac{R_2 R_3 - R_1 R_4}{(R_1 + R_3)(R_2 + R_4)}\right]$$

(b) When the differential voltage (ΔV) is 0, V_A equals V_B, so:

$$\left(\frac{R_3}{R_1 + R_3}\right)V = \left(\frac{R_4}{R_2 + R_4}\right)V$$

$$\frac{R_3}{R_1 + R_3} = \frac{R_4}{R_2 + R_4}$$

$$R_2 R_3 + R_3 R_4 = R_1 R_4 + R_3 R_4$$

$$R_2 R_3 = R_1 R_4$$

Therefore, the bridge resistance ratio is:

$$\frac{R_1}{R_2} = \frac{R_3}{R_4}$$

where R_4 is the measuring resistance element.

Current-Sensitive Bridge. A *current-sensitive bridge* creates a current flow change through the output of the bridge, that is, between point A and point B (refer to Figure 13-4). The current flow is the result of a bridge imbalance created by resistance changes in the measuring element. The other resistors in the bridge have known, fixed values.

When current changes are being measured, the detecting device D has a very low resistance R_D, allowing current to flow from point A to B through the detector. Typical devices that have very low impedance include galvanometers and low-input impedance current amplifier interfaces (PLC modules).

The following equation describes the current that flows through a current-sensitive bridge's detector as a result of a bridge imbalance:

$$I_D = \frac{VR_4}{\left[R_i\left(1 + \frac{R_2}{R_1}\right) + R_2 + R_{4B}\right]\left[R_D\left(1 + \frac{R_3}{R_1}\right) + R_3 + R_{4B}\right]}$$

The term R_{4B} is the resistance value when the bridge is balanced. The following example illustrates how this equation is used to obtain a current proportional to the change in resistance.

EXAMPLE 13-4

A bridge circuit uses a thermistor with a nominal resistance of 10 Ω to measure small changes in temperature (see Figure 13-5). An amplifier input module, which has an input impedance of 300 Ω, measures small changes in current. What is the current if a change in temperature results in a 10% change in resistance?

SOLUTION

The resistance of the thermistor (R_4) changes 10% due to temperature change, which translates into an R_4 value of 11 Ω (10 Ω + 1 Ω). The term ΔR_4 is the absolute value of the difference between R_{4B} and the new value of R_4 due to the measurement change. Therefore, the difference in thermistor resistance is calculated as:

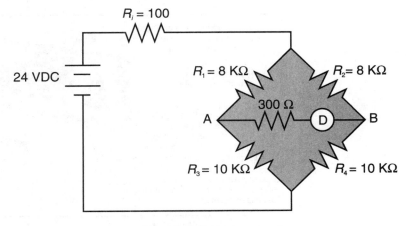

Figure 13-5. Bridge circuit.

$$\Delta R_4 = |R_4 - R_{4B}|$$
$$= |10\ \text{K}\Omega - 11\ \text{K}\Omega|$$
$$= 1\ \text{K}\Omega$$

The current measurement is defined by (values of R in KΩ):

$$I_D = \frac{V\Delta R_4}{\left[R_i\left(1 + \frac{R_2}{R_1}\right) + R_2 + R_{4B}\right]\left[R_D\left(1 + \frac{R_3}{R_1}\right) + R_3 + R_{4B}\right]}$$

$$= \frac{(24)(1)}{\left[0.1\left(1 + \frac{8}{8}\right) + 8 + 10\right]\left[0.3\left(1 + \frac{10}{8}\right) + 10 + 10\right]}$$

$$= \frac{24}{(18.2)(20.675)}$$

$$= 0.06378\ \text{mA}$$

LVDT TECHNIQUES

A **linear variable differential transformer (LVDT)** is an electromechanical mechanism that provides a voltage reference that is proportional to the displacement of a core inside a coil. Figure 13-6 illustrates a cutaway and a diagram of an LVDT, while Table 13-2 lists the types of transducers that use LVDT mechanisms.

An AC voltage, when applied to the primary coil, creates an induced voltage in the secondary coils of an LVDT. As the LVDT's core (which is made of a magnetic material) moves, the voltage at the output of the secondary coil changes. The induced voltage created by the core movement and the way the secondary coils are wound determine the value of the voltage change (see Figure 13-7). The secondary coil is wound in the opposite direction of the primary, so that the induced voltage will change polarity as the coil moves.

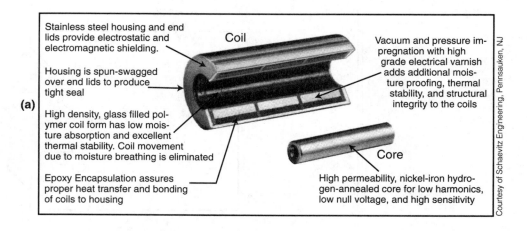

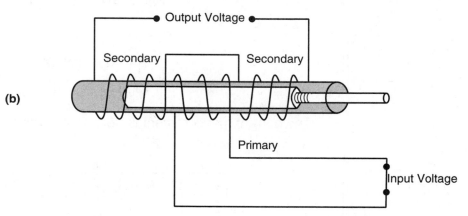

Courtesy of Schaevitz Engineering, Pennsauken, NJ

Figure 13-6. (a) Cutaway and **(b)** diagram of an LVDT.

Transducer	Uses and Measurements
Load cell	Torque, force, weight, tensile testing
Bourdon tube	Pressure, fluid pressure, volume porosity
Bellows, diaphragm	Pressure, fluid pressure, low-range pressure measurements
LVDT lineal	Lineal displacement, level measurement
Manometer	Measuring the height of a column of liquid with a known density
LVDT gauge	Mechanical dial position indicators, gauging and displacement (very small) measurements
Accelerometer	Acceleration in one or more axes, servo positioning applications, seismic systems
Inclinometer	Level sensing, incline
Variable reluctance proximity detector	Proximity detection of ferromagnetic objects

Table 13-2. Transducers that use LVDT mechanisms.

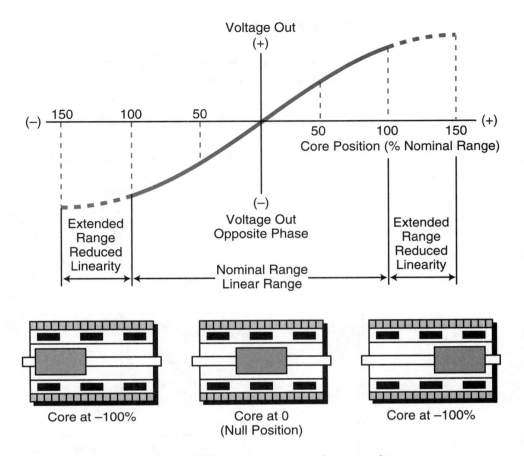

Figure 13-7. LVDT core movement and output voltage.

Modern LVDTs provide *demodulation*, or *rectification*, circuits to convert the secondary output into a DC voltage signal. This voltage signal is in linear proportion to the core movement within its range. The resultant voltage when the core is at its starting position is +*V*; when the core is at its end position the resultant voltage is –*V*. When the core is at the middle, it provides a null, or zero, voltage output. Figure 13-8 illustrates a simple demodulator circuit for an LVDT.

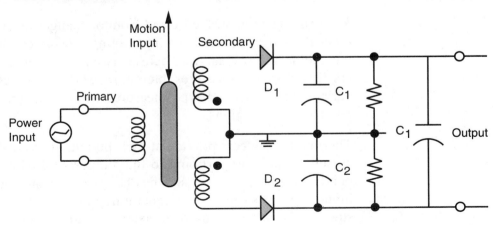

Figure 13-8. Demodulator circuit for an LVDT.

EXAMPLE 13-5

Graphically illustrate the position of an LVDT core that has a total displacement range of 20 inches and an output of ±10 VDC.

SOLUTION

Figure 13-9 shows the graph for this LVDT core.

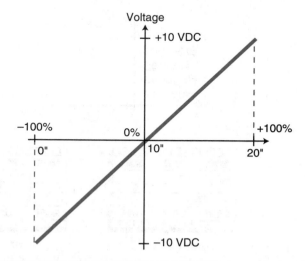

Core Position	Linear Position	Voltage Output
Core at −100%	0"	−10 VDC
Core at 0% (null)	10"	0 VDC
Core at +100%	20"	+10 VDC

Figure 13-9. Position of example LVDT core.

13-4 THERMAL TRANSDUCERS

Thermal transducers sense and monitor changes in temperature. Either the process itself or induced heating/cooling process control inputs may cause these temperature changes. There are two primary types of thermal transducers. The first type measures internal *resistance* changes due to temperature variations; the second type measures *voltage* differentials as a result of temperature variations.

Thermal transducers provide at their output, after conditioning, voltage or current signals proportional to the temperature measurement range. Depending on the transducer and the PLC used, special input modules or analog input interfaces input this temperature data into the controller. An understanding of thermal transducer operation will help you know how and where to use these transducers in a process control application.

There are many types of thermal transducers in the marketplace; however, this section only discusses the most commonly used ones: resistance temperature detectors (RTDs), thermistors, and thermocouples. RTDs and thermistors are internal resistance–type transducers, while thermocouples measure voltage differentials.

RESISTANCE TEMPERATURE DETECTORS (RTDs)

Resistance temperature detectors (RTDs) are temperature transducers made of conductive wire elements. The most common types of wires used in RTDs are platinum, nickel, copper, and nickel-iron. A protective sheath material (protecting tube) covers these wires, which are coiled around an insulator that serves as a support. Figure 13-10 shows the construction of an RTD. In an RTD, the resistance of the conductive wires increases linearly with an increase in the temperature being measured; for this reason, RTDs are said to have a *positive temperature coefficient.*

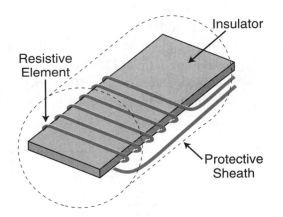

Figure 13-10. Resistance temperature detector.

RTDs are generally used in a bridge circuit configuration. Figure 13-11 illustrates an RTD in a bridge circuit. As mentioned in the previous section, a bridge circuit provides an output proportional to changes in resistance. Since the RTD is the variable resistor in the bridge (i.e., it reacts to temperature changes), the bridge output will be proportional to the temperature measured by the RTD.

As shown in Figure 13-11, an RTD element may be located away from its bridge circuit. In this configuration, the user must be aware of the lead wire resistance created by the wire connecting the RTD with the bridge circuit. The lead wire resistance causes the total resistance in the RTD arm of the bridge to increase, since the lead wire resistance adds to the RTD resistance. If the RTD circuit does not receive proper lead wire compensation, it will provide an erroneous measurement.

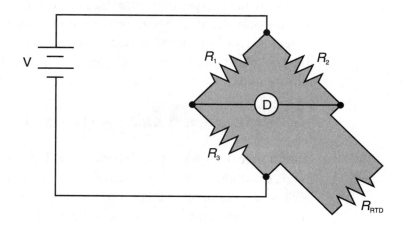

Figure 13-11. RTD in a bridge circuit.

Figure 13-12 presents a typical wire compensation method used to balance lead wire resistance. The lead resistances of wires L1 and L2 are identical because they are made of the same material. These two resistances, R_{L1} and R_{L2}, are added to R_2 and R_{RTD}, respectively. This adds the wire resistance to two adjacent sides of the bridge, thereby compensating for the resistance of the lead wire in the RTD measurement. The equations in Figure 13-12 represent the bridge before and after compensation. Note that R_{L3} has no influence on the bridge circuit since it is connected to the detector (e.g., input module, amplifier, etc.).

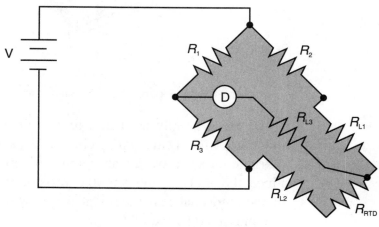

$$\frac{R_1}{R_2} = \frac{R_3}{R_{RTD}}$$ without lead wire consideration

$$\frac{R_1}{R_2} = \frac{R_3}{R_{RTD} + R_{L1} + R_{L2}}$$ taking lead wire into consideration (no compensation)

$$\frac{R_1}{R_2 + R_{L1}} = \frac{R_3}{R_{RTD} + R_{L2}}$$ taking lead wire into consideration (with compensation)

Figure 13-12. RTD bridge configuration with lead wire compensation.

As mentioned previously, the changes in RTD resistance are proportional to changes in temperature. The following equation defines these resistance changes:

$$R_T = R_{T_0}\left[1 + \alpha_1(T - T_0) + \alpha_2(T - T_0)^2\right]$$

where:

R_T = the change in resistance at temperature T

R_{T_0} = the RTD resistance at a reference temperature point T_0 (e.g., copper is 10 Ω at 25°C)

α_1 = a constant per degree Celsius that varies with the first RTD material

α_2 = a constant per degree Celsius that varies with the second RTD material

When an RTD is connected to a PLC's RTD input module, the interface determines the temperature (T) based on changes in resistance R_T. The module stores this value, which is calculated through an equation that corresponds to the RTD's type of input (e.g., copper), in a table. During this process, the input module also compensates for lead wire connections.

If an RTD is used with a standard analog input module, the user must design the bridge circuit, as well as the amplifier, so that the signal matches that of the input module range (e.g., 0 to 10 VDC). To do this, the PLC must compute the temperature by determining the temperature-versus-voltage curve. It determines this linear curve by analyzing another curve, the resistance-versus-temperature curve. It then computes the temperature using the temperature-versus-VDC equation or the linear interpolation look-up table for the input count value of the analog input voltage. This technique can be used with any transducer that uses a bridge circuit for signal detection (e.g., thermistor, strain gauge, etc.). If the transducer's temperature detection range is linear with respect to resistance, the PLC can use an equation to compute the temperature. If the transducer's temperature detection range is not linear, the PLC must perform a linear interpolation based on a look-up table. Chapter 7 explains linear equations in analog readings, while Chapter 11 provides examples of linear interpolations of analog readings.

THERMISTORS

Like RTDs, **thermistors** (see Figure 13-13) are temperature transducers that exhibit changes in internal resistance proportional to changes in temperature. Thermistors are made of semiconductor materials, such as oxides of cobalt, nickel, manganese, iron, and titanium. These semiconductor materials exhibit a temperature-versus-resistance behavior that is opposite of the behavior of RTD conducting materials. As the temperature increases, the resistance of a thermistor decreases; therefore, a thermistor is said to have a *negative temperature coefficient*. Although most thermistors have negative coefficients, some do have positive temperature coefficients.

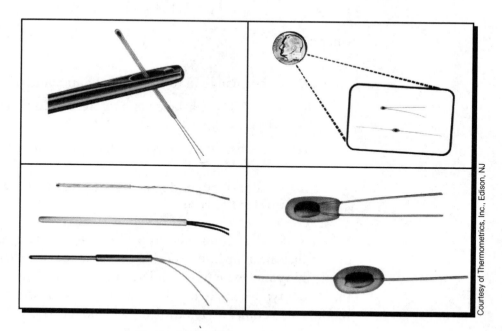

Figure 13-13. Different types of thermistors.

Thermistors can be classified into two major groups: bead thermistors and metallized-surface thermistors. Table 13-3 lists the types of thermistors that fall under these categories. Each of these two groups of thermistors offer advantages and disadvantages, as shown in Table 13-4.

Bead-Type Thermistors	Metallized-Surface Contact Thermistors
Bare beads	Discs
Glass-coated beads	Chips
Glass probes	Flakes
Glass rods (axial lead probes)	Rods
Bead-in-glass (tube or enclosure)	Washers or wafers

Table 13-3. Classification of thermistors.

Thermistors experience a much greater change in resistance than RTDs. Figure 13-14 illustrates a graph of the temperature-versus-resistance ratio for thermistors (R_T/R at 25°C, where R_T is the resistance at temperature T). The graph shows that thermistors experience a large change in resistance with relatively small increases in temperature. An advantage to these abrupt changes in resistance is that a thermistor can provide better resolution than an

Type	Advantages	Disadvantages
Bead-type thermistors	• Good to excellent stability, lead wires are strain relieved in glass hermetic seal • High operating and storage temperatures • Smaller sizes available • Fast response times	• Normally broad resistance tolerances, high cost for close tolerances • Medium to low dissipation constants • Matched pairs or resistive padding are required for interchangeability
Metallized-surface contact-type thermistors	• Normally tighter tolerances, lower cost for close tolerances • Low-cost single units for interchangeability • Medium dissipation constants	• Moderate to good stability, difficult to obtain high stability without hermetic seal • Limited operating and storage temperatures • Medium sizes available • Medium response times

Table 13-4. Advantages and disadvantages of thermistor types.

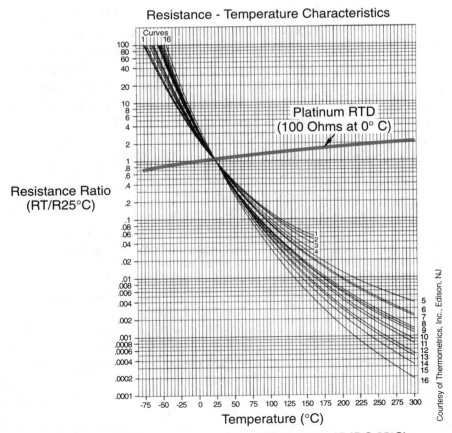

Figure 13-14. Temperature-versus-resistance curve (R_T/R @ 25°C).

RTD for certain temperature ranges. Therefore, thermistors provide more accurate readings when the span of measurement is narrow. For example, a thermistor with a 1 MΩ resistance at 25°C will have a resistance of approximately 300 KΩ at 50°C, meaning that the resistance changes by 28 KΩ for every 1°C temperature change. Energy management applications, which have narrow temperature spans and need accurate control measurements, may require this type of resolution.

A thermistor's resistance as a function of temperature can be defined by:

$$R_T = R_0 e^{\beta\left[\left(\frac{1}{T}\right) - \left(\frac{1}{T_0}\right)\right]}$$

where:

$R_T =$ the thermistor resistance at absolute temperature T

$T =$ °C plus 273° (absolute temperature)

$T_0 =$ the temperature reference (absolute °K)

$\beta =$ a constant between 3400 and 4000 absolute degrees, depending on the type of thermistor used

As indicated in the previous equation and the resistance ratio graph in Figure 13-14, the resistance change in a thermistor is proportional to the natural logarithm of the temperature, indicating rapid changes in resistance in response to changes in temperature. When using thermistors, the temperature measurement range should not exceed 100 to 150°C, so that the temperature-versus-resistance ratio is linear. Table 13-5 compares the advantages and disadvantages of thermistors and RTDs.

Type	Advantages	Disadvantages
Thermistor	• Fast response • Small size • High resistances eliminate most lead resistance problems • Rugged, not affected by shock or vibration • Inexpensive	• Nonlinear • Narrow span for any singular input • Interchangeability limited unless matched pairs are used
RTD	• Linear over wide range • Wide temperature range • High temperature range • Interchangeable over wide range • Better stability at high temperatures	• Low sensitivity • More expensive • No point sensing • Affected by shock and vibration • Requires 3- or 4-wire operation • Can be affected by contact resistance

Table 13-5. Advantages and disadvantages of thermistors and RTDs.

EXAMPLE 13-6

A thermistor has a resistance value of 100 KΩ at 3°C and a β constant of 3900°K. At what temperature (°C) will the thermistor's resistance be 20 Ω?

SOLUTION

The value of T_0 is equal to 3°C. This translates to an absolute temperature value of 276°K (3°C + 273°). So, when R_T equals 20 KΩ, the temperature will be 311.5°K:

$$R_T = R_0 e^{\beta \left[\left(\frac{1}{T} \right) - \left(\frac{1}{T_0} \right) \right]}$$

$$20 = 100 e^{3900 \left[\left(\frac{1}{T} \right) - \left(\frac{1}{276} \right) \right]}$$

$$0.2 = e^{3900 \left[\left(\frac{1}{T} \right) - (0.00362) \right]}$$

$$\ln 0.2 = 3900 \left[\left(\frac{1}{T} \right) - (0.00362) \right]$$

$$-1.609 = 3900 \left[\left(\frac{1}{T} \right) - (0.00362) \right]$$

$$-0.000413 = \frac{1}{T} - 0.00362$$

$$\frac{1}{T} = 0.00321$$

$$T = 311.5°K$$

Therefore, the temperature in °C is:

$$311.5°K - 273°K = 38.5°C$$

THERMOCOUPLES

Thermocouples are bimetallic temperature measuring devices (i.e., they are made of two different metals). In a thermocouple, the two metals are joined together at junctions with different temperatures (see Figure 13-15). This temperature differential creates a voltage across the thermocouple—a phenomenon known as the Seebeck effect. Temperature T_1, the *hot junction*, is the temperature being measured, while T_2, the *cold junction*, is the reference temperature. As temperature T_1 increases, the voltage differential (emf) between materials A and B increases in proportion to the temperature. Table 13-6 compares the characteristics of different types of thermocouples.

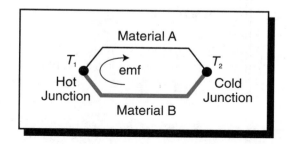

Figure 13-15. Thermocouple diagram.

The standard reference junction (cold junction) temperature for a thermocouple is 0°C; therefore, all standard tables list thermocouple voltage outputs based on a 0°C cold junction reference. Figure 13-16 graphically illustrates how to use an ice bath to keep the reference junction at 0°C. Although in theory the reference junction should be at 0°C, this is certainly not practical in industrial applications. Therefore, to implement thermocouple readings based on the standard tables (0°C reference), the user must perform a cold junction compensation. Cold junction compensation factors the effect of the nonzero reference temperature into the output voltage reading so that a standard thermocouple table can be used to find the hot junction temperature. Let's look at an example to see how this compensation is performed.

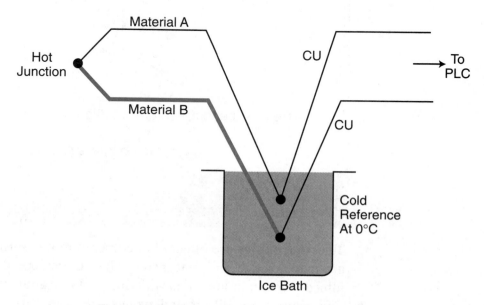

Figure 13-16. Thermocouple kept at 0°C in an ice bath.

ISA Type Designation	Positive Wire	Negative Wire	Millivolts per °F	Minimum Temp (°F)	Maximum Temp (°F)	Scale Linearity	Recommended Environment	Advantages	Disadvantages
B	Pt70-Rh30	Pt94-Rh6	0.0003–0.006	32	3380	Good at high temps; poor below 1000°F	Inert or slow oxidizing	—	—
E	Chromel	Constantan	0.015–0.042	–320	1830	Good	Oxidizing	Highest emf/°F	Larger drift than other base metal T/Cs
J	Iron	Constantan	0.014–0.035	–320	1400	Good; nearly linear from 300–800°F	Reducing	Most economical	—
K	Chromel	Alumel	0.009–0.024	–310	2500	Good; most linear of all T/Cs	Oxidizing	Most linear	More expensive than J or T
R	Pt87-Rh13	Platinum	0.003–0.008	0	3100	Same as type B	Oxidizing	Small size, fast response	More expensive than type K
S	Pt90-Rh10	Platinum	0.003–0.007	0	3200	Same as type B	Oxidizing	Same as type R	More expensive than type K
T	Copper	Constantan	0.008–0.035	–310	750	Good, but crowded at low end	Oxidizing or reducing	Good resistance to corrosion from moisture	Limited temperature
Y	Iron	Constantan	0.022–0.033	–200	1800	About the same as type J	Reducing	—	Not industrial standard
—	Tungsten	W74-Re26	0.001–0.012	0	4200	Same as type B	Inert or vacuum	High temperature	Brittle, hard to use, expensive
—	W94-Re6	W74-Re26	0.001–0.010	0	4200	Same as type B	Inert or vacuum	High temperature	Little less brittle than above
—	Copper	Gold-cobalt	0.0005–0.025	–450	0	Reasonable above 60°K	—	Good output at very low temperature	Expensive lab-type T/C
—	Ir40-Rh60	Iridium	0.001–0.004	0	3800	Same as type B	Inert	—	Brittle, expensive

(From the *Instrument Engineer's Handbook: Process Measurement*)

Table 13-6. Thermocouple comparisons.

EXAMPLE 13-7

A Chromel-Constantan (type-E) thermocouple has a reading of 16.42 mV at its output. The reference junction is at 46°F (see Figure 13-17). Find the temperature T at the hot junction (i.e., the temperature being measured). Utilize compensation to obtain the correct temperature reading.

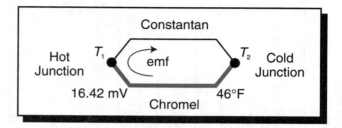

Figure 13-17. Chromel-Constantan thermocouple.

SOLUTION

Since the reference temperature is not at 0°C, we must perform a cold junction compensation. To perform this compensation, we must first find the millivolt reading for 46°F from the type-E thermocouple table (referenced at 0°C, 32°F). Table 13-7 shows relevant values from this table.

Temperature	Millivolt Output
40°F	0.26 mV
46°F	?
50°F	0.59 mV

Table 13-7. Excerpt from type-E thermocouple table.

To obtain the millivolt output at 46°F, we must perform a linear interpolation of the available table values. Referring to Figure 13-18, we interpolate this value by finding the proportional ratio of the listed values as follows:

$$\frac{X - 0.26}{0.59 - 0.26} = \frac{46 - 40}{50 - 40}$$

$$\frac{X - 0.26}{0.33} = \frac{6}{10}$$

$$X = (0.33)(0.6) + 0.26$$

$$X = 0.458 \, \text{mV}$$

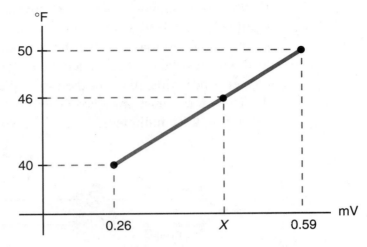

Figure 13-18. Interpolation for compensation example.

The value 0.458 mV is an offset value that must be added to the thermocouple output value to compensate for the cold junction reading. Therefore, the compensated output reading is:

$$Output = 16.42 \text{ mV} + 0.458 \text{ mV}$$
$$= 16.878 \text{ mV}$$

To obtain the temperature for a 16.878 mV output, we must return to the type-E thermocouple table and find the values closest to 16.878 mV. Table 13-8 shows these values. Again, since there is no exact value for the millivolt reading, we must perform a linear interpolation of the two known table values. The temperature T for the cold junction compensated output value is:

$$\frac{470-480}{16.68-17.10} = \frac{470-T}{16.68-16.878}$$
$$\frac{-10}{-0.42} = \frac{470-T}{-0.198}$$
$$T = (0.198)(23.809) + 470$$
$$T = 474.71°F$$

Temperature	Millivolt Output
470°F	16.68 mV
?	16.878 mV
480°F	17.10 mV

Table 13-8. Excerpt from type-E thermocouple table.

Figure 13-19 shows how thermocouples are connected using lead wires. Optimally, the lead wires should be made of the same material as the thermocouple to maintain the thermocouple's characteristics and to avoid lead wire resistance. In practice, however, lead wires are usually made of copper, since wires made of thermocouple materials are expensive and the distances they must span are generally long. Copper wires provide low resistance, thus minimizing lead wire resistance.

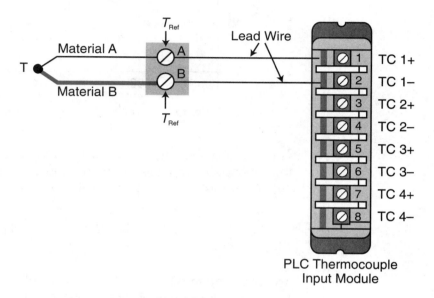

Figure 13-19. Thermocouple connection diagram.

The temperatures at the reference points of a thermocouple (A and B in Figure 13-19) must be maintained at the same value. Therefore, special shielded cable must be used so that the temperature of the cable materials remains the same all the way to the PLC input. PLCs that have thermocouple input interfaces provide cold junction compensation at the module, thus computing any necessary temperature adjustments. A thermistor usually reads the reference temperature in the module, since the span of the reference temperature is rather narrow and thermistors provide accurate readings for narrow temperature ranges. The module's memory stores all of the thermocouple tables.

To increase thermocouple resolution, the user can connect several thermocouples in series, thus forming a **thermopile**. Thermopiles generate larger output voltages than single thermocouples, which reduces the sensitivity requirements for the measuring device. When thermocouples are combined to form a thermopile, they should be clustered together as closely as possible in order to measure the temperature at a particular point. Figure 13-20 illustrates a thermopiling arrangement. The voltage of the thermopile is the average of the voltages at the three hot junctions. Note that the thermocouples must be isolated from each other to avoid thermal reaction among them, as well as to avoid a thermocouple emf short. The reference cold junctions must also be maintained at the same temperature (T_{Ref}).

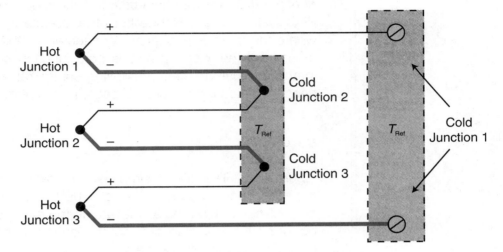

Figure 13-20. Thermopiling arrangement.

Other thermocouple group configurations allow the measurement of temperature differences, as well as direct average readings from several thermocouples (parallel thermocouple configuration). Figure 13-21 illustrates these two configurations.

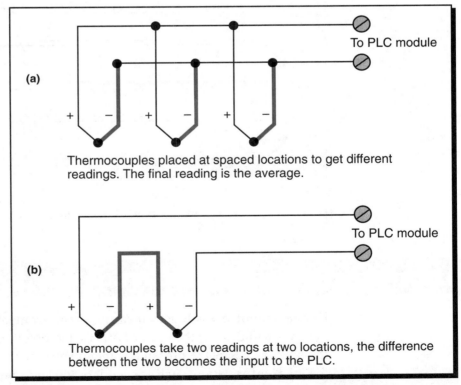

Figure 13-21. Thermocouple configurations that measure **(a)** average temperature and **(b)** temperature differential.

EXAMPLE 13-8

Three type-B thermocouples, which are connected in a thermopile configuration, will be connected to a thermocouple input module in a PLC system. The distance from the module to the 2000°F furnace where the thermocouples are connected is 300 feet. The module's measuring thermistor provides cold junction compensation at the interface. Illustrate the thermopiling configuration and interface connection, indicating the type of cable to be used.

SOLUTION

Figure 13-22 illustrates a simplified diagram of the wiring configuration. The cable used is type-B shielded cable made of the thermocouple material. It runs all the way to the module, since the module performs the cold junction compensation. Note that all reference temperatures should be the same.

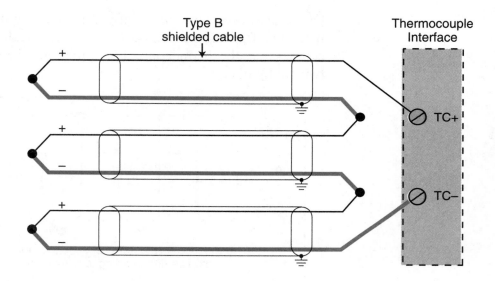

Figure 13-22. Thermopile wiring configuration.

13-5 DISPLACEMENT TRANSDUCERS

Displacement transducers measure the movement of objects. In this section, we will discuss the use of LVDTs and potentiometers as displacement transducers. Chapter 8 presented other displacement transducers that provide feedback information, such as encoders and leadscrews. It dealt with these as position- and motion-related feedback devices.

LVDTs

As mentioned previously, LVDTs (linear variable differential transformers) operate on the principle of a movable core inside a wound coil, which generates voltage changes depending on the position of the core. Therefore, the attachment of a rod or a similar element to the core provides a way to measure linear displacement of the rod along a path.

An LVDT can measure displacement of any type, whether it is induced pressure, force, or linear displacement. LVDTs are capable of measuring displacements ranging from ±0.05 inches (±1.27 mm) to ±10 inches (±254 mm). The voltage output from an LVDT is generally ±10 VDC for any displacement range.

Null repeatability is extremely stable in an LVDT due to the symmetry of the device. This makes an LVDT an excellent null (0-inch) position indicator for closed-loop control systems. LVDTs also provide excellent resolution, since even the most minute movement of the core will produce an output voltage. This exceptional movement resolution is primarily due to the very low friction inherent in an LVDT's design. Chapter 11 provided a programming and implementation example that included the use of an LVDT.

POTENTIOMETERS

A **potentiometer** is perhaps the most simple displacement transducer available. Its measurement principle is based on resistance changes due to the movement of an arm called a *wiper* (see Figure 13-23). A voltage source powers a potentiometer, causing the wiper to provide an output voltage proportional to the movement of the attached element. This voltage output corresponds to the voltage drop between the top section of the potentiometer and the resistance accumulated by the wiper position.

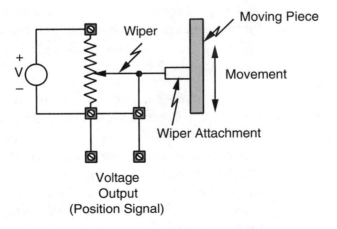

Figure 13-23. Potentiometer diagram.

Potentiometer transducers are prone to problems such as excess friction in the wiper arm, limited resolution in the wire-wound unit, and mechanical breakdown due to wear. They are also quite sensitive to vibration. On the other hand, potentiometers have a wide range of applications and are relatively inexpensive.

13-6 PRESSURE TRANSDUCERS

Pressure transducers transform the force per unit area exerted on their surroundings into a proportional electrical signal through signal conditioning. This measurement can be used simply as a pressure value or as a means of obtaining other transducer measurements, such as flow, strain, and vibration.

Three of the most common types of pressure transducers are strain gauges, Bourdon tubes, and load cells. Bridge circuits usually provide signal conditioning for these pressure transducers. Another type of pressure transducer is the piezoelectric crystal. However, we will discuss this type of transducer when we discuss vibration transducers, since it is primarily used for the detection of vibration.

STRAIN GAUGES

A **strain gauge** is a mechanical transducer that measures the body deformation, or strain, of a rigid body as a result of the force applied to the body. Strain gauges are often used in applications, such as flow measurement, that require pressure differential measurements. However, strain gauges are also used in simple direct strain measurements, where stress is directly applied to a rigid body.

A strain gauge measures pressure by sensing resistance changes in its wires due to an applied force. These wires are made of either metal, such as copper, iron, or platinum, or a semiconductor material, such as silicon or germanium. Strain gauges made of semiconductor material are more sensitive, since they provide a greater resistance change in response to the deformation caused by the applied force.

The two main categories of strain gauges are *bonded* and *unbonded*, as illustrated in Figure 13-24. A bonded strain gauge attaches directly to the area where stress is being applied to the rigid body. A thin layer of synthetic thermosetting resin (epoxy) connects the bonded strain gauge to the body. Unbonded strain gauges operate under the same principle; however, a moving part of the gauge moves with the force applied. This movement changes the resistance of the wires, creating a voltage differential (due to force) in the bridge circuit.

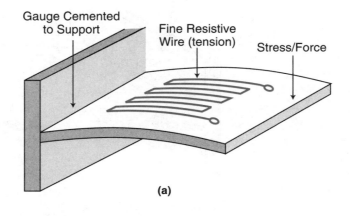

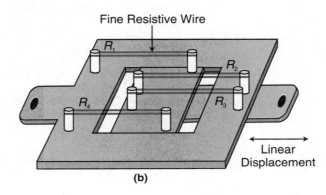

Figure 13-24. (a) Bonded and **(b)** unbonded strain gauges.

Changes in temperature affect strain gauge circuits, so these circuits require temperature compensation. A *dummy gauge*, added to the bridge-conditioning circuit, can provide this compensation (see Figure 13-25). As the temperature increases, the resistance in the measuring gauge changes as a result of both temperature and pressure. The resistance in the dummy gauge also changes due to temperature, since it is made of the same material as the measuring gauge. However, it does not change due to force because stress is measured in only one direction. Thus, the dummy gauge provides a value for resistance change based solely on temperature change. The bridge circuit uses the information from the dummy circuit to compensate for temperature resistance change in the measuring gauge value, resulting in a resistance based on force only. The resistance-equivalent signal provided by a strain gauge is usually very small, so the signal requires amplification through a transmitter to provide the proper voltage level to a PLC analog input.

BOURDON TUBES

A **Bourdon tube** is a pressure transducer that converts pressure measurements into displacement. This displacement is proportional to the pressure being measured. The different types of Bourdon tubes include spiral, helical, twisted, and C-tube.

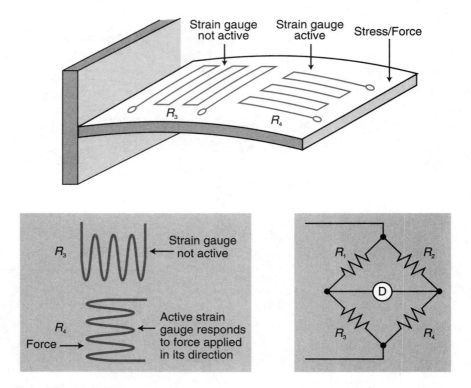

Figure 13-25. Strain gauge using a dummy gauge to compensate for changes in temperature.

Figure 13-26 illustrates the most commonly used type of Bourdon tube, the C-tube. The pressure inlet end of the tube is fixed, while the other end of the tube is free to move. Since the intermediate conversion is displacement, a linear variable differential transformer converts the pressure in the tube into a proportional electrical signal. Hence, the Bourdon tube and the LVDT perform a pressure measurement–to–linear displacement–to–electrical signal conversion.

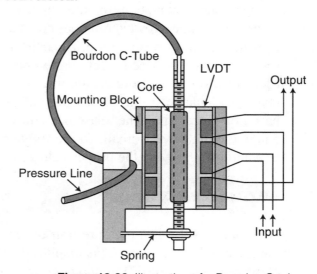

Figure 13-26. Illustration of a Bourdon C-tube.

LOAD CELLS

Load cells are force or weight transducers based on a direct application of bonded strain gauges. These devices measure deformations produced by weight. Load cells are used in many applications, especially ones that require weight measurements.

13-7 FLOW TRANSDUCERS

Flow transducers measure the flow of materials in a process. This flow of materials can be in solid, gas, or liquid form. All flow control applications utilize the term Q, or *rate of flow*, to define flow measurement in the system. In this section, we will discuss the rate of flow for each kind of material—solid, liquid, and gas—and the transducers used to measure them.

SOLID FLOW TRANSDUCERS

Solid flow is typically measured with a strain gauge–based load cell transducer, which measures the weight of the product. Solid flow measurement frequently integrates the use of a conveyor or belt product transporter with a load cell. The units generally used for this type of flow measurement are lbs/min or kg/min. Figure 13-27 illustrates an example of how a load cell measures the flow of solids. The equation that describes the flow (Q) is:

$$Q = \frac{WV}{L}$$

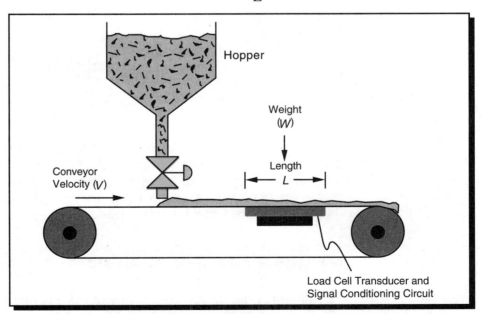

Figure 13-27. Load cell measuring the flow of solids.

where:

$Q =$ the rate of flow

$W =$ the mass/weight of the solid

$V =$ the velocity/speed of the moving transporter

$L =$ the length of the weight transducer (load cell)

If English units are used—pounds, feet, and feet/min—then Q will be expressed in lbs/min. Conversely, if metric units are used—kg, meters, and meters/min—then Q will be expressed in kg/min.

EXAMPLE 13-9

A conveyor transports material that is weighed on a platform 2 meters in length. A load cell, which is connected to an analog input module through a bridge circuit and an amplifier, must weigh 50 kgs of the material. The required flow is 1200 kgs/min. Find the speed at which the conveyor must run to obtain the required flow. Also, suggest how to control the conveyor so that the flow rate remains constant.

SOLUTION

The velocity of the conveyor is expressed as:

$$Q = \frac{WV}{L}$$

$$V = \frac{QL}{W}$$

$$= \frac{(1200)(2)}{50}$$

$$= 48 \text{ m/min}$$

To keep the flow rate constant, a PLC could control the speed of the conveyor by either computing the flow rate and making changes to a motor or by changing the drive reference speed of the motor. It could also control the analog valve, varying the hopper output according to the required flow rate and speed.

FLUID FLOW TRANSDUCERS

To measure fluid flow, you must measure one of two conditions in the process line: pressure differential or fluid motion. The two most common devices for measuring the pressure differential in a process line are Venturi tubes and orifice plates. One of the most common fluid flow transducers for detecting fluid motion is the turbine flow meter. This transducer transforms flow directly into electrical signals.

Pressure-Based Fluid Flow Meters. Both the **Venturi tube** and the **orifice plate** are based on the Bernoulli effect, which relates flow velocity to the pressure differential between two points. These fluid flow meters use pressure transducers, which transform pressure into an electrical signal to determine the pressure differential. The strain gauge and the Bourdon C-tube (see Section 13-5) are the two types of transducers most commonly used in pressure-based flow meters. These transducers use the bridge circuit and LVDT techniques, respectively, to convert measured pressure values into electrical signals. If low pressures are to be measured, a Venturi tube or orifice plate may incorporate a low-pressure transducer, such as a bellows, diaphragm, or capsule to enhance the pressure reading resolution. Figure 13-28 shows these low-pressure transducers.

Courtesy of Schaevitz Engineering, Pennsauken, NJ

Figure 13-28. Low-pressure transducers.

Figure 13-29 illustrates a diagram of a Venturi tube, while Figure 13-30 shows an orifice plate flow transducer. The pressure differential ΔP in these devices is equal to the difference in pressures P_1 and P_2. The value ΔP also relates to the velocity of the fluid through the Bernoulli effect. The velocity at point P_2 as a function of ΔP is:

$$V = k\sqrt{\Delta P}$$

where:

$V =$ the fluid velocity

$\Delta P = P_1 - P_2$

$k =$ a constant

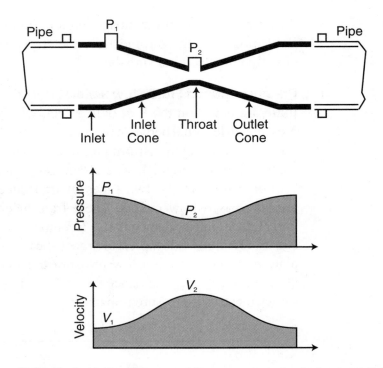

Figure 13-29. Venturi tube diagram and the pressure and velocity at points P_1 and P_2.

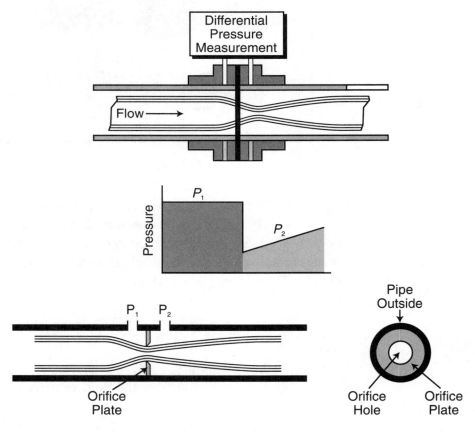

Figure 13-30. Orifice plate flow transducer.

The constant k takes into account the density of the fluid, the ratio of pipe to obstruction at cross-sectional points P_1 and P_2, temperature, and other factors. The equation to obtain the flow rate measurement is:

$$Q = VA$$
$$= Ak\sqrt{\Delta P}$$
$$= K\sqrt{\Delta P}$$

where:

$V =$ the fluid velocity

$A =$ the cross-sectional area of the pipe

$K =$ a new constant composed of k times the area A_2

The flow rate value Q gives us the volume per unit time of the flow (ft/min $\times$ ft² = ft³/min). Note that the velocity times the area at point 1 ($V_1 A_1$) is equal to the velocity times the area at point 2 ($V_2 A_2$).

EXAMPLE 13-10

Illustrate the PLC connections and functions necessary to implement the ratio control computation shown in Figure 13-31.

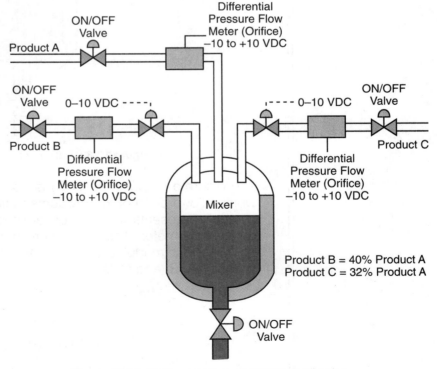

Figure 13-31. Ratio control computation application.

SOLUTION

To implement the ratio control of products B and C at the specified percentage of product A (wild flow), we must read the differential pressures (*DP*) from the orifice flow meter and control the output of the analog servo valves. Figure 13-32 illustrates this ratio control implementation using flow ratio as the process variable.

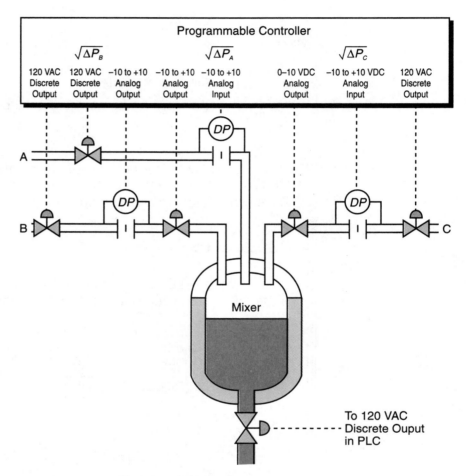

Figure 13-32. Ratio control implementation.

The discrete output (120 VAC) is connected to the ON/OFF valve, which allows each of the products to flow. Each DP instrumentation symbol represents the differential pressure measurement from the orifice flow meter. These pressure measurements are input to the analog input modules (−10 to +10 VDC). The flow rate for each product, A, B, and C, is:

$$Q_A = K_A\sqrt{\Delta P_A}$$
$$Q_B = K_B\sqrt{\Delta P_B}$$
$$Q_C = K_C\sqrt{\Delta P_C}$$

where K_A, K_B, and K_C are the given constants. The square root value of the analog input ΔP should be taken after the input count value corresponding to it has been converted to engineering units (through linearization, etc.). As product A flows, the PLC computes the flows of products B and C to maintain the proper ratios between A, B, and C (B = 0.40A and C = 0.32A). The PLC must control the output control valves for products B and C to maintain the proper ratios.

Motion Detection Fluid Flow Meters. The **turbine flow meter** is one of the most common types of motion detection flow meters. This device is used in applications that measure liquid and gas flows, as well as in applications with very low flow rates. Turbine meters are widely used in petrochemical and pipeline transfers of petroleum flows. Special types of turbine flow meters are also used in liquid oxygen and nitrogen gas-metering applications.

A turbine meter consists of a multibladed rotor, which is suspended in a liquid flow. The fluid flow passing through the blades creates a rotary motion in the turbine. This rotary motion creates a magnetic flux that is sensed by a coil inside the turbine flow meter. The coil changes the flux into a small voltage (as low as 10 to 20 mV) and then amplifies it. This design allows the turbine meter to convert the movement of its blades into output pulses that are proportional to the volume passing through the turbine. The output pulses generally provide information in gallons per minute (gpm). Some turbine meters also provide an analog output proportional to the flow rate being measured. Figure 13-33 illustrates a simple diagram of a turbine flow meter.

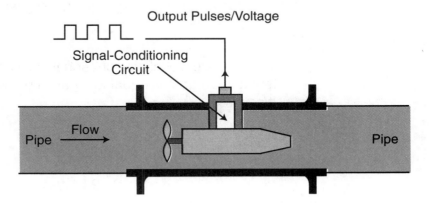

Figure 13-33. Turbine flow meter.

EXAMPLE 13-11

A programmable controller system receives an analog signal from a turbine flow meter. The flow rate is given as 60 gpm and the area of the pipe is 2 square inches. Find the velocity of the flow to be displayed in feet per second on a four-digit LED display.

SOLUTION

The flow rate of the fluid is described by :

$$Q = VA$$

where:

$Q =$ the flow rate
$V =$ the velocity of the flow
$A =$ the cross - sectional area of the pipe

The velocity of the flow is:

$$V = \frac{Q}{A}$$

Note that the units given must be converted to obtain the velocity in ft/sec. To convert gallons to cubic feet, we must first convert gallons to cubic meters and then to cubic feet:

$$1 \text{ gal} = 3.785 \times 10^{-3} \text{ m}^3$$
$$1 \text{ m}^3 = 35.31 \text{ ft}^3$$

Therefore:

$$1 \text{ gal} = \left(3.785 \times 10^{-3}\right)\left(35.31 \text{ ft}^3\right)$$
$$= 0.1336 \text{ ft}^3$$

The cross-sectional area of 2 square inches is equal to 0.0139 square feet and 60 gpm is equal to 1 gallon per second. So, the velocity in ft/sec is:

$$V = \frac{(1)(0.1336)}{0.0139}$$
$$= 9.61 \text{ ft/sec}$$

Hence, to obtain the velocity of a fluid in a pipe in feet/second when the flow rate is given in gpm and the area is given in square inches, the following equation can be used:

$$V_{(ft/sec)} = \frac{\left(Q_{gpm}\right)\left(0.3208\right)}{A_{(sq\ in)}}$$

13-8 VIBRATION TRANSDUCERS

Vibration transducers are used in system applications that require the detection of vibration, which can severely damage process control equipment. For example, a vibration detector can monitor the amount of vibration in a large motor, thereby preventing potential failure of the bearings. Vibration transducers also sense excessive machine vibration, helping to prevent catastrophic equipment damage, extensive repairs, and downtime. Some of the most common causes of system vibration failures, along with their frequencies, are:

- imbalance of a rotating member (approximately 40%)

- misalignment (15%)

- defective bearings (15%)

- defective belts (15%)

- other miscellaneous causes (15%)

Before we explain vibration transducers, let's first explore some of the basics of vibrational motion.

VIBRATION BASICS

Vibration is defined as the oscillatory movement of a mass about a reference position characterized by displacement, velocity, and acceleration. Displacement (s) is the distance that the mass moves from its reference position in meters (see Figure 13-34), velocity (v) is the speed at which the mass moves in meters per second (m/sec), and acceleration (a) is the rate of change of the mass's velocity per second (m/sec²). Table 13-9 displays the equations for these vibration motion parameters. Vibration also involves other parameters, including frequency, amplitude, and wave form. Vibration can be mathematically defined in terms of periodic motion of a mass from a reference position by:

$$s_t = s_{max} \sin \omega t$$

where:

s_t = the position and distance of movement in meters (displacement)

s_{max} = the maximum displacement in meters (peak displacement)

ω = the angular frequency in radians per second

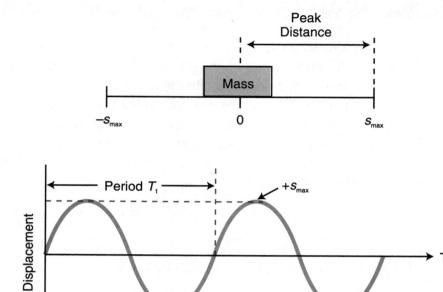

Figure 13-34. Displacement.

Parameter	Motion Equation	Description
Displacement	$s = f(t)$	Displacement as a function of time
Velocity	$v = \dfrac{ds}{dt}$	First derivative of displacement
Acceleration	$a = \dfrac{dv}{dt} = \dfrac{d^2s}{dt^2}$	Second derivative of displacement

Table 13-9. Motion parameters associated with vibration.

The angular frequency can also be expressed as angular velocity where $\omega = 2\pi f$. In vibration, velocity is the first derivative of displacement, while acceleration is the first derivative of velocity (or the second derivative of displacement):

$$v(t) = \frac{ds_t}{dt} = -s\omega \cos \omega t$$

$$a(t) = \frac{dv(t)}{dt} = \frac{d^2s_t}{dt} = -s_0\omega^2 \sin \omega t$$

All three vibration terms—displacement, velocity, and acceleration—have the same periodic frequency. Another important term in vibration monitoring is the *peak acceleration*, which is frequency squared (ω^2) times the peak displacement (s_0):

$$a_{peak} = s_0 \omega^2$$

This peak acceleration equation indicates that acceleration can become large even with very small displacement, since the displacement term is multiplied by the square of the frequency. Thus, acceleration can easily reach a level of several g values ($1g = 9.8 \ m/sec^2$), creating a potentially destructive vibration. Table 13-10 lists the characteristics of several types of vibration.

EXAMPLE 13-12

A steam pipe in a heat batching system (see Figure 13-35) vibrates at a frequency of 8 cycles per second (8 Hz) with a peak displacement of 10 mm (1 cm or 0.01 m). **(a)** Find and plot the displacement equation indicating the period, and **(b)** calculate the peak acceleration in m/sec^2 and its equivalent in g units.

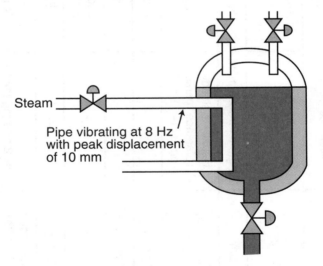

Figure 13-35. Heat batching system.

SOLUTION

(a) Figure 13-36 presents the graph of displacement versus time of vibration, which is given mathematically by the equation:

$$
\begin{aligned}
s_t &= s_{max} \sin \omega t \\
&= s_{max} \sin 2\pi f t \\
&= (0.01) \sin 2\pi (8) t \\
&= (0.01) \sin 50.265 t
\end{aligned}
$$

Cause	Frequency Relative to Machine rpm	Phase-Strobe Picture	Amplitude	Notes
Imbalance	1 × rpm	Single steady reference mark	Radial, steady, proportional to imbalance	Common cause of vibration
Defective antifriction bearing	10–100 × rpm	Unstable	Measured velocity 0.2 to 1.0 in/sec; radial	Velocity largest at defective bearing; as failure approaches, velocity signal will increase, frequency will decrease
Sleeve bearing	1 × rpm	Single reference mark	Not large	Shaft and bearing amplitude about the same
Misaligned coupling or bearing	2 × rpm (sometimes 1 or 3 × rpm)	Usually 2 steady reference marks (sometimes 1 or 3)	High axial	Axial vibration can be twice radial; use dial indicator to check
Bent shaft	1 or 2 × rpm	1 or 2	High axial	—
Defective gears	High rpm × gear teeth	—	Radial	Use velocity measurement
Mechanical looseness	1 or 2 × rpm	1 or 2	Proportional to looseness	Radial vibration largest in the direction of looseness
Defective belt	1 or 2 × belt rpm	—	Erratic	Strobe light will freeze belt
Electrical	Power line frequency × 1 or 2 (3600 or 7200 rpm)	1 or 2 rotating marks	Usually low	Vibration stops instantly when power is turned off
Oil whip	Less than rpm	Unstable	Radial; unsteady	Frequency may be as low as one-half rpm
Aerodynamic	1 × rpm or number of fan blades × rpm	—	—	May cause trouble in case of resonance
Best frequency	1 × rpm	Rotates at best rate	Variable at best rate	Caused by two machines running at close rpm
Resonance	Specific criticals	Single reference mark	High	Phase will change with speed; amplitude will decrease above and below resonant speed; resonance can be removed from operative range by stiffening

Table 13-10. Vibration identification guide.

(Courtesy of PMC/BETA Corp., Natick, MA)

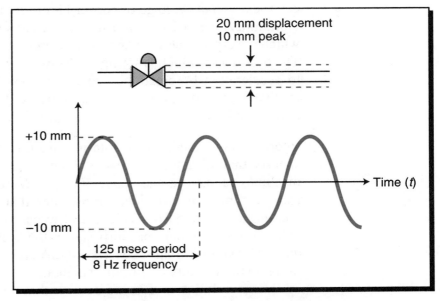

Figure 13-36. Displacement versus time of vibration.

This system has a period (*T*) equal to:

$$f = \frac{1}{T}$$

$$T = \frac{1}{f} = \frac{1}{8\,\text{Hz}} = 0.125\ \text{sec}$$

(b) The peak acceleration is:

$$a_{peak} = \omega^2 s_{max}$$
$$= (2\pi f)^2 s_{max}$$
$$= (2\pi 8)^2 (0.01)$$
$$= (50.265)^2 (0.01)$$
$$= 25.66\ \text{m/sec}^2$$

This value in g units is:

$$a_{peak} = \left(25.266\ \text{m/m}^2\right)\left(\frac{1\,\text{g}}{9.8\ \text{m/m}^2}\right)$$
$$= 2.578\ \text{g}$$

VIBRATION DETECTION

Vibration can be detected by measuring displacement, velocity, or acceleration; therefore, vibration transducers can measure any of these factors. One of the most commonly used vibration transducers, the *piezoelectric transducer*, is based on the piezoelectric accelerometer, which produces an electrical output (voltage or current) proportional to the acceleration of the vibration. A piezoelectric transducer does this using a piezoelectric crystal, which is a crystalline substance that exhibits electric polarity under pressure. The transducer, which is spring loaded with a crystal of known mass, reacts to acceleration by creating a voltage across the crystal, generally in the millivolt range. It measures acceleration by detecting the force applied to the known mass, since force is equal to mass times acceleration ($F = ma$).

International standards for rotating machinery (the ISO 2378 and 3945) specify that vibration severity is directly related to vibratory velocity for machines running at and above 500 rpm. Vibration velocity can be found using a vibration transducer/transmitter that integrates the acceleration measurement taken from a piezoelectric-based accelerometer, resulting in a velocity measurement proportional to the vibration. The vibration transducer/transmitter then sends a 4–20 mA signal to the PLC that is proportional to the velocity of vibration in inches or meters per second. Figure 13-37 shows vibration measuring devices that provide a 4–20 mA output.

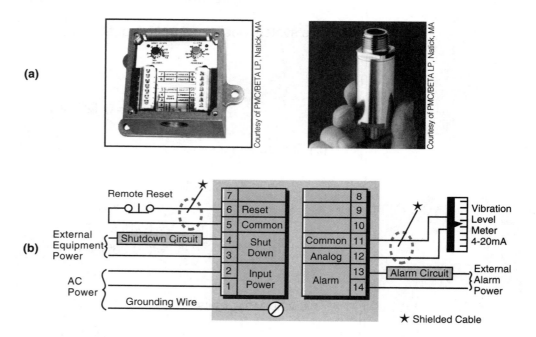

Figure 13-37. (a) Vibration measuring devices and **(b)** the connection diagram for the first device in part (a).

There are several guidelines for determining the level at which vibration becomes critical. Figure 13-38 illustrates a vibration warning level guide provided by PMC/BETA LP (Natick, MA), a vibration transducer manufacturer. A PLC can monitor the level of vibration in a machine or equipment and provide the operator with a warning indication, according to the guide, before damage occurs. Figure 13-39 illustrates a severity chart for machines with vibration warning levels of 0.2–0.4 inches/sec over a frequency range of 100 to 100,000 rpm. This chart shows the variable peak-to-peak displacements for

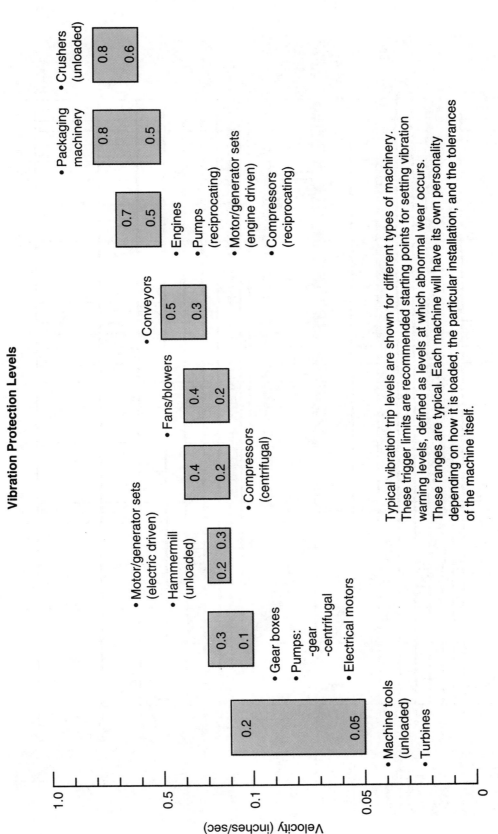

Typical vibration trip levels are shown for different types of machinery. These trigger limits are recommended starting points for setting vibration warning levels, defined as levels at which abnormal wear occurs. These ranges are typical. Each machine will have its own personality depending on how it is loaded, the particular installation, and the tolerances of the machine itself.

(Courtesy of PMC/BETA LP, Natick, MA)

Figure 13-38. Vibration warning level guide.

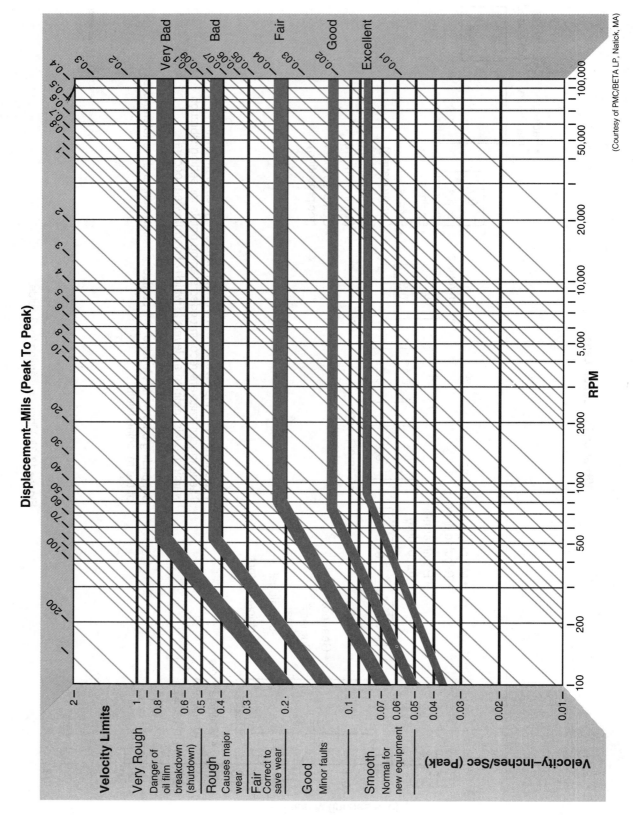

Figure 13-39. Vibration severity chart.

(Courtesy of PMC/BETA LP, Natick, MA)

smooth to very rough severity and indicates possible consequences (bold blue lines indicating very bad, bad, fair, etc.). For machines with higher or lower warning levels, the limits shown on the vibration severity chart should be increased or decreased, respectively.

If the vibration velocity surpasses the maximum allowable limit in a vibration monitoring system, the PLC can annunciate an alarm condition and initiate a shutdown of the system before a catastrophic failure occurs. Figure 13-40 illustrates a typical direct interface application of a vibration transducer (4–20 mA) to a PLC system where the PLC is responsible for a shutdown command to the machine, if necessary. A vibration transducer/

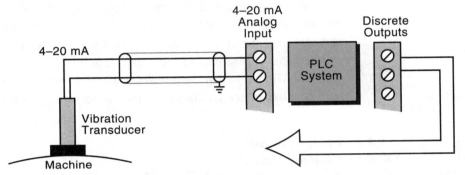

(a) Digital signal controls a shutdown if the vibration limit exceeds the maximum level.

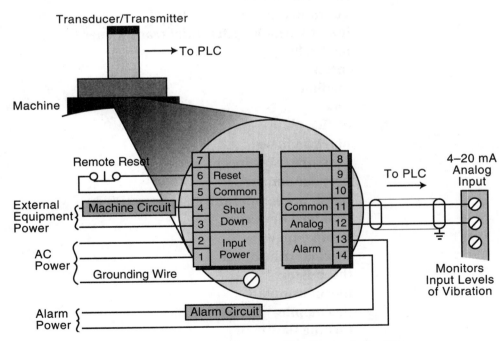

(b) Digital signal annunciates working levels of vibration.

Figure 13-40. Vibration transducers used in a PLC system to (a) control a shutdown command and (b) monitor vibration levels.

transmitter combination can also be interfaced with a programmable controller to monitor vibration. The internal contacts of the transmitter can then be used to shut down the machine or system if the vibration level surpasses the specified alarm limits.

13-9 SUMMARY

In this chapter, we introduced basic measurement concepts that explain how data and errors are interpreted and analyzed. This information, which is based on statistical analysis, is very helpful when implementing an intelligent or knowledge-based PLC system.

We also explained different techniques used to transform the physical measurement of a transducer sensor into a voltage or current signal. Transducers, in general, are composed of several intermediate measurement and connection elements, which vary depending on the type of transducer—thermal, displacement, pressure, or flow. Process control systems, which we will discuss next, use these devices to monitor system variables.

KEY TERMS

bridge circuit
Bourdon tube
displacement transducer
flow transducer
guarantee error
linear variable differential transformer (LVDT)
load cells
mean
median
mode
orifice plate
potentiometer
pressure transducer
propagation error
resistance temperature detector (RTD)
standard deviation
strain gauge
thermal transducer
thermistor
thermocouple
thermopile
turbine flow meter
Venturi tube
vibration transducer

CHAPTER FOURTEEN

PROCESS RESPONSES AND TRANSFER FUNCTIONS

Mathematics may be defined as the subject in which we never know what we are talking about, nor whether what we are saying is true.

—Bertrand Russell

 **CHAPTER HIGHLIGHTS**

As we have already discussed, PLCs control machines and processes via discrete, analog, and special I/O interfaces that communicate with the real world. With the aid of control software, a user can program a PLC to control any process through these I/O interfaces. In our discussion, however, we have not yet explained process control in its true form, as it applies to the behavior and control of a manufacturing activity. Therefore, we will dedicate this chapter to the explanation of basic process control concepts. In the next chapter, we will explain how these concepts apply to a process control operation.

14-1 PROCESS CONTROL BASICS

Process control is the regulation of designated process parameters to within a specified target range or to a set target value called the **set point**. Process control is most often used in product manufacturing, because many factors, such as color, composition, and density, must be accurate for a product to be well made. Therefore, to implement a quality product, process control is used to monitor and correct process parameters by analyzing the state of dynamic variables. Dynamic variables are process characteristics, such as temperature, flow, and pressure, that vary with time. Through its I/O interfaces, a PLC can regulate these dynamic variables to a desired set point, thus implementing process control.

Figure 14-1 illustrates the basic concept of process control using a reactor tank in which steam controls the temperature in the tank. In this case, the temperature must be maintained at a target value, or set point, of 125°C. Because it varies with time, the temperature is the dynamic variable, which is also called the **process variable** (*PV*). The steam level, which regulates the process variable (i.e., raises and lowers the temperature), is called the **control variable** (*CV*). The valve that controls the amount of steam entering the reactor tank's jacket is called the **control element**, or *final output field device*, because the more the controller opens the valve, the more the steam increases the temperature.

Figure 14-2 shows the block diagram of the process control system illustrated in Figure 14-1. The PLC reads the process variable from the system (i.e., obtains feedback) and compares it with the set point to determine how well the temperature is being regulated. This configuration is known as a closed-loop system, because the controller uses feedback to monitor the system. An open-loop system does not use feedback, so the controller does not receive process variable data. Figure 14-3 illustrates the configuration of an open-loop system.

If the temperature reading in the process in Figure 14-2 is low, the controller will adjust the control variable by opening the valve to allow steam to enter the tank, thereby raising the temperature. The controller will then recheck

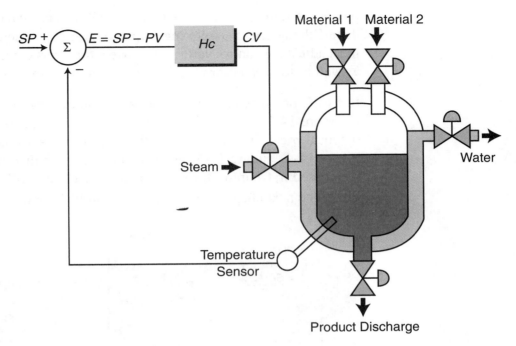

Figure 14-1. Reactor tank control system.

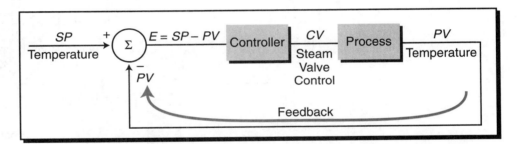

Figure 14-2. Block diagram of the closed-loop reactor tank system.

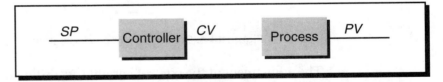

Figure 14-3. Open-loop process control system.

the process variable. If the temperature is still low, it will again open the steam valve to increase the temperature. The controller will repeat this process until the actual temperature (process variable) is as close as possible to the target temperature (set point value). The difference between the process variable and the set point is called the **error**. The error can be either positive or negative, depending on whether the process variable is too high or too low. However, regardless of the sign of the error, the controller still performs the same basic function—adjusting the process variable until it equals the set point (i.e., making the error equal to zero). Once equality is achieved, the process is said to be *regulated*. As we will discuss later, many factors can disturb the system, thus altering the process variable. Therefore, the controller must adjust the control variable to correct for errors created by these factors, as well as to correct for errors due to a change in the set point.

In a PLC-based system, a control system block diagram like the one shown in Figure 14-2 can be expanded to include interfaces that control the field output devices, as well as those that read process variable input data (see Figure 14-4). Figure 14-5 shows a control system that uses a PID interface (discussed in Chapter 8) to implement process control independent of the PLC. The next chapter will further explain PID control.

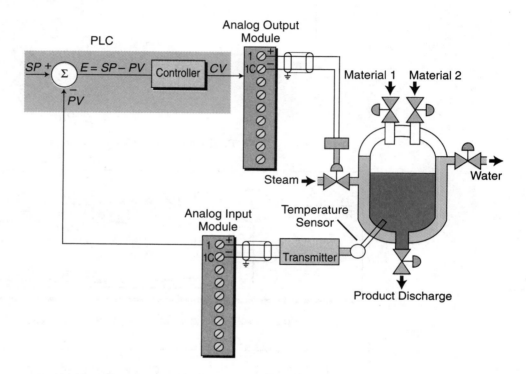

Figure 14-4. Control system block diagram including I/O interfaces.

The adjustment of the control variable according to data obtained by reading the process variable and analyzing the error between it and the set point is referred to as the **control loop**. Most control loops are affected by *disturbances*, which influence the process and alter the process variable (see Figure 14-6). To understand disturbances, let's examine a simple control loop example—a car's cruise control mechanism. As shown in Figure 14-7, once the cruise control has been set at a target speed (set point), the system will maintain that speed by keeping the accelerator (control variable) at a constant level. However, if the system experiences a disturbance, such as pavement with higher friction or an uphill climb, the system will increase the control variable (i.e., increase acceleration) to maintain the set point speed. This is demonstrated by the fact that the accelerator pedal of a car

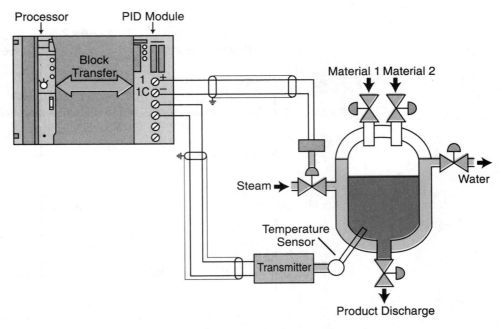

Figure 14-5. Control system using a PID interface.

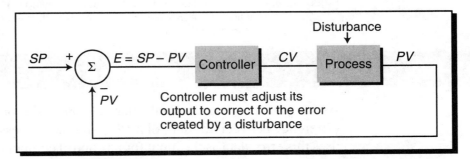

Figure 14-6. A control loop with a disturbance.

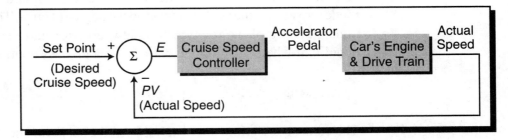

Figure 14-7. Cruise control process loop.

under cruise control is depressed further when the car is going uphill than when it is going downhill. Figure 14-8 illustrates how a cruise control system compensates for disturbances. The components that form this simple system respond to keep the process variable at the set point by adjusting the control variable to maintain the error at zero during the disturbance. This, in fact, is the main function of process control—monitoring the error signal generated by the system and adjusting the outputs accordingly.

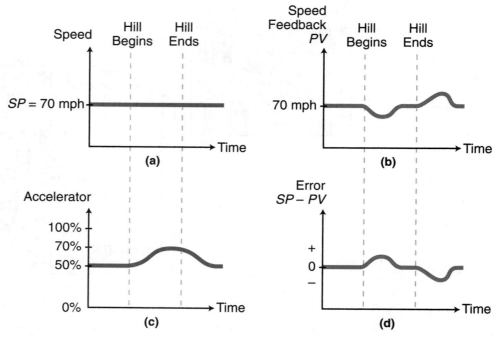

Figure 14-8. Cruise control compensation graphs showing **(a)** the set point speed, **(b)** the reaction of the process variable to the disturbance, **(c)** the reaction of the control variable to the disturbance, and **(d)** the error.

14-2 CONTROL SYSTEM PARAMETERS

As we just discussed, a controller calculates the process error (E) as the difference between the set point (SP) and the process variable (PV). It then uses this error data as the input for its control computations. It uses this input data to apply control to the process by manipulating the control variable (CV) so as to eliminate the error. The way it does this depends on the controller's mode (covered in the next chapter) and the degree of error. Therefore, a thorough understanding of the relationship between error and the control variable is beneficial when designing and applying a process control application.

ERROR

The control deviation, or error, between the set point and the process variable is given by the equation:

$$E = SP - PV$$

where:

$E =$ the error
$SP =$ the set point value
$PV =$ the process variable value

This equation can also be represented as:

$$E = PV - SP$$

where the set point is subtracted from the process variable. Both equations give the same magnitude (value) of error, but with different signs. The first error equation ($E = SP - PV$) is used in *negative feedback* control loops, where the process variable is fed back into the system and subtracted from the set point for error correction control. Negative feedback is used in closed-loop control systems instead of positive feedback, because negative feedback reduces the error in the system while positive feedback magnifies it. As shown in Figure 14-9a, positive feedback results in a feedback signal that is in phase with the error deviation, thus enhancing the system error. Negative feedback, on the other hand, produces a signal that is directly out of phase with the error deviation (see Figure 14-9b). This reversal of phase causes the system error to decrease as the control variable regulates the process. All feedback controllers produce a 180° phase shift in the feedback signal to provide negative feedback.

(a)

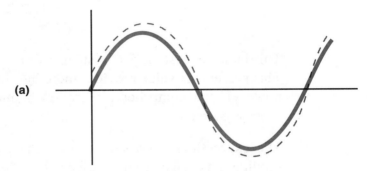

Positive Feedback

(b)

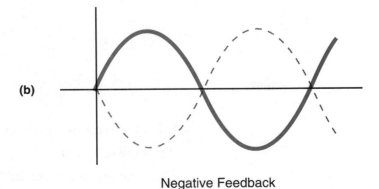

Negative Feedback

Figure 14-9. (a) Positive and **(b)** negative feedback.

INTERPRETATION OF ERROR

The representation of error as the difference between the set point and the process variable provides a "natural" value; that is, a value expressed in the units being measured. For example, a PLC that receives process variable and set point data in the form of analog counts will express the error as a function of analog counts. Likewise, a system with a set point of 125°C and a process variable of 120°C will have an error of 5°C. However, the system cannot determine if 5°C is an acceptable error because it does not know how close the error is to zero relative to the variable range. Therefore, another way for the controller to calculate error is as a percentage of the target set point. This is expressed as:

$$E = \frac{SP - PV}{SP}$$

Using this equation, the error for the previous temperature example would be:

$$E = \frac{125°C - 120°C}{125°C}$$
$$= 4\%$$

This indicates that the 5°C system error is within 4% of the set point target. This percentage value provides more information than the 5°C error value; however, the controller requires even more information to adjust the process correctly.

The expression of error as a percentage of the process variable range provides an even more indicative value of error. The range of *PV* indicates the maximum and minimum values that the process value can have. Figure 14-10 illustrates the error as a percentage of the process variable range. Mathematically, this can be expressed as:

$$E\% = \frac{SP - PV}{PV_{min} - PV_{max}}$$

where:

$E\%$ = the error as a percentage of *PV* range

SP = the set point

PV = the process variable

PV_{max} = the maximum value of *PV*

PV_{min} = the minimum value of *PV*

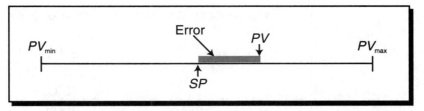

Figure 14-10. Error as a percentage of the process variable range.

Note that, in this equation, the sign of the terms PV_{min} and PV_{max} (positive and negative, respectively) are such that the error will be positive if the process variable is above the set point and negative if it is below the set point. This error representation provides additional information about the magnitude of the error.

EXAMPLE 14-1

A process with a temperature set point of 180°C has a process variable input of 168°C (see Figure 14-11). Express the error as a percentage of range given that the process variable has a range of **(a)** 100°C to 200°C and **(b)** 50°C to 350°C.

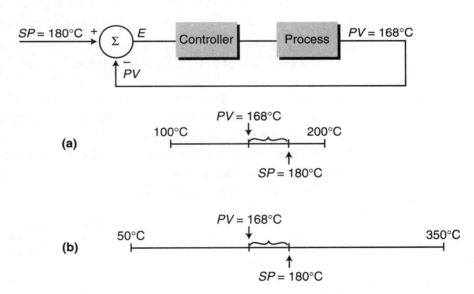

Figure 14-11. Process control loop for Example 14-1 given **(a)** a 100–200°C process variable range and **(b)** a 50–350°C range.

SOLUTION

(a) The value of the error as a percentage of range (*E%*) is expressed as:

$$E\% = \frac{SP - PV}{PV_{min} - PV_{max}}$$

For a process variable range of 100 to 200°C, the error is:

$$E\% = \frac{180°C - 168°C}{100°C - 200°C}$$
$$= \frac{12°C}{-100°C}$$
$$= -0.12 = -12\%$$

(b) For a process variable range of 50 to 350°C, the error is:

$$E\% = \frac{180°C - 168°C}{50°C - 350°C}$$
$$= \frac{12°C}{-300°C}$$
$$= -0.04 = -4\%$$

Although the actual natural value of the error is the same in both parts (a) and (b)—i.e., 12°C—the magnitude of the error in the first case (12%) is three times greater than the magnitude of the error in the second case (4%). The negative sign of the error calculations indicates that the process variable is lower than the set point.

THE CONTROL VARIABLE

During the control of a process, the controller calculates the error value and adjusts the control variable accordingly to bring the error to zero. Like the error, the value of the control variable can also be expressed as a percentage of range; however, the control variable is expressed in terms of the full range of the controller's output (i.e., the control field device). This range of the controller output is defined as:

$$CV\% = \frac{CV_{actual} - CV_{min}}{CV_{max} - CV_{min}}$$

where:

$CV\%$ = the control variable value as a percentage of its range

CV_{actual} = the actual value of the controller output

CV_{max} = the maximum value of the controllable signal

CV_{min} = the minimum value of the controllable signal

The order and the sign of the nominator and denominator terms in this equation result in a control variable percentage value that is always positive, since the value of CV_{actual} cannot be less than its minimum possible value.

EXAMPLE 14-2

The PLC system shown in Figure 14-12 has an analog output module that sends a 0–10 VDC signal to an electric-to-pneumatic (I/P) converter. The I/P converter controls a steam valve that regulates the process to a set point of 140°C. The range of the controller output is from 0 to 4095 counts, which provides a range of 20 to 220°C in steam temperature control. The process variable has a value of 130°C. Find the percentage of controller output as a function of voltage.

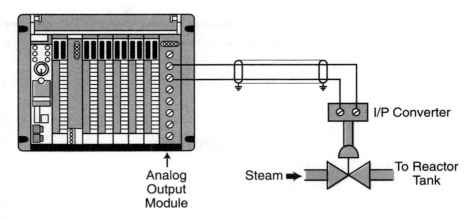

Analog
Output
Module

Steam

I/P Converter

To Reactor
Tank

Figure 14-12. Process regulated by an I/P converter.

SOLUTION

Figure 14-13 illustrates the relationship between the control variable output and the controllable range of temperature. Since the relationship between the controller output and the temperature is linear, the equation of the control variable as a function of voltage is represented by:

$$CV_{volt} = \left(\frac{V_{max} - V_{min}}{Temp_{max} - Temp_{min}} \right) T - CV_{(T=0)}$$

$$CV_{volt} = \left(\frac{10\ V - 0\ V}{220°C - 20°C} \right) T - 1$$

$$= \left(\frac{10}{200} \right) T - 1$$

$$= \left(\frac{T}{20} \right) - 1$$

where T is the given value of the temperature and CV_{volt} is the output of the controller in voltage. Note that this equation takes the form of the equation of a line, $Y = mX + b$ (see Appendix E). At a temperature of 140°C, then, the controller output in voltage would be:

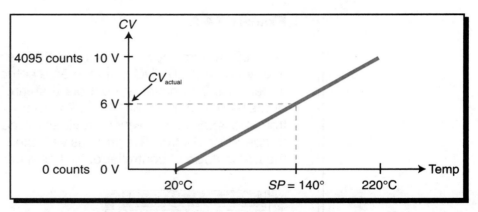

Figure 14-13. Relationship between control variable output and temperature range.

$$CV_{volt} = \left(\frac{140}{20}\right) - 1$$

$$= 6 \text{ volts}$$

So, the control variable in voltage as a percentage of total output would be:

$$CV\% = \frac{CV_{actual} - CV_{min}}{CV_{max} - CV_{min}}$$

$$= \frac{6 \text{ V} - 0 \text{ V}}{10 \text{ V} - 0 \text{ V}}$$

$$= 0.60 = 60\%$$

If the minimum temperature output of 20°C corresponded to a controller output of 1 volt instead of 0 volts, the percentage output would be different because the value of the control variable (CV_{actual}) would change, thus changing the percentage result (see Figure 14-14). A temperature of 140°C would require a 6.4 volt output, which as a percentage of the range would become:

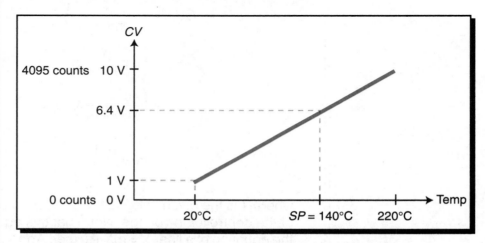

Figure 14-14. Minimum temperature output of 20°C corresponding to a 1 V control variable output.

$$CV\% = \frac{6.4\ V - 0\ V}{10\ V - 0\ V}$$

$$= \frac{6.4}{10}$$

$$= 0.64 = 64\%$$

Having a minimum control output value that is greater than zero is common in many process control systems, because most systems require that the control element be constantly on to regulate the process variable. Note, however, that the range of the control variable is still from 0 to 10 V.

ERROR AND THE CONTROL VARIABLE

The control variable can affect the error in two ways, as illustrated in Figure 14-15. If the error becomes more positive as the control variable *increases*, the action of the controller is called *direct acting*. On the other hand, if the error becomes more positive as the control variable *decreases*, the action of the controller is called *reverse acting*. Table 14-1 illustrates the relationship between error and direct- and reverse-acting controllers. The error values refer to the error as a percentage of the process variable range, meaning that a positive error corresponds to a process variable that is greater than the set point and a negative error corresponds to one that is less than the set point. Note that the equation $E = SP - PV$ yields error values that will have a sign that is opposite of those just described. Therefore, the correct way to identify the type of controller action (direct or reverse) is through the error computed as a percentage of the full range. Chapter 15 discusses reverse- and direct-acting controller modes in more detail.

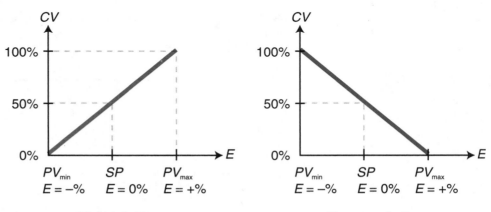

Direct Acting
As percentage of error *increases*,
the control variable *increases*

Reverse Acting
As percentage of error *increases*,
the control variable *decreases*

Figure 14-15. Direct and reverse action in process control.

	Control Variable	
	Direct Acting	Reverse Acting
Error ↑	↑	↓
Error ↓	↓	↑

Table 14-1. Relationship between error and the control variable in direct- and reverse-acting controllers.

ERROR DEADBAND

The purpose of a process control system is to keep the error value as close to zero as possible. However, all systems have an allowable fluctuation in error, meaning that the error can vary from zero by a certain amount without hampering the final product. Figure 14-16 illustrates this fluctuation allowance, which is called the **error deadband**. Within the deadband, the controller treats the error as if it were zero, meaning that it will not make any corrective actions to the control variable. If the value of the error deviates from zero by more than the deadband, the controller will initiate a correction by changing its output (*CV*). The error deadband is a user-specified value, which the PLC stores in one of its registers.

Most PLCs that offer process control loop manipulation software have upper and lower limit alarms, which engage when the process variable deviates from the error deadband. An error will trigger these alarms, as shown

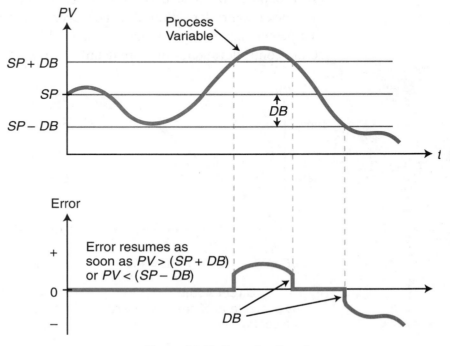

Figure 14-16. Error deadband.

in Figure 14-17, when the error condition exists for more than a specified amount of time. Once the error exceeds the upper or lower limit, a timer starts timing and triggers the alarm after it times out. The alarm indicator will remain ON until the process variable returns to within the specified range.

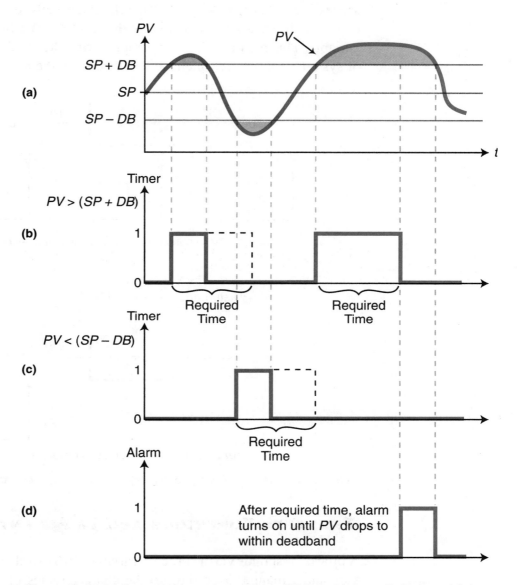

Figure 14-17. (a) Control loop process variable, **(b)** timer for upper limit alarm, **(c)** timer for lower limit alarm, and **(d)** alarm activation.

14-3 PROCESS DYNAMICS

The term *dynamics*, as used in process control, refers to the changes that occur in a process. These changes involve the response of the process system to changes in its input (*CV*), which occur when disturbances are present, or to changes in its set point (see Figure 14-18). Process dynamics does not refer to the behavior of the process when the error is zero (or within the error deadband). At this point, the process is said to be at **steady state**. Instead, process dynamics deals with the system's (process variable's) reaction to corrective actions taken by the controller to bring the error to zero after it senses that the error is too large. Therefore, the analysis of process dynamics explores the relationship between the control variable and the process variable. This relationship is important during the "tuning," or adjustment, of system parameters, which we will discuss in the next chapter.

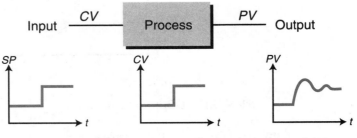

(a) Change in output due to a change in the set point

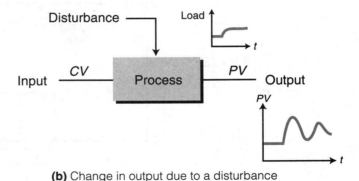

(b) Change in output due to a disturbance

Figure 14-18. Process change due to **(a)** a change in set point and **(b)** a disturbance.

TRANSFER FUNCTIONS AND TRANSIENT RESPONSES

A process responds via the process variable (PV) to a change in input (CV) in a dynamic manner according to the characteristics of the process. These process characteristics, which include factors such as delay time and inherent physical responses of the process, are defined by a **transfer function**, represented by the term H_T (see Figure 14-19). A transfer function is an equation that describes a process in terms of response over time, as well as calculates the outcome of the process variable. Therefore, the value of the term H_T equals the value of the process variable at a particular control variable value and time, given the characteristics of the process. Every process has its own unique transfer function based on its particular characteristics, and for most processes, the transfer function equation is not known. Thus, certain assumptions must be made about the process to estimate H_T. Experimentation can also be used to approximate the outcome of H_T (i.e., the process variable response) to a forced change in the process input. This experimental change in process input is called a **step test** and the response is called a **step response**.

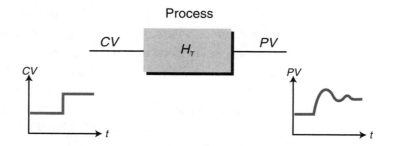

Change in *CV*...through H_T...affects change in *PV*

$$H_T = \frac{PV}{CV}$$

Figure 14-19. Transfer function.

The most important aspect of a transfer function is not so much its composition or form, but its response to sudden process input changes created by disturbances. This behavioral response of a process is called a **transient response**, and it includes the time required for the output to reach a steady-state final value given a sudden change in input. Transient responses provide much information about the dynamics of a process and, therefore, about the transfer function.

As shown in Figure 14-20, a closed-loop control system includes two transfer functions—one that defines the controller (*Hc*) and another that defines the process (*Hp*). The input to the controller's transfer function is the error signal (*E*), and its output is the control variable (*CV*). This control variable becomes the input to the process's transfer function, whose output is the process variable (*PV*). In this chapter, we will discuss the process's transfer function and its behavior. In the next chapter, we will discuss the controller's transfer function and the different forms that it can take.

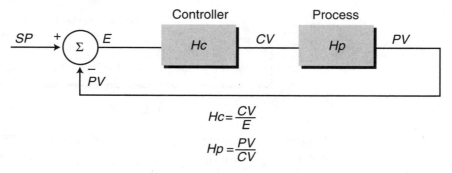

$$Hc = \frac{CV}{E}$$

$$Hp = \frac{PV}{CV}$$

Figure 14-20. A closed-loop control system with two transfer functions.

To better understand transient responses and the information we can obtain from them, let's explore a basic example of an open-loop hot-water heater system. Figure 14-21 shows this process, while Figure 14-22 shows the corresponding process block diagram. The transfer function of the process depends on many factors, such as the rate of flow of the steam, the temperature of the steam, the temperature of the incoming water at the inlet, the ambient temperature, and the inflow and outflow rates of the water. Regardless of all these process factors, the controller must maintain the temperature in the tank (the process variable) as close as possible to the user-defined set point by manipulating the control variable. For this example, let's assume that the temperature in the tank (at steady state) is at a set point of 65°C, the temperature range spans from 15°C to 93°C, and the steam control valve is at 55% of its open position.

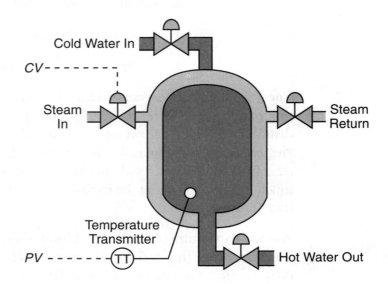

Figure 14-21. Hot-water heater system.

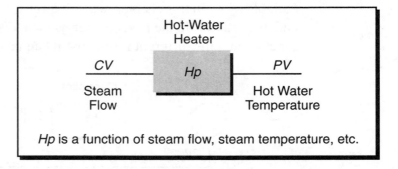

Figure 14-22. Water heater process diagram.

With a step change in valve position from 55% to 75% open, the temperature (*PV*) in the tank will begin to heat up (see Figure 14-23). After 15 minutes, the process variable will increase to 81°C. The process variable's behavior during this 15-minute period is the transient response of the transfer function *Hp*. Note that the transient response is very smooth, heating the water slowly over the 15 minutes until steady state is once again achieved.

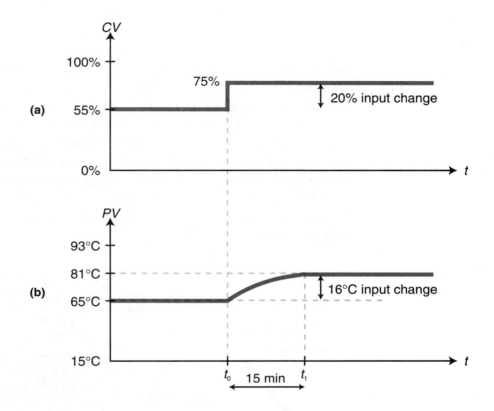

Figure 14-23. (a) Control variable step change and **(b)** its corresponding process variable change.

PROCESS GAIN

The **process gain,** represented by the term K, defines the ratio between process output and process input. This gain is another dynamic element that is observed in a transient response. It is calculated by dividing the change in process output over a period of time by the corresponding change in process input. Thus, the process gain is equal to the change in the process variable divided by the change in the control variable:

$$K = \frac{PV_{\text{final}} - PV_{\text{initial}}}{CV_{\text{final}} - CV_{\text{initial}}}$$

Hence, for the previous hot-water tank example, K would be the change in the tank temperature divided by the corresponding change in the valve output over the 15-minute period:

$$K = \frac{81°C - 65°C}{75\% - 55\%}$$

$$= \frac{16°C}{20\%}$$

$$= 0.8°C / \%$$

So, the process gain is 0.8°C/%. This means that the process variable (temperature) changes 0.8°C for every one percent of change in the control variable (steam valve). The process gain calculation of 0.8°C/% (see Figure 14-24) is only valid for the process conditions under which it was calculated (i.e., a constant inflow and outflow of water—Q_{in} and Q_{out}, respectively—at a constant water temperature). Under these conditions, the mass of water in the heater tank remains constant and the steam heating system's gain (0.8°C/%) operates linearly over the span of the temperature range. If either the input water flow Q_{in} or any other parameter changes, the gain of the system will also change.

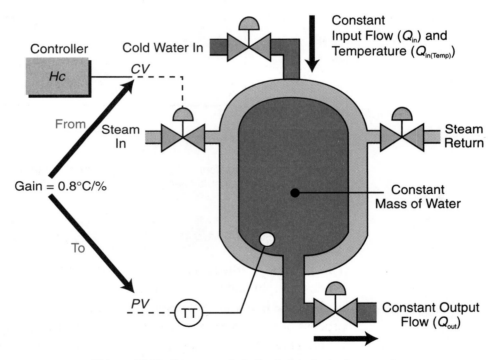

Figure 14-24. Process gain in the hot-water tank example.

DEAD TIME

The perfect response of a process variable to a step change in the control variable is instantaneous, as shown in Figure 14-25. In this type of perfect system, the process's transfer function is equal to 1, meaning that a control variable input immediately results in an equal process variable output. In

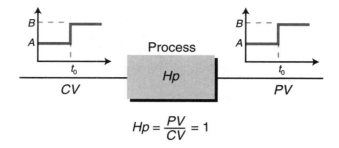

$$Hp = \frac{PV}{CV} = 1$$

Figure 14-25. Instantaneous response of the process output to a step change in input.

reality, however, characteristics distinctive to the process influence the relationship between the control variable and process variable, resulting in a transfer function that is not equal to 1. One of these process characteristics is the inherent delay associated with an output's response to an input (see Figure 14-26). This delay is called the **dead time**. A typical example of dead time occurs when you first turn on the hot-water knob in the shower. The water will not become hot until all the cool water runs out of the pipe and the water from the hot-water heater reaches the shower head. The time that elapses between turning the knob and receiving the hot water is the dead time.

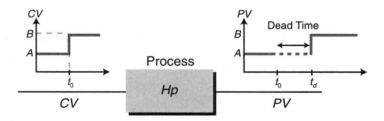

Figure 14-26. Dead time delay.

A dead time delay occurs not only when a control variable input is introduced into the system, but also when an existing input is increased or decreased. Figure 14-27 illustrates a trailing-end dead time delay for a stepped-down input. In this system, a dead time delay occurs between the decrease of the input and the response of the output to the change in input. In addition to process dead time delay, sensor measuring devices also introduce a dead time delay into a system, because a time lapse occurs between the moment the analog process reading is taken and the moment the voltage or current value is available at the transmitter's output (see Figure 14-28). Although the sensor delay is, for the most part, small in comparison to the process variable dead time, it is still an important part of the transfer function because it affects the process. This is especially important in fast-reacting, closed-loop systems, such as servo motor and other positioning applications.

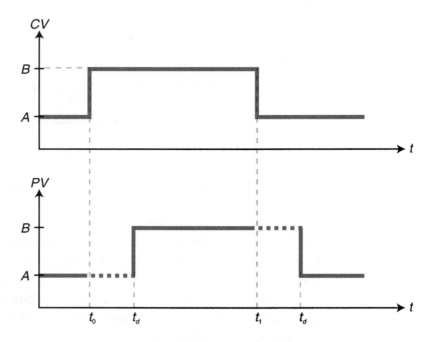

Figure 14-27. Trailing-end dead time delay.

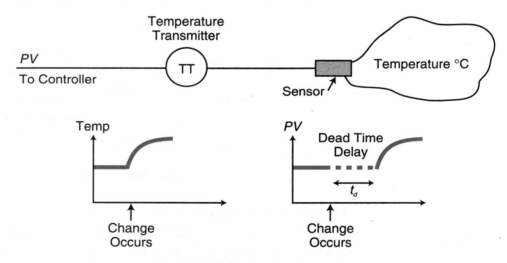

Figure 14-28. Sensor dead time delay.

LAG TIME

Dead time is not the only delay associated with a process and its transfer function. Another delay is **lag time**. Unlike a dead time delay, which is the delay between a change in input and the initial response of the process variable to the input change, a lag time delay occurs when a process variable exhibits a time lapse between its initial response to the input variable and its optimal response to it. Lag time occurs due to process characteristics that are contained in the transfer function. An example of lag time delay can be observed in a car's cruise control mechanism (see Figure 14-29). If you are driving at 70 mph under cruise control and disengage the control because you see a police car, the car's speed will decrease. Once you pass the police car and press the cruise control's resume button, the car will start to accelerate immediately; however, a delay will occur before the car reaches the set point speed (70 mph). This delay is the lag time. Note that, in this example, the dead time is minimal, because once the resume button is pressed, the control variable (accelerator) increases and the speed (process variable) begins to react almost immediately.

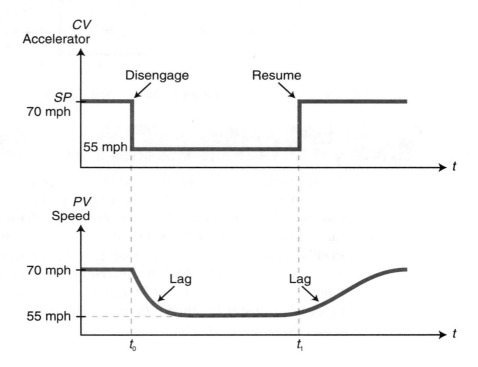

Figure 14-29. Lag time in a car's cruise control mechanism.

A process generally has one of two types of lag, which are the result of its transfer function:

• first-order lag

• second-order lag

First-order lag time is the lag a process variable exhibits in response to a rapid change in the control variable. *Second-order lag time* is the oscillating response of a process variable as it settles to its steady-state value after a step change in the input. Figure 14-30 illustrates both first- and second-order lags. Note how the two processes respond differently to the same change in input because of the different types of lags. A first-order lag is also called a **first-order response**, while a second-order lag is called a **second-order response**. Both types of responses are called *system transient responses*.

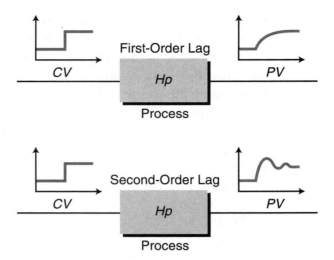

Figure 14-30. First- and second-order lags.

14-4 LAPLACE TRANSFORM BASICS

As mentioned previously, the response of a process is tied to the transfer function of the process itself. Each section of a control system has a transfer function that can be described mathematically. This includes one transfer function for the controller, one for the process, and one for the total system in either an open-loop or closed-loop configuration. As shown in Figure 14-31, the controller's transfer function (Hc) is the ratio of its output (CV) to its input, which is the error between the set point and the process variable. The process's transfer function, Hp, is also the ratio of its output (PV) to its input, the control variable.

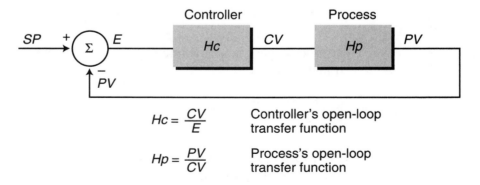

$$Hc = \frac{CV}{E}$$ Controller's open-loop transfer function

$$Hp = \frac{PV}{CV}$$ Process's open-loop transfer function

Figure 14-31. Transfer functions in a control system.

Mathematically, transfer functions are expressed through **Laplace transforms**. Laplace transforms are mathematical functions that are used to solve complex differential equations by converting them into easy-to-manage algebraic equations. It is beyond the scope of this book to explain the mathematical derivation of Laplace transforms. However, we will discuss how they are used in process control functions, to aid in the understanding of first- and second-order systems and their transient responses.

TRANSFER FUNCTIONS

As explained earlier, a system can have either a first-order response or a second-order response. These responses (transfer functions) are expressed as complex differential equations, with first-order system responses characterized by first-order differential equations and second-order systems characterized by second-order differential equations. A first-order differential equation is a mathematical statement that expresses the rate of change of a function with respect to its independent variable. A second-order differential equation expresses the rate of change of a first-order term with respect to its independent variable (i.e., the rate of change of the rate of change). The notations used to express first- and second-order differential equations are as follows:

$$\text{First-order equation} \quad y = \frac{dx}{dt} + x$$

$$\text{Second-order equation} \quad y = \frac{d^2x}{dt^2} + \frac{dx}{dt} + x$$

Most control processes found in industrial applications can be described as either first order or second order. For more complex processes with third-order responses, a second-order equation can be used to approximate the process response.

Laplace transforms use known substitutions for complex differential equations to change them into more easily solvable algebraic equations. To accomplish this, Laplace transforms convert differential equations from the time (t) domain—the process response as a function of time—to the frequency (s) domain—the process response as a function of frequency. Table 14-2 shows some of the most common Laplace transforms found in process control applications. This table also includes inverse Laplace transforms, which are used to convert Laplace equations back into time domain responses. Table 14-2 also includes the time domain process responses to a step input.

DERIVATIVE LAPLACE TRANSFORMS

Laplace transforms replace the derivative terms in both first-order and second-order differential equations with their respective frequency domain s terms. Table 14-3 shows the Laplace transforms for both first- and second-

DESCRIPTION	LAPLACE TRANSFORM $F(s)$	TIME FUNCTION $f(t)$	PROCESS RESPONSE
Any Function f			
Unit Step Input	$\dfrac{1}{s}$	$u(t)$	$u(t)=0 \quad t<t_0$ $u(t)=1 \quad t\geq t_0$
Step Input	$\dfrac{A}{s}$	$A(t)$	$A(t)=0 \quad t<t_0$ $A(t)=A \quad t\geq t_0$
Delay (Dead Time) to Step Input	$Ae^{-t_d s}$	$A(t-t_d)$	$A(t-t_d)=0 \quad t<t_d$ $A(t-t_d)=A \quad t\geq t_d$

First-Order Equations

DESCRIPTION	LAPLACE TRANSFORM	TIME FUNCTION	PROCESS RESPONSE
First-Order Response	$\dfrac{A}{s+a}$	Ae^{-at}	For $A=1$
First-Order Response with Lag	$\dfrac{A}{\tau s+1}$	$\dfrac{A}{\tau}e^{\frac{-t}{\tau}}$	For $A=1$
First-Order Response plus Dead Time	$\dfrac{A_1 A_2}{s+a}e^{-t_d s}$	$A_1 A_2 e^{-at}\quad$ for $t\geq t_d$	For $A_1 A_2=1$
First-Order Response with Lag plus Dead Time	$\dfrac{A_1 A_2}{\tau s+1}e^{-t_d s}$	$A_1 A_2 e^{\frac{-t}{\tau}}\quad$ for $t\geq t_d$	For $A_1 A_2=1$

First-Order Response (For $A=1$):

	$\tau=1$	$\tau=2$
$t=0$	1.000	1.000
$t=1$	0.368	0.135
$t=2$	0.135	0.018

First-Order Response with Lag (For $A=1$):

	$\tau=1$	$\tau=2$
$t=0$	1.000	0.500
$t=1$	0.368	0.303
$t=2$	0.135	0.184

First-Order Response plus Dead Time (For $A_1 A_2=1$):

	$a=1$	$a=2$
$t=t_d$	1.000	1.000
$t=t_d+1$	0.368	0.135
$t=t_d+2$	0.135	0.018

First-Order Response with Lag plus Dead Time (For $A_1 A_2=1$):

	$\tau=1$	$\tau=2$
$t=t_d$	1.000	0.500
$t=t_d+1$	0.368	0.135
$t=t_d+2$	0.135	0.018

Table 14-2. Laplace transforms.

DESCRIPTION	LAPLACE TRANSFORM	TIME FUNCTION	PROCESS RESPONSE
First-Order Response to Step Input (A_1/s) with Lag	$\dfrac{A_1 A_2}{s(\tau s + 1)}$	$A_1 A_2\left(1 - e^{\frac{-t}{\tau}}\right)$	For $A_1A_2=1$: $t=0$: $\tau=1$: 0.000, $\tau=2$: 0.000 $t=1$: 0.632, 0.393 $t=2$: 0.865, 0.632
First-Order Response to Step Input (A_1/s) with Lag plus Dead Time	$\dfrac{A_1 A_2 A_3}{s(\tau s + 1)}\, e^{-t_d s}$	$A_1 A_2 A_3\left(1 - e^{\frac{-t}{\tau}}\right)$ for $t \geq t_d$	For $A_1A_2A_3=1$: $t=t_d$: $\tau=1$: 1.000, $\tau=2$: 1.000 $t=t_d+1$: 0.632, 0.393 $t=t_d+2$: 0.865, 0.632

Second-Order Equations

DESCRIPTION	LAPLACE TRANSFORM	TIME FUNCTION
Second-Order Transfer Function ($Hp_{(s)}$) for $\zeta < 1$ (Underdamped)	$\dfrac{A\omega_n^2}{s^2 + 2\zeta\omega_n s + \omega_n^2}$	$\dfrac{A\omega_n e^{-\zeta\omega_n t}}{\sqrt{1-\zeta^2}}\sin\left(\omega_n\sqrt{1-\zeta^2}\,t\right)$
Second-Order Transfer Function ($Hp_{(s)}$) for $\zeta = 1$ (Critically Damped)	$\dfrac{A}{(\tau s + 1)^2}$	$\dfrac{At}{\tau^2}e^{\frac{-t}{\tau}}$
Second-Order Transfer Function ($Hp_{(s)}$) for $\zeta > 1$ (Overdamped)	$\dfrac{A}{(\tau_1 s + 1)(\tau_2 s + 1)}$	$\dfrac{A}{\tau_1 - \tau_2}\left(e^{\frac{-t}{\tau_1}} - e^{\frac{-t}{\tau_2}}\right)$
Second-Order Step Response (A_1/s) for $\zeta < 1$ (Underdamped)	$\dfrac{A_1 A_2 \omega_n^2}{s(s^2 + 2\zeta\omega_n s + \omega_n^2)}$	$A_1 A_2\left[1 + \dfrac{e^{-\zeta\omega_n t}}{\sqrt{1-\zeta^2}}\sin\left(\omega_n\sqrt{1-\zeta^2}\,t - \psi\right)\right]$ where $\psi = \tan^{-1}\dfrac{\sqrt{1-\zeta^2}}{-\zeta}$ $(0 < \psi < \pi)$
Second-Order Step Response (A_1/s) for $\zeta = 1$ (Critically Damped)	$\dfrac{A_1 A_2}{s(\tau s + 1)^2}$	$A_1 A_2\left(1 - \dfrac{\tau - t}{\tau}e^{\frac{-t}{\tau}}\right)$
Second-Order Step Response (A_1/s) for $\zeta > 1$ (Overdamped)	$\dfrac{A_1 A_2}{s(\tau_1 s + 1)(\tau_2 s + 1)}$	$A_1 A_2\left(1 + \dfrac{\tau_1 e^{\frac{-t}{\tau_1}} - \tau_2 e^{\frac{-t}{\tau_2}}}{\tau_2 - \tau_1}\right)$

$Hp_{(s)} = \dfrac{\text{Out}}{\text{In}}$

Critically damped $\zeta = 1$; Overdamped $\zeta > 1$; Underdamped $\zeta < 1$; $\dfrac{2\pi}{\omega_n}$ or $\dfrac{1}{f_n}$

Table 14-2 continued.

635

	Time (t) Domain	**Laplace (s) Domain**
Function	$x_{(t)}$	$X_{(s)}$
First Derivative	$\dfrac{dx}{dt}$	$sX_{(s)} - x_{(t\,=\,0)}$
Second Derivative	$\dfrac{d^2x}{dt^2}$	$s^2 X_{(s)} - s\dfrac{dx_{(t=0)}}{dt} - x_{(t=0)}$
Integral	$\displaystyle\int_0^t A\,dt$	$\dfrac{A}{s}$

Table 14-3. Derivative Laplace transforms.

order derivative terms. In the frequency domain, a first-order derivative term becomes an s term times the function in the frequency domain minus a constant, which is the value of the function at $t = 0$ in the time domain. A second-order derivative becomes an s^2 Laplace term times the Laplace function minus s times the value of the time domain first derivative at $t = 0$ minus the value of the function at $t = 0$ in the time domain. Therefore, a simple first-order differential equation of the form:

$$y_{(t)} = \frac{dx}{dt} + x_{(t)}$$

becomes the following equation in Laplace form:

$$Y_{(s)} = sX_{(s)} - x_{(t=0)} + X_{(s)}$$

Assuming that the value of the function $x_{(t\,=\,0)}$ is zero, the equation becomes:

$$Y_{(s)} = sX_{(s)} + X_{(s)}$$
$$= X_{(s)}(s+1)$$

If $X_{(s)}$ represents the Laplace output of the process and $Y_{(s)}$ represents the input, as shown in Figure 14-32, the equation for the transfer function of the process (output divided by input) in Laplace form becomes:

$$\frac{X_{(s)}}{Y_{(s)}} = \frac{1}{s+1}$$

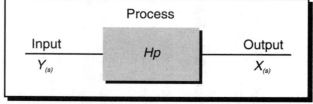

Figure 14-32. Process inputs and outputs in Laplace form.

This indicates that the Laplace output $X_{(s)}$ is equal to the Laplace input times the transfer function in Laplace form:

$$X_{(s)} = Y_{(s)}\left(\frac{1}{s+1}\right)$$

So, by working in the frequency domain instead of the time domain, the solving of the differential equation is reduced to an algebraic manipulation. In the Laplace domain, the exponent of the s term indicates the order of the transfer function—in this case, a first-order transfer function.

A second-order differential equation of the form:

$$y_{(t)} = \frac{d^2x}{dt^2} + \frac{dx}{dt} + x_{(t)}$$

becomes the following equation in Laplace form:

$$Y_{(s)} = \left[s^2 X_{(s)} - s\frac{dx_{(t=0)}}{dt} - x_{(t=0)}\right] + \left[sX_{(s)} - x_{(t=0)}\right] + X_{(s)}$$

Assuming that the values of the initial parameters are zero, this equation becomes:

$$Y_{(s)} = s^2 X_{(s)} + sX_{(s)} + X_{(s)}$$
$$= X_{(s)}\left(s^2 + s + 1\right)$$

Since $Y_{(s)}$ represents the input and $X_{(s)}$ represents the output, the transfer function for a second-order response in Laplace form is:

$$\frac{X_{(s)}}{Y_{(s)}} = \frac{1}{\left(s^2 + s + 1\right)}$$

Note that the exponent of the s term indicates a second-order transfer function. As a result of the different mathematical formats of first-order and second-order transfer functions, a second-order system will have an oscillating response, while a first-order system will have a smooth response toward its final steady-state value. We will discuss these response curves in more detail later.

The substitutions used in Laplace transforms are the result of computations performed on electrical circuit networks (e.g., resistors, capacitors, and inductors) that create transfer functions. Mechanical and hydraulic systems also use Laplace transforms to represent system transfer functions, primarily because of their similarity in mathematical representation to electrical systems. The following example illustrates the derivation of a transfer function for a resistor/capacitor (R/C) network.

EXAMPLE 14-3

Figure 14-33 represents an R/C (resistor/capacitor) electrical network. Find **(a)** the differential equation that represents this network, **(b)** the network transfer function in Laplace, and **(c)** the equivalent closed-loop block diagram. Refer to Appendix G for information about the characteristics of electrical circuit elements.

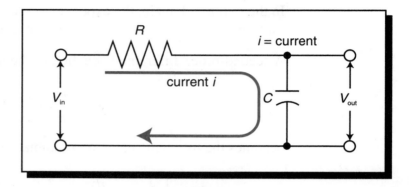

Figure 14-33. R/C electrical network diagram.

SOLUTION

The term V_C represents the voltage across the capacitor (C). This term is equivalent to V_{out} and is represented by:

$$V_{out} = V_C = \frac{1}{C} \int_0^t i \, dt$$

Kirchhoff's voltage law states that the voltage at the output of an electrical circuit is equal to the input voltage minus the voltage across the resistor (V_R), which is equal to the current (i) times the resistance (R). Therefore:

$$V_{out} = V_{in} - V_R$$
$$= V_{in} - iR$$

Solving for the current yields:

$$V_{out} = V_{in} - iR$$
$$iR = V_{in} - V_{out}$$
$$i = \left(V_{in} - V_{out}\right)\left(\frac{1}{R}\right)$$

Replacing the current term in the capacitor's output voltage equation with this value of i produces:

$$V_{out} = \frac{1}{C}\int_0^t i\, dt$$

$$= \frac{1}{C}\int_0^t (V_{in} - V_{out})\frac{1}{R}\, dt$$

$$= \frac{1}{C}\int_0^t \frac{V_{in}}{R}\, dt - \frac{1}{C}\int_0^t \frac{V_{out}}{R}\, dt$$

$$= \frac{1}{RC}\int_0^t V_{in}\, dt - \frac{1}{RC}\int_0^t V_{out}\, dt$$

$$= \frac{1}{RC}\left[\int_0^t V_{in}\, dt - \int_0^t V_{out}\, dt\right]$$

(a) To obtain the differential equation of this network, we must take the derivative of both sides of the output voltage equation to eliminate the integral terms:

$$\frac{dV_{out}}{dt} = \frac{1}{RC}\left(V_{in} - V_{out}\right)$$

$$V_{in} = RC\frac{dV_{out}}{dt} + V_{out}$$

(b) The Laplace form of this first-order differential equation is:

$$V_{in(s)} = RCsV_{out(s)} - V_{out(t=0)} + V_{out(s)}$$

Since there is no initial value at $t = 0$, this equation becomes:

$$V_{in(s)} = RCsV_{out(s)} + V_{out(s)}$$

$$= V_{out(s)}\left(RCs + 1\right)$$

The transfer function is:

$$\frac{V_{out(s)}}{V_{in(s)}} = \frac{1}{sRC + 1}$$

The equation for a first-order system with lag is represented by:

$$Hp_{(s)} = \frac{1}{\tau s + 1}$$

Therefore, this electrical network is a first-order lag system where the time constant τ is equal to RC (the resistance times the capacitance).

(c) Figure 14-34 shows the block diagram of this closed-loop system.

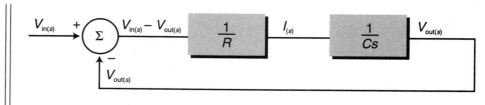

Figure 14-34. Block diagram of the R/C network.

Referring to the hot-water heater example used in Section 14-3, let's assume that the differential equation that describes the process, otherwise known as the heating system's *enthalpy balance equation*, is given by:

$$\text{Steam flow} = A\frac{dT}{dt} + BT_{(t)}$$

where A and B are constants and T is the temperature. Since the steam flow is directly related to the control variable and the temperature is the process variable, we can rewrite this equation as:

$$CV_{(t)} = A\frac{dPV}{dt} + BPV_{(t)}$$

Taking the Laplace transform of this equation, we get:

$$CV_{(s)} = AsPV_{(s)} - APV_{(t=0)} + BPV_{(s)}$$

Assuming that the value of PV at $t = 0$ is zero, this equation becomes:

$$CV_{(s)} = AsPV_{(s)} + BPV_{(s)}$$
$$= PV_{(s)}(As + B)$$

The process's transfer function, which is equal to the process's output divided by its input, is:

$$\frac{PV_{(s)}}{CV_{(s)}} = \frac{1}{As + B}$$

To obtain a standard first-order lag equation (i.e., a fraction whose denominator is $\tau s + 1$), we can divide both the numerator and the denominator by B:

$$\frac{PV_{(s)}}{CV_{(s)}} = \frac{\left(\frac{1}{B}\right)}{\left(\frac{As+B}{B}\right)}$$
$$= \frac{\left(\frac{1}{B}\right)}{\left(\frac{A}{B}s + 1\right)}$$

Previously, we calculated experimentally that the gain for this process is 0.8°C/% and that it took 15 minutes to achieve the final steady-state value. Furthermore, as we will discuss in Section 14-6, it is a given that in a first-order system with lag the output variable will be at 99.33% of the input variable when $t = 5\tau$. The observed value of the response at 15 minutes is 100%, which is close to the 99.33% at 5τ. Therefore, we can approximate the value of τ as:

$$5\tau \approx 15 \text{ min}$$

$$\tau \approx \frac{15 \text{ min}}{5} \approx 3 \text{ min}$$

Thus, we can represent the process's transfer function as:

$$\frac{PV_{(s)}}{CV_{(s)}} = \frac{0.8}{3s + 1}$$

INTEGRAL LAPLACE TRANSFORMS

Aside from a simple constant gain, an integral transfer function represents the simplest of all process responses (see Figure 14-35). It is represented by the equation:

$$\frac{\text{Output}}{\text{Input}} = \frac{PV}{CV} = \int_0^t A\,dt$$

$$= A\int_0^t dt$$

$$PV = A\int_0^t CV\,dt$$

where A is a constant gain. In Laplace, the term ($\frac{A}{s}$) replaces the time domain integral term. Therefore, the Laplace transform of this equation is:

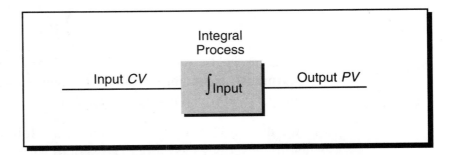

Figure 14-35. Integral transfer function.

$$PV_{(s)} = \left(\frac{A}{s}\right)CV_{(s)}$$

An integral process integrates the input with the process over time. This implies that the rate of change of the process output varies according to the input, as shown in the following equation:

$$PV = A\int_0^t CV dt$$

$$dPV = ACVdt$$

$$\frac{dPV}{dt} = ACV$$

Therefore, in an integral process, a change in the control variable will produce a rate of change in the process variable over time.

In process systems, changes in the control variable are usually simulated by producing a step change with amplitude B. In Laplace form, a step change to an integral process is represented as:

$$PV = CV\left(\frac{A}{s}\right)$$

$$= \left(\frac{B}{s}\right)\left(\frac{A}{s}\right)$$

$$= \frac{AB}{s^2}$$

where:

$A =$ the constant integral gain

$B =$ the amplitude of the step change

The inverse Laplace (time domain) response of this equation is:

$$PV_{(t)} = ABt$$

This implies that the rate of increase of the process variable is AB per second (or per minute, depending on time units). Figure 14-36 illustrates this integral response curve, which is a ramp-type function.

As we discussed briefly in Chapter 8, the following integro-differential equation describes the output (i.e., control variable) of a typical proportional-integral-derivative (PID) controller (see Figure 14-37):

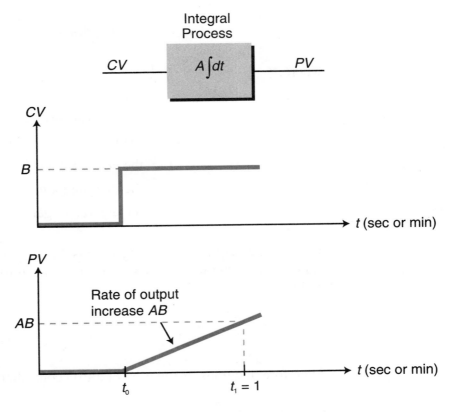

Figure 14-36. Integral response to a step change.

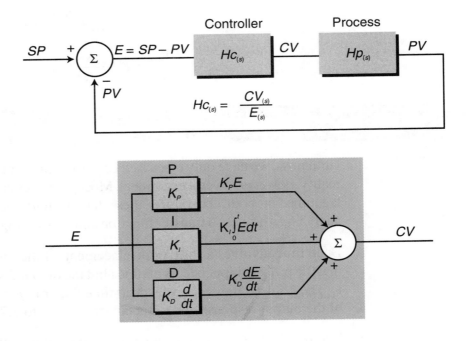

Figure 14-37. PID process control loop.

$$CV_{(t)} = K_P E + K_I \int_0^t E \, dt + K_D \frac{dE}{dt} + CV_{(t=0)}$$

where:

$$CV_{(t)} = \text{the control variable output}$$
$$K_P = \text{the proportional gain}$$
$$E = \text{the error } (SP - PV)$$
$$K_I = \text{the integral gain}$$
$$K_D = \text{the derivative gain}$$

To find the Laplace transform of this PID equation, we must make the appropriate substitutions, assuming that the initial parameters are zero, which yields:

$$CV_{(s)} = K_P E_{(s)} + \frac{K_I E_{(s)}}{s} + K_D E_{(s)} s$$

Therefore, the transfer function of a PID controller (Hc) in Laplace is:

$$Hc_{(s)} = \frac{\text{Out}}{\text{In}} = \frac{CV_{(s)}}{E_{(s)}}$$

$$CV_{(s)} = K_P E_{(s)} + \frac{K_I E_{(s)}}{s} + K_D E_{(s)} s$$

$$\frac{CV_{(s)}}{E_{(s)}} = K_P + \frac{K_I}{s} + K_D s$$

14-5 DEAD TIME RESPONSES IN LAPLACE FORM

Until now, we have only discussed Laplace transforms of ideal processes. In reality, however, no process is ideal. Most processes contain either dead time, lag time, or both. Therefore, these factors must be accounted for when analyzing a process's transfer function and performing its Laplace transform.

Dead time involves a shift, or displacement, of the time variable t, meaning that the process input occurs at time t but the output does not occur until time t_d. The Laplace transform of a dead time factor is $e^{-t_d s}$, where t_d is the delay of the output response and e is a constant equal to 2.718.

Figure 14-38 illustrates a transfer function of a simple system with dead time (and no lag), which receives a step (OFF-to-ON) input with amplitude A. Note that the value of the process gain is 1, because an input of magnitude

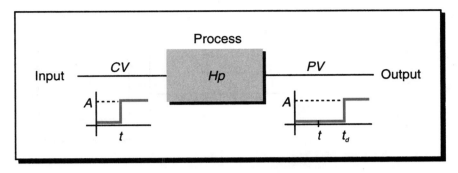

Figure 14-38. Transfer function of a system with dead time.

A yields an output of magnitude *A*. The output, however, has a dead time equal to t_d. Therefore, the transfer function *Hp* is equal to the process gain times the dead time. If the process gain equals 1, then the Laplace transform of this transfer function is:

$$Hp_{(s)} = e^{-t_d s}$$

The input (*CV*) to the process in Figure 14-38 is a step input of amplitude *A*. In the previous section, we explained that the Laplace transform of a step input is:

$$CV = \frac{A}{s}$$

Knowing these two equations, along with the fact that *Hp* is equal to the output divided by the input, we can find the value, in Laplace, of the process variable:

$$Hp = \frac{PV}{CV}$$
$$PV = (CV)(Hp)$$
$$= \left(\frac{A}{s}\right)e^{-t_d s}$$

14-6 LAG RESPONSES IN LAPLACE FORM

Now that we have explained the basics of Laplace transforms, let's examine the various responses exhibited by processes. Most processes exhibit either a first-order or second-order lag response with one or two time lags, respectively (see Figure 14-39). Note that this figure shows open-loop processes whose inputs are the control variable from the controller.

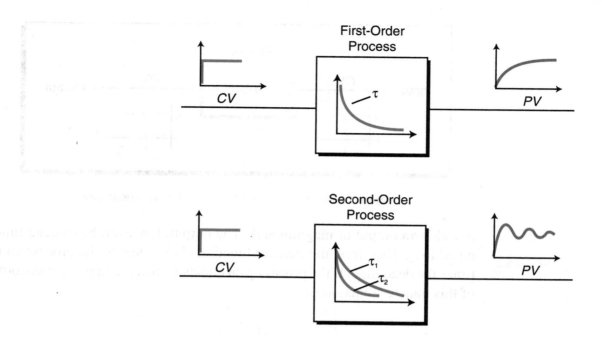

Figure 14-39. First- and second-order lags.

FIRST-ORDER LAG RESPONSES

First-order lag is one of the most common types of process responses. In this type of system, the process variable response lags behind a rapid change (step change) in the control variable. In the time domain, a first-order response to a step input is represented by the equation:

$$V_{out} = V_{in}\left(1 - e^{\frac{-t}{\tau}}\right)$$

where:

$$V_{in} = \text{the step input to the process}$$
$$V_{out} = \text{the output of the process}$$
$$t = \text{time}$$
$$\tau = \text{the time constant}$$

As shown in Figure 14-40, this response is an exponential function, meaning that the value of V_{out} increases rapidly with time to equal V_{in}. The time constant τ describes how quickly the output catches up with the input value—the smaller the value of τ, the faster V_{out} equals V_{in} and vice versa. So, first-order responses with smaller time constants have shorter lag times.

For a first-order system with a step input and lag, the value of the output, given that $t = \tau$, is 63.2% of the input is:

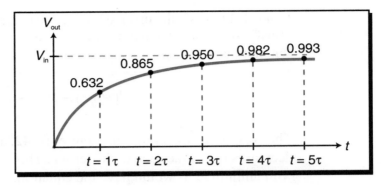

Figure 14-40. First-order lag process response.

$$V_{out} = V_{in}\left(1 - e^{\frac{-t}{\tau}}\right)$$

$$= V_{in}\left(1 - e^{\frac{-\tau}{\tau}}\right)$$

$$= V_{in}\left(1 - e^{-1}\right)$$

$$= V_{in}(1 - 0.368)$$

$$= 0.632V_{in}$$

As the value of t increases, the value of V_{out} becomes closer to 100% of V_{in}. As shown in Figure 14-40, at the time $t = 4\tau$, the value of V_{out} will be over 98% of that of V_{in}, and when the time variable reaches 5τ, V_{out} will be over 99% of V_{in}.

In the Laplace domain, a first-order system transfer function with lag is represented by the equation (see Figure 14-41):

$$Hp_{(s)} = \frac{A_1}{\tau s + 1}$$

where:

$$A_1 = \text{the process amplitude gain}$$

$$\tau = \text{the system's lag time}$$

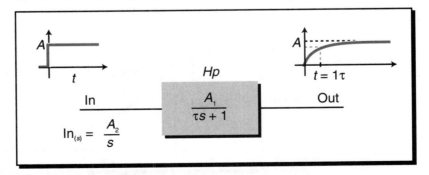

Figure 14-41. First-order transfer function with lag in the Laplace domain.

Using Table 14-2, the inverse Laplace transform (represented by $\mathscr{L}^{-1}$) of this transfer function, which turns the Laplace equation into a time-based transfer function, is:

$$\mathscr{L}^{-1}[Hp_{(s)}] = Hp_{(t)} = \left(\frac{A_1}{\tau}\right)e^{\frac{-t}{\tau}}$$

This response, as shown in Figure 14-42, has a decaying form (i.e., it decreases over time) due to the term e. This same system given a step input with amplitude A_2 would have the equation:

$$Out_{(s)} = \left(In_{(s)}\right)\left(Hp_{(s)}\right)$$

$$= \left(\frac{A_2}{s}\right)\left(\frac{A_1}{\tau s + 1}\right)$$

$$= \frac{A_1 A_2}{s(\tau s + 1)}$$

Again using Table 14-2, the inverse Laplace response, or real-time response of the system ($Out_{(t)}$), is:

$$Out_{(t)} = A_1 A_2\left(1 - e^{\frac{-t}{\tau}}\right)$$

where the value of the output will be $0.63A_1A_2$ (63% of A_1A_2) when $t = \tau$. If the input is a unit step (that is, the amplitude A_2 equals 1) and the gain of $Hp_{(s)}$ equals 1 ($A_1 = 1$), then the output will be:

$$Out_{(t)} = A_1 A_2\left(1 - e^{\frac{-t}{\tau}}\right)$$

$$= (1)(1)\left(1 - e^{\frac{-t}{\tau}}\right)$$

$$= \left(1 - e^{\frac{-t}{\tau}}\right)$$

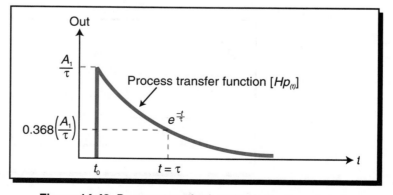

Figure 14-42. Process transfer function in the time domain.

Figure 14-43 graphs the response of the process variable $Out_{(t)}$ for $A_1 = A_2 = 1$. This curve, representing a first-order response to a step input plus lag, is a function of the system's transfer function, which is the step value (1) minus the system's curve term ($e^{\frac{-t}{\tau}}$).

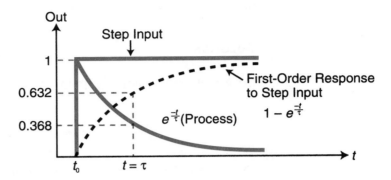

Figure 14-43. Process variable's lag response.

Adding a simple dead time term ($e^{-t_d s}$) to a first-order step response with lag generates the Laplace transfer function:

$$Out_{(s)} = \left(\frac{A_1 A_2}{s(\tau s + 1)}\right)e^{-t_d s}$$

where t_d is the dead time. Figure 14-44 shows the graph of this function in the time domain. The value of the output is:

$$Out_{(t)} = A_1 A_2 \left(1 - e^{\frac{-t}{\tau}}\right) \qquad \text{for } t \geq t_d$$

$$Out_{(t)} = 0 \qquad \text{for } t < t_d$$

Note that the first output response equation is valid for t values greater than t_d, the dead time. $Out_{(t)}$ will be zero for time values before the dead time t_d.

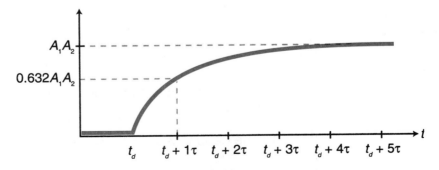

Figure 14-44. First-order step response with dead time and lag.

EXAMPLE 14-4

A process system has a first-order response with a time constant of 10.8 minutes. **(a)** Calculate how long it will take for the value of the output V_{out} to be at 90% of the input V_{in}. **(b)** Calculate the value of V_{out} at 90% of V_{in} given a 5 minute dead time.

SOLUTION

(a) A first-order system with lag has a response of:

$$V_{out} = V_{in}\left(1 - e^{\frac{-t}{\tau}}\right)$$

The value of τ is 10.8 seconds and the required ratio of output over input is 90%, or 0.90; therefore:

$$V_{out} = V_{in}\left(1 - e^{\frac{-t}{\tau}}\right)$$

$$\frac{V_{out}}{V_{in}} = 1 - e^{\frac{-t}{\tau}}$$

$$0.90 = 1 - e^{\frac{-t}{10.8}}$$

$$e^{\frac{-t}{10.8}} = 1 - 0.90$$

$$= 0.10$$

Solving for t by taking the natural logarithm (ln) of both sides of the equation yields:

$$e^{\frac{-t}{10.8}} = 0.10$$

$$\frac{-t}{10.8} = \ln 0.10$$

$$-t = (10.8)(\ln 0.10)$$

$$= (10.8)(-2.303)$$

$$= 24.87\ \text{minutes}$$

So, in 24.87 minutes, the value of the output will be at 90% of the value of the input.

(b) The dead time will simply add to the time required to achieve the 90% value. Therefore, with a lag of 5 minutes, the system will reach a value of 90% final output in 29.87 minutes (24.87 min + 5 min).

SECOND-ORDER LAG RESPONSES

A second-order lag response exhibits oscillations that occur while the output signal is settling into its final steady-state value. This type of response is caused by a step change in the input or a disturbance in the process.

A second-order transfer function with lag is characterized by a second-order differential equation that is represented in Laplace form as:

$$Hp_{(s)} = \frac{\text{Out}}{\text{In}} = \frac{A\omega_n^2}{\left(s^2 + 2\zeta\omega_n s + \omega_n^2\right)}$$

where:

A = the gain

ω_n = the resonant, or natural, frequency of oscillation in radians/second

ζ = the damping coefficient

Figure 14-45 illustrates this second-order, oscillating response to a step input. The frequency term ω_n is the factor that determines how quickly the response oscillates above and below the desired outcome. The damping coefficient ζ is the factor that suppresses the oscillation over time, so that the response finally levels off at the desired outcome value. The complete numerator term $A\omega_n^2$ represents the system gain (K_{sys}), which specifies the total amplitude of the response signal given its frequency.

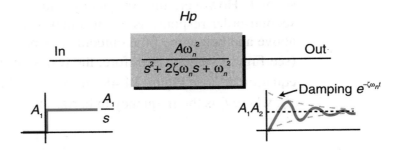

Figure 14-45. Second-order response to a step input.

The amplitude of the oscillation of a second-order response dies off exponentially due to the damping of the factor $e^{-\zeta\omega_n t}$, which is part of the inverse Laplace transform representation (time domain) of the system. If the damping coefficient (ζ) is equal to 0, then the term $e^{-\zeta\omega_n t}$ will be 1 and the response will oscillate indefinitely in a sinusoidal manner at a frequency of ω_n instead of leveling out. Thus, the damping coefficient determines the shape of the response (see Figure 14-46).

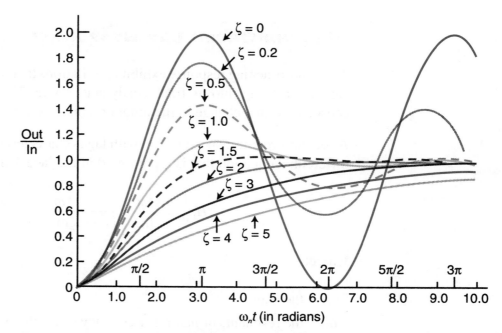

Figure 14-46. Damping coefficient effect on the oscillation of a second-order response.

Unlike a first-order system, a second-order system has two lag times (τ_1 and τ_2), which are related to the frequency of oscillation (ω_n). These two lag times combine to create a system second-order time constant τ_{sys}, which is equal to:

$$\tau_{sys} = \frac{1}{\omega_n}$$

As used in Laplace and time domain second-order response equations, the term ω_n represents frequency. This frequency is expressed in radians per second. However, this frequency can also be expressed in degrees. A second-order response is a sinusoidal response, meaning that it fluctuates above and below the final outcome (set point) value once every 2π periods (see Figure 14-47). Therefore, the response period is characterized by the equation $\frac{2\pi}{\omega_n}$ (see Figure 14-48). In degrees, this same period is expressed as $\frac{1}{f_n}$, where f_n is the frequency in hertz. Therefore:

$$\frac{2\pi}{\omega_n} = \frac{1}{f_n}$$

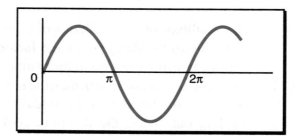

Figure 14-47. Sinusoidal response of a second-order system around the set point.

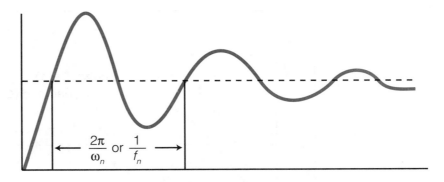

Figure 14-48. Response period of a sinusoidal curve.

Solving for ω_n yields:

$$\omega_n = 2\pi f_n$$

So, the radian/sec frequency term ω_n is equivalent to the degree frequency term $2\pi f_n$.

14-7 TYPES OF SECOND-ORDER RESPONSES

A second-order system can exhibit one of three types of responses:

- overdamped ($\zeta > 1$)

- critically damped ($\zeta = 1$)

- underdamped ($\zeta < 1$)

These responses differ in how they reach the final steady-state value, or set point, over time due to the value of their damping coefficients (see Figure 14-49). An underdamped response oscillates around the set point because its time domain transfer function contains the damping term $e^{-\zeta\omega_n t}$. Critically damped and overdamped responses do not contain this term, so they overshoot the set point and then settle back to it.

OVERDAMPED RESPONSES

An **overdamped response** is a second-order response with lag whose damping coefficient (ζ) is greater than 1. By algebraically manipulating the transfer function of a second-order system with lag (see Section 14-5), the transfer function of an overdamped response can be expressed as:

$$Hp_{(s)} = \left(\frac{A_1}{\tau_1 s + 1} \right)\left(\frac{A_2}{\tau_2 s + 1} \right)$$

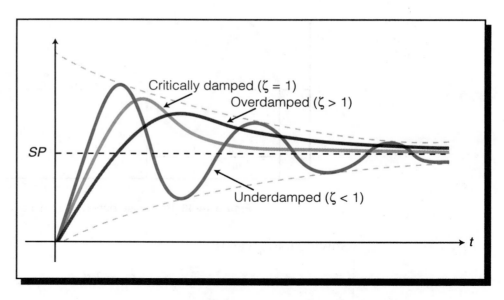

Figure 14-49. Overdamped, critically damped, and underdamped responses.

In this equation, which is a function of two first-order systems (i.e., two time lags), the terms A_1 and A_2 represent the gains. By substituting the term K_{OD} for the total overdamped system gain $A_1 A_2$, this function can be simplified to:

$$Hp_{(s)} = \frac{A_1 A_2}{(\tau_1 s + 1)(\tau_2 s + 1)}$$

$$= \frac{K_{OD}}{(\tau_1 s + 1)(\tau_2 s + 1)}$$

EXAMPLE 14-5

For an overdamped system ($\zeta > 1$), solve for **(a)** K_{sys}, **(b)** ω_n, **(c)** τ_{sys}, and **(d)** ζ using the transfer functions for a second-order response and an overdamped response.

$$Hp_{(s)} = \frac{A\omega_n^2}{\left(s^2 + 2\zeta\omega_n s + \omega_n^2\right)} \quad \text{(Second-order transfer function)}$$

$$Hp_{(s)} = \left(\frac{A_1}{\tau_1 s + 1}\right)\left(\frac{A_2}{\tau_2 s + 1}\right) \quad \text{(Overdamped response)}$$

SOLUTION

Multiplying the terms in the overdamped transfer function yields the equation:

$$Hp_{(s)} = \frac{A_1 A_2}{(\tau_1 s + 1)(\tau_2 s + 1)}$$

$$= \frac{A_1 A_2}{\left[\tau_1 \tau_2 s^2 + (\tau_1 + \tau_2)s + 1\right]}$$

Dividing by the term $\tau_1 \tau_2$ generates the following equation, which has a denominator in the form of a second-order lag transfer function:

$$Hp_{(s)} = \frac{\left(\frac{A_1 A_2}{\tau_1 \tau_2}\right)}{\left(s^2 + \frac{(\tau_1 + \tau_2)}{\tau_1 \tau_2} s + \frac{1}{\tau_1 \tau_2}\right)}$$

Therefore, this equation is equal to the second-order lag equation:

$$Hp_{(s)} = \frac{A \omega_n^2}{\left(s^2 + 2\zeta \omega_n s + \omega_n^2\right)}$$

because the denominator of the polynomial can be separated into two real factors, where τ_1 and τ_2 are the two time constants. Thus, the relationship of the terms in these two equations is:

$$Hp_{(s)} = \frac{\overbrace{\left(\frac{A_1 A_2}{\tau_1 \tau_2}\right)}^{K_{sys}}}{\left(s^2 + \underbrace{\frac{(\tau_1 + \tau_2)}{\tau_1 \tau_2}}_{2\zeta \omega_n} s + \underbrace{\frac{1}{\tau_1 \tau_2}}_{\omega_n^2}\right)}$$

(a) Knowing that the term K_{OD} is equal to the overdamped system gain $A_1 A_2$, the term K_{sys} for an overdamped system is:

$$K_{sys} = A \omega_n^2 = \frac{A_1 A_2}{\tau_1 \tau_2} = \frac{K_{OD}}{\tau_1 \tau_2}$$

(b) For an overdamped system, ω_n^2 is equal to:

$$\omega_n^2 = \frac{1}{\tau_1 \tau_2}$$

Solving for ω_n generates:

$$\omega_n^2 = \frac{1}{\tau_1 \tau_2}$$

$$\omega_n = \sqrt{\frac{1}{\tau_1 \tau_2}}$$

$$= \frac{1}{\sqrt{\tau_1 \tau_2}}$$

(c) Earlier, we explained that τ_{sys} is equal to 1 over the frequency ω_n. Using the information from part (b), τ_{sys} for an overdamped system is:

$$\tau_{sys} = \frac{1}{\omega_n}$$

$$= \frac{1}{\left(\frac{1}{\sqrt{\tau_1 \tau_2}}\right)}$$

$$= \sqrt{\tau_1 \tau_2}$$

(d) The damping coefficient terms for a second-order system with lag and an overdamped system relate as follows:

$$2\zeta\omega_n = \frac{\left(\tau_1 + \tau_2\right)}{\tau_1 \tau_2}$$

Solving for ζ yields:

$$2\zeta\omega_n = \frac{\left(\tau_1 + \tau_2\right)}{\tau_1 \tau_2}$$

$$\zeta = \frac{\left(\tau_1 + \tau_2\right)}{\left(\tau_1 \tau_2\right)\left(2\omega_n\right)}$$

$$= \frac{\left(\tau_1 + \tau_2\right)}{\left(\tau_1 \tau_2\right)\left(2\right)\left(\frac{1}{\sqrt{\tau_1 \tau_2}}\right)}$$

$$= \frac{\left(\tau_1 + \tau_2\right)}{\left(\frac{2\tau_1 \tau_2}{\sqrt{\tau_1 \tau_2}}\right)}$$

$$= \frac{\left(\tau_1 + \tau_2\right)}{\left(2\sqrt{\tau_1 \tau_2}\right)}$$

An overdamped second-order transfer function in real time (the time domain) is described by the inverse Laplace transform (see Table 14-2):

$$H_{(t)} = \frac{K_{OD}}{\tau_1 - \tau_2}\left(e^{\frac{-t}{\tau_1}} - e^{\frac{-t}{\tau_2}}\right)$$

This indicates two exponential decaying responses—one at a rate of τ_1 and the other at the rate of τ_2. Figure 14-50 illustrates the form of these two exponential responses, along with the response of $H_{(t)}$, which is a function of a combination of these two responses. Note that, as indicated in the time domain transfer function term ($e^{\frac{-t}{\tau_1}} - e^{\frac{-t}{\tau_2}}$), the curve of $H_{(t)}$ is equal to the curve of the τ_1 response minus the curve of the τ_2 response.

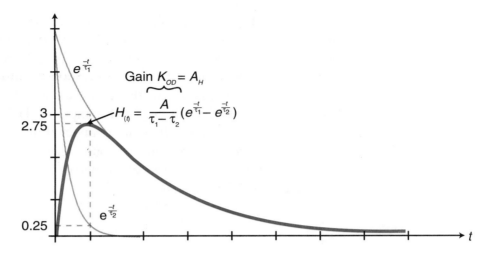

Figure 14-50. Real-time transfer function of an overdamped second-order process.

Figure 14-51 illustrates a second-order system response to a step input with amplitude *B*. As shown in this figure, the output of the time domain transfer function in response to a step input is similar in form to the second-order response curve shown in Figure 14-50. The overdamped response may also follow the shape of a first-order system response curve if one of the time

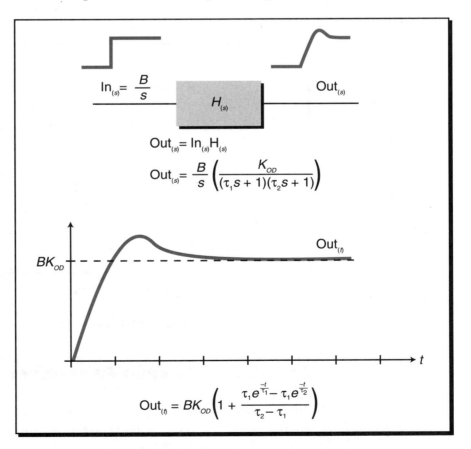

Figure 14-51. Second-order response to a step input.

constants is significantly longer than the other (i.e., $\tau_1 \gg \tau_2$). Figure 14-52 illustrates a case like this, where one of the exponential components (τ_2) dies out much more rapidly than the other (τ_1). Thus, the response to the unit step is heavily damped, causing a sluggish response similar to a first-order one. The system response is heavily damped because the value of ζ in this system becomes large (see the value of ζ in Example 14-4). Two first-order systems with different time lags, which are connected in series (or *cascaded*), will produce this type of overdamped second-order response (see Figure 14-53). In a cascaded system, the output of one part of the system depends on the input to another part of the system.

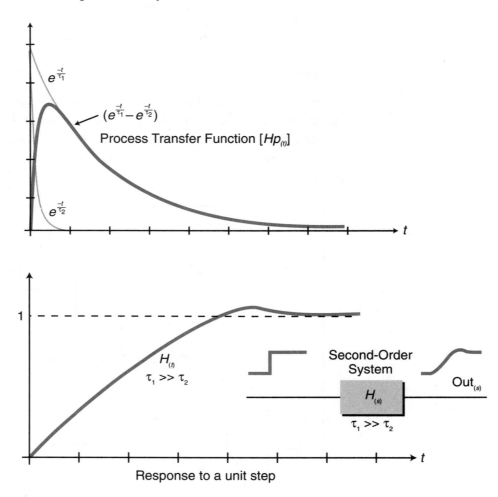

Figure 14-52. A heavily damped response.

CRITICALLY DAMPED RESPONSES

Critically damped responses, which are second-order responses where $\zeta = 1$, are the result of second-order transfer functions where $\tau_1 = \tau_2$. Therefore, the transfer function for this type of response is ($\tau = \tau_1 = \tau_2$):

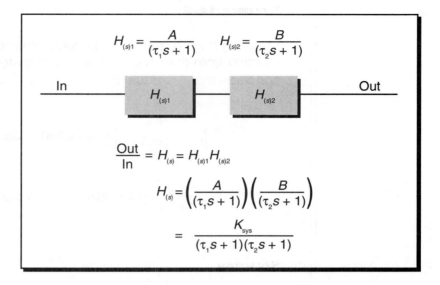

$$H_{(s)1} = \frac{A}{(\tau_1 s + 1)} \qquad H_{(s)2} = \frac{B}{(\tau_2 s + 1)}$$

In ▭ $H_{(s)1}$ ▭ $H_{(s)2}$ Out

$$\frac{Out}{In} = H_{(s)} = H_{(s)1} H_{(s)2}$$

$$H_{(s)} = \left(\frac{A}{(\tau_1 s + 1)}\right)\left(\frac{B}{(\tau_2 s + 1)}\right)$$

$$= \frac{K_{sys}}{(\tau_1 s + 1)(\tau_2 s + 1)}$$

Figure 14-53. Two first-order systems with different lag times cascaded to form an overdamped second-order system.

$$H_{(s)} = \frac{A}{(\tau_1 s + 1)(\tau_2 s + 1)}$$

$$= \frac{A}{(\tau s + 1)^2}$$

This computation comes from substituting $\zeta = 1$ in the second-order lag transfer function, making the denominator a second-order polynomial of the form $(s^2 + 2\omega_n s + \omega_n^2)$:

$$H_{(s)} = \frac{A\omega_n^2}{s^2 + 2\zeta\omega_n s + \omega_n^2}$$

$$= \frac{A\omega_n^2}{s^2 + 2\omega_n s + \omega_n^2}$$

$$= \frac{A\omega_n^2}{\left(s + \omega_n\right)^2}$$

Therefore, substituting K_{sys} for the system gain (nominator) and τ_{sys} for the system lag time produces:

$$H_{(s)} = \frac{K_{sys}}{\left(\tau_{sys} s + 1\right)^2}$$

EXAMPLE 14-6

Show **(a)** how to derive the second-order critically damped transfer function from the second-order lag transfer function and **(b)** how to conclude that ζ is equal to 1.

$$H_{(s)} = \frac{A\omega_n^2}{\left(s^2 + 2\zeta\omega_n s + \omega_n^2\right)} \quad \text{(Second-order transfer function)}$$

$$H_{(s)} = \frac{K_{sys}}{(\tau s + 1)^2} \quad \text{(Critically damped transfer function)}$$

SOLUTION

(a) Given that $\zeta = 1$ for a critically damped response, dividing the numerator and denominator of the second-order function by ω_n^2 yields:

$$H_{(s)} = \frac{A\omega_n^2}{s^2 + 2s\omega_n + \omega_n^2}$$

$$= \frac{\left(\dfrac{A\omega_n^2}{\omega_n^2}\right)}{\left(\dfrac{s^2}{\omega_n^2} + \dfrac{2s\omega_n}{\omega_n^2} + \dfrac{\omega_n^2}{\omega_n^2}\right)}$$

$$= \frac{A}{\left(\dfrac{1}{\omega_n^2}s + \dfrac{2}{\omega_n}s + 1\right)}$$

Replacing the term $\frac{1}{\omega_n}$ with $\frac{1}{\tau_{sys}}$ generates the equation:

$$H_{(s)} = \frac{A}{\dfrac{1}{\left(\dfrac{1}{\tau_{sys}^2}\right)}s + \dfrac{2}{\left(\dfrac{1}{\tau_{sys}}\right)}s + 1}$$

$$= \frac{A}{\tau_{sys}^2 s + 2\tau_{sys}s + 1}$$

Substituting K_{sys} for A, this equation forms the critically damped second-order response:

$$H_{(s)} = \frac{K_{sys}}{\tau_{sys}^2 s + 2\tau_{sys}s + 1}$$

$$= \frac{K_{sys}}{(\tau_{sys}s + 1)^2}$$

(b) From Example 14-4, we know that:

$$\tau_{sys} = \sqrt{\tau_1 \tau_2}$$

$$\zeta = \frac{\tau_1 + \tau_2}{2\sqrt{\tau_1 \tau_2}}$$

Because $\tau_1 = \tau_2$ in a critically damped system, τ_{sys} becomes:

$$\tau_{sys} = \sqrt{\tau_1^2} = \tau_1$$

or

$$\tau_{sys} = \sqrt{\tau_2^2} = \tau_2$$

Since both τ_1 and τ_2 are equal, τ_{sys} equals τ. Using the τ_1 and τ_2 equality in the damping coefficient equation proves that $\zeta = 1$:

$$\zeta = \frac{2\tau}{2\sqrt{\tau^2}}$$

$$= \frac{2\tau}{2\tau}$$

$$= 1$$

A critically damped system where $\zeta = 1$ indicates that the response of the system will be rapid as compared to a more sluggish overdamped response where $\zeta > 1$. Cascading two first-order systems with the same (or nearly the same) lag time ($\tau_1 = \tau_2$) will produce a critically damped second-order system response (see Figure 14-54).

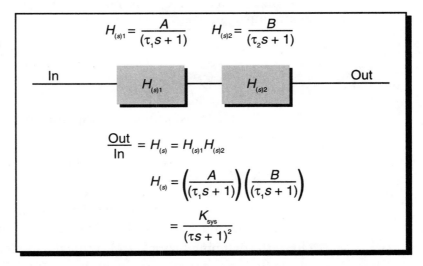

Figure 14-54. Two first-order systems with the same lag time cascaded to form a critically damped second-order system.

The inverse Laplace transform of a second-order critically damped system's response to a unit step, given that the gain K_{sys} equals $A_1 A_2$, is (from Table 14-2):

$$\mathscr{L}^{-1}\left[\frac{K_{sys}}{s(\tau s+1)^2}\right] = K_{sys}\left[1-\left(\frac{\tau-t}{\tau}\right)e^{\frac{-t}{\tau}}\right]$$

A critically damped system achieves a steady-state value quicker than the other two types of second-order systems. However, the amplitude of the overshoot of a critically damped response is larger than that of an overdamped system.

UNDERDAMPED RESPONSES

Second-order **underdamped responses** exhibit an over and undershoot signal (oscillating response) at a natural resonant frequency of ω_n in radians/second. This oscillation is the result of a damping factor (ζ) that is less than 1. This means that instead of being able to factor the denominator of the second-order lag transfer function into a polynomial (i.e., $s^2 + 2\zeta\omega_n s + \omega_n^2$), the denominator becomes a complex-root quadratic equation. The inverse Laplace transform of this equation produces an exponential, decreasing sinusoidal response to a unit step input ($\frac{1}{s}$) represented by (from Table 14-2):

$$\mathscr{L}^{-1}\left[\left(\frac{1}{s}\right)\left(\frac{A\omega_n^2}{s^2 + 2\zeta\omega_n s + \omega_n^2}\right)\right] = A\left[1 + \frac{e^{-\zeta\omega_n t}}{\sqrt{1-\zeta^2}}\sin\left(\omega_n\sqrt{1-\zeta^2}t - \psi\right)\right]$$

For small values of ζ, this response exhibits a behavior to a unit step approximate to:

$$\underset{\text{unit step}}{\text{Out}_{(t)}} = A\left[1 + e^{-\zeta\omega_n t}\sin(\omega_n t)\right]$$

This makes the transfer function response approximate to:

$$H_{(t)} \approx e^{-\zeta\omega_n t}\sin(\omega_n t) \approx \sin(\omega_n t)\big|_{\zeta=0}$$

which is the form of a sine curve. The response of this equation, shown in Figure 14-55, illustrates the damping factor ζ of the sinusoidal response. The closer the damping factor is to 1 (critical damping), the lower the frequency of oscillation and the sooner it will level off (see Figure 14-56a). Remember that if $\zeta = 0$, the response will oscillate forever as a sinusoidal response at a

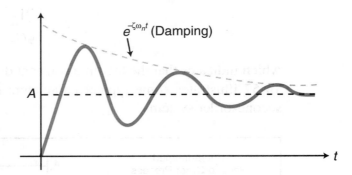

Figure 14-55. Underdamped response.

frequency ω_n; therefore, the closer the value of ζ gets to zero (see Figure 14-56b), the higher the frequency and the longer the oscillations will last (τ becomes longer).

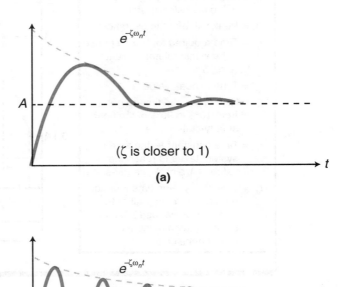

Figure 14-56. Frequency of oscillation is **(a)** lower when ζ is closer to 1 and **(b)** higher when ζ is closer to 0.

The exponential sinusoidal response of an underdamped second-order system will settle to 5% of its steady-state value within 3τ (three time constants), to 2% within 4τ, and to 0.5% within 5τ. The second-order lag response (τ_{sys}) for an underdamped system is defined as:

$$\tau_{sys} = \frac{1}{\zeta \omega_n}$$

which indicates that the lag time constant depends on the value of ζ. Figure 14-57 illustrates some typical parameters used to describe underdamped second-order systems.

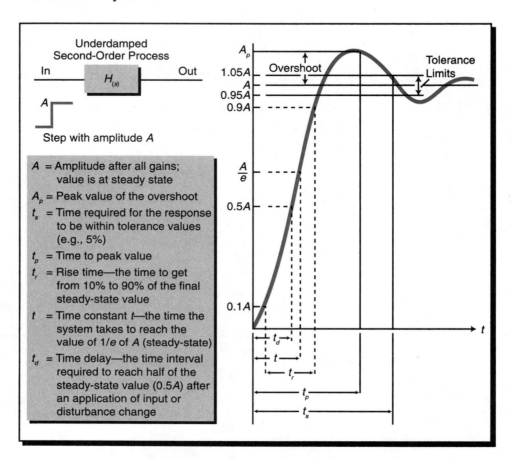

Figure 14-57. Parameters of an underdamped second-order system.

EXAMPLE 14-7

Compare the relationship between the first-order time response term $e^{\frac{-t}{\tau}}$ and the second-order, sinusoidal, exponential decay term $e^{-\zeta \omega_n t}$, which is used in the underdamped transfer function:

$$e^{-\zeta \omega_n t} \sin(\omega_n t)$$

SOLUTION

The time constant τ_{sys} for an underdamped system is equal to:

$$\tau_{sys} = \frac{1}{\zeta\omega_n}$$

or

$$\zeta\omega_n = \frac{1}{\tau_{sys}}$$

The response of a first-order exponential system is $e^{\frac{-t}{\tau}}$, where τ is the system's time constant. Therefore, the decaying term $e^{-\zeta\omega_n t}$ in the equation:

$$e^{-\zeta\omega_n t}\sin(\omega_n t)$$

is equal to:

$$e^{\frac{-t}{\tau_{sys}}}\sin(\omega_n t)$$

This indicates that, in an underdamped second-order system, the value of τ_{sys} (lag time) becomes larger as the value of ζ becomes smaller ($\tau_{sys} = \frac{1}{\zeta\omega_n}$) This is similar to the behavior of a first-order system with a long lag, because the time to reach the steady-state value will be long. In a second-order underdamped system, the oscillation continues for a longer time since the term $e^{\frac{-t}{\tau_{sys}}}$ provides less damping. If $\zeta = 1$, the value of τ_{sys} becomes $\frac{1}{\omega_n}$, which is the lag time of a critically damped system.

14-8 SUMMARY

The objective of a process control system is to maintain the process variable (process output) at a desired target value, referred to as the set point. The system provides this control by implementing a feedback loop, meaning that it reads the process variable and compares it to the set point value. The controller then uses the difference between these values, as computed by $E = SP - PV$, to determine how much corrective action it must take. The error, which the controller calculates as a percentage of the full range of the process variable, can be caused by changes in the set point or by disturbances to the process.

Open-loop systems are systems in which the process variable is not fed back into the control system for reference. Closed-loop systems, on the other hand, do receive process variable feedback. Most process control systems are closed-loop systems that receive negative feedback. In a negative feedback system, the controller determines the error by subtracting the process variable from the set point.

Process dynamics refers to changes in the process that occur due to disturbances or changes in the set point. Process gain changes are a result of gains in the process variable value created by changes in the control variable output. The dynamics of a process also includes dead time and lag time. Dead time is the delay that occurs between the moment a change is made in the control variable and the moment the process variable begins to react to the control variable change. Lag time is the delay associated with the time required by the process control loop to bring the process variable to the set point by adjusting the final control element. The lag time is a finite time required by the control system to physically adjust the final control element (e.g., a steam valve).

A transient is the process variable response to a change in set point or to the creation of a disturbance (e.g., a load change). The transient response depends not only on the dynamics of the process, but also on the characteristics of the process itself. These characteristics are the result of the transfer functions of the controller and the process. A transfer function is the mathematical representation of a system's response, where the response is computed by dividing the output by the input. Transfer functions are expressed in the frequency domain using Laplace transforms, to allow easy algebraic manipulations of the equations. The inverse Laplace transform of a transfer function converts a frequency-based Laplace response into a time-based response.

Each element in a control system loop has a transfer function associated with it—the controller has one and the process has one. The combined controller/process system also has a transfer function. Transfer functions are categorized as either first-order or second-order responses. First-order systems have one lag time associated with the process, while second-order systems have two lag times. Laplace transforms are used to mathematically represent both first- and second-order process transfer functions, as well as controller transfer functions and the combination of both process and controller functions in a closed-loop configuration. Although it is difficult to obtain the actual transfer function of a process, a knowledge of the type of transfer function expected from a process response is extremely useful, especially when tuning the controller.

First-order systems have one lag time, resulting in an exponential, decaying response. When the system receives a step input change, its open-loop output will have the following time domain response, which smoothly follows the input:

$$V_{out} = V_{in}\left(1 - e^{\frac{-t}{\tau}}\right)$$

The time constant τ specifies the time the output takes to achieve 63.2% of the final steady-state value. The time constant τ is sometimes referred to as the 63% response time. After 5τ periods have elapsed, the value of the output response will be at 99.33% of its final value. In Laplace form, the transfer function of a first-order system has the form:

$$H_{(s)} = \frac{\text{Out}}{\text{In}} = \frac{A_1}{\tau s + 1}$$

Second-order systems have two lag times and are described by the transfer equation:

$$H_{(s)} = \frac{\text{Out}}{\text{In}} = \frac{A\omega_n^2}{\left(s^2 + 2\zeta\omega_n s + \omega_n^2\right)}$$

Second-order systems can have three types of responses, depending on their damping coefficient: overdamped ($\zeta > 1$), critically damped ($\zeta = 1$), and underdamped ($\zeta < 1$). Each of these types of responses have different inverse Laplace transforms, translating into different time domain responses. Overdamped responses ($\zeta > 1$) have two different time constants (τ_1 and τ_2), and their response over time in reaching the set point is sluggish. Critically damped systems ($\zeta = 1$) have two lag times, or time constants, that are equal ($\tau_1 = \tau_2$). These systems reach the set point much faster than overdamped systems. Underdamped responses ($\zeta < 1$) produce a faster response than either overdamped or critically damped responses, resulting in an overshoot and undershoot of the final value that dies off exponentially as the steady-state value is approached. An underdamped response has two time constants that are imaginary, or mathematically speaking, that have complex roots produced by their quadratic equation.

Although some processes have more complicated responses (third- and fourth-order responses), these processes' transfer functions can be approximated by second-order system transfer functions. Most manufacturing processes can be classified as either first-order or second-order systems. In the next chapter, we will discuss how to use PID control to adjust the inputs to these complex systems to obtain a desired output.

KEY TERMS

control element
control loop
control variable
critically damped response
dead time
error
error deadband
first-order response
lag time
Laplace transforms

overdamped response
process control
process gain
process variable
second-order response
set point
steady state
step response
step test
transfer function
transient response
underdamped response

CHAPTER FIFTEEN

PROCESS CONTROLLERS AND LOOP TUNING

Confusion worse confounded.

—John Milton

CHAPTER HIGHLIGHTS In the previous chapter, we explained some important topics, such as process variable responses and transfer functions, that are elemental to the understanding of process control. In this chapter, we will continue our discussion of process control by explaining how a controller regulates a process. We will discuss the different types of controllers available, their advantages and disadvantages, and the effects that they have on the processes being controlled. We will also examine several tuning methods that are used to stabilize the process and determine the controller's tuning constants. After you finish this chapter, you will be ready to integrate PID control into a PLC application.

15-1 INTRODUCTION

The behavior of a closed-loop control system depends not only on the characteristics of the process and its transfer function, but also on the type of controller and the design decisions that occur during the selection of tuning parameters. As we explained previously, in a process control system, the process receives control information from the controller in the form of the control variable, which acts on the control element or process actuator (e.g., valve). The normal value of the control variable is usually at 50% of its range, so that it can either increase or decrease to accommodate for changes in the process variable.

The effect that a controller has on the process is the result of its *action*, or operational mode. Like the process itself, a controller also has a transfer function, which can be represented mathematically by Laplace transforms. The interaction between the controller and the process comprises the true essence of closed-loop process control.

In process control, the controller is responsible for the stability of the control system. Figure 15-1 illustrates three types of stability responses:

- stable

- conditionally stable

- unstable

Stable responses have an asymptotic characteristic, meaning that, as time increases, the response of the system approaches some finite value (see Figure 15-1a). Conversely, *conditionally stable responses* have a sinusoidal-type wave shape (see Figure 15-1b). This sinusoidal response has a low amplitude and may be acceptable in noncritical control loops, but not in the control of critical processes. *Unstable responses*, as the name implies, are system responses that are not acceptable because they create an unstable, or "runaway," condition (see Figure 15-1c). The sinusoidal amplitude of an unstable response increases as time increases.

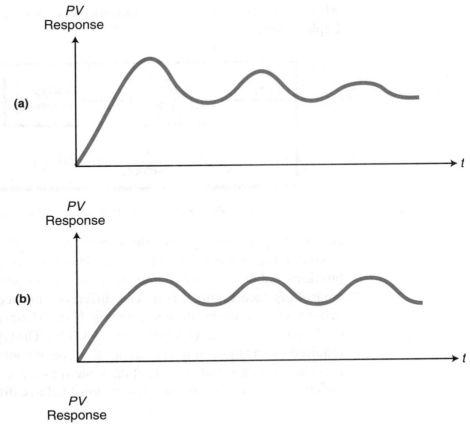

Figure 15-1. (a) Stable, **(b)** conditionally stable, and **(c)** unstable responses.

15-2 CONTROLLER ACTIONS

In the previous chapter, we explained how a process's transfer function indicates the process's behavioral response to an input change. Now, we will explain the controller's transfer function, *Hc*, along with the types of controller modes used to control the process. Figure 15-2 illustrates an open-loop configuration of the controller and process transfer functions by themselves, in which the controller's output (the control variable *CV*) is the input to the process. The transfer function for this open-loop configuration is:

$$\frac{\text{Output}}{\text{Input}} = \frac{PV_{(s)}}{\text{Input}_{(s)}} = \left(Hc_{(s)}\right)\left(Hp_{(s)}\right)$$

where $Hc_{(s)}$ and $Hp_{(s)}$ are the controller and process transfer functions in Laplace form.

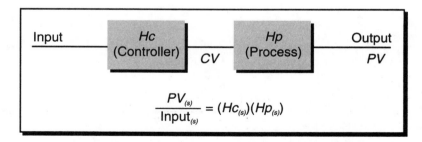

$$\frac{PV_{(s)}}{Input_{(s)}} = (Hc_{(s)})(Hp_{(s)})$$

Figure 15-2. Open-loop system configuration.

As in an open-loop system, the controller in a closed-loop system also regulates the process variable value. However, a closed-loop controller provides either a *direct action* or a *reverse action* to the process it is controlling (see Figure 15-3). The difference between these two types of actions is the effect that the control variable (CV) from the controller (Hc) has on the process variable (PV) of the process (Hp). The type of process behavior required by an application determines the type of controller action needed in the system. For instance, a heating system and a cooling system behave differently, so their controller actions must behave differently as well.

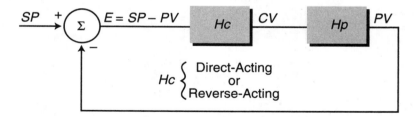

Figure 15-3. Closed-loop system configuration.

Note that, in a closed-loop control system, the key variable input to the controller is the error signal. After interpreting the error signal, the controller sends commands to the process via the control variable to bring the error to zero. In this chapter, we will refer to the error signal as the input to the controller and to the control variable as the controller output.

DIRECT-ACTING CONTROLLERS

A **direct-acting controller** is a closed-loop controller whose control variable output increases in response to an increase in the process variable. This is the type of action exhibited by a typical air-cooling system. As the temperature (PV) increases (i.e., it becomes warmer), the controller increases the value of its output (i.e., it increases the output of the air-conditioning compressor) to bring the process variable back to the set point.

Figure 15-4 illustrates another example of a direct-acting controller in which two materials are mixed in an exothermic (heat-producing) batch. Cold water flowing through the tank jacket cools the batch. A temperature sensor measures the temperature process variable, which has a cool set point.

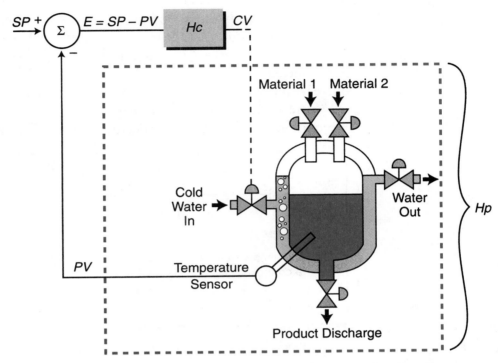

Figure 15-4. Direct-acting controller controlling the temperature in a batch-cooling process.

Figure 15-5 shows the reaction of the process in Figure 15-4. If the control variable that controls the cold water valve is open 100% (full open), the temperature of the batch will be 100°F; if the cold water valve is at 0% (closed), the temperature of the batch will be at 200°F. The desired set point of this process is 150°F, which corresponds to a 50% controller output. Therefore, in this process, if the cold water control valve opening increases, the system temperature decreases and vice versa.

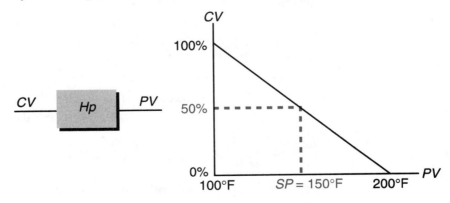

If *CV* increases, *PV* decreases in the process

Figure 15-5. Process variable's reaction to a direct-acting controller.

Figure 15-6 shows the reaction of the controller to the process variable. If the controller senses that the temperature is too hot, it opens the cold water valve more to cool off the batch. Conversely, if the temperature is too cold, the controller decreases the opening of the control valve to warm up the temperature. Therefore, the controller in this system is a direct-acting controller because, as the process variable (temperature) increases, the controller increases its control variable output (opens the valve for more cold water) to bring *PV* closer to the set point, thus bringing the error to zero. In terms of error, the equation $E = SP - PV$ indicates that a direct-acting controller will increase its output as the error value in the system becomes more negative (as *PV* increases, *E* becomes more negative) and will decrease it as the error becomes more positive (as *PV* decreases, *E* becomes more positive).

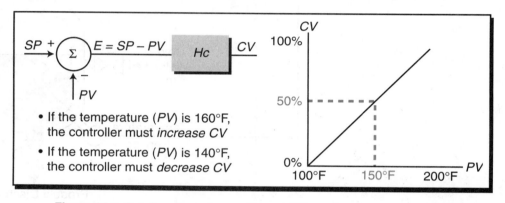

Figure 15-6. Relationship between *CV* and *PV* in a direct-acting controller.

In process control applications, a direct-acting controller is sometimes said to respond to a positive increase in error with an increase in the control variable (increase in controller output). The term "positive error," however, can be deceiving because it refers to the error change in the process variable, not to the change in the actual system error value. For example, referring to Figure 15-6, if the temperature (*PV*) increases from the set point of 150°F to 160°F, the direct-acting controller will increase the control variable because the process variable has increased by +10°F. This change in the process variable is a positive error because the actual *PV* value has changed in a positive direction. The system error (*E*), on the other hand, will become more negative due to this same change. When *PV* equaled the set point, the system error was 0 (150°F – 150°F). When the process variable increased to 160°F, however, the system error became –10°F (150°F – 160°F). Regardless of the terminology used, a direct-acting controller senses the direction of change in both the process variable and error and responds appropriately.

REVERSE-ACTING CONTROLLERS

A **reverse-acting controller** behaves oppositely of a direct-acting controller—if the controller detects an increase in the process variable, it will respond by decreasing the control variable. This behavior is typical of a heating system. As the temperature becomes warmer (*PV* increases), the controller decreases the amount of heat the furnace produces to maintain the temperature at the set point.

Figure 15-7 illustrates a heating process in which a steam control valve allows heat to enter into the tank jacket of the batch system. The graph illustrated in Figure 15-8 shows that, in this heating system, if the steam control valve (*CV*) is at 100%, the temperature (*PV*) of the batch will be 200°F. Conversely, if the steam valve is at 0%, or completely closed, the batch temperature will drop to 100°F. To maintain the set point at 150°F, the controller must maintain the control variable output at 50% of its range. Figure 15-9 shows the relationship between the control variable and the process variable for the controller in this system. Because it is a reverse-acting controller, if the process variable (temperature) increases, the controller decreases its output to bring the error closer to zero.

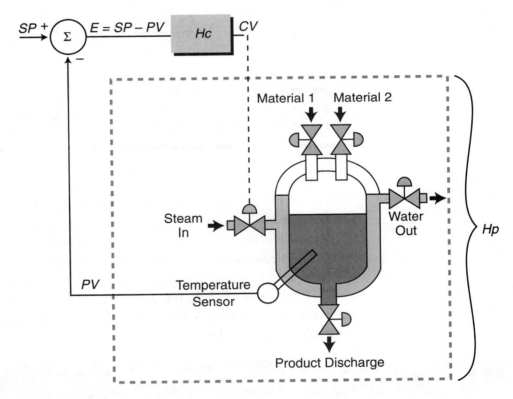

Figure 15-7. Reverse-acting controller controlling the temperature in a batch-heating process.

The selection of a direct- or reverse-acting controller depends on the behavior of the process itself. If the process reacts in a direct manner (*PV* increases as *CV* increases), the system's controller must provide reverse action, as in the case of a heating system, to control the process. If the process reacts in an inverse manner (*PV* increases as *CV* decreases), the controller must use direct action to control the process. Some single-loop controllers have a toggle

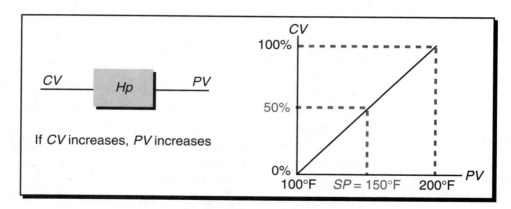

Figure 15-8. Process variable's reaction in a reverse-acting controller.

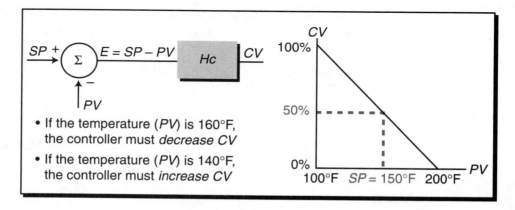

Figure 15-9. Relationship between *CV* and *PV* in a reverse-acting controller.

switch that can be used to select the desired action of the controller (direct or reverse). The control switch on a home thermostat is an example of this type of switch. During the winter, when the switch is set to heat, the system operates in a reverse-acting mode. During the summer, when the switch is set to cool, the system operates in a direct-acting mode. The closed-loop system remains the same, except for the behavior of the controller. The process behavior, which changes from winter to summer, necessitates the switch from reverse to direct action.

15-3 DISCRETE-MODE CONTROLLERS

A controller can have one of two modes that describes its output signal:

- discrete (ON/OFF) mode

- continuous (analog) mode

In discrete mode, the controller produces a discontinuous ON/OFF signal as its output (the control variable), which serves as the input to the process (see Figure 15-10a). In continuous mode, the controller emits an analog output signal (see Figure 15-10b). In this section, we will discuss discrete-mode controllers. In the next section, we will explore continuous-mode controllers.

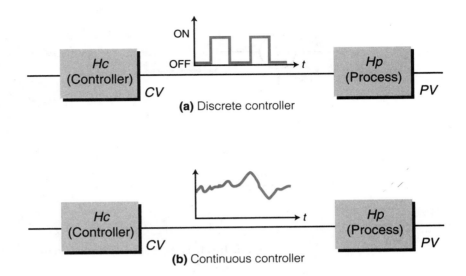

Figure 15-10. (a) Discrete- and **(b)** continuous-mode controllers.

Due to the nature of their signal, **discrete-mode controllers** produce a conditionally stable response. This means that the system error fluctuates between a predetermined deadband, creating a low-amplitude sinusoidal response. These controllers are used in systems where this type of response is acceptable. A noncritical heating system that uses an ON/OFF signal to control the heater is an example of a discrete-mode controller. A home air-conditioning and heating system is another example of a discrete-mode system, because the process variable cycles between two values on either side of the set point when the air conditioner or heater is turned ON or OFF. The two most common types of discrete-mode controllers are:

- two-position controllers
- three-position controllers

TWO-POSITION DISCRETE CONTROLLERS

A **two-position controller**, also called an *ON/OFF controller*, is the most basic type of process controller. As the name implies, it provides an ON/OFF signal to the process's control element (see Figure 15-11). A typical example of an ON/OFF controller is a home heating system. The heater turns ON when the temperature is below the set point and turns OFF when the temperature reaches an acceptable level. Ideally, if the set point temperature is 70°F, the heater will turn ON when the temperature is less than 70°F and turn OFF when it is greater than 70°F, as the heater tries to keep the error ($SP - PV$) at

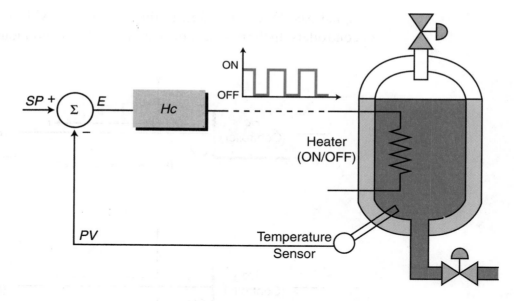

Figure 15-11. Two-position discrete controller controlling a heater.

zero. However, most heating systems have an error deadband, meaning that the heater will turn OFF at a value just above the target temperature and turn ON at a value just below it. So, if the heater in our example has a deadband of 68°F to 72°F, the heater will turn OFF when the temperature reaches 72°F and turn ON when it falls to 68°F. This deadband range avoids the constant ON/OFF action associated with trying to keep the process variable at one exact set point.

The example heating system has a reverse-acting controller, because if the controller senses that the process variable (temperature) decreases, it will increase its output to 100% (ON), as shown in Figure 15-12. As a result of the error deadband, the heater will turn ON when the temperature drops to 68°F and turn OFF when it reaches 72°F. Furthermore, note that the controller is sometimes ON and sometimes OFF within the error deadband. This depends

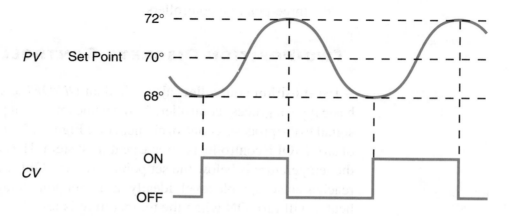

Figure 15-12. Behavior of the process and control variables in the heating system.

on the direction of the process variable. If the value of the process variable is decreasing within the deadband, then the controller is OFF, since it senses that the temperature is still at an acceptable level (reverse-acting). When the process variable reaches the lower limit of the deadband, then the controller will turn ON, causing the direction of the process variable to change (i.e., to increase). The controller will remain ON until *PV* reaches the upper limit of the deadband. At that time, the controller will turn OFF, *PV* will begin to decrease, and the cycle will repeat.

The action of an ON/OFF controller can be described by:

$$CV = 100\% \text{ (ON)} \quad \text{IF error} > -\Delta E$$

$$CV = 0\% \text{ (OFF)} \quad \text{IF error} < +\Delta E$$

where $\pm\Delta E$ represents the error deadband. Graphically, this controller output can be represented as shown in Figure 15-13, where the deadband of $\pm\Delta E$ is equal to $\pm 2°F$. If the process variable "comes" from the positive side (i.e., the error declines to a value less than $-\Delta E$, or 68°F), the controller output will turn ON at point 1 and remain ON (point 2) until the error reaches $+\Delta E$ (point 3).

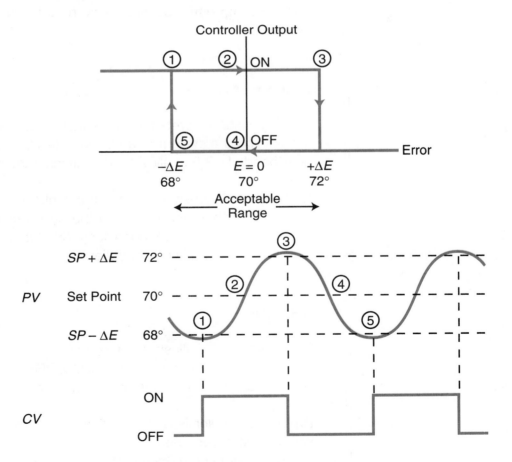

Figure 15-13. Controller output of the heating process.

At that time, the controller will turn OFF and remain OFF (through point 4) until the error drops to $-\Delta E$ (point 5), causing the cycle to repeat. This deadband curve is said to have *hysteresis*, meaning that the reaction of the system depends on its previous actions. It also produces an oscillating response, which is acceptable in this case. Also note that the curve of the ON/OFF controller signal will tend to overshoot the $SP + \Delta E$ value and undershoot the $SP - \Delta E$ value of the heater system due to finite warm up and cool off times (lag times).

ON/OFF controllers are appropriate for applications where large-scale, sudden changes are uncommon and the process reaction rate is slow. If the error deadband of the controller is reduced, then the amount of error in the system will decrease; however, the frequency of the ON/OFF and process variable cycles will increase. Conversely, if the deadband is increased, the oscillation frequency will decrease, but the error will be maximized. Thus, a trade-off exists between the desired error deadband and the frequency of the ON/OFF activation of the control element. The control element (e.g., valve, compressor, etc.) and other system components may be seriously damaged if they are turned ON and OFF too rapidly. Therefore, the system must be configured to compromise between the error allowance and the frequency of oscillation.

EXAMPLE 15-1

A two-position discrete-mode controller controls a cooling system, maintaining the system at a set point of 70°F. The controller has a deadband of ±3°F to allow for deviations from the set point.

(a) Plot the relationship between the controller's ON/OFF output, the process variable response, and the error curve, disregarding any overshoot or undershoot conditions. **(b)** Determine whether this is a direct- or reverse-acting controller.

SOLUTION

(a) Figure 15-14a illustrates the response of the process variable (temperature) to the controller's ON/OFF output. Figure 15-14b shows the hysteresis curve of the controller output versus the error.

(b) This controller is a direct-acting one, because as the process variable increases (passes $+\Delta E$ of SP), the controller will increase the control variable from 0% (OFF) to 100% (ON).

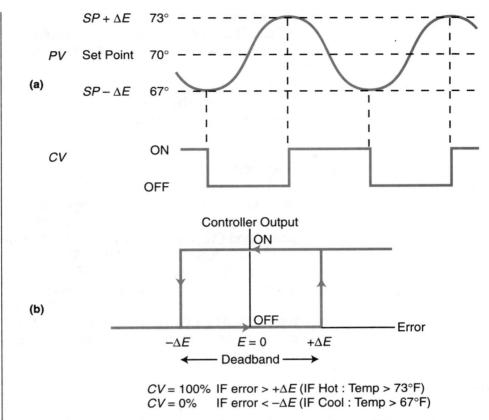

$$CV = 100\% \quad \text{IF error} > +\Delta E \text{ (IF Hot : Temp} > 73°F)$$
$$CV = 0\% \quad \text{IF error} < -\Delta E \text{ (IF Cool : Temp} > 67°F)$$

Figure 15-14. Example 15-1 **(a)** process variable response and **(b)** hysteresis curve.

EXAMPLE 15-2

Figure 15-15 shows a mixer tank that is heated by an ON/OFF heating control system. The set point temperature is 200°F with a deadband deviation of ±5% from the set point. When the heater is not on, the

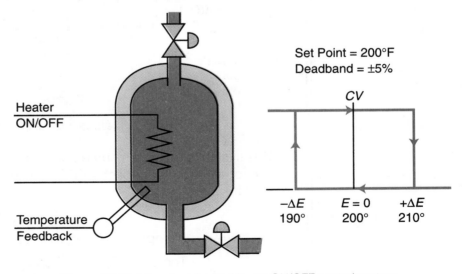

Figure 15-15. Mixer tank heated by an ON/OFF control system.

system linearly loses (cools) 4°F per minute; when the heater is applied, the system gains 8°F per minute. The system starting point is at the set point temperature with the heater in the OFF mode.

(a) Plot the oscillation response (cycle period) of the system and controller, and **(b)** calculate the response in part (a) taking into consideration a heater lag time of 30 seconds (0.5 min).

SOLUTION

(a) Figure 15-16 illustrates the response of the process variable over time, along with the controller's output status. The upper value of the deadband (ΔE = +5%) is 210°F, while the lower value (ΔE = –5%) is 190°F. This curve starts at 200°F (*SP*) and declines at a rate of 4°F/min until the temperature equals 190°F (*SP* – *ΔE*). At 190°F, the controller turns ON and starts heating the system at a rate of 8°F/min until the temperature reaches 210°F (*SP* + *ΔE*), at which point, the controller turns off the heater. The process variable starts to cool off again at the rate of 4°F/min until the temperature reaches *SP* – *ΔE*, where the cycle is repeated.

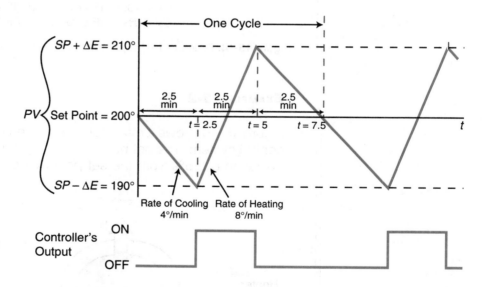

Figure 15-16. Process variable response for Example 15-2.

This system's response curve can be represented by two equations, one for when the controller is OFF and another for when the controller is ON:

$$\text{Temp1}_{(t)} = -4(t - t_1) + \text{Temp}_{(t_1)} \quad \text{(OFF mode)}$$
$$\text{Temp2}_{(t)} = 8(t - t_2) + \text{Temp}_{(t_2)} \quad \text{(ON mode)}$$

where:

$$\text{Temp1}_{(t)} = \text{the value of the } PV \text{ curve (temperature) when the controller is OFF}$$

$$\text{Temp2}_{(t)} = \text{the value of the } PV \text{ curve when the controller is ON}$$

$$\text{Temp}_{(t_1)} = \text{the value of } PV \text{ at time } t_1$$

$$\text{Temp}_{(t_2)} = \text{the value of } PV \text{ at time } t_2$$

$$t = \text{time}$$

Note that the first curve has a slope of $-4°F/min$ and that the second curve has a slope of $+8°F/min$. So, the time required to reach $SP - \Delta E$ (190°) at $t_1 = 0$ and $\text{Temp}_{(t_1)} = 200°$ is:

$$\text{Temp1}_{(t)} = -4(t - t_1) + \text{Temp}_{(t_1)}$$
$$190 = -4(t - 0) + 200$$
$$190 = -4t + 200$$
$$4t = 10$$
$$t = \frac{10}{4} = 2.5 \text{ min}$$

At 2.5 minutes, the controller will turn ON. So, knowing that $\text{Temp2}_{(t)}$ is equal to 190°F at $t = 2.5$, we need to find the time at t_2. $\text{Temp}_{(t_2)}$ is equal to 210°F, because that is the time when the controller will turn OFF again; therefore:

$$\text{Temp2}_{(t)} = 8(t - t_2) + \text{Temp}_{(t_2)}$$
$$190 = 8(2.5 - t_2) + 210$$
$$190 = 20 - 8t_2 + 210$$
$$8t_2 = 40$$
$$t_2 = \frac{40}{8} = 5 \text{ min}$$

So, the temperature value will be 210°F when $t = 5$ minutes. Thus, the time from the moment the controller turns ON the heater at $SP - \Delta E$ (190°F) to the moment the controller turns OFF the heater at $SP + \Delta E$ (210°F) is 2.5 minutes. This is the time at 210°F minus the time it took to get to 190°F (5 min – 2.5 min = 2.5 min).

To complete the calculation of the oscillation cycle period, we must find the amount of time required for the temperature to cool off to the set point value again. This is equal to half of the time it takes for the temperature to go from 210° to 190°. This value is the same as the time calculated for the curve Temp1$_{(t)}$, which is 2.5 minutes. Therefore, the frequency of oscillation will be 7.5 minutes.

Another way to calculate the time for each curve is to determine the difference between the temperatures and divide this difference by the rate required to get from one temperature to another. The time required for the first half of the curve, the OFF mode, to decline from the set point (200°F) to the lower limit of the deadband (190°F) is:

$$200°F \rightarrow 190°F @ \text{ rate} - 4°F/min$$
$$-10°F \text{ change @ rate} - 4°F/min$$
$$t = \frac{-10°F}{-4°F/min} = 2.5 \text{ min}$$

The same calculation for the OFF-to-ON state of the controller is:

$$190°F \rightarrow 210°F @ \text{ rate } 8°F/min$$
$$20°F \text{ change @ rate } 8°F/min$$
$$t = \frac{20°F}{8°F/min} = 2.5 \text{ min}$$

Finally, the time required for the next part of the curve is:

$$210°F \rightarrow 190°F @ \text{ rate} - 4°F/min$$
$$-20°F \text{ change @ rate} - 4°F/min$$
$$t = \frac{-20°F}{-4°F/min} = 5 \text{ min}$$

However, to compute the oscillation period, the system only requires half of this last time calculation, 2.5 minutes, to complete the cycle (i.e., return to the set point). Thus, the total time for the oscillation response is 7.5 min (2.5 min + 2.5 min + 2.5 min).

(b) A lag of 30 seconds, or 0.5 minutes, will cause the ON/OFF response to undershoot and overshoot the deadband, slightly affecting the frequency of oscillation (see Figure 15-17). This 0.5-minute lag may be due to the cooling off and heating up times associated with the heating element. The oscillation frequency for the first part of the curve can be calculated as follows:

$$200°F \rightarrow 190°F @ \text{ rate} - 4°F/min$$
$$t = \frac{-10°F}{-4°F/min} = 2.5 \text{ min}$$

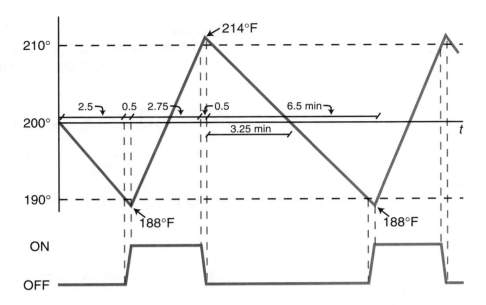

Figure 15-17. Undershoot and overshoot of the error deadband due to lag.

However, once the temperature reaches 190°F and the heater turns ON, another 0.5 minutes will elapse while the heating element heats up. Meanwhile, the temperature will continue to drop. So, during that 0.5-minute lapse, the temperature will drop another 2°F; making the final low-limit temperature 188°F:

$$t = \frac{\Delta \text{ temperature}}{\text{rate}}$$

$$0.5\,\text{min} = \frac{\Delta \text{ temperature}}{-4°F/\text{min}}$$

$$\Delta \text{ temperature} = (0.5)(-4) = -2°F$$

$$190°F - 2°F = 188°F$$

This lag will cause an undershoot of the deadband. Once the controller is ON, it will heat the tank at a rate of 8°F/min, reaching the 210°F upper temperature level in 2.75 minutes:

$$188°F \rightarrow 210°F \text{ @ rate } 8°F/\text{min}$$

$$t = \frac{22°F}{8°F/\text{min}} = 2.75\,\text{min}$$

The 0.5-minute lag will cause an overshoot of 4°F:

$$0.5\,\text{min} = \frac{\Delta \text{ temperature}}{8°F/\text{min}}$$

$$\Delta \text{ temperature} = (0.5)(8)$$

$$= 4°F$$

This will make the upper temperature 214°F (210°F + 4°F). The final period of oscillation, which is the last half of the curve, is the cooling off period between the upper temperature limit (214°F) and the lower limit (188°F):

$$214°F \rightarrow 188°F \text{ @ rate } 4°F/min$$

$$t = \frac{26°F}{4°F/min} = 6.5 \text{ min}$$

The half point of this curve, where *PV* equals the set point, will occur at 3.25 minutes. Thus, the total period of this system with lag, as shown in Figure 15-17, is:

$$Period = (2.5 \text{ min} + 0.5 \text{ min}) + (2.75 \text{ min} + 0.5 \text{ min}) + 3.25 \text{ min}$$

$$= 9.5 \text{ min}$$

The addition of a 0.5-minute lag to this system will increase the frequency from 7.5 minutes to 9.5 minutes.

THREE-POSITION DISCRETE CONTROLLERS

A **three-position controller** provides three output levels, instead of just two output levels like a two-position controller. Basically, this controller has an additional ON setting at 50% of the full ON range. The use of a three-position controller tends to reduce the cycling behavior of the process variable because it provides an intermediate output level, rather than just the two level settings of an ON/OFF controller. The controller stops at the intermediate 50% setting when the set point is achieved. In fact, a controller's output is usually designed so that its half output, or 50%, coincides with the level required by the process to maintain the process variable at the set point, minimizing the error in the system. Figure 15-18 illustrates a three-position, direct-acting controller's output (*CV*) according to the error present in the system. This output can be represented mathematically as:

$$CV = 100\% \quad \text{IF error} > +\Delta E$$
$$CV = 50\% \quad \text{IF} - \Delta E < \text{error} < +\Delta E$$
$$CV = 0\% \quad \text{IF error} < -\Delta E$$

Because three-position field devices are not as widely available as two-position ON/OFF devices, analog field devices are often used to implement three-position control. A PLC may also implement a three-position output using a contact output interface (see Figure 15-19). The incoming sides of

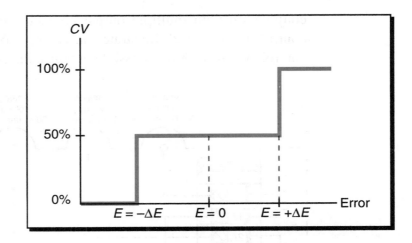

Figure 15-18. Three-position, direct-acting controller's output in response to error.

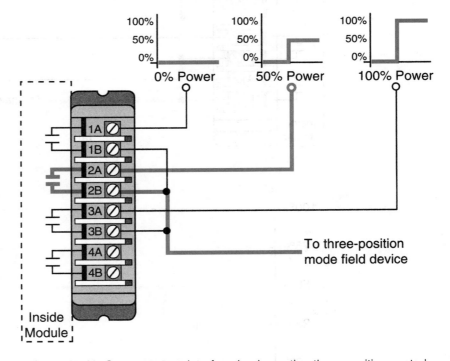

Figure 15-19. Contact output interface implementing three-position control.

three of the contact module's terminals are connected to 0%, 50%, and 100% power signals, respectively. The other sides of the contacts join together and connect to the field device (e.g., valve). In this configuration, a PLC with a contact output module can be interfaced with analog signals set at 0%, 50%, and 100% of the full range of the field control device to implement three-position control. Discrete-mode controllers with more than three positions (e.g., five positions, multipositions, etc.) may also be implemented this way in some control applications. Since output field devices with more than three positions are not readily available, analog output devices are often used in conjunction with multiple output contact cards to obtain multiposition control. Figure 15-20 illustrates an example of this type of direct-acting control system with five possible output settings.

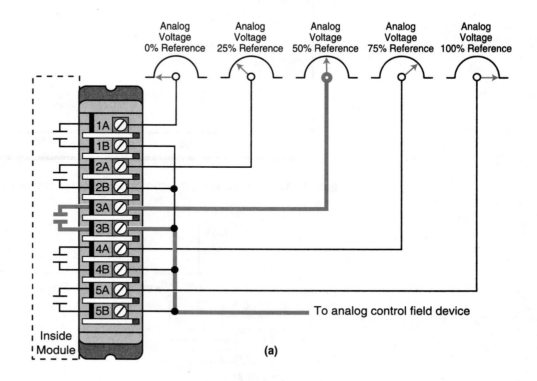

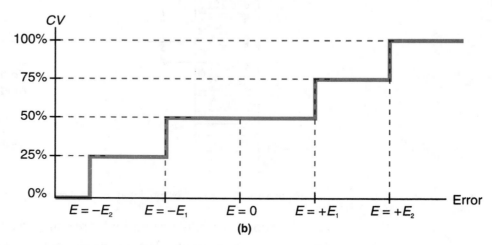

Figure 15-20. **(a)** A discrete mode, multiposition controller and **(b)** its output.

EXAMPLE 15-3

Graphically illustrate the reaction of a three-position controller output to the steam heating system shown in Figure 15-21. Include the effect of the controller's lag.

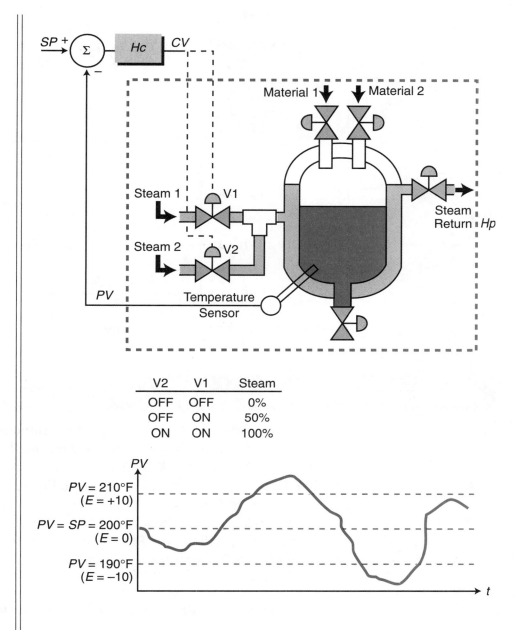

V2	V1	Steam
OFF	OFF	0%
OFF	ON	50%
ON	ON	100%

Figure 15-21. Steam heating system controlled by a three-position controller.

SOLUTION

Figure 15-22 shows the controller output (*CV*) for this three-position discrete-mode controller. Note that the response plot (see Figure 15-22a) shows that an overshoot is present, indicating a lag in steam actuation. The same lag also creates an undershoot condition when both V1 and V2 are ON. As shown, the system overshoot and undershoot pass the deadband during the lag periods.

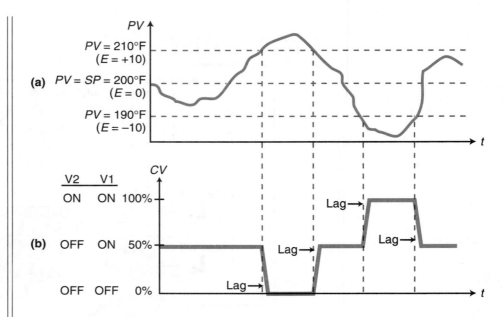

Figure 15-22. (a) Response plot and **(b)** controller output of the heating system in Figure 15-21.

15-4 CONTINUOUS-MODE CONTROLLERS

Most process control applications employ **continuous-mode controllers**, instead of discrete-mode controllers, to avoid the oscillatory system response caused by ON/OFF control. A continuous-mode controller sends an analog signal to the process control field device (see Figure 15-23) to regulate the process variable, bringing the error signal to zero in a closed-loop system. A continuous-mode controller behaves like a multiposition controller with an infinite number of positions. In a PLC-based system, the controller may be an intelligent I/O interface or software routine instructions that use standard I/O analog modules.

Continuous-mode controllers use three different modes to control the process:

- proportional control mode

- integral control mode

- derivative control mode

These three modes are also referred to as controller *actions*, each one reacting differently to the error present in the system in a direct- or reverse-acting fashion. The *proportional mode* provides a control variable adjustment that is proportional to the error deviation. The *integral mode* (or *reset mode*) provides a change in the control variable based on the time history of the error. The *derivative mode* (or *rate mode*) provides a change in the control variable based on the rate of change of the error signal. The combination of all three

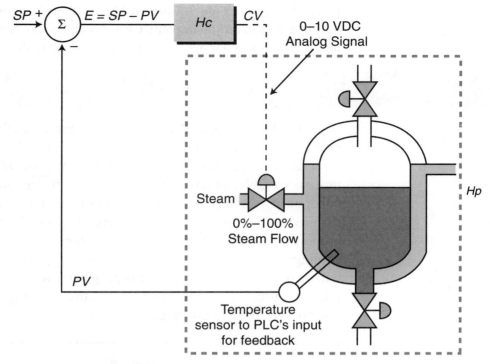

Figure 15-23. Block diagram of a continuous-mode controller.

modes in one controller forms the industry standard known as PID control. Table 15-1 shows the different possible combinations of continuous controller modes. Note that the derivative action is not used as a stand-alone mode

Controller Mode		Response	Applications
Proportional	P	*CV* changes in proportion to *E*	Systems with small load changes and/or small to moderate lag times
Integral	I	*CV* changes according to how *E* changes over time	Processes with small process lags and small capacities
Derivative	D	*CV* changes according to how fast *E* changes	Not used alone in applications
Proportional-Integral	PI	*CV* responds in a combination of P and I actions	Systems with large load changes
Proportional-Derivative	PD	*CV* responds in a combination of P and D actions	Processes with fast load changes
Proportional-Integral-Derivative	PID	*CV* responds in a combination of P, I, and D actions	Can be used in practically all process control applications

Table 15-1. Continuous controller modes.

in applications. This is due to the derivative action's response, which produces a high output but only for a short period of time. This has little effect on the process and, consequently, does not provide any process control.

15-5 PROPORTIONAL CONTROLLERS (P MODE)

A **proportional controller** adjusts the control variable output in a manner proportional to the error. As shown in Figure 15-24, the controller (Hc) receives feedback information from the process (Hp) in the form of the process variable, which is then compared to the set point. The error created, either positive or negative, tells the controller what percentage of output (CV) to provide to bring the error to zero. Figure 15-25 illustrates a typical proportional controller transfer function for a direct-acting controller (e.g., a cooling system). As the error becomes more negative ($PV > SP$), the controller will increase the control variable in proportion to the error. This will cause the process variable (from Hp) to decrease, thus pushing the error to zero. If the error becomes more positive, the opposite occurs.

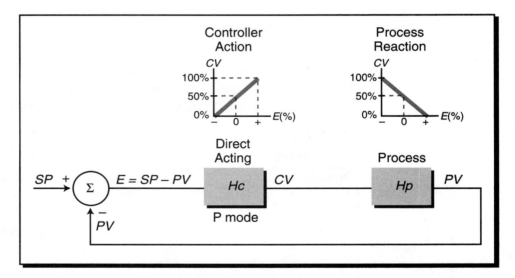

Figure 15-24. Proportional closed-loop control.

The control variable output ($CV_{(t)}$) of a proportional controller, starting from the set point value, is expressed by:

$$CV_{(t)} = K_P E + CV_{(E=0)}$$

where:

K_P = the proportional gain of the controller

E = the current error

$CV_{(E=0)}$ = the controller output when the error equals 0

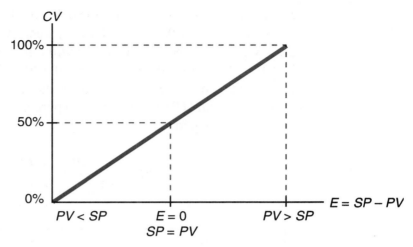

Figure 15-25. Direct-acting proportional controller transfer function.

Using this equation, a proportional controller can adjust the value of the control variable according to time and error by replacing the $CV_{(E=0)}$ term with the previous value of CV:

$$CV_{new} = K_P E + CV_{old}$$

So, if a controller with a control variable output value of 50% senses that the proportional error in the system ($K_P E$) is 20%, its new output will be 70%:

$$CV_{new} = 20\% + 50\%$$
$$= 70\%$$

This value indicates a linear correspondence between the control variable and the error, as was depicted in the graph in Figure 15-25. This graphic representation is called the *proportional band*, and it shows the error values associated with the full range of the controller output. The slope of this graph, the proportional gain K_P, is computed by dividing the percentage change in output by the percentage change in error:

$$K_P = \frac{\% \text{ change in } CV}{\% \text{ change in } E}$$

The proportional gain, therefore, is expressed in units of %/%. For example, a gain of 1 indicates that a 1% change in error will cause a 1% change in controller output. Note that the direction of the slope of the proportional gain (the positive or negative response of the control variable to a change in the error) depends on whether the controller is direct acting or reverse acting.

The proportional gain relationship between the error and the control variable depends on the width of the band upon which the controller is acting. For example, the temperature control system in Figure 15-26 has a temperature response that spans from 60°F to 180°F, **equaling a range of 120°F (180°F –** 60°F). However, if the controller only needs to exert control from 90°F to 150°F with the set point at 120°F, it will only be controlling a range of 60°F (150°F – 90°F) over the total range of 120°F. Therefore, the proportional band of the controller is 60°F over the 120°F range. Accordingly, the proportional band (*PB*) of control as a percentage of the full process variable range is represented as:

$$PB = \frac{PV_{max} - PV_{min}}{PV_{(max\ range)} - PV_{(min\ range)}}$$

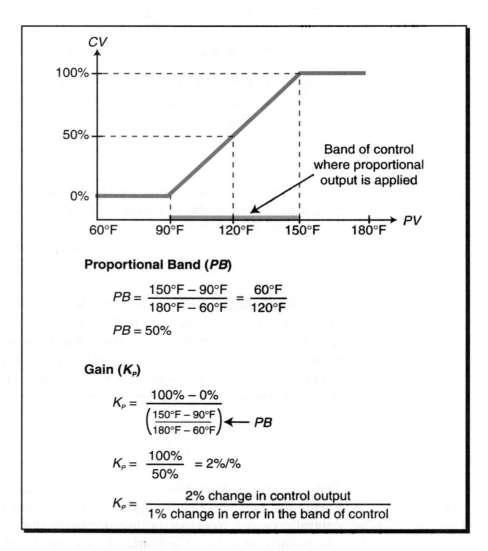

Proportional Band (*PB*)

$$PB = \frac{150°F - 90°F}{180°F - 60°F} = \frac{60°F}{120°F}$$

$$PB = 50\%$$

Gain (*K$_P$*)

$$K_P = \frac{100\% - 0\%}{\left(\frac{150°F - 90°F}{180°F - 60°F}\right)} \leftarrow PB$$

$$K_P = \frac{100\%}{50\%} = 2\%/\%$$

$$K_P = \frac{2\%\ change\ in\ control\ output}{1\%\ change\ in\ error\ in\ the\ band\ of\ control}$$

Figure 15-26. Proportional band and gain calculations.

As shown in the calculations in Figure 15-26, the proportional band of the temperature control system is 50%. The proportional gain of the system is defined by how much the control variable **output changes for each percent of** error within the control band. The error percentage range is equal to the proportional band percentage, because both express how much the process variable can deviate from the set point. The gain, according to Figure 15-26, will be 2, meaning that the controller's output will change 2% for every 1% change in error. This controller is a direct-acting controller, since the control variable will change in the same direction (+ or –) as the percentage of error. Note that the gain and proportional band are inversely related, meaning that:

$$K_P = \frac{1}{PB}$$

$$PB = \frac{1}{K_P}$$

If the process variable in the previous example is at the set point of 120%, then the controller must only maintain *CV* at 50% to keep the error at 0 (see Figure 15-27a). However, if the process variable increases to 135°F, the error incurred over the total temperature span will be:

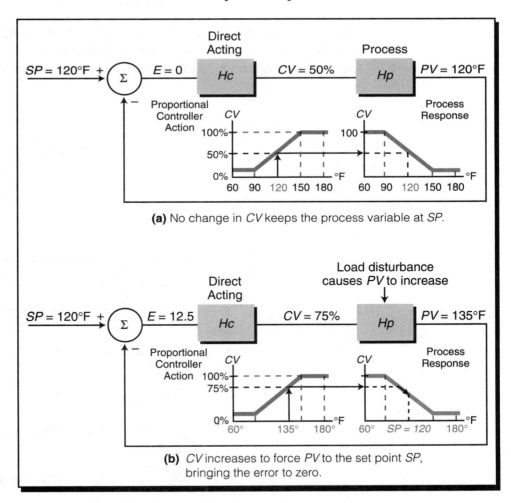

(a) No change in *CV* keeps the process variable at *SP*.

(b) *CV* increases to force *PV* to the set point *SP*, bringing the error to zero.

Figure 15-27. Process with **(a)** no change and **(b)** change in *CV*.

$$E = \frac{SP - PV}{PV_{min} - PV_{max}}$$

$$= \frac{120°F - 135°F}{60°F - 180°F}$$

$$= \frac{-15°F}{-120°F}$$

$$= 12.5\%$$

This error (12.5% above the set point over the *PV* range) in the controller equation, as well as in the graphic in Figure 15-27b, indicates that the new output of the controller will be 75%:

$$CV_{new} = K_P E + CV_{old}$$

$$= (2\%)(12.5\%) + 50\%$$

$$= 25\% + 50\%$$

$$= 75\%$$

The gain of a controller indicates how sensitive the controller is to error. The proportional band also indicates this sensitivity, since the gain and the band are related. Figure 15-28 illustrates two controllers with gains of $K_{P1} = 1$ and

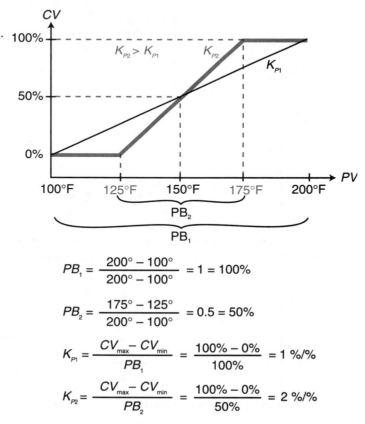

$$PB_1 = \frac{200° - 100°}{200° - 100°} = 1 = 100\%$$

$$PB_2 = \frac{175° - 125°}{200° - 100°} = 0.5 = 50\%$$

$$K_{P1} = \frac{CV_{max} - CV_{min}}{PB_1} = \frac{100\% - 0\%}{100\%} = 1 \%/\%$$

$$K_{P2} = \frac{CV_{max} - CV_{min}}{PB_2} = \frac{100\% - 0\%}{50\%} = 2 \%/\%$$

Figure 15-28. Two controllers with gains of 1 and 2.

$K_{P2} = 2$ that have proportional bands of 100% and 50%, respectively. The system with a gain of 1 will change the controller output 1% for every 1% of error, while the system with a gain of 2 will have twice the sensitivity, changing CV 2% for every 1% error.

EXAMPLE 15-4

Graph the transfer function for a proportional controller with a proportional band of 60% over a process variable range of 50°F to 150°F. The proportional band is centered around a set point of 90°F at a 50% controller output.

SOLUTION

Figure 15-29 illustrates the controller's transfer function. Note that, in this system, the set point (90°F) is not at the center of the total process variable range.

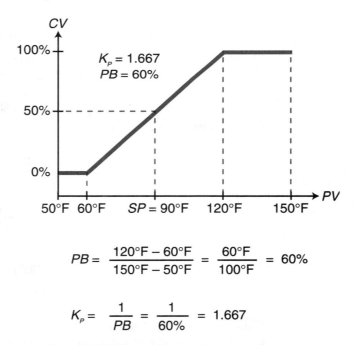

$$PB = \frac{120°F - 60°F}{150°F - 50°F} = \frac{60°F}{100°F} = 60\%$$

$$K_p = \frac{1}{PB} = \frac{1}{60\%} = 1.667$$

Figure 15-29. Example controller's transfer function.

The proportional gain for a reverse-acting controller (see Figure 15-30) is calculated using the same equations as used for a direct-acting one; however, the sign of the gain will be negative due to the slope of the curve. The proportional band for the reverse-acting controller in Figure 15-30 is:

$$PB = \frac{200°F - 100°F}{200°F - 100°F}$$
$$= 100\%$$

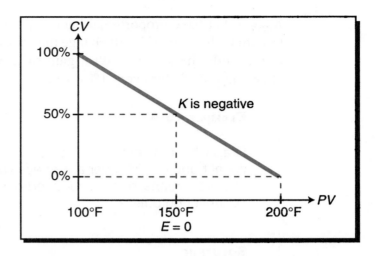

Figure 15-30. Proportional gain for a reverse-acting controller.

The gain is:

$$K_P = \frac{0\% - 100\%}{PB}$$
$$= \frac{-100\%}{100\%}$$
$$= -1\% / \%$$

If the process variable temperature is 160°F, the percentage error over the full variable range will be:

$$E = \frac{SP - PV}{PV_{min} - PV_{max}}$$
$$= \frac{150°F - 160°F}{100°F - 200°F}$$
$$= \frac{-10°F}{-100°F}$$
$$= 10\%$$

This indicates that the value of *PV* is 10% more than the *SP* value. Assuming that the previous *CV* output was at the set point (50%), the controller's new output will be:

$$CV_{new} = K_P E + CV_{old}$$
$$= (-1)(10\%) + 50\%$$
$$= -10\% + 50\%$$
$$= 40\%$$

Thus, the controller will reduce the value of its output to 40% of its range. This will affect the process by reducing the process variable to the set point value.

CLOSED-LOOP PROPORTIONAL CONTROL

Figure 15-31a illustrates a typical open-loop process control system where the process variable and transfer function in Laplace form are represented by the equation:

$$PV_{(s)} = \left(SP_{(s)}\right)\left(Hc_{(s)}\right)\left(Hp_{(s)}\right)$$

$$\frac{PV_{(s)}}{SP_{(s)}} = Hc_{(s)}Hp_{(s)}$$

Figure 15-31b shows this same system in a closed-loop configuration. For the closed-loop system, the process variable is defined by:

$$PV_{(s)} = \left(E_{(s)}\right)\left(Hc_{(s)}\right)\left(Hp_{(s)}\right)$$

Replacing $E_{(s)}$ with $SP_{(s)} - PV_{(s)}$ yields:

$$PV_{(s)} = \left(SP_{(s)} - PV_{(s)}\right)Hc_{(s)}Hp_{(s)}$$

$$PV_{(s)} = \left(SP_{(s)}Hc_{(s)}Hp_{(s)}\right) - \left(PV_{(s)}Hc_{(s)}Hp_{(s)}\right)$$

$$PV_{(s)}\left(1 + Hc_{(s)}Hp_{(s)}\right) = SP_{(s)}Hc_{(s)}Hp_{(s)}$$

Solving for $PV_{(s)}$ over $SP_{(s)}$ yields the closed-loop transfer function of this process control system:

$$\frac{PV_{(s)}}{SP_{(s)}} = \frac{Hc_{(s)}Hp_{(s)}}{Hc_{(s)}Hp_{(s)} + 1}$$

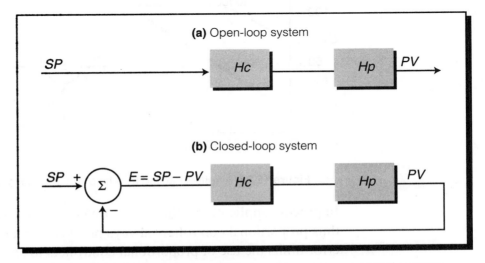

Figure 15-31. (a) Open-loop and **(b)** closed-loop process control systems.

The term $Hc_{(s)}$ represents the controller's transfer function, while the term $Hp_{(s)}$ represents the process's transfer function. The process's transfer function may take the form of a first-order response or a second-order response (overdamped, underdamped, or critically damped). As we will discuss later, the ideal controller transfer function for a system with a second-order process plus lag and dead time is one with proportional, integral, and derivative (PID) components.

A closed-loop system's response to a proportional controller creates an error that cannot be eliminated. This error is referred to as *offset*. Figure 15-32 shows the graph of a proportional controller's transfer function, in which a 50% *CV* output keeps the process variable at the set point. If a load disturbance occurs, the error will increase and the controller will change the output variable to try to bring the error back to zero. However, if the load disturbance requires a permanent output change in the controller (CV_{new}), the one-to-one relationship between the controller and error will prohibit a zero error value, because the original function curve is changed. For instance, if the process variable in Figure 15-33a increases due to a load disturbance, the control variable (assuming a direct-acting controller) will increase proportionally to try to bring the error back to zero. If the load disturbance causes a permanent change in the required controller output, the new output level (CV_{new}) will cause the process variable to return to the set point value. However, as *PV* begins to approach the set point (see Figure 15-33b), the controller will reduce its output because the error is diminishing. The reduced *CV* output will cause the error to increase, because the process requires a control variable output at the level of CV_{new} to maintain the process variable at the set point. Thus, error will always be present in the system.

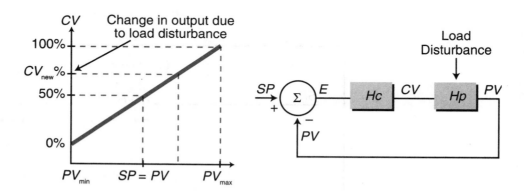

Figure 15-32. An offset caused by a load disturbance in a closed-loop system.

In process applications, the need for a permanent change in *CV* is typical, thus proportional controllers always produce a small amount of error. This error limits the use of proportional controllers to applications that include a

(a)

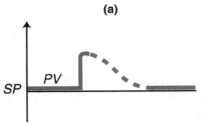

A load disturbance causes *PV*
to deviate from the set point.

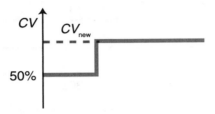

The controller increases the
control variable to *CV*~new~ to
compensate for the disturbance,
bringing the error to zero.

(b)

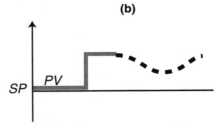

A load disturbance causes *PV*
to deviate from the set point.

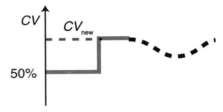

The controller increases the control
variable to *CV*~new~ to compensate for
the disturbance. The increase to *CV*~new~
causes a decrease in the value of *PV*,
instigating a decrease in the control
variable. The decrease in the control
variable causes the process variable
to increase again.

Figure 15-33. (a) Desired response of a proportional controller to a load disturbance and
(b) its actual response.

manual reset, allowing the operator to change the *bias*, or operating point, of
the controller. This changes the level of controller output associated with
$E = 0$ from *CV* to CV_{new}. Another way to minimize the system error is to
increase the gain, K_P, thus reducing the proportional band. This method is
precarious, however, because too much gain will make the system oscillate
like an ON/OFF controller with a small deadband. The reason for this is that
if a small error occurs around the set point, the controller will have a large
output swing (*CV*) to correct for the process variable (*PV*). This will push the
error in the opposite direction. When the error goes the opposite direction of
the set point, the controller will quickly respond with another large output
change, thus forcing the process variable to return to its original direction.
Hence, the system will behave like an ON/OFF system if the proportional gain
is too large.

The following example illustrates the effect of error in a proportional
controller controlling a first-order system. Note that the step change in set
point is permanent, simulating a permanent disturbance or change.

EXAMPLE 15-5

The closed-loop system shown in Figure 15-34 has a first-order process (Hp) with a gain of 5 and a time constant of $\tau = 30$ seconds. The proportional controller (Hc) has a proportional gain of 8. **(a)** Find the closed-loop transfer function, **(b)** calculate the response to a unit step $\frac{1}{s}$, and **(c)** plot the response, indicating the system time constant (τ_{sys}) and the steady-state value.

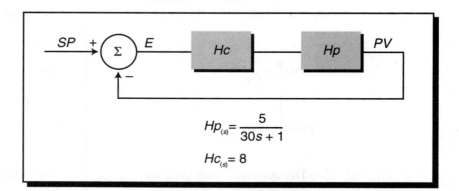

$$Hp_{(s)} = \frac{5}{30s + 1}$$

$$Hc_{(s)} = 8$$

Figure 15-34. Closed-loop system.

SOLUTION

(a) The transfer function of a closed-loop system is expressed as:

$$\frac{PV_{(s)}}{SP_{(s)}} = \frac{Hc_{(s)}Hp_{(s)}}{Hc_{(s)}Hp_{(s)} + 1}$$

As discussed in Chapter 14, the Laplace transfer function of a first-order process with lag is:

$$Hp_{(s)} = \frac{A}{\tau s + 1}$$

So the process's transfer function is:

$$Hp_{(s)} = \frac{5}{30s + 1}$$

A proportional controller's Laplace transfer function is simply the value of its gain, so:

$$Hc_{(s)} = 8$$

Therefore, the closed-loop transfer function of the entire system is:

$$\frac{PV_{(s)}}{SP_{(s)}} = \frac{(8)\left(\frac{5}{30s+1}\right)}{\left[(8)\left(\frac{5}{30s+1}\right)\right]+1}$$

$$= \frac{\left(\frac{40}{30s+1}\right)}{\left(\frac{40}{30s+1}+1\right)}$$

$$= \frac{\left(\frac{40}{30s+1}\right)}{\left(\frac{40+30s+1}{30s+1}\right)}$$

$$= \frac{40}{40+30s+1}$$

$$= \frac{40}{30s+41}$$

This transfer function indicates that this is a first-order system. To express it in the form of a first-order system, we must divide the numerator and denominator by 41 to obtain:

$$\frac{PV_{(s)}}{SP_{(s)}} = \frac{\left(\frac{40}{41}\right)}{\left(\frac{30s}{41}+\frac{41}{41}\right)}$$

$$= \frac{0.976}{0.732s+1}$$

(b) The response to a unit step change in the set point is given by:

$$SP_{(s)} = \frac{1}{s} \quad \text{(unit step)}$$

Therefore, using the previous equation, the process variable response will be:

$$PV_{(s)} = SP_{(s)}\left(\frac{0.976}{0.732s+1}\right)$$

Thus:

$$PV_{(s)} = \left(\frac{1}{s}\right)\left(\frac{0.976}{0.732s+1}\right)$$

$$PV_{(s)} = \frac{0.976}{s(0.732s+1)}$$

According to Table 14-1, this response is in the form of the inverse Laplace transform of a first-order response to a step input with lag:

$$\mathscr{L}^{-1}\left[\frac{A}{s(\tau s + 1)}\right] = A\left(1 - e^{\frac{-t}{\tau \text{sec}}}\right)$$

Hence, in the time domain, the process variable response will be equal to:

$$PV_{(t)} = 0.976\left(1 - e^{\frac{-t}{0.732\,\text{sec}}}\right)$$

(c) Figure 15-35 illustrates the time response of the closed-loop system to a unit step change in the set point. Note that the gain of the system is 0.976, meaning that the process variable in the system will not reach the value of the unit step input. Instead, the system will respond only 0.976 to the unit step change of 1. The process variable steady-state value (PV_{ss}), which is the final value of PV, can be computed using the final value theorem:

$$PV_{SS} = \lim_{t \to \infty}\left[0.976\left(1 - e^{\frac{-t}{0.732}}\right)\right]$$

$$= \lim_{t \to \infty}[0.976(1 - 0)]$$

$$= 0.976$$

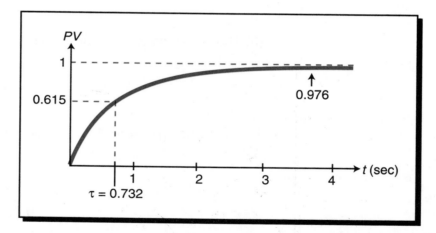

Figure 15-35. System response to step change.

Therefore, the system will always have a residual error of 2.4% (1.0 − 0.976 = 0.024). The system time constant, $\tau_{sys} = 0.732$, indicates that the system will take 0.732 seconds to reach 0.615, 63% of the steady-state value.

The open-loop response of the process to a step change would have a steady-state value of 5 (see Figure 15-36a), where 1τ would occur at 30 seconds. The open-loop response of the controller and process to a unit step would have a gain of 40 with this same τ constant (30 sec).

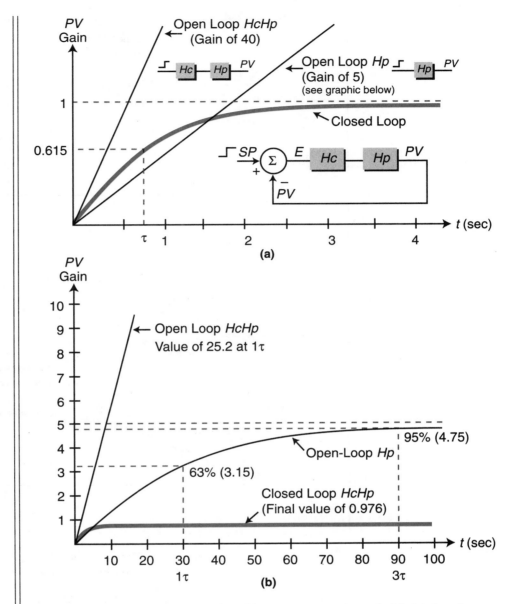

Figure 15-36. Open- and closed-loop responses shown in **(a)** detail and **(b)** 10×detail.

The closed-loop response of the controller and process, however, would have a much smaller gain (in this case, 0.976) due to the negative feedback in the control loop.

The value of the gain in a proportional controller influences the response of a second-order closed-loop system as shown in Figure 15-37. The higher the gain, the faster the process responds; however, cycling and overshoot occur. Lowering the gain makes the response much slower and the value at steady state smaller. For example, the value of $K_P = 1$ in Figure 15-37 will cause a slow, closed-loop response to a unit step change in the set point with a final steady-state value of 0.5.

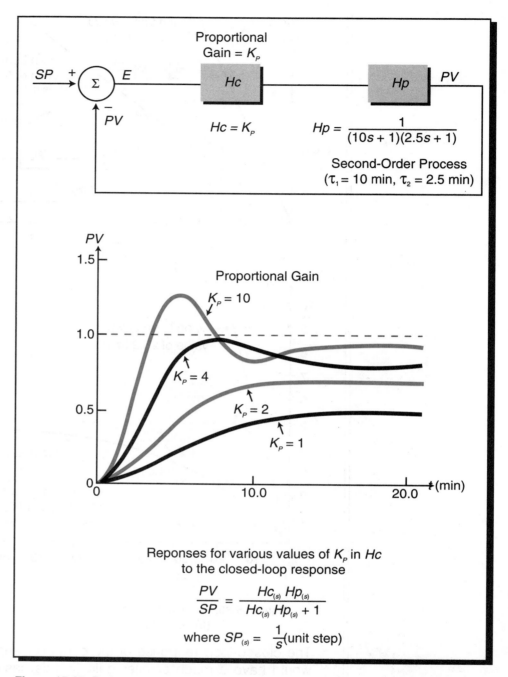

Figure 15-37. Responses of a second-order closed-loop system to different values of proportional gain.

15-6 INTEGRAL CONTROLLERS (I MODE)

An **integral controller** provides an output whose rate of change is proportional to the error deviation. This means that the larger the error, the faster the controller's output changes and vice versa. An integral controller will stop adjusting its output once the error becomes zero. When used in conjunction with a proportional controller, an integral controller will bring the system's residual error to zero. An integral controller's output (*CV*) is represented by:

$$\frac{dCV}{dt} = K_I E$$

where:

$\dfrac{dCV}{dt}$ = the rate of change in controller output in % over seconds

K_I = the integral gain in % of the controller output per second per % error

E = the error in %

This differential equation indicates that the controller's output $CV_{(t)}$ can be obtained by taking the integral over time of both sides of the equation so that:

$$\frac{dCV}{dt} = K_I E$$

$$dCV = K_I E dt$$

$$\int_0^t dCV = \int_0^t K_I E dt$$

$$CV_{(t)} - CV_{(t=0)} = K_I \int_0^t E dt$$

$$CV_{(t)} = K_I \int_0^t E dt + CV_{(t=0)}$$

where the term $CV_{(t=0)}$ is the value of the output at $t = 0$. When an integral controller is used in a closed-loop system, it calculates $CV_{(t)}$ for every change in error. So, if the value of the error changes after the controller has calculated a previous value $CV_{(t)}$, then it will use this previous value of $CV_{(t)}$ as the $CV_{(t=0)}$ value and calculate a new $CV_{(t)}$ output based on the new error.

The integral gain K_I (see Figure 15-38) indicates the sensitivity of the output's rate of change to the percentage of error that occurs over time. A large value of K_I means that a small error will produce a large rate of change in the controller output. Conversely, a small value of K_I means that a small error produces a small rate of change in the controller output. In Figure 15-38, the rate of change of K_{I1} is greater than that of K_{I2}.

Figure 15-39 illustrates the reaction of an integral controller's output to a change in the process variable due to a load disturbance. Note that, at the moment the error occurs, the controller starts the integration of the error

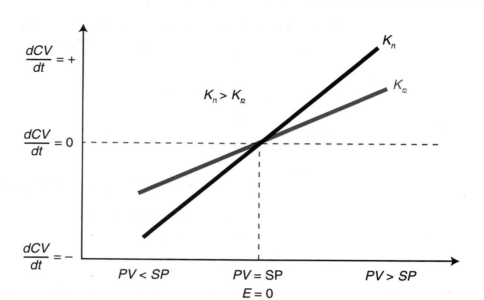

Figure 15-38. Integral gain.

value, meaning that the control variable begins to increase as a function of the magnitude of the error. The error in Figure 15-39 is constant, creating a ramp integration of the controller's output. That is, the amount of error remains constant over time, so the control variable increases at a steady rate.

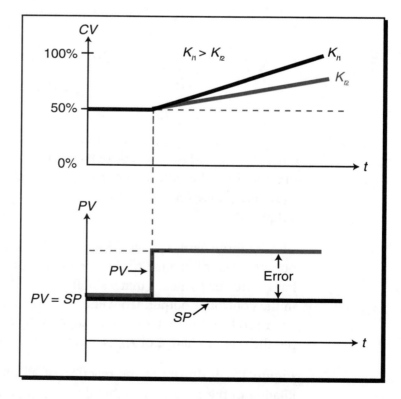

Figure 15-39. Integral controller's response to a step change in the process variable.

To further illustrate the effect of the measured error on the control variable output, let's examine Figure 15-40, which shows the graph of a direct-acting integral controller's output response to a change in error. If the error makes a large jump (1), the controller will respond with a steep increase in output. As the error begins to decrease (2), the rate of increase of the output variable will also decrease (less ramping). When the error becomes zero (3), then the controller will keep its output at its previous level. As the error increases again, but in the opposite direction (4), the output will begin to decrease. As the error decreases, but still remains negative (5), the control variable will continue to decrease but at a less rapid rate. Furthermore, if the error increases positively (6), then the output will increase again. Finally, as the error goes to zero and remains there (7), the controller will level out the control variable and make no more changes to its output level. Thus, an integral controller can adjust its output level to bring the error to zero. An integral controller does not exhibit the limitations of the linear relationship of a proportional controller; thus, it is able to keep a zero error at an output value other than 50% of the controller output.

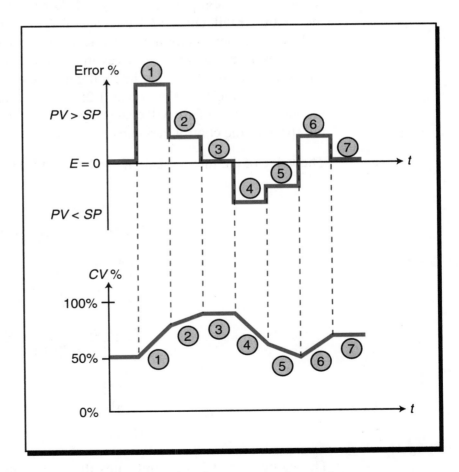

Figure 15-40. Output response to changes in error.

The gain of an integral controller (K_I) is defined by the equation:

$$K_I = \frac{\left(\frac{dCV}{dt}\right)}{E}$$

$$= \frac{\text{change in \% of } CV \text{ per second}}{\text{change in \% of error}}$$

A value of $K_I = 0.2$ indicates that the controller will change 0.2% per second for every 1% of error present in the system. So, if the 1% error in the system lasts for 2 seconds and then goes to zero, the controller will increase its output 0.4%.

As discussed earlier, different values of K_I have different curves, or slopes, associated with them. Figure 15-41a shows the curves for two integral gains, K_{I1} and K_{I2}. For both gain values, the controller will make no change to the output (CV) if the error equals 0. However, if the error increases to PV_{max}, the controller will change its output at a rate of 25% per second if the integral gain equals K_{I1}, while it will only change its output at a rate of 15% per second if the integral gain equals K_{I2}. Likewise, if the error drops to PV_{min}, the controller will change CV at a rate of –25% per second for the K_{I1} value and –15% per second for the K_{I2} value. Both K_{I1} and K_{I2} can be thought of as belonging to a "family" of curves that expresses the value of the control variable over time for given integral gain and error values (see Figure 15-41b). For example, if $K_I E$ equals 1.25 ($K_I = 0.5$ and $E = 2.5\%$), then in 8 seconds the value of CV will change by 10%. The family of curves illustrates the speed of the control variable change for different error values.

The value of the integral gain K_{I1} in Figure 15-41 can be computed as:

$$K_I = \frac{\% \text{ change in } \left(\frac{dCV}{dt}\right)}{\% \text{ error over full range}}$$

$$= \frac{\left(\frac{dCV_{max}}{dt} - \frac{dCV_{min}}{dt}\right)}{\left(\frac{PV_{max} - PV_{min}}{PV_{(max\ range)} - PV_{(min\ range)}}\right)}$$

$$= \frac{25\ \%/\sec - (-25\ \%/\sec)}{\left(\frac{200°F - 100°F}{200°F - 100°F}\right)}$$

$$= \frac{50\ \%/\sec}{100\%}$$

$$= 0.5\ \sec^{-1}$$

The value $K_I = 0.5$ sec^{-1} indicates that the controller will gain 0.5% in output per second for each percentage of error present. If the error is 50%

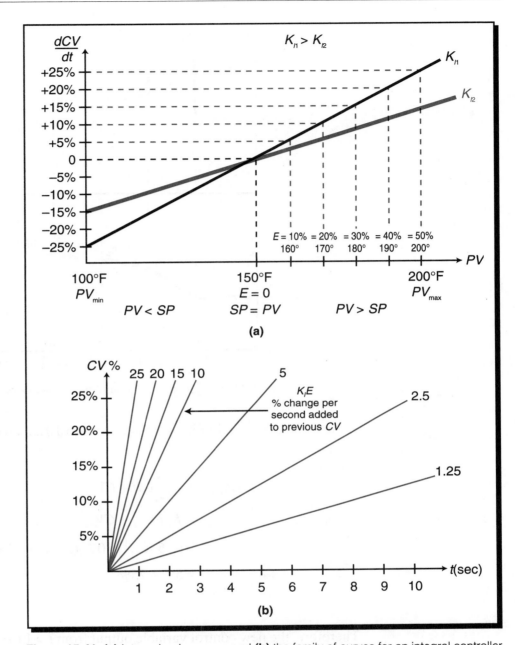

Figure 15-41. (a) Integral gain curves and **(b)** the family of curves for an integral controller.

(i.e., $PV = 200°F$), then after one second the controller's output will be 75% (see Figure 15-42):

$$CV_{(t=1)} = K_I \int_0^t Edt + CV_{(t=0)}$$

$$= K_I Et\Big|_{t=0}^{t=1} + CV_{(t=0)}$$

$$= K_I E(t - t_0) + CV_{(t=0)}$$

$$= (0.5 \text{ sec}^{-1})(50\%)(1 - 0) + 50\%$$

$$= 75\%$$

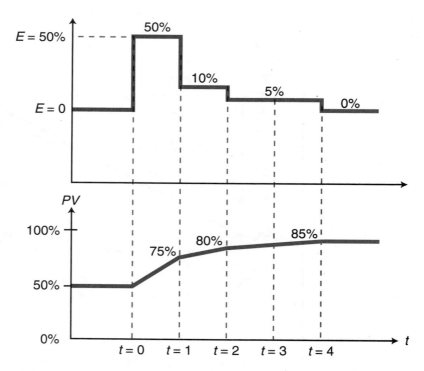

Figure 15-42. Integral controller output response to a change in error.

If the error as shown in Figure 15-42 drops to 10%, the output will be:

$$CV_{(t=2)} = K_I \int_0^t E dt + CV_{(t=1)}$$

$$= K_I E t \big|_{t=1}^{t=2} + CV_{(t=1)}$$

$$= (0.5 \text{ sec}^{-1})(10\%)(2-1) + 75\%$$

$$= 5\% + 75\%$$

$$= 80\%$$

Therefore, the new control variable output from $t = 1$ to $t = 2$ will be 80%. If the controller error drops to 5% for the next two seconds (from $t = 2$ to $t = 4$), the controller output will increase steadily from 80% (at $t = 2$) to 85% (at $t = 4$). After $t = 4$, the error is 0%, therefore, the controller output stays at 85%. Note that the family of curves shown in Figure 15-41b is the product of $K_I E$ for a particular value of error. If the error stays constant for t seconds, then the change in the value of CV over that time period will follow the $K_I E$ curve for that error value.

The inverse of the gain term K_I is referred to as the *integral time* (T_I), or *reset time*, in seconds. The integral time is the time it takes for the control variable (CV) to change 1% for a 1% change in error. It is expressed as:

$$T_I = \frac{1}{K_I}$$

The T_I variable is used by some manufacturers to allow the user to indirectly enter the integral gain into the controller. If the integral time must be specified in minutes, as is required by some manufacturers, a simple conversion can change T_I from seconds to minutes:

$$T_I = \frac{1}{K_I} \text{ (in seconds) or } \frac{1}{(K_I)\left(\frac{60 \text{ sec}}{\text{min}}\right)} \text{ (in minutes)}$$

So, for the previous example, the reset time is equal to:

$$T_I = \frac{1}{K_I}$$

$$= \frac{1}{0.5 \text{ sec}^{-1}}$$

$$= 2 \text{ seconds}$$

The integral controller mode is also referred to as *reset action*, because it automatically resets the error to zero over time.

EXAMPLE 15-6

Illustrate the transfer function of an integral controller with a gain of $K_I = 0.2 \text{ sec}^{-1}$ over a process variable range of 100°F to 200°F. Plot the response of the controller's output for an error due to a permanent load disturbance of +10% above the set point of 150°F over the full range. Two seconds after the controller increases its output, the error will drop by 5%. After 3 more seconds, the error will become zero. Find the value of *CV* after 5 seconds.

SOLUTION

Figure 15-43 shows the integral controller's transfer function. When the error is +10% above the set point, the process variable will be at 160°F, which will cause the controller's output to increase at a rate of 2% per second:

$$\frac{dCV}{dt} = K_I E$$

$$= (0.2)(10)$$

$$= 2\% \text{ per second}$$

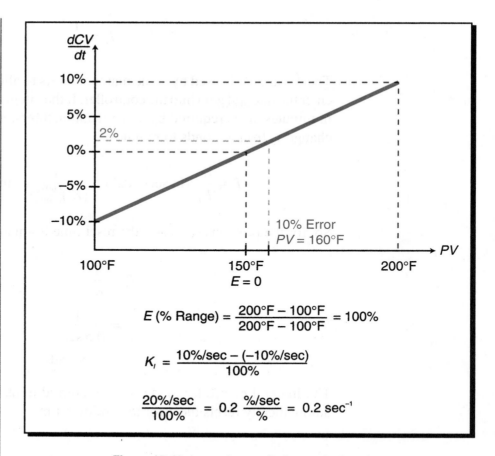

Figure 15-43. Integral controller's transfer function.

As Figure 15-44 illustrates, after 2 seconds of integral action, the controller output will be 54%:

$$CV_{(t=2)} = K_I \int_{t=0}^{t=2} E\,dt + CV_{(t=0)}$$

$$= (0.2)Et\Big|_{t=0}^{t=2} + 50\%$$

$$= [0.2][10(2-0)] + 50\%$$

$$= 54\%$$

After the 2 seconds have elapsed, the error will drop to 5% and the controller will integrate at a rate of:

$$\frac{dCV}{dt} = K_I E$$

$$= (0.2)(5\%)$$

$$= 1\% \text{ per second}$$

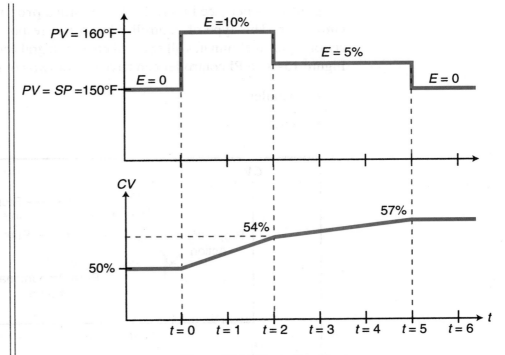

Figure 15-44. Controller output.

Therefore, at the end of the next 3 seconds, the controller output will be 57%:

$$CV_{(t=5)} = K_I \int_{t=2}^{t=5} Edt + CV_{(t=2)}$$

$$= (0.2)Et\Big|_{t=2}^{t=5} + 54\%$$

$$= [0.2][5(5-2)] + 54\%$$

$$= 57\%$$

At this point, the error will drop to zero, so the controller will stop changing the *CV*, maintaining its output at a new zero error value of 57%.

15-7 PROPORTIONAL-INTEGRAL CONTROLLERS (PI MODE)

Although an integral controller does not have the residual error at steady state that a proportional controller has, its response action to a step change in input (step in error) is often too slow to be used in real-life applications. This slow speed, as compared with the immediate response of a proportional controller, is due to the ramping effect of the integral action as the controller increases its output. Therefore, proportional action is normally added to an integral controller (see Figure 15-45) to form a **proportional-integral (PI) controller**. This type of controller has a fast response time (proportional action), plus it eliminates all residual error (integral action). As illustrated in Figure 15-46, a PI controller can have one of two configurations:

- parallel

- series

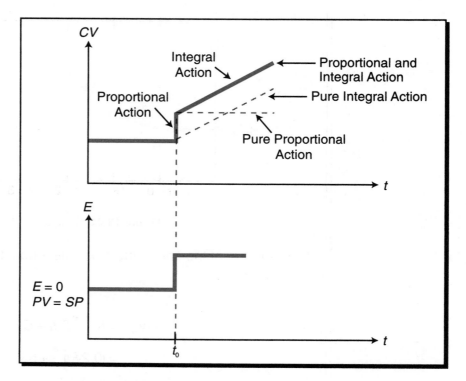

Figure 15-45. Proportional-integral action.

In a *parallel PI controller*, the proportional and integral actions occur independently of each other, so the controller's output (*CV*) is equal to the proportional action plus the integral action:

$$CV_{new} = K_P E + K_I \int_0^t E dt + CV_{old}$$

In a *series PI controller*, on the other hand, the integral action occurs after the proportional action. Therefore, the input to the integral action is not the system error *E*, but rather the result of the proportional action $K_P E$. Accordingly, a series PI controller's output is defined by:

$$CV_{new} = K_P E + K_I \int_0^t K_P E dt + CV_{old}$$

$$= K_P E + K_P K_I \int_0^t E dt + CV_{old}$$

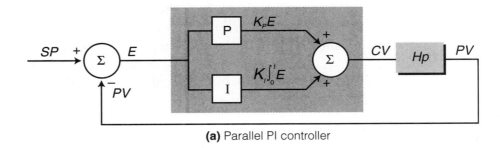

(a) Parallel PI controller

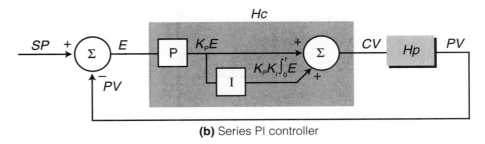

(b) Series PI controller

Figure 15-46. (a) Parallel and **(b)** series PI controllers.

Both of these types of PI controllers eliminate error offset and have a faster response time than an integral-only controller. However, series PI controllers multiply the integral gain times the proportional gain, producing an effect called *repeating*. In repeating, the effect of the proportional gain ($K_P E$) is repeated during every integral time period T_I, causing the integral action of the controller to equal that of the proportional action. This means that a series PI controller responds faster to a change in error than a parallel controller when their proportional gains are greater than one ($K_P > 1$) and their integral times are the same.

The term *repeats* is used when referring to how many times the proportional amount is repeated in one minute. If the value of T_I is less than 1 minute, then the integral gain is repeated more than one time per minute. This can be seen in the equation:

$$CV_{\text{new}} = K_P E + K_P K_I \int_0^t E \, dt + CV_{\text{old}}$$

$$= K_P E + K_P K_I E t \big|_{t=0}^{t=1} + CV_{\text{old}}$$

$$= K_P E + K_P K_I E t + CV_{\text{old}}$$

$$= K_P E + K_P E K_I t + CV_{\text{old}}$$

$$= K_P E + K_P E \left(\frac{1}{T_I}\right) t + CV_{\text{old}}$$

When $t = T_I$, the term $K_P E$ is repeated once by the integral action, in this case in the period from $t = 0$ to $t = 1$ minute. Figure 15-47a illustrates the integral gain of a PI controller with a repeat of 1 (i.e., $T_I = 1$). After 1 minute, the term $K_P E$ is repeated. Figure 15-47b illustrates a PI controller with an integral time of $T_I = 0.333$, indicating that the term $K_P E$ will be repeated three times in one minute.

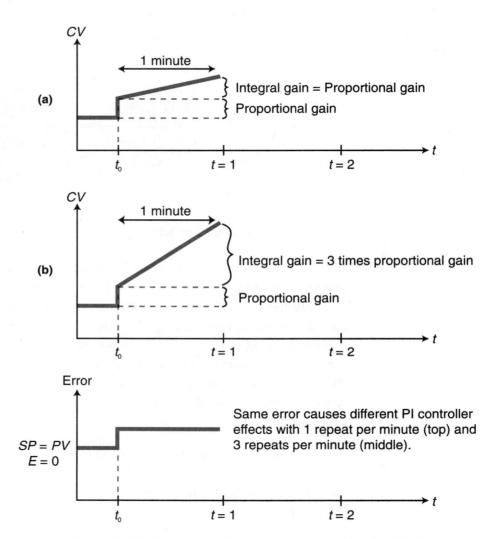

Figure 15-47. PI controller with an integral time of **(a)** 1 and **(b)** 1/3.

EXAMPLE 15-7

(a) Graph the value of the control variable after 1 minute for a series PI controller given that the proportional gain is 2 and the integral gain is 0.01 sec^{-1}. The process variable changes from the set point of 150°F to 155°F over a process variable range of 100°F to 200°F. At the set point, the controller has an output of 50%. **(b)** How long will it take for the integral gain to equal the proportional gain?

SOLUTION

(a) The error created in the system over the *PV* range is:

$$E(\%) = \frac{155°F - 150°F}{200°F - 100°F}$$
$$= 5\%$$

The given values of the proportional and integral actions are:

$$K_P = 2\%/\%$$
$$K_I = 0.01 \, \text{sec}^{-1}$$

Thus, the value K_I in minutes is:

$$K_I = \left(\frac{0.01}{\text{sec}}\right)\left(\frac{60 \, \text{sec}}{1 \, \text{min}}\right)$$
$$= \frac{0.6}{\text{min}}$$
$$= 0.6 \, \text{min}^{-1}$$

The control variable for this series PI controller is defined as:

$$CV_{new} = K_P E + K_P K_I \int_0^t E dt + CV_{(t=0)}$$
$$= \left[\left(2\tfrac{\%}{\%}\right)\left(5\%\right)\right] + \left[\left(2\tfrac{\%}{\%}\right)\left(0.6 \, \text{min}^{-1}\right)\left(5\%\big|_{t=0}^{t=1}\right)\right] + 50\%$$
$$= 10\% + \left[\left(10\%\right)\left(0.6 \, \text{min}^{-1}\right)\left(1 \, \text{min} - 0 \, \text{min}\right)\right] + 50\%$$
$$= 10\% + 6\% + 50\%$$
$$= 66\%$$

Figure 15-48 illustrates this control variable response.

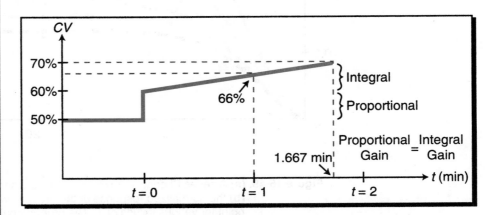

Figure 15-48. Control variable response.

(b) The integral gain will equal the proportional gain in 1.667 minutes:

$$K_P E = K_P K_I E t$$

$$(2\%)(5\%) = (2\%)(0.6\ \text{min}^{-1})(5\%)(t)$$

$$t = \frac{1}{(0.6\ \text{min}^{-1})}$$

$$= 1.667\ \text{min}$$

The integral, or reset, time of a PI controller influences the ultimate closed-loop response of the system (see Figure 15-49). As the reset time decreases, the response speed increases, creating an overshoot. The overshoot in the response will cause the proportional action to initiate a negative increase (reduction of output), producing an oscillating response.

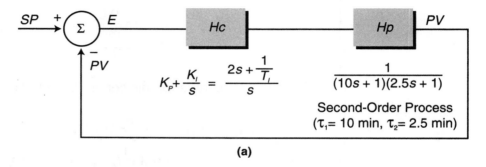

(a)

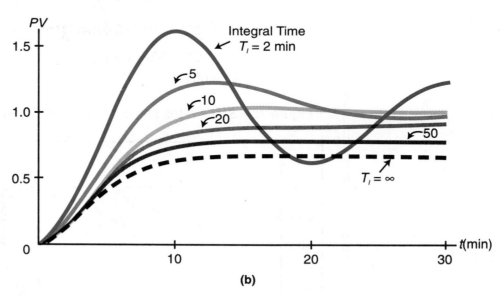

(b)

Figure 15-49. (a) A series PI controller (*Hc*) controlling a second-order process and **(b)** the normalized response of the process variable to a change in set point for various values of T_I. The proportional gain K_P is equal to 2%/% for all values of T_I.

The signs of the proportional gain ($K_P E$) and the combined proportional-integral gain ($K_P K_I \int E dt$) terms are important when determining the integral gain curves for reverse-acting and direct-acting series PI controllers. In the direct-acting mode (see Figure 15-50a), the signs of K_P and K_I are both positive. Therefore, a negative error will make both the K_P and K_I terms negative, resulting in proper controller action. Similarly, a positive error will make both terms positive, again resulting in proper direct action control.

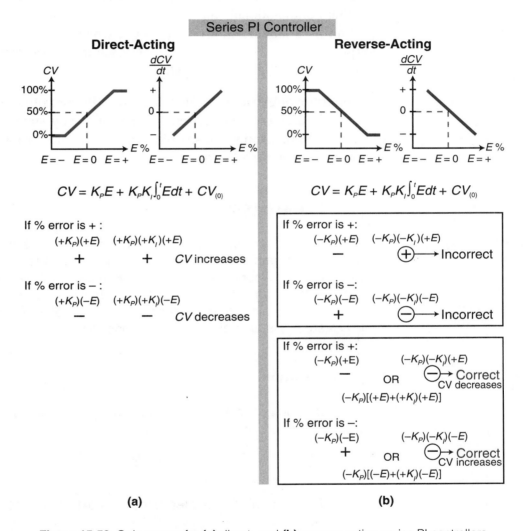

Figure 15-50. Gain curves for **(a)** direct- and **(b)** reverse-acting series PI controllers.

In a reverse-acting series PI controller, both the proportional gain and the combined proportional-integral gain ($K_P K_I$) must be negative ($-K_P K_I$) for the controller to correctly implement a reverse action (see Figure 15-50b). This means that the integral gain must be positive—a negative integral gain would result in a positive combined gain term. Since the proportional gain must be negative, the output of a reverse-acting series PI controller can be expressed as:

$$CV_{new} = -K_P \left(E + K_I \int_0^t E\,dt \right) + CV_{(t=0)}$$

The negative sign in the proportional gain term ensures that the controller will operate as reverse-acting. In a PLC system, the user enters the values for K_P and K_I; therefore, some manufacturers of series PI controllers allow the user to select a reverse-acting controller by specifying the proportional gain as a negative value. In this type of system, the controller takes care of all other computational signs, to ensure proper controller action and a proper control variable response. Otherwise, when the error is positive, one term (proportional) reduces the value of CV, while the other (integral) adds to it and vice versa if the error is negative.

The following example illustrates how a PI controller ultimately brings the error in a closed-loop system to zero at steady state. This example is an extension of Example 15-5, which used only proportional control and, as a result, had an offset error.

EXAMPLE 15-8

The closed-loop system in Example 15-5 has a first-order process with a gain of 5 and a time constant of $\tau = 30$ seconds. The controller has a proportional gain of $K_P = 8$. If the controller also has an integral action with a gain of $K_I = 0.1$ sec^{-1}, forming a PI parallel controller, find **(a)** the closed-loop transfer function of the system and **(b)** the steady-state value of the response to a unit step change in set point.

SOLUTION

(a) The process's transfer function is defined by:

$$Hp_{(s)} = \frac{5}{30s + 1}$$

The controller's transfer function is expressed as:

$$Hc_{(t)} = K_P E + K_I \int_0^t E\,dt$$

$$Hc_{(s)} = K_P + \frac{K_I}{s}$$

$$= \frac{K_P s + K_I}{s}$$

$$= \frac{8s + 0.1}{s}$$

Therefore, the closed-loop transfer function is:

$$\frac{PV_{(s)}}{SP_{(s)}} = \frac{Hc_{(s)}Hp_{(s)}}{Hc_{(s)}Hp_{(s)} + 1}$$

$$= \frac{\left[\left(\frac{8s+0.1}{s}\right)\left(\frac{5}{30s+1}\right)\right]}{\left[\left(\frac{8s+0.1}{s}\right)\left(\frac{5}{30s+1}\right) + 1\right]}$$

$$= \frac{\left(\frac{40s+0.5}{30s^2+s}\right)}{\left(\frac{40s+0.5+30s^2+s}{30s^2+s}\right)}$$

$$= \frac{40s + 0.5}{30s^2 + 41s + 0.5}$$

(b) The response of the process variable to a step change in set point is represented by:

$$PV_{(s)} = SP_{(s)}\left(\frac{40s + 0.5}{30s^2 + 41s + 0.5}\right)$$

$$= \left(\frac{1}{s}\right)\left(\frac{40s + 0.5}{30s^2 + 41s + 0.5}\right)$$

The final value of the process variable at steady state can be computed by taking the inverse Laplace transform of $PV_{(s)}$ to obtain $PV_{(t)}$ and then evaluating the response value as t approaches infinity ($t \to \infty$). However, obtaining the inverse Laplace transform of this response can be very cumbersome. So, as an alternative, we can use the final value theorem and apply it to the equation in the Laplace, or frequency, domain:

$$\lim_{t \to \infty} f(t) = \lim_{s \to 0} sF_{(s)}$$

The steady-state value of the process variable in response to a unit step input change can be found by multiplying the Laplace equation times s and evaluating it as s approaches zero. Therefore:

$$\lim_{s \to 0} sPV_{(s)} = (s)\left(\frac{1}{s}\right)\left(\frac{40s + 0.5}{30s^2 + 41s + 0.5}\right)$$

$$= \frac{40(0) + 0.5}{30(0)^2 + 41(0) + 0.5}$$

$$= \frac{0.5}{0.5}$$

$$= 1$$

Thus, the error will be zero at steady state:

$$E = SP - PV$$
$$= 1 - 1$$
$$= 0$$

A PI controller may create a situation in which it *saturates* the control variable output. Saturation occurs when the control variable output remains pegged at its maximum value (100%). The control variable will remain saturated even if the error starts to come down (see Figure 15-51). The integral action will not change direction until the percentage of error becomes negative (*PV > SP*). This situation is called **integral windup**, or *reset windup,* and it can be damaging to the process. It occurs when a large error is present in a system with a slow response (large time constant). In this situation, the controller will keep increasing the control variable value because the error remains constant due to the lag's effect on the integral corrective action. Eventually, the control variable will saturate at 100%. In other words, the controller's corrective action continues to occur when the process takes too long to respond. The start-up of a batch process is a typical example of a situation in which a reset windup can occur. As we'll discuss later, this condition can be prevented.

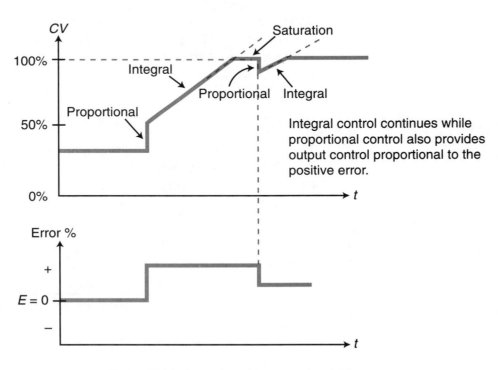

Figure 15-51. Saturation of the control variable output.

15-8 DERIVATIVE CONTROLLERS (D MODE)

STANDARD DERIVATIVE CONTROLLERS

The output of a **derivative controller** is proportional to the rate of change of the error in the system, which is expressed as $\frac{dE}{dt}$ (see Figure 15-52). This derivative action, also referred to as *rate mode*, is expressed mathematically as:

$$CV_{new} = K_D \frac{dE}{dt} + CV_{old}$$

where:

CV_{new} = the control variable

CV_{old} = the previous value of CV

K_D = the derivative gain constant in %(sec/%)

$\dfrac{dE}{dt}$ = the rate of change of error over the duration of change in %/sec

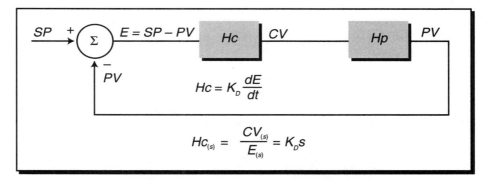

Figure 15-52. Derivative controller action.

The derivative gain constant (K_D) is also referred to as the *rate time*. It can be expressed in seconds or minutes as:

$$K_D = T_D \text{ seconds} \quad \text{(rate time)}$$

$$\text{or}$$

$$K_D = \frac{T_D}{60} \text{ minutes} \quad \text{(if } T_D \text{ is given in seconds)}$$

In Laplace form, the derivative controller transfer function takes the form:

$$Hc_{(s)} = \frac{CV_{(s)}}{E_{(s)}}$$

$$= K_D s = T_D s$$

Figure 15-53 illustrates the derivative gain transfer function in a direct-acting system by indicating the corresponding controller outputs for different rates of change ($\frac{dE}{dt}$) in error. Like in the integral mode, the rates of error change form several family curves (see Figure 15-53b). For example, if the error increases at a rate of 1.0%/sec, the controller will apply a derivative action that makes its output jump from 50% to 70% (see Figure 15-53a). If the rate of increase slows down to 0.5%/sec, the controller will decrease its output to 60%. When the rate of change of error equals zero, the controller will decrease its output to 50% again (see Figure 15-54). Note that the derivative action is based on the rate at which the error changes, not the actual value of the error.

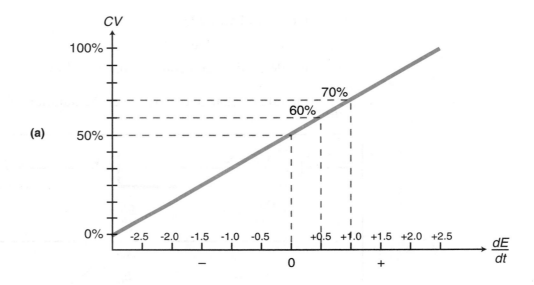

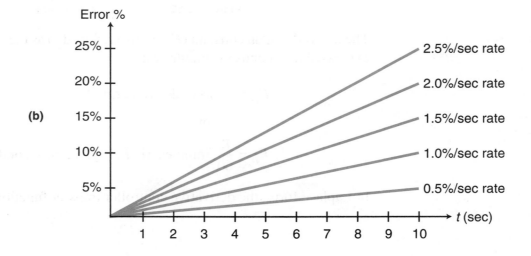

Figure 15-53. (a) Derivative controller transfer function and **(b)** its family of curves.

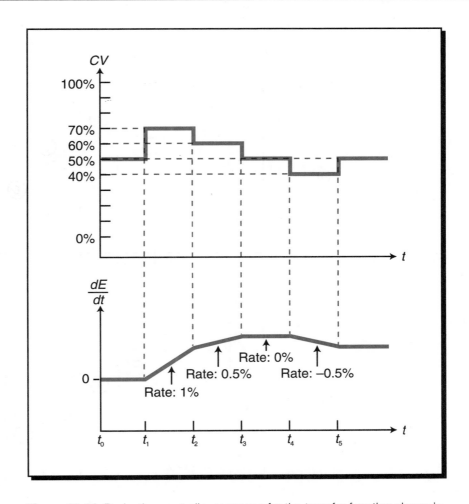

Figure 15-54. Derivative controller response for the transfer function shown in Figure 15-53a.

Derivative action is not used by itself in a controller; rather, it is used in combination with proportional and proportional-integral actions. There are several reasons for this. First, the derivative action response to a step change (see Figure 15-55a) creates an infinite change in error over time ($\frac{dE}{dt} = \infty$), causing the output of the controller to have 100% saturation for an instant (point 1 in Figure 15-55b). If the error remains at its stepped up value, the controller will sense no change and will return the control variable to 50% (between points 1 and 2). At point 2, when the error drops in a step fashion (see Figure 15-55a), the control variable will again have an infinite change over time, thus causing a 0% output (point 2 in Figure 15-55b).

The second reason why derivative action is not used alone is that it only produces a change in output if there is a change in the rate of error (points 3, 5, 6, and 7 in Figure 15-55). If a large error remains constant, the controller will maintain the control variable at 50% of its range (point 8), thus the error will not be corrected.

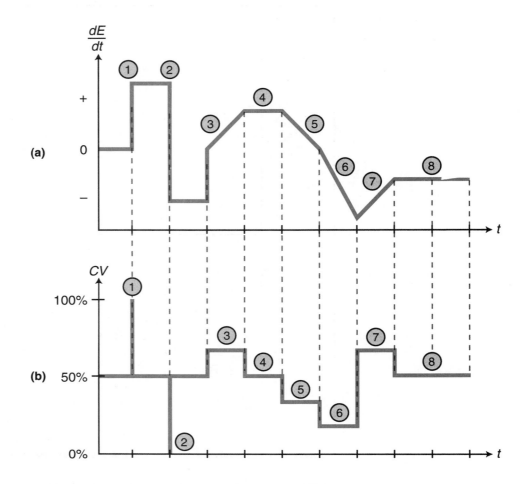

Figure 15-55. (a) Step changes and **(b)** their corresponding derivative responses.

MODIFIED DERIVATIVE CONTROLLERS

Derivative action may also be expressed in terms of the change in the process variable rate over time:

$$CV_{new} = -K_D \frac{dPV}{dt} + CV_{old}$$

This type of derivative action, used by some PLCs, avoids the saturation of the control variable in response to a step change in the set point. In this type of controller, the control variable tracks the process variable, which is very unlikely to change in a step fashion.

Note that the sign of the K_D term for a modified derivative controller is negative. It is derived from:

$$E = SP - PV$$

$$CV_{(t)} = K_D \frac{dE}{dt} + CV_{(t=0)}$$

$$= K_D \left[\left(\frac{dSP}{dt} \right) - \left(\frac{dPV}{dt} \right) \right] + CV_{(t=0)}$$

$$= \left[K_D \left(\frac{dSP}{dt} \right) - K_D \left(\frac{dPV}{dt} \right) \right] + CV_{(t=0)}$$

Since the set point is a constant value, the term $K_D(\frac{dSP}{dt})$ will equal 0. Therefore, the control variable is equal to:

$$CV_{(t)} = -K_D \frac{dPV}{dt} + CV_{(t=0)}$$

As shown in Figure 15-56, a modified derivative controller cannot be used by itself because the error signal is not fed back to the controller for error correction. Therefore, this type of derivative controller must be used in combination with either a proportional or proportional-integral controller.

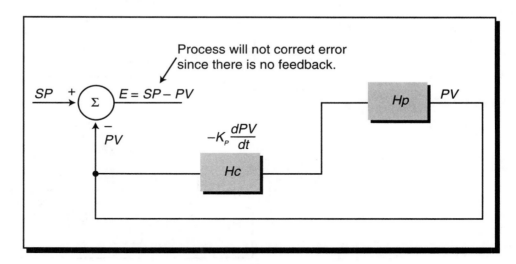

Figure 15-56. Modified derivative controller.

15-9 PROPORTIONAL-DERIVATIVE CONTROLLERS (PD MODE)

Proportional-derivative controllers are composite controllers that combine the actions of proportional and derivative controllers. A PD controller's output equation is represented as (see Figure 15-57):

$$\text{For } \frac{dE}{dt}: \quad CV_{(t)} = K_P E + K_D \frac{dE}{dt} + CV_{(t=0)} \qquad \text{(parallel)}$$

$$CV_{(t)} = K_P E + K_P K_D \frac{dE}{dt} + CV_{(t=0)} \qquad \text{(series)}$$

$$\text{For } \frac{dPV}{dt}: \quad CV_{(t)} = K_P E - K_D \frac{dPV}{dt} + CV_{(t=0)} \qquad \text{(parallel)}$$

$$CV_{(t)} = K_P E - K_P K_D \frac{dPV}{dt} + CV_{(t=0)} \qquad \text{(series)}$$

(a) Parallel PD controller

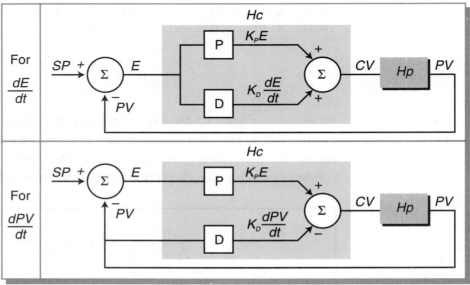

(b) Series PD controller

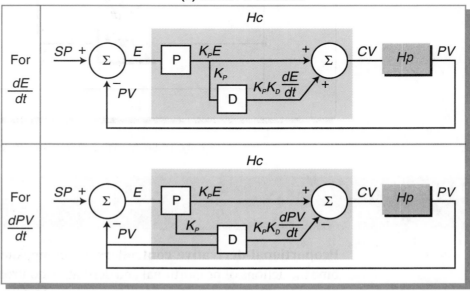

Figure 15-57. (a) Parallel and **(b)** series PD controllers.

These equations are formed by adding the equations for the proportional and derivative actions. Sometimes, the term K_D is replaced with the term T_D, since both have their units in time (seconds or minutes). The term K_D (or T_D) in a series PD controller (see Figure 15-58) indicates the time it takes for the proportional action to equal the derivative action, in other words, for the controller to *repeat* the derivative action.

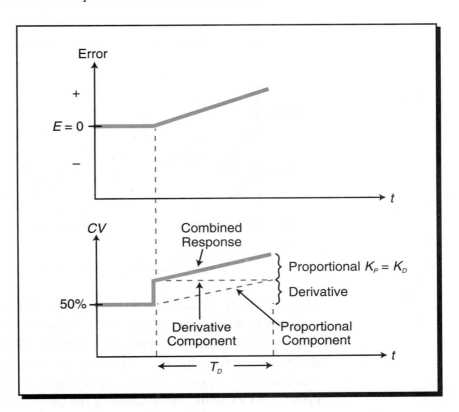

Figure 15-58. Proportional-derivative controller's response to an error.

The derivative component of a PD controller provides a faster response than just the proportional action alone, since it provides an immediate response to an error change that behaves in ramp form (see Figure 15-59). The proportional response to a ramping error is slower than the anticipatory response of a derivative action. The proportional action increases the output as it reads the error level. Since the proportional action only senses the amount of error and not its rate of change, it does not anticipate the top error value until that point is reached. A derivative action, on the other hand, anticipates the error value because it evaluates the rate at which the error is changing and, correspondingly, provides an extra amount of controller output. Therefore, when the error changes in ramp form instead of step form, the derivative gain compensates for the proportional control's delay in action. Although the derivative gain offsets the integral delay in a PD controller, it does not eliminate the offset error at steady state, which is shown in Figure 15-59 at t_5.

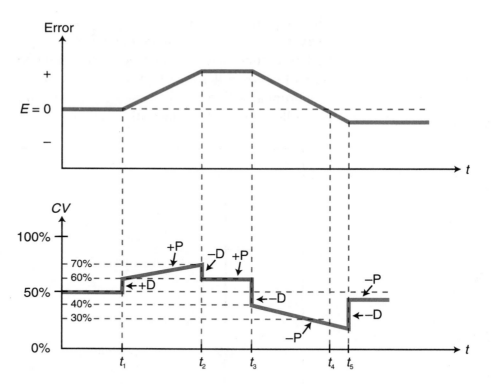

Figure 15-59. PD responses to step changes in error.

The derivative action in a PD controller adds stability to a closed-loop system by reducing the amount of overshoot and undershoot in the system's response. The derivative component acts as a "brake" in the system, slowing the proportional response as the process variable approaches its set point. The speed of response, however, also slows down. To observe this braking effect, let's examine the reaction of the closed-loop system in Figure 15-60 to a unit step. This is a second-order system with a proportional gain of $K_P = 8$ and no derivative gain (switch open). The addition of derivative action to this system will help to stabilize the overshoot and undershoot of the response to a change in error.

If the set point in Figure 15-60 changes, the proportional controller will try to bring the error to zero by making PV equal SP. The error at the start (t_0) is 1, and as PV approaches SP, this error becomes smaller. In this proportional action, the controller output is positive (direct acting), which makes the PV value become more positive. The slope of PV is also positive, as seen at point A. This positive value of $\frac{dPV}{dt}$ can be approximated as shown in Figure 15-60. If derivative action were present in this system (switch closed), then the value of $\frac{dPV}{dt}$ would be negative:

$$PV_{(t)} = K_P E - K_P K_D \frac{dPV}{dt} + PV_{(t=0)}$$

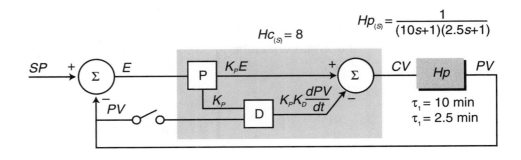

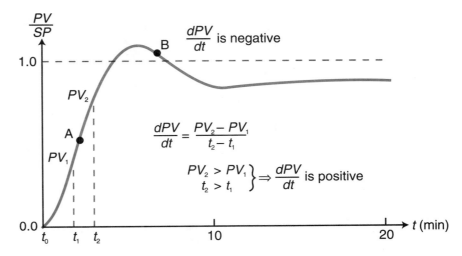

Figure 15-60. Reaction of a closed-loop system to a unit step.

thus having the opposite sign of the proportional gain. Therefore, with derivative action, the output of $PV_{(t)}$ (at point A) would be less than a pure proportional controller without derivative action. In fact, a positive $\frac{dPV}{dt}$ term (slope) would make the derivative term in the PD system equation negative. This indicates that the derivative action of the PD controller will brake the response of the proportional action, therefore reducing the amount of overshoot. The same holds true when the slope is negative, which occurs when the response of the pure proportional action starts to decrease (point B). When the proportional response becomes negative, the derivative term becomes positive, thus braking the undershoot:

$$CV_{(t)} = K_P(-E) - K_P K_D\left(-\frac{dPV}{dt}\right) + CV_{(t\,@\,b)}$$

$$= -K_P E + K_P K_D \frac{dPV}{dt} + CV_{(t\,@\,b)}$$

So, by adding the derivative action to this closed-loop system (switch closed in block diagram), it is possible to reduce the overshoot and undershoot through the braking effect of the derivative action. Figure 15-61 illustrates this closed-loop system for several values of derivative gain K_D (or T_D, derivative time). As the gain of the derivative action increases, the overshoot and undershoot decrease. However, the system response also slows down.

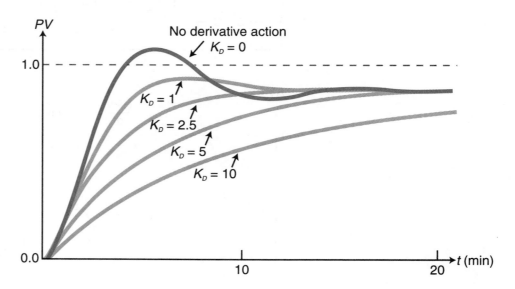

Figure 15-61. Closed-loop process response to a proportional-derivative controller for several values of K_D.

If a proportional-derivative controller has too much derivative gain, the system response will start to look like the graph in Figure 15-62. This indicates that the derivative action is no longer effective in restoring the desired stability margin.

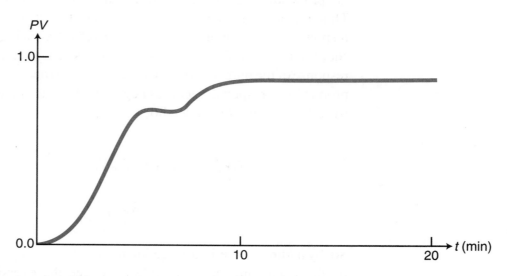

Figure 15-62. Process reponse of a proportional-derivative controller with too much derivative gain.

EXAMPLE 15-9

The closed-loop system described in Examples 15-5 and 15-8 employed proportional and proportional-integral controllers, respectively, to control a first-order system with a gain of 5 and a time constant of $\tau = 30$ seconds. Given that the system utilizes a proportional-derivative controller with a proportional gain of $K_P = 8$ and a derivative gain of 2 minutes (120 seconds), find **(a)** the closed-loop transfer function of the system and **(b)** the steady-state value of the response to a step input $(\frac{1}{s})$.

SOLUTION

(a) The transfer function of the process is:

$$Hp_{(s)} = \frac{5}{30s+1}$$

The controller's transfer function is:

$$Hc_{(s)} = K_P + K_D s$$
$$= 8 + 120s$$

Therefore, the closed-loop system transfer function is:

$$\frac{PV_{(s)}}{SP_{(s)}} = \frac{Hc_{(s)}Hp_{(s)}}{Hc_{(s)}Hp_{(s)}+1}$$

$$= \frac{\left[(8+120s)\left(\frac{5}{30s+1}\right)\right]}{\left[(8+120s)\left(\frac{5}{30s+1}\right)+1\right]}$$

$$= \frac{\left(\frac{40+600s}{30s+1}\right)}{\left(\frac{40+600s}{30s+1}\right)+1}$$

$$= \frac{\left(\frac{40+600s}{30s+1}\right)}{\left(\frac{40+600s+30s+1}{30s+1}\right)}$$

$$= \frac{\left(\frac{40+600s}{30s+1}\right)}{\left(\frac{41+630s}{30s+1}\right)}$$

$$= \frac{40+600s}{41+630s}$$

(b) The system response to a step input change is represented by:

$$PV_{(s)} = SP_{(s)} \left(\frac{40 + 600s}{41 + 630s} \right)$$

$$= \left(\frac{1}{s} \right) \left(\frac{40 + 600s}{41 + 630s} \right)$$

Applying the final value theorem to this Laplace function yields the following value:

$$\lim_{s \to 0} sPV_{(s)} = \lim_{s \to 0} s \left(\frac{1}{s} \right) \left(\frac{40 + 600s}{41 + 630s} \right)$$

$$= \frac{40 + 600(0)}{41 + 630(0)}$$

$$= \frac{40}{41}$$

$$= 0.976$$

So, the final value of the process variable at steady state will be 0.976, producing an offset error of 2.4%.

15-10 PROPORTIONAL-INTEGRAL-DERIVATIVE CONTROLLERS (PID MODE)

A **proportional-integral-derivative (PID) controller** combines the actions of all three controller modes. A PID controller, also called a *three-mode controller*, can be used to control almost any process that involves lags and dead times. A PID controller can be arranged in either a series or parallel configuration using either a standard or modified derivative action ($\frac{dE}{dt}$ and $\frac{dPV}{dt}$, respectively). The basic expression for the process variable output for a standard parallel PID controller (see Figure 15-63) is:

$$PV_{(t)} = K_P E + K_I \int_0^t E dt + K_D \frac{dE}{dt} + PV_{(t=0)}$$

In Laplace form, this controller's transfer function is represented as:

$$Hc_{(s)} = \frac{CV_{(s)}}{E_{(s)}}$$

$$= K_P + \frac{K_I}{s} + K_D s$$

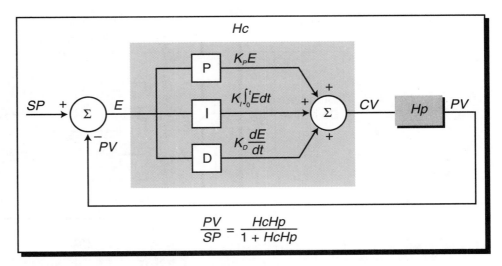

$$\frac{PV}{SP} = \frac{HcHp}{1 + HcHp}$$

Figure 15-63. Standard parallel PID controller.

Figure 15-64 illustrates the serial and parallel system configurations for a PID controller, along with their respective closed-loop equations.

PID control eliminates the offset of the proportional action through its integral action and suppresses oscillation with its derivative action. When properly tuned (see Section 15-12), a PID controller will smoothly regulate the response of a complex system or process.

ORIGINS OF PID CONTROL

In this section, we will explain why the PID controller is the perfect controller for a typical process. To illustrate the relationship between a PID controller and a process, we will examine a typical second-order process system. For computational purposes, a second-order system can be thought of as including a first-order system, in order to determine what type of controller will make the process in an open-loop system have a transfer function equal to one. We will discuss this in more detail shortly.

It is very difficult to determine the exact transfer function ($Hp_{(s)}$) of a real-life process (i.e., a manufacturing process). However, it can be approximated by a second-order system with two lag times and a dead time delay. In Laplace form, this transfer function is defined as:

$$\frac{PV_{(s)}}{CV_{(s)}} = \frac{Ae^{-t_d s}}{(\tau_1 s + 1)(\tau_2 s + 1)}$$

The $e^{-t_d s}$ term, the dead time delay, can be omitted from this equation, since we know that this term only indicates that there is a shift in time in the response. For practical purposes, the dead time will cause the response to behave in the same manner, only displaced in time by the delay. So, for

(a) Parallel PID controller

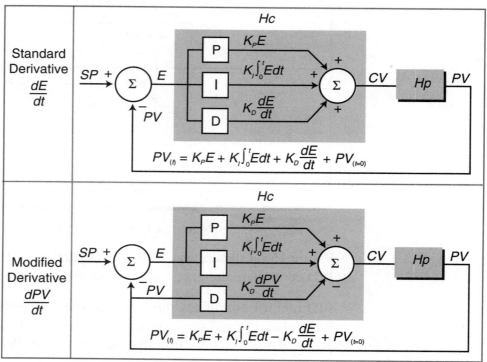

(b) Serial PID controller

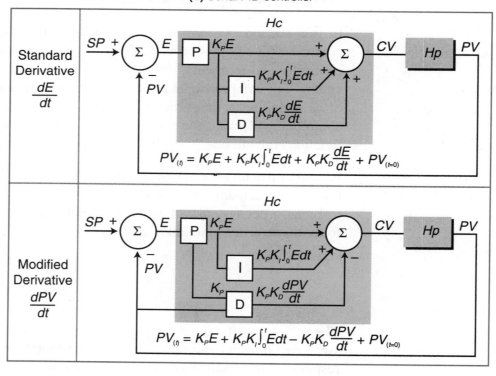

Figure 15-64. (a) Parallel and **(b)** serial PID controllers.

purposes of obtaining an equation for the best controller to govern this system, the delay can be omitted during the initial controller calculations. However, we should remember that there is a dead time response in the system.

The A term in the second-order system transfer function indicates the process gain and the τ_1 and τ_2 terms are the two lag times (see Figure 15-65). This second-order system is said to be inclusive of a first-order system, meaning that if one of the lag times is zero, the second-order equation will represent a first-order system. The ideal transfer function of a perfect process control system ($\frac{PV}{SP}$) should equal one, indicating that the output of the process (PV) immediately follows any changes in the set point without requiring negative feedback to correct the error since there is no error. In other words, if there is a step change in the set point from 0 to 1, the process will respond immediately with a change from 0 to 1. The controller-process relationship in a perfect system is such that they complement each other perfectly. Therefore, the transfer function of the process variable over the set point will be one. Accordingly, the equation for a perfect open-loop system is (see Figure 15-66):

$$\frac{PV_{(s)}}{SP_{(s)}} = Hc_{(s)}Hp_{(s)} = 1$$

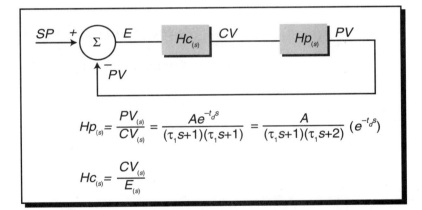

$$Hp_{(s)} = \frac{PV_{(s)}}{CV_{(s)}} = \frac{Ae^{-t_d s}}{(\tau_1 s + 1)(\tau_1 s + 1)} = \frac{A}{(\tau_1 s + 1)(\tau_1 s + 2)}\left(e^{-t_d s}\right)$$

$$Hc_{(s)} = \frac{CV_{(s)}}{E_{(s)}}$$

Figure 15-65. Second-order system.

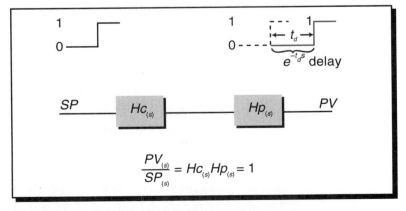

$$e^{-t_d s}\ \text{delay}$$

$$\frac{PV_{(s)}}{SP_{(s)}} = Hc_{(s)}Hp_{(s)} = 1$$

Figure 15-66. Perfect open-loop system.

So, for a perfect system, the controller's transfer function should be the inverse of the processor's transfer function. Therefore, the controller's transfer function in a perfect system using a typical process approximation is:

$$Hc_{(s)} = \frac{1}{Hp_{(s)}}$$

$$= \frac{1}{\left(\frac{A}{(\tau_1 s + 1)(\tau_2 s + 1)}\right)}$$

$$= \left(\frac{1}{A}\right)(\tau_1 s + 1)(\tau_2 s + 1)$$

$$= \left(\frac{1}{A}\right)(\tau_1 \tau_2 s^2 + \tau_1 s + \tau_2 s + 1)$$

$$= \left(\frac{1}{A}\right)\left[\tau_1 \tau_2 s^2 + (\tau_1 + \tau_2)s + 1\right]$$

The term $\frac{1}{A}$ is a constant; therefore, it can be renamed as A_1:

$$Hc_{(s)} = A_1\left[\tau_1 \tau_2 s^2 + (\tau_1 + \tau_2)s + 1\right]$$

Dividing each term in the bracket by s yields:

$$Hc_{(s)} = A_1\left[\frac{\tau_1 \tau_2 s^2}{s} + \frac{(\tau_1 + \tau_2)s}{s} + \frac{1}{s}\right]$$

$$= A_1\left[\tau_1 \tau_2 s + (\tau_1 + \tau_2) + \frac{1}{s}\right]$$

Multiplying the A_1 term and rearranging the equation produces:

$$Hc_{(s)} = \underbrace{A_1(\tau_1 + \tau_2)}_{P} + \underbrace{\frac{A_1}{s}}_{I} + \underbrace{A_1 \tau_1 \tau_2 s}_{D}$$

The terms in this equation indicate that the controller has a proportional gain, an integral action ($\frac{1}{s}$), and a derivative component (s). Therefore, the perfect system controller exhibits proportional, integral, and derivative actions. Note that the constant gain term in this equation does not imply that the gains should be the same for each of the PID actions. Rather, it indicates that these gains must be present and specified. Because a PID-type controller is the natural derivation from a perfect system, PID is considered a universal type of control for manufacturing processes. In fact, of all the PID configurations shown in Figure 15-64, perhaps the most commonly used in PLCs is the serial, modified derivative, PID configuration.

DIGITAL IMPLEMENTATION OF PID IN A PLC

A programmable controller system implements the PID control action using a discrete, or digital, algorithm to update the control variable (*CV*). For example, a modified serial PID controller may use the following digital algorithm, where the current control variable output (*CV_n*) is represented as:

$$CV_n = CV_{(n-1)} + K_P\left(E_n - E_{(n-1)}\right) + K_P K_I T_s E_n - \left(\frac{K_P K_D}{T_s}\right)\left(PV_n - 2PV_{(n-1)} + PV_{(n-2)}\right)$$

where:

$$CV_n = \text{the controller output at the } n\text{th update}$$

$$CV_{(n-1)} = \text{the controller output at the } n\text{th minus one update}$$

$$K_P = \text{the proportional gain (in seconds, where appropriate)}$$

$$K_I = \text{the integral gain (in seconds, where appropriate)}$$

$$K_D = \text{the derivative gain (in seconds, where appropriate)}$$

$$E_n = \text{the error at the } n\text{th update}$$

$$E_{(n-1)} = \text{the error at the } n\text{th minus one update}$$

$$T_s = \text{the loop sample time in seconds}$$

$$PV_n = \text{the process variable at the } n\text{th update}$$

$$PV_{(n-1)} = \text{the process variable at the } n\text{th minus one update}$$

$$PV_{(n-2)} = \text{the process variable at the } n\text{th minus two update}$$

The loop sample time (T_s) is the frequency of how often the PLC reads and executes the integration and derivative terms in the algorithm equation. In PLCs, this time can be selected from a range of 0.1 seconds to several hundred seconds (e.g., 600 seconds, or 10 minutes) Figure 15-67 illustrates several sampling rates. A small value of T_s (fast update time) is desirable in a process application where the process variable responds rapidly to control variable changes. However, because large values of T_s are necessary to evoke a stable derivative action, the trade-off between a low and high T_s value must be balanced carefully to ensure a correct system response. Otherwise, the derivative action can produce a bumpy action.

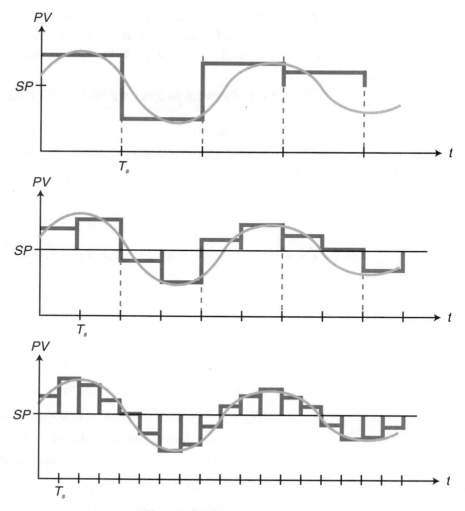

Figure 15-67. Loop sample rates.

The digital PID algorithm implemented in PLC systems calculates the error by approximating the area between the process variable and the set point (see Figure 15-68). This area calculation provides an approximate value of error.

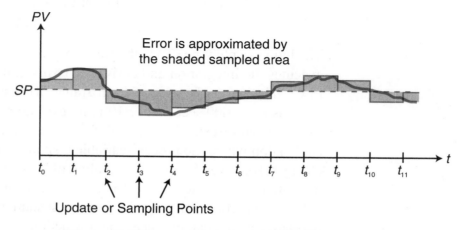

Figure 15-68. Error approximation using loop sample times.

INTEGRAL (RESET) WINDUP

As discussed in Section 15-7, integral (or reset) windup is a problematic condition that occurs in PI and PID controllers, resulting in the saturation of the controller's output ($CV = CV_{max}$). Integral windup is typical in startup situations during batch processes. To avoid integral windup, some PLC manufacturers offer PID interfaces that prevent integral action when the controller's output reaches 100%. These interfaces accomplish this by forcing the error input to the integrator section of the PID controller to zero. The block diagram in Figure 15-69 illustrates this method of integral windup prevention.

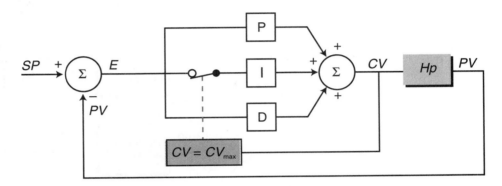

Figure 15-69. PID controller with integral windup prevention.

PID BUMPLESS AUTO/MANUAL TRANSFER

Most PLC applications that implement PID control employ automatic/manual control stations that allow the operator to switch between manual and PLC process control. To prevent a step change or "bump" during this switch, the control station must ensure that both controllers, the manual controller and the PLC (automatic), send the same output (CV) to the process. Otherwise, the process may receive a change in the control variable, which could produce a transient response in the system.

Figure 15-70 illustrates a PLC system that uses a PID controller interface with a manual control station that allows for bumpless transfer. Basically, the automatic (PLC) and manual controllers must follow each others outputs when they are operating. Figure 15-71 illustrates this configuration in block diagram form for a modified serial PID controller. When the system is in manual mode, the PID controller tracks the manual controller's output, so that when the transfer from manual to automatic occurs, both controller outputs are the same. A similar operation takes place during an automatic-to-manual transfer.

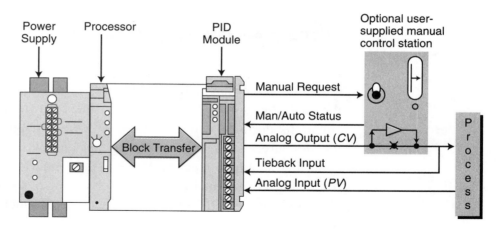

Figure 15-70. PID interface with a manual control station for bumpless transfer.

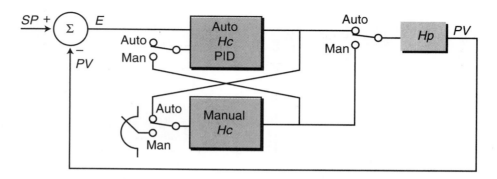

Figure 15-71. Auto/manual control station block diagram.

During a manual-to-automatic transfer in a PLC system, the PID interface processor may also set the set point equal to the process variable. This forces the system error ($SP - PV$) to zero, ensuring that a bump does not occur during the transfer. The control variable output of the PLC controller, which tracks the manual CV output, is left unchanged during the transfer. After the transfer, the PID processor returns the set point to its original value.

15-11 ADVANCED CONTROL SYSTEMS

While conventional PID control provides universal control for most processes, other techniques can increase the performance of a process control system. One of the most commonly practiced techniques used to increase process control performance is cascade control.

CASCADE CONTROL

Cascade control uses two controllers configured so that the output of one feedback loop becomes the set point for the other one. Figure 15-72a illustrates a temperature control batch system that utilizes a single PID controller, while Figure 15-72b shows the same system with cascade

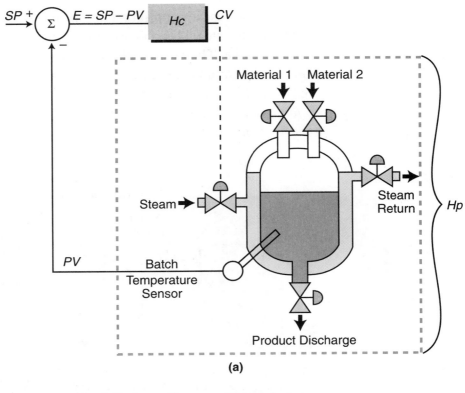

(a)

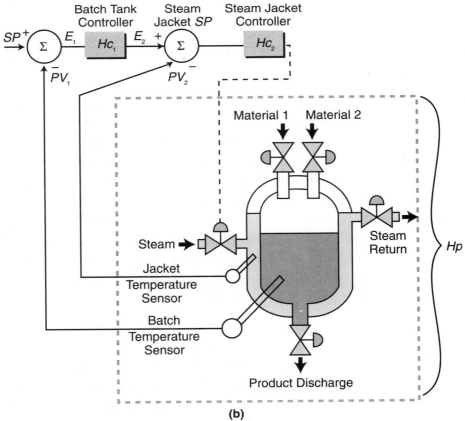

(b)

Figure 15-72. Temperature control systems with **(a)** single and **(b)** cascaded PID controllers.

control. In the cascade configuration, the batch tank controller provides the set point for the steam jacket temperature controller, which in turn, actuates the steam valve. The batch tank loop is called the *primary loop*, since the main process variable (the batch temperature) is the primary control concern (see Figure 15-73). The steam jacket temperature loop is called the *secondary loop,* or *inner loop,* since the jacket temperature is of secondary interest in the control system.

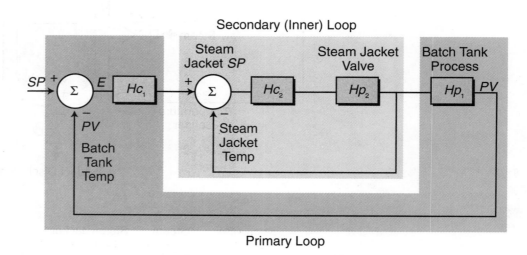

Figure 15-73. Primary and secondary loops of the temperature control system.

The greatest advantage of cascade control systems, and in fact the main reason for their use, is that they respond quicker than single-controller systems to disturbances that affect the primary loop. In cascade control, the secondary loop response to a disturbance generally occurs first, before the primary loop starts to respond. In the batch system shown in Figure 15-72b, a change in steam temperature will affect the tank's jacket temperature before it affects the main batch temperature due to the lag and dead times associated with the batch process. The steam jacket secondary control loop will respond first to this disturbance and try to correct it, thus minimizing the effect of the disturbance on the main batch system. This fast response of the secondary loop enhances the performance of cascaded control systems as compared to single-loop control systems.

Most programmable controller systems allow cascade control directly to the PID intelligent interface or analog input modules (if PID is implemented in the main PLC processor). Therefore, the user must only identify the input to the secondary loop (see Figure 15-74). The secondary loop cascade input is also referred to as the *remote input* in conventional, single-loop controllers.

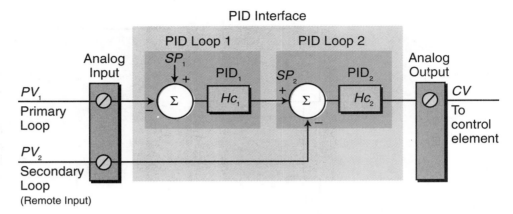

Figure 15-74. Cascade control directly to a PID interface.

BUMPLESS CASCADE CONTROL

PLC systems also provide bumpless transfer in cascade control configurations. Most often, the transfer sequence from manual to automatic is initiated in the secondary loop (see Figure 15-75). Once the secondary loop is placed in automatic mode, the secondary loop set point is set to the value of the secondary process variable PV_2. Then, the primary loop is placed in automatic mode and the primary loop set point is set to the value of the primary process variable PV. The primary loop's output is left unchanged. This is the initial operation that avoids a bump. From there, the primary loop's set point returns to the desired value, and the primary's output adjusts the set point of the secondary loop controller. In general, the primary loop cannot be activated unless the secondary loop is already active or in the AUTO mode.

The tuning of cascade controllers, which we'll explain in the next section, must be performed in a sequential, logical fashion. In most systems, the user must tune the secondary loop first, with the primary loop in manual mode. After the secondary loop is tuned, the tuning of the primary loop can begin.

15-12 CONTROLLER LOOP TUNING

For a process control system to work correctly, its control loop(s) must be tuned. **Loop tuning** involves selecting the constants [K_P, K_I (or T_I), and K_D (or T_D)] that will be used with the proportional, integral, and derivative actions of a controller. With these constants at the proper levels, the controller can effectively and efficiently regulate the process variable to the set point.

A process often experiences disturbances caused by changes in the set point or the process load (see Figure 15-76). These disturbances cause an error in the system, thereby changing the controller output, which in turn, impacts

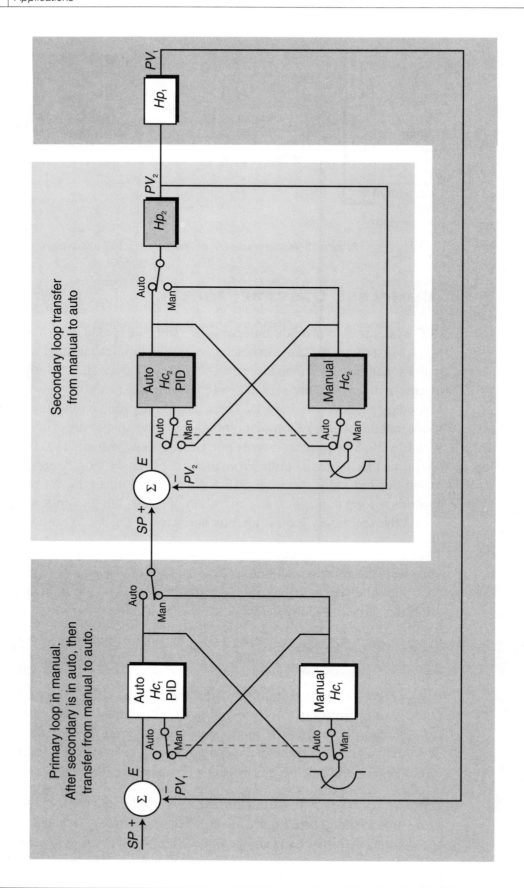

Figure 15-75. Bumpless transfer in cascaded PID controllers.

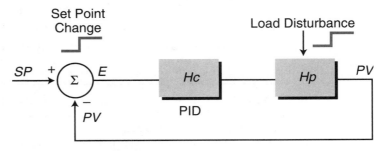

Figure 15-76. Process disturbances.

the process variable. The process variable response to both system distur-bances and the controller action must be stable; that is, it must not oscillate or grow in value without limit. This process variable response can typically be categorized as either overdamped, critically damped, or underdamped (see Figure 15-77a). However, another type of process variable response is a **quarter-amplitude response**, also called a *quarter-delay ratio response* (see Figure 15-77b). This response, which is the desired response after closed-loop tuning, reduces the *PV* overshoot by one-quarter each cycle.

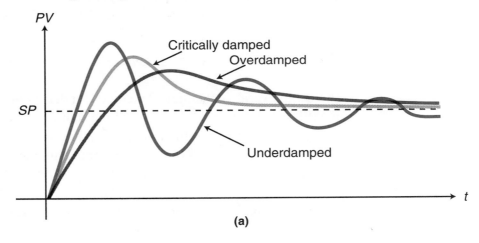

(a)

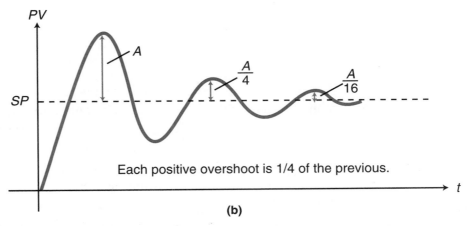

(b)

Figure 15-77. Process variable reponses: **(a)** overdamped, critically damped, underdamped, and **(b)** quarter-amplitude.

Table 15-2 shows the characteristics of each type of process variable response. The tuning parameters of the system will have a decisive impact on the type of process response exhibited by the system.

Response Type	Quality of Response
Overdamped	This response approaches the set point smoothly without oscillation. The maximum deviation from the set point is less than that of a critically damped response.
Critically Damped	This response approaches the set point smoothly, but faster than an overdamped response. This results in a larger maximum deviation from the set point. This is a relatively good response for applications in which oscillation is not acceptable.
Underdamped	This response cycles, producing several deviations over and under the set point before settling to a steady-state value. It gives a fast response to a disturbance change.
Quarter Amplitude	This response is the desired tuned system response. The closed-loop system reduces the magnitude of each overshoot by one-quarter of the previous one.

Table 15-2. Process variable response characteristics.

There are several mathematical methods for determining the tuning constants of a PID controller and for analyzing the stability of a system. One method is a *Bode plot analysis*, which analyzes amplitude (gain) and frequency response (phase shifts) to tune the system. This method is useful when the process transfer function is known. However, in continuous manufacturing processes, this is rarely the case. Therefore, we will present three other practical methods for determining the tuning constants that will produce a quality stable response. These methods are:

- the Ziegler-Nichols open-loop tuning method

- the integral of time and absolute error (ITAE) open-loop tuning method

- the Ziegler-Nichols closed-loop tuning method

The open-loop methods test the process response while the controller is in manual mode without any feedback connections (i.e., *PV* is not being fed back to the controller). Batch control processes are typical applications for open-loop tuning methods. The closed-loop technique tests the process response when the controller is in automatic mode. This tends to produce a better result, since the controller and the process are operating normally. Servo and positioning control processes are typical applications for closed-loop tuning methods. These processes cannot be tuned without feedback; therefore, they cannot use open-loop tuning techniques.

ZIEGLER–NICHOLS OPEN-LOOP TUNING METHOD

John Ziegler and Nathaniel Nichols developed the **Ziegler-Nichols open-loop tuning method** in 1942, and it remains a popular technique for tuning controllers that use proportional, integral, and derivative actions. The Ziegler-Nichols open-loop method is also referred to as a *process reaction method*, because it tests the open-loop reaction of the process to a change in the control variable output (see Figure 15-78). This basic test requires that the response of the system be recorded, preferably by a chart recorder or plotter. Once certain process response values are found, they can be plugged into the Ziegler-Nichols equation with specific multiplier constants for the gains of a controller with either P, PI, or PID actions.

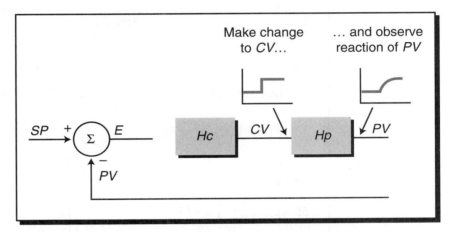

Figure 15-78. Ziegler-Nichols open-loop tuning method.

To use the Ziegler-Nichols open-loop tuning method, you must perform the following steps, which we will illustrate using the system in Figure 15-79:

1. **Bring *PV* to 50%.** With the controller in manual mode (see Figure 15-79), vary the controller's output (*CV*) so that the process variable is at 50% of its range. Turn on the chart recorder and let the system stabilize. For the system in Figure 15-79, let's assume that the control variable must be increased from 50% to 55% to increase *PV* from 40% to 50% of its range.

2. **Step change the *CV* output by 10%.** Manually step the controller's output (*CV*) by 10%. Record on the chart the time value when the step occurs. Observe the process variable response. In Figure 15-79, *CV* steps from 55% to 65% at t_1. The final value of *PV* in response to this change is 165°F, or 65% of the full range, at steady state.

   Figure 15-80 illustrates the process variable response. This response provides important information about the lag time and the rate of change of *PV*.

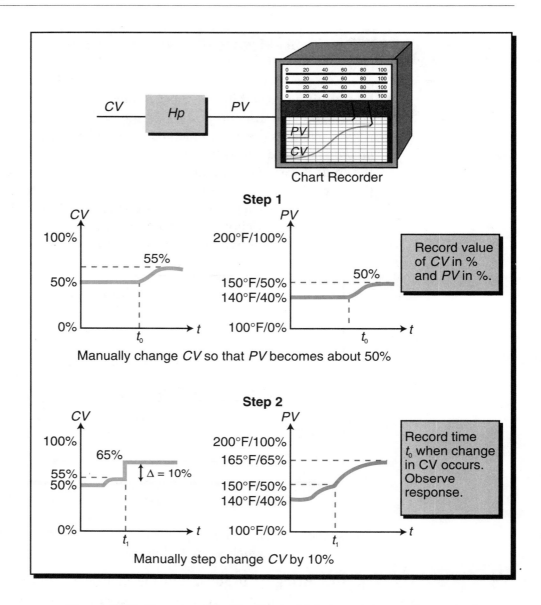

Figure 15-79. Steps 1 and 2 of the Ziegler-Nichols open-loop tuning method.

3. **Find the reaction rate.** Extend the line for the process variable response before the step change (see point A in Figure 15-80). Draw a tangent (point B) to the *PV* response to the *CV* step change at the steepest rise point on the graph to determine the *reaction rate* (*N*) of the process variable. The reaction rate is equal to the change in the process variable over the change in time. It is found by making a right triangle from the tangent line (point C) and finding the tangent of angle θ (the value of the opposite side of the triangle divided by the value of the adjacent side).

4. **Calculate the lag time.** To determine the lag time L_t, find the point at which the tangent line intersects the extension of the original

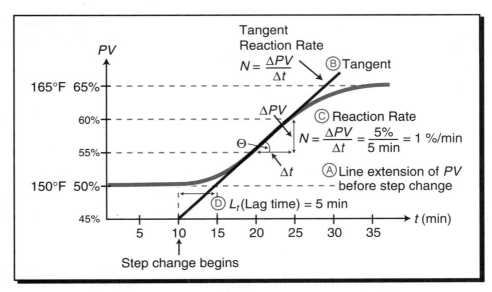

Figure 15-80. Process variable response to step change.

process variable response line (point D). Subtract the time at which the step change occurred from the time at which the tangent and response extension lines crossed.

5. **Determine the loop tuning constants.** Plug in the reaction rate and lag time values to the Ziegler-Nichols open-loop tuning equations for the appropriate type of controller—P, PI, or PID—to calculate the controller constants. Table 15-3 shows these tuning equations. For our example, the constant values for a P, PI, and PID controller would be:

$$\text{P mode}: \quad K_P = \frac{\Delta CV}{(L_t)(N)} = \frac{10\%}{(5\text{ min})(1\frac{\%}{\min})} = 2\frac{\%}{\%}$$

$$\text{PI mode}: \quad K_P = \frac{(0.9)\Delta CV}{(L_t)(N)} = \frac{(0.9)(10\%)}{(5\text{ min})(1\frac{\%}{\min})} = 1.8\frac{\%}{\%}$$
$$T_I = (3.33)(L_t) = (3.33)(5\text{ min}) = 16.65\text{ min}$$

$$\text{PID mode}: K_P = \frac{(1.2)\Delta CV}{(L_t)(N)} = \frac{(1.2)(10\%)}{(5\text{ min})(1\frac{\%}{\min})} = 2.4\frac{\%}{\%}$$
$$T_I = (2)(L_t) = (2)(5\text{ min}) = 10\text{ min}$$
$$T_D = (0.5)(L_t) = (0.5)(5\text{ min}) = 2.5\text{ min}$$

The objective of these tuning constants is to produce a quarter-amplitude response in the process variable.

Type of Controller	Loop Tuning Constant	Tuning Equation
Proportional (P)	K_P	$K_P = \dfrac{\Delta CV}{(L_t)(N)}$
Proportional-Integral (PI)	K_P	$K_P = \dfrac{(0.9)(\Delta CV)}{(L_t)(N)}$
	T_I	$T_I = (3.33)(L_t)$
Proportional-Integral-Derivative (PID)	K_P	$K_P = \dfrac{(1.2)(\Delta CV)}{(L_t)(N)}$
	T_I	$T_I = (2)(L_t)$
	T_D	$T_D = (0.5)(L_t)$

Table 15-3. Ziegler-Nichols open-loop tuning equations.

There are two problems with the Ziegler-Nichols open-loop tuning method. The first problem is that the ratio of the derivative time T_D to the integral time T_I in the equations is designed for a quarter-amplitude response:

$$\frac{T_D}{T_I} = \frac{0.5L_t}{2L_t} = \frac{1}{4} \quad \Rightarrow \quad T_I = 4T_D$$

This does not allow for small changes in T_I and/or T_D. For instance, a system with a PID controller and a process that has a large error at stabilization and a tendency to overshoot and undershoot the set point requires an increase in integral action and an increase in derivative action at the same time. In a PID controller's output equation, the derivative time and the integral time are inversely related:

$$CV_{(t)} = K_P E + \frac{K_P}{T_I} \int_0^t E \, dt - K_P T_D \frac{dPV}{dt} + CV_{(t=0)}$$

For the derivative action to increase, the derivative gain must increase; yet, for the integral action to increase, the integral gain must decrease. However, in the open-loop, quarter-amplitude tuning equations, the relationship between T_I and T_D is $T_I = 4T_D$, meaning that an increase in T_I causes an increase in T_D. Thus, this relationship causes an imbalance in the controller's PID output equation.

The second problem occurs in systems that have processes in which the lag time L_t equals the dead time D_T. In this situation, the derivative time T_D is equal to:

$$T_D = 0.5L_t$$

or

$$T_D = 0.5D_T \quad \text{(when } D_T = L_t\text{)}$$

This means that, as the dead time gets larger, the derivative time T_D also increases. In a process with a long dead time, the opposite is required. As the dead time increases, the derivative action should decrease to compensate for it. This is due to the fact that the dead time is a time delay, which changes the derivative action's effect on the overshoot from the desired negative feedback braking effect into an aggravating effect similar to an undesired positive feedback loop. This is similar to driving a car on an icy surface—the car can veer out of control due to the driver's delay in steering correction because of the slippery road.

EXAMPLE 15-10

The Ziegler-Nichols open-loop tuning method was used to obtain the process response shown in Figure 15-81. Find the tuning parameters for a serial PID controller given that the control variable change that caused this response was 11%.

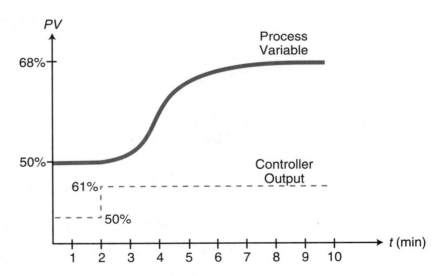

Figure 15-81. Process response obtained by the Ziegler-Nichols open-loop tuning method.

SOLUTION

Figure 15-82 shows the tangent used to determine the tuning values. The lag time L_t is estimated at 1.15 minutes (3.15 min – 2 min). The value of the reaction time N is calculated by finding the tangent of the

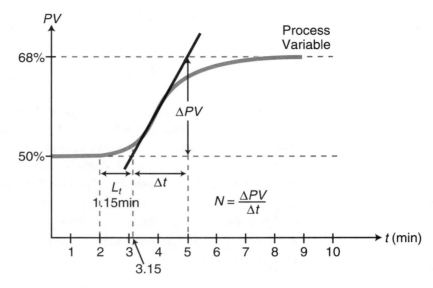

Figure 15-82. Tangent used to determine tuning values.

angle formed by the intersection of the tangent line with the *PV* line extension:

$$N = \frac{\Delta PV}{\Delta t}$$

$$= \frac{68\% - 50\%}{5\,\text{min} - 3.15\,\text{min}}$$

$$= \frac{18\%}{1.85\,\text{min}}$$

$$= 9.730$$

The PID tuning constants, using the Ziegler-Nichols open-loop equations from Table 15-3, are:

$$K_P = \frac{(1.2)\Delta CV}{(L_t)(N)} = \frac{(1.2)(11\%)}{(1.15\,\text{min})(9.73\,\tfrac{\%}{\text{min}})} = 1.18\,\tfrac{\%}{\%}$$

$$T_I = (2)(L_t) = (2)(1.15\,\text{min}) = 2.3\,\text{min}$$

$$T_D = (0.5)(L_t) = (0.5)(1.15\,\text{min}) = 0.575\,\text{min}$$

ITAE OPEN-LOOP TUNING METHOD

The **integral of time and absolute error (ITAE) open-loop tuning method** produces less response oscillation than the Ziegler-Nichols open-loop method and also minimizes the problems associated with it. This method can be used to calculate the tuning constants for processes, such as pH control, that cannot tolerate as much oscillation as produced by the quarter-amplitude response. The ITAE method, which is based on the minimization of the integral of time and the absolute error of the response, is represented by:

$$\text{ITAE} = \int_0^\infty E_{(t)} t \, dt$$

where $E_{(t)}$ is the absolute error as a function of time.

The minimization of the overshoot error of a quarter-amplitude response, such as the one achieved with the open-loop Ziegler-Nichols method, occurs during the first overshoot in an ITAE-tuned controller, bringing the system response close to the behavior of a critically damped response. The controller's ITAE tuning settings result in a response that minimizes the first and second amplitude overshoots and virtually eliminates the third (see Figure 15-83). In fact, the ratio of the damping of the second overshoot to the first overshoot is less than 1/8 (second overshoot divided by first overshoot); therefore, the response approximates critically damped behavior. Consequently, the damping in the ITAE method is much better than that in the Ziegler-Nichols open-loop method.

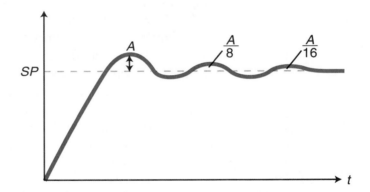

Figure 15-83. ITAE-tuned controller with minimized overshoot.

The procedure for obtaining the controller's tuning constants using the ITAE method is the same as that for the Ziegler-Nichols open-loop method, except that the data interpretation of the graphic response is more detailed. As an example, let's examine the response obtained previously with the Ziegler-Nichols open-loop method and apply the ITAE techniques to obtain the new equation values.

Figure 15-84 illustrates the same response as before with new values for the dead time (D_T) and the process lag (τ, previously called L_t). The tangent to the response is drawn at the steepest rise (like in the Ziegler-Nichols open-loop method). This tangent line determines D_T and τ. Note that τ is calculated from the intersection of the tangent and the *PV* value extension to the time

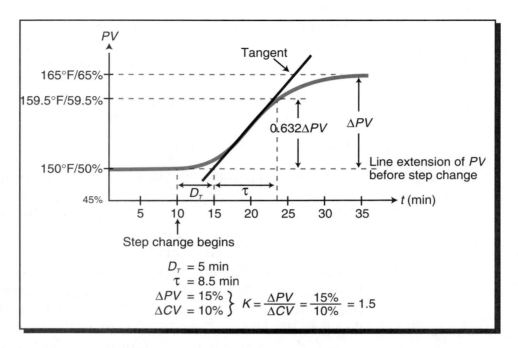

Figure 15-84. Process variable response to step change.

where the actual response, not the tangent, has a value of 63.2% of the final steady-state value. Whereas the final value of the response at steady-state was not used in the Ziegler-Nichols open-loop method, it is used in the ITAE method to determine the percentage gain in the process variable's response. This gain in the process value (ΔPV) is used to determine the gain value K, which will be used in the tuning equations. The gain K is equal to ΔPV divided by ΔCV, where the term ΔCV is the manual change (in percentage of range) executed by the controller's output (process input) over the controlling element (e.g., steam valve). Table 15-4 shows the tuning equations for the ITAE open-loop tuning method.

For the example shown in Figure 15-84, the values of the tuning constants will be:

$$\text{Process gain}: \quad K = \frac{\Delta PV}{\Delta CV} = \frac{15\%}{10\%} = 1.5$$

$$\text{P mode}: \quad K_P = \frac{0.490}{K}\left(\frac{D_T}{\tau}\right)^{-1.084} = \frac{0.490}{1.5}\left(\frac{5}{8.5}\right)^{-1.084} = 0.581$$

$$\text{PI mode}: \quad K_P = \frac{0.586}{K}\left(\frac{D_T}{\tau}\right)^{-0.916} = \frac{0.586}{1.5}\left(\frac{5}{8.5}\right)^{-0.916} = 0.635$$

$$T_I = \frac{\tau}{1.03 - 0.165\left(\frac{D_T}{\tau}\right)} = \frac{8.5}{1.03 - 0.165\left(\frac{5}{8.5}\right)} = 9.111\,\text{min}$$

Type of Controller	Loop Tuning Constant	Tuning Equation
Proportional (P)	K_P	$K_P = \dfrac{0.490}{K}\left(\dfrac{D_T}{\tau}\right)^{-1.084}$
Proportional-Integral (PI)	K_P	$K_P = \dfrac{0.586}{K}\left(\dfrac{D_T}{\tau}\right)^{-0.916}$
	T_I	$T_I = \dfrac{\tau}{1.03 - 0.165\left(\frac{D_T}{\tau}\right)}$
Proportional-Integral-Derivative (PID)	K_P	$K_P = \dfrac{0.965}{K}\left(\dfrac{D_T}{\tau}\right)^{-0.855}$
	T_I	$T_I = \dfrac{\tau}{0.796 - 0.147\left(\frac{D_T}{\tau}\right)}$
	T_D	$T_D = 0.308\tau\left(\dfrac{D_T}{\tau}\right)^{0.929}$

Note: $K = \dfrac{\Delta PV}{\Delta CV}$

Table 15-4. ITAE open-loop tuning equations.

PID mode : $K_P = \dfrac{0.965}{K}\left(\dfrac{D_T}{\tau}\right)^{-0.855} = \dfrac{0.965}{1.5}\left(\dfrac{5}{8.5}\right)^{-0.855} = 1.013$

$T_I = \dfrac{\tau}{0.796 - 0.147\left(\frac{D_T}{\tau}\right)} = \dfrac{8.5}{0.796 - 0.147\left(\frac{5}{8.5}\right)} = 11.980 \text{ min}$

$T_D = 0.308\tau\left(\dfrac{D_T}{\tau}\right)^{0.929} = (0.308)(8.5)\left(\dfrac{5}{8.5}\right)^{0.929} = 1.599 \text{ min}$

In the ITAE loop tuning method, the controller settings ensure a damping ratio of less than 1/8 for the P and PI modes. The PID mode, however, still presents a problem in systems with large dead times, although this problem is not as severe as it is in the Ziegler-Nichols open-loop method. This problem stems from the fact that the exponent of the derivative action (0.929) term T_D is close to the value of 1, which makes an approximate value of T_D be 0.308 times the value of the dead time:

$$T_D \approx 0.308\tau\left(\dfrac{D_T}{\tau}\right)^{\approx 1}$$

$$\approx 0.308\tau\left(\dfrac{D_T}{\tau}\right)$$

$$\approx 0.308 D_T$$

As discussed earlier, a derivative term proportional to the dead time can cause an aggravating response in the system that affects the overshoot. This problem in the ITAE method, however, is less pronounced than it is in the Ziegler-Nichols open-loop method where $T_D = 0.5D_T$ when $L_t = D_T$. Note also that the ITAE method does not contain a fixed ratio constant of T_D/T_I, which is the case in the Ziegler-Nichols open-loop method ($T_D/T_I = 1/4$). Therefore, the ITAE method cannot cause a potential imbalance in the PID controller equation.

Derivative control action should not be used in processes with large dead times. As a de facto rule of thumb, a large dead time is one in which the ITAE open-loop test produces a value of D_T that is greater than τ. If this is the case, then T_D should be set to zero, implementing control without derivative action.

ZIEGLER-NICHOLS CLOSED-LOOP TUNING METHOD

The **Ziegler-Nichols closed-loop tuning method** is used to obtain the controller constants [K_P, K_I (or T_I), and K_D (or T_D)] in a system with feedback. This technique allows for the tuning of processes, such as servo positioning systems, that cannot run in an open-loop environment.

The main objective of the Ziegler-Nichols closed-loop method is to find the value of the proportional-only gain that causes the control loop to oscillate indefinitely at a constant amplitude (see Figure 15-85). This gain, which causes steady-state oscillations, is called the *ultimate proportional gain* (K_{PU}). Another important value associated with this proportional-only control tuning method is the *ultimate period* (T_U). The ultimate period is the time required to complete one full oscillation once the response begins to oscillate at a constant amplitude. These two parameters, K_{PU} and T_U, are used to find the loop-tuning constants of the controller (P, PI, or PID). To find the values of these parameters and to calculate the tuning constants, you must do the following:

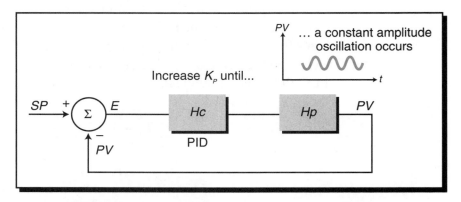

Figure 15-85. Ziegler-Nichols closed-loop tuning method.

1. **Implement proportional-only control.** Remove all integral and derivative actions. In most controllers, the removal of integral time (T_I) is done by setting T_I equal to 999 (or its largest number) or by setting K_I equal to 0. To remove the derivative action, set K_D (or T_D) to 0. Place the controller in automatic mode with the control variable and process variable at 50%.

2. **Create a disturbance in the system.** Create a small disturbance in the loop by slightly changing the set point (see point A in Figure 15-86). Start increasing the proportional gain, or lowering the percentage proportional band (%PB), until the process variable begins to oscillate (point B). Continue to increase and decrease the gain until the oscillations have a constant amplitude (point C). Record this response and determine the ultimate proportional gain and ultimate period.

   In the example system in Figure 15-86, the set point of 150°F is slightly changed to 155°F while the gain is increased to $K_D = 3$ (point A). Once the oscillation starts, the set point is returned to 150°F. The oscillation begins to decay at t_2, so the gain is increased again to $K_P = 4$ (point B), However, the response starts to grow in amplitude at t_3, so the gain is reduced to $K_P = 3.5$. At this point, the response exhibits a constant amplitude oscillation (point C). Therefore, the ultimate gain (K_{PU}) is 3.5 and the ultimate period (T_U) is 10 minutes.

3. **Calculate the constants.** Plug the K_{PU} and T_U values into the Ziegler-Nichols closed-loop tuning equations to determine the settings for the controller to be used. Table 15-5 provides the tuning equations for this closed-loop method.

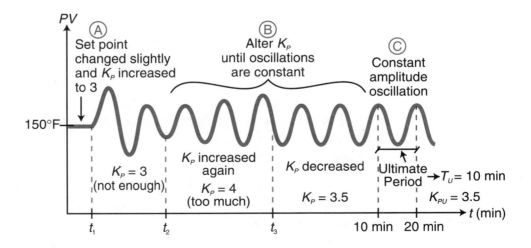

Figure 15-86. System tuned using the Ziegler-Nichols closed-loop tuning method.

Type of Controller	Loop Tuning Constant	Tuning Equation
Proportional (P)	K_P	$K_P = (0.5)(K_{PU})$
Proportional-Integral (PI)	K_P	$K_P = (0.45)(K_{PU})$
	T_I	$T_I = \dfrac{T_U}{1.2}$
Proportional-Integral-Derivative (PID)	K_P	$K_P = (0.6)(K_{PU})$
	T_I	$T_I = \dfrac{T_U}{2}$
	T_D	$T_D = \dfrac{T_U}{8}$

Note: $\%PB = \dfrac{1}{K_P}$; $K_I = \dfrac{1}{T_I}$; $K_D = T_D$

Table 15-5. Ziegler-Nichols closed-loop tuning equations.

For the example system in Figure 15-86, the tuning constants for each controller mode will be:

$$P \text{ mode}: \quad K_P = (0.5)(K_{PU}) = (0.5)(3.5) = 1.75$$

$$PI \text{ mode}: \quad K_P = (0.45)(K_{PU}) = (0.45)(3.5) = 1.575$$

$$T_I = \frac{T_U}{1.2} = \frac{10 \text{ min}}{1.2} = 8.33 \text{ min}$$

$$PID \text{ mode}: \quad K_P = (0.6)(K_{PU}) = (0.6)(3.5) = 2.10$$

$$T_I = \frac{T_U}{2} = \frac{10 \text{ min}}{2} = 5 \text{ min}$$

$$T_D = \frac{T_U}{8} = \frac{10 \text{ min}}{8} = 1.25 \text{ min}$$

The magnitude of the constant oscillation amplitude is not important in the Ziegler-Nichols closed-loop tuning equations; however, all the elements in the loop must be within operating range. For example, the control variable must not vary from fully open to fully closed to create the oscillation.

The Ziegler-Nichols closed-loop method provides a quarter-amplitude response. This response is acceptable for P and PI modes; however, in PID mode, it presents the same equation imbalance as experienced in the Ziegler-Nichols open-loop technique. Again, this is due to the fixed ratio of the derivative time to the reset time.

Another problem with this closed-loop technique is that the majority of process control loops in manufacturing operations cannot tolerate oscillations for long periods of time, especially if many trials are necessary. The time required to obtain a steady-state oscillating response is typically 30–60 minutes, but it can take up to several hundred minutes. The Ziegler-Nichols closed-loop method can be slightly altered to avoid this time problem.

ALTERED ZIEGLER-NICHOLS CLOSED-LOOP TUNING METHOD

The altered version of the Ziegler-Nichols closed-loop method provides an approximation of the desired quarter-amplitude response. The test procedure in this method reduces the time required to observe the process variable's steady-state response to a change in controller output. Instead of changing the value of *CV* (through the inverse of the proportional gain) until *PV* exhibits a constant amplitude oscillation, this altered method compares the responses of several trials until an approximate quarter-amplitude response is obtained. The procedures and steps used in this altered method are the same as in the standard Ziegler-Nichols closed-loop method, where several trials are performed by changing the proportional gain until the response approximates a quarter-amplitude response.

Once the process variable response approximates a quarter-amplitude response, the chart recorder records the period of decaying oscillation (see Figure 15-87), as well as the gain of the proportional action. The gain and decaying oscillation period for this quarter-amplitude test are called $K_{1/4}$ (quarter-amplitude gain) and $T_{1/4}$ (quarter-amplitude period), instead of K_{PU} and T_U. These $K_{1/4}$ and $T_{1/4}$ parameters are then plugged into the standard Ziegler-Nichols closed-loop tuning equations (shown previously in Table 15-5) in place of K_{PU} and T_U, respectively. The values of K_{PU} and T_U will be:

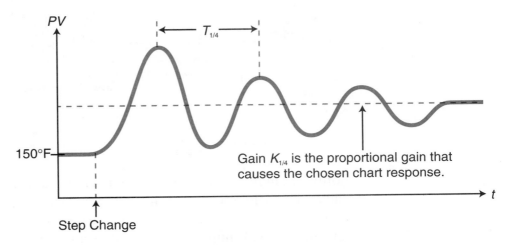

Figure 15-87. Quarter-amplitude test.

$$K_{PU} = 2K_{1/4}$$

$$T_U = T_{1/4}$$

This is due to the fact that the proportional gain in the standard Ziegler-Nichols method is:

$$K_P = 0.5K_{PU}$$

$$= 0.5(2K_{1/4})$$

$$= K_{1/4}$$

The values that will be obtained for T_I in the integral action of the PI mode and for T_I and T_D in the integral and derivative actions of the PID mode will be slightly larger than those values calculated through the unaltered Ziegler-Nichols closed-loop method, since the $T_{1/4}$ reading will be slightly larger than the T_U reading. Although this method avoids the problem of long test periods, it does present another problem—the determination of the period of decaying oscillation. Reading a decaying oscillation period is more difficult than reading a period of constant amplitude oscillation, because it is harder to ascertain where the decaying sinusoidal curve's middle section is.

SOFTWARE TUNING METHODS

Another method for tuning PID controllers is the use of software tuning systems. These software packages run on personal computers using Unix, Windows, or another platform (see Figure 15-88). They connect to the controller or PLC either directly or via a DDE (dynamic data exchange) interface. These software systems reduce the tuning time and, at the same time, optimize control loop performance.

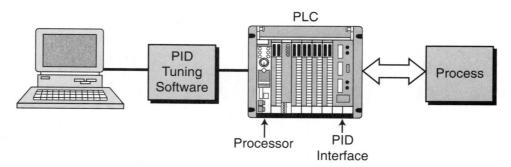

Figure 15-88. Software loop tuning.

Software tuning programs provide numerous viewing selections, windows, and on-line help screens that show process characteristics including simulations, modeling, plots, and frequency responses (see Figure 15-89). Additionally, they provide important information about the process itself that can be extremely difficult, if not impossible, to obtain manually. For example,

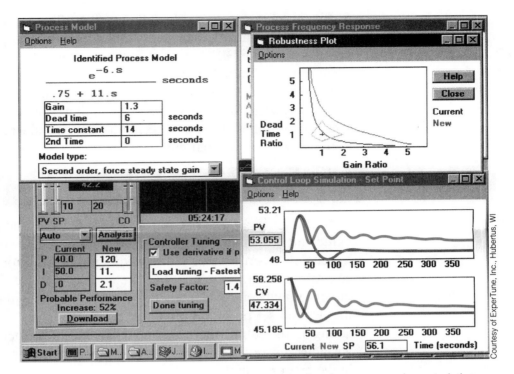

Figure 15-89. Software tuning program screens showing process characteristics.

ExperTune®, by ExperTune, Inc., identifies the transfer function of the process during the tuning test (see Figure 15-90), thus providing information such as process gain, dead time, and lag time constants. This software optimizes the PID tuning parameters automatically for even the most complicated of processes. It also performs tuning for cascade PID loops. Additionally, it provides a "robustness" plot, which shows how a change in the process dead time and/or gain will affect the closed-loop system's stability given the current loop tuning constants. In fact, the robustness plot shows all the gain and dead time values that will allow for stable closed-loop operation. For example, it shows how a change in gain will affect the stability of the closed-loop system as it is tuned. Thus, the robustness plot provides a quick look at the trade-offs between tuning and stability.

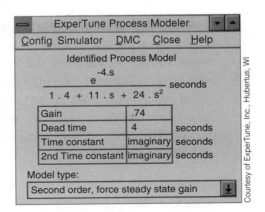

Figure 15-90. Process transfer function obtained through software loop tuning.

Software tuning systems also allow "what if" analysis, meaning that they suggest new controller constants given hypothetical process values. They can also provide PID tuning parameters from ASCII data files about the plant process. These features reduce the amount of time required to tune the system because they can produce a database of tuning constants for a variety of process scenarios. These software tuning systems greatly benefit users wanting to simplify their process tuning efforts while reducing the amount of time required for manual tuning.

15-13 SUMMARY

A controller in a process control system receives data about the set point value and the actual process variable value and then compares these values to generate an error value. The controller uses this error value according to a control algorithm to manipulate the control variable. The control variable directs a final control element (e.g., a valve) to bring the process variable to the desired set point, eliminating the system error. A controller is direct acting if its output increases in response to an increase in the process variable; it is reverse acting if its output decreases in response to an increase in the process variable.

There are two types of controller modes: discrete and continuous. Some of the most commonly used discrete-mode controllers are the two-position, or ON/OFF, mode and the three-position mode. Two-position controllers turn the output ON (100% open) or OFF (0% open) once the process variable crosses an error deadband around the set point. Three-position controllers are an extension of two-position ones in the sense that they have one more output level. This type of discrete controller provides 0%, 50%, and 100% controller output levels.

Continuous-mode controllers include proportional controllers, integral controllers, and derivative controllers. These controllers can also be combined to provide proportional-integral (PI), proportional-derivative (PD), and proportional-integral-derivative (PID) controllers.

In proportional control, the corrective output action is proportional to the size of the error deviation ($E = SP - PV$). This type of control provides a fast response and is relatively simple to implement. Proportional control, however, always leaves some offset error between the desired and actual values of the process variable. If the proportional band in a proportional controller is set too wide, the offset error will be larger than if the proportional band is narrow. However, too narrow of a proportional band will create oscillation and, thus, system instability.

Integral control provides corrective action as a function of the integral of the error (i.e., the sum of the error over time). It provides its highest gain, or corrective action, at low frequencies (i.e., in a slowly changing process).

Integral action tends to ignore high-frequency changes, such as noise or rapid transients, in the process. Although this control mode eliminates the inherent offset error present in a proportional controller, it adversely affects stability. If the integration time is reduced, the response during the slow period of the process will become faster, inducing cycling.

Derivative control provides corrective action as a function of either the rate of change of the error or the process variable. A derivative controller provides its highest gain, or corrective action, at high frequencies. Hence, it provides an anticipatory response to the process variable change. The derivative mode cannot be used alone and does not eliminate residual error.

A proportional-integral, or PI, controller combines the fast response of proportional action with the offset error elimination of integral action. A PI controller with an integral time that is too long will exhibit a response that will take a long time to return to the set point. Conversely, a shorter integral time will cause the process variable to cross the set point faster, resulting in damped oscillations. An integral time that is too short, however, will produce continuous oscillations.

A proportional-derivative, or PD, controller provides better response stability than a PI controller, but it does not provide offset error elimination. A PD controller is useful in applications where the process has a long lag time delay in its recovery from a disturbance. The derivative action in this controller provides a lead function, which cancels some of the process lag and allows the proportional band to become narrower. This improves response and stability. A PD controller does not eliminate offset error, but a narrower proportional band can reduce the amount of residual error in the system. A derivative time constant (K_D or T_D) that is too long will cause the process variable to change too rapidly and overshoot the set point with damped oscillation. Conversely, if the derivative constant is too short, the process variable will take too long to reach the set point.

A proportional-integral-derivative, PID, controller combines the increased stability of a PD controller with the eliminated offset feature of a PI controller. A PID controller can be used to control almost any type of process, including those with long lag times. The gains for each of the control actions in a PID controller can be derived experimentally utilizing several tuning methods.

An integral, or reset, windup situation occurs when an integral action saturates a controller's output at 100%. Integral windup usually happens during the start-up of a process. This condition occurs when the error in a slow-responding process system is large. The proportional action tries to bring the process variable closer to the set point, but the slow speed of the process response and the presence of error induces the integral action to continue, keeping the controller at 100% output. This condition can be prevented by disabling the integral action once the controller's output reaches 100%.

Bumpless transfer refers to a controller's ability to switch from manual to automatic control and vice versa without a step change in the input to the process. In a bumpless transfer, the manual controller station tracks the automatic controller's output and vice versa to keep the control variable output constant.

Cascade control refers to an advanced control technique where the output of one controller is the set point input to another controller. Cascade control provides more precise control than noncascaded control, since the secondary (or inner) loop will react quickly to a disturbance before it starts to affect the primary (or outer) loop.

Controller loop tuning is the process of manipulating the parameters (gains) in a PID controller so that the response of the process system is satisfactory. A satisfactory response is one that exhibits the desired speed of response, yet meets the required accuracy and stability criteria. Control processes are generally tuned under operating conditions, as opposed to start-up conditions, so that the process variable is stable at an operating point. Since the transfer function of a process is rarely known, experimental measurements and tests can be made to obtain parameters that will help determine the desired controller gains for PID control. These experimental measurements are known as the "modeling" of the system. Some of the most popular experimental tuning methods are the Ziegler-Nichols open-loop method, the integral of time and absolute error (ITAE) open-loop method, and the Ziegler-Nichols closed-loop method.

During the modeling of a process, a known disturbance is created and the resulting response is observed and recorded. The disturbance should be one that actually occurs during process operation (e.g., a change in load, flow rate, or speed of the system). However, the creation of this type of disturbance is impractical to implement in a real-life situation; therefore, a change in set point is most often used as the disturbance to the process. The values obtained from this disturbance are then plugged into the tuning method's equations to obtain the values for the proportional, integral, and derivative gain terms. These values are then used as the starting parameters for the controller, which will provide process control by minimizing the error in the system.

KEY   cascade control
TERMS continuous-mode controller
derivative controller
direct-acting controller
discrete-mode controller
integral controller
integral of time and absolute error open-loop tuning method (ITAE)
integral windup
loop tuning

proportional controller
proportional-derivative controller
proportional-integral controller
proportional-integral-derivative controller
quarter-amplitude response
reverse-acting controller
three-position controller
two-position controller
Ziegler-Nichols closed-loop tuning method
Ziegler-Nichols open-loop tuning method

ADVANCED PLC TOPICS AND NETWORKS

- Artificial Intelligence and PLC Systems
- Fuzzy Logic
- Local Area Networks
- I/O Bus Networks

CHAPTER SIXTEEN

ARTIFICIAL INTELLIGENCE AND PLC SYSTEMS

Computers can figure out all kinds of problems, except the things in the world that just don't add up.

—James Magary

CHAPTER HIGHLIGHTS In previous chapters, we highlighted both simple and complex PLC applications. In this chapter, we will present an area of PLC applications that goes one step beyond these—artificial intelligence. We will explain the basics of artificial intelligence (AI) systems by explaining their organization and methodology. We will also discuss how each of the three types of artificial intelligence systems—diagnostic, knowledge, and expert—work. Finally, we will present an example of an AI application to further explain how these complex systems operate. After you finish learning about artificial intelligence, you will be ready to explore fuzzy logic, another advanced application that involves PLCs.

16-1 INTRODUCTION TO AI SYSTEMS

Artificial intelligence (AI) is an area of computer science that has been around for some time. In fact, the conceptual design of AI was first developed in the early 1960s. The definition of artificial intelligence varies among people in the computer industry, making the concept somewhat difficult to perceive and understand. In general, AI can be defined as the subfield of computer science that encompasses the creation of computer programs to solve tasks requiring extensive knowledge.

The software programs that form an AI system are developed using the knowledge of an expert person (or persons) in the field where the system will be applied. For instance, a food-processing AI system that involves the making and packaging of a food product will consist of knowledge obtained from chemists, food technologists, packaging experts, maintenance personnel, and others closely associated with the operation.

In this chapter, we will present AI techniques that can be implemented through a PLC-based process control system. These techniques will define the methods for implementing AI into the process. The result will be a system that can successfully diagnose, control, and predict outcomes based on resident knowledge and program sophistication.

16-2 TYPES OF AI SYSTEMS

An exact classification of the types of artificial intelligence systems is very difficult to obtain because of the varying definitions of AI applications. For the purposes of this text, however, we will divide artificial intelligence into three types of systems:

- diagnostic
- knowledge
- expert

Each of these types of AI systems have similar characteristics, and in fact, the systems evolve sequentially. As the systems become more sophisticated, the size of the database grows and the extent of how the process data is compiled and interpreted increases.

DIAGNOSTIC SYSTEMS

Diagnostic AI systems are the lowest level of artificial intelligence implementation. These systems primarily detect faults within an application, but they do not try to solve them. For example, a diagnostic system can diagnose a pump fault by detecting a loss of tank pressure or by reading flow meter values.

A diagnostic system reaches a fault conclusion through inferring techniques based on known facts (knowledge) introduced into its detection system. This type of AI is used in applications that have a small knowledge and database structure. Diagnostic systems typically make GO or NO GO decisions and sometimes provide information about the fault's probable cause.

KNOWLEDGE SYSTEMS

A **knowledge AI system** is, in reality, an enhanced diagnostic system. Knowledge systems not only detect faults and process behaviors based on resident knowledge, but also make decisions about the process and/or the probable cause of a fault.

In the batching system example mentioned in the diagnostic system section, a knowledge system would go beyond just diagnosing the fault. It would also provide suggestions about probable faulty devices, as well as make a decision about whether to continue the process (if the fault is noncritical) or to shut down (if the fault is critical). The system bases these decisions on its programmed knowledge and a set of rules that defines each fault condition.

It is possible that the detection of a fault in the previous example could have been a false alarm. As part of its enhanced features, a knowledge system checks whether the elements signaling the fault condition (i.e., flow meter, pressure transducer) are operating correctly. It then compares these observations (process feedback) with the procedures and measures based on this information. For example, if a fault does occur and it is a valid noncritical fault, the control system may issue *continue process, stop after finished*, and *alert personnel* commands.

EXPERT SYSTEMS

An **expert AI system** is the top of the line in AI-type applications; it has all of the capabilities of a knowledge system and more. An expert system provides an additional capability for examining process data using statistical analysis. The use of statistical data analysis lets the system predict outcomes based on current process assessments. The outcome prediction may be a decision to continue a process in spite of a fault detection.

For the example used in the other two types of AI systems, an expert system may decide to continue the batching operation until the noncritical fault generates another fault. The system might arrive at this decision because the average pressure sensed in the mixing reactor tank is within tolerance limits (i.e., readings observed about the mean). Thus, the system continues the batching operation in spite of the fact that the flow meter reported a loss of flow. The system then continues production and alerts personnel that pump and flow meter feedback may have been lost.

The knowledge introduced into an expert system is more complex than in the other types of AI systems; therefore, expert systems generate more data verification (feedback information). The decisions made by expert systems also require more sophisticated software programming, since their decision trees involve more options and attributes.

The implementation of an expert AI system requires not only extra programming effort but also more hardware capability. The total system will need more transducers to check other transducers and field devices. Moreover, the PLC will require the use of two or more processors to implement the control and intelligence programs. The speed of the system must also be fast so that it can operate in real time. Furthermore, the system's memory requirements will be larger, since knowledge data must be incorporated and stored into the AI system.

16-3 ORGANIZATIONAL STRUCTURE OF AN AI SYSTEM

A typical artificial intelligence system consists of three primary elements:

- a global database

- a knowledge database

- an inference engine

Figure 16-1 shows a block diagram of an AI system's architecture. As the figure illustrates, the AI system must receive its knowledge from a person who thoroughly understands the process or machine being controlled. This individual, called the *expert*, must communicate all information about system maintenance, fault causes, etc. to the *knowledge engineer*, the person responsible for system implementation. The process of gathering data from the expert and transmitting it to the knowledge engineer is known as *knowledge acquisition*.

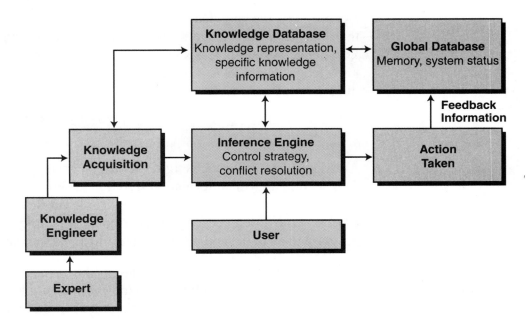

Figure 16-1. Artificial intelligence system architecture.

GLOBAL DATABASE

The **global database** section of an AI system contains all of the available information about the system being controlled. This information mainly deals with the input and output data flow from the process. The global database resembles a storage area where information about the process is stored and updated. The AI system can access the data in this area at any time to perform statistical analysis on historical process control data, which in turn can be used to implement AI decisions.

The global database resides in the memory of the control system implementing the artificial intelligence. If a PLC is used to implement a diagnostic AI system, the global database will most likely be located in the storage area of the PLC's data table. If a PLC is used in conjunction with a computer or computer module to implement an AI system, then the global database will probably be located in the computer, the computer module's memory, or a hard disk storage subsystem.

KNOWLEDGE DATABASE

The **knowledge database** section of an AI system stores the information extracted from the expert. Like the global database, this database includes information about the process; however, it also stores information about faults, along with their probable causes and possible solutions. Moreover, the knowledge database stores all of the rules governing the AI decisions to be made. The more involved the AI system, the larger the knowledge database.

Accordingly, the knowledge database of a diagnostic system is less complex than that of a knowledge system; likewise, the knowledge database of a knowledge system is less sophisticated than that of an expert system. The knowledge database is stored in the section of the system memory that implements the AI techniques.

INFERENCE ENGINE

An AI system's **inference engine** is the place where all decisions are made. This section uses the information stored in the knowledge database to arrive at a decision and then execute all applicable rules and decisions about the process. The inference engine also constantly interacts with the global database to examine and test real-time and historical data about the process.

The inference engine usually resides in the main CPU (i.e., the one that performs the AI computations). However, in a PLC-based system, the inference engine may or may not be stored in the main CPU, depending upon the system's complexity (i.e., diagnostic, knowledge, or expert).

16-4 KNOWLEDGE REPRESENTATION

Knowledge representation is the way the complete artificial intelligence system strategy is organized—that is, how the knowledge engineer represents the expert's input. This representation is stored in the knowledge database of the AI system. In rule-based knowledge representation, the expert's knowledge is transformed into IF and THEN/ELSE statements, which facilitate actions and decisions.

All control systems that implement artificial intelligence, whether diagnostic, knowledge, or expert, execute the control strategy (via the software control program) in the inference engine. Whenever a decision must be made due to a fault or another situation, the inference engine refers to the knowledge representation to obtain a decision about the probable cause. This decision is the result of a group of software subroutines. Once the knowledge database reaches an AI decision, the inference engine will determine the appropriate course of action. Depending on the control strategy formulation (main program), the inference engine may, at this time, refer to the global database to verify data or obtain more information.

RULE-BASED KNOWLEDGE REPRESENTATION

Rule-based knowledge representation defines how the expert's knowledge is used to make a decision. The rules used are either *antecedent* (IF something happens) or *consequent* (THEN take this action). For example, to the question, What causes the volume in the tank to drop?, the expert may respond with the answer, a malfunctioning tank system. The knowledge engineer may implement this information as the following rule: IF the volume is less than the set point, THEN annunciate a system malfunction due to a loss of volume.

Rules can be as long and complex as needed for the process, and they usually define the involvement of the AI system. For instance, a simple rule-based system (few rules, not very complex) may formulate a simple diagnostic rule, such as:

IF the temperature is less than the set point, THEN open the steam valve

A more complex diagnostic formula would involve rules that depend on parent rules:

IF case 1, THEN — ELSE nothing
|
IF case 2, THEN — ELSE something
|
IF case 3 — THEN nothing
•
•
•

where each of the case conditions represents a particular measurement, comparison, or situation. Figure 16-2 illustrates a decision tree for forming AI rules.

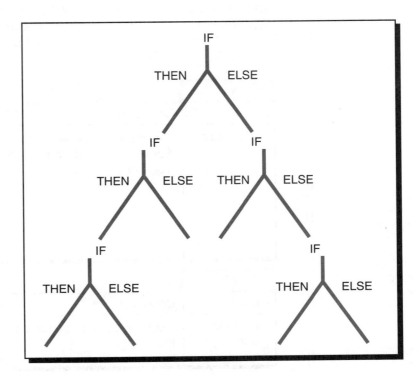

Figure 16-2. Decision tree.

A slightly different degree of complexity occurs in a rule-based knowledge representation when the rule has several probable causes. For example:

$$\text{IF volume drops, THEN} \begin{cases} \text{valve failure} \\ \text{or} \\ \text{pump failure} \\ \text{or} \\ \text{feedback failure} \end{cases}$$

In this case, the consequents must be further investigated to arrive at a complete formal rule. The inference engine can use the consequents derived in the knowledge representation to obtain a better definition of the problem's cause. Knowledge and expert AI systems use this process to provide advanced decision-making capabilities.

EXAMPLE 16-1

A PLC-controlled box conveyor transports two sizes of boxes that are diverted to different palletizer operations according to their size. A solenoid activates the diverter that sorts the boxes. Write the rules that a knowledge database could use to detect a possible cause for the solenoid's malfunction.

SOLUTION

One of two factors can result in solenoid failure according to the situation presented: coil burnout or mechanical damage. The conditions and causes in Figure 16-3 describe these two possible factors that could lead to a fault.

Rule #1

Result	Condition	Cause
Burned-out coil	Excessive temperature is developed due to continuous high-current input	—Low line voltage causes failure to pull plunger —High ambient temperature —Mechanically blocked plunger —Operations too rapid

Rule #2

Result	Condition	Cause
Mechanical damage	Excessive force exerted on the plunger	—Overvoltage —Reduced load

Figure 16-3. Knowledge database rules for conveyor fault.

16-5 KNOWLEDGE INFERENCE

Knowledge inference is the methodology used for gathering and analyzing data to draw conclusions. Knowledge inference occurs in the inference engine during the execution of the main control strategy program. It also occurs in the knowledge database during the comparison and computation of rule solutions.

The system's software program determines the approach used to derive AI solutions. Operator interaction on control problems can enhance the solution-finding process. For example, if the system detects a failure due to a misreading in an inspection system, it may alert the operator to the problem and advise him/her of probable causes. Furthermore, the system may wait for the operator's input (e.g., check for laser intensity in the receiver side to determine if the laser beam is reflecting at the correct angle) and then use the operator's input to develop more intelligent solutions to the problem.

In small systems, knowledge inference occurs on a local basis. That is, the control system houses the resident software for the inference engine. In large, distributed, intelligent systems, knowledge inference often occurs at a main host in the hierarchical system.

Remember that the degree of AI involvement in the system will determine how much hardware is required (e.g., computer modules, powerful PLCs, small PLCs with personal computers, etc.). When all global databases are in constant network communication, allowing knowledge inference information to be passed from one controller to another, the intelligent system is said to have a *blackboard architectural structure*.

In all types of intelligent systems, certain methods of rule evaluation are used to implement knowledge inference. These methods include *forward chaining* and *backward chaining*. Intelligent systems also analyze statistical information as part of knowledge inferencing to obtain predictions about outcomes.

BLACKBOARD ARCHITECTURE

Large, complex, distributed control systems involve the interaction of several subsystems, which continuously communicate with each other either directly or over a local area network. When artificial intelligence is added to these large systems, system elements, such as knowledge inferencing and the global and knowledge databases, are distributed throughout the architecture of the control system. Whether or not each of the controllers in the network has a local inference engine, global database, and knowledge database depends on the degree of inferencing that occurs on a local basis. **Blackboard architecture** is the name given to this type of large system, which utilizes several subsystems containing local global and knowledge databases.

Figure 16-4 illustrates a blackboard configuration of an intelligent control system. The PLCs at the subsystem level may contain computer modules, which help them perform inference engine computations. The hierarchy of the control system allows the supervisory PLC controller to poll each of the subsystems and obtain all or part of their local global database information. The host computer element in this control structure holds the blackboard, the area that stores all of the information obtained from the subsystems by the supervisory PLC. The inference engine of the host element then implements the complex AI solution according to its knowledge inference about the total control system.

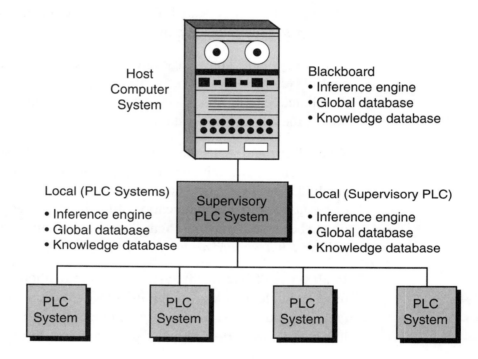

Figure 16-4. Example of blackboard architecture.

FORWARD CHAINING

Forward chaining is a method used to determine possible outcomes for given data inputs. Forward chaining inference engines typically receive process information via the global database and monitor specific inputs to the control system to determine the outcomes. For instance, in Example 16-1, forward chaining specifies the following consequences for a failed solenoid: a jammed conveyor or misplaced boxes in the two palletizers.

Two different types of fact searching occur within the forward chaining method: *depth first* and *breadth first*. Both searches deal with how the outcome is obtained. A depth-first search, shown in Figure 16-5, evaluates the rules that form the knowledge database (A, B, C, etc.) on a priority basis going *down* the tree. In the conveyor example mentioned earlier, when the control system detects the solenoid failure (A), it will evaluate a new rule to see if jamming has occurred (B). If the conveyor has jammed, then the system will evaluate the consequences that can occur (e.g., material inside box may break or material could spill (D)).

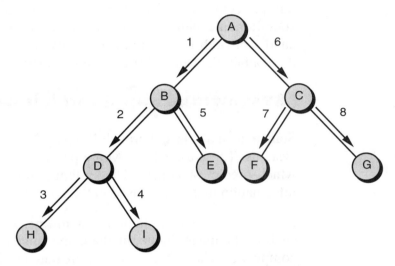

Figure 16-5. Forward chaining depth-first search.

In contrast, the breadth-first method evaluates each rule in the *same level* of the tree before proceeding to the next level down (see Figure 16-6). In our conveyor example, a breadth-first evaluation of the rules means that after the solenoid failure (A) the system will check for a possible jam (B), then it will check for palletizer misplacement (C), and so on.

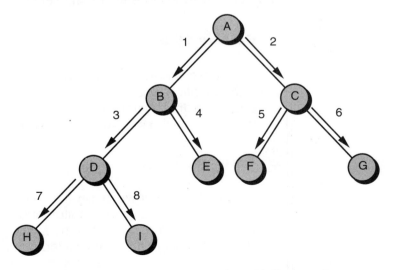

Figure 16-6. Forward chaining breadth-first search.

BACKWARD CHAINING

Backward chaining is a method for finding the causes of an outcome. Referring to Example 16-1, the rule tables present backward chaining information—that is, causes for the solenoid failure outcome. Basically, backward chaining analyzes the consequences to obtain the antecedents.

Similar to forward chaining, backward chaining uses both the depth-first and breadth-first search methods. In our conveyor example, after the solenoid failure occurs, a backward chaining depth-first search will first check one condition rule then check each possible cause of that condition. On the other hand, a breadth-first search will first examine both of the condition rules and then obtain the causes for each of the conditions.

STATISTICAL AND PROBABILITY ANALYSIS

Statistical analysis and probability play a large role in artificial intelligence systems. These aspects of AI are particularly important in expert systems, which predict outcomes. The system's global database stores the process information that will be used in the AI statistical analysis.

In Chapter 13, we explained how to interpret and obtain statistical data, such as the mean, mode, median, and standard deviation. These statistical computations help determine a future outcome based on what is happening in the current process. Decisions based on statistics can be related to the consequences of the rules described in the knowledge representation. For example, just because a system detects an error fault does not mean that the fault actually occurred, even though the feedback data transducer devices may be operating correctly. Using statistical analysis, the inference engine may decide not to advise personnel or apply the corresponding control to the fault, but instead to continue monitoring the situation more closely.

EXAMPLE 16-2

A control system monitors and controls a cooker in a temperature loop with specifications as shown in Figure 16-7. Indicate how AI can be added to the system to detect real temperature problems. Also, indicate how the system can screen out false temperature faults.

SOLUTION

Figure 16-7a shows a profile of temperature readings from the system. The PLC can monitor and accumulate temperature data continuously from time t_0 to time t_1 using FIFO instructions, storing this data in a storage area with a fixed number of registers (see Figure 16-7b). The program can also compute the mean, median, and standard deviation of the current temperature readings.

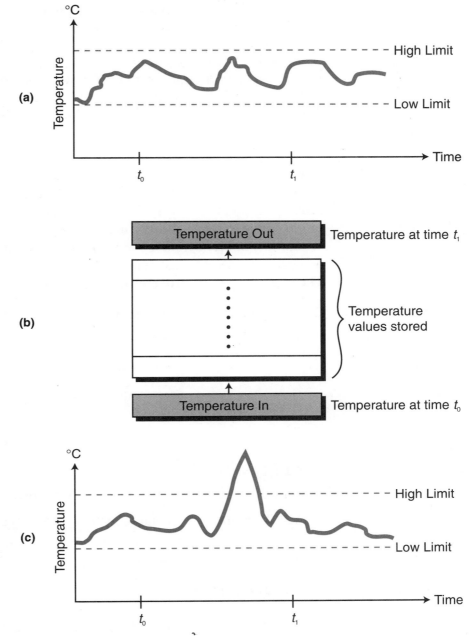

Figure 16-7. (a) Profile of temperature readings, **(b)** FIFO storage method, and **(c)** high-limit alarm value in the example process.

If a high-limit alarm occurs (see Figure 16-7c), a normal system would control the cooker by adjusting its temperature loop. However, the temperature fault may not have been caused by a temperature loop malfunction; a noise spike near the temperature transducer could have caused it.

An intelligent system would detect this sudden temperature increase by recognizing that it is well beyond the mean and median of the readings for the t_0 to t_1 period, therefore exhibiting a large standard deviation. By implementing a rule that considers the statistics of the process, the AI system will ignore the false alarm and not add the temperature reading value to the average calculations report. Furthermore, the global database of the system will receive information, for future use, about the time of day, location, and level of the spike reading. The system will closely analyze the temperature increase in case it is a true alarm. It will use temperature rate of change computations to help determine if it is a true fault.

Probability can be useful when determining or approximating the possible cause of a fault in a diagnostic ruling. One of the most commonly used probability methods is **Baye's theorem** of conditional probability. The use of this type of probability in an AI system is known as *conditional probability inferencing*. To employ probability computations in any system, however, the system must maintain historical information about the process. The expert generally provides this type of data.

Baye's theorem defines the probability of X event occurring based on the fact that Y has already occurred [$P(X/Y)$] as:

$$P(X/Y) = \frac{[P(Y/X)][P(X)]}{[P(Y/X)][P(X)] + [P(Y/\overline{X})][P(\overline{X})]}$$

where:

$P(Y/X) =$ the probability that Y occurs when X has occurred

$P(X) =$ the prior probability that X has occurred

$P(Y/\overline{X}) =$ the conditional probability that Y occurs if X does not occur

$P(\overline{X}) =$ the prior probability that X has not occurred

EXAMPLE 16-3

Part of a conveyor system controls a solenoid-operated diverter, which sends two types of boxes to two different repackaging areas. The system uses several photoelectric eyes to determine which box goes where.

The material-handling expert indicates that, due to the size and type of the boxes and environment, the following probabilities exist for conveyor faults:

For a solenoid-caused fault:

- The prior probability of a solenoid fault is 20% (80% probability that it does not fault).

- The probability that the boxes will go to the right place when the solenoid is faulty is 35%.

- The probability that the boxes will go to the right place when the solenoid is good is 60%.

For a photoeye-caused fault:

- The probability of a photoeye fault is 35% (65% probability that it does not fault).

- The probability that the boxes go to the right place when the eye is faulty is 25%.

- The probability that the boxes go to the right place when the eye is good is 45%.

Find the most probable cause of a conveyor fault when the boxes are going to the right place.

SOLUTION

We can include the expert's data in the knowledge representation by calculating which element has a higher percentage probability of having occurred. Using Baye's theorem, the probability that the solenoid is faulty (S) even though the boxes are going to the right place (B) is:

$$P(S/B) = \frac{[P(B/S)][P(S)]}{[P(B/S)][P(S)] + [P(B/\overline{S})][P(\overline{S})]}$$
$$= \frac{(0.35)(0.20)}{(0.35)(0.20) + (0.60)(0.80)}$$
$$= 12.73\%$$

The probability that the photoeye is faulty (E) even though the boxes are going to the right place (B) is:

$$P(E/B) = \frac{[P(B/E)][P(E)]}{[P(B/E)][P(E)] + [P(B/\overline{E})][P(\overline{E})]}$$
$$= \frac{(0.25)(0.35)}{(0.25)(0.35) + (0.45)(0.65)}$$
$$= 23.03\%$$

The computations indicate that a photoeye fault is most likely to have occurred in the conveyor system. In this event, the operator should be alerted and the system temporarily halted. Also, the global database should be updated with the statistics of the fault occurrence, so that this information can be used in the future.

CONFLICT RESOLUTION

A conflict occurs when more than one rule is triggered at the same time in an AI system. Normally, a system starts executing rules based on the order of occurrence of the situation. However, when situations happen at the same time, a conflict may occur in the system. For example, a system may receive information indicating that a high temperature, a low pressure, and a flow obstruction have all occurred. These three situations, on their own or in combination, can trigger the following rule consequents:

> Rule 1: IF high temperature, THEN start cooling procedure.
>
> Rule 2: IF low pressure and flow obstruction, THEN open relief valve in main supply pipe.
>
> Rule 3: IF high temperature and low pressure, THEN open relief valve in main supply pipe and alert personnel in the area.

Therefore, the system must make a decision about which of these three rules to implement. It must select the rule that exhibits the greater priority—in this case, rule 3. The expert provides the system with this information about the priority of rule execution.

16-6 AI FAULT DIAGNOSTICS APPLICATION

The example presented in this section illustrates the use of the methodology described in the previous sections. For simplicity, we will not elaborate on the PLC program coding for the application, but we will describe the rules used to define the knowledge representation.

The AI setup in this example is a diagnostic-level system implemented by a PLC-based control system. The method of rule evaluation is backward chaining (i.e., once the system detects a fault, it searches for the cause of the fault). In the batching system, the control program implements AI fault detection for only one of the two ingredients. The rules for the second ingredient are similar to the first ingredient.

DEFINITION OF THE PROCESS

Two ingredients, A and B, are to be mixed in the tank of the batching system shown in Figure 16-8. Figures 16-9 through 16-12 show the flowcharts of the process, as well as the steam valve–versus–temperature relationships. The process is as follows:

- A flow meter counts the number of pulses to monitor the amount of the ingredients in the tank (in gallons).

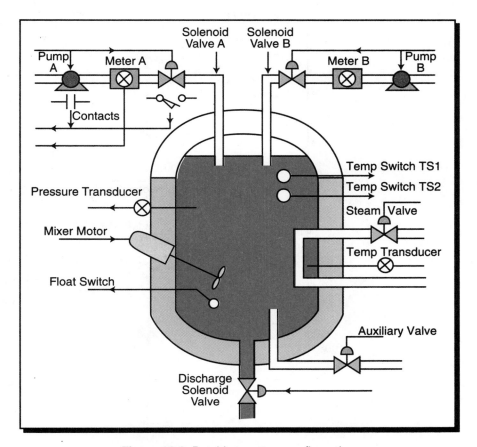

Figure 16-8. Batching system configuration.

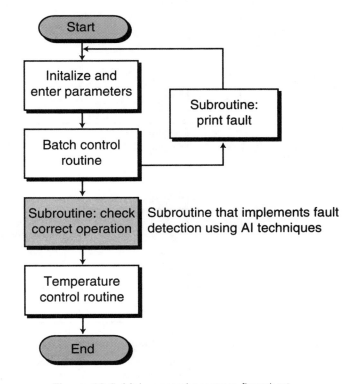

Figure 16-9. Main control program flowchart.

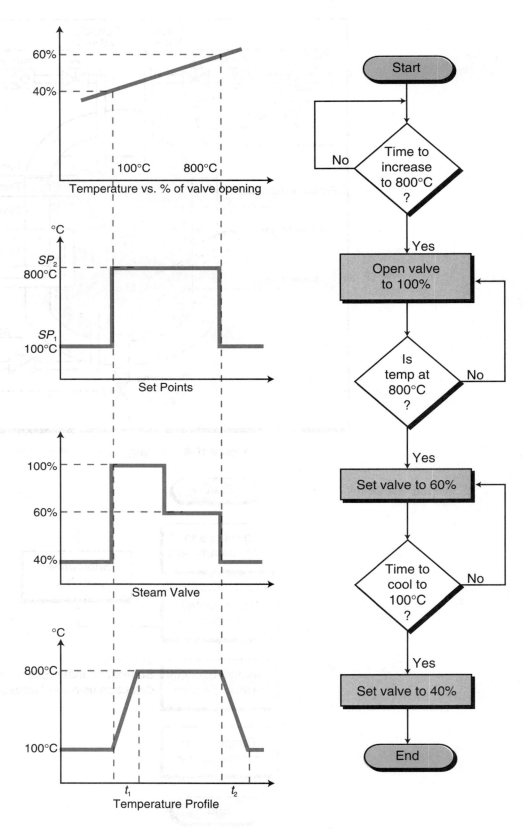

Figure 16-10. Temperature and steam valve relationship.

Figure 16-11. Temperature control subroutine.

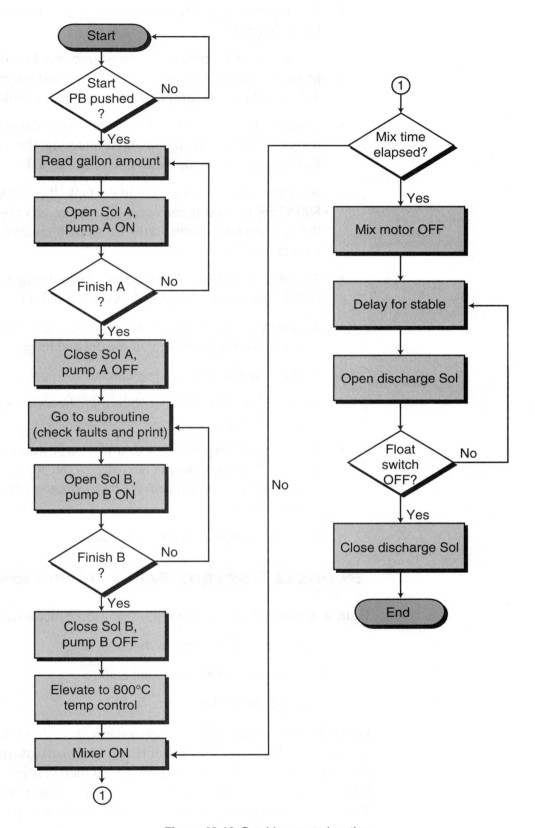

Figure 16-12. Batching control routine.

- A pump motor provides the necessary pressure to send the ingredients through the line.

- Before any of the ingredients are poured into the tank, the temperature inside the tank should be 100°C. A solenoid opens a steam valve to 40% to achieve the proper temperature in the tank.

- A load cell pressure transducer reads the volume inside the tank. It detects whether an ingredient is entering the tank, serving as a feedback device in the event of a faulty signal.

- After the two ingredients are in the tank, the temperature must be at 800°C before mixing can occur. The steam valve opens to 100% until the temperature reaches 800°C, then it remains at 60% open to maintain 800°C.

- Two thermoswitches detect the two desired temperatures (100°C and 800°C) and serve as feedback in case of a fault.

- A steam valve heats up the tank. A temperature transducer controls the temperature, maintaining it at the desired level.

- A motor agitates the two ingredients.

- An auxiliary valve disposes of the ingredients in the event that they are not mixed properly.

- When the mixing is finished, a discharge valve drains the desired solution (mixture) into the next step of the process. The steam valve returns to 40% open to cool the temperature in the tank to 100°C for the next batch.

- A float switch detects an empty tank.

PROCESS CONTROL FAULT DETECTION

Fault detection in the system occurs during three major stages of the process:

1. when the ingredients are being poured

2. during the elevation of temperature

3. during the cooling of the tank

For each of these stages, the system can provide fault–versus–possible cause information. It detects the fault through feedback information from each of the controlling and measuring devices. It then verifies this fault information by comparing it with feedback data from additional control devices. Table 16-1 shows the control and feedback devices used to perform the system check.

Control Devices	Feedback	Purpose
Valve	Limit switch and pressure transducer	Check solenoid actuation in valve
Pump	Contacts and pressure transducer	Check pump operation
Flow meter	Pressure transducer	Check ingredient flow
Steam valve	Temperature switch	Check steam valve

Table 16-1. Control and feedback devices used in batching system.

RULE DEFINITIONS

Based on the process control description and the possible failures, the system has the rules described in Table 16-2. These rules specify actions based on process occurrences and measurements.

Given the AI system's rules, we can define a set of faults *F*, representing the possible malfunctions, as:

$$F_{n,i} \text{ for } n = 0 \text{ to } 9, \ i = 1 \text{ to } 2$$

where:

$$n = \text{ rule number}$$
$$i = \text{ type of fault } (1 = \text{critical}, \ 2 = \text{noncritical})$$

We can divide the set of faults ($F_{n,i}$) into two subsets—critical faults ($F_{n,1}$) and noncritical faults ($F_{n,2}$):

$$F_{n,i} \in F_{n,1} \text{ or } F_{n,2} \text{ for } n = 1 \text{ to } 9$$

The actions taken for critical faults are abort batch process, alert operator of critical fault, open auxiliary valve, and inform operator of possible faulty devices. The actions taken for noncritical faults are alert the operator, continue process and stop at end of batch, and inform operator of possible faulty devices.

APPLICATION SUMMARY

Applying AI techniques to a control system usually involves adding hardware and software to the system. The complexity of the AI program varies depending on how much fault detection is desired. The previous example presented only the rules for one ingredient. Although the rules for the second ingredient would be similar, the control system would still have to be programmed with them, and this could be time consuming.

	Rules
1	IF there is no fault, THEN operation OK. Proceed with entire process.
2	IF there is a fault in the valve, THEN check pressure. IF pressure is OK, THEN there is a fault in the limit switch. Continue process and stop. IF pressure is not OK, THEN there is a fault in the valve. Stop process and alert operator; open auxiliary valve; fix fault.
3	IF there is a fault in the pump, THEN read pressure. IF pressure is OK, THEN there is a fault in the pump's contact. Continue process and stop. IF pressure is not OK, THEN there is a fault in the pump. Stop process and alert operator; open auxiliary valve; fix fault.
4	IF there is a fault in the meter, THEN read pressure. IF pressure is OK, THEN there is a fault in the flow meter. Continue process and stop. IF pressure is not OK, THEN there is a fault in the pump and/or valve. Stop process and alert operator; open auxiliary valve; fix fault.
5	IF there is a fault in the meter and valve, THEN read pressure. IF pressure is OK, THEN there is a fault in the flow meter and/or limit switch. Continue process and stop. IF pressure is not OK, THEN there is a fault in the valve and/or pump. Stop process and alert operator; open auxiliary valve; fix fault.
6	IF there is a fault in the meter and pump, THEN read pressure. IF pressure is OK, THEN there is a fault in the meter and/or pump contacts. Continue process and stop; fix fault. IF pressure is not OK, THEN there is a fault in the pump and/or valve. Stop process and alert operator; open auxiliary valve; fix fault.
7	IF there is a fault in the pump and valve and flow meter, THEN read pressure. IF pressure is OK, THEN there is a fault in the flow meter and/or pump and/or limit switch. Continue process and stop. IF pressure is not OK, THEN there is a fault in the valve and/or pump. Stop process and alert operator; open auxiliary valve; fix fault.
8	IF there is a fault during the heating of the tank to 800°C and TS2 is ON in a set time, THEN there is a fault in the temperature transducer. Continue process and stop. IF TS2 does not respond in the set time period, THEN there is a fault in the steam valve. Stop process and alert operator; open auxiliary valve; fix fault.
9	IF there is a fault during the cooling of the tank to 100°C and TS1 is ON in a set time, THEN there is a fault in the temperature transducer. IF TS1 does not respond in the set time period, THEN there is a fault in the steam valve. Both faults are noncritical, since the batch is finished; they do not require abort batch commands.

Table 16-2. Batching system rules.

We could add intelligence to the system by storing data from the process (e.g., how many times the pump has been turned ON, the contact status feedback to the system, how many times the valve has been turned ON and OFF, which limit switch responded, etc). This data, in conjunction with information about the last time and type of failure, when and how it was fixed, and when the last maintenance was performed, would allow the system to identify whether two possible causes generated a single fault. The global database would store this additional information, allowing the system to make decisions based on the probabilities assigned or calculated throughout several past process performances. Undoubtedly, the more intelligent a system is, the more productive it will be. Additional intelligence means less downtime and a safer process environment.

KEY TERMS

artificial intelligence (AI)
backward chaining
Baye's theorem
blackboard architecture
diagnostic AI system
expert AI system
forward chaining
global database
inference engine
knowledge AI system
knowledge database
knowledge inference
knowledge representation
rule-based knowledge representation

CHAPTER SEVENTEEN

FUZZY LOGIC

Slumber not in the tents of your fathers. The world is advancing. Advance with it.

—Giuseppe Mazzini

**CHAPTER**
HIGHLIGHTS
Fuzzy logic provides PLCs with the ability to make "reasoned" decisions about a process. In this chapter, we will introduce you to the basics of fuzzy logic, including fundamental concepts and historical origins. We will demonstrate how fuzzy logic can be used in practical applications to provide real-time, logical control of a process. When you finish this chapter, you will have learned about the advanced applications of PLCs. You will then be ready to learn how to connect PLCs through local area networks.

17-1 INTRODUCTION TO FUZZY LOGIC

Fuzzy logic is a branch of artificial intelligence that deals with reasoning algorithms used to emulate human thinking and decision making in machines. These algorithms are used in applications where process data cannot be represented in binary form. For example, the statements "the air feels cool" and "he is young" are not discrete statements. They do not provide concrete data about the air temperature or the person's age (i.e., the air is at 65°F or the boy is 12 years old). Fuzzy logic interprets vague statements like these so that they make logical sense. In the case of the cool air, a PLC with fuzzy logic capabilities would interpret both the level of coolness and its relationship to warmth to ascertain that "cool" means somewhere between hot and cold. In straight binary logic, hot would be one discrete value (e.g., logic 1) and cold would be the other (e.g., logic 0), leaving no value to represent a cool temperature (see Figure 17-1).

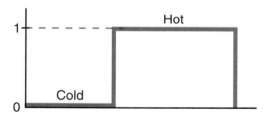

Figure 17-1. Binary logic representation of a discrete temperature value.

In contrast to binary logic, fuzzy logic can be thought of as *gray logic*, which creates a way to express in-between data values. Fuzzy logic associates a **grade**, or *level,* with a data range, giving it a value of 1 at its maximum and 0 at its minimum. For example, Figure 17-2a illustrates a representation of a cool air temperature range, where 70°F indicates perfectly cool air (i.e., a grade value of 1). Any temperature over 80°F is considered hot, and any temperature below 60°F is considered cold. Thus, temperatures above 80°F and below 60°F have a value of 0 cool, meaning they are not cool at all. Figure 17-2b shows another representation of the cool temperature range, where the dotted line shows that hot and cold temperatures are not cool. At 65°F, the fuzzy logic algorithm considers the temperature to be 50% cool and 50% cold, indicating a level of coolness. Below 60°F, the fuzzy logic algorithm considers the temperature to be cold.

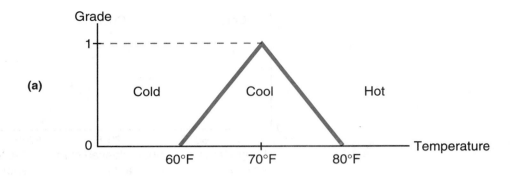

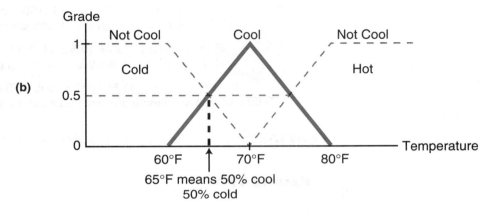

65°F means 50% cool
50% cold

Figure 17-2. (a) Cool air temperature range with **(b)** dotted lines showing not cool range.

In real life, this fuzzy logic temperature algorithm can be associated with the decision you make about the type of clothing you wear at different times of the year. The type of clothing is based on the temperature (input) and its grade representation. As shown in Figure 17-3, at 70°F, you may only need a short-sleeved shirt and pants. However, as the temperature drops to 65°F, you may decide to wear a long-sleeved shirt instead of a short-sleeved one. Moreover, if the input is 25% cool and 75% cold (62.5°F), then you may decide to add another layer, a jacket, based on the temperature and its value of coolness. As we will explain later, a fuzzy system's output may be based on several inputs, not just one, like temperature. In this situation, the output decision is made using the knowledge base represented in the fuzzy logic graph.

Fuzzy logic requires knowledge in order to reason. This knowledge, which is provided by a person who knows the process or machine (the expert), is stored in the fuzzy system. For example, if the temperature rises in a temperature-regulated batch system, the expert may say that the steam valve needs to be turned clockwise a "little bit." A fuzzy system may interpret this expression as a 10-degree clockwise rotation that closes the current valve opening by 5%. As the name implies, a description such as a "little bit" is a fuzzy description, meaning that it does not have a definite value.

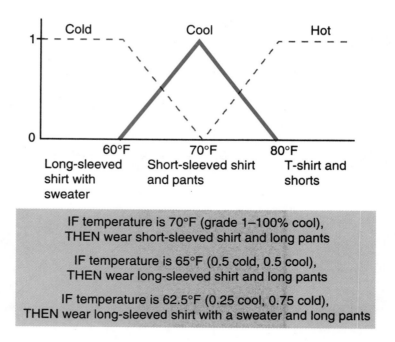

IF temperature is 70°F (grade 1–100% cool),
THEN wear short-sleeved shirt and long pants

IF temperature is 65°F (0.5 cold, 0.5 cool),
THEN wear long-sleeved shirt and long pants

IF temperature is 62.5°F (0.25 cool, 0.75 cold),
THEN wear long-sleeved shirt with a sweater and long pants

Figure 17-3. Fuzzy logic graph illustrating clothing choices based on temperature.

EXAMPLE 17-1

Figure 17-4 illustrates one representation of age (i.e., young, middle age, and old) based on the number of years a person has been alive. In this representation, the exact moment that someone passes the age of 35, he or she is considered middle-aged. Illustrate **(a)** a fuzzy logic representation of this same set of ages, and **(b)** how the representation would change if the age was divided into four ranges: young (up to 35 years), middle age (35–55 years), mature (45–65 years), and old (more than 65 years).

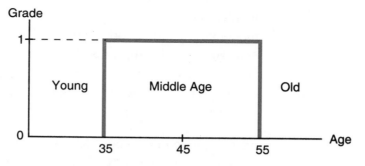

Figure 17-4. Age representation graph.

SOLUTION

(a) Figure 17-5 shows a triangular fuzzy representation that describes the age ranges. In this graph, a person who is 45 years old is perfectly middle-aged, while a person who is 50 years old is 50% middle-aged and 50% old.

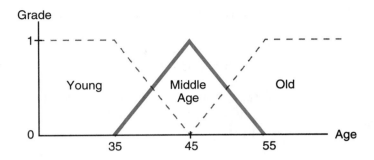

Figure 17-5. Fuzzy logic age ranges.

(b) Figure 17-6 illustrates the fuzzy logic representation for the four age groups: young, middle age, mature, and old. In this chart, a person who is 50 years old is 50% middle-aged and 50% mature.

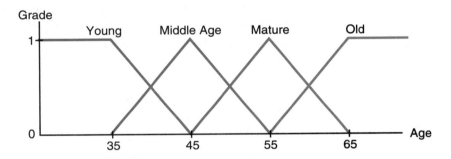

Figure 17-6. Fuzzy logic graph using four age groups.

17-2 HISTORY OF FUZZY LOGIC

Fuzzy logic has existed since the ancient times, when Aristotle developed the *law of the excluded middle*. In this law, Aristotle pointed out that the middle ground is lost in the art of logical reasoning—statements are either true or false, never in-between. When PLCs were developed, their discrete logic was based on the ancient reasoning techniques. Thus, inputs and outputs could belong to only one set (i.e., ON or OFF); all other values were excluded. Fuzzy logic breaks the law of the excluded middle in PLCs by allowing elements to belong to more than just one set. In the cool air example, the 65°F temperature input belonged to two sets, the cool set and the cold set, with grade levels indicating how well it fit into each set.

The origins of fuzzy logic date back to the early part of the twentieth century when Bertrand Russell discovered an ancient Greek paradox that states:

> A Cretan asserts that all Cretans lie. So, is he lying? If he lies, then he is telling the truth and does not lie. If he does not lie, then he tells the truth and, therefore, he lies.

In either case—that all Cretans lie or that all Cretans do not lie—a contradiction exists, because both statements are true and false. Russell found that this same paradox applied to the set theory used in discrete logic. Statements must either be totally true or totally false, leading to areas of contradiction.

Fuzzy logic surmounted this problem in classical logic by allowing statements to be interpreted as both true and false. Therefore, applying fuzzy logic to the Greek paradox yields a statement that is both true and false: Cretans tell the truth 50% of the time and lie 50% of the time. This interpretation is very similar to the idea of a glass of water being half empty or half full. In fuzzy logic the glass is both—50% full and 50% empty. Even as the amount of water decreases, the glass still retains percentages of both conditions.

Around the 1920s, independent of Bertrand Russell, a Polish logician named Jan Lukasiewicz started working on multivalued logic, which created fractional binary values between logic 1 and logic 0. In a 1937 article in *Philosophy of Science,* Max Black, a quantum philosopher, applied this multivalued logic to lists (or sets) and drew the first set of fuzzy curves, calling them *vague sets.* Twenty-eight years later, Dr. Lofti Zadeh, the Electrical Engineering Department Chair at the University of California at Berkeley, published a landmark paper entitled "Fuzzy Sets," which gave the name to the field of fuzzy logic. In this paper, Dr. Zadeh applied Lukasiewicz's logic to all objects in a set and worked out a complete algebra for fuzzy sets. Due to this groundbreaking work, Dr. Zadeh is considered to be the father of modern fuzzy logic.

Around 1975, Ebrahim Mamdani and S. Assilian of the Queen Mary College of the University of London (England) published a paper entitled "An Experiment in Linguistic Synthesis with a Fuzzy Logic Controller," where the feasibility of fuzzy logic control was proven by applying fuzzy control to a steam engine. Since then, the term *fuzzy logic* has come to mean mathematical or computational reasoning that utilizes fuzzy sets.

17-3 FUZZY LOGIC OPERATION

Figure 17-7 illustrates a fuzzy logic control system. The input to the fuzzy system is the output of the process, which is entered into the system via input interfaces. For example, in a temperature control application, the input data would be entered using an analog input module. This input information would

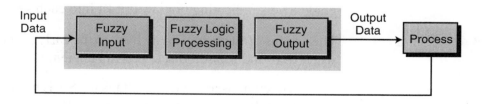

Figure 17-7. Fuzzy logic control system.

then go through the fuzzy logic process, where the processor would analyze a database to obtain an output. Fuzzy processing involves the execution of IF...THEN rules, which are based on the input conditions. An input's grade specifies how well it fits into a particular graphic set (e.g., too little, normal, too much). Note that input data, as shown in Figure 17-8, may also be represented as a count value ranging from 0 to 4095 or as a percentage of error deviation. If the fuzzy logic system utilizes an analog input that has a count range from 0 to 4095, the graphs representing the input will cover the span from 0 to 4095 counts. Furthermore, the analog input information (0–4095 counts) may represent an error range, from –50% to +50%, of a process.

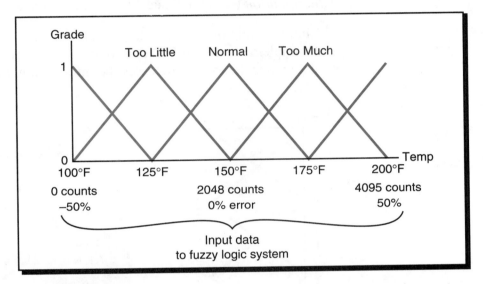

Figure 17-8. Input data to a fuzzy logic system represented as counts and percentages.

The output of a fuzzy controller is also defined by grades, with the grade determining the appropriate output value for the control element. The output of the fuzzy logic system in Figure 17-9, for example, controls a steam valve,

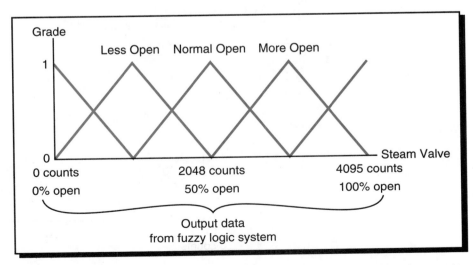

Figure 17-9. Output data from a fuzzy logic system represented as counts and percentages.

which opens or closes according to its grade on the output chart. Figure 17-10 illustrates a fuzzy logic cooling system chart with both input and output grades, where the horizontal axis is the input condition (temperature) and the vertical axis is the output (air-conditioner motor speed). In this chart, a single input can trigger more than one output condition. For example, if the input temperature is 137.5°F, then the temperature is part of two input curves—it is 50% too cool and 50% normal. Consequently, the input will trigger two outputs—the too cool input condition will trigger a less speed output, while the normal input will trigger a normal speed output condition. Since the fuzzy logic controller can have only one output, it completes a process called *defuzzification* (explained later) to determine the actual final output value.

The implementation and operation of a fuzzy logic control system is similar to the implementation of PID control using intelligent interfaces, where the module reads the input, processes the information, and provides an output.

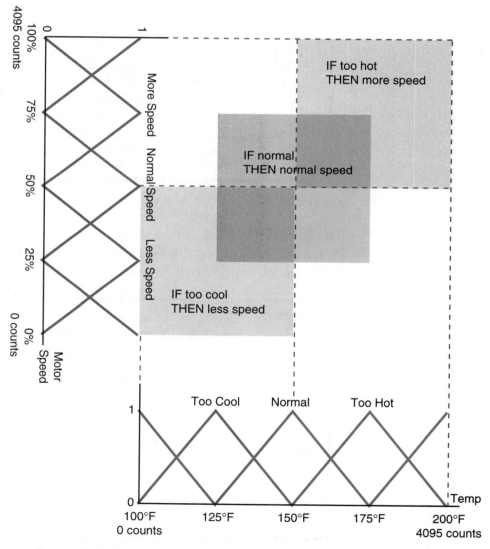

Figure 17-10. Fuzzy logic system chart showing both input and output grades.

However, fuzzy controllers are usually independent interfaces, which plug into the PLC rack and use the PLC's I/O system to communicate with the process under fuzzy control. In Chapter 8, we discussed the operation and interfacing of intelligent fuzzy logic modules.

17-4 FUZZY LOGIC CONTROL COMPONENTS

In this section, we will explain the main components of a fuzzy logic controller and also implement a simple fuzzy control program. The three main actions performed by a fuzzy logic controller are:

- fuzzification
- fuzzy processing
- defuzzification

As shown in Figure 17-11, when the fuzzy controller receives the input data, it translates it into a fuzzy form. This process is called **fuzzification**. The controller then performs **fuzzy processing**, which involves the evaluation of the input information according to IF…THEN rules created by the user during the fuzzy control system's programming and design stages. Once the fuzzy controller finishes the rule-processing stage and arrives at an outcome conclusion, it begins the **defuzzification** process. In this final step, the fuzzy controller converts the output conclusions into "real" output data (e.g., analog counts) and sends this data to the process via an output module interface. If the fuzzy logic controller is located in the PLC rack and does not have a direct or built-in I/O interface with the process, then it will send the defuzzification output to the PLC memory location that maps the process's output interface module.

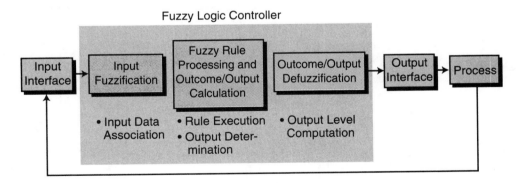

Figure 17-11. Fuzzy logic controller operation.

FUZZIFICATION COMPONENTS

The fuzzification process is the interpretation of input data by the fuzzy controller. Fuzzification consists of two main components:

- membership functions

- labels

Membership Functions. During fuzzification, a fuzzy logic controller receives input data, also known as the fuzzy variable, and analyzes it according to user-defined charts called **membership functions** (see Figure 17-12). Membership functions group input data into sets, such as temperatures that are too cold, motor speeds that are acceptable, etc. The controller assigns the input data a grade from 0 to 1 based on how well it fits into each membership function (e.g., 0.45 too cold, 0.7 acceptable speed). Membership functions can have many shapes, depending on the data set, but the most common are the S, Z, Λ, and Π shapes shown in Figure 17-13. Note that these membership functions are made up of connecting line segments defined by the lines' end points. Each membership function can have up to three line segments with a maximum of four end points. The grade

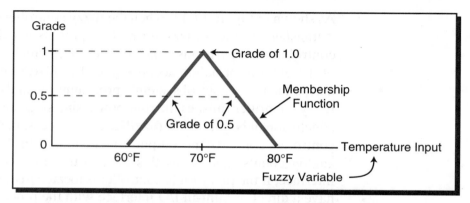

Figure 17-12. Membership function chart.

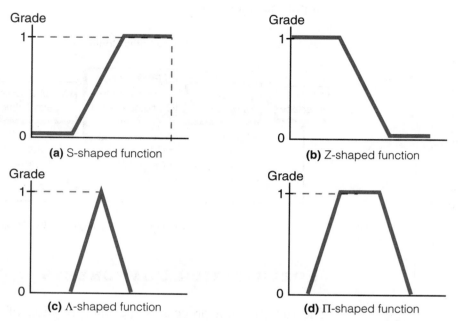

Figure 17-13. Membership function shapes: **(a)** S, **(b)** Z, **(c)** Λ, and **(d)** Π.

at each end point must have a value of 0 or 1. As shown in Figure 17-14, a membership function's shape does not have to be symmetrical; however, it must comply with the previously discussed specifications. Figure 17-15 illustrates some incorrect membership function shapes.

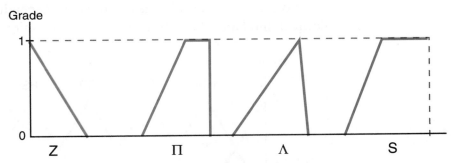

Figure 17-14. Asymmetrical membership functions.

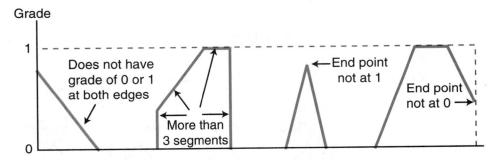

Figure 17-15. Incorrect membership function shapes.

Labels. Each fuzzy controller input can have several membership functions, with seven being the maximum and the norm, that define its conditions. Each membership function is defined by a name called a **label**. For example, an input variable such as temperature might have five membership functions labeled as cold, cool, normal, warm, and hot. Generically, the seven membership functions have the following labels, which span from the data range's minimum point (negative large) to its maximum point (positive large):

- NL (negative large)
- NM (negative medium)
- NS (negative small)
- ZR (zero)
- PS (positive small)
- PM (positive medium)
- PL (positive large)

Figure 17-16 illustrates an example of an input variable with seven Λ-shaped membership functions using all of the possible labels. A group of membership functions forms a **fuzzy set**. Figure 17-17 shows a fuzzy set with five membership functions. Although most fuzzy sets have an odd number of labels, a set can also have an even number of labels. For example, a fuzzy set may have four or six labels in any shape, depending on how the inputs are defined in relationship to the membership function.

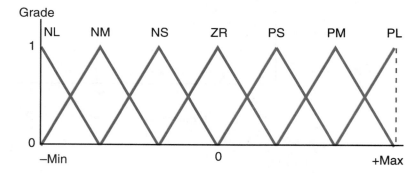

Figure 17-16. Fuzzy logic input using seven membership function labels.

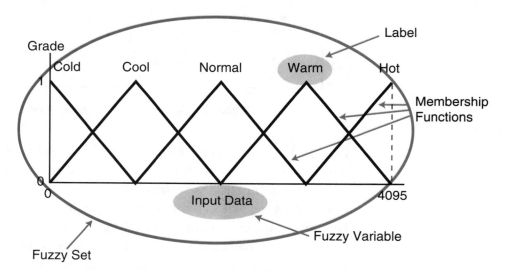

Figure 17-17. Fuzzy set with five membership functions.

FUZZY PROCESSING COMPONENTS

During fuzzy processing, the controller analyzes the input data, as defined by the membership functions, to arrive at a control output. During this stage, the processor performs two actions:

- rule evaluation

- fuzzy outcome calculation

Rule Evaluation. Fuzzy logic is based on the concept that most complicated problems are formed by a collection of simple problems and can, therefore, be easily solved. Fuzzy logic uses a reasoning, or *inferencing*, process composed of IF...THEN **rules**, each providing a response or outcome. Basically, a rule is activated, or *triggered*, if an input condition satisfies the IF part of the rule statement. This results in a control output based on the THEN part of the rule statement. In a fuzzy logic system, many rules may exist, corresponding to one or more IF conditions (see Figure 17-18). A rule may also have several input conditions, which are logically linked in either an AND or an OR relationship to trigger the rule's outcome (see Figure 17-19).

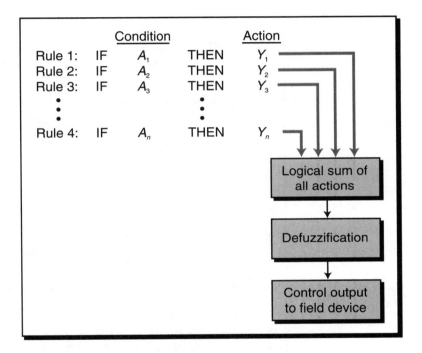

Figure 17-18. Multiple rules in a fuzzy system.

		Condition		Action
Rule 1:	IF	A_1 AND B_1 AND C_1	THEN	Y_1
Rule 2:	IF	A_2 OR B_2	THEN	Y_2
Rule 3:	IF	(A_3 AND B_3) OR C_3	THEN	Y_3

Figure 17-19. Rules with multiple input conditions linked in AND and OR relationships.

Sometimes, more than one rule is triggered at a time in a fuzzy control process. In this case, the controller evaluates all the rules to arrive at a single outcome value and then proceeds to the defuzzification process. For instance, if two inputs are logically ANDed or ORed in several rules, then they will produce several outcomes, of which only one will be logically added to determine the final outcome. Figure 17-20a illustrates an example of two fuzzy inputs, X_1 and X_2, and one fuzzy output, Y_1. The rules shown in Figure 17-20b represent four of nine possible rules that cover the two inputs. The four shown, however, cover the four possible triggering points for the two input readings, X_1 and X_2. Given the input values in Figure 17-20a, the inputs will trigger rule 1 because $X_1 = \text{ZR}$ AND $X_2 = \text{NL}$. This will generate two outputs for $Y_1 = \text{NL}$, one at a grade of 0.6 (due to the input value of X_1) and the other at a grade of 0.75 (due to the value of X_2). In a fuzzy logic situation where a two-input rule with an AND relationship produces two outcome values, the controller will choose the outcome with the *lowest* grade, in this case 0.6NL. If the rule utilizes OR logic, the chosen outcome will be the one with the

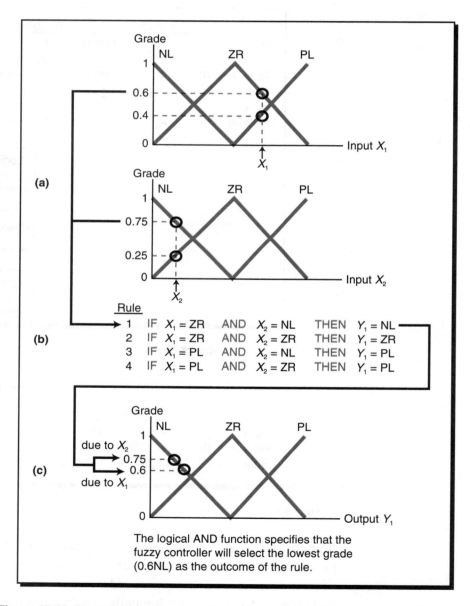

Figure 17-20. Fuzzy processing example showing **(a)** two fuzzy input values, **(b)** the four rules that they trigger, and **(c)** the resulting output.

largest grade. If rule 1 in Figure 17-20 had used an OR function instead of an AND function, then the controller would have selected the $Y_1 = 0.75\text{NL}$ outcome, the largest of the two outcomes.

Different fuzzy logic controllers have different rule evaluation capabilities. The fuzzy logic controller from Omron Electronics shown in Figure 17-21, for example, is capable of handling eight inputs and four outputs, where each input can be represented by a maximum of seven membership functions for a total of 56 membership functions (8×7). The controller also allows a maximum of 128 programmed rules. Each rule can have up to eight input conditions (which can be logically ANDed or ORed) and two outcomes.

Figure 17-21. Omron's fuzzy logic controller in a PLC system.

Fuzzy logic rules with two inputs are often represented in matrix form to represent AND conditions. For example, Figure 17-22 illustrates a 3×3 matrix (9 rules) that uses two inputs, X_1 and X_2, and one output Y_1. One advantage of this matrix representation is that it makes it easy to represent all the rules for a system. A five-label system translates into a 5×5 matrix with 25 rules, while a seven-label system produces a 7×7 matrix with 49

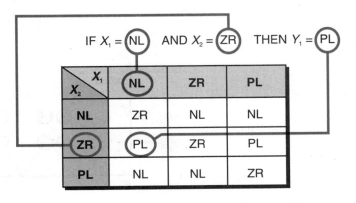

Figure 17-22. Fuzzy logic rule matrix.

rules. An even membership function combination (e.g., a system with 6 labels for one input and 4 labels for another) will have a 24-rule matrix. When more than three inputs are used, the matrix becomes more difficult to represent, since it becomes a three-dimensional matrix resembling a cube (three inputs). In this type of complicated system, the rules would be broken down into several two-dimensional matrices.

Fuzzy Outcome Calculations. Once a rule is triggered, meaning that the input data belongs to a membership function that satisfies the rule's IF statement, the rule will generate an output outcome. This fuzzy output is composed of one or more membership functions (with labels), which have grades associated with them. The outcome's membership function grade is affected by the grade level of the input data in its input membership function. In Figure 17-23a, the fuzzy input *FI* of 60% belongs to two membership functions, ZR and PS, corresponding to the grades of 0.6 and 0.4, respectively. These two grades will have an impact on the amount of the output (see Figure 17-23b) by intersecting the output membership functions at the same grade levels (0.6 and 0.4). However, the output membership function that is selected for the final output value depends on the user's programming of the IF...THEN rules.

For example, in Figure 17-24, the input triggers rules 3 and 4 because the input *FI* belongs to membership functions ZR and PS. These rules indicate that both fuzzy output action ZR and action PL must be applied to the process. These output actions will be applied at a value that corresponds to the grades generated in the input membership functions (i.e., output 0.6ZR and

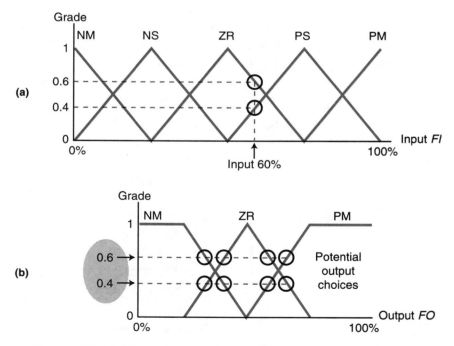

Figure 17-23. (a) Fuzzy input grades and **(b)** the resulting output grades.

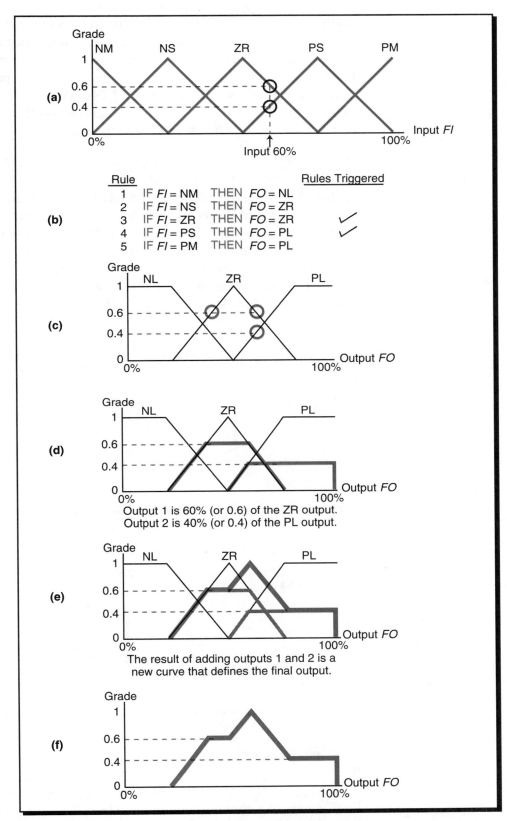

Figure 17-24. Fuzzy logic process: **(a)** inputs, **(b)** rules, **(c)** outputs, **(d)** output curves, **(e)** combined output curve, and **(f)** the output signal for the field device.

0.4PL). Note that the 0.6 grade is applied to output ZR and the 0.4 grade is applied to output PL because the user programmed the rules that way. Figure 17-24c shows these two outputs. To arrive at a final outcome value, the fuzzy logic controller logically adds both fuzzy outcomes to produce an aggregate outcome curve, illustrated in Figure 17-24e. The controller then generates an output signal (during defuzzification) that controls the process's field device (e.g., valve, motor, etc.) according to the input data (see Figure 17-24f).

A fuzzy logic controller may implement its output membership functions as noncontinuous functions that resemble spikes rather than geometrical shapes. Figure 17-25 illustrates an example of seven output membership functions represented as spikes and described by labels. Each label has a relationship to the output interface. For example, each label shown in Figure 17-25 corresponds to a value between 0 and 4095 counts. As another example, the three output membership functions presented in Figure 17-24 can be represented as noncontinuous spikes (see Figure 17-26), where the outcome grade levels specify 0.6 of ZR and 0.4 of PM.

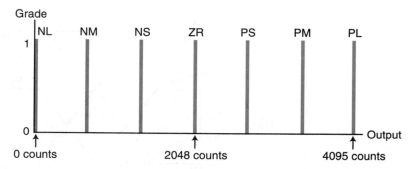

Figure 17-25. Output membership functions represented as noncontinuous functions.

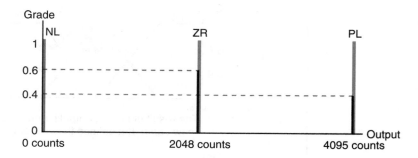

Figure 17-26. The three output membership functions from Figure 17-24 shown as spikes.

EXAMPLE 17-2

Figure 17-27 illustrates a three–membership function fuzzy set and the three rules that dictate the outcomes. For a fuzzy input (*FI*) of 37.5%, **(a)** indicate the triggered rules and the outcome membership functions selected and **(b)** illustrate the logical sum of the selected outputs.

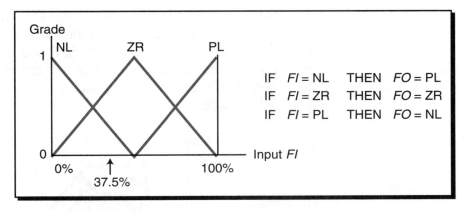

Figure 17-27. Three–membership function fuzzy set and its rules.

SOLUTION

(a) Figure 17-28a shows the two rules triggered (rules 1 and 2) by the 37.5% *FI* input, where *FI* intercepts the input membership functions at 0.25NL and 0.75ZR. Consequently, these rules trigger output values of 0.25PL and 0.75ZR, as shown in Figure 17-28b.

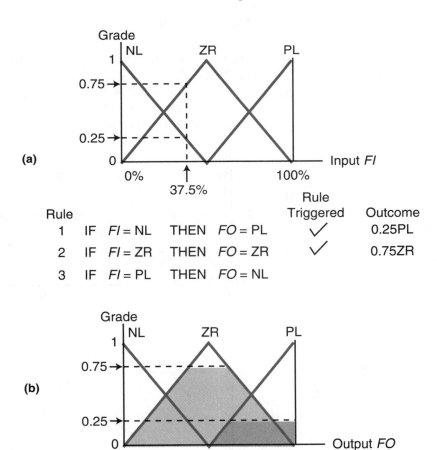

Figure 17-28. (a) Inputs triggered by the rules and **(b)** the resulting outputs.

(b) Figure 17-29 shows the logical sum that the fuzzy controller will perform. This logical sum is the result of geometrically adding the areas of the two outcomes (0.75ZR and 0.25PL) to form one graphic output, from which a final output (fuzzy output *FO*) will be selected during defuzzification. This output value will then be sent to the control field device.

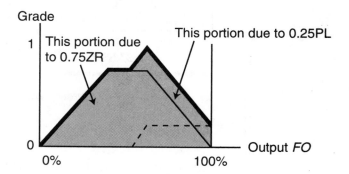

Figure 17-29. Outcome curve for Example 17-2.

EXAMPLE 17-3

Figure 17-30 illustrates two input fuzzy sets, one with five labels and the other with two labels, while Figure 17-31 shows one fuzzy output set with five labels. The rules that govern the system (as defined by the expert) are shown in Figure 17-31a in matrix form for a maximum of 10 possible combinations.

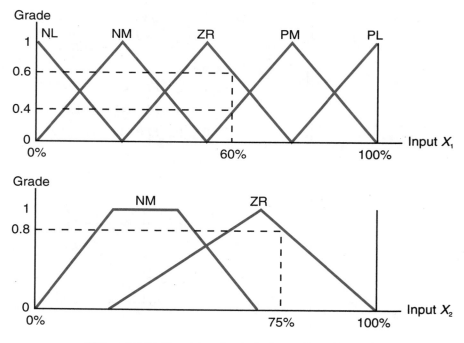

Figure 17-30. Fuzzy input sets for input X_1 and input X_2.

Output Rule Matrix

X_2 \ X_1	NL	NS	ZR	PS	PL
(a) **NM**	NL	NM	NM	ZR	PM
ZR	NM	ZR	ZR	PM	PL

(b)

Grade

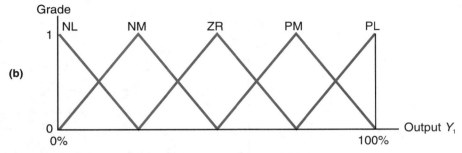

Figure 17-31. (a) Rule matrix and **(b)** fuzzy set.

(a) Indicate the rules that are triggered for the two input conditions X_1 = 60% and X_2 = 75%, as well as all the possible outcomes of the rules' triggered inputs. Also, indicate the outputs that will be selected. **(b)** Illustrate the selected outcomes and the logical outcome summation that will be used for defuzzification.

SOLUTION

(a) Figure 17-32 illustrates the two rules that will be triggered due to inputs X_1 and X_2. Input X_1 will intercept membership functions ZR and PM at grades 0.6 and 0.4, respectively. Input X_2 will intercept ZR at a

Output Rule Matrix

X_2 \ X_1	NL	NM	ZR	PM	PL
NM	NL	NM	NM	ZR	PM
ZR	NM	ZR	ZR	PM	PL

X_1 = 0.6ZR and 0.4PM
X_2 = 0.8ZR

IF X_1 = ZR AND X_2 = ZR   THEN Y_1 = ZR
IF X_1 = PM AND X_2 = ZR   THEN Y_1 = PM

Outcome

0.6ZR (due to X_1)
0.8ZR (due to X_2)
0.4PM (due to X_1)
0.8PM (due to X_2)

Select Y_1 = ZR outcome of 0.6
Select Y_1 = PM outcome of 0.4

Figure 17-32. Triggered rules.

grade of 0.8. Figure 17-30 presented these grade levels. Because the rules are linked with AND functions, each rule will have two outputs, of which the one with the lowest value will be chosen for the logical sum of the outputs.

(b) Figure 17-33 illustrates the two selected outputs from the two triggered rules and the resulting output after the two rule outcomes are logically added.

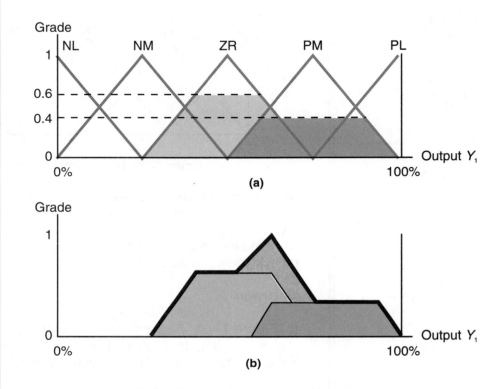

Figure 17-33. (a) Triggered outputs and **(b)** outcome curve.

DEFUZZIFICATION COMPONENTS

The final output value from the fuzzy controller depends on the defuzzification method used to compute the outcome values corresponding to each label. The defuzzification process examines all of the rule outcomes after they have been logically added and then computes a value that will be the final output of the fuzzy controller. The PLC then sends this value to the output module. Thus, during defuzzification, the controller converts the fuzzy output into a real-life data value (e.g., 1720 counts).

There are many defuzzification methods, but all are based on mathematical algorithms. The two most common defuzzification methods are:

- maximum value

- center of gravity

Maximum Value Method. The **maximum value method** bases the final output value on the rule output with the highest membership function grade. This method is mainly used with discrete output membership functions. Referring to Figure 17-26 (shown again in Figure 17-34), the maximum value defuzzification method would specify that the output value of 2048 counts be chosen as the final output value because it has the largest grade value. If two or more outcomes from two or more rules have the same grade level, then the controller will select the outcome that will be the final value based on criteria supplied by the user during the fuzzy application programming setup or system definition. Such criteria is determined by choosing either the left-most or right-most grade value of the two equal labels and their corresponding number of counts. The left-most criteria selects the lowest output (counts), while the right-most criteria selects the highest output value (highest counts).

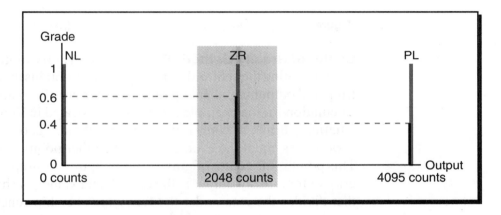

Figure 17-34. The maximum value method selects the largest output grade level.

Figure 17-35a illustrates the outcome of three rules. If the maximum value defuzzification method is used, ZR will be the final output value, meaning that the output of the controller will be 2340 counts. If the rules triggered two equal maximum grade values, as is the case in Figure 17-35b, then the controller would use the programmed criteria to select the appropriate output value. If this criteria specified the left-most maximum value, then the controller would chose label NM, which would provide an output of 1170 counts. Note that, during the defuzzification process, the fuzzy controller sends the actual output value (e.g., counts), not the grade value, to the output device. So, in Figure 17-35a, the output will be approximately 2340 counts. In Figure 17-35b, the left-most output will be approximately 1170 counts and the right-most output, if chosen, will be 3510 counts.

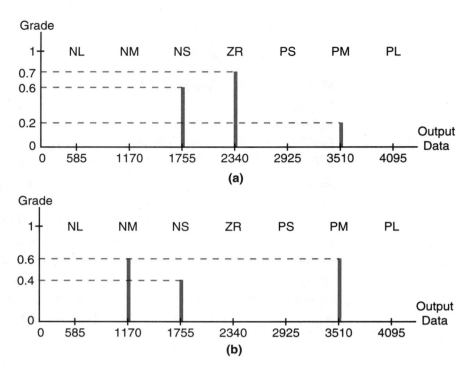

Figure 17-35. (a) Single maximum output value and **(b)** multiple maximum output values.

Center of Gravity Method. The **center of gravity method**, also referred to as "calculating the centroid," mathematically obtains the center of mass of the triggered output membership functions. Figure 17-36 illustrates the centroid calculation for the example previously illustrated in Figure 17-24. In mathematical terms, a **centroid** is the point in a geometrical figure whose coordinates equal the average of all the other points comprising the figure. This point is the center of gravity of the figure. In simple terms, the center of gravity for a fuzzy output is the output data value (as shown on the X-axis), that divides the area under the fuzzy membership function curve into two equal parts. The center of gravity method is the most commonly used defuzzification method because it provides an accurate result based on the weighted values of several output membership functions. The output value that is sent to the output interface module is the output data value at the intersection of the horizontal axis and the centroid.

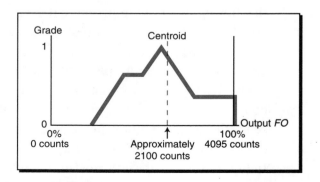

Figure 17-36. Centroid calculation of the output from Figure 17-24.

The center of gravity method applies to noncontinuous, or discrete, output membership functions, as well as continuous ones. In noncontinuous functions, the final output value that will be obtained for a seven-label output membership function (labels A through G) is expressed by the formula:

$$\text{Output data} = \frac{\sum\limits_{n=A}^{n=G}\left[(FO_n)(FGrade_n)\right]}{\sum\limits_{n=A}^{n=G}FGrade_n}$$

where:

Output data = the number of counts to be used for the output

FO = the fuzzy output in counts for labels n = A through G

$FGrade$ = the fuzzy grade level for levels n = A through G

Referring to our previous noncontinuous membership example, now shown in Figure 17-37, this equation implies that the final value of the output will be equal to the sum of each rule outcome's grade times its actual output data

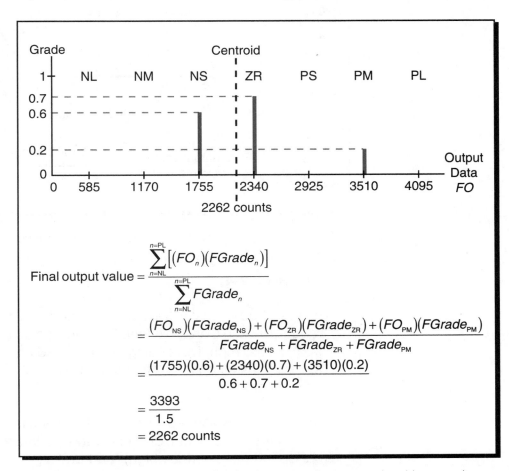

Figure 17-37. Centroid calculation for the noncontinuous membership example.

value (i.e., counts) divided by the sum of the rule outcome grades. In this case, the fuzzy logic controller will decide to send an output of 2262 counts to the output interface after completing the center of gravity calculation. The fuzzy controller's output is less than it would have been using the maximum value method (the maximum output label is ZR, which is 2340 counts), indicating that the weighted value of the 0.6NS label pulls the value to the left (less counts). However, the output value is slightly balanced by the right-pulling action of the 0.2PM label.

Fuzzy controllers utilizing continuous membership functions and the center of gravity defuzzification method also use the previous summation equation to approximate the centroid value (see Figure 17-38). However, in this case, the controller uses approximate digitized values for each membership function to compute each of the points in the summation.

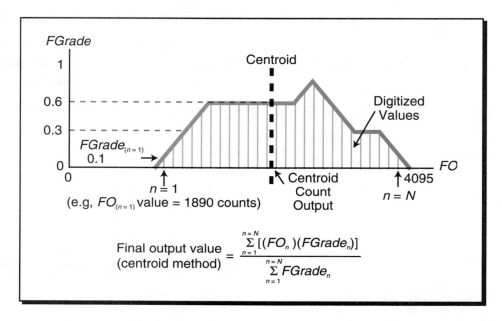

Figure 17-38. Centroid value approximation.

EXAMPLE 17-4

Figure 17-39 illustrates two cars separated by a distance *d*, which can range between 0 and 120 feet. Car 1 travels at a speed of *v*, ranging from 0 to 80 mph. Depending on the speed and distance, car 2 has several braking options (*B*) ranging from light to hard if car 1 slows down or stops.

(a) Create a Λ-shaped fuzzy set that contains three membership functions for each input: distance between cars (short, normal, long), car speed (low, normal, high), and braking strength (light, normal, hard). **(b)** Establish a set of rules for the braking output as a function of the speed and distance. Illustrate these rules in matrix form. **(c)** Using the center of gravity method, calculate the value of the outcome if car 1 is traveling at 65 mph and the distance between car 1 and car 2 is 45 feet.

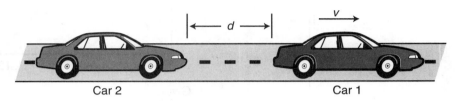

d = distance between cars
v = velocity (speed) of car 1
B = braking strength (function of d and v)

Figure 17-39. Example 17-4.

SOLUTION

(a) Figure 17-40 illustrates the three fuzzy sets, each with three membership functions. Figures 17-40a and 17-40b illustrate the two input fuzzy sets, distance and speed. The distance fuzzy set ranges from 0 to 120 feet and the speed fuzzy set ranges from 0 to 80 mph. Figure 17-40c shows the output fuzzy set (braking strength), whose output ranges from light braking to hard braking.

(b) The fuzzy system's rules are composed of IF...THEN statements defining all possible outcomes. For example:

> IF the distance between the two cars is long
> and the speed is normal, THEN brake lightly.

This rule implies that normal braking will be applied if the speed is normal and the distance between the cars is long. Using the fuzzy sets illustrated in Figure 17-40, this rule can be expressed as:

$$\text{IF } d = \text{PL AND } v = \text{ZR THEN } B = \text{NL}$$

Table 17-1 lists the rules for this fuzzy system, based on distance and speed. Remember that the user defines the rules of the system based on the desired outcome, according to his or her experience and knowledge. For example, a regular driver would tend to brake either normally or hard if the distance was short and the speed was high. However, a NASCAR driver might brake very lightly or not at all if the front car slows down, even though the NASCAR driver's distance may be very short and speed very high.

Figure 17-41 illustrates these rules in matrix form. This matrix configuration allows you to see a large number of rules and their outcomes at a glance.

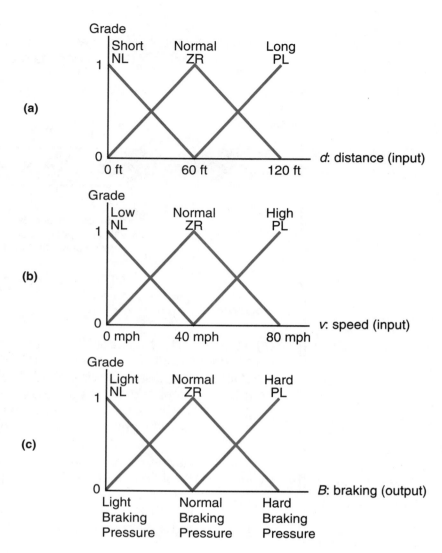

Figure 17-40. The fuzzy input sets—**(a)** distance and **(b)** speed—and **(c)** the fuzzy output set, braking.

Fuzzy Rules	
1	IF d = NL AND v = NL THEN B = ZR
2	IF d = NL AND v = ZR THEN B = PL
3	IF d = NL AND v = PL THEN B = PL
4	IF d = ZR AND v = NL THEN B = NL
5	IF d = ZR AND v = ZR THEN B = ZR
6	IF d = ZR AND v = PL THEN B = PL
7	IF d = PL AND v = NL THEN B = NL
8	IF d = PL AND v = ZR THEN B = NL
9	IF d = PL AND v = PL THEN B = ZR

Table 17-1. The fuzzy system's rules.

Distance *d* / Speed *v*	Short NL	Normal ZR	Long PL
Low NL	ZR (Brake Normal)	NL (Brake Light)	NL (Brake Light)
Normal ZR	PL (Brake Hard)	ZR (Brake Normal)	NL (Brake Light)
High PL	PL (Brake Hard)	PL (Brake Hard)	ZR (Brake Normal)

Figure 17-41. Fuzzy logic rule matrix.

(c) Figures 17-42a and 17-42b illustrate the graphs for the inputs $d =$ 45 ft and $v =$ 65 mph. Each input triggers (crosses) two membership functions—input d crosses membership functions ZR and NL; input v crosses membership functions PL and ZR. Thus, these inputs trigger four rules, rules 2, 3, 5, and 6, as were shown in Table 17-1.

Note that the inputs to these rules are connected logically by AND functions, meaning that the rules' outputs will correspond to the smallest input grade value. For example, rule 2 will be triggered because the 45-foot distance input crosses the NL (short distance) membership function (IF $d =$ NL...) and the 65 mph speed input crosses the ZR (normal speed) membership function (...AND $v =$ ZR). The grades for each input to rule 2 are as follows: distance = 0.25NL and speed = 0.375ZR. In other words, the 45-foot distance is 25%

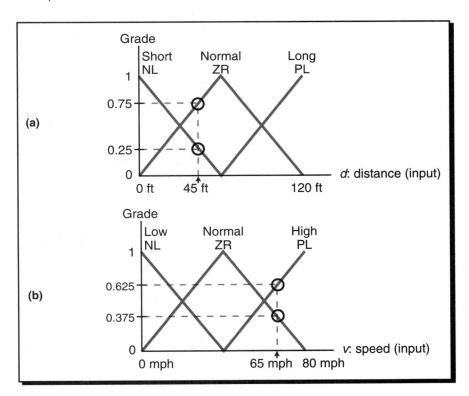

Figure 17-42. Graphs for **(a)** distance and **(b)** speed inputs.

short and the 65 mph speed is 37.5% normal. The rule implies two braking outputs (...THEN B = PL) based on these inputs, one at 0.25 (due to distance) and the other at 0.375 (due to speed). Because of the AND condition, however, the fuzzy controller will select the smallest outcome, 0.25 (see Figure 17-43). Figure 17-44a shows the outcome summary for all four rules, including the rule outcome selected (i.e., the smallest outcome because of the AND rule logic). Figure 17-44b illustrates all four triggered outcomes. Note that both rules 2 and 3 have 0.25PL outcomes; the dotted line represents the addition of these two outcomes. Figure 17-44c shows the logical sum of all the outcomes and the approximate output result, which was obtained using the center of gravity defuzzification method. By visually inspecting the output, the braking strength will be at approximately 70% normal and 30% hard.

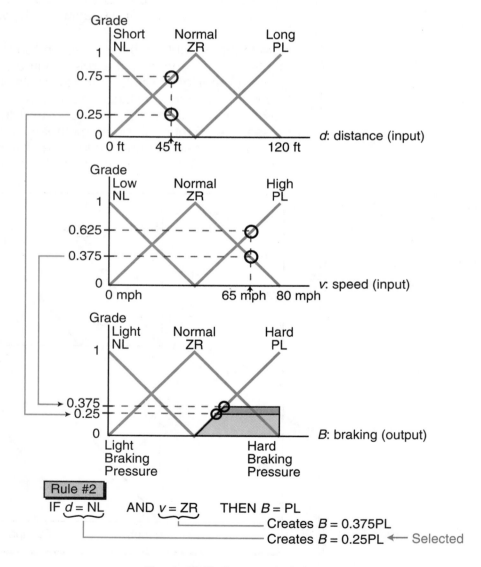

Figure 17-43. Output calculation.

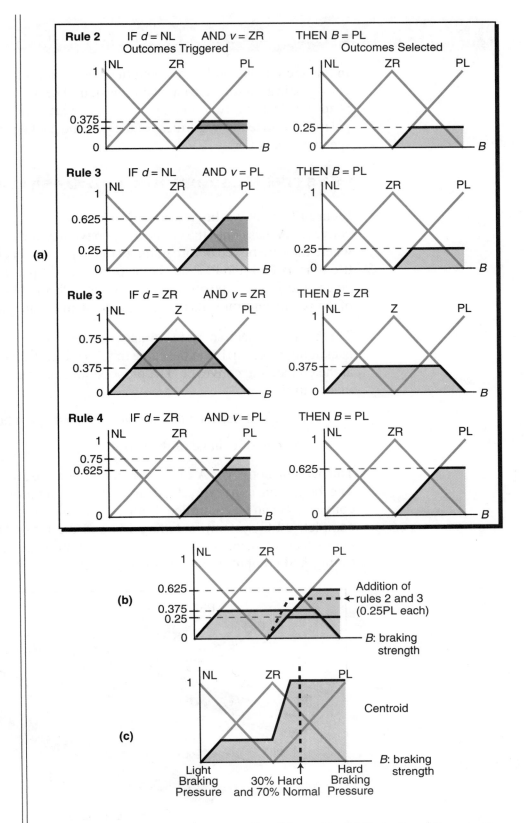

Figure 17-44. (a) Outcome summary for all four rules, **(b)** illustration of the outputs, and **(c)** defuzzification of the result.

17-5 FUZZY LOGIC CONTROL EXAMPLE

In this section, we will implement fuzzy logic rules to control the speed of a conveyor in an automated packaging system. The objective of this application is to synchronize two conveyors so that parts and packaging boxes are positioned correctly, regardless of the part and package box positions and the speed of conveyor.

SYSTEM DESCRIPTION AND OPERATION

Figure 17-45 shows the two-conveyor packaging system. The parts travel on conveyor A, pass onto the connecting conveyor, and then go to conveyor B, where they are boxed before going to the wrapping machine. The photoelectric sensors PE1 and PE2 detect the presence of a part and initiate a count to determine the part's position from encoder 1. PE3 and PE4 detect the presence of a box and determine its position based on the count inputs from encoder 2.

The control objective is to adjust the speed of conveyor B so that the packaging boxes arrive at the same time as the parts, meaning that they meet at the connecting conveyor. The process information required to implement this control is:

- the offset between the part and the packaging box

- the rate of change of the offset

The parts on conveyor A travel at random intervals, but at a constant speed. The boxes on conveyor B occur at regular intervals, and the speed of conveyor B can be controlled. Figure 17-46 shows the block diagram of the complete PLC system, including I/O and the fuzzy logic controller. The photoelectric sensors will be used in the PLC program to detect when to start timing and computing the data from the encoders. A section of the PLC's

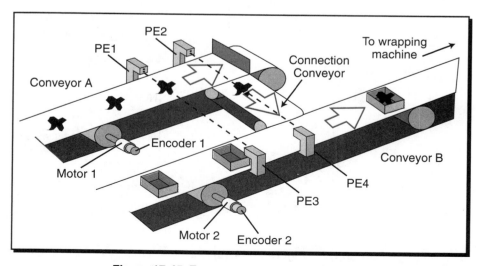

Figure 17-45. Two-conveyor packaging system.

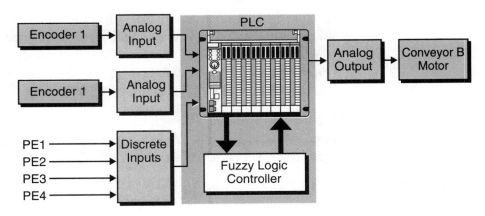

Figure 17-46. Block diagram of the PLC system.

main program must adjust the two fuzzy inputs, the part/box offset and the rate of change of the offset, so that the data is centered around a value of 2048 counts. The reason for this is that the range of the fuzzy set will be 0 to 4095 analog counts; the count value of 2048 is the middle membership function.

If a box is present at PE3 and a part is present at PE1, conveyor B should run at the same speed as conveyor A (the reference speed set initially by the operator). If the box is at PE3 but the part is behind PE1 (see Figure 17-47),

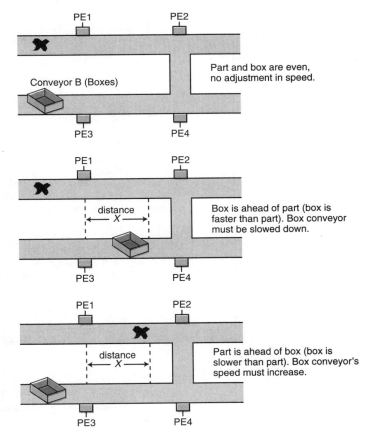

Figure 17-47. Conveyor B may slow down or speed up.

the system will slow conveyor B until the part is at PE1, at which time the fuzzy controller will indicate an increase in the speed of conveyor B so that it will catch up with conveyor A. The distance traveled by the box is calculated, using the input data from encoder 2, as the difference between the time the box passes PE3 and the time part passes PE1.

Figure 17-48 shows a flowchart of the steps that must occur to pass correct input information to the fuzzy processor. The value at position X provides the part/box offset data. This value is calculated as:

$$X = (\text{Encoder 1 counts}) - (\text{Encoder 2 counts})$$

The rate of change of the offset is calculated as the difference between the current offset reading (X_n) and the previous one ($X_{(n-1)}$):

$$\Delta X = X_n - X_{(n-1)}$$

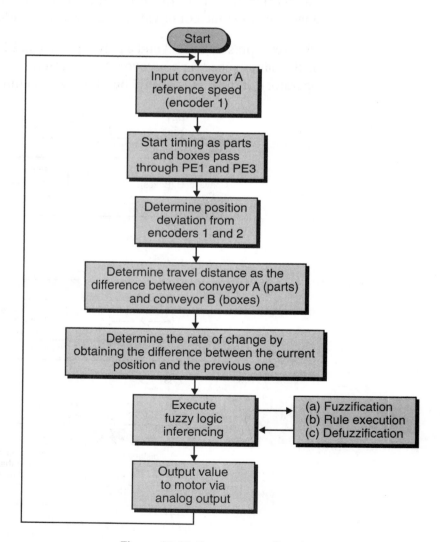

Figure 17-48. Fuzzy system flowchart.

MEMBERSHIP FUNCTIONS AND RULE CREATION

To provide enhanced resolution and accuracy, this system uses a five–membership function (five-label) fuzzy sets for the two inputs and a seven–membership function fuzzy set for the output. Figure 17-49 shows these three fuzzy sets. The offset input is named X (deviation between part and box) and the offset rate of change input is named ΔX (rate of change of deviation). The fuzzy set for the output is named S (speed), which corresponds to the motor speed of conveyor B. Note that the range of each fuzzy input and output

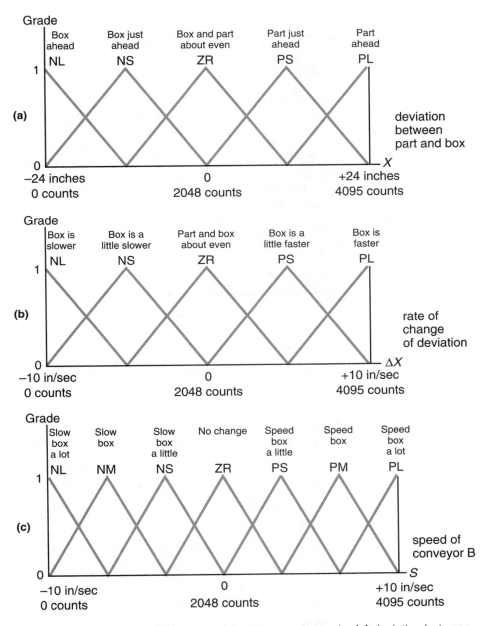

Figure 17-49. Three fuzzy sets used for the conveyor example: **(a)** deviation between part and box (input), **(b)** rate of change of deviation (input), and **(c)** speed of conveyor B (output).

variable is from 0 to 4095 counts. This corresponds to a range of ±24 inches for the deviation between the part and box positions, a range of ±10 inches/second for the rate of change of the offset, and a range of ±10 inches/second for the speed of the box conveyor.

The fuzzy logic database for this system contains 25 rules. Figure 17-50 shows a matrix of the rules, describing the desired output according to the deviation between the part and the box and the rate of change of deviation. This matrix includes a description of the rule inputs and outputs, as well as their respective membership function labels. Table 17-2 lists the actual rules that will be entered into the fuzzy controller in IF…THEN form.

Deviation X / Rate ΔX	NL Box is ahead	NS Box is just ahead	ZR About even	PS Part is just ahead	PL Part is ahead
NL Box is slower	NM Slow the box	NS Slow the box a little	PS Speed up the box a little	PM Speed up the box	PL Speed up the box a lot
NS Box is a little slower	NM Slow the box	NS Slow the box a little	PS Speed up the box a little	PM Speed up the box	PL Speed up the box a lot
ZR About even	NM Slow the box	NS Slow the box a little	ZR No change	PS Speed up the box a little	PM Speed up the box
PS Box is a little faster	NL Slow the box a lot	NM Slow the box	NS Slow the box a little	PS Speed up the box a little	PM Speed up the box
PL Box is faster	NL Slow the box a lot	NL Slow the box a lot	NS Slow the box a little	PS Speed up the box a little	PM Speed up the box

Figure 17-50. Fuzzy logic rule matrix.

Conveyor System Rules

1 IF X = NL AND ΔX = NL THEN S = NM
2 IF X = NL AND ΔX = NS THEN S = NM
3 IF X = NL AND ΔX = ZR THEN S = NM
4 IF X = NL AND ΔX = PS THEN S = NL
5 IF X = NL AND ΔX = PL THEN S = NL
6 IF X = NS AND ΔX = NL THEN S = NS
7 IF X = NS AND ΔX = NS THEN S = NS
8 IF X = NS AND ΔX = ZR THEN S = NS
9 IF X = NS AND ΔX = PS THEN S = NM
10 IF X = NS AND ΔX = PL THEN S = NL
11 IF X = ZR AND ΔX = NL THEN S = PS
12 IF X = ZR AND ΔX = NS THEN S = PS
13 IF X = ZR AND ΔX = ZR THEN S = ZR
14 IF X = ZR AND ΔX = PS THEN S = NS
15 IF X = ZR AND ΔX = PL THEN S = NS
16 IF X = PS AND ΔX = NL THEN S = PM
17 IF X = PS AND ΔX = NS THEN S = PM
18 IF X = PS AND ΔX = ZR THEN S = PS
19 IF X = PS AND ΔX = PS THEN S = PS
20 IF X = PS AND ΔX = PL THEN S = PS
21 IF X = PL AND ΔX = NL THEN S = PL
22 IF X = PL AND ΔX = NS THEN S = PL
23 IF X = PL AND ΔX = ZR THEN S = PM
24 IF X = PL AND ΔX = PS THEN S = PM
25 IF X = PL AND ΔX = PL THEN S = PM

Table 17-2. Fuzzy system rules.

Once the fuzzy controller receives the inputs, it will determine the final output value based on a logical addition of the selected outcomes. The outcome calculation may be very complex, due to the large number of rules. Remember, however, that the entire fuzzy logic analysis—fuzzification, rule execution, and defuzzification—is based on user-specified criteria for desired outputs based on photoelectric and encoder input data.

EXAMPLE 17-5

Figure 17-51 illustrates the part/box input membership functions for a conveyor system with two input readings, where the deviation between the part and the box is 9 inches and the rate of change is about –3.75 inches per second. This means that the part is just ahead of the box because the box is a little slower than the part. Figure 17-52 illustrates this situation.

(a) Determine which rules are triggered, indicate which outcomes are selected, and plot the output membership functions. **(b)** Illustrate the logical sum of the selected outputs and indicate an approximate output using the center of gravity method.

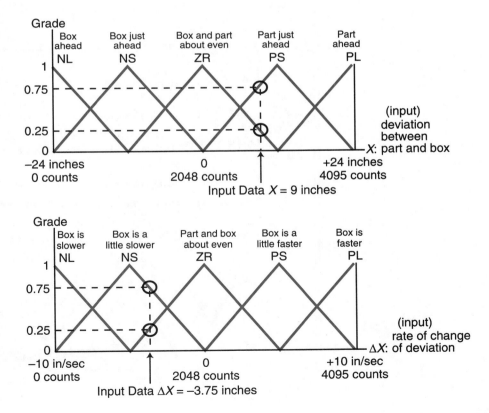

Figure 17-51. Part/box input membership functions.

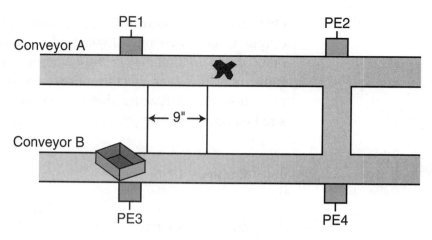

Figure 17-52. Part/box configuration.

SOLUTION

(a) Figure 17-53 shows the four rules that will be triggered by the input reading, along with the selected outcomes and their graphical representation. Note that the outcomes selected are the lowest of the two possible outcomes due to the AND logical link. Note that two of the rules generate a 0.25PS output. The bold line in Figure 17-53b indicates the sum of both outputs (0.25PS × 2 = 0.5PS).

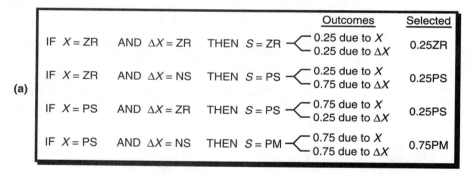

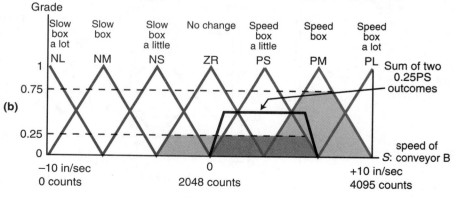

Figure 17-53. (a) Triggered rules and **(b)** their output graphic.

(b) Figure 17-54 shows the logical sum of all the rules' actions. The centroid for this output is located at approximately 2990 counts, which increases the conveyor speed approximately 4.6 inches/second, so that the box can catch up with the part.

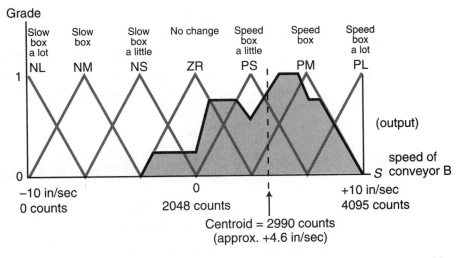

Figure 17-54. Logical sum of the rules' actions and the corresponding centroid.

17-6 FUZZY LOGIC DESIGN GUIDELINES

The guidelines presented in this section provide you with the proper procedures for designing an effective fuzzy logic control system. Although some of these design guidelines are similar to those used in standard PLC systems, others are design requirements that are specific to fuzzy logic systems. The basic elements for the successful implementation of a fuzzy logic control system include:

- control objectives
- control system configuration
- input/output determination
- fuzzy inference engine design

CONTROL OBJECTIVES

Fuzzy logic can be applied to virtually any type of control system, but it is especially suited for applications that rely heavily on human intuition and experience. The primary objective of applying fuzzy logic to an existing process is to improve the overall process and to automate tasks that previously required human judgment. In a new system, the primary objective of using fuzzy logic is usually to implement control that cannot be implemented using standard control methods. A system designer should not use fuzzy logic control just because it is available. Rather, he or she should use it because it will enhance the system. Otherwise, the outcome may not be enhanced; it may just become confusing.

Typical applications of fuzzy logic involve batching systems and temperature control loops, where process control involves "tweaking" the output based on judgments about input conditions. For example, a temperature control loop application typically requires a knowledgeable operator who can regulate the control element based on decisions such as "if the temperature is a little high but all other inputs are OK, then turn the steam valve a little clockwise." This rationale lends itself to fuzzy logic control.

CONTROL SYSTEM CONFIGURATION

The control objective may lead you to one of several types of system configurations where fuzzy logic can be implemented. Fuzzy logic does not have to be applied only in dedicated fuzzy control applications. It can also be used as a complementary system that supports other more conventional control methods. When used in this manner, the system is said to be a conventional, fuzzy, hybrid control system.

Figure 17-55 illustrates a typical process control system controlled by a PID loop. A fuzzy logic controller could enhance this PID system by regulating the steam volume based on the tank jacket temperature and the batch temperature (see Figure 17-56). If the jacket temperature decreases

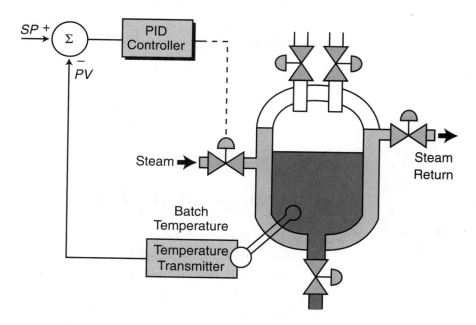

Figure 17-55. PID-controlled heating system.

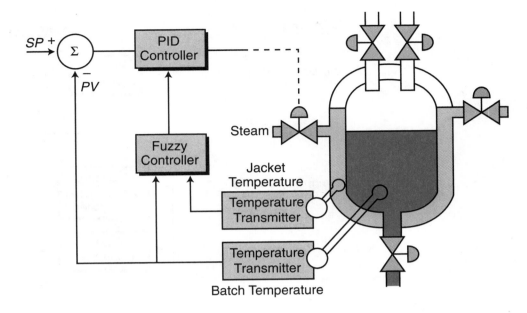

Figure 17-56. PID-controlled heating system with a fuzzy logic controller.

before the batch temperature, the fuzzy controller can take corrective action by suggesting an increase in the steam volume going into the jacket. This operation is similar to cascade control utilizing a fuzzy controller for the inner loop (secondary loop). The fuzzy controller's responsibility is to maintain a proper ratio between the jacket and batch temperatures. Figure 17-57 illustrates several other fuzzy logic system block diagrams, including a pure fuzzy control system.

INPUT/OUTPUT DETERMINATION

Once you have selected the fuzzy system configuration that is appropriate for the control objective, you must determine which inputs and outputs will be used in the fuzzy logic controller. The input conditions, or fuzzy input variables, must be able to be expressed by IF...THEN statements. That is, the input conditions to the fuzzy controller must be able to trigger conditional rules, meaning that they specify one or more output conditions. Inputs should be selected according to the process situations they describe. In other words, if you select two inputs that have little to do with each other, the outcomes that they generate will not be as precise or intuitive as the outcomes generated by inputs that deal with the same process element. For example, referring to Figure 17-56, the batch temperature and jacket temperature both relate to the regulation of the steam valve output. By analyzing these two inputs together, the fuzzy controller can make a precise decision about how much to adjust the steam valve. An analysis of two other unrelated inputs, such as batch temperature and liquid level, would not provide such an informed decision.

Fuzzy Logic System   **System Block Diagram**

Pure Fuzzy Control

Parallel Fuzzy Control

Fuzzy Logic—Refined Control
(Modified with fuzzy logic)

Man-Machine Fuzzy Interface

Fuzzy Logic—Tuned Control

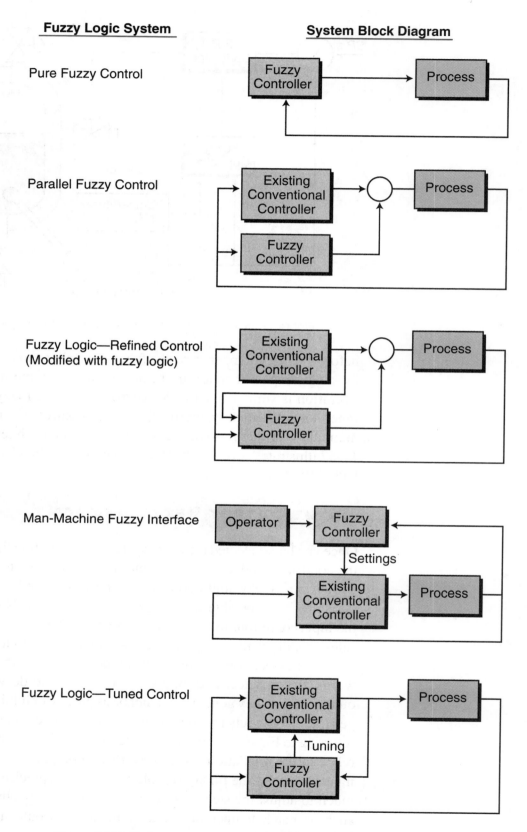

Figure 17-57. Various types of fuzzy logic systems and their block diagrams.

FUZZY INFERENCE ENGINE

The selection of the fuzzy inference engine encompasses the determination of how the fuzzification process will take place (e.g., the number and form of membership function, etc.), how the rules determine an outcome, and how the fuzzy controller implements the defuzzification.

Fuzzification. The fuzzification process, which utilizes the membership functions defined by the user, assigns a grade to each fuzzy input received. This grade determines the level of outcome that will be triggered. Therefore, the shape of the fuzzy set's membership functions is important, since the shape determines the input signals' grades, which are mapped on the output membership function.

Some fuzzy controllers allow the user to choose the shape of the membership functions by trial and error, while others have predefined membership function shapes. When using trial and error to determine the function shapes in a closed-loop fuzzy control system, the input membership functions should begin with overlapped Λ-shaped labels (see Figure 17-58). This ensures smoother control for the first trial due to the coverage provided by the Λ shape and the overlapping at the minimum and maximum points, which creates a balance (i.e., when one label grade is 1, the other is 0). The number of labels, or membership functions, that will form the fuzzy set is also an important part of the system design. For example, if a fuzzy set has five

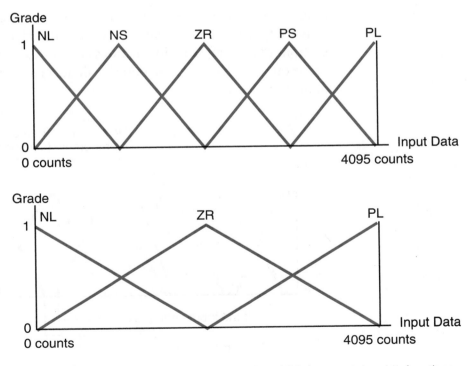

Figure 17-58. Fuzzy input sets using **(a)** five and **(b)** three membership functions.

labels covering the same input data range as a three-label fuzzy set (refer to Figure 17-58), the one with five labels will provide more fine-tuned control, especially if the output membership function also has five labels.

Although membership functions do not have to be symmetrical (see Figure 17-59), asymmetrical fuzzy sets should be carefully designed to ensure that they describe the fuzzy variable input properly. In Figure 17-59, the inner membership functions provide more sensitivity near the zero label (from NS to PS) than at the NL and PL labels (from NL to NS and from PS to PL). Asymmetrical membership functions are typically used in open-loop system applications.

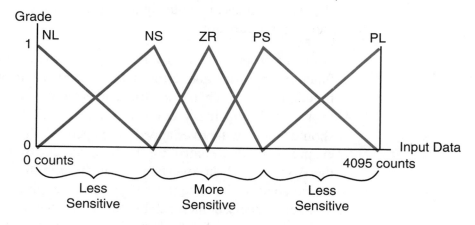

Figure 17-59. Asymmetrical fuzzy input set.

Sometimes, a membership function in a fuzzy set may not provide any sensitivity between two labels. As illustrated in Figure 17-60, the flat sections of the membership functions do not influence neighboring functions or the output. Therefore, the output will not change if the input variable falls in these regions.

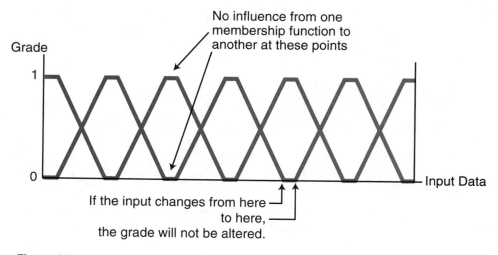

Figure 17-60. Fuzzy input set with no sensitivity at the flat sections between the labels.

Rule Decision Making and Outcome Determination. The easiest way to formulate the rules for a fuzzy logic controller is to first write them as IF…THEN statements that describe how the inputs affect the outcome. Some fuzzy controllers are capable of handling two outputs at the same time, thus allowing two rules to be combined. For example, the rules:

$$\text{IF } A = \text{PS AND } B = \text{NS THEN } C = \text{ZR}$$

$$\text{IF } A = \text{PS AND } B = \text{NS THEN } D = \text{NS}$$

can be combined into one rule:

$$\text{IF } A = \text{PS AND } B = \text{NS THEN } C = \text{ZR and } D = \text{NS}$$

This rule gives two outcomes, thus invoking two defuzzification processes, one for each controlling output. It is easiest, however, to create each rule individually (with only one outcome) and then combine them later. If at any point during the rule definition you are uncertain of the operational knowledge required for that particular rule, you should consult a knowledgeable operator so that he/she can provide you with more input information.

As mentioned earlier, you may or may not have a choice of output membership function shapes (Λ, Π, S, or Z). You also may or may not have a choice about whether the functions are continuous or discrete (see Figure 17-61).

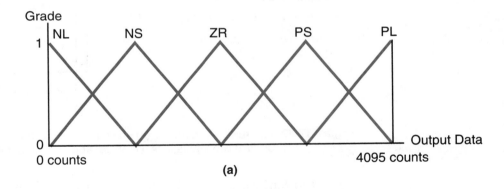

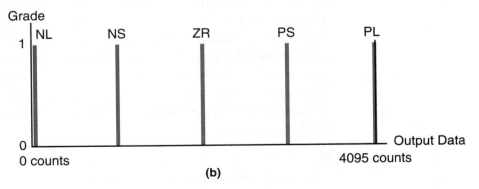

Figure 17-61. Fuzzy output sets with **(a)** continuous and **(b)** noncontinuous membership functions.

Remember that, before defuzzification occurs, the fuzzy controller adds all the outcomes based on the appropriate logic. If the rule contains a logical AND function, the controller will select the *lowest* output value; if the rule contains an OR function, the controller will select the *highest* output value.

If an application requires a highly accurate or smooth output, the rules should be designed so that an input condition triggers two or more rules. To do this, either the input membership functions must overlap or two input conditions must influence the same output (see Figure 17-62).

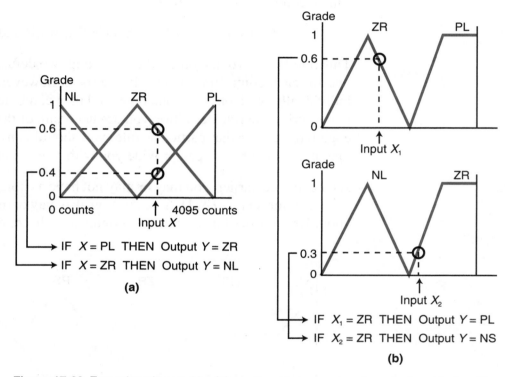

Figure 17-62. Two rules triggered by **(a)** one input in an overlapping membership function and **(b)** two inputs in two nonoverlapping membership functions.

Defuzzification. During the design of a fuzzy logic system, you may be required to choose a defuzzification method, especially if the output membership function is noncontinuous. Defuzzification methods include the center of gravity (centroid), the left-most maximum, and the right-most maximum (see Figure 17-63). If the selected defuzzification method is the center of gravity approach, the triggering rules must be arranged so that at least one rule is triggered at all times. Thus, there must always be an output from a rule. The controller will generate an error if there is no output due to a gap in input condition coverage.

Figure 17-64 illustrates two fuzzy input sets with four rules, which have a potential error condition due to improper coverage of the inputs by the rules defined. For instance, if the X_1 input intersects label ZR at the point where only ZR, and not PL, is triggered (shown as the gap in Figure 17-64a) and

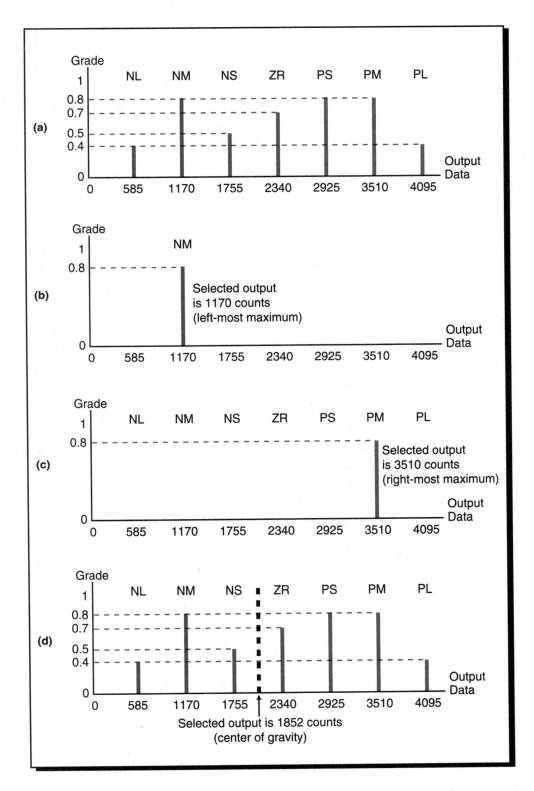

Figure 17-63. (a) Seven outputs with the final output selected using **(b)** the left-most maximum, **(c)** the right-most maximum, and **(d)** the center of gravity methods.

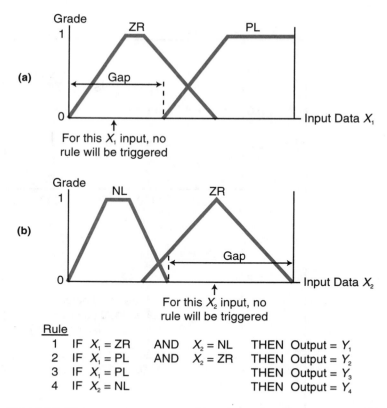

Figure 17-64. Improper coverage of inputs leading to an error condition.

input X_2 intersects label ZR anywhere in the gap area shown in Figure 17-64b, no rule will be triggered. Therefore, no output will be generated and an error will occur. Figure 17-65 shows another gap situation where a region with no sensitivity has no label (membership function); thus, no rule can be triggered. To avoid these potential error conditions, the rules should be designed so that there are no gaps in the rules.

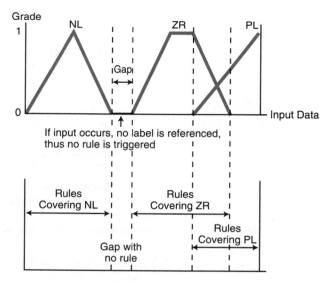

Figure 17-65. A gap in a fuzzy input set.

KEY TERMS
center of gravity method
centroid
defuzzification
fuzzification
fuzzy logic
fuzzy processing
fuzzy set
grade
label
maximum value method
membership function
rule

CHAPTER EIGHTEEN

LOCAL AREA NETWORKS

Synergy means behavior of whole systems unpredicted by the behavior of their parts.

—Richard Buckminster Fuller

CHAPTER HIGHLIGHTS
As control systems become more complex, they require more effective communication schemes between the system components. Some machine and process control systems require that programmable controllers be interconnected, so that data can be passed among them easily to accomplish the control task. Other systems require a plantwide communication system that centralizes functions, such as data acquisition, system monitoring, maintenance diagnostics, and management production reporting, thus providing maximum efficiency and productivity. This chapter presents one type of PLC communication scheme—the local area network—and the role it plays in achieving factory integration. The next chapter will discuss I/O bus networks, a type of communication scheme in which I/O field devices are connected directly to a network.

18-1 HISTORY OF LOCAL AREA NETWORKS

The proliferation of electronic and computer technologies in the 1970s made it feasible to place small personal computers at locations where users needed them. Before this, computational tasks had been performed by large computers in centralized locations. The widespread use of personal computers prompted the need for a communication method that could link this equipment. This led to the creation of **local area networks (LANs)**. These networks facilitated the decentralization of computing tasks by allowing network-connected computers to exchange information among themselves, without having to go through a central location.

Local area networks soon made their way to the industrial arena, where control had previously been exercised through a central PLC or main control system. LANs allowed many PLCs to be placed at different locations, each having its own intelligence to implement control. They also allowed PLCs to communicate system information with other PLCs performing other control tasks throughout the plant. This wave of industrial technology created further networking developments, including a special type of network—the I/O bus network—which allows intelligent field devices to communicate information to PLCs without standard PLC input/output interfaces. The next chapter explains I/O bus networks in detail.

18-2 PRINCIPLES OF LOCAL AREA NETWORKS

DEFINITION

A local area network is a high-speed, medium-distance communication system. For most LANs, the maximum distance between two nodes in the network is at least one mile, and the transmission speed ranges from 1 to 20 megabaud. Also, most local networks support at least 100 stations, or nodes. A special type of local area network, the industrial network, is one which meets the following criteria:

- capable of supporting real-time control

- high data integrity (error detection)

- high noise immunity

- high reliability in harsh environments

- suitable for large installations

Two other common types of local area networks are business system networks (e.g., Ethernet) and parallel-bus networks (e.g., Cluster/One). Business networks do not require as much noise immunity as industrial networks, since they are used in office environments. They also have less stringent access time requirements. The user of a business work station can wait a few seconds for information without problem, but a machine being controlled by a PLC may require information within milliseconds to operate correctly. Parallel-bus networks have requirements similar to business networks and are intended for microcomputers and minicomputers used in office environments over short distances.

Different types of networks have different allowable distances between connected devices. Figure 18-1 illustrates the distances at which different types of networks and buses can be used. Note that long-distance communication still relies on public networks, such as telephone systems, which have long-range data-channeling capabilities. However, developments in cable TV data transmission are enabling data exchange of information via TV cables at distances of up to 200 miles. Figure 18-2 illustrates a cable TV network, developed by LANcity (Cable Modem Division of Bay Networks), that allows connection between manufacturing plants and other locations.

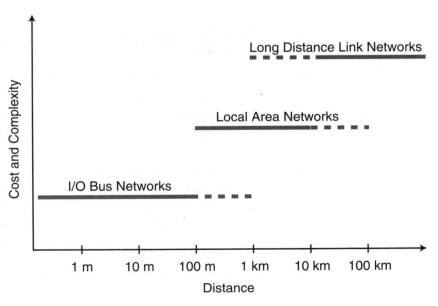

Figure 18-1. Network distance ranges.

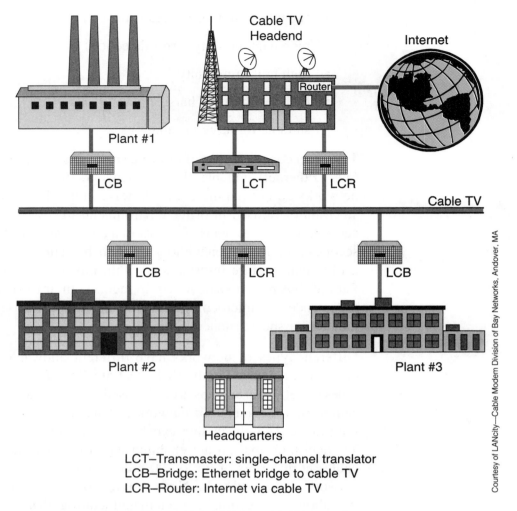

LCT–Transmaster: single-channel translator
LCB–Bridge: Ethernet bridge to cable TV
LCR–Router: Internet via cable TV

Figure 18-2. LANcity cable TV network.

Courtesy of LANcity—Cable Modem Division of Bay Networks, Andover, MA

ADVANTAGES OF LANS

Before local area networks came into use, two other methods were employed to implement communication between PLCs. The first method used a pair of wires to connect the output card of one PLC to the input card of a second PLC. This method, which transmitted only one bit of information per pair of wires, was expensive to install and very cumbersome to use. In the second method, PLCs communicated through their programming ports via a central computer, which was customer-supplied and programmed. The disadvantages of this method were that it limited the data throughput rate to the baud rate of the PLC's programming port and that the network became unusable if the central computer failed due to the system's star topology.

The local area network offers distinct advantages over its predecessors because it greatly reduces the cost of wiring for large installations. It also uses a dedicated communication link to efficiently exchange large amounts of usable data among PLCs and other hosts. Moreover, because each PLC in the network can communicate independently with the others (without the use of a central computer), a LAN does not have the disadvantage of depending solely on one computer.

LAN APPLICATIONS OF THE PLC

Centralized data acquisition and distributed control are the most common applications of local area networks. Data collection and processing, when performed by an individual controller, can burden the processor's scan time, consume large amounts of memory, and complicate the control logic program. A data highway configuration, in which all data is passed to a host computer that performs all data processing, eliminates these problems. Also, distributed control applications allocate control functions, once performed by a single controller, among several controllers. This eliminates dependence on a single controller and improves performance and reliability. To use the distributed processing approach, a local area network and the PLCs attached to it must provide the following functions:

- communication between programmable controllers

- upload capability to a host computer from any PLC

- download capability from a host computer to any PLC

- reading/writing of I/O values and registers to any PLC

- monitoring of PLC status and control of PLC operation

18-3 NETWORK TOPOLOGIES

The *topology* of a local area network is the geometry of the network, or how individual nodes are connected to it. A network's topology greatly affects its throughput rate, implementation cost, and reliability. The basic network topologies used today are star, common bus, and ring (see Figure 18-3). We should note, however, that a large network, such as the one shown in Figure 18-4, may consist of a number of interconnected topologies.

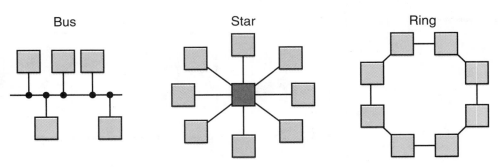

Figure 18-3. Bus, star, and ring topologies.

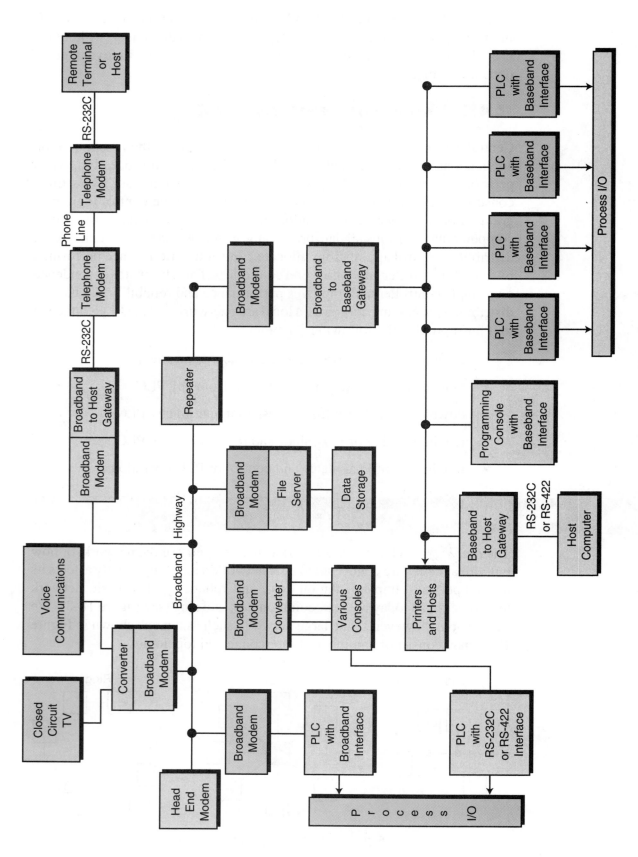

Figure 18-4. Large network using many different topologies.

STAR

As mentioned previously, the first PLC networks consisted of a multiport host computer with each port connected to the programming port of a PLC. Figure 18-5 shows this arrangement, known as **star topology**. The network controller can be either a computer, a PLC, or another intelligent host.

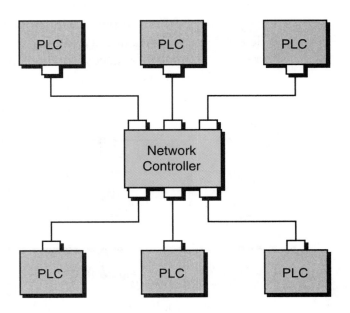

Figure 18-5. Star topology.

Most commercial computer installations are star networks, in which many terminals are tied to a central computer. This star topology is the same as the one used in telephone networks, where the central node has the task of establishing connections between the various network stations. The main advantage of this topology is that it can be implemented with a simple point-to-point protocol—that is, each node can transmit whenever necessary. If error checking is not required or if a simple parity bit per character check will suffice, then a *dumb terminal*, a terminal without network intelligence (e.g., a display monitor), can be a node. Star topology, however, has the following disadvantages:

- It does not lend itself to distributed processing due to its dependence on a central node.

- The wiring costs are high for large installations.

- Messages between two nodes must pass through the central node, resulting in low throughput.

- There is no broadcast mode, which lowers throughput even more.

- Failure of the central node will crash the network.

COMMON BUS

The **common bus topology** has a main trunkline to which individual PLC nodes are connected in a multidrop fashion (see Figure 18-6). A coaxial cable with proper terminators is typically the communication medium for the trunkline. In contrast to the star topology, communication in a common bus network can occur between any two nodes without passing information through a network controller. An inherent problem of this scheme, however, is determining which node may transmit at which time, to avoid data collision. Several communication access methods have been developed to solve this problem. We will discuss these later.

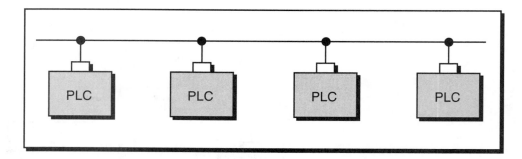

Figure 18-6. Common bus topology.

Common bus topologies are very useful in distributed control applications, since each station has equal independent control capability and can exchange information at any given time. Also, this topology requires little reconfiguration to add or remove stations from the network. The main disadvantage of this topology is that all of the nodes depend on a common bus trunkline. A break in this trunkline can affect many nodes.

Another configuration of the bus topology is the **master/slave bus topology**, consisting of several slave controllers and one master network controller (see Figure 18-7). In this configuration, the master sends data to the slaves; if the master needs data from a slave, it will *poll* (address) the slave and wait for a response. No communication takes place without the master initiating it. The implementation of a master/slave bus topology uses two pairs of wires. Through one pair of wires, the master transmits data and the slaves receive it. Through the other pair of wires, the slaves transmit data and the master receives it.

RING

Ring topology, shown in Figure 18-8, is not used in industrial environments because failure of any node (not just the master) will crash the network, unless the failed node is bypassed. We mention it here because it does not require multidropping due to its point-to-point connection restriction (see

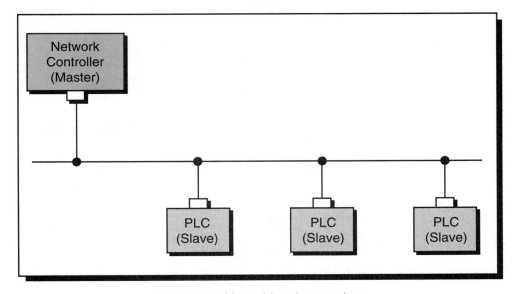

Figure 18-7. Master/slave bus topology.

Section 18-5). Thus, it is a good candidate for fiber-optic networks, since fiber-optic transmission media allows fast communication speed and long-distance connectivity.

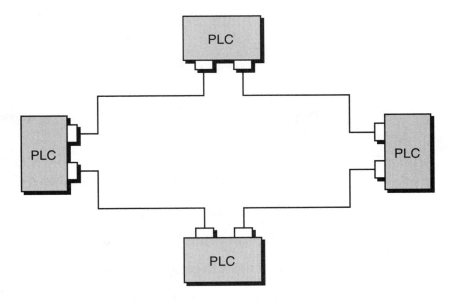

Figure 18-8. Ring topology.

Some LAN manufacturers have overcome the problem of node failure in a ring topology by using a wire center. The wire center, shown in Figure 18-9, automatically bypasses failed nodes in the ring. This **star-shaped ring topology**, however, requires twice as much wire as standard ring topology. Therefore, it must offer some other significant advantage (such as use in fiber optics) to be practical for large installations.

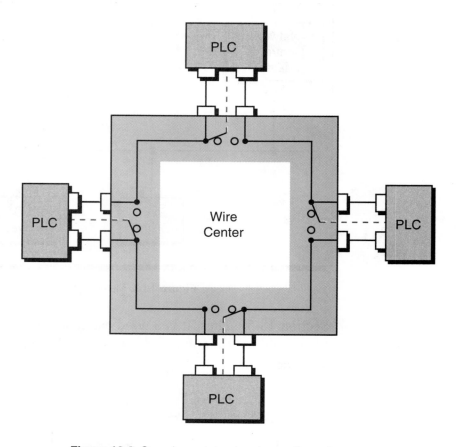

Figure 18-9. Star-shaped ring topology with a wire center.

DATA TRANSMISSION TECHNIQUES

Several transmission techniques are used to send data through a network (see Figure 18-10). Among the most common are:

- Manchester encoding

- frequency shift keying (FSK)

- nonreturn to zero invert on ones (NRZI)

Manchester encoding, also referred to as *baseband transmission encoding*, changes the signal polarity to positive for every logic 1 and to negative for every logic 0. During normal operation, the DC voltage on the cable is zero. *Frequency shift keying* (FSK) utilizes two frequencies to transmit logical values of 1 and 0. The *nonreturn to zero invert on ones* (NRZI) transmission technique involves a signal change whenever the next transmitted value is a 1. Ethernet networks use Manchester coding as their data transmission method.

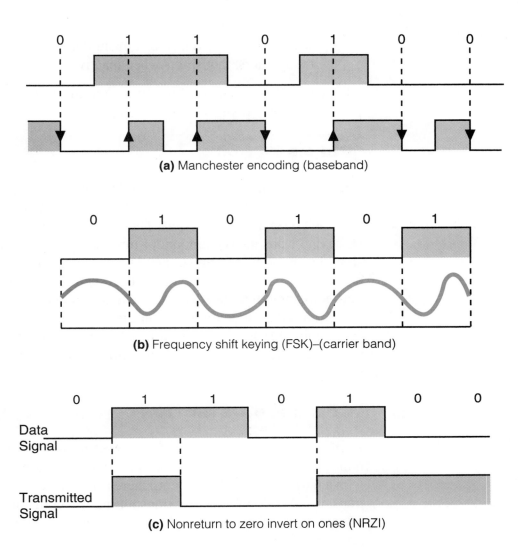

(a) Manchester encoding (baseband)

(b) Frequency shift keying (FSK)–(carrier band)

(c) Nonreturn to zero invert on ones (NRZI)

Figure 18-10. Data transmission techniques: **(a)** Manchester encoding, **(b)** frequency shift keying, and **(c)** nonreturn to zero invert on ones.

18-4 NETWORK ACCESS METHODS

An access method is the manner in which a PLC accesses the network to transmit information. In other words, it defines the method used by the node to talk through the network. As mentioned in the previous section, a bus topology requires that the nodes take turns transmitting on the medium. This process requires that each node be able to shut down its transmitter without interfering with the network's operation. This can be done in one of the following ways:

- with a modem that can turn off its carrier

- with a transmitter that can be set to a high independence state

- with a passive current-loop transmitter, wired in series with the other transmitters, that shorts when inactive

Although many access methods exist, the most commonly used ones are polling, collision detection, and token passing.

POLLING

The access method most often used in master/slave protocols is **polling**. In polling, the master interrogates, or polls, each station (slave) in sequence to see if it has data to transmit. The master sends a message to a specific slave and waits a fixed amount of time for the slave to respond. The slave should respond by sending either data or a short message saying that it has no data to send. If the slave does not respond within the allotted time, the master assumes that the slave is dead and continues polling the other slaves. Interslave communication in a master/slave configuration is inefficient, since polling requires that data first be sent to the master and then to the receiving slave. Since master/slave configurations use this technique, polling is often referred to as the *master/slave access method*.

COLLISION DETECTION

Collision detection is generally referred to as CSMA/CD (carrier sense multiple access with collision detection). In this access method, each node with a message to send waits until there is no traffic on the network and then transmits. While the node is transmitting, its collision detection circuitry checks for the presence of another transmitter. If the circuit detects a collision (two nodes transmitting at the same time), the node disables its transmitter and waits a random amount of time before trying again. This method works well as long as the network does not have an excessive amount of traffic.

Each collision and retry uses time that cannot be used for transmission of data; therefore, the network's throughput decreases and access time increases as traffic increases. For this reason, collision detection is not popular in control networks, but it is popular in business applications. In industrial applications, collision detection can be used for data gathering and program maintenance in large systems and real-time distributed control applications with a relatively small number of nodes.

TOKEN PASSING

Token passing is an access technique that eliminates contention among the PLC stations trying to gain access to the network. In this technique, the PLCs pass a token, which is a message granting a polled station the exclusive, but temporary, right to control the network (i.e., transmit information). The station with the token has the exclusive right to transmit on the network; however, it must relinquish this right to the next designated node upon termination of transmission. Thus, token passing is actually a distributed form of polling. The token-passing access method is preferred in distributed control applications that have many nodes or stringent response time requirements.

In a common bus network configuration using the token-passing technique, each station is identified by an address. During operation, the token passes from one station to the next sequentially. The node that is transmitting the token also knows the address of the next station that will receive the token. The network circulates transmitted data in one or more information packets containing source, destination, and control data. Each node receives this information and uses it, if needed. If the node has information to send, it sends it in a new packet.

In the token-passing scenario shown in Figure 18-11, station 10 passes the token to station 15 (the next address), which in turn passes the token to station 18 (the next address after 15). If the next station does not transmit the token to its successor within a fixed amount of time (token pass timeout), then the token-passing station assumes that the receiving station has failed. In this case, the originating station starts polling addresses until it finds a station that will accept the token. For instance, if station 15 fails, station 10 will poll stations 16 and 17 without response, since they are not present in the network, and then poll station 18, which will respond to the token. This receiving station will become the new successor and the failed station will be removed from, or patched out of, the network (i.e., station 18 will become station 10's next address). The time required to pass the token around the entire network depends on the number of nodes in the network. This time can be approximated by multiplying the token holding time by the number of nodes in the network.

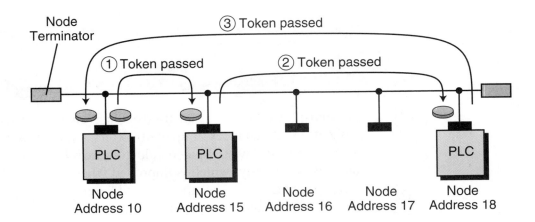

Figure 18-11. Example of token passing.

18-5 COMMUNICATION MEDIA

This section discusses the communication media (i.e., cables) used to implement local area networks. If installed properly, most local area networks can interface using any of these media. Proper installation includes the appropriate physical connectors and the correct electrical terminations. Media types commonly used for PLC networks include twisted-pair conductors, coaxial cables, and fiber optics. The type of media used and the number of nodes installed will affect the performance of the network (i.e., speed and distance). Figure 18-12 shows a comparison of different communication methods used with these media.

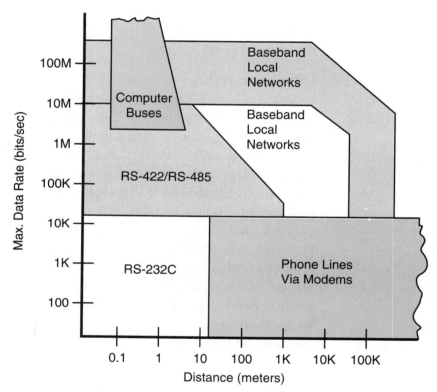

Figure 18-12. Comparison of the data transmission speeds and distances of various communication methods.

TWISTED-PAIR CONDUCTORS

Twisted-pair conductors are used extensively in industry for point-to-point applications at distances of up to 4000 feet and at transmission rates as high as 250 kilobaud. Twisted-pair conductors are relatively inexpensive and have fair noise immunity, which is improved when shielded. Performance, however, drops off rapidly as nodes are added to a twisted-pair bus. Moreover, nonuniformity also compromises the performance of these conductors. Characteristic impedance varies throughout the cable, making reflections difficult to reduce because there is no "right" value for termination resistance.

BASEBAND COAXIAL CABLE

Baseband coaxial cable, which can send one signal at a time at its original frequency, can transmit data in a local area network at speeds of up to 2 megabaud and distances of up to 18,000 feet. Unlike twisted-pair conductors, coaxial cable is extremely uniform, thus eliminating problematic reflections. The limiting factor for this type of cable is capacitive and resistive loss. Baseband cable is usually 3/8 inches in diameter.

BROADBAND COAXIAL CABLE

Broadband coaxial cable is thicker than baseband cable, ranging from 1/2 to 1 inch in diameter. Broadband cable, which has been used for years to carry cable television signals, can support a transmission rate of up to 150 megabaud. Although this type of coaxial cable can be used to increase distance in a baseband network, it is intended for use with a broadband network. Baseband networks use frequency division multiplexing to provide many simultaneous channels, each with a different RF carrier frequency. Broadband networks, on the other hand, use just one of these channels and one of the access methods previously discussed. The transmission rate on the channel is typically 1, 5, or 10 megabaud. Broadband local area networks can support thousands of nodes and are capable of spanning many miles through the use of bidirectional repeaters. One advantage of using broadband cable is that network communication can be implemented with just one of the broadband channels. The other channels can be used for video, computer access, and various monitoring and control functions.

Each broadband channel consists of two channels—a high-frequency forward channel and a low-frequency return channel. If only two nodes need to communicate, one can transmit on the forward channel and the other can transmit on the return channel. In a multidrop network, a head-end modem is required to retransmit the return channel signal on its corresponding forward channel in order for proper transmission and propagation to occur. The repeaters amplify the forward channel signals in one direction and the return channel signals in the other direction. Figure 18-4 presented an example of a broadband network with a baseband subnetwork.

FIBER-OPTIC CABLE

Fiber-optic cable consists of thin fibers of glass or plastic enclosed in a material with low refraction. This type of cable transmits signals through pulses of reflected light. The main shortcoming of fiber optics is that a low-loss terminal access point, also called a *tap* or *T-connector*, has yet to be perfected. Currently, T-connectors in fiber-optic cable only pick up a small percentage of the light energy that transmits the information through the cable. This deficiency eliminates fiber optics from use in large bus topologies, but not from use in star or ring topologies. In addition, fiber-optic cable is three to four times more expensive than baseband coaxial cable, and optical couplers are several times more expensive than strictly electrical interfaces.

Fiber optics does, however, have some impressive advantages. First, it is totally immune to all kinds of electrical interference. Second, it is small and lightweight. Finally, it can sustain transmission rates of up to 800 megabaud at distances of up to 30,000 feet. In light of these qualities, the use of fiber optics should increase in industrial applications as the technology develops.

18-6 UNDERSTANDING NETWORK SPECIFICATIONS

This section explains how to determine if a particular network can support a given application. The designer should examine all aspects of the network, including device specifications, response time, maximum length, throughput, and interface, when choosing a network for an application.

DEVICE SPECIFICATIONS

When selecting a network, the system designer must analyze the application to determine how many nodes are required and what type of device—PLC, vendor-supplied network programmer, host computer, or intelligent terminal—will be used at each node. The designer must determine if the network will support each type of device used and examine how that device will interface with the network (hardware and software). For network PLCs, the designer must also choose the model, because some PLC models are not capable of interfacing with a network. The network must be capable of supporting the number of nodes required for the current application, plus a reasonable number of nodes for future expansion.

MAXIMUM LENGTH

The maximum length of a network consists of two parts: the maximum length of the main cable and the maximum length of each drop cable used between a node and the main cable. Maximum drop lengths usually range between 30 and 100 feet; however, drop lengths should be kept as short as possible, since drops introduce reflection into the network. The ideal case is to run the main cable straight to the device and back again, even though this procedure increases wiring costs.

Another important piece of information that the designer should obtain from the vendor is the type of cable that must be used to achieve the specified transmission distance. If the system requires the maximum network transmission distance, the designer must use the proper type of cable. If the system requires a much shorter transmission distance, the designer can save money by using a less expensive cable.

RESPONSE TIME

Response time (RT), as used in this book, is the time between an input transition at one node and the corresponding output transition at another node. Response time, then, is the sum of the time required to detect the input transition, transmit the information to the output node, and operate the output. It is expressed as:

$$RT = IT + 2ST_1 + PT_1 + AT + TT + PT_2 + 2ST_2 + OT$$

where:

IT = the input delay time (the electrical delay involved in detecting the input transmission)

ST_1 = the scan time for the sending node

ST_2 = the scan time for the receiving node

PT_1 = the processing time for the sending node (the time between solving the program logic and becoming ready to transmit the data)

PT_2 = the processing time for the receiving node (the time between receiving the data and having data ready to be operated on by the program logic)

AT = the access time (the time involved in both becoming ready to transmit and in transmission)

TT = the transmission time (the time required to transmit data—this is the only time that is directly proportional to baud rate)

OT = the output delay time (the electrical delay involved in creating the output transition)

The scan time includes the I/O update time and any other overhead time, as well as the program logic execution time. It can be defined as the time between I/O updates. In the previous equation, the scan time is doubled to include the case where the input signal changes just after the I/O update. In this case, the network first executes the logic with the old information, then performs an I/O update, and finally executes the logic with the new information. This causes a two-scan delay.

I/O delay times and scan times are readily available values. Transmission time can be determined once the data rate and frame length are known. The data rate is sometimes equal to the baud rate, but it is usually less. Synchronous systems, which use Manchester encoding, have a data rate that is half of the baud rate. These systems utilize a transmission method in which the data characters and bits are transmitted at a fixed rate with the transmitter and receiver in synchronization. The data rate of asynchronous systems is

80% of the baud rate due to the start and stop bits that accompany each 8 data bits. In these systems, the time intervals between transmitted characters may be unequal in length. Transmission in an asynchronous system is controlled by start and stop signals at the beginning and end of each character.

The access time and the two processing times depend on the particular installation and generally must be obtained from the manufacturer. If the equipment is available, it is much easier and more accurate to determine the overall response time through actual measurements than through specifications. Section 18-8 presents a procedure for performing this measurement.

The parameter that should be determined by the response time equation is not the average response time, but rather the maximum response time. Therefore, the designer should take steps to create a worst-case environment during response time measurements. Creating this scenario involves performing tasks such as downloading programs and monitoring points while taking the measurements, because this sort of activity increases PLC scan times and network access times.

THROUGHPUT

Some manufacturers specify the LAN throughput value. This value represents the number of I/O points that can be updated per second through the network. The throughput value does not provide enough information to derive actual values for access time and data rate, although it gives the system designer some idea of these values. In addition, throughput varies with system loading as a result of each node's processing time. Therefore, to obtain an accurate value for throughput, the designer must know the conditions under which the measurement was taken.

DEVICES SUPPORTED

When considering each device in the system, the designer must ask not only, Will the local area network support this device?, but also, What is involved in connecting the device to the network? For user-supplied devices, the designer must also determine what support software will be required.

Programmable Controllers. All of the standard networks support at least some PLCs. A separately purchased interface unit usually connects a PLC to a network. The interface unit is connected to the PLC through either a high-speed parallel bus or the PLC's serial programming port. In the latter case, two additional terms must be added to the response time equation: the programming port transmission time and the programming port processing time.

Programming Devices. Most manufacturers offer some type of personal computer as a programming device that can be connected to a network. A PC unit connected to a network provides centralized programming of any

PLC on the network, along with various monitoring and control functions, if available. If a network-compatible programming device is not available, all programming must be done through the programming port of the individual PLCs.

Hosts. Host support means that a user-supplied host computer can perform programming functions, provided that its programming conforms to the network manufacturer's protocol. The host computer is usually connected to the network through a device called a **gateway**. The gateway contains a network port and another port (usually RS-232), which is connected to the host. A gateway greatly simplifies the software that the user must write for the host, because a host-to-gateway link requires only a simple point-to-point protocol rather than a masterless multidrop protocol, as is required by a network. A gateway also provides the appropriate electrical interfaces for the network. Since most computers have an RS-232 port, additional hardware is seldom required.

Intelligent Terminals. The type of intelligent terminal referred to here is actually a small host computer complete with an operating system and mass storage. It can interface with the network in exactly the same way as a large host computer. Anyone considering using one of these terminals on a network should investigate the software requirements closely to determine if the terminal's operating system will support the network's requirements. Some operating systems, for instance, provide for the transmission of only ASCII data, not binary data.

Gateways. In addition to the host gateway mentioned previously, some manufacturers provide gateways to other multidrop networks. They also provide other types of host gateways, for example, a high-speed, RS-422, synchronous host interface. In this case, the gateway would use a protocol designed for synchronous use, such as HDLC.

APPLICATION INTERFACE

When developing an application interface, the designer must determine how each PLC's application program allows it to share information with other PLCs. Most manufacturers provide at least one of the following methods:

- reading of registers in other PLCs

- writing to registers in other PLCs

- reading and writing of network points or registers

For example, a PLC can detect the input status of another PLC on the network through the use of a network coil and a network contact. Figure 18-13 illustrates this configuration. When the network coil (Net 200) in PLC #1 is energized, the network contact, Net 200 (⊣⊢), in PLC #2 will close. PLC #2 can use this contact like any other contact in its ladder program. The user, however, must ensure that only one PLC on its network uses each network coil.

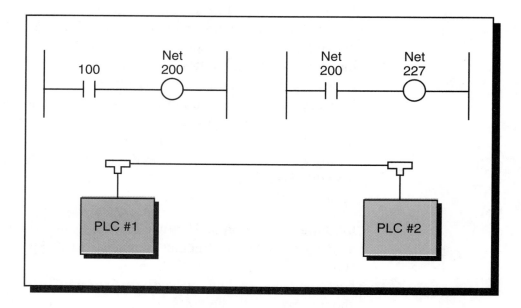

Figure 18-13. Network coils and contacts.

Network PLCs read to and write from registers through functional blocks. The designer must ensure that the capabilities of the network and PLCs are sufficient to support the communication needs of the application. Chapter 9 shows some of the typical network instructions found in PLCs.

18-7 NETWORK PROTOCOLS

A protocol is a set of rules that two or more devices must follow if they are to communicate with each other. Protocol includes everything from the meaning of data to the voltage levels on connection wires. A network protocol defines how a network will handle the following problems and tasks:

- communication line errors
- flow control (to keep buffers from overflowing)
- access by multiple devices
- failure detection
- data translation
- interpretation of messages

OSI REFERENCE MODEL

Networks follow a protocol to implement the transmission and reception of data over the network medium (e.g., coaxial cable). In 1979, the International Standards Organization (ISO) published the Open Systems Interconnection (OSI) reference model, also known as the ISO IS 7498, to provide guidelines for network protocols. This model divides the functions that protocols must perform into seven hierarchical layers (see Figure 18-14). Each layer interfaces only with its adjacent layers and is unaware of the existence of the other layers. Table 18-1 describes the seven layers of the OSI. The OSI model further subdivides the second layer into two sublayers, 2A and 2B, called *medium access control* (MAC) and *logical link control* (LLC), respectively. In network protocols, the physical layer (layer 1) and the medium access control sublayer (layer 2A) are usually implemented with hardware, while the remaining layers are implemented using software. The hardware components of layers 1 and 2A are generally referred to as modems (or transceivers) and drivers (or controllers), respectively.

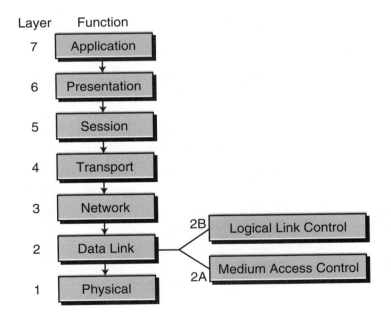

Figure 18-14. OSI reference model.

Strictly speaking, a network requires only layers 1, 2, and 7 of the protocol model to operate. In fact, many device bus networks, which we will cover in the next chapter, use only these three layers. The other layers are added only as more services are required (e.g., error-free delivery, routing, session control, data conversion, etc.). Most of today's local area networks contain all or most of the OSI layers to allow connection to other networks and devices.

Layer	Layer Name	Function
Layer 7	Application	The level seen by users; the user interface
Layer 6	Presentation	Control functions requested by the user; data is restructured from other standard formats; code and data conversion
Layer 5	Session	System-to-system connection; log-in and log-off controlled here; establishes connections and disconnections
Layer 4	Transport	Provides reliable data transfer between end devices; network connections for a given transmission are established by protocol
Layer 3	Network	Outgoing messages are divided into packets; incoming packets are assembled into messages for higher levels, establishing connections between equipment on the network
Layer 2	Data link	Outgoing messages are assembled into frame and acknowledgements; error detection or error correction is perfomed
Layer 1	Physical	Parameters, such as signal voltage swing, bit duration, and electrical connections, are established in this layer

Table 18-1. Seven layers of the OSI reference model.

To understand this seven-layer architecture, let's examine a familiar every-day example, an interoffice memo (see Figure 18-15). Imagine that two offices form two network nodes at two separate locations. If the manager of one office wants to send a memo to the manager of the other office, he/she must write the message with pencil and paper. After the message is written, the manager passes it to the secretary to be properly typed, addressed, stamped, and mailed. The pencil and paper corresponds to the seventh layer (application layer), which is the level that concerns the manager (i.e., the network user). He/she "applies" the pencil and paper to send the message. After that, it is no longer his/her responsibility; however, the memo remains in the system, meaning that the memo is still in the manager's office, having passed to the next steps that must occur before it enters the postal mail system. These other steps are the next six layers of the OSI model:

- The secretary types the memo and puts both the correct sender and receiver addresses on the envelope (layer 6—coding and conversion).

- He/she puts the memo in an envelope, affixes the correct amount of postage, and takes it to the mail room (layers 5, 4, and 3, respectively).

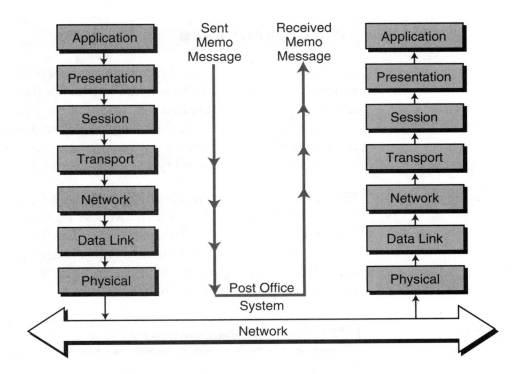

Figure 18-15. Seven-layer architecture.

- The mail room clerk takes the memo, makes sure that it has the right postage and address, and puts it in the outgoing mail basket with the other mail (layer 2).

- The mail carrier then physically picks up the memo (layer 1) and sends it through the postal mail system, or in other words, through the network.

After a couple of days, the other office receives the memo and a similar operation takes place, but in reverse order. The mail carrier delivers the memo, the clerk checks to see if the receiver works there and in which department. Then, the clerk sends the memo through the internal company delivery system and it arrives at the receiving manager's secretary. The secretary passes the memo to the manager, who reads it and interprets the message. This seven-step method ensures proper creation, implementation, and delivery of the message, since a protocol of orderly operations takes place.

The ISO's OSI model embraces an architecture that is followed by most protocol standards. Each standard is intended to be open so that network devices from different manufacturers can be interconnected. Specialized technical organizations, as opposed to standards committees such as the ISO, have made the largest efforts towards the standardization of network protocols. The ISO, however, will accept and validate a network standard as long as it complies with the protocol architecture defined by the OSI model.

IEEE STANDARDS

The Institute of Electrical and Electronic Engineers (IEEE) computer society established the Standards Project 802 in 1980 for the purpose of developing a local area network standard that would allow equipment from different manufacturers to communicate through a local area network. After studying all the users' and manufacturers' requirements, the committee developed standards that define several types of local networks.

IEEE 802.3. The IEEE, in accordance with the ISO, agreed to be responsible for the specifications of local area networks whose transmission speeds range between 1 and 20 megabaud (megabits/sec). The IEEE 802.3 standard, which the ISO accepted as its own standard (ISO 8802), regulates layers 1 and 2A of the OSI model. Figure 18-16 illustrates the different parts of the IEEE 802 standard and its relationship to the OSI model.

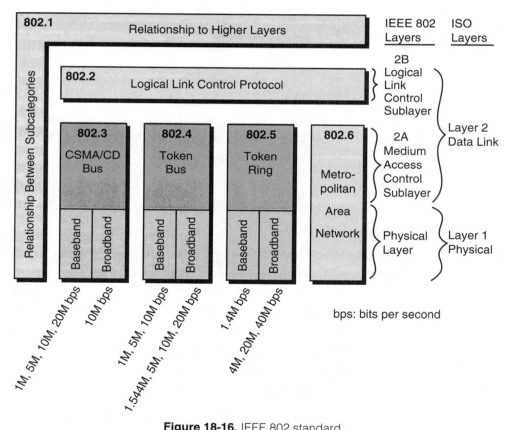

Figure 18-16. IEEE 802 standard.

The IEEE 802.3 standard specifies that network access should occur through CSMA/CD using a bus topology at a rate of 1 to 20 Mbaud (baseband) or 10 Mbaud (broadband). The widely used Ethernet network complies with the IEEE 802.3 standard. In fact, when Ethernet was first developed in the early 1980s through a joint effort of Digital Equipment Corporation (DEC), Xerox, and Intel, the IEEE accepted it with only a few modifications to make it comply with the 802.3 (CSMA/CD bus). The ISO has also taken Ethernet as a standard, the ISO 8802.3. In control systems, the Ethernet (802.3) network is primarily suited for noncritical applications, such as supervisory monitoring and PLC program management.

IEEE 802.4 and 802.5. The IEEE 802.4 standard specifies a token bus network at different baseband and broadband transmission rates than the IEEE 802.3 standard. The 802.4 standard is used by many PLC manufacturers as the protocol structure of the lower layers of their local area networks. Furthermore, another IEEE standard, the IEEE 802.5, specifies a token ring network with lower transmission rates for baseband cables (1.4 Mbaud). IBM adopted the 802.5 standard for their token-passing protocol with ring topology. Figures 18-17a, b, and c illustrate the general characteristics of the IEEE 802.3, 802.4, and 802.5 standards, respectively.

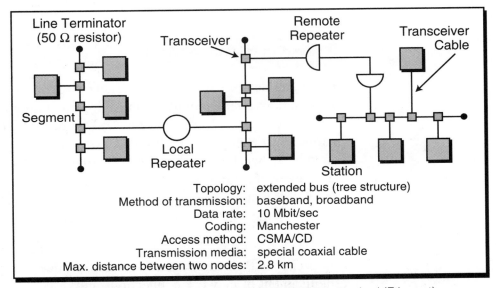

Topology:	extended bus (tree structure)
Method of transmission:	baseband, broadband
Data rate:	10 Mbit/sec
Coding:	Manchester
Access method:	CSMA/CD
Transmission media:	special coaxial cable
Max. distance between two nodes:	2.8 km

Figure 18-17a. Characteristics of the IEEE 802.3 standard (Ethernet).

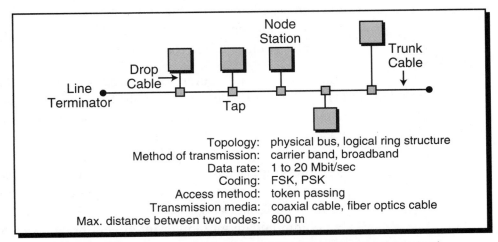

Topology:	physical bus, logical ring structure
Method of transmission:	carrier band, broadband
Data rate:	1 to 20 Mbit/sec
Coding:	FSK, PSK
Access method:	token passing
Transmission media:	coaxial cable, fiber optics cable
Max. distance between two nodes:	800 m

Figure 18-17b. Characteristics of the IEEE 802.4 standard (token bus).

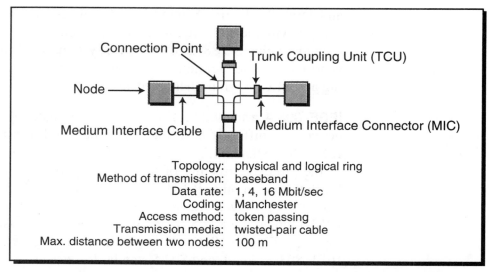

Topology:	physical and logical ring
Method of transmission:	baseband
Data rate:	1, 4, 16 Mbit/sec
Coding:	Manchester
Access method:	token passing
Transmission media:	twisted-pair cable
Max. distance between two nodes:	100 m

Figure 18-17c. Characteristics of the IEEE 802.5 standard (token ring network).

TCP/IP PROTOCOL

Most manufacturers who offer Ethernet compatibility to implement supervisory functions over equipment controlling plant floor functions use a TCP/IP protocol for layers 3 and 4 of the OSI model. The **transmission control protocol/internet protocol (TCP/IP)** was initially developed for Arpanet, a computer network created in the early 1970s in the United States. The U.S. Department of Defense established this protocol to communicate information in a reliable manner from one computer to another over the Arpanet network. Nowadays, the TCP/IP protocol is utilized in the Internet data network.

In the TCP/IP protocol, the TCP guarantees control of end-to-end connections. The TCP makes several services available to the user, such as the establishment of network connections and disconnections, guaranteed data sequencing, protection against loss of sequence, connection time control, and transparent multiplexing and transport of data. The IP (internet protocol) performs complementary functions such as addressing network data, distributing data packages, and routing data in multinetwork systems.

Some PLC manufacturers offer programmable controllers with TCP/IP-over-Ethernet protocol built into the PLC processor (see Figure 18-18). This allows the PLC to connect directly to a supervisory Ethernet network (see Figure 18-19). Note that the PLC in Figure 18-19 can also have a control network with other PLCs. Sometimes, the TCP/IP section in a supervisory network is replaced by another protocol, the *manufacturing message specification* (MMS) protocol, which is used by plant floor devices to communicate through 802.3 networks (see Figure 18-20). In this configuration, a PLC can communicate with other intelligent systems, such as robots and CNC machining centers.

Figure 18-18. Allen-Bradley's PLC-5 controllers with built-in TCP/IP-over-Ethernet protocol.

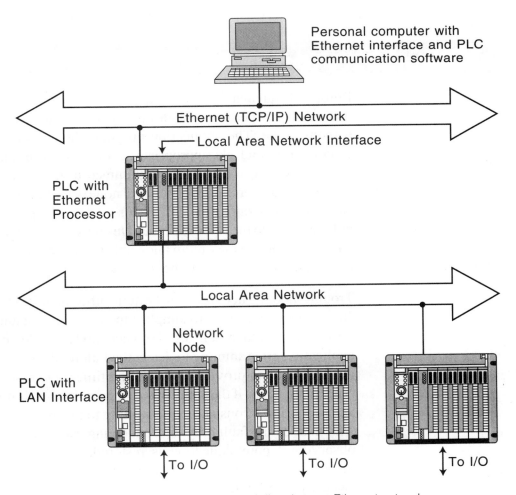

Figure 18-19. PLC connected directly to an Ethernet network.

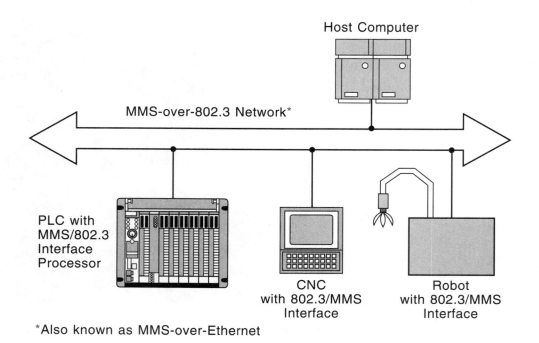

Host Computer

MMS-over-802.3 Network*

PLC with
MMS/802.3
Interface
Processor

CNC
with 802.3/MMS
Interface

Robot
with 802.3/MMS
Interface

*Also known as MMS-over-Ethernet

Figure 18-20. MMS protocol.

18-8 NETWORK TESTING AND TROUBLESHOOTING

Before a local area network is installed, the designer should test it to ensure that it not only performs the desired function, but also provides the required response time. The application program should continuously monitor the response time and take appropriate action if it exceeds the maximum time that the process will tolerate. A programmed buzzer circuit, which passes the contact closure through every critical node in the network before it is returned, can test the response time. Using this process, the pulse width of the created pulse is equal to the response time. This pulse width can be applied to a timer, which is set to the maximum allowable response time. When the timer times out, the circuit knows that the response time has been exceeded.

Troubleshooting networks can be quite difficult unless both the manufacturer and the user take steps to simplify the task. The manufacturer can provide error counts and a self-test for each node, while the user can provide application programming to detect the failure of a node. An extreme case of this would be to provide a buzzer and timer between each node and every other node. Thus, if the entire network goes down, it is probably due to a node with a short or a constantly transmitting transmitter. The user can determine which node is faulty by disconnecting each node one at a time and observing if communication is restored. Some manufacturers provide a network monitor that can detect a failed node, an open cable, or excessive electrical interference.

18-9 NETWORK COMPARISON AND SELECTION CRITERIA

NETWORK COMPARISON

The most distinctive differences among local area networks are the transmission or communication medium and the network access method. Table 18-2 shows the advantages and disadvantages of each type of communication medium and access method. The communication medium directly affects the cost of a LAN installation from the outset due to the price difference between the types of network cables. For instance, baseband cable is cheaper to install and maintain, as well as troubleshoot. Broadband cable is more expensive to install but has the capability for multiple transmission through the same cable, which is the case in cable TV (multiple channel transmission). Depending on the type of network, the troubleshooting of voice, process data, and other information parameters may be more difficult with broadband cable.

	Transmission Medium	
	Broadband	Baseband
Multiple transmission	+	−
Installation cost	−	+
Maintenance cost	−	+
Troubleshooting	−	+

	Medium Access	
	CSMA/CD (Ethernet, 802.3)	Baseband (802.4)
Response time	even	+
Operation with multiple nodes (50% of max)	−	+
Safety against failure	+	even
Network expandibilty	+	even

Table 18-2. Advantages and disadvantages of transmission media and access methods.

The network's access method also influences the manner in which nodes communicate with each other and the time required for that communication. CSMA/CD, for example, has the disadvantage of not being able to accurately predict the response time of a message transmission due to the delay caused by too many nodes trying to communicate at the same time. This short delay may be acceptable in an office environment using Ethernet (IEEE 802.3), where the information transfer speed is not of vital importance. However, in an industrial control environment, this type of delay could cause a major process breakdown. Token passing, on the other hand, has a predictable response time, even when the network has a large number of nodes.

SELECTION CRITERIA

Table 18-3 lists some of the criteria that should be evaluated during the selection of a local area network. These criteria cover four important areas: the speed and capacity of the network, the reliability of the network, the flexibility of the network, and the overall cost associated with network configuration.

Local Area Network Considerations	
Speed and capacity	• Data rate/throughput • Possible delays due to error transmission • Response time based on network load (number of nodes)
Reliability	• Safe transmission • Total failure protection • Data protection against unauthorized access
Flexibility	• Changes • Expansions • Compatibility with other networks
Costs	• Initial installation • Expansion • Maintenance • Network hardware/software cost

Table 18-3. Criteria to evaluate when choosing a LAN.

Most industrial networks can transfer information fast enough to suit the majority of applications; therefore, it is not necessary to obtain a very high-speed network unless the application specifically requires it. The processing speed of the PLCs connected to the network and the total scan requirements of the system determine the required network speed. If a supervisory system is being used to monitor a PLC network, however, speed may not be a factor. In this situation, an Ethernet or 802.3 network (CSMA/CD) may be appropriate because compatibility may already exist between the supervisory equipment (e.g., nonprocess automation computers) and the PLC network. A supervisory network like Ethernet ensures the support of many devices, since most PLC manufacturers can provide Ethernet compatibility either through a gateway or directly through the PLC using the local area network. In contrast, PLC manufacturers' proprietary networks may not have as many compatible peripherals and field equipment as an Ethernet network.

Reliability, flexibility, and cost are all as important as speed in network selection. Reliability of a network deals with the detection and correction of system errors. A network must have a reliable way of automatically detecting any system errors, and it must also provide a way for the user/ programmer to shut down a machine or process. The flexibility of a network deals with the ease of adding a node to the network, as well as the addressability of each network node. Many manufacturers of PLC local area networks provide network management software that gives the user flexibility when programming the network. Finally, the cost of a network must be analyzed not only for the initial installation costs, but also for maintenance and expansion costs. A network that is initially inexpensive to implement may turn out to be expensive due to restrictions on the addition of nodes and the lack of flexibility for changes.

KEY TERMS

baseband coaxial cable
broadband coaxial cable
collision detection
common bus topology
fiber-optic cable
gateway
local area network (LAN)
master/slave bus topology
polling
response time
ring topology
star-shaped ring topology
star topology
token passing
transmission control protocol/internet protocol (TCP/IP)
twisted-pair conductor

CHAPTER NINETEEN

I/O Bus
Networks

Necessity is the mother of invention.

—Latin Proverb

CHAPTER HIGHLIGHTS

Advances in large-scale electronic integration and surface-mount technology, coupled with trends towards decentralized control and distributed intelligence to field devices, have created the need for a more powerful type of network—the I/O bus network. This new network lets controllers better communicate with I/O field devices, to take advantage of their growing intelligence. In this chapter, we will introduce the I/O bus concept and describe the two types of I/O bus networks—device-level bus and process bus. In our discussion, we will explain these network's standards and features. We will also list the specifications for I/O bus networks and summarize their uses in control applications. When you finish this chapter, you will have learned about all the aspects of a PLC control system—hardware, software, and communication schemes—and you will be ready to apply this knowledge to the installation and maintenance of a PLC system.

19-1 INTRODUCTION TO I/O BUS NETWORKS

I/O bus networks allow PLCs to communicate with I/O devices in a manner similar to how local area networks let supervisory PLCs communicate with individual PLCs (see Figure 19-1). This configuration decentralizes control in the PLC system, yielding larger and faster control systems. The topology, or physical architecture, of an I/O bus network follows the bus or extended bus (tree) configuration, which lets field devices (e.g., limit, photoelectric, and proximity switches) connect directly to either a PLC or to a local area network bus. Remember that a bus is simply a collection of lines that transmit data and/or power. Figure 19-2 illustrates a typical connection between a PLC, a local area network, and an I/O bus network.

The basic function of an I/O bus network is to communicate information with, as well as supply power to, the field devices that are connected to the bus (see Figure 19-3). In an I/O bus network, the PLC drives the field devices directly, without the use of I/O modules; therefore, the PLC connects to and communicates with each field I/O device according to the bus's protocol. In essence, PLCs connect with I/O bus networks in a manner similar to the way they connect with remote I/O, except that PLCs in an I/O bus use an **I/O bus network scanner**. An I/O bus network scanner reads and writes to each field device address, as well as decodes the information contained in the network information packet. A large, **tree topology** bus network (i.e., a network with many branches) may have up to 2048 or more connected discrete field devices.

The field devices that connect to I/O bus networks contain intelligence in the form of microprocessors or other circuits (see Figure 19-4). These devices communicate not only the ON/OFF state of input and output controls, but also diagnostic information about their operating states. For example, a photoelectric sensor (switch) can report when its internal gain starts to

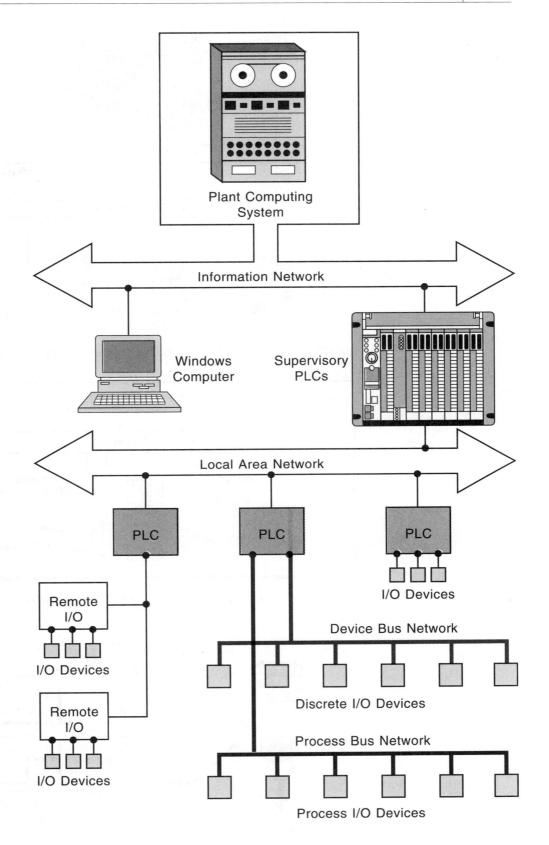

Figure 19-1. I/O bus network block diagram.

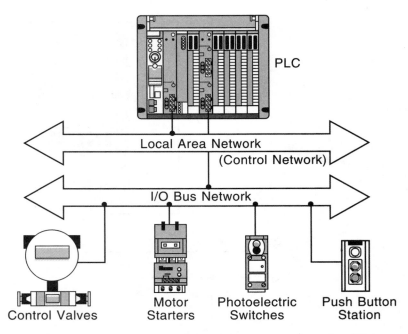

Figure 19-2. Connection between a PLC, a local area network, and an I/O bus network.

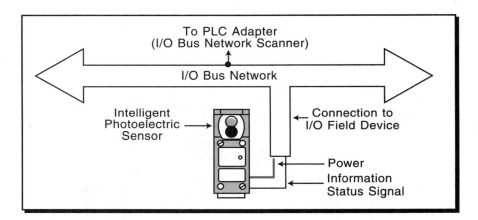

Figure 19-3. Connections for an I/O bus network.

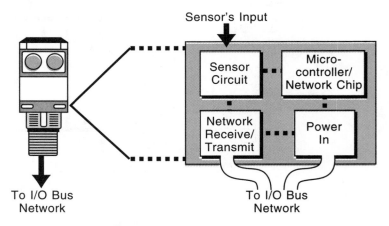

Figure 19-4. Intelligent field device.

decrease because of a dirty lens, or a limit switch can report the number of motions it has performed. This type of information can prevent I/O device malfunction and can indicate when a sensor has reached the end of its operating life, thus requiring replacement.

19-2 TYPES OF I/O BUS NETWORKS

I/O bus networks can be separated into two different categories—one that deals with low-level devices that are typical of discrete manufacturing operations and another that handles high-level devices found in process industries. These bus network categories are:

- device bus networks

- process bus networks

Device bus networks interface with low-level information devices (e.g., push buttons, limit switches, etc.), which primarily transmit data relating to the state of the device (ON/OFF) and its operational status (e.g., operating OK). These networks generally process only a few bits to several bytes of data at a time. **Process bus networks**, on the other hand, connect with high-level information devices (e.g., smart process valves, flow meters, etc.), which are typically used in process control applications. Process bus networks handle large amounts of data (several hundred bytes), consisting of information about the process, as well as the field devices themselves. Figure 19-5 illustrates a classification diagram of the two types of I/O bus networks.

The majority of devices used in process bus networks are analog, while most devices used in device bus networks are discrete. However, device bus networks sometimes include analog devices, such as thermocouples and variable speed drives, that transmit only a few bytes of information. Device

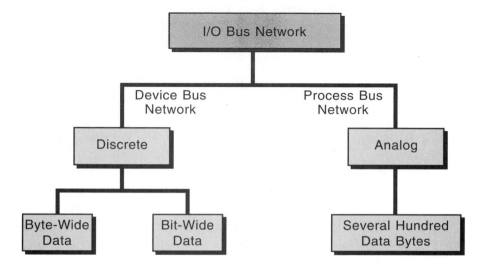

Figure 19-5. I/O bus network classification diagram.

bus networks that include discrete devices, as well as small analog devices, are called **byte-wide bus networks**. These networks can transfer between 1 and 50 or more bytes of data at a time. Device bus networks that only interface with discrete devices are called **bit-wide bus networks**. Bit-wide networks transfer less than 8 bits of data from simple discrete devices over relatively short distances.

The primary reason why device bus networks interface mainly with discrete devices and process bus networks interface mainly with analog devices is the different data transmission requirements for these devices. The size of the information packet has an inverse effect on the speed at which data travels through the network. Therefore, since device bus networks transmit only small amounts of data at a time, they can meet the high speed requirements for discrete implementations. Conversely, process bus networks work slower because of their large data packet size, so they are more applicable for the control of analog I/O devices, which do not require fast response times. The transmission speeds for both types of I/O bus networks can be as high as 1 to 2.5 megabits per second. However, a device bus network can deliver many information packets from many field devices in the time that it takes a process bus network to deliver one large packet of information from one device.

Since process bus networks can transmit several hundred bytes of data at a time, they are suitable for applications requiring complex data transmission. For example, an intelligent, process bus network–compatible pressure transmitter can provide the controller with much more information than just pressure; it can also transmit information about temperature flow rate and internal operation. Thus, this type of pressure transmitter requires a large data packet to transmit all of its process information, which is why a process bus network would be appropriate for this application. This amount of information just would not fit on a device bus network.

PROTOCOL STANDARDS

Neither of the two I/O bus networks have established protocol standards; however, many organizations are working towards developing both discrete and process bus network specifications. In the process bus area, two main organizations, the Fieldbus Foundation (which is the result of a merger between the Interoperable Systems Project, ISP, Foundation and the World FIP North American group) and the Profibus (Process Field Bus) Trade Organization, are working to establish network and protocol standards. Other organizations, such as the Instrument Society of America (ISA) and the European International Electronics Committee (IEC), are also involved in developing these standards. This is the reason why some manufacturers specify that their analog products are compatible with Profibus, Fieldbus, or another type of protocol communication scheme. Figure 19-6 illustrates a block diagram of available network and protocol standards.

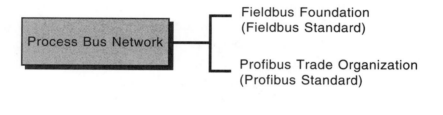

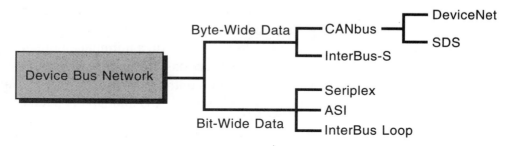

Figure 19-6. Network and protocol standards.

Although no proclaimed standards exist for device bus network applications, several de facto standards are emerging due to the availability of company-specific protocol specifications from device bus network manufacturers. These network manufacturers or associations provide I/O field device manufacturers with specifications in order to develop an open network architecture, (i.e., a network that can interface with many types of field devices). In this way, each manufacturer hopes to make its protocol the industry standard. One of these de facto standards for the byte-wide device bus network is DeviceNet, originally from PLC manufacturer Allen-Bradley and now provided by an independent spin-off association called the Open DeviceNet Vendor Association. Another is SDS (Smart Distributed System) from Honeywell. Both of these device bus protocol standards are based on the control area network bus (CANbus), developed for the automobile industry, which uses the commercially available CAN chip in its protocol. InterBus-S from Phoenix Contact is another emerging de facto standard for byte-wide device bus network.

The de facto standards for low-end, bit-wide device bus networks include Seriplex, developed by Square D, and ASI (Actuator Sensor Interface), a standard developed by a consortium of European companies. Again, this is why I/O bus network and field device manufacturers will specify compatibility with a particular protocol (e.g., ASI, Seriplex, InterBus-S, SDS, or DeviceNet) even though no official protocol standard exists.

19-3 ADVANTAGES OF I/O BUS NETWORKS

Although device bus networks interface mostly with discrete devices and process bus networks interface mostly with complex analog devices, they both transmit information the same way—digitally. In fact, the need for digital communication was one of the major reasons for the establishment of I/O bus networks. Digital communication allows more than one field device to be connected to a wire due to addressing capabilities and the device's ability to recognize data. In digital communication, a series of 1s and 0s is serially transmitted through a bus, providing important process, machine, and field device information in a digital format. These digital signals are less susceptible than other types of signals to signal degradation caused by electromagnetic interference (EMI) and radio frequencies generated by analog electronic equipment in the process environment. Additionally, PLCs in an I/O bus perform a minimal amount of analog-to-digital and digital-to-analog conversions, since the devices pass their data digitally through the bus to the controller. This, in turn, eliminates the small, but cumulative, errors caused by A/D and D/A conversions.

Another advantage of digital I/O bus communication is that, because of their intelligence, process bus–compatible field devices can pass a digital value proportional to a real-world value to the PLC, thus eliminating the need to linearize or scale the process data. For example, a flow meter can pass data about a 535.5 gallons per minute flow directly to the PLC instead of sending an analog value to an analog module that will then scale the value to engineering units. Thus, the process bus is an attempt to eliminate the need for the interpretation of analog voltages and 4–20 mA current readings from process field devices.

The advantages of digital communication in I/O bus networks are enormous. However, I/O bus networks have physical advantages as well. The reduction in the amount of wiring in a plant alone can provide incredible cost savings for manufacturing and process applications.

19-4 DEVICE BUS NETWORKS

BYTE-WIDE DEVICE BUS NETWORKS

The most common byte-wide device bus networks are based on the InterBus-S network and the CANbus network. As mentioned previously, the CANbus network includes the DeviceNet and SDS bus networks.

InterBus-S Byte-Wide Device Bus Network. InterBus-S is a sensor/actuator device bus network that connects discrete and analog field devices to a PLC or computer (soft PLC) via a ring network configuration. The InterBus-S has built-in I/O interfaces in its 256 possible node components, which also include terminal block connections for easy I/O interfacing (see Figure 19-7). This network can handle up to 4096 field I/O devices (depending on the configuration) at a speed of 500 kbaud with cyclic redundancy check (CRC) error detection.

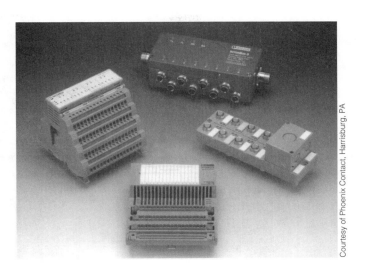

Figure 19-7. InterBus-S I/O block interfaces.

A PLC or computer in an InterBus-S network communicates with the bus in a master/slave method via a host controller or module (see Figure 19-8). This host controller has an additional RS-232C connector, which allows a laptop computer to be interfaced to the network to perform diagnostics. The laptop computer can run CMD (configuration, monitoring, and diagnostics) software while the network is operating to detect any transmission problems. The software detects any communication errors and stores them in a time-stamped file, thus indicating when possible interference might have taken place. Figure 19-9 illustrates a typical InterBus-S network with a host controller interface to a PLC.

Figure 19-8. InterBus-S I/O network interface connected to a Siemens PLC.

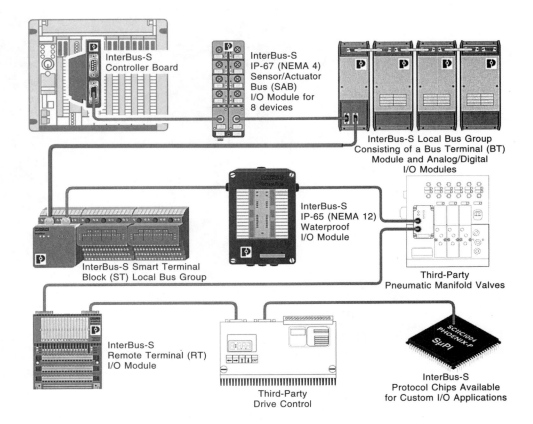

Figure 19-9. An InterBus-S network with a host controller interface to a PLC.

I/O device addresses in an InterBus-S network are automatically determined by their physical location, thus eliminating the need to manually set addresses. The host controller interface continuously scans data from the I/O devices, reading all the inputs in one scan and subsequently writing output data. The network transmits this data in *frames*, which provide simultaneous updates to all devices in the network. The InterBus-S network ensures the validity of the data transmission through the CRC error-checking technique. Table 19-1 lists some of the features and benefits of the InterBus-S device bus network. Note that this network uses the first, second, and seventh layers—the physical, data link, and application layers, respectively—of the ISO OSI reference model.

CANbus Byte-Wide Device Bus Networks. CANbus networks are byte-wide device bus networks based on the widely used CAN electronic chip technology, which is used inside automobiles to control internal components, such as brakes and other systems. A CANbus network is an open protocol system featuring variable length messages (up to 8 bytes), nondestructive arbitration, and advanced error management. A four-wire cable plus shield—two wires for power, two for signal transmission, and a "fifth" shield wire—

Network Characteristics	InterBus-S Features	User Benefits
Physical Layer (Layer 1):		
Protocol structure	Hardware ring network	Self-configuring, no network addresses to set
Distance	Inherently distributed up to 42,000 feet	Significantly lowers system installation cost
Physical media	Cabeling options allow for twisted-pair, fiber-optic, slipring, infrared, or SMG connections	Network connections can be made in all types of industrial environments
Data Link Layer (Layer 2):		
Protocol transmission	Full-duplex, total frame transmission	All network I/O updated simultaneously
Protocol arbitration	No arbitration	All data is transmitted continuously without any interruptions
Throughput	Read and write up to 4096 digital inputs and outputs in under 14 msec	Updates I/O many times faster than the application logic can be solved
Error checking	CRC error checking between every network connection	Accurate, reliable data transmissions
Application Layer (Layer 3):		
Diagnostics	Pinpoints the cause and location of network problems	More uptime, less downtime, reduced maintenance cost, improved reliability
Protocol flexibility	Supports high-speed digital data, analog data, and client-server messaging	Achieves maximum control
I/O expandability	Connects up to 256 I/O drops for a total of 4096 digital input and 4096 digital output points or a combination of digital and analog signal types	Provides greater system flexibility
Connectivity:		
Openness	Over 300 third-party manufacturers provide compatible products	Standard analog and digital I/O signal types and the widest variety of form factors available to provide optimum system flexibility for tomorrow's manufacturing requirements
Standards	DIN standard 19258 with profiles for robotics, drives, process controllers, encoders, and operator interfaces	Greater system integrity

Table 19-1. Features and benefits of the InterBus-S network.

provides the communication link with field devices (see Figure 19-10). This communication can either be master/slave or peer to peer. The speed of the network (data transmission rate) depends on the length of the trunk cable. Table 19-2 illustrates speed-versus-length tables for the DeviceNet and SDS CANbus networks.

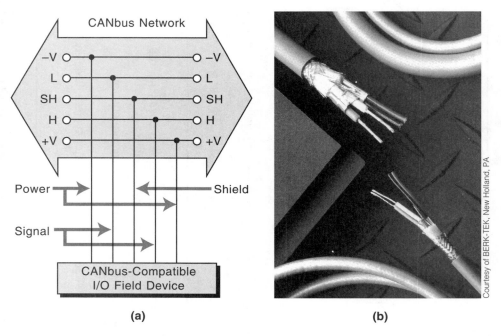

Figure 19-10. (a) A CANbus communication link and **(b)** a CANbus four-wire cable.

	Distance		
	Meters	**Feet**	**Transmission Rate**
(a)	500	1640	125K bits/sec
	200	656	250K bits/sec
	100	328	500K bits/sec

	Maximum Total Cable Trunk Length (ft.)	**Data Rate (bits/second)**
	1600	125 kbaud
(b)	800	250 kbaud
	400	500 kbaud
	100	1000 kbaud (1Mb)

Table 19-2. Speed-versus-length tables for **(a)** DeviceNet and **(b)** SDS CANbus networks.

The DeviceNet byte-wide network can support 64 nodes and a maximum of 2048 field I/O devices. The SDS network can also support 64 nodes; however, this number increases to 126 addressable locations when multiport I/O interfaces are used to multiplex the nodes. Using a 4-to-1 multiport I/O interface module, an SDS network can connect to up to 126 nonintelligent I/O devices in any combination of inputs and outputs. Figure 19-11 shows this multiplexed configuration. This multiport interface to nonintelligent field devices contains a slave CAN chip inside the interface, which provides status information about the nodes connected to the interface. In a DeviceNet network, the PLC connects to the field devices in a trunkline configuration, with either single drops off the trunk or branched drops through multiport interfaces at the device locations.

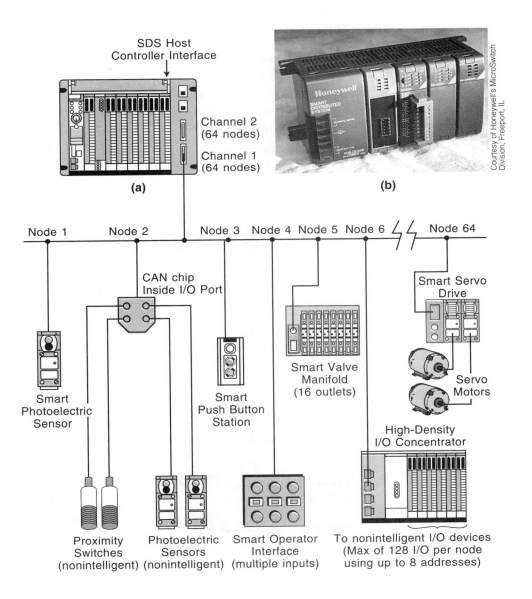

Figure 19-11. (a) A multiplexed SDS network and (b) a high-density I/O concentrator.

Because an SDS network can transmit many bytes of information in the form of variable length messages, it can also support many intelligent devices that can translate one, two, or more bytes of information from the network into 16 or 32 bits of ON/OFF information. An example of this type of intelligent device is a solenoid valve manifold. This kind of manifold can have up to 16 connections, thereby receiving 16 bits (two bytes) of data from the network and controlling the status of 16 valve outputs. However, this device uses only one address of the 126 possible addresses. Thus, in this configuration, the SDS network can actually connect to more than just 126 addressable devices.

The CANbus device bus network uses three of the ISO layers (see Figure 19-12) and defines both the media access control method and the physical signaling of the network, while providing cyclic redundancy check (CRC) error detection. The media access control function determines when each device on the bus will be enabled.

A CANbus scanner or an I/O processor provides the interface between a PLC and a CANbus network. Figure 19-13 illustrates a CANbus scanner designed for Allen-Bradley's DeviceNet network, which has two channels with up to 64 connected devices per channel. Block transfer instructions in the control program pass information to and from the scanner's processor (see Figure 19-14). The scanner converts the serial data from the CANbus network to a form usable by the PLC processor.

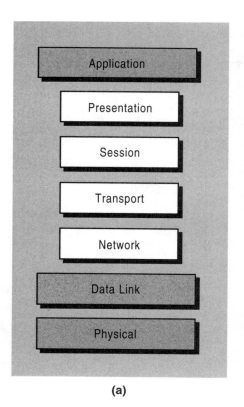

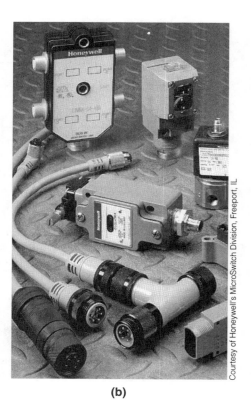

Courtesy of Honeywell's MicroSwitch Division, Freeport, IL

(a)   (b)

Figure 19-12. (a) CANbus ISO layers and **(b)** typical components and devices that connect and support the CANbus (SDS) layers.

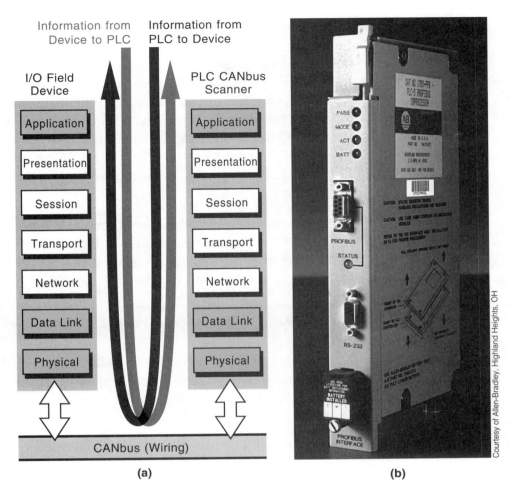

Figure 19-13. (a) Information transfer through a CANbus network and **(b)** Allen-Bradley's CANbus DeviceNet scanner.

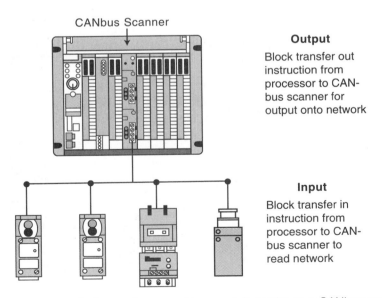

CANbus Scanner

Output

Block transfer out instruction from processor to CANbus scanner for output onto network

Input

Block transfer in instruction from processor to CANbus scanner to read network

Figure 19-14. Block transfer instructions used to pass information to a CANbus scanner.

As mentioned earlier, the SDS CANbus network can handle 126 addressable I/O devices per network per channel. To increase the number of connectable devices, a PLC or computer bus interface module with two channels can be used to link two independent networks for a total of 252 I/O addresses. Moreover, each address can be multiplexed, making the network capable of more I/O connections. If the application requires even more I/O devices, another I/O bus scanner can be connected to the same PLC or computer to implement another set of networks. The SDS CANbus network connects the PLC and field devices in the same way as a DeviceNet network—in a trunkline configuration.

Some manufacturers provide access to remote I/O systems via a CANbus with an I/O rack/CANbus remote processor. Figure 19-15 illustrates an example of this type of configuration using Allen-Bradley's Flex I/O system with a DeviceNet processor, thus creating a DeviceNet I/O subsystem.

BIT-WIDE DEVICE BUS NETWORKS

Bit-wide device bus networks are used for discrete applications with simple ON/OFF devices (e.g., sensors and actuators). These I/O bus networks can only transmit 4 bits (one nibble) of information at a time, which is sufficient to transmit data from these devices. The smallest discrete sensors and

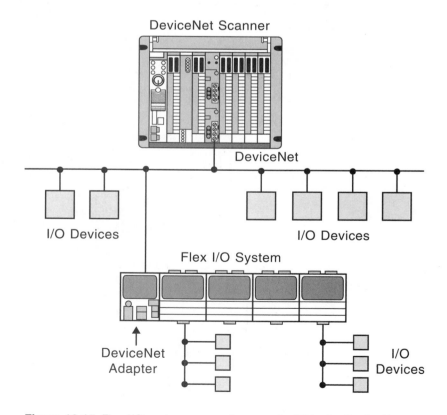

Figure 19-15. Flex I/O system connecting remote I/O to the DeviceNet processor.

actuators require only one bit of data to operate. By minimizing their data transmission capabilities, bit-wide device bus networks provide optimum performance at economical costs. The most common bit-wide device bus networks are ASI, InterBus Loop, and Seriplex.

ASI Bit-Wide Device Bus Network. The ASI network protocol is used in simple, discrete network applications requiring no more than 124 I/O field devices. These 124 input and output devices can be connected to up to 31 nodes in either a tree, star, or ring topology. The I/O devices connect to the PLC or personal computer via the bus through a host controller interface. Figure 19-16 illustrates an ASI bit-wide device bus network.

The ASI network protocol is based on the ASI protocol chip, thus the I/O devices connected to this type of network must contain this chip. Typical ASI-compatible devices include proximity switches, limit switches, photoelectric sensors, and standard off-the-shelf field devices. However, in an application using an off-the-shelf device, the ASI chip is located in the node (i.e., an intelligent node with a slave ASI chip), instead of in the device.

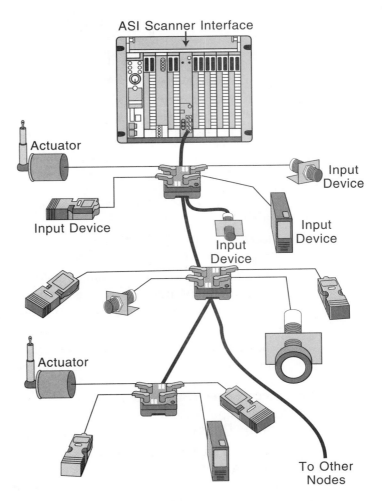

Figure 19-16. ASI bit-wide device bus network.

ASI networks require a 24-VDC power supply connected through a two-wire, unshielded, untwisted cable. Both data and power flow through the same two wires. The cycle time is less than 5 msec with a transfer rate of 167K bits/ second. The maximum cable length is 100 meters (330 ft) from the master controller. Figure 19-17 illustrates an I/O bus network that uses both the ASI bit-wide network and the byte-wide CANbus network. Note that the ASI network connects to the byte-wide CANbus network through a gateway.

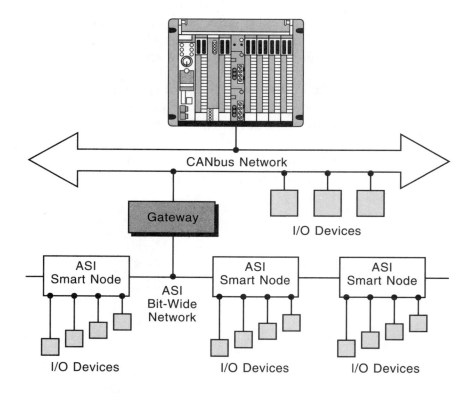

Figure 19-17. I/O bus network using the CANbus and ASI networks.

InterBus Loop Bit-Wide Device Bus Network. The InterBus Loop from Phoenix Contact Inc. is another bit-wide device bus network used to interface a PLC with simple sensor and actuator devices. The InterBus Loop uses a power and communications technology called PowerCom to send the InterBus-S protocol signal through the power supply wires (i.e., the protocol is modulated onto the power supply lines). This reduces the number of cables required by the network to only two conductors, which carry both the power and communication signals to the field devices.

Since the InterBus-S and InterBus Loop networks use the same protocol, they can communicate with each other via an InterBus Loop terminal module (see Figure 19-18). The InterBus Loop connects to the bus terminal module, located in the InterBus-S network, which attaches to the field devices via two wires. An InterBus Loop network can also interface with nonintelligent, off-the-shelf devices by means of module interfaces containing an intelligent slave network chip.

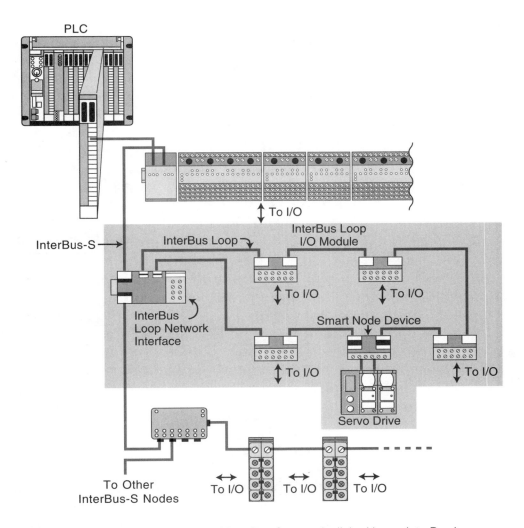

Figure 19-18. InterBus Loop and InterBus-S networks linked by an InterBus Loop terminal module.

Seriplex Bit-Wide Device Bus Network. The Seriplex device bus network can connect up to 510 field devices to a PLC in either a master/slave or peer-to-peer configuration. The Seriplex network is based on the application-specific integrated circuit, or ASIC chip, which must be present in all I/O field devices that connect to the network. I/O devices that do not have the ASIC chip embedded in their circuitry (i.e., off-the-shelf devices) can connect to the network via a Seriplex I/O module interface that contains a slave ASIC chip. The ASIC I/O interface contains 32 built-in Boolean logic function used to create logic that will provide the communication, addressability, and intelligence necessary to control the field devices connected to the network bus (see Figure 19-19).

A Seriplex network can span distances of up to 5,000 feet in a star, loop, tree, or multidrop configuration. This bit-wide bus network can also operate without a host controller. Unlike the ASI network, the Seriplex device bus

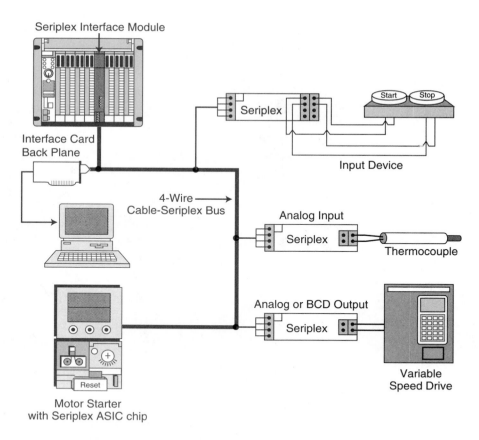

Figure 19-19. Seriplex bus network with a controller.

network can interface with analog I/O devices; however, the digitized analog signal is read or written one bit at a time in each scan cycle. Figure 19-20 illustrates a typical Seriplex bus network without a controller.

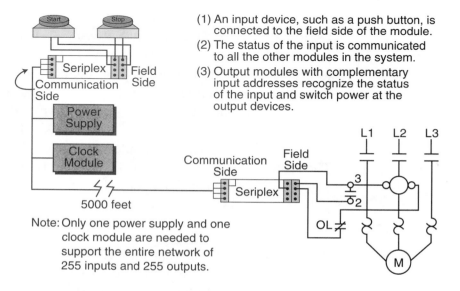

(1) An input device, such as a push button, is connected to the field side of the module.

(2) The status of the input is communicated to all the other modules in the system.

(3) Output modules with complementary input addresses recognize the status of the input and switch power at the output devices.

Note: Only one power supply and one clock module are needed to support the entire network of 255 inputs and 255 outputs.

Figure 19-20. Seriplex I/O module interface without a controller.

19-5 PROCESS BUS NETWORKS

A process bus network is a high-level, open, digital communication network used to connect analog field devices to a control system. As mentioned earlier, a process bus network is used in process applications, where the analog input/output sensors and actuators respond slower than those in discrete bus applications (device bus networks). The size of the information packets delivered to and from these analog field devices is large, due to the nature of the information being collected at the process level.

The two most commonly used process bus network protocols are Fieldbus and Profibus (see Section 19-2). Although these network protocols can transmit data at a speed of 1 to 2 megabits/sec, their response time is considered slow to medium because of the large amount of information that is transferred. Nevertheless, this speed is adequate for process applications, because analog processes do not respond instantaneously, as discrete controls do. Figure 19-21 illustrates a typical process bus configuration.

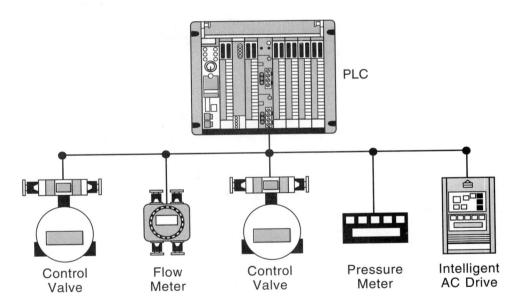

Figure 19-21. Process bus configuration.

Process bus networks can transmit enormous amounts of information to a PLC system, thus greatly enhancing the operation of a plant or process. For example, a smart, process bus–compatible motor starter can provide information about the amount of current being pulled by the motor, so that, if current requirements increase or a locked-rotor current situation occurs, the system can alert the operator and avoid a potential motor failure in a critical production line. Implementation of this type of system without a process bus network would be too costly and cumbersome because of the amount of wire runs necessary to transmit this type of process data.

Process bus networks will eventually replace the commonly used analog networks, which are based on the 4–20 mA standard for analog devices. This will provide greater accuracy and repeatability in process applications, as well as add bidirectional communication between the field devices and the controller (e.g., PLC). Figure 19-22 illustrates an intelligent valve/manifold system that can be used in a process bus network.

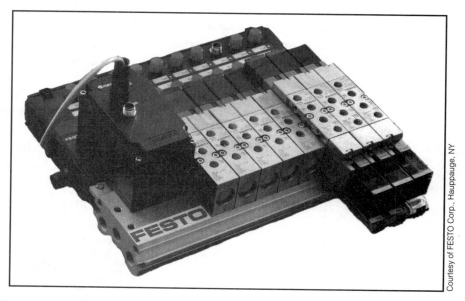

Courtesy of FESTO Corp., Hauppauge, NY

Figure 19-22. Intelligent valve/manifold system compatible with the Fieldbus protocol.

A PLC or computer communicates with a process bus network through a host controller interface module using either Fieldbus or Profibus protocol format. Block transfer instructions relay information between the PLC and the process bus processor. The process bus processor is generally inserted inside the rack enclosure of the PLC. Figure 19-23 shows a PLC with a Profibus processor communication interface.

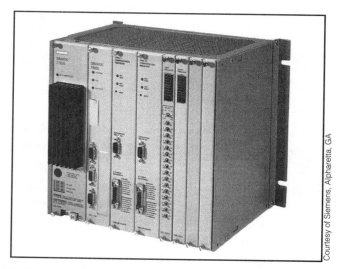

Courtesy of Siemens, Alpharetta, GA

Figure 19-23. Siemens' Simatic 505 PLC with an integrated Profibus-DP interface.

FIELDBUS PROCESS BUS NETWORK

The Fieldbus process bus network from the Fieldbus Foundation (FF) is a digital, serial, multiport, two-way communication system that connects field equipment, such as intelligent sensors and actuators, with controllers, such as PLCs. This process bus network offers the desirable features inherent in 4–20 mA analog systems, such as:

- a standard physical wiring interface

- bus-powered devices on a single pair of wires

- intrinsic safety options

However, the Fieldbus network technology offers the following additional advantages:

- reduced wiring due to multidrop devices

- compatibility among Fieldbus equipment

- reduced control room space requirements

- digital communication reliability

Fieldbus Protocol. The Fieldbus network protocol is based on three layers of the ISO's seven-layer model (see Figure 19-24). These three layers are layer 1 (physical interface), layer 2 (data link), and layer 7 (application). The section comprising layers 2 and 7 of the model are referred to as the Fieldbus *communication stack*. In addition to the ISO's model, Fieldbus adds an extra

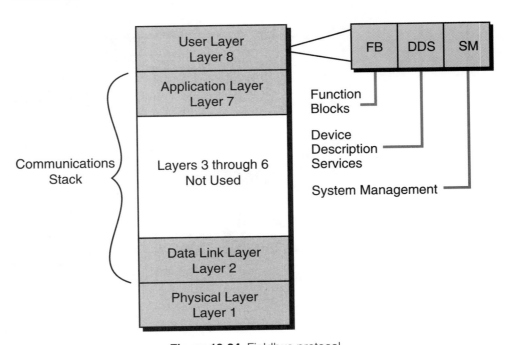

Figure 19-24. Fieldbus protocol.

layer on top of the application layer called the *user layer*. This user layer provides several key functions, which are function blocks, device description services, and system management.

Physical Layer (Layer 1). The physical layer of the Fieldbus process bus network conforms with the ISA SP50 and IEC 1152-2 standards. These standards specify the type of wire that can be used in this type of network, as well as how fast data can move through the network. Moreover, these standards define the number of field devices that can be on the bus at different network speeds, with or without being powered from the bus with *intrinsic safety* (IS). Intrinsically safe equipment and wiring does not emit enough thermal or electrical energy to ignite materials in the surrounding atmosphere. Thus, intrinsically safe devices are suitable for use in hazardous environments (e.g., those containing hydrogen or acetylene).

Table 19-3 lists the specifications for the Fieldbus network's physical layer, including the type of wire (bus), speed, number of devices, and wiring characteristics. The Fieldbus has two speeds—a low speed of 31.25 kbaud, referred to as H1, and a high speed of 1 Mbaud or 2.5 Mbaud (depending on the mode—AC current or DC voltage mode), called H2. Figure 19-25 illustrates how a bridge can connect an H1 Fieldbus network to an H2 Fieldbus network.

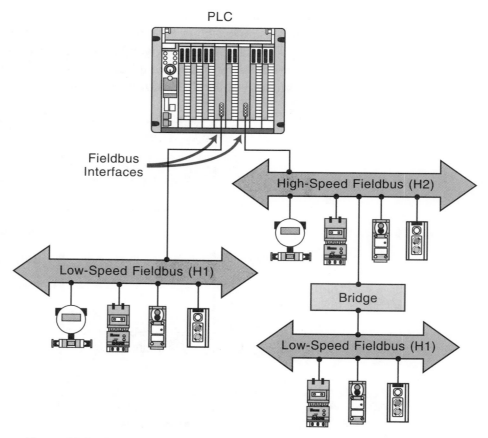

Figure 19-25. Bridge connecting low-speed and high-speed Fieldbus networks.

Bus Type	Speed	Number of Devices/ Fieldbus Segment	Type of Wire	
			New Bus Segment (Shielded/Twisted-Pair)	Existing Bus Segment (Shielded/Multitwisted Pair)
Low-speed bus (H1)	31.25 kbaud	2–32 devices that are not bus powered 2–12 devices that are bus powered 2–6 devices that are bus powered in an intrinsically safe (IS) area	#18 AWG, up to 1900 m*	#22 AWG, up to 1200 m
High-speed bus (H2)	1 Mbaud	127 devices, AC current mode (16 KHz frequency), powered from bus in IS area	#22 AWG, up to 750 m	
	1 Mbaud	127 devices, DC voltage mode, not powered from bus, and no IS devices	#22 AWG, up to 750 m	
	2.5 Mbaud	127 devices, DC voltage mode, not powered from bus, and no IS devices	#22 AWG, up to 500 m	

*A Fieldbus low-speed bus (H1) can also be implemented using unshielded, multitwisted wired with #26 AWG at up to 400 m. The low-speed bus can also use unshielded, multicore wire with #16 AWG at up to 200 m.

Table 19-3. Fieldbus physical layer specifications.

At a speed of 31.25 kbaud, the physical layer of the Fieldbus process network can support existing 4–20 mA wiring. This increases cost-effectiveness when upgrading a plant or process's network communication scheme. At this H1 speed, the Fieldbus network can also support intrinsically safe network segments with bus-powered devices.

Communication Stack (Layers 2 and 7). The communication stack portion of the Fieldbus process bus network consists of layer 2 (the data link layer) and layer 7 (the application layer). The data link layer controls the transmission of messages onto the Fieldbus through the physical layer. It manages access to the bus through a *link active scheduler*, which is a deterministic, centralized bus transmission regulator based on IEC and ISA standards. The application layer contains the *Fieldbus messaging specification* (FMS) standard, which encodes and decodes commands from the user layer, Fieldbus's additional 8th layer. The FMS is based on the Profibus process bus standard. Layer 7 also contains an *object dictionary*, which allows Fieldbus network data to be retrieved by either tag name or index record.

The Fieldbus process network uses two types of message transmissions: **cyclic** (scheduled) and **acyclic** (unscheduled). Cyclic message transmissions occur at regular, scheduled times. The master network device monitors how busy the network is and then grants the slave devices permission to send network transmissions at specified times. Other network devices can listen to and receive these messages if they are subscribers.

Acyclic, or unscheduled, messages occur between cyclic, scheduled messages, when the master device sends an unscheduled informational message to a slave device. Typically, acyclic messages involve alarm acknowledgment signals or special retrieving commands designed to obtain diagnostic information from the field devices.

User Layer (Layer 8). The user layer implements the Fieldbus network's distributed control strategy. It contains three key elements, which are function blocks, device description services, and system management. The user layer, a vital segment of the Fieldbus network, also defines the software model for user interaction with the network system.

Function Blocks. Function blocks are encapsulated control functions that allow the performance of input/output operations, such as analog inputs, analog outputs, PID control, discrete inputs/outputs, signal selectors, manual loaders, bias/gain stations, and ratio stations. The function block capabilities of Fieldbus networks allow Fieldbus-compatible devices to be programmed with blocks containing any of the instructions available in the system. Through these function blocks, users can configure control algorithms and implement them directly through field devices. This gives these intelligent field devices the capability to store and execute software routines right at their connection to the bus. The process information gathered through these function block programs can then be passed to the host through the network, either cyclically or acyclically.

Figure 19-26 illustrates an example of a process control loop that is executed directly on the Fieldbus network. In this loop, the analog input function block reads analog process information from the meter/transmitter, executes a PID function block, and then outputs analog control data to an intelligent process valve. This configuration creates an independent, self-regulating loop, which obtains its own analog input data from the flow meter. Information about the required flow parameters is passed from the host controller to the intelligent valve system, so that it can properly execute its function blocks. The function blocks allow the field device to be represented in the network as a collection of software block instructions, rather than just as an instrument.

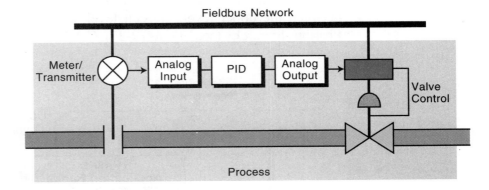

Figure 19-26. Process control loop executed on the Fieldbus network.

<u>Device Description Services.</u> *Device descriptions* (DD) are Fieldbus software mechanisms that let a host obtain message information, such as vendor name, available function blocks, and diagnostic capabilities, from field devices. Device descriptions can be thought of as "drivers" for field devices connected to the network, meaning that they allow the device to communicate with the host and the network. The network's host computer uses a device description services, or DDS, interpreter to read the desired information from each device. All devices connected to a Fieldbus process network must have a device description. When a new field device is added to the network, the host must be supplied with its device description. Device descriptions eliminate the need to revise the whole control system software when revisions are made to existing field device software or when new devices are added to the process control system.

<u>System Manager.</u> The system management portion of the user layer schedules the execution of function blocks at precisely defined intervals. It also controls the communication of all the Fieldbus network parameters used by the function blocks. Moreover, the system manager automatically assigns field device addresses.

PROFIBUS PROCESS BUS NETWORK

Profibus is a digital process bus network capable of communicating information between a master controller (or host) and an intelligent, slave process field device, as well as from one host to another. Profibus actually consists of three intercompatible networks with different protocols designed to serve distinctive application requirements. The three types of Profibus networks are:

- Profibus-FMS

- Profibus-DP

- Profibus-PA

The Profibus-FMS network is the universal solution for communicating between the upper level, the cell level, and the field device level of the Profibus hierarchy (see Figure 19-27). Cell level control occurs at individual

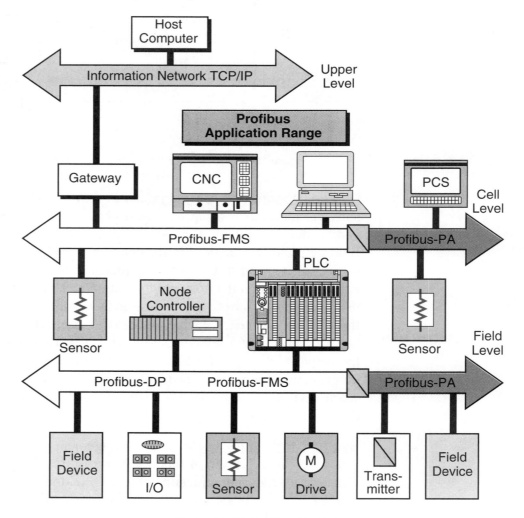

Figure 19-27. Profibus hierarchy.

(or cell) areas, which exercise the actual control during production. The controllers at the cell level must communicate with other supervisory systems. The Profibus-FMS utilizes the Fieldbus message specification (FMS) to execute its extensive communication tasks between hierarchical levels. This communication is performed through cyclic or acyclic messages at medium transmission speeds.

The Profibus-DP network is a performance-optimized version of the Profibus network. It is designed to handle time-critical communications between devices in factory automation systems. The Profibus-DP is a suitable replacement for 24-V parallel and 4–20 mA wiring interfaces.

The Profibus-PA network is the process automation version of the Profibus network. It provides bus-powered stations and intrinsic safety according to the transmission specifications of the IEC 1158-2 standard. The Profibus-PA network has device description and function block capabilities, along with field device interoperability.

Profibus Network Protocol. The Profibus network follows the ISO model; however, each type of Profibus network contains slight variations in the model's layers. The Profibus-FMS does not define layers 3 through 6; rather, it implements their functions in a lower layer interface (LLI) that forms part of layer 7 (see Figure 19-28). The Profibus-FMS implements the Fieldbus message specification (FMS), which provides powerful network communication services and user interfaces, in layer 7 as well.

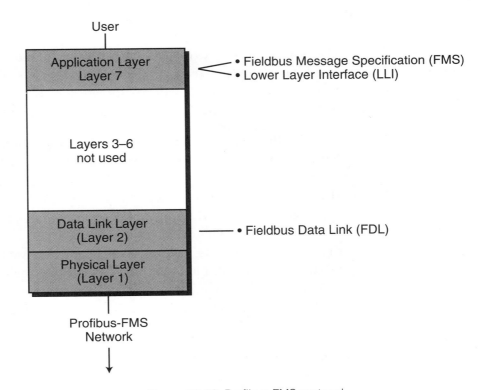

Figure 19-28. Profibus-FMS protocol.

The Profibus-DP network, on the other hand, does not define layers 3 through 7 (see Figure 19-29). It omits layer 7 primarily to achieve the high operational speed required for its applications. A *direct data link mapper* (DDLM), located in layer 2, provides the mapping between the user interface and layer 2 of the Profibus-DP network.

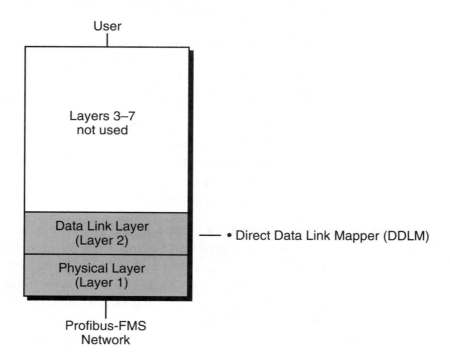

Figure 19-29. Profibus-DP protocol.

The Profibus-PA network uses the same type of model as the Profibus-FMS (see Figure 19-30), except its seventh layer differs slightly. Layer 7 implements the function block control software and also contains a device description language used for field device identification and addressing.

The data link layer, designated in the Profibus network as the *fieldbus data link layer* (FDL), executes all message and protocol transmissions. This data layer is equivalent to layer 2 of the ISO model. The fieldbus data link layer also provides **medium access control (MAC)** and data integrity. Medium access control ensures that only one station has the right to transmit data at any time. Because Profibus can communicate between masters with equal access rights (e.g., two PLCs), medium access control is used to provide each of the master stations with the opportunity to execute their communication tasks within precisely defined time intervals. For communication between a master and slave field devices, cyclic, real-time data exchange is achieved as quickly as possible through the network.

The Profibus's medium access protocol is a hybrid communication method that includes a token-passing protocol for use between masters and a master-slave protocol for communication between a master and a field device.

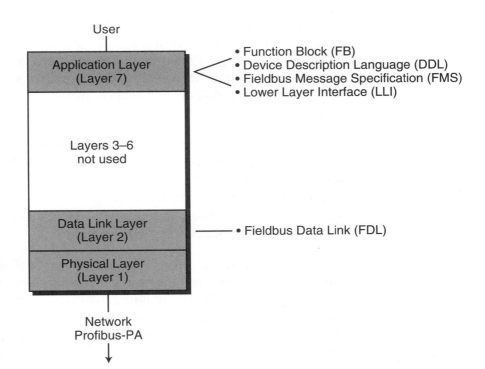

User

Application Layer
(Layer 7)

• Function Block (FB)
• Device Description Language (DDL)
• Fieldbus Message Specification (FMS)
• Lower Layer Interface (LLI)

Layers 3–6
not used

Data Link Layer
(Layer 2)

• Fieldbus Data Link (FDL)

Physical Layer
(Layer 1)

Network
Profibus-PA

Figure 19-30. Profibus-PA protocol.

Through this hybrid medium access protocol, a Profibus network can function as a master-slave system, a master-master system (token passing), or a combination of both systems (see Figure 19-31).

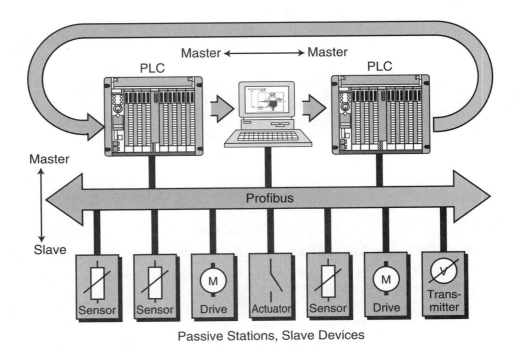

Master ←——→ Master

PLC   PLC

Master

Profibus

Slave

Sensor Sensor Drive Actuator Sensor Drive Trans-mitter

Passive Stations, Slave Devices

Figure 19-31. Master-slave and master-master Profibus communications.

As mentioned earlier, layer 2 of the Profibus network is responsible for data integrity, which is ensured through the Hamming Distance HD = 4 error detection method. The Hamming distance method can detect errors in the transmission medium, as well as in the transceivers. As defined by the IEC 870-5-1 standard, this error detection method uses special start and end delimiters, along with slip-free synchronization and a parity bit for 8 bits.

Profibus networks support both peer-to-peer and multipeer communication in either broadcast or multicast configurations. In broadcast communication, an active station sends an unconfirmed message to all other stations. Any of these stations (including both masters and slaves) can take this information. In multicast communication, an active station sends an unconfirmed message to a particular group of master or slave stations.

The physical layer, or layer 1, of the ISO model defines the network's transmission medium and the physical bus interface. The Profibus network adheres to the EIA RS-485 standard, which uses a two-conductor, twisted-pair wire bus with optional shielding. The bus must have proper terminations at both ends. Figure 19-32 illustrates the pin assignment used in the Profibus. The maximum number of stations or device nodes per segment is 32 without repeaters and 127 with repeaters. The network transmission speed is selectable from 9.6 kbaud to 12 Mbaud, depending on the distance and cable type. Without repeaters, the maximum bus length is 100 m at 12 Mbaud. With conventional type-A copper bus cable, the maximum distance is 200 m at 1.5 Mbaud. This distance can be increased to up to 1.2 km if the speed of the network is reduced to 93.75 kbaud. With type-B cable, the maximum distance is 200 m at 500 kbaud and up to 1.2 km at 93.75 kbaud. The type of connector used is a 9-pin, D-sub connector.

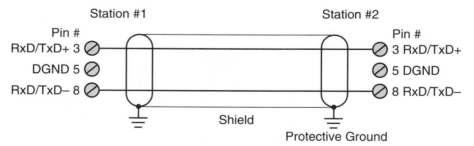

Figure 19-32. Profibus pin assignment.

19-6 I/O BUS INSTALLATION AND WIRING CONNECTIONS

INSTALLATION GUIDELINES

One of the most important aspects of an I/O bus network installation is the use of the correct type of cable, number of conductors, and type of connectors for the network being used. In device bus networks, the number of conductors and type of communication standard (i.e., RS-485, RS-422, etc.) varies depending on the specific network (e.g., DeviceNet, Seriplex, ASI, Profibus, Fieldbus, etc.). The connector ports (see Figure 19-33), which connect the I/O field devices to the I/O bus network, can be implemented in either an open or an enclosed configuration. Figure 19-34 illustrates the port connections for a DeviceNet I/O bus network.

Figure 19-33. Connector ports from a DeviceNet bus network (left: enclosed, right: open).

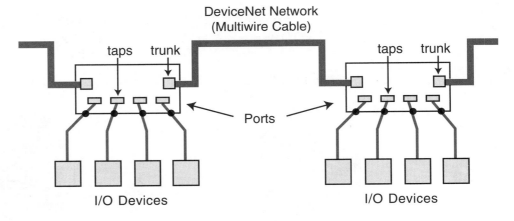

Figure 19-34. DeviceNet I/O bus port connections.

In general, an enclosed configuration can connect from 4 to 8 I/O field devices in one drop, while an open configuration can accommodate two to four I/O devices. Enclosed connector ports are used when the network must be protected from the environment, as in a NEMA 4–type enclosure. Open ports are used when replacing I/O connections in a system that already has a DIN rail installation, where the open ports can be easily mounted onto the rail.

DEVICE BUS NETWORK WIRING GUIDELINES

Figure 19-35 illustrates a typical wiring diagram connection for a DeviceNet CANbus network. Note that the two trunk connections constitute the main cable of the network, with the five wires providing signal, power, and shielding. A printed circuit board assembly internally connects the two trunk connectors, or ports, and the I/O device taps. Most manufacturers of device bus networks provide "plug-and-play" connectors and wiring systems, which facilitate installation and system modifications (see Figure 19-36).

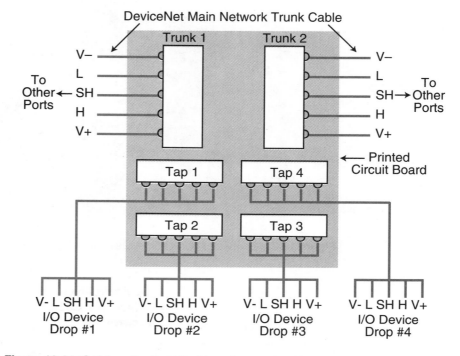

Figure 19-35. CANbus DeviceNet wiring diagram for the multiport tap in Figure 19-34.

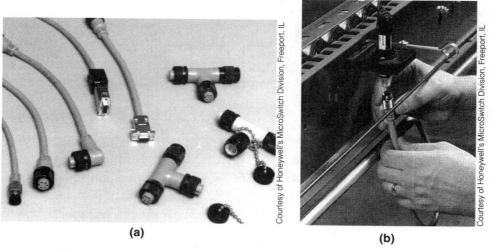

Figure 19-36. (a) Plug-and-play connectors and **(b)** their installation.

The majority of device bus networks require that a terminator resistor be present at the end of the main trunk line for proper operation and transmission of network data. Each network may also specify the number of nodes that can be connected to the network, the speed of transmission depending on the trunk length, and the maximum drop length at which field devices can be installed. The network may also limit the cumulative drop length, meaning that the combined lengths of all the drops cannot exceed a particular specification. Table 19-4 shows the specifications for Allen-Bradley's DeviceNet communication link network.

Data Transmission Rate	Trunk Length	Max. Drop Length	Max. # of Nodes	Cumulative Drop Length
125K bits/sec	500 m (1640 ft)			156 m (512 ft)
250K bits/sec	200 m (656 ft)	3 m (10 ft)	64	78 m (256 ft)
500K bits/sec	100 m (328 ft)			39 m (128 ft)

Table 19-4. DeviceNet specifications.

PROCESS BUS NETWORK WIRING GUIDELINES

Cable criteria similar to device bus networks apply to process bus networks. Depending on the network protocol specifications, specifically those of layer 1 (physical) of the OSI model, the conductor may be twisted pair or coaxial, operating at different network transmission speeds. Table 19-5 shows the wiring and network speed characteristics of the Fieldbus Foundation network (Fieldbus protocol). Figure 19-37 shows the process bus interface for Allen-Bradley's family of PLCs, which is compatible with the Profibus protocol. This Profibus interface can work at network speeds of 9.6, 19.2, 93.75, 187.5,

	Data Rate		
	Slow	**Standard**	**High**
Speed	31.25K bps	1M bps	2 Mbps
Cable	twisted-pair	twisted-pair	twisted-pair
Distance	1900 m	750 m	500 m

Table 19-5. Fieldbus network characteristics.

and 500 kbits/sec. Process bus wiring installations may also require a termination block at the end of the wiring. T-junction connectors provide the connections to different I/O field devices (see Figure 19-38).

(a)   (b)

Figure 19-37. (a) Allen-Bradley's Profibus process bus interface and **(b)** the wiring installation of a Fieldbus network using two sets of shielded twisted-pair wire.

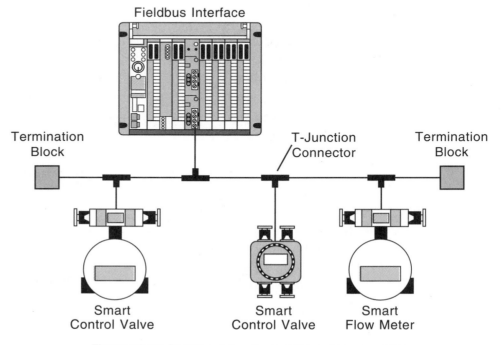

Figure 19-38. Fieldbus network using T-junction connectors.

I/O BUS NETWORK ADDRESSING

Addressing of the I/O devices in an I/O bus network occurs during the configuration, or programming, of the devices in the system. Depending on the PLC, this addressing can be done either directly on the bus network via a PC and a gateway (see Figure 19-39a) or through a PC connected directly to the bus network interface (see Figure 19-39b). It can also be done through the PLC's RS-232 port (see Figure 19-40). Some I/O bus networks have switches that can be used to define device addresses, while others have a predefined address associated with each node drop.

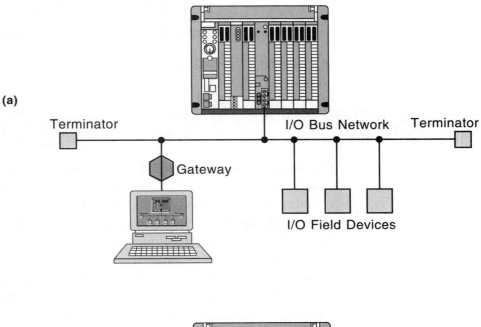

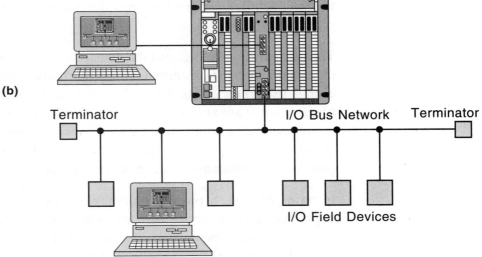

Figure 19-39. I/O addresses assigned using **(a)** a PC connected to the network through a gateway and **(b)** a PC connected directly to the network.

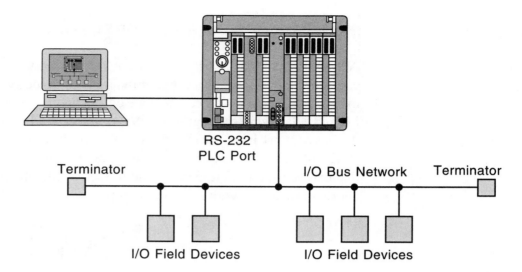

Figure 19-40. I/O addresses assigned using a PC connected to the PLC's RS-232 port.

19-7 SUMMARY OF I/O BUS NETWORKS

The device and process types of I/O bus networks provide incredible potential system cost savings, which are realized during installation of a control system. These two types of I/O networks can also form part of a larger, networked operation, as shown in Figure 19-41. In this operation, the information network communicates via Ethernet between the main computer system (or a personal computer) and a supervisory PLC. In turn, these PLCs communicate with other processors through a local area control network. The PLCs may also have remote I/O, device bus, and process bus subnetworks. The addition of field devices to this type of I/O network is relatively easy, as long as each field device is compatible with its respective I/O bus network protocol.

The main difference between the device bus and the process bus networks is the amount of data transmitted. This is due to the type of application in which each is used. Device bus networks are used in discrete applications, which transmit small amounts of information, while process bus networks are used in process/analog applications, which transmit large amounts of data. Figure 19-42 shows a graphic representation of these networks based on the potential amount of information that can be transmitted through them.

In terms of cost, a process bus network tends to be more expensive to implement than a device bus network, simply because analog I/O field devices are more expensive. Also, the intelligence built into a process bus network is more costly than the technology incorporated into a device bus network. For example, the CAN, SDS, ASI, ASIC, and InterBus-S chips used in device networks are readily available, standard, off-the-shelf chips, which

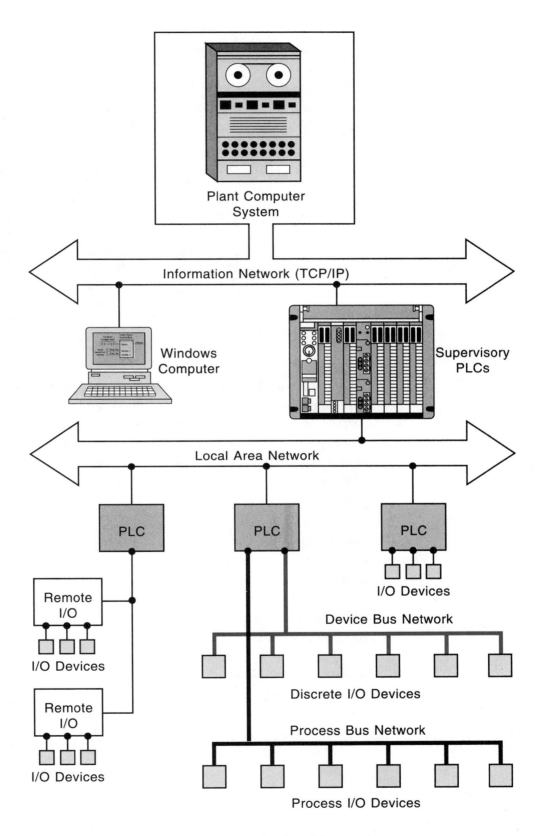

Figure 19-41. Large plantwide network.

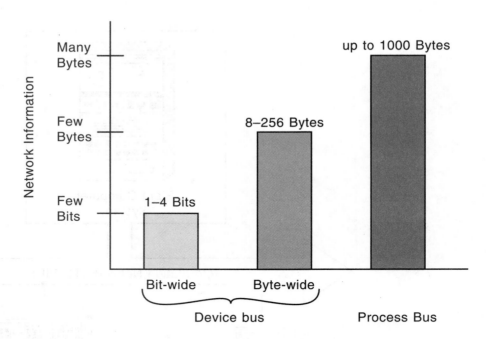

Figure 19-42. Network data transmission comparison.

can be purchased at a relatively low cost. Process bus networks, on the other hand, require devices with more sophisticated electronics, such as microprocessors, memory chips, and other supporting electronic circuitry, which makes process network I/O devices more expensive. This expense, however, is more than offset by the total savings for system wiring and installation, especially in the modernization of existing operations where wire runs may already be in place.

KEY TERMS

acyclic message
bit-wide bus network
byte-wide bus network
cyclic message
device bus network
I/O bus network
I/O bus network scanner
medium access control (MAC)
process bus network
tree topology

INSTALLATION AND START-UP

- PLC Start-Up and Maintenance
- System Selection Guidelines

CHAPTER TWENTY

PLC START-UP
AND MAINTENANCE

If I had been present at the Creation, I would have given some useful hints for the better arrangement of the Universe.

—Alfonso the Wise, King of Castille

 **CHAPTER HIGHLIGHTS** The design of programmable controllers includes a number of rugged features that allow PLCs to be installed in almost any industrial environment. Although programmable controllers are tough machines, a little foresight during their installation will ensure proper system operation. In this chapter, we will explore PLC installation, explaining the specifications for proper PLC component placement and environment. We will also explain other factors that affect PLC operation, such as noise, heat, and voltage. In addition, we will discuss wiring guidelines and safety precautions. Although proper PLC installation leads to good system operation, no programmable controller system is without faults. Therefore, we will investigate proactive maintenance techniques, as well as reactive troubleshooting processes. When you finish this chapter, you will understand the fundamentals of PLC start-up and operation.

20-1 PLC SYSTEM LAYOUT

System layout is the conscientious approach to placing and interconnecting components not only to satisfy the application, but also to ensure that the controller will operate trouble free in its environment. In addition to programmable controller equipment, the system layout also encompasses the other components that form the total system. These components include isolation transformers, auxiliary power supplies, safety control relays, and incoming line noise suppressors. In a carefully constructed layout, these components are easy to access and maintain.

PLCs are designed to work on a factory floor; thus, they can withstand harsh environments. Nevertheless, careful installation planning can increase system productivity and decrease maintenance problems. The best location for a programmable controller is near the machine or process that it will control, as long as temperature, humidity, and electrical noise are not problems. Placing the controller near the equipment and using remote I/O where possible will minimize wire runs and simplify start-up and maintenance. Figure 20-1 shows a programmable controller installation and its wiring connections.

PANEL ENCLOSURES AND SYSTEM COMPONENTS

PLCs are generally placed in a NEMA-12 **panel enclosure** or another type of NEMA enclosure, depending on the application. A panel enclosure holds the PLC hardware, protecting it from environmental hazards. Table 20-1 describes the different types of NEMA enclosures. The enclosure size depends on the total space required. Mounting the controller components in

Courtesy of Siemens, Alpharetta, GA, and Phoenix Contact, Harrisburg, PA

Figure 20-1. Installation of a PLC-based system using modular I/O terminal blocks.

an enclosure is not always required, but it is recommended for most applications to protect the components from atmospheric contaminants, such as conductive dust, moisture, and other corrosive and harmful airborne substances. Metal enclosures also help minimize the effects of electromagnetic radiation, which may be generated by surrounding equipment.

The enclosure layout should conform to NEMA standards, and component placement and wiring should take into consideration the effects of heat, electrical noise, vibration, maintenance, and safety. Figure 20-2 illustrates a typical enclosure layout, which can be used for reference during the following layout guideline discussion.

NEMA Panel Enclosures

Type 1 (Surface mount)
For indoor use to protect against contact with the enclosed equipment in applications where unusual service conditions do not exist

Type 1 (Flush mount)
Used for the same types of applications as Type 1 surface-mounted enclosures in situations where installation in a machine frame or plaster wall is desired

Type 3
For outdoor use to protect against windblown dust, rain, sleet, and external ice formation

Type 3R
For outdoor use to protect against falling rain, sleet, and external ice formation

Type 3R, 7, and 9 (Unilock enclosure for hazardous locations)
Used for the same types of applications as Type 3R, 7, and 9 enclosures but provides a copper-free aluminum, bronze-chromated housing

Type 4
For indoor or outdoor use to protect against windblown dust and rain, splashing water, and hose-directed water

Type 4X (Nonmetallic, corrosion-resistant, fiberglass-reinforced polyester)
For indoor and outdoor use to protect against corrosion, windblown dust and rain, splashing water, and hose-directed water

Type 6P
For indoor and outdoor use to protect against the entry of water during prolonged submersion at a limited depth

Type 7 (Hazardous gas locations bolted enclosure)
For indoor use in applications using hazardous gases; capable of withstanding an internal explosion of specified gases and containing such an explosion to prevent the ignition of the surrounding atmosphere

Type 9 (Hazardous dust locations)
For indoor use in applications where hazardous dust is present; designed to prohibit the entry of dust as well as prevent the ignition of dust by enclosed heat-generating devices

Type 12
For indoor use to protect against dust, falling dirt, and dripping noncorrosive liquids

Type 13
For indoor use to protect against dust, spraying of water, oil, and noncorrosive coolants

Table 20-1. NEMA panel enclosure descriptions.

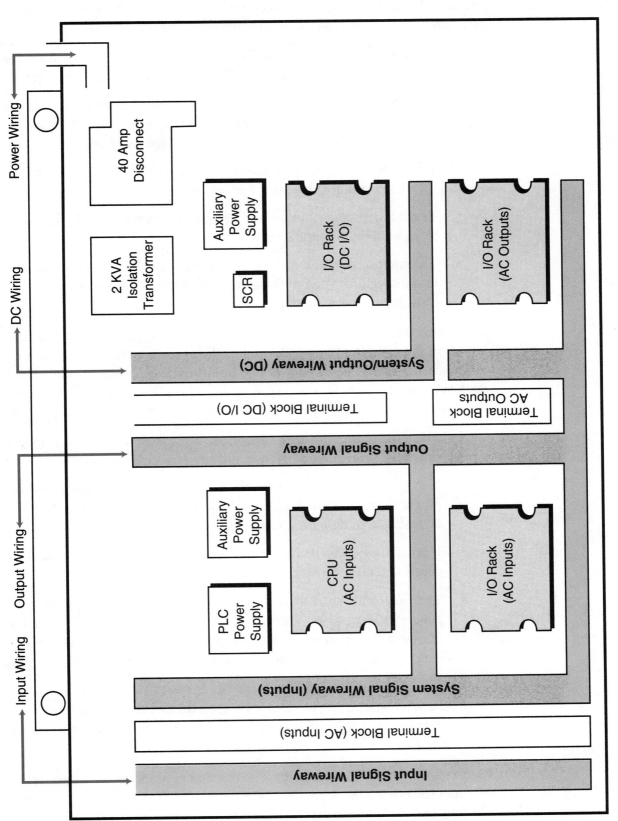

Figure 20-2. Enclosure layout.

General. The following recommendations address preliminary considerations for the location and physical aspects of a PLC enclosure:

- The enclosure should be located so that the doors can fully open for easy access when testing or troubleshooting wiring and components.

- The enclosure depth should provide adequate clearance between the closed enclosure door (including any print pockets mounted on the door) and the enclosed components and related cables.

- The enclosure's back panel should be removable to facilitate mounting of the components and other assemblies.

- The cabinet should contain an emergency disconnect device installed in an easily accessible location.

- The enclosure should include accessories, such as AC power outlets, interior lighting, and a gasketed, clear acrylic viewing window, for installation and maintenance convenience.

Environmental. The effects of temperature, humidity, electrical noise, and vibration are important when designing the system layout. These factors influence the actual placement of the controller, the inside layout of the enclosure, and the need for other special equipment. The following considerations help to ensure favorable environmental conditions for the controller:

- The temperature inside the enclosure must not exceed the maximum operating temperature of the controller (typically 60°C).

- If the environment contains "hot spots," such as those generated by power supplies or other electrical equipment, a fan or blower should be installed to help dissipate the heat.

- If condensation is likely, the enclosure should contain a thermostat-controlled heater.

- The enclosure should be placed well away from equipment that generates excessive electromagnetic interference (EMI) or radio frequency interference (RFI). Examples of such equipment include welding machines, induction heating equipment, and large motor starters.

- In cases where the PLC enclosure must be mounted on the controlled equipment, the vibrations caused by that equipment should not exceed the PLC's vibration specifications.

Placement of PLC Components. The placement of the major components of a specific controller depends on the number of system components and the physical design or modularity of each component (see Figure 20-3).

Figure 20-3. Placement of PLC components.

Although different controllers have different mounting and spacing requirements, the following considerations and precautions apply when placing any PLC inside an enclosure:

- To allow maximum convection cooling, all controller components should be mounted in a vertical (upright) position. Some manufacturers may specify that the controller components can be mounted horizontally. However, in most cases, components mounted horizontally will obstruct air flow.

- The power supply (main or auxiliary) has a higher heat dissipation than any other system component; therefore, it should not be mounted directly underneath any other equipment. The power supply should be installed at the top of the enclosure above all other equipment, with adequate spacing (at least ten inches) between the power supply and the top of the enclosure. The power supply may also be placed adjacent to other components, but with sufficient spacing.

- The CPU should be located at a comfortable working level (e.g., at sitting or standing eye level) that is either adjacent to or below the power supply. If the CPU and power supply are contained in a single PLC unit, then the PLC unit should be placed toward the top of the enclosure with no other components directly above it, unless there is sufficient space.

- Local I/O racks (in the same panel enclosure as the CPU) can be arranged as desired within the distance allowed by the I/O rack interconnection cable. Typically, the racks are located below or adjacent to the CPU, but not directly above the CPU or power supply.

- Remote I/O racks and their auxiliary power supplies are generally placed inside an enclosure at the remote location, following the same placement practices as described for local racks.

- Spacing of the controller components (to allow proper heat dissipation) should adhere to the manufacturer's specifications for vertical and horizontal spacing between major components.

Placement of Other Components. In general, other equipment inside the enclosure should be located away from the controller components, to minimize the effects of noise and heat generated by these devices. The following list outlines some common practices for locating other equipment inside the enclosure:

- Incoming line devices, such as isolation and constant voltage transformers, local power disconnects, and surge suppressors, should be located near the top of the enclosure and beside the power supply. This placement assumes that the incoming power enters at the top of the panel. The proper placement of incoming line devices keeps power wire runs as short as possible, minimizing the transmission of electrical noise to the controller components.

- Magnetic starters, contactors, relays, and other electromechanical components should be mounted near the top of the enclosure in an area segregated from the controller components. A good practice is to place a six-inch barrier between the magnetic area and the controller area. Typically, magnetic components are adjacent and opposite to the power supply and incoming line devices.

- If fans or blowers are used to cool the components inside the enclosure, they should be located close to the heat-generating devices (generally power supply heat sinks). When using fans, outside air should not be brought inside the enclosure unless a fabric or other reliable filter is used. Filtration prevents conductive particles and other harmful contaminants from entering the enclosure.

Grouping Common I/O Modules. The grouping of I/O modules allows signal and power lines to be routed properly through the ducts, thus minimizing crosstalk interference. Following are recommendations concerning the grouping of I/O modules:

- I/O modules should be segregated into groups, such as AC input modules, AC output modules, DC input modules, DC output modules, analog input modules, and analog output modules, whenever possible.

- If possible, a separate I/O rack should be reserved for common input or output modules. If this is not possible, then the modules should be separated as much as possible within the rack. A suitable partitioning would involve placing all AC modules or all DC modules together and, if space permits, allowing an unused slot between the two groups.

Duct and Wiring Layout. The duct and wiring layout defines the physical location of wireways and the routing of field I/O signals, power, and controller interconnections within the enclosure. The enclosure's duct and wiring layout depends on the placement of I/O modules within each I/O rack. The placement of these modules occurs during the design stage, when the I/O assignment takes place. Prior to defining the duct and wiring layout and assigning the I/O, the following guidelines should be considered to minimize electrical noise caused by crosstalk between I/O lines:

- All incoming AC power lines should be kept separate from low-level DC lines, I/O power supply cables, and I/O rack interconnection cables.

- Low-level DC I/O lines, such as TTL and analog, should not be routed in parallel with AC I/O lines in the same duct. Whenever possible, keep AC signals separate from DC signals.

- I/O rack interconnection cables and I/O power cables can be routed together in a common duct not shared by other wiring. Sometimes, this arrangement is impractical or these cables cannot be separated from all other wiring. In this case, the I/O cables can either be routed with low-level DC lines or routed externally to all ducts and held in place using tie wraps or some other fastening method.

- If I/O wiring must cross AC power lines, it should do so only at right angles (see Figure 20-4). This routing practice minimizes the possibility of electrical noise pickup. I/O wiring coming from the conduits should also be at right angles (see Figure 20-5).

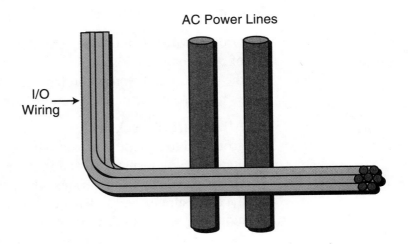

Figure 20-4. I/O wiring must cross AC power lines at a right angle.

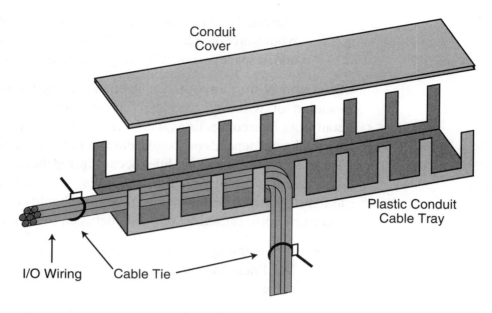

Figure 20-5. I/O wiring from a conduit.

- When designing the duct layout, the separation between the I/O modules and any wire duct should be at least two inches. If terminal strips are used, then the terminal strip and wire duct, as well as the terminal strip and I/O modules, should be at least two inches apart.

Grounding. Proper grounding is an important safety measure in all electrical installations. When installing electrical equipment, users should refer to National Electric Code (NEC) Article 250, which provides data about the size and types of conductors, color codes, and connections necessary for safe grounding of electrical components. The code specifies that a grounding path must be permanent (no solder), continuous, and able to safely conduct the ground-fault current in the system with minimal impedance. The following grounding practices have significant impacts on the reduction of noise caused by electromagnetic induction:

- Ground wires should be separated from the power wiring at the point of entry to the enclosure. To minimize the ground wire length within the enclosure, the ground reference point should be located as close as possible to the point of entry of the plant power supply.

- All electrical racks/chassis and machine elements should be grounded to a central ground bus, normally located in the magnetic area of the enclosure. Paint and other nonconductive materials should be scraped away from the area where the chassis makes contact with the enclosure. In addition to the ground connection made through the mounting bolt or stud, a one-inch metal braid or size #8 AWG wire (or the manufacturer's recommended wire size) should be used to connect each chassis to the enclosure at the mounting bolt or stud.

- The enclosure should be properly grounded to the ground bus, which should have a good electrical connection at the point of contact with the enclosure.

- The machine ground should be connected to the enclosure and to the earth ground.

20-2 POWER REQUIREMENTS AND SAFETY CIRCUITRY

The source for a PLC power supply is generally single-phase and 120 or 240 VAC. If the controller is installed in an enclosure, the two power leads (L1 hot and L2 common) normally enter the enclosure through the top part of the cabinet to minimize interference with other control lines. The power line should be as clean as possible to avoid problems due to line interference in the controller and I/O system.

POWER REQUIREMENTS

Common AC Source. The system power supply and I/O devices should have a common AC source (see Figure 20-6). This minimizes line interference and prevents faulty input signals stemming from a stable AC source to the power supply and CPU, but an unstable AC source to the I/O devices. By keeping both the power supply and the I/O devices on the same power source, the user can take full advantage of the power supply's line monitoring feature.

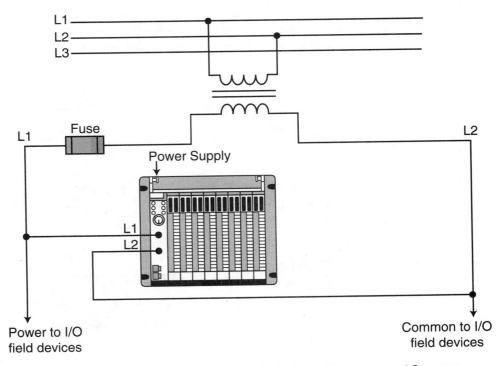

Figure 20-6. System power supply and I/O devices with a common AC source.

If line conditions fall below the minimum operating level, the power supply will detect the abnormal condition and signal the processor, which will stop reading input data and turn off all outputs.

Isolation Transformers. Another good practice is to use an isolation transformer on the AC power line going to the controller. An isolation transformer is especially desirable when heavy equipment is likely to introduce noise into the AC line. An isolation transformer can also serve as a step-down transformer to reduce the incoming line voltage to a desired level. The transformer should have a sufficient power rating (in units of volt-amperes) to supply the load, so users should consult the manufacturer to obtain the recommended transformer rating for their particular application.

SAFETY CIRCUITRY

The PLC system should contain a sufficient number of emergency circuits to either partially or totally stop the operation of the controller or the controlled machine or process (see Figure 20-7). These circuits should be routed outside the controller, so that the user can manually and rapidly shut down the system in the event of total controller failure. Safety devices, like emergency pull rope switches and end-of-travel limit switches, should bypass the controller to operate motor starters, solenoids, and other devices directly. These emergency circuits should use simple logic with a minimum number of highly reliable, preferably electromechanical, components.

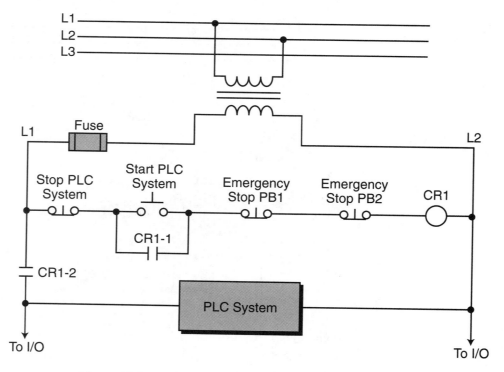

Figure 20-7. Emergency circuits hardwired to the PLC system.

Emergency Stops. The system should have emergency stop circuits for every machine directly controlled by the PLC. To provide maximum safety, these circuits should not be wired to the controller, but instead should be left hardwired. These emergency switches should be placed in locations that the operator can easily access. Emergency stop switches are usually wired into master control relay or safety control relay circuits, which remove power from the I/O system in an emergency.

Master or Safety Control Relays. Master control relay (MCR) and **safety control relay (SCR)** circuits provide an easy way to remove power from the I/O system during an emergency situation (see Figure 20-8). These control relay circuits can be de-energized by pushing any emergency stop

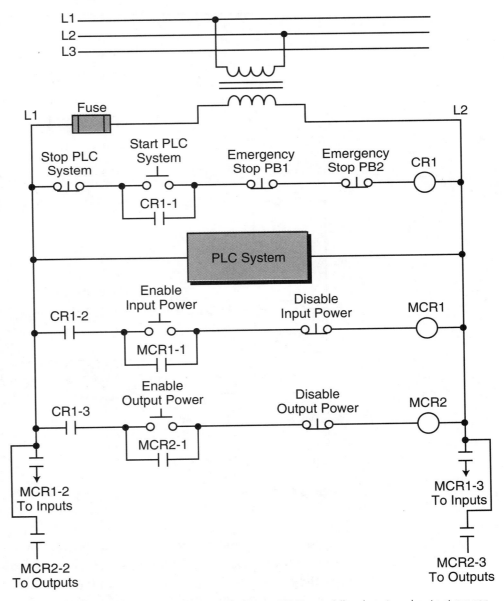

Figure 20-8. Master start control for a PLC with MCRs enabling input and output power.

switch connected to the circuit. De-energizing the control relay coil removes power to the input and output devices. The CPU, however, continues to receive power and operate even though all of its inputs and outputs are disabled.

An MCR circuit may be extended by placing a PLC fault relay (closed during normal PLC operation) in series with any other emergency stop condition. This enhancement will cause the MCR circuit to cut the I/O power in the case of a PLC failure (memory error, I/O communications error, etc.). Figure 20-9 illustrates the typical wiring of a master control relay circuit.

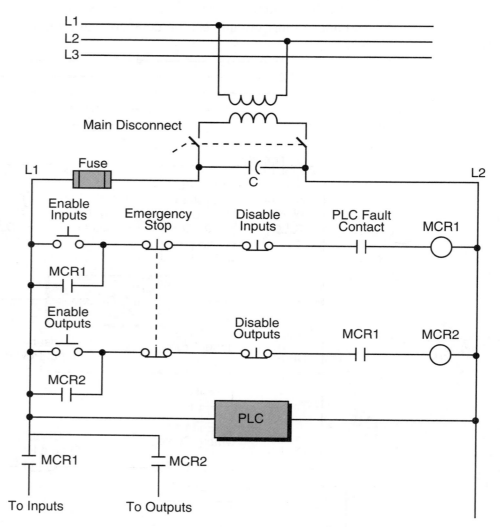

Figure 20-9. Circuit that enables/disables I/O power through MCRs and PLC fault contact detection.

Emergency Power Disconnect. The power circuit feeding the power supply should use a properly rated emergency power disconnect, thus providing a way to remove power from the entire programmable controller system (refer to Figure 20-9). Sometimes, a capacitor (0.47 µF for 120 VAC,

0.22 µF for 220 VAC) is placed across the disconnect to protect against an *outrush* condition. Outrush occurs when the power disconnect turns off the output triacs, causing the energy stored in the inductive loads to seek the nearest path to ground, which is often through the triacs.

20-3 NOISE, HEAT, AND VOLTAGE REQUIREMENTS

Implementation of the previously outlined recommendations should provide favorable operating conditions for most programmable controller applications. However, in certain applications, the operating environment may have extreme conditions that require special attention. These adverse conditions include excessive noise and heat and nuisance line fluctuations. This section describes these conditions and provide measures to minimize their effects.

Excessive Noise. Electrical noise seldom damages PLC components, unless extremely high energy or high voltage levels are present. However, temporary malfunctions due to noise can result in hazardous machine operation in certain applications. Noise may be present only at certain times, or it may appear at widespread intervals. In some cases, it may exist continuously. The first case is the most difficult to isolate and correct.

Noise usually enters a system through input, output, and power supply lines. Noise may also be coupled into these lines electrostatically through the capacitance between them and the noise signal carrier lines. The presence of high-voltage or long, closely spaced conductors generally produces this effect. The coupling of magnetic fields can also occur when control lines are located close to lines carrying large currents. Devices that are potential noise generators include relays, solenoids, motors, and motor starters, especially when operated by hard contacts, such as push buttons and selector switches.

Analog I/O and transmitters are very susceptible to noise from electromechanical sources, causing jumps in counts during the reading of analog data. Therefore, motor starters, transformers, and other electromechanical devices should be kept away from analog signals, interfaces, and transmitters.

Although the design of solid-state controls provides a reasonable amount of noise immunity, the designer must still take special precautions to minimize noise, especially when the anticipated noise signal has characteristics similar to the desired control input signals. To increase the operating noise margin, the controller must be installed away from noise-generating devices, such as large AC motors and high-frequency welding machines. Also, all inductive loads must be suppressed. Three-phase motor leads should be grouped together and routed separately from low-level signal leads. Sometimes, if the noise level situation is critical, all three-phase motor leads must be suppressed (see Figure 20-10). Figure 20-11 illustrates line-filtering configurations used for removing input power noise to a controller or transmitter.

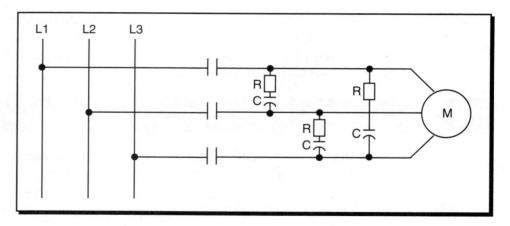

Figure 20-10. Suppression of a three-phase motor lead.

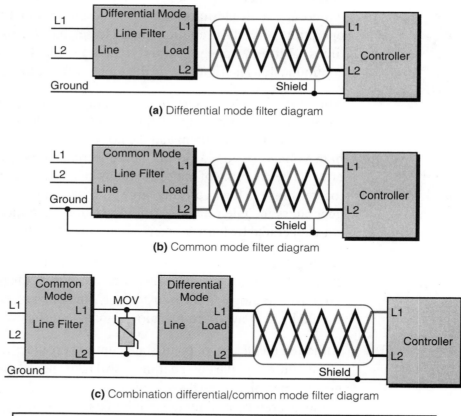

(a) Differential mode filter diagram

(b) Common mode filter diagram

(c) Combination differential/common mode filter diagram

Note 1: Keep line filters 12 inches or less from the controller. Minimize the line distance where noise can be introduced into the controller.

Note 2: To prevent ground loops, do not tie the common mode line metal case filters with other metal that is at ground potential. Doing so will reduce the filters' effectiveness.

Courtesy of Watlow Electric Co., St. Louis, MO

Figure 20-11. Power noise reduction using one of three line-filtering configurations.

Excessive Heat. Programmable controllers can withstand temperatures ranging from 0 to 60°C. They are normally cooled by *convection*, meaning that a vertical column of air, drawn in an upward direction over the surface of the components, cools the PLC. To keep the temperature within limits, the cooling air at the base of the system must not exceed 60°C.

The PLC components must be properly spaced when they are installed to avoid excess heat. The manufacturer can provide spacing recommendations, which are based on typical conditions for most PLC applications. Typical conditions are as follows:

- 60% of the inputs are ON at any one time

- 30% of the outputs are ON at any one time

- the current supplied by all of the modules combined meets manufacturer-provided specifications

- the air temperature is around 40°C

Situations in which most of the I/O are ON at the same time and the air temperature is higher than 40°C are not typical. In these situations, spacing between components must be larger to provide better convection cooling. If equipment inside or outside of the enclosure generates substantial amounts of heat and the I/O system is ON continuously, the enclosure should contain a fan that will reduce hot spots near the PLC system by providing good air circulation. The air being brought in by the fan should first pass through a filter to prevent dirt or other contaminants from entering the enclosure. Dust obstructs the components' heat dissipation capacity, as well as harms heat sinks when thermal conductivity to the surrounding air is lowered. In cases of extreme heat, the enclosure should be fitted with an air-conditioning unit or cooling control system that utilizes compressed air (see Figures 20-12 and 20-13). Leaving enclosure doors open to cool off the system is not a good practice, since this allows conductive dust to enter the system.

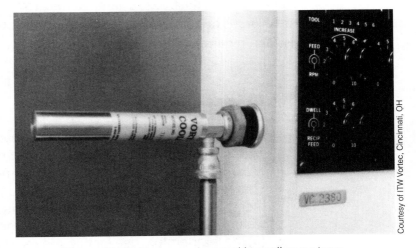

Figure 20-12. Vortex cooler used in cooling systems.

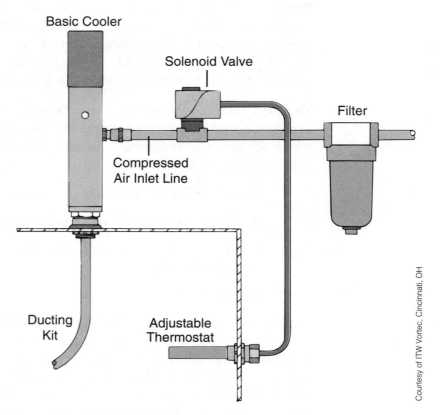

Basic Cooler

Solenoid Valve

Filter

Compressed
Air Inlet Line

Ducting
Kit

Adjustable
Thermostat

Courtesy of ITW Vortec, Cincinnati, OH

Figure 20-13. Compressed air cooling system.

There are methods available to calculate the temperature rise and heat dissipation requirements of an enclosure based on its size and equipment contents. Temperature rise is the temperature difference between the air inside an enclosure and the outside air temperature (ambient air temperature). Hoffman Engineering Co., a manufacturer of control system enclosures, has developed temperature rise graphs for use with their panels and enclosures. Figure 20-14 illustrates a temperature rise graph for a NEMA 12–type enclosure. The following example illustrates how to calculate temperature rise and required airflow using the graph.

EXAMPLE 20-1

The NEMA 12 enclosure shown in Figure 20-15 contains a programmable controller with a power supply transformer, power supplies for an analog transmitter and other equipment, and various electromechanical equipment. The combined power dissipation of the equipment, found by adding each element's power dissipation, is 1011 watts. The ambient temperature of the enclosure is 90°F (32.2°C). Find **(a)** the temperature rise for this enclosure and **(b)** the required airflow.

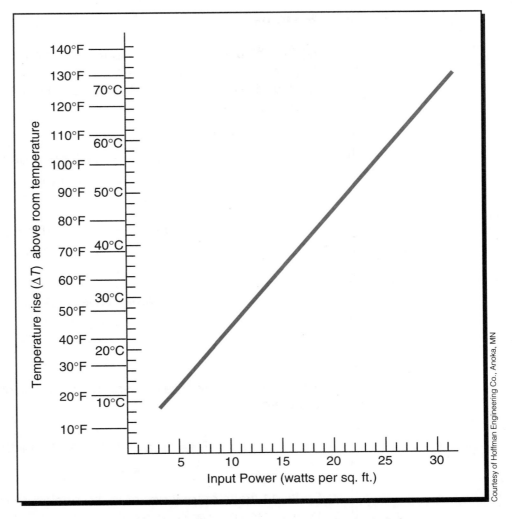

Figure 20-14. Temperature rise graph for a NEMA 12 enclosure.

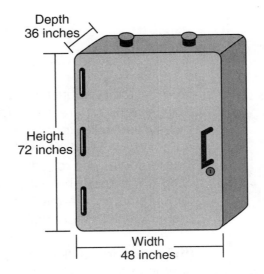

Figure 20-15. NEMA 12 enclosure.

SOLUTION

(a) To calculate the temperature rise, first calculate the total area (square feet) of the exposed sides of the enclosure. Assuming that the back and bottom sides of the enclosure are not exposed, the area of each exposed side equals:

$$
\begin{aligned}
\text{Front area} &= \text{(Height)(Width)} \\
&= \text{(6 ft)(4 ft)} \\
&= 24 \text{ ft}^2
\end{aligned}
$$

$$
\begin{aligned}
\text{Side area} &= \text{(Height)(Depth)} \\
&= \text{(6 ft)(3 ft)} \\
&= 18 \text{ ft}^2
\end{aligned}
$$

$$
\begin{aligned}
\text{Top area} &= \text{(Depth)(Width)} \\
&= \text{(3 ft)(4 ft)} \\
&= 12 \text{ ft}^2
\end{aligned}
$$

Therefore, the total area for heat dissipation, taking into account that there are two sides, is:

$$
\begin{aligned}
\text{Total area} &= 24 \text{ ft}^2 + 2(18 \text{ ft}^2) + 12 \text{ ft}^2 \\
&= 72 \text{ ft}^2
\end{aligned}
$$

So, 1011 watts of total power in the enclosure is distributed over a total surface area of 72 ft², resulting in a power dissipation per square foot of 14.04 watts:

$$
\begin{aligned}
\text{Power dissipation} &= \frac{1011 \text{ watts}}{72 \text{ ft}^2} \\
&= 14.04 \text{ watts/ft}^2
\end{aligned}
$$

From the temperature rise curve for a NEMA 12 enclosure, we can find that the temperature rise is approximately 32°C or 57.5°F. Therefore, this system will experience a final temperature (ambient + rise) of approximately 64.2°C (32.2°C + 32°C) or 147.5°F (90°F + 57.5°F). This temperature exceeds the PLC's maximum operating temperature of 60°C, meaning that a malfunction could occur because of the high temperature inside the enclosure. This system, therefore, requires proper ventilation or cooling.

(b) The required airflow inside the enclosure is based on the maximum operating temperature of the components (e.g., 60°C for a PLC).

Assuming that all inside components can withstand up to 60°C (140°F), the permissible temperature rise (ΔT) in °F of the cooling air is:

$$\Delta T = \text{Max temp of enclosure} - \text{Max temp of components}$$
$$= 179.6°F - 140°F$$
$$= 39.6°F$$

The required airflow Q_{air} is given by the equation:

$$Q_{air} = \frac{(3160)(\text{KW of enclosure})}{\Delta T}$$

where the term 3160 is a constant, KW is the kilowatt heat of the enclosure (in this case 1.011 KW) and ΔT is the permissible temperature. Therefore, the airflow requirement is:

$$Q_{air} = \frac{(3160)(1.011)}{39.6}$$
$$= 80.68 \text{ ft}^3/\text{min}$$

Thus, a minimum airflow of 80.68 ft³/min is required to dissipate the heat in the enclosure.

Excessive Line Voltage Variation. The power supply section of a PLC system can sustain line fluctuations and still allow the system to function within its operating margin. As long as the incoming voltage is adequate, the power supply provides all the logic voltages necessary to support the processor, memory, and I/O. However, if the voltage drops below the minimum acceptable level, the power supply will alert the processor, which will then execute a system shutdown.

In applications that are subject to "soft" AC lines and unusual line variations, the first step towards a solution is to correct any possible feeder problem in the distribution system. If this correction does not solve the problem, then a constant voltage transformer can be used to prevent the system from shutting down too often (see Figure 20-16). The constant voltage transformer stabilizes the input voltage to the power supply and input field devices by compensating for voltage changes at the primary to maintain a steady voltage in the secondary. When using a constant voltage transformer, the user should check that its power rating is sufficient to supply the input devices and the power supply. Also, the user should connect the output devices in front of the constant voltage transformer, rather than behind it, so that the transformer is not providing power to the outputs. This arrangement will lessen the load supported by the transformer, allowing a smaller transformer to be used. The manufacturer can provide information regarding power rating requirements.

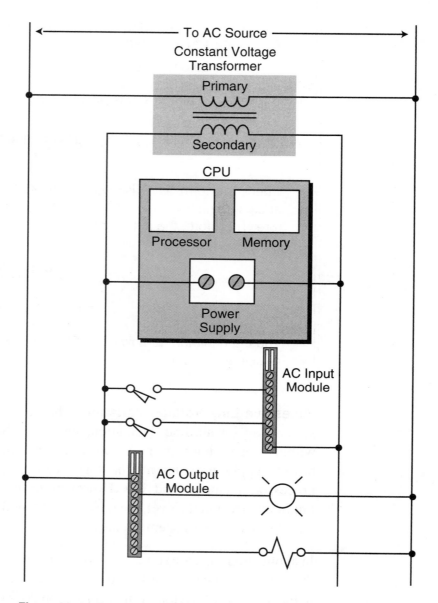

Figure 20-16. Constant voltage transformer used to stabilize input voltage.

20-4 I/O INSTALLATION, WIRING, AND PRECAUTIONS

Input/output installation is perhaps the biggest and most critical job when installing a programmable controller system. To minimize errors and simplify installation, the user should follow predefined guidelines. All of the people involved in installing the controller should receive these I/O system installation guidelines, which should have been prepared during the design phase. A complete set of documents with precise information regarding I/O placement and connections will ensure that the system is organized properly. Furthermore, these documents should be constantly updated during every stage of the installation. The following considerations will facilitate an orderly installation.

I/O Module Installation

Placement and installation of the I/O modules is simply a matter of inserting the correct modules in their proper locations. This procedure involves verifying the type of module (115 VAC output, 115 VDC input, etc.) and the slot address as defined by the I/O address assignment document. Each terminal in the module is then wired to the field devices that have been assigned to that termination address. The user should remove power to the modules (or rack) before installing and wiring any module.

Wiring Considerations

Wire Size. Each I/O terminal can accept one or more conductors of a particular wire size. The user should check that the wire is the correct gauge and that it is the proper size to handle the maximum possible current.

Wire and Terminal Labeling. Each field wire and its termination point should be labeled using a reliable labeling method. Wires should be labeled with shrink-tubing or tape, while tape or stick-on labels should identify each terminal block. Color coding of similar signal characteristics (e.g., AC: red, DC: blue, common: white, etc.) can be used in addition to wire labeling. Typical labeling nomenclature includes wire numbers, device names or numbers, and the input or output address assignment. Good wire and terminal identification simplifies maintenance and troubleshooting.

Wire Bundling. Wire bundling is a technique commonly used to simplify the connections to each I/O module. In this method, the wires that will be connected to a single module are bundled, generally using a tie wrap, and then routed through the duct with other bundles of wire with the same signal characteristics. Input, power, and output bundles carrying the same type of signals should be kept in separate ducts, when possible, to avoid interference.

Wiring Procedures

Once the I/O modules are in place and their wires have been bundled, the wiring to the modules can begin. The following are recommended procedures for I/O wiring:

- Remove and lock out input power from the controller and I/O before any installation and wiring begins.

- Verify that all modules are in the correct slots. Check module type and model number by inspection and on the I/O wiring diagram. Check the slot location according to the I/O address assignment document.

- Loosen all terminal screws on each I/O module.

- Locate the wire bundle corresponding to each module and route it through the duct to the module location. Identify each of the wires in the bundle and check that they correspond to that particular module.

- Starting with the first module, locate the wire in the bundle that connects to the lowest terminal. At the point where the wire is at a vertical height equal to the termination point, bend the wire at a right angle towards the terminal.

- Cut the wire to a length that extends 1/4 inch past the edge of the terminal screw. Strip approximately 3/8 inch of insulation from the end of the wire. Insert the uninsulated end of the wire under the pressure plate of the terminal and tighten the screw.

- If two or more modules share the same power source, jumper the power wiring from one module to the next.

- If shielded cable is being used, connect only one end to ground, preferably at the rack chassis. This connection will avoid possible **ground loops**. A ground loop condition exists when two or more electrical paths are created in a ground line or when one or more paths are created in a shield (Section 20-7 explains how to identify a ground loop). Leave the other end cut back and unconnected, unless otherwise specified.

- Repeat the wiring procedure for each wire in the bundle until the module wiring is complete.

- After all of the wires are terminated, check for good terminations by gently pulling on each wire.

SPECIAL I/O CONNECTION PRECAUTIONS

Chapters 6, 7, and 8 presented typical connection diagrams for the various types of I/O modules. Certain field device wiring connections, however, may need special attention. These connections include leaky inputs, inductive loads, output fusing, and shielded cable.

Connecting Leaky Inputs. Some field devices have a small leakage current even when they are in the OFF state. Both triac and transistor outputs exhibit this leakage characteristic, although transistor leakage current is much lower. Most of the time, the leaky input will only cause the module's input indicator to flicker; but sometimes, the leakage can falsely trigger an input circuit, resulting in misoperation. A typical device that exhibits this leakage situation is a proximity switch. This type of leakage may also occur when an output module drives an input module when there is no other load.

Figure 20-17 illustrates two leakage situations, along with their corrective actions. A leaky input can be corrected by placing a bleeding (or loading) resistor across the input. A bleeding resistor introduces resistance into the circuit, causing the voltage to drop on the line between the leaky field device

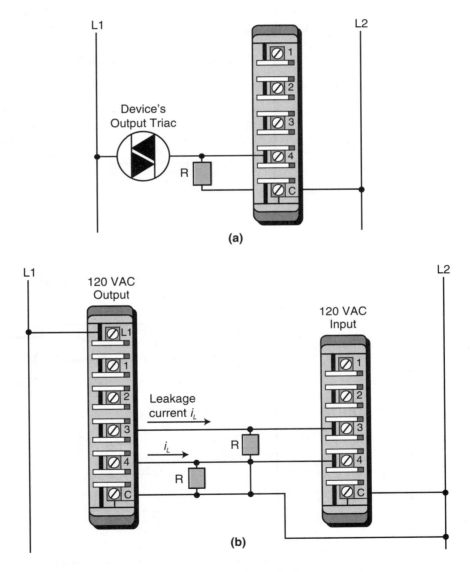

Figure 20-17. (a) A connection for a leaky input device and **(b)** the connection of an output module to an input module.

and the input circuit. This causes a shunt on the input's terminals. Consequently, the leakage current is routed through the bleeding resistor, minimizing the amount of current to the input module (or to the output device). This prevents the input or output from turning ON when it should be OFF.

Suppression of Inductive Loads. The interruption of current caused by turning an inductive load's output OFF generates a very high voltage spike. These spikes, which can reach several thousands of volts if not suppressed, can occur either across the leads that feed power to the device or between both power leads and the chassis ground, depending on the physical construction of the device. This high voltage causes erratic operation and, in some cases, may damage the output module. To avoid this situation, a snubber circuit, typically a resistor/capacitor network (RC) or metal oxide varistor (MOV), should be installed to limit the voltage spike, as well as control the rate of current change through the inductor (see Figure 20-18).

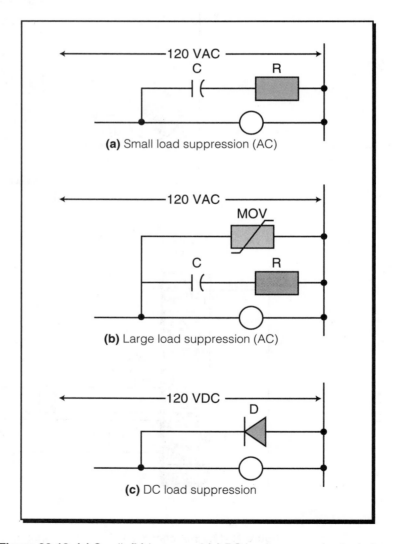

Figure 20-18. (a) Small, (b) large, and (c) DC load suppression techniques.

Most output modules are designed to drive inductive loads, so they typically include suppression networks. Nevertheless, under certain loading conditions, the triac may be unable to turn OFF as current passes through zero (commutation), thus requiring additional external suppression in the system.

An RC snubber circuit placed across the device can provide additional suppression for small AC devices, such as solenoids, relays, and motor starters up to size 1. Larger contactors (size 2 and above) require an MOV in addition to the RC network. A free-wheeling diode placed across the load can provide DC suppression. Figure 20-19 presents several examples of inductive load suppression.

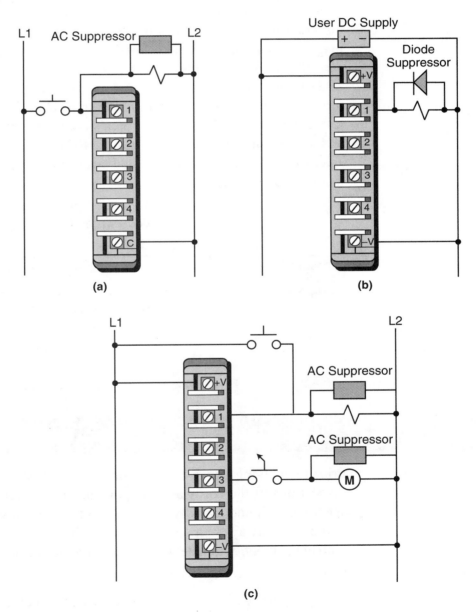

Figure 20-19. Suppression of **(a)** a load in parallel with a PLC input module, **(b)** a DC load, and **(c)** loads with switches in parallel and series with a PLC output module.

Fusing Outputs. Solid-state outputs normally have fusing on the module, to protect the triac or transistor from moderate overloads. If the output does not have internal fuses, then fuses should be installed externally (normally at the terminal block) during the initial installation. When adding fuses to an output circuit, the user should adhere to the manufacturer's specifications for the particular module. Only a properly rated fuse will ensure that the fuse will open quickly in an overload condition to avoid overheating of the output switching device.

Shielding. Control lines, such as TTL, analog, thermocouple, and other low-level signals, are normally routed in a separate wireway, to reduce the effects of signal coupling. For further protection, shielded cable should be used for the control lines, to protect the low-level signals from electrostatic and magnetic coupling with both lines carrying 60 Hz power and other lines carrying rapidly changing currents. The twisted, shielded cable should have at least a one-inch lay, or approximately twelve twists per foot, and should be protected on both ends by shrink-tubing or a similar material. The shield should be connected to control ground at only one point (see Figure 20-20), and shield continuity must be maintained for the entire length of the cable. The shielded cable should also be routed away from high noise areas, as well as insulated over its entire length.

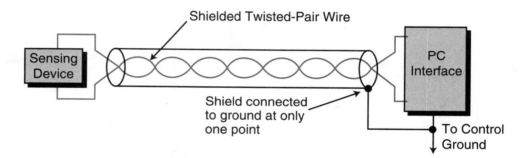

Figure 20-20. Shielded cable ground connection.

20-5 PLC START-UP AND CHECKING PROCEDURES

Prior to applying power to the system, the user should make several final inspections of the hardware components and interconnections. These inspections will undoubtedly require extra time. However, this invested time will almost always reduce total start-up time, especially for large systems with many input/output devices. The following checklist pertains to prestart-up procedures:

• Visually inspect the system to ensure that all PLC hardware components are present. Verify correct model numbers for each component.

- Inspect all CPU components and I/O modules to ensure that they are installed in the correct slot locations and placed securely in position.

- Check that the incoming power is correctly wired to the power supply (and transformer) and that the system power is properly routed and connected to each I/O rack.

- Verify that the I/O communication cables linking the processor to the individual I/O racks correspond to the I/O rack address assignment.

- Verify that all I/O wiring connections at the controller end are in place and securely terminated. Use the I/O address assignment document to verify that each wire is terminated at the correct point.

- Check that the output wiring connections are in place and properly terminated at the field device end.

- Ensure that the system memory has been cleared of previously stored control programs. If the control program is stored in EPROM, remove the chips temporarily.

STATIC INPUT WIRING CHECK

A **static input wiring check** should be performed with power applied to the controller and input devices. This check will verify that each input device is connected to the correct input terminal and that the input modules or points are functioning properly. Since this test is performed before other system tests, it will also verify that the processor and the programming device are in good working condition. Proper input wiring can be verified using the following procedures:

- Place the controller in a mode that will inhibit the PLC from any automatic operation. This mode will vary depending on the PLC model, but it is typically *stop, disable, program,* etc.

- Apply power to the system power supply and input devices. Verify that all system diagnostic indicators show proper operation. Typical indicators are *AC OK, DC OK, processor OK, memory OK,* and *I/O communication OK.*

- Verify that the emergency stop circuit will de-energize power to the I/O devices.

- Manually activate each input device. Monitor the corresponding LED status indicator on the input module and/or monitor the same address on the programming device, if used. If properly wired, the indicator will turn ON. If an indicator other than the expected one turns ON when the input device is activated, the input device may be wired to the wrong input terminal. If no indicator turns ON, then a fault may exist in either the input device, field wiring, or input module (see Section 20-4).

- Take precautions to avoid injury or damage when activating input devices that are connected in series with loads that are external to the PLC.

STATIC OUTPUT WIRING CHECK

A **static output wiring check** should be performed with power applied to the controller and the output devices. A safe practice is to first locally disconnect all output devices that involve mechanical motion (e.g., motors, solenoids, etc.). When performed, the static output wiring check will verify that each output device is connected to the correct terminal address and that the device and output module are functioning properly. The following procedures should be used to verify output wiring:

- Locally disconnect all output devices that will cause mechanical motion.

- Apply power to the controller and to the input/output devices. If an emergency stop can remove power to the outputs, verify that the circuit does remove power when activated.

- Perform the static check of the outputs one at a time. If the output is a motor or another device that has been locally disconnected, reapply power to that device only prior to checking. The output operation check can be performed using *one* of the following methods:

  - Assuming that the controller has a forcing function, test each output, with the use of the programming device, by forcing the output ON and setting the corresponding terminal address (point) to 1. If properly wired, the corresponding LED indicator will turn ON and the device will energize. If an indicator other than the expected one turns ON when the terminal address is forced, then the output device may be wired to the wrong output terminal (Inadvertent machine operation does not occur because rotating and other motion-producing outputs are disconnected). If no indicator turns ON, then a fault may exist in either the output device, field wiring, or output module (see Section 20-4).

  - Program a dummy rung, which can be used repeatedly for testing each output, by programming a single rung with a single normally open contact (e.g., a conveniently located push button) controlling the output. Place the CPU in either the RUN, single-scan, or a similar mode, depending on the controller. With the controller in the RUN mode, depress the push button to perform the test. With the controller in single-scan mode, depress and maintain the push button while the controller executes the single-scan. Observe the output device and LED indicator, as described in the first procedure.

CONTROL PROGRAM REVIEW

The **control program checkout** is simply a final review of the control program. This check can be performed at any time, but it should be done prior to loading the program into memory for the dynamic system checkout.

A complete documentation package that relates the control program to the actual field devices is required to perform the control program checkout. Documents, such as address assignments and wiring diagrams, should reflect any modifications that may have occurred during the static wiring checks. When performed, this final program review will verify that the final hardcopy of the program, which will be loaded into memory, is either free of error or at least agrees with the original design documents. The following is a checklist for the final control program checkout:

- Using the I/O wiring document printout, verify that every controlled output device has a programmed output rung of the same address.

- Inspect the hardcopy printout for errors that may have occurred while entering the program. Verify that all program contacts and internal outputs have valid address assignments.

- Verify that all timer, counter, and other preset values are correct.

DYNAMIC SYSTEM CHECKOUT

The **dynamic system checkout** is a procedure that verifies the logic of the control program to ensure correct operation of the outputs. This checkout assumes that all static checks have been performed, the wiring is correct, the hardware components are operational and functioning correctly, and the software has been thoroughly reviewed.

During the dynamic checkout, it is safe to gradually bring the system under full automatic control. Although small systems may be started all at once, a large system should be started in sections. Large systems generally use remote subsystems that control different sections of the machine or process. Bringing one subsystem at a time on-line allows the total system to start up with maximum safety and efficiency. Remote subsystems can be temporarily disabled either by locally removing their power or by disconnecting their communications link with the CPU. The following practices outline procedures for the dynamic system checkout:

- Load the control program into the PLC memory.

- Test the control logic using *one* of the following methods:

  - Switch the controller to the TEST mode, if available, which will allow the execution and debugging of the control program while the outputs are disabled. Check each rung by observing the status of the output LED indicators or by monitoring the corresponding output rung on the programming device.

  - If the controller must be in the RUN mode to update outputs during the tests, locally disconnect the outputs that are not being tested, to avoid damage or harm. If an MCR or similar instruction is available, use it to bypass execution of the outputs that are not being tested, so that disconnection of the output devices is not necessary.

- Check each rung for correct logic operation, and modify the logic if necessary. A useful tool for debugging the control logic is the single scan. This procedure allows the user to observe each rung as every scan is executed.

- When the tests indicate that all of the logic properly controls the outputs, remove all of the temporary rungs that may have been used (MCRs, etc.). Place the controller in the RUN mode, and test the total system operation. If all procedures are correct, the full automatic control should operate smoothly.

- Immediately document all modifications to the control logic, and revise the original documentation. Obtain a reproducible copy (e.g., 3.5" disk, etc.) of the program as soon as possible.

The start-up recommendations and practices presented in this section are good procedures that will aid in the safe, orderly start-up of any programmable control system. However, some controllers may have specific start-up requirements, which are outlined in the manufacturer's product manual. The user should be aware of these specific requirements before starting up the controller.

20-6 PLC SYSTEM MAINTENANCE

Programmable controllers are designed to be easy to maintain, to ensure trouble-free operation. Still, several maintenance aspects should be considered once the system is in place and operational. Certain maintenance measures, if performed periodically, will minimize the chance of system malfunction. This section outlines some of the practices that should be followed to keep the system in good operating condition.

PREVENTIVE MAINTENANCE

Preventive maintenance of programmable controller systems includes only a few basic procedures, which will greatly reduce the failure rate of system components. Preventive maintenance for the PLC system should be scheduled with the regular machine or equipment maintenance, so that the equipment and controller are down for a minimum amount of time. However, the schedule for PLC preventive maintenance depends on the controller's environment—the harsher the environment, the more frequent the maintenance. The following are guidelines for preventive measures:

- Periodically clean or replace any filters that have been installed in enclosures at a frequency dependent on the amount of dust in the area. Do not wait until the scheduled machine maintenance to check the filter. This practice will ensure that clean air circulation is present inside the enclosure.

- Do not allow dirt and dust to accumulate on the PLC's components; the central processing unit and I/O system are not designed to be dust proof. If dust builds up on heat sinks and electronic circuitry, it can obstruct heat dissipation, causing circuit malfunction. Furthermore, if conductive dust reaches the electronic boards, it can cause a short circuit, resulting in possible permanent damage to the circuit board.

- Periodically check the connections to the I/O modules to ensure that all plugs, sockets, terminal strips, and modules have good connections. Also, check that the module is securely installed. Perform this type of check more often when the PLC system is located in an area that experiences constant vibrations, which could loosen terminal connections.

- Ensure that heavy, noise-generating equipment is not located too close to the PLC.

- Make sure that unnecessary items are kept away from the equipment inside the enclosure. Leaving items, such as drawings, installation manuals, or other materials, on top of the CPU rack or other rack enclosures can obstruct the airflow and create hot spots, which can cause system malfunction.

- If the PLC system enclosure is in an environment that exhibits vibration, install a vibration detector that can interface with the PLC as a preventive measure. This way, the programmable controller can monitor high levels of vibration, which can lead to the loosening of connections.

SPARE PARTS

It is a good idea to keep a stock of replacement parts on hand. This practice will minimize downtime resulting from component failure. In a failure situation, having the right spare in stock can mean a shutdown of only minutes, instead of hours or days. As a rule of thumb, the amount of a spare part stocked should be 10% of the number of that part used. If a part is used infrequently, then less than 10% of that particular part can be stocked.

Main CPU board components should have one spare each, regardless of how many CPUs are being used. Each power supply, whether main or auxiliary, should also have a backup. Certain applications may require a complete CPU rack as a standby spare. This extreme case exists when a downed system must be brought into operation immediately, leaving no time to determine which CPU board has failed.

REPLACEMENT OF I/O MODULES

If a module must be replaced, the user should make sure that the replacement module being installed is the correct type. Some I/O systems allow modules to be replaced while power is still applied, but others may require that power be removed. If replacing a module solves the problem, but the failure reoccurs in a relatively short period, the user should check the inductive loads. The inductive loads may be generating voltage and current spikes, in which case, external suppression may be necessary. If the module's fuse blows again after it is replaced, the problem may be that the module's output current limit is being exceeded or that the output device is shorted.

20-7 TROUBLESHOOTING THE PLC SYSTEM

TROUBLESHOOTING GROUND LOOPS

As mentioned earlier, a ground loop condition occurs when two or more electrical paths exist in a ground line. For example, in Figure 20-21, the transducers and transmitter are connected to ground at the chassis (or device enclosure) and connected to an analog input card through a shielded cable. The shield connects to both chassis grounds, thereby creating a path for current to flow from one ground to another since both grounds have different potentials. The current flowing through the shield could be as high as several amperes, which would induce significant magnetic fields in the signal transmission. This could create interference that would result in a possible misreading of the analog signal. To avoid this problem, the shield should be connected to ground on only one side of the chassis, preferably the PLC side. In the example shown in Figure 20-21, the shield should only be connected to ground at the analog input interface.

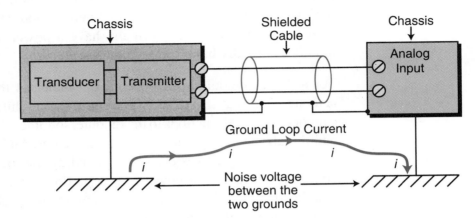

Figure 20-21. Ground loop created by shielded cable grounded at both ends.

To check for a ground loop, disconnect the ground wire at the ground termination and measure the resistance from the wire to the termination point where it is connected (see Figure 20-22). The meter should read a large ohm value. If a low ohm value occurs across this gap, circuit continuity exists, meaning that the system has at least one ground loop.

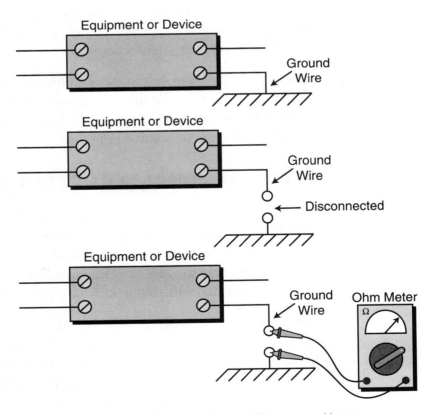

Figure 20-22. Procedure for identifying ground loops.

DIAGNOSTIC INDICATORS

LED status indicators can provide much information about field devices, wiring, and I/O modules. Most input/output modules have at least a single indicator—input modules normally have a power indicator, while output modules normally have a logic indicator.

For an input module, a lit power LED indicates that the input device is activated and that its signal is present at the module. This indicator alone cannot isolate malfunctions to the module, so some manufacturers provide an additional diagnostic indicator, a logic indicator. An ON logic LED indicates that the input signal has been recognized by the logic section of the input circuit. If the logic and power indicators do not match, then the module is unable to transfer the incoming signal to the processor correctly. This indicates a module malfunction.

An output module's logic indicator functions similarly to an input module's logic indicator. When it is ON, the logic LED indicates that the module's logic circuitry has recognized a command from the processor to turn ON. In addition to the logic indicator, some output modules incorporate either a blown fuse indicator or a power indicator or both. A blown fuse indicator indicates the status of the protective fuse in the output circuit, while a power indicator shows that power is being applied to the load. Like the power and logic indicators in an input module, if both LEDs are not ON simultaneously, the output module is malfunctioning.

LED indicators greatly assist the troubleshooting process. With both power and logic indicators, the user can immediately pinpoint a malfunctioning module or circuit. LED indicators, however, cannot diagnose all possible problems; instead, they serve as preliminary signs of system malfunctions.

TROUBLESHOOTING PLC INPUTS

If the field device connected to an input module does not seem to turn ON, a problem may exist somewhere between the L1 connection and the terminal connection to the module. An input module's status indicators can provide information about the field device, the module, and the field device's wiring to the module that will help pinpoint the problem.

The first step in diagnosing the problem is to place the PLC in standby mode, so that it is not activating the output. This allows the field device to be manually activated (e.g., a limit switch can be manually closed). When the field device is activated, the module's power status indicator should turn ON, indicating that power continuity exists. If the indicator is ON, then wiring is not the cause of the problem.

The next step is to evaluate the PLC's reading of the input module. This can be accomplished using the PLC's test mode, which reads the inputs and executes the program but does not activate the outputs. In this mode, the PLC's display should either show a 1 in the image table bit corresponding to the activated field device or the contact's reference instruction should become highlighted when the device provides continuity (see Figure 20-23). If the PLC is reading the device correctly, then the problem is not located in the input module. If it does not read the device correctly, then the module could be faulty. The logic side of the module may not be operating correctly, or its optical isolator may be blown. Moreover, one of the module's interfacing channels could be faulty. In this case, the module must be replaced.

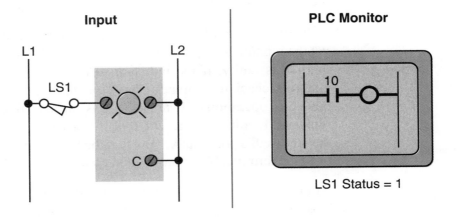

Figure 20-23. Highlighted contact indicating power continuity.

If the module does not read the field device's signal, then further tests are required. Bad wiring, a faulty field device, a faulty module, or an improper voltage between the field device and the module could be causing the problem. First, close the field device and measure the voltage to the input module. The meter should display the voltage of the signal (e.g., 120 volts AC). If the proper voltage is present, the input module is faulty because it is not recognizing the signal. If the measured voltage is 10–15% below the proper signal voltage, then the problem lies in the source voltage to the field device. If no voltage is present, then either the wiring or the field device is the cause of the problem. Check the wiring connection to the module to ensure that the wire is secured at the terminal or terminal blocks.

To further pinpoint the problem, check that voltage is present at the field device. With the device activated, measure the voltage across the device using a voltmeter. If no voltage is present on the load side of the device (the side that connects to the module), then the input device is faulty. If there is power, then the problem lies in the wiring from the input device to the module. In this case, the wiring must be traced to find the problem.

TROUBLESHOOTING PLC OUTPUTS

PLC output interfaces also contain status indicators that provide useful troubleshooting information. Like the troubleshooting of PLC inputs, the first step in troubleshooting outputs is to isolate the problem to either the module, the field device, or the wiring.

At the output module, ensure that the source power for switching the output is at the correct level. In a 120 VAC system, this value should be within 10% of the rated value (i.e., between 108 and 132 volts AC). Also, examine the output module to see if it has a blown fuse. If it does have a blown fuse, check the fuse's rated value. Furthermore, check the output device's current requirements to determine if the device is pulling too much current.

If the output module receives the command to turn ON from the processor yet the module's output status does not turn ON accordingly, then the output module is faulty. If the indicator turns ON but the field device does not energize, check for voltage at the output terminal to ensure that the switching device is operational. If no voltage is present, then the module should be replaced. If voltage is present, then the problem lies in the wiring or the field device. At this point, make sure that the field wiring to the module's terminal or to the terminal block has a good connection and that no wires are broken.

After checking the module, check that the field device is working properly. Measure the voltage coming to the field device while the output module is ON, making sure that the return line is well connected to the device. If there is power yet the device does not respond, then the field device is faulty.

Another method for checking the field device is to test it without using the output module. Remove the output wiring and connect the field device directly to the power source. If the field device does not respond, then it is faulty. If the field device responds, then the problem lies in the wiring between the device and the output module. Check the wiring, looking for broken wires along the wire path.

TROUBLESHOOTING THE CPU

PLCs also provide diagnostic indicators that show the status of the PLC and the CPU. Such indicators include *power OK*, *memory OK*, and *communications OK* conditions. First, check that the PLC is receiving enough power from the transformer to supply all the loads. If the PLC is still not working, check for voltage supply drop in the control circuit or for blown fuses. If the PLC does not come up even with proper power, then the problem lies in the CPU. The diagnostic indicators on the front of the CPU will show a fault in either memory or communications. If one of these indicators is lit, the CPU may need to be replaced.

SUMMARY OF TROUBLESHOOTING METHODS

In conclusion, the best method for diagnosing input/output malfunctions is to isolate the problem to the module, the field device, or the wiring. If both power and logic indicators are available, then module failures become readily apparent. The first step in solving the problem is to take a voltage measurement to determine if the proper voltage level is present at the input or output terminal. If the voltage is adequate at the terminal and the module is not responding, then the module should be replaced. If the replacement module has no effect, then field wiring may be the problem. A proper voltage level at the output terminal while the output device is OFF also indicates an error in the field wiring. If an output rung is activated but the LED indicator is OFF, then the module is faulty. If a malfunction cannot be traced to the I/O module, then the module connectors should be inspected for poor contact or misalignment. Finally, check for broken wires under connector terminals and cold solder joints on module terminals.

KEY TERMS

control program checkout
dynamic system checkout
ground loop
master control relay (MCR)
panel enclosure
safety control relay (SCR)
static input wiring check
static output wiring check
system layout
wire bundling

SYSTEM SELECTION
GUIDELINES

This is not the end. It is not even the beginning of the end. But it is, perhaps, the end of the beginning,

—Winston Churchill

CHAPTER HIGHLIGHTS In this chapter, we will explain the procedures for selecting the proper programmable controller for an application. We will explain how to determine application requirements, as well as how to evaluate PLC capabilities. We will also provide several guidelines for defining and configuring the control system, along with other factors that will affect the final selection. After finishing this chapter, you will be able to select the PLC system that is right for your application.

21-1 INTRODUCTION TO PLC SYSTEM SELECTION

As you have seen in this book, programmable controllers are available in all shapes and sizes, covering a wide spectrum of capabilities. On the low end are "relay replacers," with minimum I/O and memory capability. At the high end are large supervisory controllers, which play an important role in hierarchical systems by performing a variety of control and data acquisition functions. In between these two extremes are multifunctional controllers with both communication capabilities, which allow integration with various peripherals, and expansion capabilities, which allow the product to grow as the application requirements change.

Deciding on the right controller for a given application has become increasingly more difficult. With the explosion of new products, including general- and special-purpose programmable controllers, system selection now places an even greater demand on the designer to take a system approach to selecting the best product for each task. Programmable controller selection affects many factors, so the designer must determine which characteristics are desirable in the control system and which controller best fits the present and future needs of the application.

21-2 PLC SIZES AND SCOPES OF APPLICATIONS

Prior to evaluating the system requirements, the designer should understand the different ranges of programmable controller products and the typical features found within each range. This understanding will enable the designer to quickly identify the type of product that comes closest to matching the application's requirements.

Figure 21-1, previously presented in Chapter 1, illustrates PLC product ranges divided into five major areas with overlapping boundaries. The basis for this product segmentation is the number of possible inputs and outputs the system can accommodate (I/O count), the amount of memory available for the application program, and the system's general hardware and software structure. As the I/O count increases, the complexity and cost of the system also increase. Similarly, as the system complexity increases, the memory capacity, variety of I/O modules, and capabilities of the instruction set increase as well.

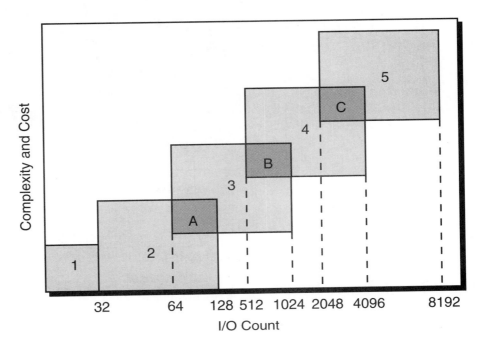

Figure 21-1. PLC product ranges.

The shaded areas in Figure 21-1, labeled A, B, and C, reflect the possibility of controllers with enhanced (not standard) features for a particular range. These enhancements place the product in a gray area that overlaps the next higher range. For example, because of its I/O count, a small PLC would fall into area 2, but it could have analog control functions that are standard in medium-sized controllers. Thus, this type of product would belong in area A. Products that fall into these overlapping areas allow the user to select the product that best matches the application's requirements, without having to select the larger product, unless it is necessary. The following discussion presents information about the five PLC categories, as well as the overlapping categories.

SEGMENT 1: MICRO PLCs

Micro PLCs are used in applications that require the control of a few discrete I/O devices, such as small conveyor controls. Some micro PLCs can perform limited analog I/O monitoring functions (e.g., monitoring a temperature set point or activating an output). Figure 21-2 shows a typical microcontroller, while Table 21-1 lists the standard features of micro PLCs.

Figure 21-2. PLC Direct's micro PLC DL105.

Courtesy of PLC Direct, Cumming, GA

Micro PLCs
• Up to 32 I/O
• 16-bit processor
• Relay replacer
• Memory up to 1K
• Digital I/O
• Built-in I/Os in a compact unit
• Master control relays
• Timers and counters
• Programmed with handheld programmer

Table 21-1. Standard features of micro PLCs.

SEGMENT 2: SMALL PLCs

Small controllers are mostly used in applications that require ON/OFF control for logic sequencing and timing functions. These PLCs, along with microcontrollers, are widely used for the individual control of small machines. Often, these products are single-board controllers. Table 21-2 lists the standard features of small PLCs.

Area A. Area A includes controllers that are capable of having up to 64 or 128 I/O, along with products that have features normally found in medium-sized controllers. The enhanced capabilities of these small controllers allow them to be used effectively in applications that need only a small number of I/O, yet require analog control, basic math, I/O bus network interfaces, LANs, remote I/O, and/or limited data-handling capabilities (see Figure 21-3). A typical application of an area A controller is a transfer line in which several small machines, under individual control, must be interlocked through a LAN.

Small PLCs
• Up to 128 I/O
• 16-bit processor
• Relay replacer
• Memory up to 2K
• Digital I/O
• Local I/O only
• Ladder or Boolean language only
• Master control relays
• Timers/counters/shift registers
• Drum timers or sequencers
• Programmed with handheld programmer

Table 21-2. Standard features of small PLCs.

Courtesy of Allen-Bradley, Highland Heights, OH

Figure 21-3. Area A (SLC500) controller capable of handling up to 72 discrete and 4 analog I/O.

SEGMENT 3: MEDIUM PLCs

Medium PLCs (see Figure 21-4) are used in applications that require more than 128 I/O, as well as analog control, data manipulation, and arithmetic capabilities. In general, the controllers in segment 3 have more flexible hardware and software features than the controllers previously mentioned. Table 21-3 lists these features.

Area B. Area B contains medium PLCs that have more memory, table-handling, PID, and subroutine capabilities than typical medium-sized PLCs, as well as more arithmetic and data-handling instructions. Figure 21-5 shows a PLC that falls into this category.

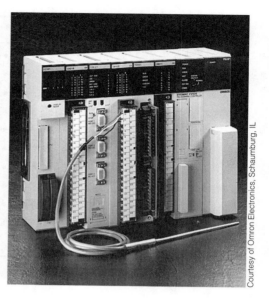

Figure 21-4. Medium-sized PLC 5/11 (left) and PLC 5/20 (right) processors with up to 512 I/O capacity.

Figure 21-5. Omron's area B CV500 PLC with a temperature control module (up to 1024 I/O).

Medium PLCs	
• Up to 1024 I/O	• Math capabilities
• 16- or 32-bit processor	–Addition
• Relay replacer and analog control	–Subtraction
• Memory up to 4K words	–Multiplication
• Expandable to 16K	–Division
• Digital I/O	• Limited data handling
• Analog I/O	–Compare
• Local and remote I/O	–Data conversion
• Ladder or Boolean language	–Move register/file
• Function block/high-level language	–Matrix functions
• Master control relays	• Special function I/O modules
• Timers/counters/shift registers	• RS-232 communication port
• Drum timers and sequencers	• Local area networks
• Jump	• Support I/O bus networks

Table 21-3. Standard features of medium PLCs.

SEGMENT 4: LARGE PLCs

Large controllers (see Figure 21-6) are used for more complicated control tasks, which require extensive data manipulation, data acquisition, and reporting. Further software enhancements allow these products to perform complex numerical computations. Table 21-4 summarizes the standard features of large PLCs.

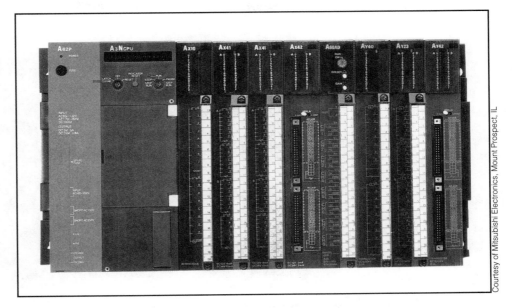

Figure 21-6. Large Mitsubishi A3NCPU controller with 2048 I/O capacity.

Large PLCs	
• Up to 4096 I/O	• Math capabilities
• 16- or 32-bit processor	–Addition
• Relay replacer and analog control	–Subtraction
• Memory up to 12K words	–Multiplication
• Expandable to 128K	–Division
• Digital I/O	–Square root
• Analog I/O	–Double precision
• Local and remote I/O	• Extended data handling
• Ladder or Boolean language	–Compare
• Function block/high-level language	–Data conversion
• Master control relays	–Move register/file
• Timers/counters/shift registers	–Matrix functions
• Drum timers and sequencers	–Block transfer
• Jump	–Binary tables
• Subroutines, interrupts	–ASCII tables
• PID modules or system software PID	• Local area networks
• One or more RS-232 communication ports	• Special function I/O modules
• Host computer communication modules	• Support I/O bus networks

Table 21-4. Standard features of large PLCs.

Area C. Area C includes the segment 4 PLCs that have a large amount of application memory and I/O capacity. The PLCs in this area also have greater math and data-handling capabilities than other large PLCs. Figure 21-7 shows an example of this type of controller.

Courtesy of Giddings & Lewis, Fond du Lac, WI

Figure 21-7. Giddings & Lewis's area C PIC900 with up to 3168 I/O and motion I/O, IEC programming, and floating-point math capabilities.

SEGMENT 5: VERY LARGE PLCS

Very large PLCs (see Figure 21-8) are utilized in sophisticated control and data acquisition applications that require large memory and I/O capacities. Remote and special I/O interfaces are also standard requirements for this type of controller. Typical applications for very large PLCs include steel mills and refineries. These PLCs usually serve as supervisory controllers in large, distributed control applications. Table 21-5 lists standard features found in segment 5 PLCs.

Courtesy of Allen-Bradley, Highland Heights, OH

Figure 21-8. Very large PLC-3 from Allen-Bradley with 8190 I/O capability.

Very Large PLCs	
• Up to 8192 I/O	• Math capabilities
• 32-bit processor or multiprocessors	–Addition
• Relay replacer and analog control	–Subtraction
• Memory up to 64K words	–Multiplication
• Expandable to 1 meg	–Division
• Digital I/O	–Square root
• Analog I/O	–Double precision
• Remote analog I/O	–Floating point
• Remote special modules	–Cosine functions
• Local and remote I/O	• Powerful data handling
• Ladder or Boolean language	–Compare
• Function block/high-level language	–Data conversion
• Master control relays	–Move register/file
• Timers/counters/shift registers	–Matrix functions
• Drum timers and sequencers	–Block transfer
• Jump	–Binary tables
• Subroutines, interrupts	–ASCII tables
• Special function I/O modules	–LIFO
• Local area networks	–FIFO
• PID modules or system software PID	• Machine diagnostics
• Two or more RS-232 communication ports	• Support I/O bus networks
• Host computer communication modules	

Table 21-5. Standard features of very large PLCs.

21-3 PROCESS CONTROL SYSTEM DEFINITION

Selecting the right programmable controller for a machine or process involves evaluating not only current needs, but future requirements as well. If present and future goals are not properly evaluated, the control system may quickly become inadequate and obsolete.

Keeping the future in mind when choosing a programmable controller will minimize the costs of changes and additions to the system. For example, with proper planning, future memory expansion may only require the installation of a memory module; furthermore, the addition of a peripheral may be as easy as connecting the device to the communication port. A local area network can also ease the future integration of programmable controllers into a plantwide communication scheme.

Once the basic control application has been defined, the user should begin evaluating the controller requirements, including:

- input/output

- type of control

- memory

- software

- peripherals

- physical and environmental

INPUT/OUTPUT CONSIDERATIONS

Determining the amount of I/O required is typically the first step in selecting a controller. Once the decision has been made to automate a machine or process, determining the amount of I/O is simply a matter of counting the discrete and/or analog devices that will be monitored or controlled. This count will help to identify the minimum size constraints for the controller. Remember that the controller should allow for future expansion and spares (typically 10% to 20% spares), although spares do not affect the choice of PLC size.

Discrete Inputs/Outputs. Input/output interfaces with standard ratings are available for accepting signals from sensors and switches (e.g., push buttons, limit switches, etc.), as well as ON/OFF control devices (e.g., pilot lights, alarms, motor starters, etc.). If these input/output devices receive power from separate sources, then the discrete interface circuits must have isolated commons (return lines). Typical discrete AC inputs/outputs range from 24 to 240 V, and typical DC inputs/outputs range from 5 to 240 V.

Input circuits vary from one manufacturer to another. Nevertheless, characteristics like debouncing circuitry, which protects against false signals, and surge protection, which guards against large transients, are desirable in any input circuit. Another good input circuit quality is optical or transformer isolation between the high-power input and the interface's control logic circuitry.

When evaluating discrete outputs, the following are key characteristics: fuses, transient surge protection, and isolation between the power and logic circuits. Fused circuits cost more initially, but they usually cost less than having a fuse installed externally. These circuits should also have easily accessible fuses, so that replacing fuses does not require shutting down several other devices for a long period of time. Moreover, fused output circuits should have blown fuse indicators, as well as an output current rating and a specified operating temperature (typically 60°F) that fits the application's requirements.

Analog Inputs/Outputs. Analog input/output interfaces sense signals generated by transducers. These interfaces measure quantity values, such as flow, temperature, and pressure, and are used to control voltage or current output devices. Typical interface ratings include -10 to $+10$ V, 0 to $+10$ V, 4 to 20 mA, and 10 to 50 mA.

Some manufacturers provide special analog interfaces that accept low-level signals (e.g., RTD, thermocouple). Typically, these interface modules accept a mix of thermocouple or RTD signals on a single module. Users should consult the vendor concerning specific requirements.

Special Function Inputs/Outputs. Sometimes an application requires a special type of I/O conditioning (e.g., positioning, fast input, frequency, etc.) that may be impossible to implement using standard I/O modules. Special function I/O modules and smart modules, a type of special interface, can perform this task. Typically, these interfaces process all of the field data within the module itself, thus relieving the CPU from performing this time-consuming duty. For example, PID, three-axis positioning, and stepper motor modules are special function I/O modules that make control implementation much easier. These modules reduce programming and implementation time.

Remote Inputs/Outputs. Remote I/O modules are convenient, cost-effective processing devices, especially when used in large systems. Remote I/O subsystems, which are located away from the CPU and connected to it by twisted-pair cables, can dramatically reduce wiring costs, both from a labor and a material standpoint. Another advantage of remote I/O subsystems is that inputs and outputs can be strategically grouped to control separate machines or sections of a machine or process. This grouping provides easy maintenance and allows start-up without involving the entire system.

Most controllers that have remote I/O have remote digital I/O. However, users who require remote analog I/O should check to see if this feature is available in the products being considered.

I/O Bus Networks. I/O bus networks, which include device bus and process bus networks, should be considered in applications requiring decentralized control within the PLC system. I/O bus networks provide a topology that allows the direct connection of field devices to a bus network, thereby simplifying wiring. At the same time, these networks let the PLC directly receive I/O field device information about the status of the device. However, the system's I/O field devices must be compatible with the I/O bus network to take advantage of these enhanced communications features.

CONTROL SYSTEM ORGANIZATION

With the advent of new, smarter programmable controllers, the decision about the type of control has become a very important consideration. Questions such as, What type of control should I use? are now asked more often when automating a process. Knowing process application and future automation requirements will help the user to decide what type of control, and thus PLC, is required. Possible control configurations include individual control, centralized control, and distributed control. Figure 21-9 illustrates these configurations.

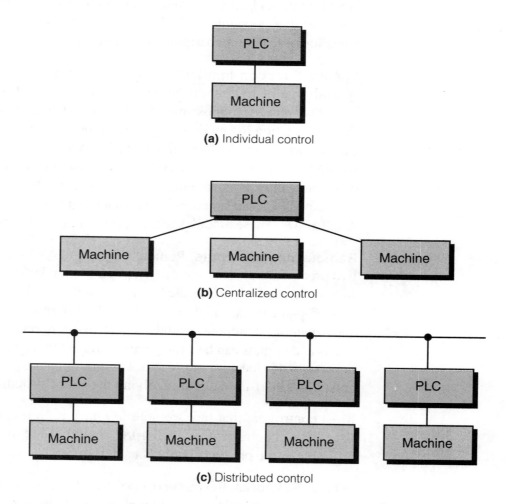

(a) Individual control

(b) Centralized control

(c) Distributed control

Figure 21-9. (a) Individual, **(b)** centralized, and **(c)** distributed control configurations.

Individual Control. Individual control, or *segregated control*, is used when a PLC controls a single machine with only local I/O or with local and a few remote I/O. This type of control does not normally require communication with any other controllers or computers. Individual control is primarily applied to OEM and end-user equipment, such as injection-molding machines, small machine tools, and small, dedicated batching processes. When deciding on this approach, the user should consider whether future intercontroller communication will be desired. If so, the user can choose the appropriate controller for the initial installation to avoid extra design expenses at a later date.

Centralized Control. Centralized control is used when a central PLC controls several machines or processes. This type of control can have many subsystems spread throughout the factory. Each of these subsystems may interface with specific I/O devices that may or may not be related to the same control. A centralized programmable controller communicates only with its subsystems and/or peripherals; it does not exchange data with other PLCs.

The flexibility and potential advantages of a centralized application depend on the PLC used and the system designer's design philosophy. For example, centralized control can be implemented as the large, individual control of a large process or the centralized control of a number of highly complex, small processes.

One distinct disadvantage of centralized control is that, if the main PLC fails, the whole process stops. Redundant systems can be used to overcome this problem in large, critical, central controls that require a backup. Several manufacturers offer this redundancy option.

Distributed Control. The need to have several main PLCs communicating with each other has brought about **distributed control**. This type of control employs local area networks (LANs), which allow several PLCs to control different stages or processes locally while constantly interchanging information about the process. Communication among PLCs occurs at very high speeds (up to 1 megabaud) through single coaxial or fiber-optic cables. Despite this powerful configuration, communication between two different manufacturers' LAN systems can be difficult. Therefore, the user should properly define the process application's functional requirements from the beginning.

MEMORY CONSIDERATIONS

The two main factors to consider when choosing memory are the type and the amount. An application may require two types of memory: nonvolatile memory and volatile memory with a battery backup. A nonvolatile memory, such as EPROM, can provide a reliable, permanent storage medium once the program has been created and debugged. If the application will require on-line changes, then it should probably be stored in read/write memory supported by a battery. Some controllers offer both of these options, which can be used individually or in conjunction with each other.

Small PLCs normally have a fixed (nonexpandable) memory capacity of 1/2K to 2K words. Therefore, the amount of memory is not a major concern when selecting a small controller. In medium and large controllers, however, memory expands incrementally in units of 1K, 2K, 4K, etc. Although there are no fixed rules for determining the amount of memory required, certain guidelines can provide an estimate of memory needs.

The amount of memory required for a given application is a function of the total number of inputs and outputs to be controlled and the complexity of the control program. The complexity refers to the amount and type of arithmetic and data manipulation functions that the PLC will perform. For each of their products, manufacturers have a rule-of-thumb formula that helps to approximate the memory requirement. This formula involves multiplying the total number of I/O by a constant (usually a number between 3 and 8). If the program involves arithmetic or data manipulation, this memory approximation should be increased by 25 to 50%.

Although memory requirement formulas do a good job of estimating memory needs, the best way to obtain memory requirement data is to create the program and count the number of words used. Knowledge of the number of words required to store each instruction will allow the user to determine exact memory requirements.

SOFTWARE CONSIDERATIONS

During system implementation, the user must program the PLC. Because the programming is so important, the user should be aware of the software capabilities of the product they choose. Generally, the software capability of a system is tailored to handle the control hardware that is available with the controller. However, some applications require special software functions that are beyond the control of the hardware components. For instance, an application may involve special control or data acquisition functions that require complex numerical calculations and data-handling manipulations. The instruction set selected will determine the ease with which these software tasks can be implemented. It will also directly affect the time required to implement and execute the control program.

PERIPHERALS

The programming device is the key peripheral in a PLC system. It is of primary importance because it must provide all of the capabilities necessary to accurately and easily enter the control program into the system. The two most common types of programming devices are handheld units and personal computers. Handheld units, which are small and low cost, are typically used to program relatively small control programs in small PLCs. The amount of information that can be displayed on a handheld unit is normally a single program element or, in some cases, a single program rung.

Personal computers provide a better way to program a system if the control program is large. Many PLC manufacturers provide software that allows their PLCs to be programmed using a standard PC. However, expansion boards or special interfacing cables may be required to link the personal computer with the programmable controller. Using a PC as a programming device becomes even more advantageous when the same program development software can be used in same-model PLCs or those of the same family. Laptop PCs equipped with programming and documentation software provide even more programming flexibility by joining the ease of PC programming with the transportability of handheld programming devices.

In addition to the programming device, a system may require other types of peripherals at certain control stations to provide an interface between the controller and the operator. The most common peripheral is the line printer, used for obtaining a hardcopy printout of the program and for sending report information about the process. Other peripherals include color displays and alphanumeric displays, which can be used to send messages or alarms about the process, as well as diskette drives, which can be used for storing hourly or monthly production reports on a floppy diskette. If a PC is used as a graphic interface to a PLC system, both systems must have compatible DDE (dynamic data exchange) drivers to properly interface with peripherals.

Peripheral requirements should be evaluated along with the CPU, since the CPU will determine the type and number of peripherals that can be interfaced to the system. The CPU also influences the method of interfacing, as well as the distance that peripherals can be placed from the PLC.

PHYSICAL AND ENVIRONMENTAL

The physical and environmental characteristics of the various controller components will significantly impact total system reliability and maintenance. Ambient conditions, such as temperature, humidity, dust level, and corrosion, can affect the controller's ability to operate properly. The user should determine operating conditions (i.e., temperature, vibration, EMI/RFI, etc.), and packaging requirements (i.e., dustproof, dripproof, ruggedness, type of connections, etc.) before selecting the controller and its I/O system. Most programmable controller manufacturers provide products that have undergone certain environmental and physical tests (e.g., temperature, EMI/RFI, shock, etc.). Users should be aware of the tests performed and whether or not the results meet the demands of the operating environment.

Table 21-6 provides a checklist of the features a user should look for when evaluating PLC requirements. The table also provides typical specifications for these features. Note that the list covers all product ranges, from small to very large; therefore, some PLCs may not have all of these features due to their range characteristics.

I/O System Checklist	Typical Specifications
I/O Count	
Digital count	Maximum of 128 I/O (mixable)
Analog count	Maximum of 16 I/O (mixable)
Digital I/O	
Inputs	
Points/module	4 points/module
Input type	AC, DC, nonvoltage, etc.
Input ratings	110 VAC, 220 VAC, 5–24 VDC, etc.
Maximum inputs/channel	64 points/channel
Input status indicators	Power, logic
Isolation	1500 volts optical
Outputs	
Points/module	16 points/module
Output type	AC, DC, contact, etc.
Output ratings	110 VAC, 220 VAC, contact
Output current (amps/point)	1 amp/point with all outputs ON at 115 VAC
Maximum outputs/channel	64 points/channel
Output status indicators	Power, individual blown fuse, logic
Output protection	Fuses, suppression on contact output
Analog I/O	
Inputs	
Points/module	4 analog inputs/module
Resolution	11 bits
Input type	Current, voltage
Input ratings	4–20 mA, 0–5 volts, 0–10 volts
Built-in transducer	Thermocoupler input
Maximum inputs/channel	32 points/channel
Power supply	Internal to PLC
Outputs	
Points/module	2 analog outputs/module
Output type	Current, voltage
Output ratings	4–20 mA, 10–50 mA, 0–10 volts
Maximum outputs/channel	16 points/channel
Power supply	+15 VDC and –15 VDC
Remote I/O	
Digital	
Distance	1500 ft
I/O per remote	32 I/O per remote
Communication link	Twisted-pair, 100 ohms impedance
Analog	
Distance	5000 ft with receiver/transmitter
I/O per remote	16 I/O per remote
Communication link	Coaxial

Table 21-6. PLC requirement checklist.

I/O System Checklist	Typical Specifications
Special I/O	
High-speed pulse counter	Local and remote, 50 kHz
Fast electronic input	5 microsecond pulse width minimum
Interrupt module	Yes
Absolute encoder	Direct connection to encoder
Incremental encoder	Not available
BCD I/O module	Remote, 4 and 8 BCD digits
Stepper motor	Yes
ASCII communications module	Full ASCII 300–4800 baud
Host computer	Yes, protocol decode on module
LAN I/O module	Extra board in CPU
PID module	Local and remote, 2 loops per module
Language module	Basic interpreter module
I/O Bus Network	
Device bus	
Networks supported	DeviceNet, InterBus-S, and SDS
Number of nodes	64 nodes, 2048 devices
Structure	Trunkline
Process bus	
Networks supported	Profibus and Fieldbus foundation
Media	Coaxial
Speed	2 Mbaud
Physical	
Wire size to I/O	20 AWG with 2 wires per I/O
Separate commons	Yes for 4 pts/module, no for 16 pts/module
Removable under power	Yes
Disturb wiring to remove I/O	No, disconnect screw from I/O module

CPU Checklist	Typical Specifications
Processor	
Microprocessor	16-bit microprocessor and multiprocessor board
Scan time	10 msec/K of memory
Communication ports	Two RS-232C ports
Memory	
Memory type	RAM, EEPROM
Total system memory	64K
Application memory size	8K for user
Word size	8 bits
Memory utilization	1 word per element (coil or contact)
Memory protection	Yes, key switch
Power Supply	
Incoming power	120/240 VAC, 24 VDC
Frequency	50/60 Hz

Table 21-6 continued.

CPU Checklist	Typical Specifications
Power Supply	
Voltage variation	+15%, −10%
Overvoltage protection	Yes
Current limiting	Yes
Maximum current supply	100 mA at 14 VDC, 2.5 amps at 5 VDC
Isolation	1500 volts
Location	Built-in CPU
Environmental	
Operating temperature	0–60°C
Humidity	5–90% relative humidity, noncondensing
EMI/RFI	Satisfies NEMA and IEEE tests
Noise	1000 volts peak-peak, 1 microsecond
Vibration	Withstands 16.7 Hz, double amplitude
Shock	10g each direction

Software Checklist	Typical Specifications
Language	
Ladder or Boolean	Ladder language
High-level	Functional blocks
IEC 1131-3	Conforms to IEC 1131-3 with SFCs
Software Coils	
Number of internals	128
Number of timers	32 with maximum count of 9999 sec BCD
Number of counters	166
Number of shift registers	32, 16 bits each
Number of drum timers	16
Timer's time base	0.1, 1.0 seconds
Timer type	ON delay and OFF delay
Counter type	Up/down counter
Latch coil	32
Transitional coil	16, OFF/ON and ON/OFF
Master control relays	8
Global coil	256 in LAN
Global register	128 in LAN
Fault coil	Yes, detection of CPU failure
Interrupt coil	Yes
Math	
Addition	Yes, double precision
Subtraction	Yes, double precision
Multiplication	Yes
Division	Yes
Square root	Yes
Floating point	Yes, 1E + 38, 1E − 38
Trigonometric functions	Yes, sine and cosine in IEC 1131-3 ST language

Table 21-6 continued.

Software Checklist	Typical Specifications
Data Handling	
Number of registers	128, 16 bits each
Data size in registers	+32767, −32767, and 9999 BCD
Compare	Yes, <, >, and =
Conversions	Binary-BCD, BCD-binary
Move	Registers and single files
Matrix	AND, OR, XOR, NAND
Tables	Move to ASCII or binary tables
PID	Software functional block, 20 loops
LIFO	Yes
FIFO	Yes
Jump	Conditional and direct
Subroutines	Yes
ASCII instructions	Yes, print and read
Sort	No
Machine diagnostics	Yes

Programming and Storage Device Checklist	Typical Specifications
Personal Computer	
Physical	
Computer type	Desktop and laptop
Display size	14 to 21" screen
Graphics	Yes
DDE driver	Yes
Ladder matrix size	10 × 7 elements, scrolling for IEC 1131-3
Built-in storage	Yes
Local area network	Yes
Communication	RS-232C and 20 mA current loop
Incoming power	115/230 VAC
Operating temperature	0–40°C
Keyboard type	Mylar or standard keys
Functional	
Intelligent	Yes
Single scan	No
Power flow	Yes, element intensified on screen
Off-line programming	Yes
Monitor function	Yes
Modify function	Yes
Force I/O	Yes, indicates forcing on mainframe
Search	No
Mnemonics	Yes
Documentation	Built-in software module
Manual Programmer	
Physical	
Display type	LCD or LED
Ladder matrix size	7 × 4 elements

Table 21-6 continued.

Programming and Storage Device Checklist	Typical Specifications
Manual Programmer	
Physical	
Communication	RS-232C
Incoming power	From unit
Operating temperature	0–40°C
Keyboard type	Mylar
Functional	
Intelligent	No
Single scan	No
Power flow	Yes
Monitor function	Yes
Modify function	Yes
Force I/O	Yes
Search	No
Mnemonics	Yes, also messages
Storage Devices	
Floppy disk	In personal computer
Computer	Yes, through computer module
Electronic memory module	Yes, for a small PLC

System Diagnostics Checklist	Typical Specifications
Power Supply	
Power loss detection	Yes, after 3 cycles
Voltage level detection	Yes, DC levels for CPU
Diagnostic monitoring	Continuously
Memory	
Memory OK	Yes, checksum and LED indicator
Battery OK	Yes, LED indicator
Diagnostic monitoring	At power-up only
Processor	
Local	Yes, watchdog timer and LED indicator
Remote	Yes, indicator in CPU
Diagnostic monitoring	Continuously
Communication	
Local I/O	Yes
Remote I/O	Yes, checksum
Programming device	CRT port OK and RS-232C OK
Diagnostic monitoring	During transmission
Fault Indications	
CPU	Yes, external relay contacts
Remote	Yes, external contacts at remote driver
LAN	Yes, internal coil
I/O	Yes, detects presence of I/O module

Table 21-6 continued.

21-4 OTHER CONSIDERATIONS

An evaluation of the previously discussed hardware and software requirements will narrow the selection of the PLC down to one of a few possible candidates. More than likely, two or more products will meet all of the requirements of the preliminary system design, meaning that a final decision must still be made. At this point, the user should evaluate a few more factors, which can lead to the selection of the product that best fits the system specifications and the application requirements. The user should discuss these factors with the potential vendors.

PROVEN PRODUCT RELIABILITY

The reliability of the controller plays an important role in overall system performance. Lack of reliability usually translates into downtime, poor quality products, and higher scrap levels.

The user can investigate several factors to determine the proven reliability of a particular product. **Mean-time-between-failures (MTBF) studies** can be helpful if the user knows how to evaluate the data. These studies provide information about the average time between equipment malfunctions and how long the equipment will operate without a failure. Knowledge of a similar application in which the product has been successfully applied is also useful. A sales representative can provide this information and even, on occasion, arrange a site visit. Moreover, the user should ensure that the vendor can truly satisfy any unique or peculiar specifications (e.g., EMI and vibration requirements). Finally, the user should research the **burn-in procedures** for the product (e.g., the total system burn-in process or the parts burn-in process). The burn-in process involves operating the product at an elevated temperature to simulate extended operation in order to force an electronic board or part to fail. If a part passes the burn-in procedure, it will have an extremely high probability for proper operation. Usually, the vendor can provide MTBF and burn-in information upon request.

STANDARDIZATION OF PLC EQUIPMENT

A last consideration when making the final decision on a PLC is the possibility of future plans to standardize machinery—that is, to use only products from a given manufacturer and product line. Many companies are adopting this practice for good reasons. Several vendors are creating complete product families of PLCs that cover the entire range of capabilities, thus making standardization more feasible. Another current trend by manufacturers is to build completely intercompatible product families, with products ranging from very small to very large PLCs. These families share the same I/O structure, programming device, and elementary instruction set. They also have similar memory organization and structure. Because of their similarities, these product families can be linked in a network configuration. PLC families also provide the following important benefits:

- Training on a new PLC family member is a progression of current knowledge, rather than the development of a totally new set of skills.

- Standardized products can result in better plant maintenance in emergency situations.

- I/O spares can be used for all family products, resulting in a smaller spare inventory.

- An outgrown product can be replaced with the next larger product by simply removing the smaller CPU, installing the larger CPU, and reloading the old program.

21-5 SUMMARY

This chapter has presented a general approach for selecting a programmable controller. PLC selection relates not only to obvious factors, such as I/O capacity, memory capacity, and sophistication of control, but also to intangible factors that have a significant impact on final system results. Selecting the appropriate PLC for an application is important because the right PLC can make a process more efficient, more effective, and less expensive. Table 21-7 lists a summary of the major steps involved in selecting a PLC.

KEY TERMS
burn-in procedures
centralized control
distributed control
individual control
mean-time-between-failures study (MTBF)

Step	Action
1	Know the process to be controlled.
2	Determine the type of control. • Distributed control • Centralized control • Individual control
3	Determine the I/O interface requirements. • Number of digital and analog inputs and outputs • Input/output specifications • Remote I/O requirements • Special I/O requirements • I/O bus network applications • Future expansion plans
4	Determine the software language and functions. • Ladder, Boolean, and/or high level • Basic instructions (timers, counters, etc.) • Enhanced instructions/functions (math, PID, etc.) • IEC 1131-3 language
5	Consider the type of memory. • Volatile (R/W) • Nonvolatile (EEPROM, EPROM, etc.) • Combination of volatile and nonvolatile
6	Consider memory capacity. • Memory requirements based on memory usage per instruction • Extra memory for complex programming and future expansion
7	Evaluate processor scan time requirements.
8	Define programming and storage device requirements. • Personal computer • Disk storage • Manual programmer • Functional capabilities of the programming device
9	Define peripheral requirements. • Graphic display • Operator interface • Line printers • Documentation system • Report generation system
10	Determine any physical and environmental constraints. • Available space for system • Ambient conditions
11	Evaluate other factors that may affect selection. • Vendor support • Proven product reliability • Plant goals for standardization

Table 21-7. Steps for selecting a PLC.

APPENDICES

- Logic Symbols, Truth Tables, and Equivalent Ladder/Logic Diagrams

- ASCII Reference

- Electrical Relay Diagram Symbols

- P&ID Symbols

- Equation of a Line and Number Tables

- Abbreviations and Acronyms

- Voltage-Current Laplace Transfer Function Relationships

APPENDIX A: LOGIC SYMBOLS, TRUTH TABLES, AND EQUIVALENT LADDER/LOGIC DIAGRAMS

LOGIC SYMBOLS AND TRUTH TABLES

AND	OR	A	B	Y
A — ⊐D— Y B —	A —o⊐D o— Y B —o	1 1 0 0	1 0 1 0	1 0 0 0
A —o⊐D— Y B —	A —⊐Do— Y B —o	1 1 0 0	1 0 1 0	0 0 1 0
A —⊐D— Y B —o	A —o⊐Do— Y B —	1 1 0 0	1 0 1 0	0 1 0 0
A —o⊐D— Y B —o	A —⊐Do— Y B —	1 1 0 0	1 0 1 0	0 0 0 1
A —o⊐Do— Y B —o	A —⊐D— Y B —	1 1 0 0	1 0 1 0	1 1 1 0
A —⊐Do— Y B —o	A —⊐D— Y B —	1 1 0 0	1 0 1 0	1 0 1 1
A —o⊐Do— Y B —	A —⊐D— Y B —o	1 1 0 0	1 0 1 0	1 1 0 1
A —⊐Do— Y B —	A —o⊐D— Y B —o	1 1 0 0	1 0 1 0	0 1 1 1

EQUIVALENT LADDER/LOGIC DIAGRAMS

Logic Diagram	Truth Table	Ladder Diagram
AND Gate	A B C 0 0 0 0 1 0 1 0 0 1 1 1	AND Equivalent Circuit
OR Gate	A B C 0 0 0 0 1 1 1 0 1 1 1 1	OR Equivalent Circuit
Exclusive-OR Gate	A B C 0 0 0 0 1 1 1 0 1 1 1 0	Exclusive-OR Equivalent Circuit
NAND Gate	A B C 0 0 1 0 1 1 1 0 1 1 1 0	NAND Equivalent Circuit
NOR Gate	A B C 0 0 1 0 1 0 1 0 0 1 1 0	NOR Equivalent Circuit

APPENDIX B: ASCII REFERENCE

OCT	DEC	PARITY	HEX	ASCII CHAR	CTRL KEYBD EQUIV	ALTERNATE CODE NAMES
000	1	EVEN	00	NUL	@	NULL, CTRL SHIFT P, TAPE LEADER
001	2	ODD	01	SOH	A	START OF HEADER, SOM
002	3	ODD	02	STX	B	START OF TEXT, EOA
003	4	EVEN	03	ETX	C	END OF TEXT, EOM
004	5	ODD	04	EOT	D	END OF TRANSMISSION, END
005	6	EVEN	05	ENQ	E	ENQUIRY, WRU, WHO ARE YOU
006	7	EVEN	06	ACK	F	ACKNOWLEDGE, RU, ARE YOU
007	8	ODD	07	BEL	G	BELL
010	9	ODD	08	BS	H	BACKSPACE, FE0
011	10	EVEN	09	HT	I	HORIZONTAL TAB, TAB
012	11	EVEN	0A	LF	J	LINE FEED, NEW LINE, NL
013	12	ODD	0B	VT	K	VERTICAL TAB, VTAB
014	13	EVEN	0C	FF	L	FORM FEED, FORM, PAGE
015	14	ODD	0D	CR	M	CARRIAGE RETURN, EOL
016	15	ODD	0E	SO	N	SHIFT OUT, RED SHIFT
017	16	EVEN	0F	SI	O	SHIFT IN, BLACK SHIFT
020	17	ODD	10	DLE	P	DATA LINK ESCAPE, DC0
021	18	EVEN	11	DC1	Q	XON, READER ON
022	19	EVEN	12	DC2	R	TAPE, PUNCH ON
023	20	ODD	13	DC3	S	XOFF, READER OFF
024	21	EVEN	14	DC4	T	TAPE, PUNCH OFF
025	22	ODD	15	NAK	U	NEGATIVE ACKNOWLEDGE, ERR
026	23	ODD	16	SYN	V	SYNCHRONOUS IDLE, SYNC
027	24	EVEN	17	ETB	W	END OF TEXT BUFFER, LEM
030	25	EVEN	18	CAN	X	CANCEL, CNCL
031	26	ODD	19	EM	Y	END OF MEDIUM
032	27	ODD	1A	SUB	Z	SUBSTITUTE
033	28	EVEN	1B	ESC	[	ESCAPE, PREFIX
034	29	ODD	1C	FS	\	FILE SEPARATOR
035	30	EVEN	1D	GS	]	GROUP SEPARATOR
036	31	EVEN	1E	RS		RECORD SEPARATOR
037	32	ODD	1F	US	_	UNIT SEPARATOR
040	33	ODD	20	SP		SPACE, BLANK
041	34	EVEN	21	!		
042	35	EVEN	22	"		
043	36	ODD	23	#		
044	37	EVEN	24	$		
045	38	ODD	25	%		
046	39	ODD	26	&		
047	40	EVEN	27	'		APOSTROPHE
050	41	EVEN	28	(		
051	42	ODD	29	)		
052	43	ODD	2A	*		ASTERISK
053	44	EVEN	2B	+		
054	45	ODD	2C	,		COMMA
055	46	EVEN	2D	-		MINUS
056	47	EVEN	2E	.		PERIOD
057	48	ODD	2F	/		
060	49	EVEN	30	0		NUMBER ZERO
061	50	ODD	31	1		NUMBER ONE
062	51	ODD	32	2		
063	52	EVEN	33	3		
064	53	ODD	34	4		
065	54	EVEN	35	5		
066	55	EVEN	36	6		
067	56	ODD	37	7		
070	57	ODD	38	8		
071	58	EVEN	39	9		
072	59	EVEN	3A	:		COLON
073	60	ODD	3B	;		SEMICOLON

To transmit control codes, depress CTRL then the desired keyboard equivalent character.

OCT	DEC	PARITY	HEX	ASCII CHAR	CTRL KEYBD EQUIV	ALTERNATE CODE NAMES
074	61	EVEN	3C	<		LESS THAN
075	62	ODD	3D	=		
076	63	ODD	3E	>		GREATER THAN
077	64	EVEN	3F	?		
100	65	ODD	40	@		SHIFT P
101	66	EVEN	41	A		
102	67	EVEN	42	B		
103	68	ODD	43	C		
104	69	EVEN	44	D		
105	70	ODD	45	E		
106	71	ODD	46	F		
107	72	EVEN	47	G		
110	73	EVEN	48	H		
111	74	ODD	49	I		LETTER I
112	75	ODD	4A	J		
113	76	EVEN	4B	K		
114	77	ODD	4C	L		
115	78	EVEN	4D	M		
116	79	EVEN	4E	N		
117	80	ODD	4F	O		LETTER O
120	81	EVEN	50	P		
121	82	ODD	51	Q		
122	83	ODD	52	R		
123	84	EVEN	53	S		
124	85	ODD	54	T		
125	86	EVEN	55	U		
126	87	EVEN	56	V		
127	88	ODD	57	W		
130	89	ODD	58	X		
131	90	EVEN	59	Y		
132	91	EVEN	5A	Z		
133	92	ODD	5B	[		SHIFT K
134	93	EVEN	5C	\		SHIFT L
135	94	ODD	5D	]		SHIFT M
136	95	ODD	5E	^		SHIFT N
137	96	EVEN	5F	_		SHIFT O, UNDERSCORE
140	97	EVEN	60	`		ACCENT GRAVE
141	98	ODD	61	a		
142	99	ODD	62	b		
143	100	EVEN	63	c		
144	101	ODD	64	d		
145	102	EVEN	65	e		
146	103	EVEN	66	f		
147	104	ODD	67	g		
150	105	ODD	68	h		
151	106	EVEN	69	i		
152	107	EVEN	6A	j		
153	108	ODD	6B	k		
154	109	EVEN	6C	l		
155	110	ODD	6D	m		
156	111	ODD	6E	n		
157	112	EVEN	6F	o		
160	113	ODD	70	p		
161	114	EVEN	71	q		
162	115	EVEN	72	r		
163	116	ODD	73	s		
164	117	EVEN	74	t		
165	118	ODD	75	u		
166	119	ODD	76	v		
167	120	EVEN	77	w		
170	121	EVEN	78	x		
171	122	ODD	79	y		
172	123	ODD	7A	z		
173	124	EVEN	7B	[		
174	125	ODD	7C	!		VERTICAL SLASH
175	126	EVEN	7D	]		ALT MODE
176	127	EVEN	7E	~		ALT MODE)
177	128	ODD	7F	DEL		DELETE, RUBOUT

APPENDIX C: ELECTRICAL RELAY DIAGRAM SYMBOLS

SWITCHES

Disconnect	Circuit Interrupter	Circuit Breaker	Limit		
			Normally Open	Normally Closed	Neutral Position

					Actuated
DISC	C1	CB	LS / Held Closed / LS	LS / Held Open / LS	LS / NP	NP / LS

Limit (cont.)			Liquid Level		Vacuum & Pressure		Temperature	
Maintained Position	Proximity Switch		Normally Open	Normally Closed	Normally Open	Normally Closed	Normally Open	Normally Closed
	Closed	Open						
LS	PRS	PRS	FS	FS	PS	PS	TS	TS

Flow (Air, Water)		Foot		Toggle	Cable Operated (Emerg.) Switch	Plugging		Nonplug
Normally Open	Normally Closed	Normally Open	Normally Closed					
FLS	FLS	FTS	FTS	TGS / PB	COS	PLS F	PLS F	PLS F / R

Plugging w/Lockout Coil	Selector		Rotary Selector	
	2-Position	3-Position	Nonbridging Contacts	Bridging Contacts
PLS F / 1 LO	SS / 1 2	SS / 1 2 3	RSS	RSS
			OR	OR
			RSS	RSS
			Total Contacts To Suit Needs	

Thermocouple Switch	Push Buttons				Connections, Etc.	
	Single Circuit	Double Circuit		Maintained Contact	Conductors	
		PB	Mushroom Head		Not Connected	Connected
TCS / OFF / 1 / 2	Normally Open PB / Normally Closed PB	PB	PB	PB		

Connections, Etc. (cont.)			Contacts						
Ground	Chassis Or Frame Not Necessarily Grounded	Plug and Recp.	Time Delay After Coil				Relay, Etc.		Thermal Over-Load
			Normally Open	Normally Closed	Normally Open	Normally Closed	Normally Open	Normally Closed	
GRD	CH	PL	TR	TR	TR	TR	CR M	CR M	OL
							CON	CON	IDL
		RECP							

Coils						
Relays, Timers, Etc.	Solenoids, Brakes, Etc.				Thermal Overload Element	Control Circuit Transformer
	General	2-Position Hydraulic	3-Position Pneumatic	2-Position Lubrication		
CR TR M CON	SOL	SOL 2-H	SOL 3-P	SOL 2-L	OL IOL	H1 H3 H2 H4 T X1 X2

Coils (cont.)			Motors	
Reactors (cont.)			3-Phase Motor	DC Motor Armature
Adjustable Iron Core	Air Core	Magnetic Amplifier Winding		
		MAX	MTR	MTR A
	X			

Pilot Lights		Horns, Siren, Etc.	Buzzer	Bell
	Push to Test			
PL	PL	AH	ABU	ABE

APPENDIX D: P&ID SYMBOLS

INSTRUMENT LINE SYMBOLS

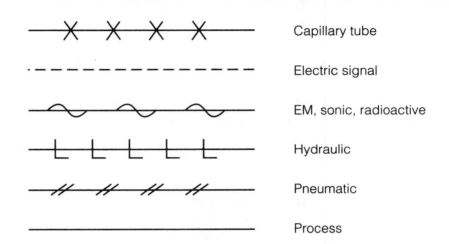

Capillary tube

Electric signal

EM, sonic, radioactive

Hydraulic

Pneumatic

Process

SYMBOLS FOR TRANSDUCERS AND ELEMENTS

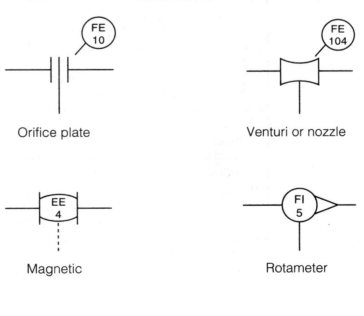

Orifice plate

Venturi or nozzle

Magnetic

Rotameter

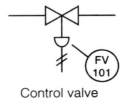

Control valve

INSTRUMENT IDENTIFICATION LETTERING

	First Letter	Second Letter
A	Analysis	Alarm
B	Burner, combustion	User's choice*
C	User's choice	Control
D	User's choice	
E	Voltage	Sensory (primary element)
F	Flow rate	
G	User's choice	Glass (sight tube)
H	Hand (manually initiated)	
I	Current (electric)	Indicate
J	Power	
K	Time or time schedule	Control station
L	Level	Light (pilot)
M	User's choice	
N	User's choice	User's choice
O	User's choice	Orifice, restriction
P	Pressure, vacuum	Point (test connection)
Q	Quantity	
R	Radiation	Record or print
S	Speed or frequency	Switch
T	Temperature	Transmit
U	Multivariable	Multifunction
V	Vibration, mechanical analysis	Valve, damper, louver
W	Weight, force	Well
X	Unclassified**	Unclassified
Y	Event, state, or presence	Relay, compute
Z	Position, dimension	Driver, actuator, unclassified

* User's choice may be used to denote a particular meaning, having one meaning as a first letter and another meaning as a second letter. The user must describe the particular meaning(s) in the legend. This letter can be used repetitively in a particular project.

** Unclassified letters may be used only once or to a limited extent. If used, the letter may have one meaning as a first letter and another meaning as a second letter. The user must specify the meaning(s) in the legend.

Reference: ANSI/ISA-S5.1-1984, *Instrumentation Symbols and Identification*, ISBN 0-87664-844-8

APPENDIX E: EQUATION OF A LINE AND NUMBER TABLES

EQUATION OF A LINE

The equation that describes the line that passes through two given points $P_1(X_1, Y_1)$ and $P_2(X_2, Y_2)$ can be calculated by:

$$Y = mX + b$$

where: m = the slope of the line
b = the value of Y when $X = 0$

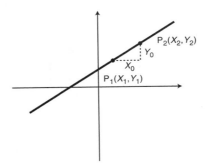

The value of m can be calculated as:

$$m = \frac{Y_0}{X_0} = \frac{Y_2 - Y_1}{X_2 - X_1}$$

If the value of b is not given but the values for points P_1 and P_2 are known, then Y can be obtained by:

$$Y - Y_1 = \frac{Y_2 - Y_1}{X_2 - X_1}(X - X_1)$$

where: X_1 = X value at point P_1
Y_1 = Y value at point P_1
X_2 = X value at point P_2
Y_2 = Y value at point P_2

For example:

$$Y - 2 = \frac{1 - 2}{-1 - 3}(X - 3)$$

$$Y - 2 = \frac{-1}{-4}(X - 3)$$

$$Y = \frac{1}{4}X - \frac{3}{4} + 2$$

$$Y = \frac{1}{4}X + 1\frac{1}{4}$$

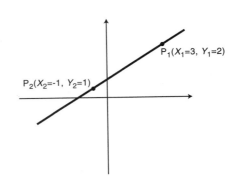

NUMBER TABLES

Powers of Two		Powers of Eight	
2^n	n	8^n	n
1	0	1	0
2	1	8	1
4	2	64	2
8	3	512	3
16	4	4,096	4
32	5	32,768	5
64	6	262,144	6
128	7	2,097,152	7
256	8	16,777,216	8
512	9	134,217,728	9
1,024	10	1,073,741,824	10
2,048	11	8,589,934,592	11
4,096	12	68,719,476,736	12
8,192	13	549,755,813,888	13
16,384	14	4,398,046,511,104	14
32,768	15	35,184,372,088,832	15

Powers of Sixteen	
16^n	n
1	0
16	1
256	2
4,096	3
65,536	4
1,048,576	5
16,777,216	6
268,435,456	7
4,294,967,296	8
68,719,476,736	9
1,099,511,627,776	10
17,592,186,044,416	11
281,474,976,710,656	12
4,503,599,627,370,496	13
72,057,594,037,927,936	14
1,152,921,504,606,846,976	15

APPENDIX F: ABBREVIATIONS AND ACRONYMS

A	ampere
AC	alternating current
A/D	analog-to-digital converter
AI	artificial intelligence
ANSI	American National Standards Institute
ASCII	American Standard Code for Information Exchange
ASI	actuator sensor interface
AWG	American Wire Gauge
BCC	block check character
BCD	binary coded decimal
°C	degrees Celsius
CAN	control area network
CIM	computer-integrated manufacturing
CNC	computer numerical control
CPU	central processing unit
CRC	cyclic redundancy check
CRT	cathode ray tube
CSMA/CD	carrier sense multiple access with collision detection
CX-ORC	cyclic exclusive-OR checksum
D/A	digital-to-analog converter
DC	direct current
DCE	data communication equipment
DIN	Deutsch Industrie Norm (German equivalent of the EIA)
DR	derivative register
DTE	data terminal equipment
EAROM	electrically alterable read-only memory
EEPROM	electrically erasable programmable read-only memory
EIA	Electronic Industries Association
emf	electromotive force
EMI	electromagnetic interference
EPROM	erasable programmable read-only memory
°F	degrees Fahrenheit
FIFO	first-in first-out
FMS	flexible manufacturing system
GUI	graphic user interface
IEC	International Electrotechnical Commission
IEEE	Institute of Electrical and Electronics Engineers
I/O	input/output
IR	integral register
ISA	Instrument Society of America
ISO	International Standards Organization
ISP	Interoperable Systems Project Foundation
IVR	input variable register
JIC	Joint Industrial Council
K	1024 bytes
°K	degrees Kelvin
LAN	local area network

LCD	liquid crystal display
LED	light-emitting diode
LIFO	last-in first-out
LRC	longitudinal redundancy check
LSB	least significant bit/byte
LVDT	linear variable differential transformer
mA	milliampere
MAP	Manufacturing Automation Protocol
MCR	master control relay
MOV	metal oxide varistor
MSB	most significant bit/byte
MTBF	mean time between failures
mF	microfarads
msec	microsecond
msec	millisecond
mV	millivolts
NEC	National Electric Code
NEMA	National Electrical Manufacturers Association
NOVRAM	nonvolatile random-access memory
OEM	original equipment manufacturer
OSI	Open Systems Interconnection reference model
OVR	output variable register
PC	personal computer
pF	picofarad
PID	proportional-integral-derivative
PLC	programmable logic controller
PR	proportional register
PROM	programmable read-only memory
RAM	random-access memory
RC	resistor-capacitor network
RFI	radio frequency interference
ROM	read-only memory
RTD	resistance temperature detector
R/W	read/write
SCR	safety control relay
SDS	smart distributed system
SPR	set point register
TTL	transistor-transistor logic
TTY	teletype
TWS	thumbwheel switch
V	volts
VA	volt-ampere
VAC	volts AC
VDC	volts DC
VP	virtual position
VRC	vertical redundancy check
VS	variable speed
XOR	exclusive OR

APPENDIX G: VOLTAGE-CURRENT LAPLACE TRANSFER FUNCTION RELATIONSHIPS

Table G-1 presents resistor, inductor, and capacitor voltage-current relationships and their corresponding Laplace equations. Note that the inductor voltage is a derivative term of the current, while the capacitor voltage is an integral term of the current. Table G-2 illustrates these same relationships in system block diagram form. In this table, if the current is the input to the resistor block transfer function, the result is the voltage ($I_{R(s)} \times R = V_{R(s)}$). The same holds true for the derivative block transfer function of an inductor and the integral block transfer function of a capacitor. If the voltage is the input, the output of each transfer function will be the current.

Element and Representation	Voltage-Current Relationship	Laplace Transform Relationship
Resistor $\quad$ i_R $\quad R$ $\quad v_R$	$v_{R(t)} = Ri_{R(t)}$	$V_{R(s)} = RI_{R(s)}$
Inductor $\quad$ i_L $\quad L$ $\quad v_L$	$v_{L(t)} = L\dfrac{di_{L(t)}}{dt}$	$V_{L(s)} = LsI_{L(s)}$
Capacitor $\quad$ i_C $\quad C$ $\quad v_C$	$v_{C(t)} = \dfrac{1}{C}\displaystyle\int_0^t i_{C(t)}\,dt$	$V_{C(s)} = \dfrac{1}{Cs}I_{C(s)}$

Table G-1. Voltage-current Laplace transfer function relationships.

	Current Input	Voltage Input
Resistor	$I_{R(s)} \rightarrow \boxed{R} \rightarrow V_{R(s)}$	$V_{R(s)} \rightarrow \boxed{\dfrac{1}{R}} \rightarrow I_{R(s)}$
Inductor	$I_{L(s)} \rightarrow \boxed{Ls} \rightarrow V_{L(s)}$	$V_{L(s)} \rightarrow \boxed{\dfrac{1}{Ls}} \rightarrow I_{L(s)}$
Capacitor	$I_{C(s)} \rightarrow \boxed{\dfrac{1}{Cs}} \rightarrow V_{C(s)}$	$V_{C(s)} \rightarrow \boxed{Cs} \rightarrow I_{C(s)}$

Table G-2. Block diagrams of transfer functions.

GLOSSARY

A

AC/DC I/O interface. A discrete interface that converts alternating current (AC) voltages from field devices into direct current (DC) signals that the processor can use. It can also convert DC signals into proportional AC voltages.

action. A set of control instructions prompting a PLC to perform a certain control function during the execution of a sequential function chart step.

acyclic message. An unscheduled message transmission.

A/D. *See* **analog-to-digital converter**.

address. (1) The location in a computer's memory where particular information is stored. (2) The alphanumeric value used to identify a specific I/O rack, module group, and terminal location.

addressability. The total number of devices that can be connected to a network.

address field. The sequence of eight (or any multiple of eight) bits immediately following the opening flag sequence of a frame, which identifies the secondary station that is sending (or is designated to receive) the frame.

AI. *See* **artificial intelligence**.

algorithm. A set of procedures used to solve a problem.

alphanumeric code. A character string consisting of a combination of letters, numbers, and/or special characters used to represent text, commands, numbers, and/or code groups.

ambient temperature. The temperature of the air surrounding a device.

American National Standards Institute (ANSI). A clearinghouse and coordinating agency for voluntary standards in the United States.

American Wire Gauge (AWG). A standard system used to designate the size of electrical conductors. Gauge numbers have an inverse relationship to size; larger gauges have a smaller diameter.

analog device. An apparatus that measures continuous information signals (i.e., signals that have an infinite number of values). The only limitation on resolution is the accuracy of the measuring device.

analog input interface. An input circuit that uses an analog-to-digital converter to translate a continuous analog signal, measured by an analog device, into a digital value that can be used by the processor.

analog output interface. An output circuit that uses a digital-to-analog converter to translate a digital value, sent from the processor, into an analog signal that can control a connected analog device.

analog signal. A continuous signal that changes smoothly over a given range, rather than switching suddenly between certain levels as discrete signals do.

analog-to-digital converter (A/D). A device that translates analog signals from field devices into binary numbers that can be read by the processor.

AND. A logical operator that requires all input conditions to be logic 1 for the output to be logic 1. If any input is logic 0, then the output will be logic 0.

ANSI. *See* **American National Standards Institute**.

application. (1) A machine or process monitored and controlled by a PLC. (2) The use of computer or processor-based routines for specific purposes.

application memory. The part of the total system memory devoted to storing the application program and its associated data.

application program. The set of instructions that provides control, data acquisition, and report generation capabilities for a specific process.

arithmetic instructions. Computer programming codes that give a PLC the ability to perform mathematical functions, such as addition, subtraction, multiplication, division, and square root, on data.

artificial intelligence (AI). A subfield of computer science dealing with the development of computer programs that solve tasks requiring extensive knowledge.

ASCII. For *American Standard Code for Information Interchange*. A seven-bit code with an optional parity bit used to represent alphanumeric, punctuation, and control characters.

ASCII I/O interface. A special function interface that transmits alphanumeric data between peripheral equipment and a PLC.

assembly language. A symbolic programming language that can be directly translated into machine language instructions.

asynchronous. Recurrent or repeated operations that occur in unrelated patterns over time.

AWG. *See* **American Wire Gauge**.

B

back plane. A printed circuit board, located in the back of a chassis, that contains a data bus, power bus, and mating connectors for modules that will be inserted into the chassis.

backup. A device or system that is kept on hand to replace a device or system that fails.

backward chaining. A method of finding the causes of an outcome by analyzing its consequents to obtain its antecedents.

bandwidth. The range of frequencies expressed in Hertz over which a system is designed to operate.

base. The maximum number of digits used to represent values in a number system.

baseband coaxial cable. A communication medium that can send one transmission signal at a time at its original frequency.

BASIC module. An intelligent I/O interface capable of performing computational tasks without affecting the PLC processor's computing time.

battery backup. A battery or set of batteries that will provide power to the processor's memory in the event of a power outage.

baud. (1) The reciprocal of the shortest pulse width in a data communication stream. (2) The number of binary bits transmitted per second during a serial data transmission.

Baye's theorem. An equation that defines the probability of one event occurring based on the fact that another event has already occurred.

BCC. *See* **block check character**.

BCD. *See* **binary coded decimal**.

binary coded decimal (BCD). A binary number system in which each decimal digit from 0 to 9 is represented by four binary digits (bits). The four positions have a weighted value of 1, 2, 4, and 8, respectively, starting from the least significant (right-most) bit.

binary number system. A base 2 number system that uses only the numbers 0 and 1 to express all values. Each digit position of a binary number has a weighted value of 1, 2, 4, 8, 16, 32, 64, and so on, starting with the least significant (right-most) digit.

bit. For *binary digit*. The smallest unit of binary information. A bit can have a value of 1 or 0.

bit rate. *See* **baud**.

bit-wide bus network. An I/O bus network that interfaces with discrete devices that transmit less than 8 bits of data at a time.

blackboard architecture. The distribution of knowledge inferencing, as well as global and knowledge databases, in a control system through the use of several subsystems containing local, global, and knowledge databases that work independently of each other.

block. A group of words transmitted as a unit.

block check character (BCC). A character, placed at the end of a data block, that corresponds to the characteristics of the block.

block diagram. A schematic drawing.

block length. The total number of words transmitted at one time.

block transfer. A programming technique used to transfer up to 64 words of data to or from an intelligent I/O module.

Boolean action. A set of control instructions that assigns a discrete value to a variable during a sequential function chart step.

Boolean language. A PLC programming language, based primarily on the Boolean logic operators, that implements all of the functions of the basic ladder diagram instruction set.

Boolean operators. Logical operators, such as AND, OR, NAND, NOR, NOT, and exclusive-OR, that can be used singly or in combination to form logical statements that have output responses of TRUE or FALSE.

Boolean variable. A single-bit variable whose value is transmitted in the form of 1s and 0s.

Bourdon tube. A pressure transducer available in spiral, helical, twisted, and C-tube configurations that converts pressure measurements into displacement.

branch. A parallel logic path within a rung.

breadth-first search. A method of rule evaluation that evaluates each rule in the same level of a decision tree before proceeding downward.

bridge circuit. A mechanism found in transducer circuits that uses resistors to change the parameters (e.g., voltage and current) of an incoming signal.

broadband coaxial cable. A communication medium that can transmit two or more transmission signals at one time via frequency division multiplexing.

burn-in procedure. The process of operating a device at an elevated temperature to identify early-failing parts.

bus. (1) A group of lines used for data transmission or control. (2) Power distribution conductors.

bus topology. A network configuration in which all stations are connected in parallel with the communication medium and all stations can receive information from any other station on the network.

bypass/control station. A device that allows a process to be switched to either PLC or manual control.

byte. A group of eight adjacent bits that are operated on as a unit, such as when moving data to or from memory.

byte-wide bus network. An I/O bus network, which interfaces with discrete and small analog devices, that can transmit between 1 and 50 or more bytes of data at a time.

C

cascade control. The use of two controllers to regulate a process so that the feedback loop of one controller is the set point of the other controller.

center of gravity method. A method of calculating the final output value of a fuzzy logic controller by finding the value that corresponds to the center of the mass under the control output curve.

centralized control. A PLC control system organization in which a central PLC controls several machines or processes.

central processing unit (CPU). The part of a programmable controller responsible for reading inputs, executing the control program, and updating outputs. Sometimes referred to as the *processor*, the CPU consists of the arithmetic logic unit, timing/control circuitry, accumulator, scratch pad memory, program counter, address stack, and instruction register.

centroid. The point in a geometrical figure whose coordinates equal the average of all the other points comprising the figure.

channel. A designated path for a signal.

channel capacity. The amount of information that can be transmitted per second on a given communication channel depending on the medium, line length, and modulation rate.

character. One symbol of a set of elementary symbols, such as a letter of the alphabet or a number.

chassis. A hardware assembly that houses PLC devices, such as I/O modules, adapter modules, processor modules, power supplies, and processors.

checksum. A transmission verification algorithm that adds the binary values of all the characters in a data block and places the sum in the block check character position.

chip. A very small piece of semiconductor material that holds electronic components. Chips are normally made of silicon and are typically less than 1/4 inch square and 1/100 inch thick.

closed loop. A control system that uses feedback from the process to maintain outputs at a desired level.

coaxial cable. A transmission medium, consisting of a central conductor surrounded by dielectric materials and an external conductor, that possesses a predictable characteristic impedance.

code. (1) A binary representation of numbers, letters, or symbols that have some meaning. (2) A set of programmed instructions.

coil. A ladder diagram symbol that represents an output instruction.

cold junction compensation. A compensation factor that allows a thermocouple to operate as though it has an ice-point reference.

collision detection (CSMA/CD). A network access method in which each node waits until there is no traffic on the network then transmits its message. If the node detects another transmission on the network, it will disable its transmitter and wait until the network clears before retransmitting the message.

combined error. *See* **propagation error.**

common bus topology. A network configuration in which individual PLCs connect to a main trunkline in a multidrop fashion.

compatibility. (1) The ability of various specified units to replace one another with little or no reduction in capability. (2) The ability of units to be interconnected and used without modification.

complement. A logical operation that inverts a signal or bit.

conditional probability inferencing. The conditional probability of an event happening in an artificial intelligence system.

constant voltage transformer. A transformer that maintains a steady output voltage (secondary) regardless of input voltage (primary) fluctuations.

contact. A ladder diagram symbol that represents an input condition.

contact output interface. A discrete interface, which does not require an external power source, that is triggered by the change in state of a normally open or normally closed contact.

contact symbology. A set of symbols used to express a control program through conventional relay symbols (e.g., normally open contacts, normally closed contacts, etc.).

continuous-mode controller. A process controller that sends an analog signal to a process control field device.

control element. The output field device that regulates the actual control variable level in a process control system.

control logic. The control plan for a given system.

control loop. The method of adjusting the control variable in a process control system by analyzing process variable data and then comparing it to the set point to determine the amount of error in the system.

control panel. A panel that contains instruments used to control devices.

control program checkout. A final review of a PLC's control program prior to starting up the system.

control program printout. A hard copy of the control logic program stored in a PLC's memory.

control strategy. The sequence of steps that must occur during a process or PLC program to produce the desired output control.

control task. The desired results of a control program.

control variable. The independent variable in a process control system that is used to adjust the dependent variable, the process variable.

convergence. A point in a sequential function chart where many elements flow into one element.

counter. An electromechanical device that counts the number of times an event occurs.

counter instructions. Computer programming codes that allow a PLC to perform the counting functions (count up, count down, counter reset) of a hardware counter.

CPU. *See* **central processing unit**.

CRC. *See* **cyclic redundancy check**.

critically damped response. A second-order control system response in which the damping coefficient equals 1, causing the response to overshoot the set point and then quickly settle back to it.

CSMA/CD. *See* **collision detection**.

current loop. A two-wire communication link in which the presence of a 20 milliamp current level indicates a binary 1 (mark) and its absence indicates no data, a binary 0 (space).

CX-ORC. *See* **cyclic exclusive-OR checksum**.

cyclic exclusive-OR checksum (CX-ORC). An error detection method in which the words in the data block are exclusive-ORed with the checksum word and then rotated to the left. This action is repeated until all of the words in the block have been operated on.

cyclic message. A scheduled message transmission.

cyclic redundancy check (CRC). An error detection method in which all the bits in a block are divided by a predetermined binary number. The remainder becomes the block check character.

D

D/A. *See* **digital-to-analog converter**.

data. A general term for any type of information.

data link layer. Layer 2 of the OSI network protocol. This layer provides functional and procedural means for establishing, maintaining, and releasing data link connections among network entities.

data manipulation instructions. Computer codes that provide a PLC with the ability to compare, convert, shift, examine, and operate on data in multiple registers.

data table. The part of a processor's memory, containing I/O values and files, where data is monitored, manipulated, and changed for control purposes.

data transfer instructions. Computer codes that allow a PLC to move numerical data within a controller, either in single register units or in blocks of registers.

DC I/O interface. A discrete module that links a processor with direct current field devices.

dead time. The delay between the time a control system's control variable changes and the time the process variable begins to respond to the change.

debouncing. The act of removing intermediate noise from a mechanical switch.

decimal number system. A base 10 number system that uses ten numbers—0, 1, 2, 3, 4, 5, 6, 7, 8, and 9—to represent all values. Each digit position has a weighted value of 1, 10, 100, 1000, and so on, beginning with the least significant (right-most) digit.

defuzzification. The process of converting a fuzzy logic controller's output conclusions into real output data and sending the data to the field device.

depth-first search. A rule evaluation method that evaluates all the rules in a downward branch of a decision tree before proceeding to the next branch.

derivative controller. A continuous-mode controller whose output to the control field device is proportional to the rate of change of error in the system.

device bus network. A network that allows low-level input/output devices that transmit relatively small amounts of information to communicate directly with a PLC.

diagnostic AI system. The lowest level of artificial intelligence system. This type of system primarily detects faults within an application but does not provide information about possible solutions.

diagnostics. The detection and isolation of an error or malfunction.

differential input/output. A signal transmission system where inputs and outputs have individual return lines for each channel, as opposed to all data running through one line.

digital device. A device that processes and sends discrete (two-state) electrical signals.

digital signal. A noncontinuous signal that has a finite number of values.

digital-to-analog converter (D/A). A device that translates binary numbers from a processor into analog signals that field devices can understand.

direct-acting controller. A closed-loop controller whose control variable output increases in response to an increase in the process variable.

direct action I/O interface. A special I/O interface that detects, preprocesses, and transmits low-level and fast-speed signals.

discrete input interface. An input circuit that allows a PLC to receive data from digital field devices.

discrete-mode controller. A controller that sends a noncontinuous signal to the field device controlling a process.

discrete output interface. An output circuit that allows a PLC to send data to digital field equipment.

displacement transducer. A device that measures the movement of an object.

distributed control. A PLC control system organization in which factory or machine control is divided into several subsystems, each managed by a separate PLC, yet all interconnected to form a single entity.

distributed I/O processing. The allocation of various control tasks to several intelligent I/O interfaces.

divergence. A point in a sequential function chart where one element flows into many elements.

documentation. An orderly collection of recorded hardware and software information about a control system. These records provide valuable reference data for installing, debugging, and maintaining the PLC.

double-precision arithmetic. Arithmetic instructions that use double the number of registers than single-precision arithmetic to hold the operands and result (i.e., two registers each for the operands and two or four registers for the result).

downtime. The time when a system is not available for use.

dynamic system checkout. The process of verifying the correct operation of a control program by actually implementing it.

E

EAROM. *See* **electrically alterable read-only memory**.

EEPROM. *See* **electrically erasable programmable read-only memory**.

EIA. *See* **Electronic Industries Association.**

electrically alterable read-only memory (EAROM). A type of nonvolatile, programmable, read-only memory that can be erased completely by applying the proper voltage to the memory chip.

electrically erasable programmable read-only memory (EEPROM). A type of nonvolatile, programmable, read-only memory that can be erased by electrical pulses.

Electronic Industries Association (EIA). An agency that sets electrical/electronic standards.

encoder/counter module. An interface, which is used in positioning applications, that links encoders and high-speed counter devices with programmable logic controllers.

enhanced ladder language. A PLC language that implements basic ladder language instructions, as well as more sophisticated functional block instructions, which can perform multiple operations in a single instruction.

EPROM. *See* **erasable programmable read-only memory**.

erasable programmable read-only memory (EPROM). A type of nonvolatile, programmable, read-only memory that can be erased with ultraviolet light.

error. The difference between the set point and the process variable in a control system.

error-correcting code. A code in which each acceptable expression conforms to specific rules of construction that also define one or more nonacceptable expressions, so that if certain errors occur, they can be detected and corrected.

error deadband. The amount that the process variable can fluctuate from the set point before the control system provides corrective action.

error-detecting code. A code in which each expression conforms to specific rules of construction, so that if an error occurs in an expression, it can be detected.

exclusive-OR (XOR). A logical operation, which has only two inputs, that yields a logic 1 output if only one of the two inputs is logic 1 and a logic 0 output if both inputs are the same, either logic 1 or logic 0.

execute. To perform a specific operation by processing either one instruction, a series of instructions, or a complete program.

execution time. The time required to perform one specific instruction, a series of instructions, or a complete program.

executive memory. The part of the system memory that permanently stores a system's supervisory programs, as well as instruction software. This area of memory is not accessible to the user.

expert AI system. The highest level of AI systems. This type of system detects process faults, provides information about possible causes of the faults, and makes complex decisions about resulting actions based on statistical analysis.

F   **FALSE.** As related to PLC instructions, a reset logic state associated with a binary 0.

fast-input interface. An intelligent I/O module that functions as a pulse stretcher, detecting very fast input pulses that regular I/O modules cannot read.

fast-response interface. A special I/O module designed to detect fast inputs and respond with an output.

FBD. *See* **function block diagram**.

feedback. The signal or data transmitted to a PLC from a controlled machine or process to denote its response to the command signal.

fiber-optic cable. A communication medium composed of thin fibers of glass or plastic enclosed in a material with low refraction.

first-order response. A process response to a rapid change in the control variable characterized by one lag time and a process response curve that slowly approaches the set point.

floating-point math. A data manipulation format, which is used to express a number by expressing the power of the base, that usually involves the use of two sets of digits. For example, in a floating decimal notation where the base is 10, the number 8,700,000 would be expressed as 8.7(10)6 or 8.7E6.

flowchart. A graphical representation of the definition or solution of a task or problem.

flowcharting. A method of pictorially representing the operation of a process in a sequential manner.

flow transducer. A device that measures the amount of solid, liquid, or gaseous materials flowing through a process by measuring either weight, differential pressure, or fluid motion.

forward chaining. A method for determining all possible outcomes for a given set of inputs.

frequency shift keying (FSK). A signal modulation technique that offers a high amount of noise immunity in which a carrier frequency is shifted to high or low to represent a binary 1 or 0, respectively.

FSK. *See* **frequency shift keying**.

full-duplex line. A communication line used to simultaneously transmit data in two directions.

function block diagram (FBD). A graphical PLC programming language in which instructions are programmed as blocks that are then used as needed to control process elements.

fuzzification. The translation of input data into fuzzy logic membership sets.

fuzzy logic. The branch of artificial intelligence that deals with reasoning algorithms used to simulate human judgment.

fuzzy logic interface. A special I/O interface that provides intelligent, closed-loop process control by analyzing input data according to specified mathematical algorithms and then providing a correlating output response.

fuzzy processing. The interpretation of fuzzy input data to determine an appropriate outcome based on user-programmed IF...THEN rules.

fuzzy set. A group of membership functions.

G

gate. A circuit having two or more input terminals and one output terminal, where an output is present only when the prescribed inputs are present.

gateway. A device or pair of devices that connects two or more communication networks. This device may act as a host to each network and may transfer messages between the networks by translating their protocols.

global database. The section of an AI system that stores data measurements from the controlled process.

grade. A measure of how well a value fits into a given membership function.

Grafcet. A PLC programming language that uses an object-oriented, flowchart-like framework, along with steps, transitions, and actions, to define the control program.

Gray code. A cyclic code, similar to a binary code, in which only one bit changes as the counting number increases.

gross error. An error resulting from human miscalculation.

ground loop. A condition in which two or more electrical paths exist within a ground line.

guarantee error. A value of error derived from a known specification that defines the amount that a product or material will arithmetically deviate from the mean.

H

half-duplex line. A communication line that can transmit data in two directions, but in only one direction at a time.

Hamming code. An error-detecting code that combines parity and data bits to generate a byte containing a value that identifies the erroneous bit.

hard copy. A printed document.

hardware. All the physical components of a programmable controller, including peripherals, as opposed to the software components that control its operation.

hardwired logic. Logic control functions that are determined by the way a system's devices are physically interconnected.

hexadecimal number system (hex). A base 16 number system that uses the numbers 0, 1, 2, 3, 4, 5, 6, 7, 8, and 9 and the letters A, B, C, D, E, and F to represent numbers and codes.

host. A central computer in a network system.

I

IEC 1131 programming standard. A standardized set of PLC programming guidelines, set forth by the International Electrotechnical Commission, that includes general PLC information, equipment and test requirements, programming languages, user guidelines, and communication standards.

IEEE 802. A family of standards specified by the Institute of Electrical and Electronic Engineers for data communication over local and metropolitan area networks.

IL. *See* **instruction list**.

image table. An area in a PLC's memory dedicated to I/O data where 1s and 0s represent ON and OFF conditions, respectively.

individual control. A PLC control system organization in which a PLC controls a single machine or process.

inference engine. The section of an AI system where all decisions are made using the knowledge stored in the knowledge database.

input. Information sent to the processor from connected devices.

input device. Any connected equipment, such as control devices (e.g., switches, buttons, and sensors) or peripheral devices (e.g., cathode ray tubes and manual programmers), that supply information to the central processing unit. Each type of input device has a unique interface to the processor.

input/output system. A collection of plug-in modules that transmit control data between a PLC and field devices.

input table. The area of a PLC's memory where information about the status of input devices is stored.

instruction list (IL). A low-level, text-based PLC programming language that uses assembly language–like mnemonics to represent the control program.

integer variable. A nondiscrete variable whose value is transmitted in the form of a whole number.

integral controller. A continuous-mode controller whose output to the control field device changes according to how the error signal changes over time.

integral of time and absolute error open-loop tuning method (ITAE). A method used to determine the proper tuning constants for a controller based on the minimization of the integral of time and the absolute error of the response.

integral windup. The situation in which the control variable in a system remains at its maximum level even though the amount of error in the system starts to decrease.

intelligent I/O interface. A microprocessor-based module that can perform sophisticated processing functions independently of the central processing unit.

interface. A circuit that permits communication between a central processing unit and a field input or output device. Different devices require different interfaces.

interlock. A device actuated by the operation of another device to which it is linked to govern the succeeding operation of the same or allied devices.

internal output. A program output that does not drive a field device and is used for internal purposes only. It provides interlocking functions like a hardwired control relay. An internal output may also be referred to as an *internal storage bit* or an *internal coil*.

internal storage address assignment document. A document that identifies the address, type, and function of every internal used in a control program.

International Standards Organization (ISO). An organization established to promote the development of international standards.

interrupt. The act of redirecting a program's execution to perform a more urgent task.

I/O address. A unique number, assigned to each input/output device, that corresponds to the device's location in the rack enclosure. The address number is used when programming, monitoring, or modifying a specific input or output.

I/O address assignment document. A document that identifies every field device by address, type of input/output module, type of field device, and the function the field device performs.

I/O bus network. A network that lets input and output devices communicate directly to a PLC through digital communication.

I/O bus network scanner. A device connected to a PLC that reads and writes to field devices connected to an I/O bus network, as well as decodes the data in the network information packet.

I/O module. A plug-in assembly, containing two or more identical input or output circuits, that provides the connection between a processor and connected devices. Normal I/O module capacities are 2, 4, 8, and 16 circuits.

I/O scan time. The time required to update all local and remote I/O.

I/O update scan. The process of revising the bits in a PLC's I/O tables based on the latest results from reading the inputs and processing the outputs according to the control program.

I/O wiring connection diagram. A drawing that shows the actual connections of the field I/O devices to a PLC, including power supplies and subsystem connections.

ISO. *See* **International Standards Organization.**

isolated I/O interface. An input module in which each input has a separate return line. Isolated I/O interfaces can connect field devices powered from different sources to one module.

isolation transformer. A transformer that protects its connected devices from surrounding electromagnetic interference.

ITAE. *See* **integral of time and absolute error open-loop tuning method.**

K. 2^{10}. Used to denote memory size in either bits, bytes, or words.

knowledge AI system. A mid-level AI system that detects faults based on resident knowledge and also makes decisions about the cause of the fault and ensuing process actions.

knowledge database. The section of an AI system that stores information extracted from the expert.

knowledge inference. A decision-making methodology used to gather and analyze process data in order to draw conclusions.

knowledge representation. The way an artificial intelligence system strategy is organized.

L

label. A name given to a membership function.

ladder diagram. An industry standard for representing relay logic control systems.

ladder diagram language (LD). A graphical set of instructions that implements basic relay ladder functions in a PLC.

ladder relay instructions. Computer codes that implement relay coils and contacts and their corresponding functions in a PLC.

ladder rung matrix. A rectangular array that defines the maximum number of contacts that can be programmed in a ladder rung, along with the maximum number of parallel branches allowed in the rung.

lag time. The delay between the initial response of the process variable to a change in the control variable and the process variable's optimal response to it.

LAN. *See* **local area network.**

language. A set of symbols and rules for representing and communicating information between people and machines.

Laplace transform. A mathematical function used to convert differential equations from the time domain into the frequency domain so that they become easy-to-manage algebraic equations.

LCD. *See* **liquid crystal display.**

LD. *See* **ladder diagram language.**

lead resistance compensation. A factor that compensates for signal loss due to resistance present in electrical wires.

least significant bit (LSB). The bit representing the smallest value in a nibble, byte, or word.

least significant digit (LSD). The digit representing the smallest value in a byte or word.

LED. *See* **light-emitting diode.**

light-emitting diode (LED). A semiconductor diode whose junction emits light when current passes through it in a forward direction.

limit switch. An electrical switch actuated by the motion of a machine or equipment.

linear variable differential transformer (LVDT). An electromechanical mechanism that provides a voltage reference that is proportional to the movement or displacement of a core inside a coil.

liquid crystal display (LCD). A display device consisting of a liquid crystal hermetically sealed between two glass plates.

load. The power used by a machine or apparatus.

load cell. A force or weight transducer that is based on a direct application of a bonded strain gauge.

local area network (LAN). An ensemble of interconnected processing elements (nodes), which are typically located within a few miles of each other.

local rack. An enclosure, placed in the same area as the master rack, that contains a local I/O processor, which sends data to and from the central processing unit.

location. A storage position or register in memory identified by a unique address.

logic. The process of solving complex problems through the use of simple functions that can be either true or false.

logic diagram. A drawing that uses interconnected AND, OR, and NOT logic symbols to graphically describe a system's operation or control.

longitudinal redundancy check (LRC). An error-checking technique based on an accumulated exclusive-OR of transmitted characters. LRC characters are accumulated at both the sending and receiving stations.

loop tuning. The process of determining the proportional, integral, and derivative constants that will allow a PID controller to perform optimally.

LRC. *See* **longitudinal redundancy check.**

LSB. *See* **least significant bit.**

LSD. *See* **least significant digit.**

LVDT. *See* **linear variable differential transformer.**

M MAC. *See* **medium access control.**

macrostep. A small sequential function chart program embedded as an action within a larger sequential function chart.

mask. A logical function used to set certain bits in a word to an established state.

master. A device used to control other devices.

master control relay (MCR). A hardwired or softwired relay instruction that will de-energize its associated I/O devices when the instruction is de-energized.

master rack. The enclosure containing the CPU or processor module.

master/slave bus topology. A network configuration in which one master controller manages several slave controllers.

maximum value method. A method of calculating the final output value of a fuzzy logic controller by finding the rule output value with the highest membership function grade.

MCR. *See* **master control relay**.

mean. The average value of a set of data readings.

mean-time-between-failures study. A study, which contains data about the average time between equipment failures, that provides information about the reliability of a product.

median. The middle value of a set of data readings organized in ascending order.

medium access control (MAC). A technique that ensures that only one device is transmitting on a network at any given time.

membership function. A group of fuzzy logic rules used to divide input data into sets, which are then analyzed to provide reasoned control of a field device.

memory. The part of a programmable controller that stores data, instructions, and the control program either temporarily or semipermanently.

memory map. A diagram showing a system's memory addresses, as well as which programs and data are assigned to each section of memory.

message. A group of data and control bits transferred as an entity from a data source.

microprocessor. A digital, electronic logic package (usually on a single chip) capable of performing the program execution, control, and data-processing functions of a central processing unit. A microprocessor usually contains an arithmetic logic unit, temporary storage registers, instruction decoder circuitry, a program counter, and bus interface circuitry.

miniprogrammer. A portable device used for programming, changing, and monitoring a PLC's control logic.

mode. The most frequently occurring value in a set of data readings.

module. An interchangeable, plug-in item containing electronic components.

most significant bit (MSB). The bit representing the greatest value of a nibble, byte, or word.

most significant digit (MSD). The digit representing the greatest value of a byte or word.

MSB. *See* **most significant bit**.

MSD. *See* **most significant digit**.

multidrop link. A cable that terminates at more than one point.

multiplexing. The act of channeling two or more signals to one source using the same channel.

multiprocessing. Concurrent execution of two or more tasks residing in memory.

N **NAND.** A logical operator that yields a logic 1 output if any input is logic 0 and a logic 0 output if all inputs are logic 1. This operator is a negated AND function, the result of negating the output of an AND gate by following it with a NOT symbol.

negative logic. The use of binary logic so that logic 0 represents the voltage level normally associated with logic 1 (i.e., logic 0 = +5 V, logic 1 = 0 V).

network. A series of points (or devices) connected by some type of communication medium.

network communications instructions. Computer codes that allow a PLC to share data with other PLCs connected to a local area network.

network interface module. A special function interface that allows PLCs and other intelligent devices to communicate and transfer data over a high-speed local area communication network.

network layer. Layer 3 of the OSI protocol. This layer routes information in the network.

nibble. A group of four bits.

node. A station, such as a personal computer or a PLC, that is connected to a network and can thereby send and receive messages through the network.

nonreturn to zero invert on ones (NRZI). A self-clocking pulse code used to establish reliable synchronous transmission.

nonvolatile memory. A type of memory whose contents are not lost or disturbed if operating power is lost.

NOR. A logical operator that yields a logic 1 output if all inputs are logic 0 and a logic 0 output if any input is logic 1. This operator is a negated OR function, the result of negating the output of an OR gate by following it with a NOT symbol.

normal action. A set of IEC 1131-3 instructions that is executed continuously for the duration of an SFC step's activity.

normally closed contact. (1) A relay contact pair that is closed when the coil of the relay is not activated and open when the coil is activated. (2) A ladder program symbol that allows logic continuity (flow) if the referenced input is logic 0 when evaluated.

normally open contact. (1) A relay contact pair that is open when the coil of the relay is not activated and closed when the coil is activated. (2) A ladder program symbol that allows logic continuity (flow) if the referenced input is logic 1 when evaluated.

NOT. A logical operator that yields a logic 1 output if a logic 0 is entered at the input and a logic 0 output if a logic 1 is entered at the input. The NOT function, also called an *inverter*, is normally used in conjunction with AND and OR functions.

NRZI. *See* **nonreturn to zero invert on ones.**

O **octal number system.** A base 8 number system that uses eight numbers—0, 1, 2, 3, 4, 5, 6, and 7—to represent all values.

off-line. The state of not being in continuous direct communication with the processor.

one's complement. An operation that represents the negative value of a binary word by assigning the most significant bit of the word with a value equal to its normal value minus one.

one shot. A programming technique that sets a storage bit or output to a certain state for only one scan.

on-line. The state of being in continuous communication with the processor.

open loop. A control system that does not receive process feedback in order to perform self-correcting actions.

optical coupler. A device that couples signals from one circuit to another by means of electromagnetic radiation.

OR. A logical operator that yields a logic 1 output if any input is logic 1 and a logic 0 output if all inputs are logic 0.

orifice plate. A transducer that measures fluid flow by measuring the pressure differential between two points.

OSI model. A description of network communications functions organized in seven layers to promote open system interconnections.

output. Information sent from the processor to connected field devices.

output device. Any connected equipment, such as control devices (e.g., motors, solenoids, and alarms) or peripheral devices (e.g., line printers, disk drives, and color displays), that receives information or instructions from the central processing unit. Each type of output device has a unique interface to the processor.

output table. The area of a PLC's memory where information about the status of output devices is stored.

overdamped response. A second-order control system response in which the damping coefficient is greater than 1, causing the response to overshoot the set point and then slowly settle back to it.

P

packet. Data and sequences of control bits arranged in a specified format and transferred as an entity during data transmission.

panel enclosure. The physical enclosure that houses a PLC's hardware and components.

parallel circuit. A circuit in which two or more of the connected components or contact symbols in a ladder program are connected to the same pair of terminals so that current may flow through all the branches.

parity. The even or odd characteristic of the number of 1s in a byte or word of memory.

parity bit. A bit added to a memory word as a means of error detection.

parity check. A check for a certain number of 1s and 0s in a memory word to ensure data integrity.

peripherals. External devices, such as line printers, disk drives, recorders, etc., that are connected to a PLC.

PID interface. *See* **proportional-integral-derivative interface**.

PLC. *See* **programmable logic controller**.

polling. A network access method where a master controller manages the communication process by interrogating each slave controller under it to determine whether the slave has any information to send.

positive logic. The conventional use of binary logic in which logic 1 represents a positive logic level (e.g., logic 1 = +5 V, logic 0 = 0 V).

potentiometer. A simple transducer that measures displacement based on resistance changes due to the movement of a wiper arm.

power supply. The unit that supplies the necessary voltage and current to a system's circuitry.

presentation layer. Layer 6 of the OSI protocol. This layer communicates data while resolving syntax differences between network devices.

pressure transducer. A transducer that measures pressure by transforming exerted force into an electrical signal.

process. (1) Continuous and regular production executed in a definite, uninterrupted manner. (2) One or more entities threaded together to perform a requested service.

process bus network. A network that allows high-level analog input/output devices that transmit large amounts of information to communicate directly with a PLC.

process control. The regulation of process parameters to within specified target parameters through the manipulation of the control variable.

process gain. The ratio between a process's output and its input. In an ideal process control situation, the process gain equals one.

process variable. A process control system's dependent variable, which is controlled by its independent variable, the control variable.

program. A planned set of instructions stored in memory and executed in an orderly fashion by the central processing unit.

program coding. The process of translating a logic or relay diagram into PLC ladder program form.

program/flow control instructions. Computer codes that give a PLC the ability to direct the flow of operation and alter the order of execution of a control program.

programmable logic controller (PLC). A solid-state control device that can be programmed to control process or machine operations. It consists of five basic components: the processor, memory, input/output modules, the power supply, and the programming device.

programmable read-only memory (PROM). A read-only memory that can be programmed once and never altered again.

programming device. A device that is used to enter the control program into memory and make changes to the stored program.

program scan. The time required by the processor to evaluate and execute the control logic. This time does not include the I/O update time. The program scan repeats continuously while the processor is in the run mode.

PROM. *See* **programmable read-only memory**.

propagation error. A combined error caused by the interaction of two or more independent variables, each causing a different error.

proportional controller. A continuous-mode controller whose output to the control field device in proportional to the change in error.

proportional-derivative controller. A continuous-mode controller that uses both proportional and derivative actions to determine the control variable output based on both the amount of error and its rate of change.

proportional-integral controller. A continuous-mode controller that uses both proportional and integral actions to determine the control variable output based on the amount of error and its change over time.

proportional-integral-derivative controller. A continuous-mode controller that uses proportional, integral, and derivative actions to determine the control variable output based on the amount of error, its change over time, and its rate of change. This type of controller provides the optimum type of control in most process applications.

proportional-integral-derivative (PID) interface. An intelligent I/O module that provides automatic, closed-loop control of multiple, continuous-process control loops.

protocol. A formal definition of how communication will occur in a network.

pulse action. A set of IEC 1131-3 instructions that is executed only once after a step becomes active.

Q

quarter-amplitude response. A process variable response whose amplitude diminishes by one-fourth during each cycle.

R

rack enclosure. The location in a PLC that physically houses plug-in devices, such as I/O modules and supplementary power supplies.

RAM. *See* **random-access memory**.

random-access memory (RAM). A volatile, alterable memory that provides storage for the application program and data.

random error. An error resulting from an unexpected action in a process line.

read. (1) To acquire data from a storage device. (2) The transfer of data between devices, such as a peripheral device and a computer.

read-only memory (ROM). A type of memory that permanently stores an unalterable program or set of instructions.

real variable. A nondiscrete variable whose value is transmitted in the form of fractional and floating-point data.

register. A temporary storage device for data and information (e.g., timer/counter preset values). A PLC register is normally 16 bits wide.

register/BCD I/O interface. A multibit module that uses thumbwheel switches to interface between discrete devices and a programmable controller.

relay. An electrically operated device that mechanically switches electrical circuits.

relay logic. The representation of a control program or other logic in relay form (i.e., using electrically operated devices to mechanically switch electrical circuits).

remote I/O subsystem. A system where some or all of the I/O racks are mounted away from the PLC.

remote rack. An enclosure, containing I/O modules and a remote I/O processor, located away from the CPU.

resistance temperature detector (RTD). A temperature transducer composed of conductive wire elements typically made of platinum, nickel, copper, or nickel-iron.

resistance temperature detector (RTD) interface. An intelligent I/O module that interprets temperature information from RTD devices.

resolution. The smallest detectable increment of measurement.

response time. The time, including terminal delay, network delay, and service node delay, between the transmission of the last character of a network node's message and its receipt of the first character of the reply.

reverse-acting controller. A closed-loop controller whose control variable output decreases in response to an increase in the process variable.

ring topology. A network architecture where signals from one node are relayed through all the other nodes in the network.

ROM. *See* **read-only memory**.

RTD. *See* **resistance temperature detector**.

RTD interface. *See* **resistance temperature detector interface**.

rule. An algorithm consisting of IF conditions and THEN actions that a fuzzy logic module uses to interpret input data and respond with a corresponding output value.

rule-based knowledge representation. A method of expressing an expert's knowledge in an AI system using IF...THEN rules that determine the actions and decisions to be made.

rung. A ladder program term that refers to the programmed instructions that drive one output. A complete control program may have several rungs.

safety control relay (SCR). A hardwired or softwired relay instruction that will de-energize its associated I/O devices when de-energized.

scaling. Changing analog output data to reflect engineering units.

scan. The process of reading all inputs, executing the control program, and updating all outputs.

scan time. The time required to complete the scan. Effectively, this is the time required to activate an output that is controlled by programmed logic.

SCR. *See* **safety control relay**.

scratch pad memory. A temporary storage area used by the CPU to store a relatively small amount of data used for interim calculations or control. Data that is needed quickly is stored in this area to avoid the extra access time involved in retrieving data from the main memory.

second-order response. A process response to a rapid change in the control variable characterized by two lag time and a process response curve that either oscillates around the set point or overshoots the set point before settling to it.

sequential function charts (SFC). An object-oriented programming framework that organizes actions written in IEC 1131-3 programming languages (ladder diagram, instruction list, function block diagram, and structured text) into a unified sequential control program.

series circuit. A circuit in which the components or contact symbols are connected end to end. All components must be closed to permit current flow.

servo motor interface. An intelligent I/O module used in applications requiring position control via a servo drive controller, which translates the rotational movement of a servo motor into linear displacement.

set point. The target process variable value in a process control system.

SFC. *See* **sequential function charts**.

SFC action. A set of IEC 1131-3 instructions, organized as an SFC program, that is activated when a certain step in the main SFC program becomes active.

single-ended input/output. An analog I/O connection in which the commons are electrically tied together resulting in only one return line.

single-precision arithmetic. Arithmetic instructions that use one register each to hold the operands and one or two registers to hold the result of the operation.

sinking configuration. An electrical configuration that causes a device to receive current when the device is ON.

slave. A remote system or terminal whose function is controlled by a master device.

software. The programs that control the processing of data in a system.

solenoid. A transducer that converts a current into linear motion through the use of one or more electromagnets that move a metal plunger.

solid-state. Circuitry designed using only integrated circuits, transistors, diodes, etc., without any electromechanical devices, such as relays.

sourcing configuration. An electrical configuration that causes a device to provide current when the device is ON.

special function instructions. Computer codes that allow a PLC to perform special operations, such as sequencing, diagnostics, and PID control.

ST. *See* **structured text**.

stand-alone action. A set of IEC 1131-3 programming instructions, not attached to the SFC program itself, that directs the program to jump to a particular step when the action's logical conditions are satisfied.

standard deviation. A measure of the dispersion of a set of data readings about the mean.

star-shaped ring topology. A network architecture in which signals from one node are relayed through all the other nodes in the network, yet a node can be bypassed in the event of its failure to avoid a break in the ring.

star topology. A network architecture in which all network nodes are connected to a central device that routes the nodes' messages.

static input wiring check. A procedure performed with power applied to the PLC and input devices that verifies that each input device is connected to the proper input terminal and is operating properly.

static output wiring check. A procedure performed with power applied to the PLC and output devices that verifies that each output device is connected to the proper output terminal and is operating properly.

steady state. The situation in which the error in a process control system is at zero or within the error deadband.

step. A stage in a control process as defined by the process's sequential function chart.

stepper motor interface. A positioning interface that controls a stepper motor, which translates incoming pulses into mechanical motion, by generating a pulse train indicating distance, rate, and direction commands to the motor.

step response. The process variable's response to a sudden change in the process input (i.e., the control variable).

step test. A forced, sudden change in the control variable used to elicit a response from the process.

storage area. The area of a PLC's memory that stores blocks of input/output data, as well as data about the status of internal bits.

storage register assignment document. A document that lists the storage registers used in a control program, including their contents and a description of their function.

strain gauge. A mechanical transducer that measures body deformation (or strain) due to the force applied to a rigid body.

structured text (ST). A high-level, text-based PLC programming language, resembling the BASIC and PASCAL computer languages, that allows a control program or any other complex task to be broken down into smaller tasks.

subprogram. A semi-independent program, embedded in a larger, main control program, that executes a specialized control sequence when activated by the main program.

subroutine. A program segment in a ladder diagram that performs a separate task.

subsystem. A part of a larger system having the properties of a system in its own right.

sum-of-the-weights method. A method of changing values from other number systems into their decimal equivalents by multiplying each digit by the weighted value of its position and then summing the results.

synchronous. A type of serial transmission that maintains a constant time interval between successive events.

syntax. Rules governing the structure of a language.

system. A set of one or more PLCs, I/O devices and modules, computers, associated software, peripherals, terminals, and communication networks that together provide a means of performing information processing to control a machine or process.

system abstract. A definition of the process to be controlled including a clear statement of the control problem, a description of the design strategy, and a statement of objectives.

system configuration diagram. A drawing of the PLC control system that shows the location, simplified connections, and minimum details of the system's major hardware components.

system error. An error resulting from an instrument or from the environment.

system layout. The planned approach to placing and connecting PLC components to satisfy the control strategy and to provide system reliability and ease of maintenance.

T

tap. A device that provides mechanical and electrical connections to a trunk cable. A tap allows the signals on the trunk to be passed to a station and the signals transmitted by the stations to be passed to the trunk.

task. A set of instructions, data, and control information capable of being executed by a CPU to accomplish a specific purpose.

TCP/IP. *See* **transmission control protocol/internet protocol.**

termination. (1) The load connected to the output end of a transmission line. (2) A provision for ending a transmission line and connecting to a bus bar or other terminating device.

thermal transducer. A device that measures changes in temperature.

thermistor. A temperature transducer made of semiconductor material, such as oxides of cobalt, nickel, manganese, iron, and titanium, that exhibits changes in internal resistance proportional to changes in temperature.

thermocouple. A bimetallic temperature transducer that provides a temperature value by measuring the voltage differential caused by joining together two different metals at different temperatures.

thermocouple input module. A module that amplifies, digitizes, and converts the input signal from a thermocouple into a digital signal equivalent to the temperature reading.

thermopile. The connection of several thermocouples in series to enhance their resolution.

three-position controller. A discrete-mode controller that provides three output levels—ON, 50% ON, and OFF.

throughput. The speed at which an application or part of an application is performed. Throughput depends on the transmission speed, medium, protocol, packet size, and amount of data handled by a network.

thumbwheel switch. A rotating switch used to input numeric information into a controller.

time base. A unit of time generated by the system clock and used by software timer instructions. Typical time bases are 0.01, 0.1, and 1.0 seconds.

timer instructions. Computer codes that allow a PLC to perform the timing functions (ON-delay energize/de-energize, OFF-delay energize/de-energize, reset) of a hardware timer.

token. (1) A signal that grants bus transmission rights to a node on a network. (2) A signal that enables a transition or action in a sequential function chart.

token passing. A network transmission technique in which a token is passed along the bus and each node has a set amount of time to receive it and respond to it.

topology. The way in which a network or system is physically structured.

transducer. A device used to convert physical parameters, such as temperature, pressure, and weight, into electrical signals.

transfer function. The unique characteristics of a process that determine its output due to changes over time.

transient response. The behavioral response of a process.

transistor-transistor logic (TTL). A semiconductor logic family characterized by high speed and medium power dissipation in which the basic logic element is a multiple-emitter transistor.

transition. A variable input, action result, conditional statement, or other program element that signals a sequential function chart to progress from one step to another.

transmission control protocol/internet protocol (TCP/IP). A network protocol developed by the U.S. Department of Defense.

transmission medium. The physical device used to transfer data in a transmission system (e.g., coaxial cable, fiber-optic cable, etc.).

transmitter. A device that amplifies a voltage signal.

tree topology. A network architecture in which the network has many nodes located in many branches of the network.

triac. A semiconductor device that functions as an electrically controlled switch for AC loads.

TRUE. As related to PLC instructions, a set logic state associated with a binary 1.

truth table. A table that shows the state of a given output as a function of all possible input combinations.

TTL. *See* **transistor-transistor logic.**

TTL I/O interface. A discrete interface that allows a controller to accept signals from TTL field devices, which are 5 VDC–level semiconductor devices.

turbine flow meter. A flow transducer that measures fluid flow by measuring the fluid's motion through the meter's multibladed rotor.

twisted-pair conductor. A communication medium used mainly for point-to-point applications that can transmit data up to 4000 feet at transmission rates as high as 250 kbaud.

two-position controller. A discrete-mode controller that provides two output levels—ON and OFF.

two's complement. A numbering system, used to express negative binary numbers, in which all numbers from right to left are inverted after the first 1 is detected.

U   **underdamped response.** A second-order control system response in which the damping coefficient is less than 1, causing the response to oscillate around the set point before settling to it.

user program memory. The memory section where the application control program is stored.

V   **variable.** A factor that can be altered, measured, and controlled.

Venturi tube. A transducer that measures fluid flow by measuring the pressure differential between two points.

vertical redundancy check (VRC). An error-detecting method in which a parity bit is added to each character in a message so that the number of bits in each character, including the parity bit, is either odd or even.

vibration transducer. A device that measures the vibration of a body by measuring its displacement, velocity, or acceleration.

volatile memory. A type of memory whose contents are irretrievable after operating power is lost.

VRC. *See* **vertical redundancy check**.

W   **watchdog timer.** A timer that monitors the logic circuits controlling a PLC. If a watchdog timer ever times out, it will disconnect the processor from the process because it will assume that the processor is faulty.

weighted value. The numerical value assigned to any single bit as a function of its position in a word.

weight input module. A special analog interface designed to read data from load cells, which convert force and weight values into electrical signals.

wire bundling. The technique of grouping an I/O module's wires according to their characteristics (e.g., input, output, power).

---

wire input module. A special input interface designed to detect short-circuit or open-circuit connections between a module and its input devices.

word. The number of bits that the central processing unit operates on at one time when it is performing an instruction or operating on data. A word is usually composed of a fixed number of bits.

write. The process of putting information into a storage location.

X XOR. *See* **exclusive-OR**.

Z **Ziegler-Nichols closed-loop tuning method.** A method for determining a controller's tuning constants by finding the value of the proportional gain that will cause the control loop to oscillate indefinitely at a constant amplitude when it is in a closed-loop system.

Ziegler-Nichols open-loop tuning method. A method for determining the tuning constants for a controller by testing the process variable's response to a change in the control variable output in an open-loop system.

INDEX